D0780973

Statistical Mechanics

Third Edition

Statistical Mechanics

Third Edition

R. K. Pathria

Department of Physics
University of California at San Diego

Paul D. Beale

Department of Physics
University of Colorado at Boulder

ELSEVIER

AMSTERDAM • BOSTON • HEIDELBERG • LONDON
NEW YORK • OXFORD • PARIS • SAN DIEGO
SAN FRANCISCO • SINGAPORE • SYDNEY • TOKYO
Butterworth-Heinemann is an imprint of Elsevier

Butterworth-Heinemann is an imprint of Elsevier
The Boulevard, Langford Lane, Kidlington, Oxford, 0X5 1GB, UK
30 Corporate Drive, Suite 400, Burlington, MA 01803, USA

First published 1972; Second edition 1996

© 2011 Elsevier Ltd. All rights reserved

The right of R. K. Pathria to be identified as the author of this work has been asserted in accordance with the Copyright, Designs and Patents Act 1988.

No part of this publication may be reproduced or transmitted in any form or by any means, electronic or mechanical, including photocopying, recording, or any information storage and retrieval system, without permission in writing from the publisher. Details on how to seek permission, further information about the Publisher's permissions policies and our arrangements with organizations such as the Copyright Clearance Center and the Copyright Licensing Agency, can be found at our website: *www.elsevier.com/permissions*.

This book and the individual contributions contained in it are protected under copyright by the Publisher (other than as may be noted herein).

Notices
Knowledge and best practice in this field are constantly changing. As new research and experience broaden our understanding, changes in research methods, professional practices, or medical treatment may become necessary.

Practitioners and researchers must always rely on their own experience and knowledge in evaluating and using any information, methods, compounds, or experiments described herein. In using such information or methods they should be mindful of their own safety and the safety of others, including parties for whom they have a professional responsibility.

To the fullest extent of the law, neither the Publisher nor the authors, contributors, or editors, assume any liability for any injury and/or damage to persons or property as a matter of products liability, negligence or otherwise, or from any use or operation of any methods, products, instructions, or ideas contained in the material herein.

British Library Cataloguing in Publication Data
A catalogue of this book is available from the British Library.

Library of Congress Cataloging in Publication Data
Pathria, R. K.
Statistical mechanics–3rd ed. / R. K. Pathria, Paul D. Beale.
p. cm.
Includes bibliographical references and index.
ISBN 978-0-12-382188-1 (pbk.)
1. Statistical mechanics. I. Beale, Paul D. II. Title.
QC174.8.P38 2011
530.13–dc22 2010048955

Cover: The image was created using the opensource software CMBview (*http://www.jportsmouth.com/code/CMBview/cmbview.html*) written by Jamie Portsmouth and used with permission. It was created using the WMAP seven-year Internal Linear Combination Map courtesy of the WMAP Science Team (*http://lambda.gsfc.nasa.gov/product/map/dr4/ilc_map_get.cfm*).

For information on all Butterworth-Heinemann publications, visit our Web site at *www.elsevierdirect.com*

Printed in the United States
11 12 13 14 15 10 9 8 7 6 5 4 3 2 1

Working together to grow libraries in developing countries
www.elsevier.com | www.bookaid.org | www.sabre.org
ELSEVIER BOOK AID International Sabre Foundation

Contents

Preface to the Third Edition

The second edition of *Statistical Mechanics* was published in 1996. The new material added at that time focused on phase transitions, critical phenomena, and the renormalization group — topics that had undergone vast transformations during the years following the publication of the first edition in 1972. In 2009, R. K. Pathria (R.K.P.) and the publishers agreed it was time for a third edition to incorporate the important changes that had occurred in the field since the publication of the second edition and invited Paul B. Beale (P.D.B.) to join as coauthor. The two authors agreed on the scope of the additions and changes and P.D.B. wrote the first draft of the new sections except for Appendix F which was written by R.K.P. Both authors worked very closely together editing the drafts and finalizing this third edition.

The new topics added to this edition are:

- *Bose–Einstein condensation and degenerate Fermi gas behavior in ultracold atomic gases:* Sections 7.2, 8.4, 11.2.A, and 11.9. The creation of Bose–Einstein condensates in ultracold gases during the 1990s and in degenerate Fermi gases during the 2000s led to a revolution in atomic, molecular, and optical physics, and provided a valuable link to the quantum behavior of condensed matter systems. Several of P.D.B.'s friends and colleagues in physics and JILA at the University of Colorado have been leaders in this exciting new field.
- *Finite-size scaling behavior of Bose–Einstein condensates:* Appendix F. We develop an analytical theory for the behavior of Bose–Einstein condensates in a finite system, which provides a rigorous justification for singling out the ground state in the calculation of the properties of the Bose–Einstein condensate.
- *Thermodynamics of the early universe:* Chapter 9. The sequence of thermodynamic transitions that the universe went though shortly after the Big Bang left behind mileposts that astrophysicists have exploited to look back into the universe's earliest moments. Major advances in astronomy over the past 20 years have provided a vast body of observational data about the early evolution of the universe. These include the Hubble Space Telescope's deep space measurements of the expansion of the universe, the Cosmic Background Explorer's precise measurements of the temperature of the cosmic microwave background, and the Wilkinson Microwave Anisotropy Probe's mapping of the angular variations in the cosmic microwave background. These data sets have led to precise determinations of the age of the universe, its composition and early evolution. Coincidentally, P.D.B.'s faculty office is located in the tower named after George Gamow, a member of the faculty at the University of Colorado in the 1950s and 1960s and a leader in the theory of nucleosynthesis in the early universe.
- *Chemical equilibrium:* Section 6.6. Chemical potentials determine the conditions necessary for chemical equilibrium. This is an important topic in its own right, but also plays a critical role in our discussion of the thermodynamics of the early universe in Chapter 9.

- *Monte Carlo and molecular dynamics simulations:* Chapter 16. Computer simulations have become an important tool in modern statistical mechanics. We provide here a brief introduction to Monte Carlo and molecular dynamics techniques and algorithms.
- *Correlation functions and scattering:* Section 10.7. Correlation functions are central to the understanding of thermodynamic phases, phase transitions, and critical phenomena. The differences between thermodynamic phases are often most conspicuous in the behavior of correlation functions and the closely related static structure factors. We have collected discussions from the second edition into one place and added new material.
- *Fluctuation–dissipation theorem and the dynamical structure factor:* Sections 15.3.A., 15.6.A, and 15.6.B. The fluctuation–dissipation theorem describes the relation between natural equilibrium thermodynamic fluctuations in a system and the response of the system to small disturbances from equilibrium, and it is one of the cornerstones of nonequilibrium statistical mechanics. We have expanded the discussion of the fluctuation–dissipation theorem to include a derivation of the key results from linear response theory, a discussion of the dynamical structure factor, and analysis of the Brownian motion of harmonic oscillators that provides useful practical examples.
- *Phase equilibrium and the Clausius–Clapeyron equation:* Sections 4.6 and 4.7. Much of the text is devoted to using statistical mechanics methods to determine the properties of thermodynamic phases and phase transitions. This brief overview of phase equilibrium and the structure of phase diagrams lays the groundwork for later discussions.
- *Exact solutions of one-dimensional fluid models:* Section 13.1. One-dimensional fluid models with short-range interactions do not exhibit phase transitions but they do display short-range correlations and other behaviors typical of dense fluids.
- *Exact solution of the two-dimensional Ising model on a finite lattice:* Section 13.4.A. This solution entails an exact counting of the microstates of the microcanonical ensemble and provides analytical results for the energy distribution, internal energy, and heat capacity of the system. This solution also describes the finite-size scaling behavior of the Ising model near the transition point and provides an exact framework that can be used to test Monte Carlo methods.
- *Summary of thermodynamic assemblies and associated statistical ensembles:* Appendix H. We provide a summary of thermodynamic relations and their connections to statistical mechanical ensembles. Most of this information can be found elsewhere in the text, but we thought it would be helpful to provide a rundown of these important connections in one place.
- *Pseudorandom number generators:* Appendix I. Pseudorandom number generators are indispensable in computer simulations. We provide simple algorithms for generating uniform and Gaussian pseudorandom numbers and discuss their properties.
- *Dozens of new homework problems.*

The remainder of the text is largely unchanged.

The completion of this task has left us indebted to many a friend and colleague. R.K.P. has already expressed his indebtedness to a good number of people on two previous occasions — in 1972 and in 1996 — so, at this time, he will simply reiterate the many words of gratitude he has already written. In addition though, he would like to thank Paul Beale for his willingness to be a partner in this project and for his diligence in carrying out the task at hand both arduously and meticulously.

On his part, P.D.B. would like to thank his friends at the University of Colorado at Boulder for the many conversations he has had with them over the years about research and pedagogy of statistical mechanics, especially Noel Clark, Tom DeGrand, John Price, Chuck Rogers, Mike

Dubson, and Leo Radzihovsky. He would also like to thank the faculty of the Department of Physics for according him the honor of serving as the chair of this outstanding department.

Special thanks are also due to many friends and colleagues who have read sections of the manuscript and have offered many valuable suggestions and corrections, especially Tom DeGrand, Michael Shull, David Nesbitt, Jamie Nagle, Matt Glaser, Murray Holland, Leo Radzihovsky, Victor Gurarie, Edmond Meyer, Matthew Grau, Andrew Sisler, Michael Foss-Feig, Allan Franklin, Shantha deAlwis, Dmitri Reznik, and Eric Cornell.

P.D.B. would like to take this opportunity to extend his thanks and best wishes to Professor Michael E. Fisher whose graduate statistical mechanics course at Cornell introduced him to this elegant field. He would also like to express his gratitude to Raj Pathria for inviting him to be part of this project, and for the fun and engaging discussions they have had during the preparation of this new edition. Raj's thoughtful counsel always proved to be valuable in improving the text.

P.D.B.'s greatest thanks go to Matthew, Melanie, and Erika for their love and support.

R.K.P.
P.D.B.

Preface to the Second Edition

The first edition of this book was prepared over the years 1966 to 1970 when the subject of phase transitions was undergoing a complete overhaul. The concepts of scaling and universality had just taken root but the renormalization group approach, which converted these concepts into a calculational tool, was still obscure. Not surprisingly, my text of that time could not do justice to these emerging developments. Over the intervening years I have felt increasingly conscious of this rather serious deficiency in the text; so when the time came to prepare a new edition, my major effort went toward correcting that deficiency.

Despite the aforementioned shortcoming, the first edition of this book has continued to be popular over the last 20 years or so. I, therefore, decided not to tinker with it unnecessarily. Nevertheless, to make room for the new material, I had to remove some sections from the present text which, I felt, were not being used by the readers as much as the rest of the book was. This may turn out to be a disappointment to some individuals but I trust they will understand the logic behind it and, if need be, will go back to a copy of the first edition for reference. I, on my part, hope that a good majority of the users will not be inconvenienced by these deletions. As for the material retained, I have confined myself to making only editorial changes. The subject of phase transitions and critical phenomena, which has been my main focus of revision, has been treated in three new chapters that provide a respectable coverage of the subject and essentially bring the book up to date. These chapters, along with a collection of more than 60 homework problems, will, I believe, enhance the usefulness of the book for both students and instructors.

The completion of this task has left me indebted to many. First of all, as mentioned in the Preface to the first edition, I owe a considerable debt to those who have written on this subject before and from whose writings I have benefited greatly. It is difficult to thank them all individually; the bibliography at the end of the book is an obvious tribute to them. As for definitive help, I am most grateful to Dr Surjit Singh who advised me expertly and assisted me generously in the selection of the material that comprises Chapters 11 to 13 of the new text; without his help, the final product might not have been as coherent as it now appears to be. On the technical side, I am very thankful to Mrs. Debbie Guenther who typed the manuscript with exceptional skill and proof read it with extreme care; her task was clearly an arduous one but she performed it with good cheer — for which I admire her greatly.

Finally, I wish to express my heartfelt appreciation for my wife who let me devote myself fully to this task over a rather long period of time and waited for its completion ungrudgingly.

R.K.P.

Preface to the First Edition

This book has arisen out of the notes of lectures that I gave to the graduate students at the McMaster University (1964–1965), the University of Alberta (1965–1967), the University of Waterloo (1969–1971), and the University of Windsor (1970–1971). While the subject matter, in its finer details, has changed considerably during the preparation of the manuscript, the style of presentation remains the same as followed in these lectures.

Statistical mechanics is an indispensable tool for studying physical properties of matter "in bulk" on the basis of the dynamical behavior of its "microscopic" constituents. Founded on the well-laid principles of *mathematical statistics* on one hand and *Hamiltonian mechanics* on the other, the formalism of statistical mechanics has proved to be of immense value to the physics of the last 100 years. In view of the universality of its appeal, a basic knowledge of this subject is considered essential for every student of physics, irrespective of the area(s) in which he/she may be planning to specialize. To provide this knowledge, in a manner that brings out the essence of the subject with due rigor but without undue pain, is the main purpose of this work.

The fact that *the dynamics of a physical system is represented by a set of quantum states* and the assertion that *the thermodynamics of the system is determined by the multiplicity of these states* constitute the basis of our treatment. The fundamental connection between the microscopic and the macroscopic descriptions of a system is uncovered by investigating the conditions for equilibrium between two physical systems in thermodynamic contact. This is best accomplished by working in the spirit of the quantum theory right from the beginning; the entropy and other thermodynamic variables of the system then follow in a most natural manner. After the formalism is developed, one may (if the situation permits) go over to the limit of the classical statistics. This message may not be new, but here I have tried to follow it as far as is reasonably possible in a textbook. In doing so, an attempt has been made to keep the level of presentation fairly uniform so that the reader does not encounter fluctuations of too wild a character.

This text is confined to the study of the *equilibrium states* of physical systems and is intended to be used for a *graduate course* in statistical mechanics. Within these bounds, the coverage is fairly wide and provides enough material for tailoring a good two-semester course. The final choice always rests with the individual instructor; I, for one, regard Chapters 1 to 9 (*minus* a few sections from these chapters *plus* a few sections from Chapter 13) as the "essential part" of such a course. The contents of Chapters 10 to 12 are relatively advanced (not necessarily difficult); the choice of material out of these chapters will depend entirely on the taste of the instructor. To facilitate the understanding of the subject, the text has been illustrated with a large number of graphs; to assess the understanding, a large number of problems have been included. I hope these features are found useful.

I feel that one of the most essential aspects of teaching is to arouse the curiosity of the students in their subject, and one of the most effective ways of doing this is to discuss with them (in a reasonable measure, of course) the circumstances that led to the emergence of the subject. One would, therefore, like to stop occasionally to reflect upon the manner in which the various developments really came about; at the same time, one may not like the flow of the text to be hampered by the discontinuities arising from an intermittent addition of historical material. Accordingly, I decided to include in this account a historical introduction to the subject which stands separate from the main text. I trust the readers, especially the instructors, will find it of interest.

For those who wish to continue their study of statistical mechanics beyond the confines of this book, a fairly extensive bibliography is included. It contains a variety of references — old as well as new, experimental as well as theoretical, technical as well as pedagogical. I hope that this will make the book useful for a wider readership.

The completion of this task has left me indebted to many. Like most authors, I owe considerable debt to those who have written on the subject before. The bibliography at the end of the book is the most obvious tribute to them; nevertheless, I would like to mention, in particular, the works of the Ehrenfests, Fowler, Guggenheim, Schrödinger, Rushbrooke, ter Haar, Hill, Landau and Lifshitz, Huang, and Kubo, which have been my constant reference for several years and have influenced my understanding of the subject in a variety of ways. As for the preparation of the text, I am indebted to Robert Teshima who drew most of the graphs and checked most of the problems, to Ravindar Bansal, Vishwa Mittar, and Surjit Singh who went through the entire manuscript and made several suggestions that helped me unkink the exposition at a number of points, to Mary Annetts who typed the manuscript with exceptional patience, diligence and care, and to Fred Hetzel, Jim Briante, and Larry Kry who provided technical help during the preparation of the final version.

As this work progressed I felt increasingly gratified toward Professors F. C. Auluck and D. S. Kothari of the University of Delhi with whom I started my career and who initiated me into the study of this subject, and toward Professor R. C. Majumdar who took keen interest in my work on this and every other project that I have undertaken from time to time. I am grateful to Dr. D. ter Haar of the University of Oxford who, as the general editor of this series, gave valuable advice on various aspects of the preparation of the manuscript and made several useful suggestions toward the improvement of the text. I am thankful to Professors J. W. Leech, J. Grindlay, and A. D. Singh Nagi of the University of Waterloo for their interest and hospitality that went a long way in making this task a pleasant one.

The final tribute must go to my wife whose cooperation and understanding, at all stages of this project and against all odds, have been simply overwhelming.

R.K.P.

Historical Introduction

Statistical mechanics is a formalism that aims at explaining the physical properties of matter *in bulk* on the basis of the dynamical behavior of its *microscopic* constituents. The scope of the formalism is almost as unlimited as the very range of the natural phenomena, for in principle it is applicable to matter in any state whatsoever. It has, in fact, been applied, with considerable success, to the study of matter in the solid state, the liquid state, or the gaseous state, matter composed of several phases and/or several components, matter under extreme conditions of density and temperature, matter in equilibrium with radiation (as, for example, in astrophysics), matter in the form of a biological specimen, and so on. Furthermore, the formalism of statistical mechanics enables us to investigate the *nonequilibrium* states of matter as well as the *equilibrium* states; indeed, these investigations help us to understand the manner in which a physical system that happens to be "out of equilibrium" at a given time t approaches a "state of equilibrium" as time passes.

In contrast with the present status of its development, the success of its applications, and the breadth of its scope, the beginnings of statistical mechanics were rather modest. Barring certain primitive references, such as those of Gassendi, Hooke, and so on, the real work on this subject started with the contemplations of Bernoulli (1738), Herapath (1821), and Joule (1851) who, in their own individual ways, attempted to lay a foundation for the so-called *kinetic theory of gases* — a discipline that finally turned out to be the forerunner of statistical mechanics. The pioneering work of these investigators established the fact that the pressure of a gas arose from the motion of its molecules and could, therefore, be computed by considering the dynamical influence of the molecular bombardment on the walls of the container. Thus, Bernoulli and Herapath could show that, if temperature remained constant, the pressure P of an ordinary gas was inversely proportional to the volume V of the container (Boyle's law), and that it was essentially independent of the shape of the container. This, of course, involved the explicit assumption that, *at a given temperature* T, the (mean) speed of the molecules was independent of both pressure and volume. Bernoulli even attempted to determine the (first-order) correction to this law, arising from the *finite* size of the molecules, and showed that the volume V appearing in the statement of the law should be replaced by $(V - b)$, where b is the "actual" volume of the molecules.[1]

Joule was the first to show that the pressure P was directly proportional to the square of the molecular speed c, which he had initially assumed to be the same for all molecules. Krönig (1856) went a step further. Introducing the "quasistatistical" assumption that, *at any time t*,

[1]As is well known, this "correction" was correctly evaluated, much later, by van der Waals (1873) who showed that, for large V, b is *four times* the "actual" volume of the molecules; see Problem 1.4.

one-sixth of the molecules could be assumed to be flying in each of the six "independent" directions, namely $+x, -x, +y, -y, +z$, and $-z$, he derived the equation

$$P = \frac{1}{3} n m c^2, \tag{1}$$

where n is the number density of the molecules and m the molecular mass. Krönig, too, assumed the molecular speed c to be the same for all molecules; so from (1), he inferred that the kinetic energy of the molecules should be directly proportional to the absolute temperature of the gas.

Krönig justified his method in these words: "The path of each molecule must be so irregular that it will defy all attempts at calculation. However, according to the laws of probability, one could assume a completely regular motion in place of a completely irregular one!" It must, however, be noted that it is only because of the special form of the summations appearing in the calculation of the pressure that Krönig's argument leads to the same result as the one following from more refined models. In other problems, such as the ones involving diffusion, viscosity, or heat conduction, this is no longer the case.

It was at this stage that Clausius entered the field. First of all, in 1857, he derived the ideal-gas law under assumptions far less stringent than Krönig's. He discarded both leading assumptions of Krönig and showed that equation (1) was still true; of course, c^2 now became the *mean square speed* of the molecules. In a later paper (1859), Clausius introduced the concept of the *mean free path* and thus became the first to analyze transport phenomena. It was in these studies that he introduced the famous "Stosszahlansatz" — the hypothesis on the number of collisions (among the molecules) — which, later on, played a prominent role in the monumental work of Boltzmann.[2] With Clausius, the introduction of the microscopic and statistical points of view into the physical theory was definitive, rather than speculative. Accordingly, Maxwell, in a popular article entitled "Molecules," written for the *Encyclopedia Britannica*, referred to Clausius as the "principal founder of the kinetic theory of gases," while Gibbs, in his Clausius obituary notice, called him the "father of statistical mechanics."[3]

The work of Clausius attracted Maxwell to the field. He made his first appearance with the memoir "Illustrations in the dynamical theory of gases" (1860), in which he went much farther than his predecessors by deriving his famous law of the "distribution of molecular speeds." Maxwell's derivation was based on elementary principles of probability and was clearly inspired by the Gaussian law of "distribution of random errors." A derivation based on the requirement that "the equilibrium distribution of molecular speeds, once acquired, should remain invariant under molecular collisions" appeared in 1867. This led Maxwell to establish what is known as *Maxwell's transport equation* which, if skilfully used, leads to the same results as one gets from the more fundamental equation due to Boltzmann.[4]

Maxwell's contributions to the subject diminished considerably after his appointment, in 1871, as the Cavendish Professor at Cambridge. By that time Boltzmann had already made his first strides. In the period 1868–1871 he generalized Maxwell's distribution law to polyatomic gases, also taking into account the presence of external forces, if any; this gave rise to the famous *Boltzmann factor* $\exp(-\beta\varepsilon)$, where ε denotes the *total* energy of a molecule. These investigations also led to the *equipartition theorem*. Boltzmann further showed that, just

[2]For an excellent review of this and related topics, see Ehrenfest and Ehrenfest (1912).

[3]For further details, refer to Montroll (1963) where an account is also given of the pioneering work of Waterston (1846, 1892).

[4]This equivalence has been demonstrated in Guggenheim (1960) where the coefficients of viscosity, thermal conductivity, and diffusion of a gas of hard spheres have been calculated on the basis of Maxwell's transport equation.

like the original distribution of Maxwell, the generalized distribution (which we now call the *Maxwell–Boltzmann distribution*) is stationary with respect to molecular collisions.

In 1872 came the celebrated *H-theorem*, which provided a molecular basis for the natural tendency of physical systems to approach, and stay in, a state of equilibrium. This established a connection between the microscopic approach (which characterizes statistical mechanics) and the phenomenological approach (which characterized thermodynamics) much more transparently than ever before; it also provided a direct method for computing the entropy of a given physical system from purely microscopic considerations. As a corollary to the H-theorem, Boltzmann showed that the Maxwell–Boltzmann distribution is the *only* distribution that stays invariant under molecular collisions and that any other distribution, under the influence of molecular collisions, will ultimately go over to a Maxwell–Boltzmann distribution. In 1876 Boltzmann derived his famous transport equation, which, in the hands of Chapman and Enskog (1916–1917), has proved to be an extremely powerful tool for investigating macroscopic properties of systems in nonequilibrium states.

Things, however, proved quite harsh for Boltzmann. His H-theorem, and the consequent *irreversible* behavior of physical systems, came under heavy attack, mainly from Loschmidt (1876–1877) and Zermelo (1896). While Loschmidt wondered how the consequences of this theorem could be reconciled with the reversible character of the basic equations of motion of the molecules, Zermelo wondered how these consequences could be made to fit with the *quasiperiodic* behavior of closed systems (which arose in view of the so-called Poincaré cycles). Boltzmann defended himself against these attacks with all his might but, unfortunately, could not convince his opponents of the correctness of his viewpoint. At the same time, the energeticists, led by Mach and Ostwald, were criticizing the very (molecular) basis of the kinetic theory,[5] while Kelvin was emphasizing the "nineteenth-century clouds hovering over the dynamical theory of light and heat."[6]

All this left Boltzmann in a state of despair and induced in him a persecution complex.[7] He wrote in the introduction to the second volume of his treatise *Vorlesungen über Gastheorie* (1898):[8]

> *I am convinced that the attacks (on the kinetic theory) rest on misunderstandings and that the role of the kinetic theory is not yet played out. In my opinion it would be a blow to science if contemporary opposition were to cause kinetic theory to sink into the oblivion which was the fate suffered by the wave theory of light through the authority of Newton. I am aware of the weakness of one individual against the prevailing currents of opinion. In order to insure that not too much will have to be rediscovered when people return to the study of kinetic theory I will present the most difficult and misunderstood parts of the subject in as clear a manner as I can.*

We shall not dwell any further on the kinetic theory; we would rather move on to the development of the more sophisticated approach known as the *ensemble theory*, which may in fact be regarded as the statistical mechanics proper.[9] In this approach, the dynamical state of a

[5]These critics were silenced by Einstein whose work on the Brownian motion (1905b) established atomic theory *once and for all*.

[6]The first of these clouds was concerned with the mysteries of the "aether," and was dispelled by the theory of relativity. The second was concerned with the inadequacy of the "equipartition theorem," and was dispelled by the quantum theory.

[7]Some people attribute Boltzmann's suicide on September 5, 1906 to this cause.

[8]Quotation from Montroll (1963).

[9]For a review of the historical development of kinetic theory leading to statistical mechanics, see Brush (1957, 1958, 1961a,b, 1965–1966).

given system, as characterized by the generalized coordinates q_i and the generalized momenta p_i, is represented by a *phase point* $G(q_i, p_i)$ in a *phase space* of appropriate dimensionality. The evolution of the dynamical state in time is depicted by the *trajectory* of the G-point in the phase space, the "geometry" of the trajectory being governed by the equations of motion of the system and by the nature of the physical constraints imposed on it. To develop an appropriate formalism, one considers the given system along with an infinitely large number of "mental copies" thereof; that is, an *ensemble* of similar systems under identical physical constraints (though, at any time t, the various systems in the ensemble would differ widely in respect of their dynamical states). In the phase space, then, one has a swarm of infinitely many G-points (which, at any time t, are widely dispersed and, with time, move along their respective trajectories). The fiction of a host of infinitely many, identical but independent, systems allows one to replace certain dubious assumptions of the kinetic theory of gases by readily acceptable statements of statistical mechanics. The explicit formulation of these statements was first given by Maxwell (1879) who on this occasion used the word "statistico-mechanical" to describe the study of ensembles (of gaseous systems) — though, eight years earlier, Boltzmann (1871) had already worked with essentially the same kind of ensembles.

The most important quantity in the ensemble theory is the *density function*, $\rho(q_i, p_i; t)$, of the G-points in the phase space; a stationary distribution ($\partial\rho/\partial t = 0$) characterizes a *stationary ensemble*, which in tum represents a system *in equilibrium*. Maxwell and Boltzmann confined their study to ensembles for which the function ρ depended solely on the energy E of the system. This included the special case of *ergodic* systems, which were so defined that "the undisturbed motion of such a system, if pursued for an unlimited time, would ultimately traverse (the neighborhood of) each and every phase point compatible with the *fixed* value E of the energy." Consequently, the *ensemble average*, $\langle f \rangle$, of a physical quantity f, taken at *any* given time t, would be the same as the *long-time average*, $\bar{f}$, pertaining to *any* given member of the ensemble. Now, $\bar{f}$ is the value we expect to obtain for the quantity in question when we make an appropriate measurement on the system; the result of this measurement should, therefore, agree with the theoretical estimate $\langle f \rangle$. We thus acquire a recipe to bring about a direct contact between theory and experiment. At the same time, we lay down a rational basis for a microscopic theory of matter as an alternative to the empirical approach of thermodynamics!

A significant advance in this direction was made by Gibbs who, with his *Elementary Principles of Statistical Mechanics* (1902), turned ensemble theory into a most efficient tool for the theorist. He emphasized the use of "generalized" ensembles and developed schemes which, in principle, enabled one to compute a complete set of thermodynamic quantities of a given system from purely mechanical properties of its microscopic constituents.[10] In its methods and results, the work of Gibbs turned out to be much more general than any preceding treatment of the subject; it applied to any physical system that met the simple-minded requirements that (i) it was mechanical in structure and (ii) it obeyed Lagrange's and Hamilton's equations of motion. In this respect, Gibbs's work may be considered to have accomplished for thermodynamics as much as Maxwell's had accomplished for electrodynamics.

These developments almost coincided with the great revolution that Planck's work of 1900 brought into physics. As is well known, Planck's *quantum hypothesis* successfully resolved the essential mysteries of the black-body radiation — a subject where the three best-established disciplines of the nineteenth century, namely mechanics, electrodynamics, and thermodynamics, were all focused. At the same time, it uncovered both the strengths and the weaknesses of these disciplines. It would have been surprising if statistical mechanics, which linked thermodynamics with mechanics, could have escaped the repercussions of this revolution.

[10]In much the same way as Gibbs, but quite independently of him, Einstein (1902, 1903) also developed the theory of ensembles.

The subsequent work of Einstein (1905a) on the photoelectric effect and of Compton (1923a,b) on the scattering of x-rays established, so to say, the "existence" of the *quantum of radiation*, or the *photon* as we now call it.[11] It was then natural for someone to derive Planck's radiation formula by treating black-body radiation as a *gas of photons* in the same way as Maxwell had derived his law of distribution of molecular speeds for a gas of conventional molecules. But, then, does a gas of photons differ so radically from a gas of conventional molecules that the two laws of distribution should be so different from one another?

The answer to this question was provided by the manner in which Planck's formula was derived by Bose. In his historic paper of 1924, Bose treated black-body radiation as a gas of photons; however, instead of considering the allocation of the "individual" photons to the various energy states of the system, he fixed his attention on the number of states that contained "a particular number" of photons. Einstein, who seems to have translated Bose's paper into German from an English manuscript sent to him by the author, at once recognized the importance of this approach and added the following note to his translation: "Bose's derivation of Planck's formula is in my opinion an important step forward. The method employed here would also yield the quantum theory of an ideal gas, which I propose to demonstrate elsewhere."

Implicit in Bose's approach was the fact that in the case of photons what really mattered was "the set of numbers of photons in various energy states of the system" and not the specification as to "which photon was in which state"; in other words, photons were *mutually indistinguishable*. Einstein argued that what Bose had implied for photons should be true for material particles as well (for the property of indistinguishability arose essentially from the wave character of these entities and, according to de Broglie, material particles also possessed that character).[12] In two papers, which appeared soon after, Einstein (1924, 1925) applied Bose's method to the study of an ideal gas and thereby developed what we now call *Bose–Einstein statistics*. In the second of these papers, the fundamental difference between the new statistics and the classical *Maxwell–Boltzmann* statistics comes out so transparently in terms of the indistinguishability of the molecules.[13] In the same paper, Einstein discovered the phenomenon of *Bose–Einstein condensation* which, 13 years later, was adopted by London (1938a,b) as the basis for a microscopic understanding of the curious properties of liquid He^4 at low temperatures.

Following the enunciation of Pauli's exclusion principle (1925), Fermi (1926) showed that certain physical systems would obey a different kind of statistics, namely the *Fermi–Dirac statistics*, in which not more than one particle could occupy the same energy state ($n_i = 0, 1$). It seems important to mention here that Bose's method of 1924 leads to the Fermi–Dirac distribution as well, provided that one limits the occupancy of an energy state to *at most* one particle.[14]

[11]Strictly speaking, it might be somewhat misleading to cite Einstein's work on the photoelectric effect as a proof of the existence of photons. In fact, many of the effects (including the photoelectric effect), for which it seems necessary to invoke photons, can be explained away on the basis of a wave theory of radiation. The only phenomena for which photons seem indispensable are the ones involving *fluctuations*, such as the Hanbury Brown–Twiss effect or the Lamb shift. For the relevance of fluctuations to the problem of radiation, see ter Haar (1967, 1968).

[12]Of course, in the case of material particles, the total number N (of the particles) will also have to be conserved; this had not to be done in the case of photons. For details, see Section 6.1.

[13]It is here that one encounters the *correct* method of counting "the number of distinct ways in which g_i energy states can accommodate n_i particles," depending on whether the particles are (i) distinguishable or (ii) indistinguishable. The occupancy of the individual states was, in each case, *unrestricted*, that is, $n_i = 0, 1, 2, \ldots$.

[14]Dirac, who was the first to investigate the connection between statistics and wave mechanics, showed, in 1926, that the wave functions describing a system of identical particles obeying Bose–Einstein (or Fermi–Dirac) statistics must be symmetric (or antisymmetric) with respect to an interchange of two particles.

Soon after its appearance, the Fermi–Dirac statistics were applied by Fowler (1926) to discuss the equilibrium states of white dwarf stars and by Pauli (1927) to explain the weak, temperature-independent paramagnetism of alkali metals; in each case, one had to deal with a "highly degenerate" gas of electrons that obey Fermi–Dirac statistics. In the wake of this, Sommerfeld produced his monumental work of 1928 that not only put the electron theory of metals on a physically secure foundation but also gave it a fresh start in the right direction. Thus, Sommerfeld could explain practically all the major properties of metals that arose from conduction electrons and, in each case, obtained results that showed much better agreement with experiment than the ones following from the classical theories of Riecke (1898), Drude (1900), and Lorentz (1904–1905). Around the same time, Thomas (1927) and Fermi (1928) investigated the electron distribution in heavier atoms and obtained theoretical estimates for the relevant binding energies; these investigations led to the development of the so-called *Thomas–Fermi model* of the atom, which was later extended so that it could be applied to molecules, solids, and nuclei as well.[15]

Thus, the whole structure of statistical mechanics was overhauled by the introduction of the concept of indistinguishability of (identical) particles.[16] The statistical aspect of the problem, which was already there in view of the large number of particles present, was now augmented by another statistical aspect that arose from the probabilistic nature of the wave mechanical description. One had, therefore, to carry out a *two-fold* averaging of the dynamical variables over the states of the given system in order to obtain the relevant expectation values. That sort of a situation was bound to necessitate a reformulation of the ensemble theory itself, which was carried out step by step. First, Landau (1927) and von Neumann (1927) introduced the so-called *density matrix*, which was the quantum-mechanical analogue of the *density function* of the classical phase space; this was elaborated, both from statistical and quantum-mechanical points of view, by Dirac (1929–1931). Guided by the classical ensemble theory, these authors considered both *microcanonical* and *canonical* ensembles; the introduction of *grand canonical* ensembles in quantum statistics was made by Pauli (1927).[17]

The important question as to which particles would obey Bose–Einstein statistics and which Fermi–Dirac remained theoretically unsettled until Belinfante (1939) and Pauli (1940) discovered the vital connection between spin and statistics.[18] It turns out that those particles whose spin is an integral multiple of $\hbar$ obey Bose–Einstein statistics while those whose spin is a half-odd integral multiple of $\hbar$ obey Fermi–Dirac statistics. To date, no third category of particles has been discovered.

Apart from the foregoing milestones, several notable contributions toward the development of statistical mechanics have been made from time to time; however, most of those contributions were concerned with the development or perfection of mathematical techniques that make application of the basic formalism to actual physical problems more fruitful. A review of these developments is out of place here; they will be discussed at their appropriate place in the text.

[15] For an excellent review of this model, see March (1957).

[16] Of course, in many a situation where the wave nature of the particles is not so important, classical statistics continue to apply.

[17] A detailed treatment of this development has been given by Kramers (1938).

[18] See also Lüders and Zumino (1958).

1 The Statistical Basis of Thermodynamics

In the annals of thermal physics, the 1850s mark a very definite epoch. By that time the science of thermodynamics, which grew essentially out of an experimental study of the macroscopic behavior of physical systems, had become, through the work of Carnot, Joule, Clausius, and Kelvin, a secure and stable discipline of physics. The theoretical conclusions following from the first two laws of thermodynamics were found to be in very good agreement with the corresponding experimental results.[1] At the same time, the kinetic theory of gases, which aimed at explaining the macroscopic behavior of gaseous systems in terms of the motion of their molecules and had so far thrived more on speculation than calculation, began to emerge as a real, mathematical theory. Its initial successes were glaring; however, a real contact with thermodynamics could not be made until about 1872 when Boltzmann developed his H-theorem and thereby established a direct connection between entropy on one hand and molecular dynamics on the other. Almost simultaneously, the conventional (kinetic) theory began giving way to its more sophisticated successor — the ensemble theory. The power of the techniques that finally emerged reduced thermodynamics to the status of an "essential" consequence of the get-together of the *statistics* and the *mechanics* of the molecules constituting a given physical system. It was then natural to give the resulting formalism the name *Statistical Mechanics*.

As a preparation toward the development of the formal theory, we start with a few general considerations regarding the statistical nature of a macroscopic system. These considerations will provide ground for a statistical interpretation of thermodynamics. It may be mentioned here that, unless a statement is made to the contrary, the system under study is supposed to be in one of its equilibrium states.

1.1 The macroscopic and the microscopic states

We consider a physical system composed of N identical particles confined to a space of volume V. In a typical case, N would be an extremely large number — generally, of order 10^{23}. In view of this, it is customary to carry out analysis in the so-called *thermodynamic limit*, namely $N \to \infty, V \to \infty$ (such that the ratio N/V, which represents the *particle density n*, stays fixed at a preassigned value). In this limit, the *extensive* properties of the system

[1]The third law, which is also known as *Nernst's heat theorem*, did not arrive until about 1906. For a general discussion of this law, see Simon (1930) and Wilks (1961); these references also provide an extensive bibliography on this subject.

© 2011 Elsevier Ltd. All rights reserved.

become directly proportional to the size of the system (i.e., proportional to N or to V), while the *intensive* properties become independent thereof; the particle density, of course, remains an important parameter for all physical properties of the system.

Next we consider the total energy E of the system. If the particles comprising the system could be regarded as noninteracting, the total energy E would be equal to the sum of the energies ε_i of the individual particles:

$$E = \sum_i n_i \varepsilon_i, \tag{1}$$

where n_i denotes the number of particles each with energy ε_i. Clearly,

$$N = \sum_i n_i. \tag{2}$$

According to quantum mechanics, the single-particle energies ε_i are discrete and their values depend crucially on the volume V to which the particles are confined. Accordingly, the possible values of the total energy E are also discrete. However, for large V, the spacing of the different energy values is so small in comparison with the total energy of the system that the parameter E might well be regarded as a *continuous* variable. This would be true even if the particles were mutually interacting; of course, in that case the total energy E cannot be written in the form (1).

The specification of the actual values of the parameters N, V, and E then defines a *macrostate* of the given system.

At the molecular level, however, a large number of possibilities still exist because at that level there will *in general* be a large number of different ways in which the macrostate (N, V, E) of the given system can be realized. In the case of a noninteracting system, since the total energy E consists of a simple sum of the N single-particle energies ε_i, there will obviously be a large number of different ways in which the individual ε_i can be chosen so as to make the total energy equal to E. In other words, there will be a large number of different ways in which the total energy E of the system can be distributed among the N particles constituting it. Each of these (different) ways specifies a *microstate*, or *complexion*, of the given system. In general, the various microstates, or complexions, of a given system can be identified with the independent solutions $\psi(\boldsymbol{r}_1, \ldots, \boldsymbol{r}_N)$ of the Schrödinger equation of the system, corresponding to the eigenvalue E of the relevant Hamiltonian. In any case, to a given macrostate of the system there does in general correspond a large number of microstates and it seems natural to assume, when there are no other constraints, that at any time t the system is *equally likely* to be in any one of these microstates. This assumption forms the backbone of our formalism and is generally referred to as the postulate of "equal *a priori* probabilities" for all microstates consistent with a given macrostate.

The actual number of all possible microstates will, of course, be a function of N, V, and E and may be denoted by the symbol $\Omega(N, V, E)$; the dependence on V comes in because the possible values ε_i of the single-particle energy ε are themselves a function of this parameter.[2] Remarkably enough, it is from the magnitude of the number Ω, and from its dependence on the parameters N, V, and E, that complete thermodynamics of the given system can be derived!

We shall not stop here to discuss the ways in which the number $\Omega(N,V,E)$ can be computed; we shall do that only after we have developed our considerations sufficiently so that we can carry out further derivations from it. First we have to discover the manner in which this number is related to any of the leading thermodynamic quantities. To do this, we consider the problem of "thermal contact" between two given physical systems, in the hope that this consideration will bring out the true nature of the number Ω.

1.2 Contact between statistics and thermodynamics: physical significance of the number $\Omega(N,V,E)$

We consider two physical systems, A_1 and A_2, which are separately in equilibrium; see Figure 1.1. Let the macrostate of A_1 be represented by the parameters N_1, V_1, and E_1 so that it has $\Omega_1(N_1, V_1, E_1)$ possible microstates, and the macrostate of A_2 be represented by the parameters N_2, V_2, and E_2 so that it has $\Omega_2(N_2, V_2, E_2)$ possible microstates. The mathematical form of the function Ω_1 may not be the same as that of the function Ω_2, because that ultimately depends on the nature of the system. We do, of course, believe that all thermodynamic properties of the systems A_1 and A_2 can be derived from the functions $\Omega_1(N_1, V_1, E_1)$ and $\Omega_2(N_2, V_2, E_2)$, respectively.

We now bring the two systems into thermal contact with each other, thus allowing the possibility of exchange of energy between the two; this can be done by sliding in a conducting wall and removing the impervious one. For simplicity, the two systems are still separated by a rigid, impenetrable wall, so that the respective volumes V_1 and V_2 and the respective particle numbers N_1 and N_2 remain fixed. The energies E_1 and E_2, however, become variable and the only condition that restricts their variation is

$$E^{(0)} = E_1 + E_2 = \text{const.} \tag{1}$$

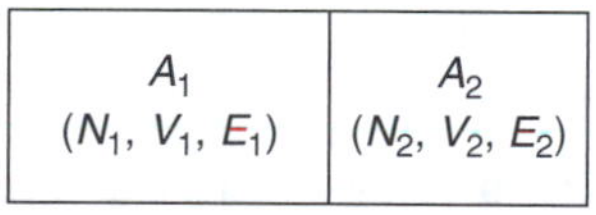

FIGURE 1.1 Two physical systems being brought into thermal contact.

[2]It may be noted that the manner in which the ε_i depend on V is itself determined by the nature of the system. For instance, it is not the same for relativistic systems as it is for nonrelativistic ones; compare, for instance, the cases dealt with in Section 1.4 and in Problem 1.7. We should also note that, *in principle*, the dependence of Ω on V arises from the fact that it is the *physical dimensions* of the container that appear in the boundary conditions imposed on the wave functions of the system.

Here, $E^{(0)}$ denotes the energy of the composite system $A^{(0)} (\equiv A_1 + A_2)$; the energy of interaction between A_1 and A_2, if any, is being neglected. Now, at any time t, the subsystem A_1 is equally likely to be in any one of the $\Omega_1(E_1)$ microstates while the subsystem A_2 is equally likely to be in any one of the $\Omega_2(E_2)$ microstates; therefore, the composite system $A^{(0)}$ is equally likely to be in any one of the

$$\Omega_1(E_1)\Omega_2(E_2) = \Omega_1(E_1)\Omega_2(E^{(0)} - E_1) = \Omega^{(0)}(E^{(0)}, E_1) \tag{2}$$

microstates.[3] Clearly, the number $\Omega^{(0)}$ itself varies with E_1. The question now arises: at what value of E_1 will the composite system be in equilibrium? In other words, how far will the energy exchange go in order to bring the subsystems A_1 and A_2 into mutual equilibrium?

We assert that this will happen at that value of E_1 which *maximizes* the number $\Omega^{(0)}(E^{(0)}, E_1)$. The philosophy behind this assertion is that a physical system, left to itself, proceeds naturally in a direction that enables it to assume an ever-increasing number of microstates until it finally settles down in a macrostate that affords the *largest possible* number of microstates. Statistically speaking, we regard a macrostate with a larger number of microstates as a more probable state, and the one with the largest number of microstates as the most probable one. Detailed studies show that, for a typical system, the number of microstates pertaining to any macrostate that departs even slightly from the most probable one is "orders of magnitude" smaller than the number pertaining to the latter. Thus, the most probable state of a system is *the* macrostate in which the system spends an "overwhelmingly" large fraction of its time. It is then natural to identify this state with the *equilibrium* state of the system.

Denoting the equilibrium value of E_1 by $\overline{E}_1$ (and that of E_2 by $\overline{E}_2$), we obtain, on maximizing $\Omega^{(0)}$,

$$\left(\frac{\partial \Omega_1(E_1)}{\partial E_1}\right)_{E_1=\overline{E}_1} \Omega_2(\overline{E}_2) + \Omega_1(\overline{E}_1)\left(\frac{\partial \Omega_2(E_2)}{\partial E_2}\right)_{E_2=\overline{E}_2} \cdot \frac{\partial E_2}{\partial E_1} = 0.$$

Since $\partial E_2/\partial E_1 = -1$, see equation (1), the foregoing condition can be written as

$$\left(\frac{\partial \ln \Omega_1(E_1)}{\partial E_1}\right)_{E_1=\overline{E}_1} = \left(\frac{\partial \ln \Omega_2(E_2)}{\partial E_2}\right)_{E_2=\overline{E}_2}.$$

Thus, our condition for equilibrium reduces to the equality of the parameters β_1 and β_2 of the subsystems A_1 and A_2, respectively, where β is defined by

$$\beta \equiv \left(\frac{\partial \ln \Omega(N, V, E)}{\partial E}\right)_{N,V,E=\overline{E}}. \tag{3}$$

[3] It is obvious that the macrostate of the composite system $A^{(0)}$ has to be defined by two energies, namely E_1 and E_2 (or else $E^{(0)}$ and E_1).

We thus find that when two physical systems are brought into thermal contact, which allows an exchange of energy between them, this exchange continues *until* the equilibrium values $\overline{E}_1$ and $\overline{E}_2$ of the variables E_1 and E_2 are reached. Once these values are reached, there is no more net exchange of energy between the two systems; the systems are then said to have attained a state of thermal equilibrium. According to our analysis, this happens only when the respective values of the parameter β, namely β_1 and β_2, become equal.[4] It is then natural to expect that the parameter β is somehow related to the *thermodynamic temperature* T of a given system. To determine this relationship, we recall the thermodynamic formula

$$\left(\frac{\partial S}{\partial E}\right)_{N,V} = \frac{1}{T}, \tag{4}$$

where S is the entropy of the system in question. Comparing equations (3) and (4), we conclude that an intimate relationship exists between the thermodynamic quantity S and the statistical quantity Ω; we may, in fact, write for any physical system

$$\frac{\Delta S}{\Delta(\ln \Omega)} = \frac{1}{\beta T} = \text{const.} \tag{5}$$

This correspondence was first established by Boltzmann who also believed that, since the relationship between the thermodynamic approach and the statistical approach seems to be of a fundamental character, the constant appearing in (5) must be a *universal constant*. It was Planck who first wrote the explicit formula

$$S = k \ln \Omega, \tag{6}$$

without any additive constant S_0. As it stands, formula (6) determines the *absolute* value of the entropy of a given physical system in terms of the total number of microstates accessible to it in conformity with the given macrostate. The zero of entropy then corresponds to the special state for which only one microstate is accessible ($\Omega = 1$) — the so-called "unique configuration"; the statistical approach thus provides a theoretical basis for the third law of thermodynamics as well. Formula (6) is of fundamental importance in physics; it provides a bridge between the microscopic and the macroscopic.

Now, in the study of the second law of thermodynamics we are told that the law of increase of entropy is related to the fact that the energy content of the universe, in its natural course, is becoming less and less available for conversion into work; accordingly, the entropy of a given system may be regarded as a measure of the so-called disorder or chaos prevailing in the system. Formula (6) tells us how disorder arises microscopically. Clearly, disorder is a manifestation of the largeness of the number of microstates the system can have. The larger the choice of microstates, the lesser the degree of predictability and hence the increased level of disorder in the system. Complete order prevails when and

[4] This result may be compared with the so-called "zeroth law of thermodynamics," which stipulates the existence of a *common* parameter T for two or more physical systems in thermal equilibrium.

only when the system has no other choice but to be in a unique state ($\Omega = 1$); this, in turn, corresponds to a state of vanishing entropy.

By equations (5) and (6), we also have

$$\beta = \frac{1}{kT}. \tag{7}$$

The universal constant k is generally referred to as the *Boltzmann constant*. In Section 1.4 we shall discover how k is related to the gas constant R and the Avogadro number N_A; see equation (1.4.3).[5]

1.3 Further contact between statistics and thermodynamics

In continuation of the preceding considerations, we now examine a more elaborate exchange between the subsystems A_1 and A_2. If we assume that the wall separating the two subsystems is movable as well as conducting, then the respective volumes V_1 and V_2 (of subsystems A_1 and A_2) also become variable; indeed, the total volume $V^{(0)} (= V_1 + V_2)$ remains constant, so that effectively we have only one more independent variable. Of course, the wall is still assumed to be impenetrable to particles, so the numbers N_1 and N_2 remain fixed. Arguing as before, the state of equilibrium for the composite system $A^{(0)}$ will obtain when the number $\Omega^{(0)}(V^{(0)}, E^{(0)}; V_1, E_1)$ attains its largest value; that is, when not only

$$\left(\frac{\partial \ln \Omega_1}{\partial E_1}\right)_{N_1, V_1;\, E_1 = \bar{E}_1} = \left(\frac{\partial \ln \Omega_2}{\partial E_2}\right)_{N_2, V_2;\, E_2 = \bar{E}_2}, \tag{1a}$$

but also

$$\left(\frac{\partial \ln \Omega_1}{\partial V_1}\right)_{N_1, E_1;\, V_1 = \bar{V}_1} = \left(\frac{\partial \ln \Omega_2}{\partial V_2}\right)_{N_2, E_2;\, V_2 = \bar{V}_2}. \tag{1b}$$

Our conditions for equilibrium now take the form of an equality between the pair of parameters (β_1, η_1) of the subsystem A_1 and the parameters (β_2, η_2) of the subsystem A_2 where, by definition,

$$\eta \equiv \left(\frac{\partial \ln \Omega(N, V, E)}{\partial V}\right)_{N, E, V = \bar{V}}. \tag{2}$$

Similarly, if A_1 and A_2 came into contact through a wall that allowed an exchange of particles as well, the conditions for equilibrium would be further augmented by the equality

[5] We follow the notation whereby equation (1.4.3) means equation (3) of Section 1.4. However, while referring to an equation in the same section, we will omit the mention of the section number.

of the parameter ζ_1 of subsystem A_1 and the parameter ζ_2 of subsystem A_2 where, by definition,

$$\zeta \equiv \left(\frac{\partial \ln \Omega(N, V, E)}{\partial N} \right)_{V, E, N=\overline{N}}. \tag{3}$$

To determine the physical meaning of the parameters η and ζ, we make use of equation (1.2.6) and the basic formula of thermodynamics, namely

$$dE = T\,dS - P\,dV + \mu\,dN, \tag{4}$$

where P is the thermodynamic pressure and μ the chemical potential of the given system. It follows that

$$\eta = \frac{P}{kT} \quad \text{and} \quad \zeta = -\frac{\mu}{kT}. \tag{5}$$

From a physical point of view, these results are completely satisfactory because, thermodynamically as well, the conditions of equilibrium between two systems A_1 and A_2, if the wall separating them is both conducting and movable (thus making their respective energies and volumes variable), are indeed the same as the ones contained in equations (1a) and (1b), namely

$$T_1 = T_2 \quad \text{and} \quad P_1 = P_2. \tag{6}$$

On the other hand, if the two systems can exchange particles as well as energy but have their volumes fixed, the conditions of equilibrium, obtained thermodynamically, are indeed

$$T_1 = T_2 \quad \text{and} \quad \mu_1 = \mu_2. \tag{7}$$

And finally, if the exchange is such that all three (macroscopic) parameters become variable, then the conditions of equilibrium become

$$T_1 = T_2, \quad P_1 = P_2, \quad \text{and} \quad \mu_1 = \mu_2. \tag{8}$$ [6]

It is gratifying that these conclusions are identical to the ones following from statistical considerations.

Combining the results of the foregoing discussion, we arrive at the following recipe for deriving thermodynamics from a statistical beginning: determine, for the macrostate (N, V, E) of the given system, the number of all possible microstates accessible to the system; call this number $\Omega(N, V, E)$. Then, the entropy of the system in that state follows from

[6] It may be noted that the same would be true for any two parts of a single thermodynamic system; consequently, in equilibrium, the parameters T, P, and μ would be constant *throughout* the system.

the fundamental formula

$$S(N, V, E) = k \ln \Omega(N, V, E), \tag{9}$$

while the leading *intensive fields*, namely temperature, pressure, and chemical potential, are given by

$$\left(\frac{\partial S}{\partial E}\right)_{N,V} = \frac{1}{T}; \quad \left(\frac{\partial S}{\partial V}\right)_{N,E} = \frac{P}{T}; \quad \left(\frac{\partial S}{\partial N}\right)_{V,E} = -\frac{\mu}{T}. \tag{10}$$

Alternatively, we can write[7]

$$P = \left(\frac{\partial S}{\partial V}\right)_{N,E} \Big/ \left(\frac{\partial S}{\partial E}\right)_{N,V} = -\left(\frac{\partial E}{\partial V}\right)_{N,S} \tag{11}$$

and

$$\mu = -\left(\frac{\partial S}{\partial N}\right)_{V,E} \Big/ \left(\frac{\partial S}{\partial E}\right)_{N,V} = \left(\frac{\partial E}{\partial N}\right)_{V,S}, \tag{12}$$

while

$$T = \left(\frac{\partial E}{\partial S}\right)_{N,V}. \tag{13}$$

Formulae (11) through (13) follow equally well from equation (4). The evaluation of P, μ, and T from these formulae indeed requires that the energy E be expressed as a function of the quantities N, V, and S; this should, in principle, be possible once S is known as a function of N, V, and E.

The rest of the thermodynamics follows straightforwardly; see Appendix H. For instance, the Helmholtz free energy A, the Gibbs free energy G, and the enthalpy H are given by

$$A = E - TS, \tag{14}$$

$$G = A + PV = E - TS + PV$$
$$= \mu N \tag{15}$$
[8]

[7]In writing these formulae, we have made use of the well-known relationship in partial differential calculus, namely that "if three variables x, y, and z are mutually related, then (see Appendix H)

$$\left(\frac{\partial x}{\partial y}\right)_z \left(\frac{\partial y}{\partial z}\right)_x \left(\frac{\partial z}{\partial x}\right)_y = -1."$$

[8]The relation $E - TS + PV = \mu N$ follows directly from (4). For this, all we have to do is to regard the given system as having grown to its present size in a gradual manner, such that the intensive parameters, T, P, and μ stayed constant throughout the process while the extensive parameters N, V, and E (and hence S) grew *proportionately* with one another.

and

$$H = E + PV = G + TS. \tag{16}$$

The specific heat at constant volume, C_V, and the one at constant pressure, C_P, would be given by

$$C_V \equiv T\left(\frac{\partial S}{\partial T}\right)_{N,V} = \left(\frac{\partial E}{\partial T}\right)_{N,V} \tag{17}$$

and

$$C_P \equiv T\left(\frac{\partial S}{\partial T}\right)_{N,P} = \left(\frac{\partial (E+PV)}{\partial T}\right)_{N,P} = \left(\frac{\partial H}{\partial T}\right)_{N,P}. \tag{18}$$

1.4 The classical ideal gas

To illustrate the approach developed in the preceding sections, we shall now derive the various thermodynamic properties of a classical ideal gas composed of monatomic molecules. The main reason why we choose this highly specialized system for consideration is that it affords an explicit, though asymptotic, evaluation of the number $\Omega(N,V,E)$. This example becomes all the more instructive when we find that its study enables us, in a most straightforward manner, to identify the Boltzmann constant k in terms of other physical constants; see equation (3). Moreover, the behavior of this system serves as a useful reference with which the behavior of other physical systems, especially real gases (with or without quantum effects), can be compared. And, indeed, in the limit of high temperatures and low densities the ideal-gas behavior becomes typical of most real systems.

Before undertaking a detailed study of this case it appears worthwhile to make a remark that applies to all *classical* systems composed of *noninteracting* particles, irrespective of the internal structure of the particles. This remark is related to the explicit dependence of the number $\Omega(N,V,E)$ on V and hence to the *equation of state* of these systems. Now, if there do not exist any spatial correlations among the particles, that is, if the probability of any one of them being found in a particular region of the available space is completely independent of the location of the other particles,[9] then the total number of ways in which the N particles can be spatially distributed in the system will be simply equal to the product of the numbers of ways in which the individual particles can be accommodated in the same space independently of one another. With N and E fixed, each of these numbers will be directly proportional to V, the volume of the container; accordingly, the total number of ways will be directly proportional to the Nth power of V:

$$\Omega(N,E,V) \propto V^N. \tag{1}$$

[9]This will be true if (i) the mutual interactions among particles are negligible, and (ii) the wave packets of individual particles do not significantly overlap (or, in other words, the quantum effects are also negligible).

Combined with equations (1.3.9) and (1.3.10), this gives

$$\frac{P}{T} = k\left(\frac{\partial \ln \Omega(N,E,V)}{\partial V}\right)_{N,E} = k\frac{N}{V}. \tag{2}$$

If the system contains n moles of the gas, then $N = nN_A$, where N_A is the *Avogadro number*. Equation (2) then becomes

$$PV = NkT = nRT \quad (R = kN_A), \tag{3}$$

which is the famous *ideal-gas law*, R being the *gas constant* per mole. Thus, for any classical system composed of noninteracting particles the ideal-gas law holds.

For deriving other thermodynamic properties of this system, we require a detailed knowledge of the way Ω depends on the parameters N, V, and E. The problem essentially reduces to determining the total number of ways in which equations (1.1.1) and (1.1.2) can be mutually satisfied. In other words, we have to determine the total number of (independent) ways of satisfying the equation

$$\sum_{r=1}^{3N} \varepsilon_r = E, \tag{4}$$

where ε_r are the energies associated with the various degrees of freedom of the N particles. The reason why this number should depend on the parameters N and E is quite obvious. Nevertheless, this number also depends on the "spectrum of values" that the variables ε_r can assume; it is through this spectrum that the dependence on V comes in. Now, the energy eigenvalues for a *free, nonrelativistic* particle confined to a cubical box of side L ($V = L^3$), under the condition that the wave function $\psi(\boldsymbol{r})$ vanishes everywhere on the boundary, are given by

$$\varepsilon(n_x, n_y, n_z) = \frac{h^2}{8mL^2}(n_x^2 + n_y^2 + n_z^2); \quad n_x, n_y, n_z = 1, 2, 3, \ldots, \tag{5}$$

where h is Planck's constant and m the mass of the particle. The number of distinct eigenfunctions (or microstates) for a particle of energy ε would, therefore, be equal to the number of independent, positive-integral solutions of the equation

$$\left(n_x^2 + n_y^2 + n_z^2\right) = \frac{8mV^{2/3}\varepsilon}{h^2} = \varepsilon^*. \tag{6}$$

We may denote this number by $\Omega(1, \varepsilon, V)$. Extending the argument, it follows that the desired number $\Omega(N, E, V)$ would be equal to the number of independent, positive-integral solutions of the equation

$$\sum_{r=1}^{3N} n_r^2 = \frac{8mV^{2/3}E}{h^2} = E^*, \quad \text{say.} \tag{7}$$

An important result follows straightforwardly from equation (7), even before the number $\Omega(N,E,V)$ is explicitly evaluated. From the nature of the expression appearing on the right side of this equation, we conclude that the volume V and the energy E of the system enter into the expression for Ω in the form of the combination $(V^{2/3}E)$. Consequently,

$$S(N,V,E) \equiv S\left(N, V^{2/3}E\right). \tag{8}$$

Hence, for the constancy of S and N, which defines a *reversible adiabatic* process,

$$V^{2/3}E = \text{const.} \tag{9}$$

Equation (1.3.11) then gives

$$P = -\left(\frac{\partial E}{\partial V}\right)_{N,S} = \frac{2}{3}\frac{E}{V}, \tag{10}$$

that is, the pressure of a system of nonrelativistic, noninteracting particles is precisely equal to two-thirds of its energy density.[10] It should be noted here that, since an explicit computation of the number Ω has not yet been done, results (9) and (10) hold for *quantum* as well as *classical* statistics; equally general is the result obtained by combining these, namely

$$PV^{5/3} = \text{const.}, \tag{11}$$

which tells us how P varies with V during a *reversible adiabatic* process.

We shall now attempt to evaluate the number Ω. In this evaluation we shall explicitly assume the particles to be *distinguishable*, so that if a particle in state i gets interchanged with a particle in state j the resulting microstate is counted as distinct. Consequently, the number $\Omega(N,V,E)$, or better $\Omega_N(E^*)$ (see equation (7)), is equal to the number of positive-integral lattice points lying on the surface of a $3N$-dimensional sphere of radius $\sqrt{E^*}$.[11] Clearly, this number will be an extremely irregular function of E^*, in that for two given values of E^* that may be very close to one another, the values of this number could be very different. In contrast, the number $\Sigma_N(E^*)$, which denotes the number of positive-integral lattice points lying *on or within* the surface of a $3N$-dimensional sphere of radius $\sqrt{E^*}$, will be much less irregular. In terms of our physical problem, this would correspond to the number, $\Sigma(N,V,E)$, of microstates of the given system consistent with *all* macrostates characterized by the specified values of the parameters N and V but having energy *less*

[10] Combining (10) with (2), we obtain for the classical ideal gas: $E = \frac{3}{2}NkT$. Accordingly, equation (9) reduces to the well-known thermodynamic relationship: $V^{\gamma-1}T = \text{const.}$, which holds during a *reversible adiabatic* process, with $\gamma = \frac{5}{3}$.

[11] If the particles are regarded as *indistinguishable*, the evaluation of the number Ω by counting lattice points becomes quite intricate. The problem is then solved by having recourse to the theory of "partitions of numbers"; see Auluck and Kothari (1946).

than or equal to E; that is,

$$\Sigma(N,V,E) = \sum_{E' \leq E} \Omega(N,V,E') \tag{12}$$

or

$$\Sigma_N(E^*) = \sum_{E^{*'} \leq E^*} \Omega_N(E^{*'}). \tag{13}$$

Of course, the number Σ will also be somewhat irregular; however, we expect that its asymptotic behavior, as $E^* \to \infty$, will be a lot smoother than that of Ω. We shall see in the sequel that the thermodynamics of the system follows equally well from the number Σ as from Ω.

To appreciate the point made here, let us digress a little to examine the behavior of the numbers $\Omega_1(\varepsilon^*)$ and $\Sigma_1(\varepsilon^*)$, which correspond to the case of a single particle confined to the given volume V. The *exact* values of these numbers, for $\varepsilon^* \leq 10{,}000$, can be extracted from a table compiled by Gupta (1947). The wild irregularities of the number $\Omega_1(\varepsilon^*)$ can hardly be missed. The number $\Sigma_1(\varepsilon^*)$, on the other hand, exhibits a much smoother asymptotic behavior. From the geometry of the problem, we note that, *asymptotically*, $\Sigma_1(\varepsilon^*)$ should be equal to the volume of an octant of a three-dimensional sphere of radius $\sqrt{\varepsilon^*}$, that is,

$$\lim_{\varepsilon^* \to \infty} \frac{\Sigma_1(\varepsilon^*)}{(\pi/6)\varepsilon^{*3/2}} = 1. \tag{14}$$

A more detailed analysis shows that, to the next approximation (see Pathria, 1966),

$$\Sigma_1(\varepsilon^*) \approx \frac{\pi}{6}\varepsilon^{*3/2} - \frac{3\pi}{8}\varepsilon^*; \tag{15}$$

the correction term arises from the fact that the volume of an octant somewhat overestimates the number of desired lattice points, for it includes, partly though, some points with one or more coordinates equal to zero. Figure 1.2 shows a histogram of the actual values of $\Sigma_1(\varepsilon^*)$ for ε^* lying between 200 and 300; the theoretical estimate (15) is also shown. In the figure, we have also included a histogram of the actual values of the corresponding number of microstates, $\Sigma_1'(\varepsilon^*)$, when the quantum numbers n_x, n_y, and n_z can assume the value zero as well. In the latter case, the volume of an octant somewhat underestimates the number of desired lattice points; we now have

$$\Sigma_1'(\varepsilon^*) \approx \frac{\pi}{6}\varepsilon^{*3/2} + \frac{3\pi}{8}\varepsilon^*. \tag{16}$$

Asymptotically, however, the number $\Sigma_1'(\varepsilon^*)$ also satisfies equation (14).

Returning to the N-particle problem, the number $\Sigma_N(E^*)$ should be *asymptotically* equal to the "volume" of the "positive compartment" of a $3N$-dimensional sphere of

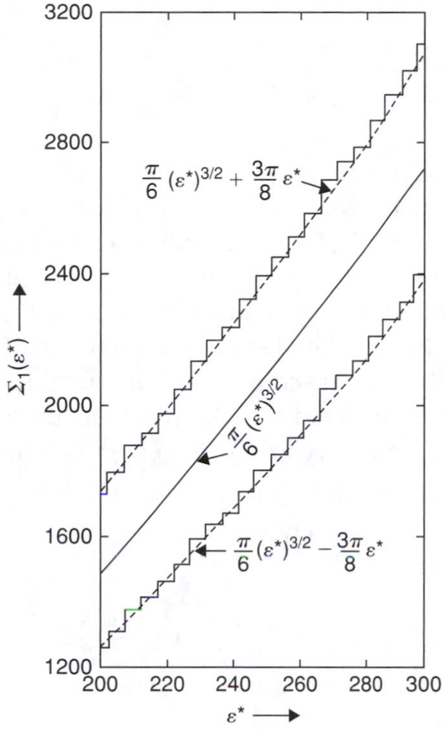

FIGURE 1.2 Histograms showing the actual number of microstates available to a particle in a cubical enclosure; the lower histogram corresponds to the so-called Dirichlet boundary conditions, while the upper one corresponds to the Neumann boundary conditions (see Appendix A). The corresponding theoretical estimates, (15) and (16), are shown by dashed lines; the customary estimate, equation (14), is shown by a solid line.

radius $\sqrt{E^*}$. Referring to equation (C.7a) of Appendix C, we obtain

$$\Sigma_N(E^*) \approx \left(\frac{1}{2}\right)^{3N} \left\{ \frac{\pi^{3N/2}}{(3N/2)!} E^{*3N/2} \right\}$$

which, on substitution for E^*, gives

$$\Sigma(N, V, E) \approx \left(\frac{V}{h^3}\right)^N \frac{(2\pi mE)^{3N/2}}{(3N/2)!}. \tag{17}$$

Taking logarithms and applying Stirling's formula, (B.29) in Appendix B,

$$\ln(n!) \approx n \ln n - n \quad (n \gg 1), \tag{18}$$

we get

$$\ln \Sigma(N,V,E) \approx N \ln\left[\frac{V}{h^3}\left(\frac{4\pi mE}{3N}\right)^{3/2}\right] + \frac{3}{2}N. \tag{19}$$

For deriving the thermodynamic properties of the given system we must somehow fix the precise value of, or limits for, the energy of the system. In view of the extremely irregular nature of the function $\Omega(N,V,E)$, the specification of a precise value for the energy of the system cannot be justified on physical grounds, for that would never yield well-behaved expressions for the thermodynamic functions of the system. From a practical point of view, too, an absolutely isolated system is too much of an idealization. In the real world, almost every system has some contact with its surroundings, however little it may be; as a result, its energy cannot be defined sharply.[12] Of course, the effective width of the range over which the energy may vary would, in general, be small in comparison with the mean value of the energy. Let us specify this range by the limits $\left(E - \frac{1}{2}\Delta\right)$ and $\left(E + \frac{1}{2}\Delta\right)$ where, by assumption, $\Delta \ll E$; typically, $\Delta/E = O(1/\sqrt{N})$. The corresponding number of microstates, $\Gamma(N,V,E;\Delta)$, is then given by

$$\Gamma(N,V,E;\Delta) \simeq \frac{\partial \Sigma(N,V,E)}{\partial E}\Delta \approx \frac{3N}{2}\frac{\Delta}{E}\Sigma(N,V,E), \tag{17a}$$

which gives

$$\ln \Gamma(N,V,E;\Delta) \approx N \ln\left[\frac{V}{h^3}\left(\frac{4\pi mE}{3N}\right)^{3/2}\right] + \frac{3}{2}N + \left\{\ln\left(\frac{3N}{2}\right) + \ln\left(\frac{\Delta}{E}\right)\right\}. \tag{19a}$$

Now, for $N \gg 1$, the first term in the curly bracket is negligible in comparison with any of the terms outside this bracket, for $\lim_{N\to\infty} (\ln N)/N = 0$. Furthermore, for any reasonable value of Δ/E, the same is true of the second term in this bracket.[13] Hence, for all practical purposes,

$$\ln \Gamma \approx \ln \Sigma \approx N \ln\left[\frac{V}{h^3}\left(\frac{4\pi mE}{3N}\right)^{3/2}\right] + \frac{3}{2}N. \tag{20}$$

We thus arrive at the baffling result that, for all practical purposes, the actual width of the range allowed for the energy of the system does not make much difference; the energy could lie between $\left(E - \frac{1}{2}\Delta\right)$ and $\left(E + \frac{1}{2}\Delta\right)$ or equally well between 0 and E. The reason underlying this situation is that the rate at which the number of microstates of the system

[12]Actually, the very act of making measurements on a system brings about, inevitably, a contact between the system and its surroundings.

[13]It should be clear that, while Δ/E is much less than 1, it must not tend to 0, for that would make $\Gamma \to 0$ and $\ln \Gamma \to -\infty$. A situation of that kind would be too artificial and would have nothing to do with reality. Actually, in most physical systems, $\Delta/E = O(N^{-1/2})$, whereby $\ln(\Delta/E)$ becomes of order $\ln N$, which again is negligible in comparison with the terms outside the curly bracket.

increases with energy is so fantastic, see equation (17), that even if we allow *all* values of energy between zero and a particular value E, it is only the "immediate neighborhood" of E that makes an overwhelmingly dominant contribution to this number! And since we are finally concerned only with the logarithm of this number, even the "width" of that neighborhood is inconsequential!

The stage is now set for deriving the thermodynamics of our system. First of all, we have

$$S(N,V,E) = k\ln\Gamma = Nk\ln\left[\frac{V}{h^3}\left(\frac{4\pi mE}{3N}\right)^{3/2}\right] + \frac{3}{2}Nk, \tag{21}$$ [14]

which can be inverted to give

$$E(S,V,N) = \frac{3h^2N}{4\pi mV^{2/3}}\exp\left(\frac{2S}{3Nk} - 1\right). \tag{22}$$

The temperature of the gas then follows with the help of formula (1.3.10) or (1.3.13), which leads to the energy–temperature relationship

$$E = N\left(\frac{3}{2}kT\right) = n\left(\frac{3}{2}RT\right), \tag{23}$$

where n is the number of moles of the gas. The specific heat at constant volume now follows with the help of formula (1.3.17):

$$C_V = \left(\frac{\partial E}{\partial T}\right)_{N,V} = \frac{3}{2}Nk = \frac{3}{2}nR. \tag{24}$$

For the equation of state, we obtain

$$P = -\left(\frac{\partial E}{\partial V}\right)_{N,S} = \frac{2}{3}\frac{E}{V}, \tag{25}$$

which agrees with our earlier result (10). Combined with (23), this gives

$$P = \frac{NkT}{V} \quad \text{or} \quad PV = nRT, \tag{26}$$

which is the same as (3). The specific heat at constant pressure is given by, see (1.3.18),

$$C_P = \left(\frac{\partial(E+PV)}{\partial T}\right)_{N,P} = \frac{5}{2}nR, \tag{27}$$

[14] Henceforth, we shall replace the sign $\approx$, which characterizes the *asymptotic* character of a relationship, by the sign of equality because for most physical systems the asymptotic results are as good as exact.

so that, for the ratio of the two specific heats, we have

$$\gamma = C_P/C_V = \frac{5}{3}. \tag{28}$$

Now, suppose that the gas undergoes an *isothermal* change of state ($T = $ const. and $N = $ const.); then, according to (23), the total energy of the gas would remain constant while, according to (26), its pressure would vary inversely with volume (Boyle's law). The change in the entropy of the gas, between the initial state i and the final state f, would then be, see equation (21),

$$S_f - S_i = Nk \ln(V_f/V_i). \tag{29}$$

On the other hand, if the gas undergoes a *reversible adiabatic* change of state ($S = $ const. and $N = $ const.), then, according to (22) and (23), both E and T would vary as $V^{-2/3}$; so, according to (25) or (26), P would vary as $V^{-5/3}$. These results agree with the conventional thermodynamic ones, namely

$$PV^{\gamma} = \text{const.} \quad \text{and} \quad TV^{\gamma-1} = \text{const.}, \tag{30}$$

with $\gamma = \frac{5}{3}$. It may be noted that, thermodynamically, the change in E during an adiabatic process arises solely from the external work done by the gas on the surroundings or vice versa:

$$(dE)_{\text{adiab}} = -PdV = -\frac{2E}{3V}dV; \tag{31}$$

see equations (1.3.4) and (25). The dependence of E on V follows readily from this relationship.

The considerations of this section have clearly demonstrated the manner in which the thermodynamics of a macroscopic system can be derived from the multiplicity of its microstates (as represented by the number Ω or Γ or Σ). The whole problem then hinges on an asymptotic enumeration of these numbers, which unfortunately is tractable only in a few idealized cases, such as the one considered in this section; see also Problems 1.7 and 1.8. Even in an idealized case like this, there remains an inadequacy that could not be detected in the derivations made so far; this relates to the *explicit* dependence of S on N. The discussion of the next section is intended not only to bring out this inadequacy but also to provide the necessary remedy for it.

1.5 The entropy of mixing and the Gibbs paradox

One thing we readily observe from expression (1.4.21) is that, contrary to what is logically desired, the entropy of an ideal gas, as given by this expression, is *not* an extensive

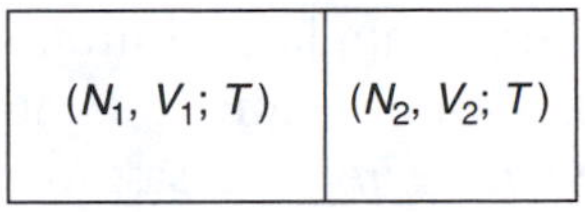

FIGURE 1.3 The mixing together of two ideal gases 1 and 2.

property of the system! That is, if we increase the size of the system by a factor α, keeping the intensive variables unchanged,[15] then the entropy of the system, which should also increase by the same factor α, does not do so; the presence of the $\ln V$ term in the expression affects the result adversely. This in a way means that the entropy of this system is different from the sum of the entropies of its parts, which is quite unphysical. A more common way of looking at this problem is to consider the so-called *Gibbs paradox*.

Gibbs visualized the mixing of two ideal gases 1 and 2, both being initially at the same temperature T; see Figure 1.3. Clearly, the temperature of the mixture would also be the same. Now, before the mixing took place, the respective entropies of the two gases were, see equations (1.4.21) and (1.4.23),

$$S_i = N_i k \ln V_i + \frac{3}{2} N_i k \left\{1 + \ln\left(\frac{2\pi m_i kT}{h^2}\right)\right\}; \quad i = 1, 2. \tag{1}$$

After the mixing has taken place, the total entropy would be

$$S_T = \sum_{i=1}^{2} \left[N_i k \ln V + \frac{3}{2} N_i k \left\{1 + \ln\left(\frac{2\pi m_i kT}{h^2}\right)\right\}\right], \tag{2}$$

where $V = V_1 + V_2$. Thus, the net increase in the value of S, which may be called the *entropy of mixing*, is given by

$$(\Delta S) = S_T - \sum_{i=1}^{2} S_i = k\left[N_1 \ln \frac{V_1 + V_2}{V_1} + N_2 \ln \frac{V_1 + V_2}{V_2}\right]; \tag{3}$$

the quantity ΔS is indeed positive, as it must be for an *irreversible* process like mixing. Now, in the special case when the initial particle densities of the two gases (and, hence, the particle density of the mixture) are also the same, equation (3) becomes

$$(\Delta S)^* = k\left[N_1 \ln \frac{N_1 + N_2}{N_1} + N_2 \ln \frac{N_1 + N_2}{N_2}\right], \tag{4}$$

which is again positive.

[15]This means an increase of the parameters N, V, and E to αN, αV, and αE, so that the energy per particle and the volume per particle remain unchanged.

So far, it seems all right. However, a paradoxical situation arises if we consider the mixing of two samples of the *same* gas. Once again, the entropies of the individual samples will be given by (1); of course, now $m_1 = m_2 = m$, say. And the entropy after mixing will be given by

$$S_T = Nk \ln V + \frac{3}{2} Nk \left\{1 + \ln\left(\frac{2\pi mkT}{h^2}\right)\right\}, \tag{2a}$$

where $N = N_1 + N_2$; note that this expression is numerically the same as (2), with $m_i = m$. Therefore, the entropy of mixing in this case will also be given by expression (3) and, if $N_1/V_1 = N_2/V_2 = (N_1 + N_2)/(V_1 + V_2)$, by expression (4). The last conclusion, however, is unacceptable because the mixing of two samples of the same gas, with a common initial temperature T and a common initial particle density n, is clearly a *reversible* process, for we can simply reinsert the partitioning wall into the system and obtain a situation that is in no way different from the one we had before mixing. Of course, we tacitly imply that in dealing with a system of identical particles we cannot track them down individually; all we can reckon with is their numbers. When two dissimilar gases, even with a common initial temperature T, and a common initial particle density n, mixed together the process was irreversible, for by reinserting the partitioning wall one would obtain two samples of the mixture and not the two gases that were originally present; to that case, expression (4) would indeed apply. However, in the present case, the corresponding result should be

$$(\Delta S)^*_{1\equiv 2} = 0. \tag{4a}$$ [16]

The foregoing result would also be consistent with the requirement that the entropy of a given system is equal to the sum of the entropies of its parts. Of course, we had already noticed that this is not ensured by expression (1.4.21). Thus, once again we are led to believe that there is something basically wrong with that expression.

To see how the above paradoxical situation can be avoided, we recall that, for the entropy of mixing of two samples of the same gas, with a common T and a common n, we were led to result (4), which can also be written as

$$(\Delta S)^* = S_T - (S_1 + S_2) \approx k[\ln\{(N_1 + N_2)!\} - \ln(N_1!) - \ln(N_2!)], \tag{4}$$

instead of the logical result (4a). A closer look at this expression shows that we would indeed obtain the correct result if our original expression for S were diminished by an *ad hoc* term, $k\ln(N!)$, for that would diminish S_1 by $k\ln(N_1!)$, S_2 by $k\ln(N_2!)$ and S_T by $k\ln\{(N_1 + N_2)!\}$, with the result that $(\Delta S)^*$ would turn out to be zero instead of the expression appearing in (4). Clearly, this would amount to an *ad hoc* reduction of the statistical numbers Γ and Σ by a factor $N!$. This is precisely the remedy proposed by Gibbs to avoid the paradox in question.

[16] In view of this, we fear that expression (3) may also be inapplicable to this case.

If we agree with the foregoing suggestion, then the modified expression for the entropy of a classical ideal gas would be

$$S(N,V,E) = Nk\ln\left[\frac{V}{Nh^3}\left(\frac{4\pi mE}{3N}\right)^{3/2}\right] + \frac{5}{2}Nk \tag{1.4.21a}$$

$$= Nk\ln\left(\frac{V}{N}\right) + \frac{3}{2}Nk\left\{\frac{5}{3} + \ln\left(\frac{2\pi mkT}{h^2}\right)\right\}, \tag{1a}$$

which indeed is truly extensive! If we now mix two samples of the same gas at a common initial temperature T, the entropy of mixing would be

$$(\Delta S)_{1\equiv 2} = k\left[(N_1+N_2)\ln\left(\frac{V_1+V_2}{N_1+N_2}\right) - N_1\ln\left(\frac{V_1}{N_1}\right) - N_2\ln\left(\frac{V_2}{N_2}\right)\right] \tag{3a}$$

and, if the initial particle densities of the samples were also equal, the result would be

$$(\Delta S)^*_{1\equiv 2} = 0. \tag{4a}$$

It may be noted that for the mixing of two dissimilar gases, the original expressions (3) and (4) would continue to hold even when (1.4.21) is replaced by (1.4.21a).[17] The paradox of Gibbs is thereby resolved.

Equation (1a) is generally referred to as the *Sackur–Tetrode* equation. We reiterate the fact that, by this equation, the entropy of the system does indeed become a truly extensive quantity. Thus, the very root of the trouble has been eliminated by the recipe of Gibbs. We shall discuss the physical implications of this recipe in Section 1.6; here, let us jot down some of its immediate consequences.

First of all, we note that the expression for the energy E of the gas, written as a function of N, V, and S, is also modified. We now have

$$E(N,V,S) = \frac{3h^2N^{5/3}}{4\pi mV^{2/3}}\exp\left(\frac{2S}{3Nk} - \frac{5}{3}\right), \tag{1.4.22a}$$

which, unlike its predecessor (1.4.22), makes energy too a truly extensive quantity. Of course, the thermodynamic results (1.4.23) through (1.4.31), derived in the previous section, remain unchanged. However, there are some that were intentionally left out, for they would come out correct only from the modified expression for $S(N,V,E)$ or $E(S,V,N)$. The most important of these is the chemical potential of the gas, for which we obtain

$$\mu \equiv \left(\frac{\partial E}{\partial N}\right)_{V,S} = E\left[\frac{5}{3N} - \frac{2S}{3N^2k}\right]. \tag{5}$$

[17] Because, in this case, the entropy S_T of the mixture would be diminished by $k\ln(N_1!N_2!)$, rather than by $k\ln\{(N_1+N_2)!\}$.

In view of equations (1.4.23) and (1.4.25), this becomes

$$\mu = \frac{1}{N}[E + PV - TS] \equiv \frac{G}{N}, \tag{6}$$

where G is the Gibbs free energy of the system. In terms of the variables N, V, and T, expression (5) takes the form

$$\mu(N, V, T) = kT \ln \left\{ \frac{N}{V} \left(\frac{h^2}{2\pi mkT} \right)^{3/2} \right\}. \tag{7}$$

Another quantity of importance is the Helmholtz free energy:

$$A = E - TS = G - PV = NkT \left[\ln \left\{ \frac{N}{V} \left(\frac{h^2}{2\pi mkT} \right)^{3/2} \right\} - 1 \right]. \tag{8}$$

It will be noted that, while A is an extensive property of the system, μ is intensive.

1.6 The "correct" enumeration of the microstates

In the preceding section we saw that an *ad hoc* diminution in the entropy of an N-particle system by an amount $k\ln(N!)$, which implies an *ad hoc* reduction in the number of microstates accessible to the system by a factor $(N!)$, was able to correct the unphysical features of some of our former expressions. It is now natural to ask: why, *in principle*, should the number of microstates, computed in Section 1.4, be reduced in this manner? The physical reason for doing so is that the particles constituting the given system are not only identical but also *indistinguishable*; accordingly, it is unphysical to label them as No. 1, No. 2, No. 3, and so on and to speak of their being *individually* in the various single-particle states ε_i. All we can sensibly speak of is their distribution over the states ε_i by *numbers*, that is, n_1 particles being in the state ε_1, n_2 in the state ε_2, and so on. Thus, the correct way of specifying a microstate of the system is through the distribution numbers $\{n_j\}$, and not through the statement as to "which particle is in which state." To elaborate the point, we may say that if we consider two microstates that differ from one another merely in an interchange of two particles in different energy states, then according to our original mode of counting we would regard these microstates as distinct; in view of the indistinguishability of the particles, however, these microstates are not distinct (for, physically, there exists no way whatsoever of distinguishing between them).[18]

[18]Of course, if an interchange took place among particles in the same energy state, then even our original mode of counting did not regard the two microstates as distinct.

Now, the total number of permutations that can be effected among N particles, distributed according to the set $\{n_i\}$, is

$$\frac{N!}{n_1!n_2!\ldots}, \tag{1}$$

where the n_i must be consistent with the basic constraints (1.1.1) and (1.1.2).[19] If our particles were distinguishable, then all these permutations would lead to "distinct" microstates. However, in view of the indistinguishability of the particles, these permutations must be regarded as leading to one and the same thing; consequently, for *any* distribution set $\{n_i\}$, we have one, and only one, distinct microstate. As a result, the total number of distinct microstates accessible to the system, consistent with a given macrostate (N, V, E), would be severely cut down. However, since factor (1) itself depends on the numbers n_i constituting a particular distribution set and for a given macrostate there will be many such sets, there is no straightforward way to "correct down" the number of microstates computed on the basis of the classical concept of "distinguishability" of the particles.

The recipe of Gibbs clearly amounts to disregarding the details of the numbers n_i and slashing the whole sequence of microstates by a *common* factor $N!$; this is correct for situations in which all N particles happen to be in different energy states but is certainly wrong for other situations. We must keep in mind that by adopting this recipe we are still using a spurious *weight factor*,

$$w\{n_i\} = \frac{1}{n_1!n_2!\ldots}, \tag{2}$$

for the distribution set $\{n_i\}$ whereas in principle we should use a factor of *unity*, irrespective of the values of the numbers n_i.[20] Nonetheless, the recipe of Gibbs does correct the situation in a gross manner, though in matters of detail it is still inadequate. In fact, it is only by taking $w\{n_i\}$ to be equal to unity (or zero) that we obtain true *quantum statistics*!

We thus see that the recipe of Gibbs corrects the enumeration of the microstates, as necessitated by the indistinguishability of the particles, only in a gross manner. Numerically, this would approach closer and closer to reality as the probability of the n_i being greater than 1 becomes less and less. This in turn happens when the given system is at a sufficiently high temperature (so that many more energy states become accessible) and has a sufficiently low density (so that there are not as many particles to accommodate). It follows that the "corrected" classical statistics represents truth more closely if the expectation values of the *occupation numbers* n_i are much less than unity:

$$\langle n_i \rangle \ll 1, \tag{3}$$

[19] The presence of the factors $(n_i!)$ in the denominator is related to the comment made in the preceding note.

[20] Or a factor of *zero* if the distribution set $\{n_i\}$ is disallowed on certain physical grounds, such as the Pauli exclusion principle.

that is, if the numbers n_i are generally 0, occasionally 1, and rarely greater than 1. Condition (3) in a way defines the *classical limit*. We must, however, remember that it is because of the application of the correction factor $1/N!$, which replaces (1) by (2), that our results agree with reality *at least* in the classical limit.

In Section 5.5 we shall demonstrate, in an independent manner, that the factor by which the number of microstates, as computed for the "labeled" molecules, be reduced so that the formalism of classical statistical mechanics becomes a true limit of the formalism of quantum statistical mechanics is indeed $N!$.

Problems

1.1. **(a)** Show that, for two *large* systems in thermal contact, the number $\Omega^{(0)}(E^{(0)}, E_1)$ of Section 1.2 can be expressed as a Gaussian in the variable E_1. Determine the root-mean-square deviation of E_1 from the mean value $\overline{E}_1$ in terms of other quantities pertaining to the problem.

(b) Make an explicit evaluation of the root-mean-square deviation of E_1 in the special case when the systems A_1 and A_2 are ideal classical gases.

1.2. Assuming that the entropy S and the statistical number Ω of a physical system are related through an arbitrary functional form

$$S = f(\Omega),$$

show that the additive character of S and the multiplicative character of Ω *necessarily* require that the function $f(\Omega)$ be of the form (1.2.6).

1.3. Two systems A and B, of identical composition, are brought together and allowed to exchange both energy and particles, keeping volumes V_A and V_B constant. Show that the minimum value of the quantity (dE_A/dN_A) is given by

$$\frac{\mu_A T_B - \mu_B T_A}{T_B - T_A},$$

where the μ's and the T's are the respective chemical potentials and temperatures.

1.4. In a classical gas of hard spheres (of diameter D), the spatial distribution of the particles is no longer uncorrelated. Roughly speaking, the presence of n particles in the system leaves only a volume $(V - nv_0)$ available for the $(n+1)$th particle; clearly, v_0 would be proportional to D^3. Assuming that $Nv_0 \ll V$, determine the dependence of $\Omega(N, V, E)$ on V (compare to equation (1.4.1)) and show that, as a result of this, V in the ideal-gas law (1.4.3) gets replaced by $(V - b)$, where b is four times the actual volume occupied by the particles.

1.5. Read Appendix A and establish formulae (1.4.15) and (1.4.16). Estimate the importance of the linear term in these formulae, relative to the main term $(\pi/6)\varepsilon^{*3/2}$, for an oxygen molecule confined to a cube of side 10 cm; take $\varepsilon = 0.05$ eV.

1.6. A cylindrical vessel 1 m long and 0.1 m in diameter is filled with a monatomic gas at $P = 1$ atm and $T = 300$ K. The gas is heated by an electrical discharge, along the axis of the vessel, which releases an energy of 10^4 joules. What will the temperature of the gas be immediately after the discharge?

1.7. Study the statistical mechanics of an extreme relativisitic gas characterized by the single-particle energy states

$$\varepsilon(n_x, n_y, n_z) = \frac{hc}{2L}\left(n_x^2 + n_y^2 + n_z^2\right)^{1/2},$$

instead of (1.4.5), along the lines followed in Section 1.4. Show that the ratio C_P/C_V in this case is 4/3, instead of 5/3.

1.8. Consider a system of quasiparticles whose energy eigenvalues are given by

$$\varepsilon(n) = nh\nu; \quad n = 0, 1, 2, \ldots.$$

Obtain an asymptotic expression for the number Ω of this system for a given number N of the quasiparticles and a given total energy E. Determine the temperature T of the system as a function of E/N and $h\nu$, and examine the situation for which $E/(Nh\nu) \gg 1$.

1.9. Making use of the fact that the entropy $S(N, V, E)$ of a thermodynamic system is an extensive quantity, show that

$$N\left(\frac{\partial S}{\partial N}\right)_{V,E} + V\left(\frac{\partial S}{\partial V}\right)_{N,E} + E\left(\frac{\partial S}{\partial E}\right)_{N,V} = S.$$

Note that this result implies that $(-N\mu + PV + E)/T = S$, that is, $N\mu = E + PV - TS$.

1.10. A mole of argon and a mole of helium are contained in vessels of equal volume. If argon is at 300 K, what should the temperature of helium be so that the two have the same entropy?

1.11. Four moles of nitrogen and one mole of oxygen at $P = 1$ atm and $T = 300\,\text{K}$ are mixed together to form air at the same pressure and temperature. Calculate the entropy of mixing per mole of the air formed.

1.12. Show that the various expressions for the entropy of mixing, derived in Section 1.5, satisfy the following relations:

(a) For all N_1, V_1, N_2, and V_2,

$$(\Delta S)_{1\equiv 2} = \{(\Delta S) - (\Delta S)^*\} \geq 0,$$

the equality holding when and only when $N_1/V_1 = N_2/V_2$.

(b) For a given value of $(N_1 + N_2)$,

$$(\Delta S)^* \leq (N_1 + N_2)k \ln 2,$$

the equality holding when and only when $N_1 = N_2$.

1.13. If the two gases considered in the mixing process of Section 1.5 were initially at different temperatures, say T_1 and T_2, what would the entropy of mixing be in that case? Would the contribution arising from this cause depend on whether the two gases were different or identical?

1.14. Show that for an ideal gas composed of monatomic molecules the entropy change, between any two temperatures, when the pressure is kept constant is 5/3 times the corresponding entropy change when the volume is kept constant. Verify this result *numerically* by calculating the actual values of $(\Delta S)_P$ and $(\Delta S)_V$ per mole of an ideal gas whose temperature is raised from 300 K to 400 K.

1.15. We have seen that the (P, V)-relationship during a reversible adiabatic process in an ideal gas is governed by the exponent γ, such that

$$PV^\gamma = \text{const.}$$

Consider a mixture of two ideal gases, with mole fractions f_1 and f_2 and respective exponents γ_1 and γ_2. Show that the *effective* exponent γ for the mixture is given by

$$\frac{1}{\gamma - 1} = \frac{f_1}{\gamma_1 - 1} + \frac{f_2}{\gamma_2 - 1}.$$

1.16. Establish thermodynamically the formulae

$$V\left(\frac{\partial P}{\partial T}\right)_\mu = S \quad \text{and} \quad V\left(\frac{\partial P}{\partial \mu}\right)_T = N.$$

Express the pressure P of an ideal classical gas in terms of the variables μ and T, and verify the above formulae.

2 Elements of Ensemble Theory

In the preceding chapter we noted that, for a given *macrostate* (N, V, E), a statistical system, at any time t, is equally likely to be in any one of an extremely large number of distinct *microstates*. As time passes, the system continually switches from one microstate to another, with the result that, over a reasonable span of time, all one observes is a behavior "averaged" over the variety of microstates through which the system passes. It may, therefore, make sense if we consider, at a *single* instant of time, a rather large number of systems — all being some sort of "mental copies" of the given system — which are characterized by the same macrostate as the original system but are, naturally enough, in all sorts of possible microstates. Then, under ordinary circumstances, we may expect that the average behavior of any system in this collection, which we call an *ensemble*, would be identical to the time-averaged behavior of the given system. It is on the basis of this expectation that we proceed to develop the so-called *ensemble theory*.

For classical systems, the most appropriate framework for developing the desired formalism is provided by the *phase space*. Accordingly, we begin our study of the various ensembles with an analysis of the basic features of this space.

2.1 Phase space of a classical system

The microstate of a given classical system, at any time t, may be defined by specifying the *instantaneous* positions and momenta of all the particles constituting the system. Thus, if N is the number of particles in the system, the definition of a microstate requires the specification of $3N$ position coordinates $q_1, q_2, \ldots, q_{3N}$ and $3N$ momentum coordinates $p_1, p_2, \ldots, p_{3N}$. Geometrically, the set of coordinates (q_i, p_i), where $i = 1, 2, \ldots, 3N$, may be regarded as a point in a space of $6N$ dimensions. We refer to this space as the *phase space*, and the phase point (q_i, p_i) as a *representative point*, of the given system.

Of course, the coordinates q_i and p_i are functions of the time t; the precise manner in which they vary with t is determined by the canonical equations of motion,

$$\left.\begin{aligned} \dot{q}_i &= \frac{\partial H(q_i, p_i)}{\partial p_i} \\ \dot{p}_i &= -\frac{\partial H(q_i, p_i)}{\partial q_i} \end{aligned}\right\} i = 1, 2, \ldots, 3N, \qquad (1)$$

where $H(q_i, p_i)$ is the *Hamiltonian* of the system. Now, as time passes, the set of coordinates (q_i, p_i), which also defines the microstate of the system, undergoes a continual change. Correspondingly, our representative point in the phase space carves out a

© 2011 Elsevier Ltd. All rights reserved.

trajectory whose direction, at any time t, is determined by the *velocity vector* $\boldsymbol{v} \equiv (\dot{q}_i, \dot{p}_i)$, which in turn is given by the equations of motion (1). It is not difficult to see that the trajectory of the representative point must remain within a limited region of the phase space; this is so because a finite volume V directly limits the values of the coordinates q_i, while a finite energy E limits the values of both the q_i and the p_i [through the Hamiltonian $H(q_i, p_i)$]. In particular, if the total energy of the system is known to have a *precise* value, say E, the corresponding trajectory will be restricted to the "hypersurface"

$$H(q_i, p_i) = E \tag{2}$$

of the phase space; on the other hand, if the total energy may lie anywhere in the range $(E - \frac{1}{2}\Delta, E + \frac{1}{2}\Delta)$, the corresponding trajectory will be restricted to the "hypershell" defined by these limits.

Now, if we consider an ensemble of systems (i.e., the given system, along with a large number of mental copies of it) then, at any time t, the various members of the ensemble will be in all sorts of possible microstates; indeed, each one of these microstates must be consistent with the given macrostate that is supposed to be common to all members of the ensemble. In the phase space, the corresponding picture will consist of a swarm of representative points, one for each member of the ensemble, all lying within the "allowed" region of this space. As time passes, every member of the ensemble undergoes a continual change of microstates; correspondingly, the representative points constituting the swarm continually move along their respective trajectories. The overall picture of this movement possesses some important features that are best illustrated in terms of what we call a *density function* $\rho(q, p; t)$.[1] This function is such that, at any time t, the number of representative points in the "volume element" $(d^{3N}q\, d^{3N}p)$ around the point (q, p) of the phase space is given by the product $\rho(q, p; t) d^{3N}q\, d^{3N}p$. Clearly, the density function $\rho(q, p; t)$ symbolizes the manner in which the members of the ensemble are distributed over all possible microstates at different instants of time. Accordingly, the *ensemble average* $\langle f \rangle$ of a given physical quantity $f(q, p)$, which may be different for systems in different microstates, would be given by

$$\langle f \rangle = \frac{\int f(q, p) \rho(q, p; t) d^{3N}q\, d^{3N}p}{\int \rho(q, p; t) d^{3N}q\, d^{3N}p}. \tag{3}$$

The integrations in (3) extend over the whole of the phase space; however, it is only the populated regions of the phase space ($\rho \neq 0$) that really contribute. We note that, in general, the ensemble average $\langle f \rangle$ may itself be a function of time.

An ensemble is said to be *stationary* if ρ does not depend explicitly on time, that is, at all times

$$\frac{\partial \rho}{\partial t} = 0. \tag{4}$$

[1]Note that (q, p) is an abbreviation of $(q_i, p_i) \equiv (q_1, \ldots, q_{3N}, p_1, \ldots, p_{3N})$.

Clearly, for such an ensemble the average value $\langle f \rangle$ of *any* physical quantity $f(q,p)$ will be independent of time. Naturally, a stationary ensemble qualifies to represent a system in *equilibrium*. To determine the circumstances under which equation (4) may hold, we have to make a rather detailed study of the movement of the representative points in the phase space.

2.2 Liouville's theorem and its consequences

Consider an arbitrary "volume" ω in the relevant region of the phase space and let the "surface" enclosing this volume be denoted by σ; see Figure 2.1. Then, the rate at which the number of representative points in this volume increases with time is written as

$$\frac{\partial}{\partial t} \int_{\omega} \rho \, d\omega, \tag{1}$$

where $d\omega \equiv (d^{3N}q \, d^{3N}p)$. On the other hand, the *net* rate at which the representative points "flow" out of ω (across the bounding surface σ) is given by

$$\int_{\sigma} \rho \boldsymbol{v} \cdot \hat{\boldsymbol{n}} \, d\sigma; \tag{2}$$

here, $\boldsymbol{v}$ is the velocity vector of the representative points in the region of the surface element $d\sigma$ while $\hat{\boldsymbol{n}}$ is the (outward) unit vector normal to this element. By the divergence theorem, (2) can be written as

$$\int_{\omega} \operatorname{div}(\rho \boldsymbol{v}) d\omega; \tag{3}$$

of course, the operation of divergence here means

$$\operatorname{div}(\rho \boldsymbol{v}) \equiv \sum_{i=1}^{3N} \left\{ \frac{\partial}{\partial q_i}(\rho \dot{q}_i) + \frac{\partial}{\partial p_i}(\rho \dot{p}_i) \right\}. \tag{4}$$

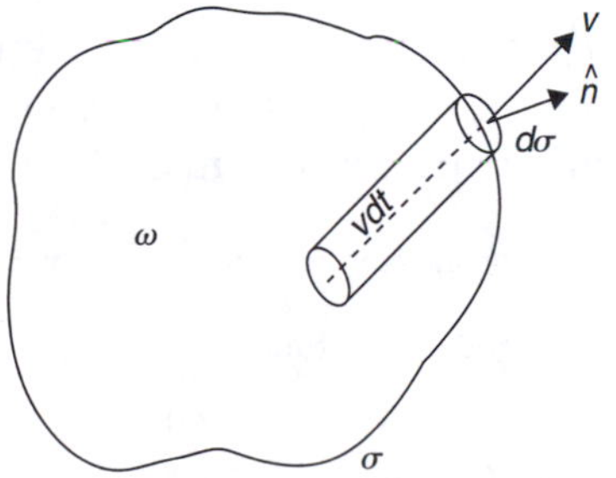

FIGURE 2.1 The "hydrodynamics" of the representative points in the phase space.

In view of the fact that there are no "sources" or "sinks" in the phase space and hence the total number of representative points remains conserved,[2] we have, by (1) and (3),

$$\frac{\partial}{\partial t}\int_{\omega} \rho\, d\omega = -\int_{\omega} \operatorname{div}(\rho \boldsymbol{v})\, d\omega, \tag{5}$$

that is,

$$\int_{\omega} \left\{ \frac{\partial \rho}{\partial t} + \operatorname{div}(\rho \boldsymbol{v}) \right\} d\omega = 0. \tag{6}$$

Now, the necessary and sufficient condition that integral (6) vanish for all arbitrary volumes ω is that the integrand itself vanish *everywhere* in the relevant region of the phase space. Thus, we must have

$$\frac{\partial \rho}{\partial t} + \operatorname{div}(\rho \boldsymbol{v}) = 0, \tag{7}$$

which is the *equation of continuity* for the swarm of the representative points.

Combining (4) and (7), we obtain

$$\frac{\partial \rho}{\partial t} + \sum_{i=1}^{3N} \left(\frac{\partial \rho}{\partial q_i} \dot{q}_i + \frac{\partial \rho}{\partial p_i} \dot{p}_i \right) + \rho \sum_{i=1}^{3N} \left(\frac{\partial \dot{q}_i}{\partial q_i} + \frac{\partial \dot{p}_i}{\partial p_i} \right) = 0. \tag{8}$$

The last group of terms vanishes identically because, by the equations of motion, we have, for all i,

$$\frac{\partial \dot{q}_i}{\partial q_i} = \frac{\partial^2 H(q_i, p_i)}{\partial q_i \partial p_i} \equiv \frac{\partial^2 H(q_i, p_i)}{\partial p_i \partial q_i} = -\frac{\partial \dot{p}_i}{\partial p_i}. \tag{9}$$

Further, since $\rho \equiv \rho(q, p; t)$, the remaining terms in (8) may be combined to form the "total" time derivative of ρ, with the result that

$$\frac{d\rho}{dt} = \frac{\partial \rho}{\partial t} + [\rho, H] = 0. \tag{10}^3$$

Equation (10) embodies *Liouville's theorem* (1838). According to this theorem, the "local" density of the representative points, *as viewed by an observer moving with a representative point*, stays constant in time. Thus, the swarm of the representative points moves in

[2]This means that in the ensemble under consideration neither are any new members being added nor are any old ones being removed.

[3]We recall that the *Poisson bracket* $[\rho, H]$ stands for the sum

$$\sum_{i=1}^{3N} \left(\frac{\partial \rho}{\partial q_i} \frac{\partial H}{\partial p_i} - \frac{\partial \rho}{\partial p_i} \frac{\partial H}{\partial q_i} \right),$$

which is identical to the group of terms in the middle of (8).

the phase space in essentially the same manner as an incompressible fluid moves in the physical space!

A distinction must be made, however, between equation (10) on one hand and equation (2.1.4) on the other. While the former derives from the basic mechanics of the particles and is therefore *quite generally* true, the latter is only a requirement for equilibrium which, in a given case, may or may not be satisfied. The condition that ensures simultaneous validity of the two equations is clearly

$$[\rho, H] = \sum_{i=1}^{3N} \left(\frac{\partial \rho}{\partial q_i} \dot{q}_i + \frac{\partial \rho}{\partial p_i} \dot{p}_i \right) = 0. \tag{11}$$

Now, one possible way of satisfying (11) is to assume that ρ, which is already assumed to have no explicit dependence on time, is *independent* of the coordinates (q, p) as well, that is,

$$\rho(q, p) = \text{const.} \tag{12}$$

over the relevant region of the phase space (and, of course, is zero everywhere else). Physically, this choice corresponds to an ensemble of systems that at *all* times are *uniformly* distributed over all possible microstates. The ensemble average (2.1.3) then reduces to

$$\langle f \rangle = \frac{1}{\omega} \int_{\omega} f(q, p) d\omega; \tag{13}$$

here, ω denotes the total "volume" of the relevant region of the phase space. Clearly, in this case, *any* member of the ensemble is equally likely to be in *any* one of the various possible microstates, inasmuch as *any* representative point in the swarm is equally likely to be in the neighborhood of *any* phase point in the allowed region of the phase space. This statement is usually referred to as the postulate of "equal *a priori* probabilities" for the various possible microstates (or for the various volume elements in the allowed region of the phase space); the resulting ensemble is referred to as the *microcanonical ensemble.*

A more general way of satisfying (11) is to assume that the dependence of ρ on (q, p) comes only through an explicit dependence on the Hamiltonian $H(q, p)$, that is,

$$\rho(q, p) = \rho[H(q, p)]; \tag{14}$$

condition (11) is then identically satisfied. Equation (14) provides a class of density functions for which the corresponding ensemble is stationary. In Chapter 3 we shall see that the most natural choice in this class of ensembles is the one for which

$$\rho(q, p) \propto \exp[-H(q, p)/kT]. \tag{15}$$

The ensemble so defined is referred to as the *canonical ensemble.*

2.3 The microcanonical ensemble

In this ensemble the macrostate of a system is defined by the number of molecules N, the volume V, and the energy E. However, in view of the considerations expressed in Section 1.4, we may prefer to specify a range of energy values, say from $(E - \frac{1}{2}\Delta)$ to $(E + \frac{1}{2}\Delta)$, rather than a sharply defined value E. With the macrostate specified, a choice still remains for the systems of the ensemble to be in *any one* of a large number of possible microstates. In the phase space, correspondingly, the representative points of the ensemble have a choice to lie *anywhere* within a "hypershell" defined by the condition

$$\left(E - \frac{1}{2}\Delta\right) \leq H(q,p) \leq \left(E + \frac{1}{2}\Delta\right). \tag{1}$$

The volume of the phase space enclosed within this shell is given by

$$\omega = \int' d\omega \equiv \int' \left(d^{3N}q\, d^{3N}p\right), \tag{2}$$

where the primed integration extends only over that part of the phase space which conforms to condition (1). It is clear that ω will be a function of the parameters N, V, E, and Δ.

Now, the microcanonical ensemble is a collection of systems for which the density function ρ is, at all times, given by

$$\left.\begin{array}{lll} \rho(q,p) = \text{const.} & \text{if} & \left(E - \frac{1}{2}\Delta\right) \leq H(q,p) \leq \left(E + \frac{1}{2}\Delta\right) \\ \qquad 0 & \text{otherwise} & \end{array}\right\}. \tag{3}$$

Accordingly, the expectation value of the number of representative points lying in a volume element $d\omega$ of the relevant hypershell is simply proportional to $d\omega$. In other words, the *a priori* probability of finding a representative point in a given volume element $d\omega$ is the same as that of finding a representative point in an equivalent volume element $d\omega$ located *anywhere* in the hypershell. In our original parlance, this means an equal *a priori* probability for a given member of the ensemble to be in *any one* of the various possible microstates. In view of these considerations, the ensemble average $\langle f \rangle$, as given by equation (2.2.13), acquires a simple physical meaning. To see this, we proceed as follows.

Since the ensemble under study is a stationary one, the ensemble average of any physical quantity f will be independent of time; accordingly, taking a time average thereof will not produce any new result. Thus

$$\begin{aligned} \langle f \rangle &\equiv \text{the ensemble average of } f \\ &= \text{the time average of (the ensemble average of } f). \end{aligned}$$

Now, the processes of time averaging and ensemble averaging are completely independent, so the order in which they are performed may be reversed without causing any change in the value of $\langle f \rangle$. Thus

$$\langle f \rangle = \text{the ensemble average of (the time average of } f\text{)}.$$

Now, the time average of any physical quantity, taken over a sufficiently long interval of time, must be the same for *every* member of the ensemble, for after all we are dealing with only *mental copies* of a given system.[4] Therefore, taking an ensemble average thereof should be inconsequential, and we may write

$$\langle f \rangle = \text{the long-time average of } f,$$

where the latter may be taken over *any* member of the ensemble. Furthermore, the long-time average of a physical quantity is all one obtains by making a measurement of that quantity on the given system; therefore, it may be identified with the value one expects to obtain through experiment. Thus, we finally have

$$\langle f \rangle = f_{\text{exp}}. \tag{4}$$

This brings us to the most important result: *the ensemble average of any physical quantity f is identical to the value one expects to obtain on making an appropriate measurement on the given system.*

The next thing we look for is the establishment of a connection between the mechanics of the microcanonical ensemble and the thermodynamics of the member systems. To do this, we observe that there exists a direct correspondence between the various microstates of the given system and the various locations in the phase space. The volume ω (of the allowed region of the phase space) is, therefore, a direct measure of the multiplicity Γ of the microstates accessible to the system. To establish a numerical correspondence between ω

[4]To provide a *rigorous* justification for this assertion is not trivial. One can readily see that if, for any particular member of the ensemble, the quantity f is averaged only over a *short* span of time, the result is bound to depend on the relevant "subset of microstates" through which the system passes during that time. In the phase space, this will mean an averaging over only a "part of the allowed region." However, if we employ instead a sufficiently long interval of time, the system may be expected to pass through *almost* all possible microstates "without fear or favor"; consequently, the result of the averaging process would depend only on the macrostate of the system, and not on a subset of microstates. Correspondingly, the averaging in the phase space would go over *practically* all parts of the allowed region, again "without fear or favor." In other words, the representative point of our system will have traversed each and every part of the allowed region *almost* uniformly. This statement embodies the so-called *ergodic theorem* or *ergodic hypothesis*, which was first introduced by Boltzmann (1871). According to this hypothesis, the trajectory of a representative point passes, in the course of time, through *each and every* point of the relevant region of the phase space. A little reflection, however, shows that the statement as such requires a qualification; we better replace it by the so-called *quasiergodic hypothesis*, according to which the trajectory of a representative point traverses, in the course of time, *any neighborhood of any point* of the relevant region. For further details, see ter Haar (1954, 1955), Farquhar (1964).

Now, when we consider an ensemble of systems, the foregoing statement should hold for every member of the ensemble; thus, *irrespective of the initial (and final) states* of the various systems, the long-time average of any physical quantity f should be the same for every member system.

and Γ, we need to discover a *fundamental volume* ω_0 that could be regarded as "equivalent to one microstate." Once this is done, we may say that, asymptotically,

$$\Gamma = \omega/\omega_0. \tag{5}$$

The thermodynamics of the system would then follow in the same way as in Sections 1.2–1.4, namely through the relationship

$$S = k \ln \Gamma = k \ln(\omega/\omega_0), \quad \text{etc.} \tag{6}$$

The basic problem then consists in determining ω_0. From dimensional considerations, see (2), ω_0 must be in the nature of an "angular momentum raised to the power $3N$." To determine it exactly, we consider certain simplified systems, both from the point of view of the phase space and from the point of view of the distribution of quantum states.

2.4 Examples

We consider, first of all, the problem of a classical ideal gas composed of monatomic particles; see Section 1.4. In the microcanonical ensemble, the volume ω of the phase space accessible to the representative points of the (member) systems is given by

$$\omega = \int'\cdots\int'\left(d^{3N}q\,d^{3N}p\right), \tag{1}$$

where the integrations are restricted by the conditions that (i) the particles of the system are confined in physical space to volume V, and (ii) the total energy of the system lies between the limits $(E - \frac{1}{2}\Delta)$ and $(E + \frac{1}{2}\Delta)$. Since the Hamiltonian in this case is a function of the p_i alone, integrations over the q_i can be carried out straightforwardly; these give a factor of V^N. The remaining integral is

$$\int_{\left(E-\frac{1}{2}\Delta\right)\le\sum_{i=1}^{3N}\left(p_i^2/2m\right)\le\left(E+\frac{1}{2}\Delta\right)}\cdots\int d^{3N}p = \int_{2m\left(E-\frac{1}{2}\Delta\right)\le\sum_{i=1}^{3N}y_i^2\le 2m\left(E+\frac{1}{2}\Delta\right)}\cdots\int d^{3N}y,$$

which is equal to the volume of a $3N$-dimensional hypershell, bounded by hyperspheres of radii

$$\sqrt{\left[2m\left(E+\frac{1}{2}\Delta\right)\right]} \quad \text{and} \quad \sqrt{\left[2m\left(E-\frac{1}{2}\Delta\right)\right]}.$$

For $\Delta \ll E$, this is given by the thickness of the shell, which is almost equal to $\Delta(m/2E)^{1/2}$, multiplied by the surface area of a $3N$-dimensional hypersphere of radius $\sqrt{(2mE)}$. By equation (7) of Appendix C, we obtain for this integral

$$\Delta\left(\frac{m}{2E}\right)^{1/2}\left\{\frac{2\pi^{3N/2}}{[(3N/2)-1]!}(2mE)^{(3N-1)/2}\right\},$$

which gives

$$\omega \simeq \frac{\Delta}{E}V^N\frac{(2\pi mE)^{3N/2}}{[(3N/2)-1]!}. \tag{2}$$

Comparing (2) with (1.4.17 and 1.4.17a), we obtain the desired correspondence, namely

$$(\omega/\Gamma)_{\text{asymp}} \equiv \omega_0 = h^{3N};$$

see also Problem 2.9. Quite generally, if the system under study has $\mathcal{N}$ degrees of freedom, the desired conversion factor is

$$\omega_0 = h^{\mathcal{N}}. \tag{3}$$

In the case of a single particle, $\mathcal{N} = 3$; accordingly, the number of microstates available would asymptotically be equal to the volume of the allowed region of the phase space divided by h^3. Let $\Sigma(P)$ denote the number of microstates available to a free particle confined to volume V of the physical space, its momentum p being less than or equal to a specified value P. Then

$$\Sigma(P) \approx \frac{1}{h^3}\int_{p\le P}\!\!\cdots\!\int\left(d^3q\,d^3p\right) = \frac{V}{h^3}\frac{4\pi}{3}P^3, \tag{4}$$

from which we obtain for the number of microstates with momentum lying between p and $p+dp$

$$g(p)dp = \frac{d\Sigma(p)}{dp}dp \approx \frac{V}{h^3}4\pi p^2 dp. \tag{5}$$

Expressed in terms of the particle energy, these expressions assume the form

$$\Sigma(E) \approx \frac{V}{h^3}\frac{4\pi}{3}(2mE)^{3/2} \tag{6}$$

and

$$a(\varepsilon)d\varepsilon = \frac{d\Sigma(\varepsilon)}{d\varepsilon}d\varepsilon \approx \frac{V}{h^3}2\pi(2m)^{3/2}\varepsilon^{1/2}d\varepsilon. \tag{7}$$

The next case we consider here is that of a one-dimensional *simple harmonic oscillator*. The classical expression for the Hamiltonian of this system is

$$H(q,p) = \frac{1}{2}kq^2 + \frac{1}{2m}p^2, \tag{8}$$

where k is the spring constant and m the mass of the oscillating particle. The space coordinate q and the momentum coordinate p of the system are given by

$$q = A\cos(\omega t + \phi), \quad p = m\dot{q} = -m\omega A \sin(\omega t + \phi), \tag{9}$$

A being the amplitude and ω the (angular) frequency of vibration:

$$\omega = \sqrt{(k/m)}. \tag{10}$$

The energy of the oscillator is a constant of the motion, and is given by

$$E = \frac{1}{2} m\omega^2 A^2. \tag{11}$$

The phase-space trajectory of the representative point (q, p) of this system is determined by eliminating t between expressions (9) for $q(t)$ and $p(t)$; we obtain

$$\frac{q^2}{(2E/m\omega^2)} + \frac{p^2}{(2mE)} = 1, \tag{12}$$

which is an ellipse, with axes proportional to $\sqrt{E}$ and hence area proportional to E; to be precise, the area of this ellipse is $2\pi E/\omega$. Now, if we restrict the oscillator energy to the interval $\left(E - \frac{1}{2}\Delta, E + \frac{1}{2}\Delta\right)$, its representative point in the phase space will be confined to the region bounded by elliptical trajectories corresponding to the energy values $\left(E + \frac{1}{2}\Delta\right)$ and $\left(E - \frac{1}{2}\Delta\right)$. The "volume" (in this case, the area) of this region will be

$$\int_{\left(E-\frac{1}{2}\Delta\right) \le H(q,p) \le \left(E+\frac{1}{2}\Delta\right)} \cdots \int (dq\, dp) = \frac{2\pi\left(E + \frac{1}{2}\Delta\right)}{\omega} - \frac{2\pi\left(E - \frac{1}{2}\Delta\right)}{\omega} = \frac{2\pi\Delta}{\omega}. \tag{13}$$

According to quantum mechanics, the energy eigenvalues of the harmonic oscillator are given by

$$E_n = \left(n + \frac{1}{2}\right)\hbar\omega; \quad n = 0, 1, 2, \ldots \tag{14}$$

In terms of phase space, one could say that the representative point of the system must move along one of the "chosen" trajectories, as shown in Figure 2.2; the area of the phase space between two consecutive trajectories, for which $\Delta = \hbar\omega$, is simply $2\pi\hbar$.[5] For arbitrary values of E and Δ, such that $E \gg \Delta \gg \hbar\omega$, the number of eigenstates within the allowed

[5]Strictly speaking, the very concept of phase space is invalid in quantum mechanics because there, in principle, it is wrong to assign to a particle the coordinates q and p *simultaneously*. Nevertheless, the ideas discussed here are tenable in the correspondence limit.

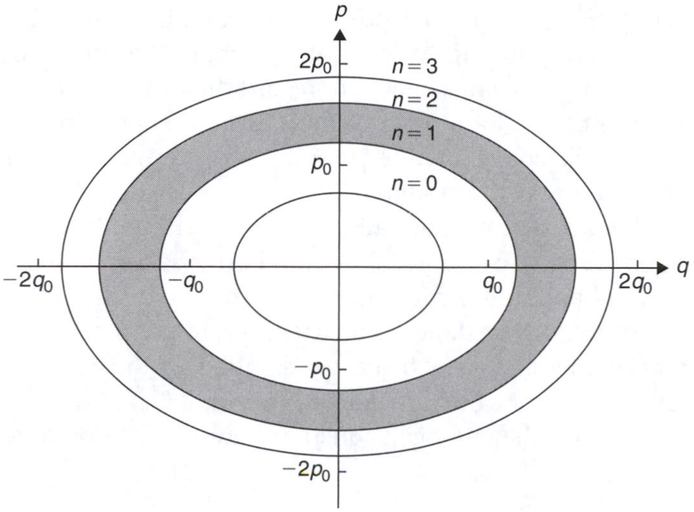

FIGURE 2.2 Eigenstates of a linear harmonic oscillator, in relation to its phase space.

energy interval is very nearly equal to $\Delta/\hbar\omega$. Hence, the area of the phase space equivalent to one eigenstate is, asymptotically, given by

$$\omega_0 = (2\pi\Delta/\omega)/(\Delta/\hbar\omega) = 2\pi\hbar = h. \tag{15}$$

If, on the other hand, we consider a system of N harmonic oscillators along the same lines as above, we arrive at the result: $\omega_0 = h^N$ (see Problem 2.7). Thus, our findings in these cases are consistent with our earlier result (3).

2.5 Quantum states and the phase space

At this stage we would like to say a few words on the central role played here by the Planck constant h. The best way to appreciate this role is to recall the implications of the Heisenberg uncertainty principle, according to which we cannot specify *simultaneously* both the position and the momentum of a particle exactly. An element of uncertainty is inherently present and can be expressed as follows: assuming that all conceivable uncertainties of measurement are eliminated, even then, by the very nature of things, the product of the uncertainties Δq and Δp in the *simultaneous* measurement of the canonically conjugate coordinates q and p would be of order $\hbar$:

$$(\Delta q \Delta p)_{\min} \sim \hbar. \tag{1}$$

Thus, it is impossible to define the position of a representative point in the phase space of the given system more accurately than is allowed by condition (1). In other words, around any point (q, p) in the (two-dimensional) phase space, there exists an area of order $\hbar$ within which the position of the representative point cannot be pinpointed. In a phase space of $2\mathcal{N}$ dimensions, the corresponding "volume of uncertainty" around any point would be of order $\hbar^{\mathcal{N}}$. Therefore, it seems reasonable to regard the phase space as made up of elementary cells, of volume $\sim \hbar^{\mathcal{N}}$, and to consider the various positions within such a cell as *nondistinct*. These cells could then be put into one-to-one correspondence with the quantum-mechanical states of the system.

It is, however, obvious that considerations of uncertainty alone cannot give us the *exact* value of the conversion factor ω_0. This could only be done by an *actual* counting of microstates on one hand and a computation of volume of the relevant region of the phase space on the other, as was done in the examples of the previous section. Clearly, a procedure along these lines could not be possible until after the work of Schrödinger and others. Historically, however, the first to establish the result (2.4.3) was Tetrode (1912) who, in his well-known work on the chemical constant and the entropy of a monatomic gas, assumed that

$$\omega_0 = (zh)^{\mathcal{N}}, \tag{2}$$

where z was supposed to be an unknown numerical factor. Comparing theoretical results with the experimental data on mercury, Tetrode found that z was very nearly equal to unity; from this he concluded that "it seems rather plausible that z is *exactly* equal to unity, as has already been taken by O. Sackur (1911)."[6]

In the extreme relativistic limit, the same result was established by Bose (1924). In his famous treatment of the photon gas, Bose made use of Einstein's relationship between the momentum of a photon and the frequency of the associated radiation, namely

$$p = \frac{h\nu}{c}, \tag{3}$$

and observed that, for a photon confined to a three-dimensional cavity of volume V, the relevant "volume" of the phase space,

$$\int' (d^3q\, d^3p) = V 4\pi p^2 dp = V(4\pi h^3 \nu^2/c^3) d\nu, \tag{4}$$

would correspond exactly to the Rayleigh expression,

$$V(4\pi \nu^2/c^3) d\nu, \tag{5}$$

for the number of normal modes of a radiation oscillator, *provided that* we divide phase space into elementary cells of volume h^3 and put these cells into one-to-one correspondence with the vibrational modes of Rayleigh. It may, however, be added that a two-fold multiplicity of these states ($g = 2$) arises from the spin orientations of the photon

[6]For a more satisfactory proof of this result, see Section 5.5, especially equation (5.5.22).

(or from the states of polarization of the vibrational modes); this requires a multiplication of both expressions (4) and (5) by a factor of 2, leaving the conversion factor h^3 unchanged.

Problems

2.1. Show that the volume element

$$d\omega = \prod_{i=1}^{3N}(dq_i\, dp_i)$$

of the phase space remains *invariant* under a canonical transformation of the (generalized) coordinates (q, p) to any other set of (generalized) coordinates (Q, P).

[*Hint*: Before considering the most general transformation of this kind, which is referred to as a *contact* transformation, it may be helpful to consider a *point* transformation — one in which the new coordinates Q_i and the old coordinates q_i transform only among themselves.]

2.2. **(a)** Verify *explicitly* the invariance of the volume element $d\omega$ of the phase space of a single particle under transformation from the Cartesian coordinates (x, y, z, p_x, p_y, p_z) to the spherical polar coordinates $(r, \theta, \phi, p_r, p_\theta, p_\phi)$.

(b) The foregoing result seems to contradict the intuitive notion of "equal weights for equal solid angles," because the factor $\sin\theta$ is invisible in the expression for $d\omega$. Show that if we average out any physical quantity, whose dependence on p_θ and p_ϕ comes only through the kinetic energy of the particle, then as a result of integration over these variables we do indeed recover the factor $\sin\theta$ to appear with the subelement $(d\theta\, d\phi)$.

2.3. Starting with the line of zero energy and working in the (two-dimensional) phase space of a classical rotator, draw lines of constant energy that divide phase space into cells of "volume" h. Calculate the energies of these states and compare them with the energy eigenvalues of the corresponding quantum-mechanical rotator.

2.4. By evaluating the "volume" of the relevant region of its phase space, show that the number of microstates available to a rigid rotator with angular momentum $\le M$ is $(M/\hbar)^2$. Hence determine the number of microstates that may be associated with the quantized angular momentum $M_j = \sqrt{\{j(j+1)\}}\hbar$, where $j = 0, 1, 2, \ldots$ or $\frac{1}{2}, \frac{3}{2}, \frac{5}{2}, \ldots$. Interpret the result physically.

[*Hint*: It simplifies to consider motion in the variables θ and φ, with $M^2 = p_\theta^2 + (p_\phi/\sin\theta)^2$.]

2.5. Consider a particle of energy E moving in a one-dimensional potential well $V(q)$, such that

$$m\hbar\left|\frac{dV}{dq}\right| \ll \{m(E - V)\}^{3/2}.$$

Show that the allowed values of the momentum p of the particle are such that

$$\oint p\, dq = \left(n + \frac{1}{2}\right)h,$$

where n is an integer.

2.6. The generalized coordinates of a simple pendulum are the angular displacement θ and the angular momentum $ml^2\dot{\theta}$. Study, both mathematically and graphically, the nature of the corresponding trajectories in the phase space of the system, and show that the area A enclosed by a trajectory is equal to the product of the total energy E and the time period τ of the pendulum.

2.7. Derive (i) an *asymptotic* expression for the number of ways in which a given energy E can be distributed among a set of N one-dimensional harmonic oscillators, the energy eigenvalues of the oscillators being $\left(n + \frac{1}{2}\right)\hbar\omega; n = 0, 1, 2, \ldots$, and (ii) the corresponding expression for the "volume" of the relevant region of the phase space of this system. Establish the correspondence between the two results, showing that the conversion factor ω_0 is precisely h^N.

2.8. Following the method of Appendix C, replacing equation (C.4) by the integral

$$\int_0^\infty e^{-r} r^2 dr = 2,$$

show that

$$V_{3N} = \int_{0 \le \sum_{i=1}^{N} r_i \le R} \cdots \int \prod_{i=1}^{N} \left(4\pi r_i^2 dr_i\right) = (8\pi R^3)^N/(3N)!.$$

Using this result, compute the "volume" of the relevant region of the phase space of an extreme relativistic gas ($\varepsilon = pc$) of N particles moving in three dimensions. Hence, derive expressions for the various thermodynamic properties of this system and compare your results with those of Problem 1.7.

2.9. **(a)** Solve the integral

$$\int_{0 \le \sum_{i=1}^{3N} |x_i| \le R} \cdots \int (dx_1 \ldots dx_{3N})$$

and use it to determine the "volume" of the relevant region of the phase space of an extreme relativistic gas ($\varepsilon = pc$) of $3N$ particles moving in one dimension. Determine, as well, the number of ways of distributing a given energy E among this system of particles and show that, asymptotically, $\omega_0 = h^{3N}$.

(b) Compare the thermodynamics of this system with that of the system considered in Problem 2.8.

3 The Canonical Ensemble

In the preceding chapter we established the basis of ensemble theory and made a somewhat detailed study of the microcanonical ensemble. In that ensemble the macrostate of the systems was defined through a fixed number of particles N, a fixed volume V, and a fixed energy E [or, preferably, a fixed energy range $(E - \frac{1}{2}\Delta, E + \frac{1}{2}\Delta)$]. The basic problem then consisted in determining the number $\Omega(N, V, E)$, or $\Gamma(N, V, E; \Delta)$, of *distinct* microstates accessible to the system. From the asymptotic expressions of these numbers, complete thermodynamics of the system could be derived in a straightforward manner. However, for most physical systems, the mathematical problem of determining these numbers is quite formidable. For this reason alone, a search for an alternative approach within the framework of the ensemble theory seems necessary.

Practically, too, the concept of a fixed energy (or even an energy range) for a system belonging to the real world does not appear satisfactory. For one thing, the total energy E of a system is hardly ever measured; for another, it is hardly possible to keep its value under strict physical control. A far better alternative appears to be to speak of a fixed temperature T of the system — a parameter that is not only directly observable (by placing a "thermometer" in contact with the system) but also controllable (by keeping the system in contact with an appropriate "heat reservoir"). For most purposes, the precise nature of the reservoir is not very relevant; all one needs is that it should have an infinitely large heat capacity, so that, irrespective of energy exchange between the system and the reservoir, an overall constant temperature can be maintained. Now, if the reservoir consists of an infinitely large number of mental copies of the given system we have once again an ensemble of systems — this time, however, it is an ensemble in which the macrostate of the systems is defined through the parameters N, V, and T. Such an ensemble is referred to as a *canonical* ensemble.

In the canonical ensemble, the energy E of a system is variable; in principle, it can take values anywhere between zero and infinity. The question then arises: what is the probability that, at any time t, a system in the ensemble is found to be in one of the states characterized by the energy value E_r?[1] We denote this probability by the symbol P_r. Clearly, there are two ways in which the dependence of P_r on E_r can be determined. One consists of regarding the system as in equilibrium with a heat reservoir at a *common* temperature T and studying the statistics of the energy exchange between the two. The other consists of regarding the system as a member of a canonical ensemble (N, V, T), in which an energy $\mathcal{E}$ is being shared by $\mathcal{N}$ identical systems constituting the ensemble, and studying the

[1]In what follows, the energy levels E_r appear as purely *mechanical* quantities — independent of the temperature of the system. For a treatment involving "temperature-dependent energy levels," see Elcock and Landsberg (1957).

© 2011 Elsevier Ltd. All rights reserved.

statistics of this sharing process. We expect that in the thermodynamic limit the final result in either case would be the same. Once P_r is determined, the rest follows without difficulty.

3.1 Equilibrium between a system and a heat reservoir

We consider the given system A, immersed in a very large heat reservoir A'; see Figure 3.1. On attaining a state of mutual equilibrium, the system and the reservoir would have a *common* temperature, *say* T. Their energies, however, would be variable and, in principle, could have, at any time t, values lying anywhere between 0 and $E^{(0)}$, where $E^{(0)}$ denotes the energy of the composite system $A^{(0)}(\equiv A+A')$. If, at any particular instant of time, the system A happens to be in a state characterized by the energy value E_r, then the reservoir would have an energy E'_r, such that

$$E_r + E'_r = E^{(0)} = \text{const.} \tag{1}$$

Of course, since the reservoir is supposed to be much larger than the given system, any *practical* value of E_r would be a very small fraction of $E^{(0)}$; therefore, for all practical purposes,

$$\frac{E_r}{E^{(0)}} = \left(1 - \frac{E'_r}{E^{(0)}}\right) \ll 1. \tag{2}$$

With the state of the system A having been specified, the reservoir A' can still be in *any one* of a large number of states compatible with the energy value E'_r. Let the number of these states be denoted by $\Omega'(E'_r)$. The prime on the symbol Ω emphasizes the fact that its functional form will depend on the nature of the reservoir; of course, the details of this dependence are not going to be of any particular relevance to our final results. Now, the larger the number of states available to the reservoir, the larger the probability of the reservoir assuming that particular energy value E'_r (and, hence, of the system A assuming the corresponding energy value E_r). Moreover, since the various possible states (with a given energy value) are *equally likely* to occur, the relevant probability would be directly proportional to this number; thus,

$$P_r \propto \Omega'(E'_r) \equiv \Omega'(E^{(0)} - E_r). \tag{3}$$

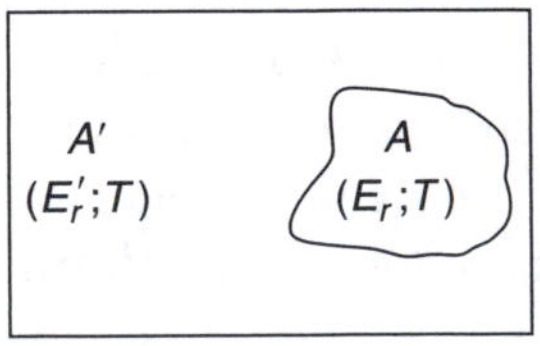

FIGURE 3.1 A given system A immersed in a heat reservoir A'; in equilibrium, the two have a common temperature T.

In view of (2), we may carry out an expansion of (3) around the value $E'_r = E^{(0)}$, that is, around $E_r = 0$. However, for reasons of convergence, it is essential to effect the expansion of its logarithm instead:

$$\begin{aligned}\ln \Omega'(E'_r) &= \ln \Omega'(E^{(0)}) + \left(\frac{\partial \ln \Omega'}{\partial E'}\right)_{E'=E^{(0)}} (E'_r - E^{(0)}) + \cdots \\ &\simeq \text{const} - \beta' E_r,\end{aligned} \tag{4}$$

where use has been made of formula (1.2.3), whereby

$$\left(\frac{\partial \ln \Omega}{\partial E}\right)_{N,V} \equiv \beta; \tag{5}$$

note that, in equilibrium, $\beta' = \beta = 1/kT$. From (3) and (4), we obtain the desired result:

$$P_r \propto \exp(-\beta E_r). \tag{6}$$

Normalizing (6), we get

$$P_r = \frac{\exp(-\beta E_r)}{\sum_r \exp(-\beta E_r)}, \tag{7}$$

where the summation in the denominator goes over *all* states accessible to the system A. We note that our final result (7) bears no relation whatsoever to the physical nature of the reservoir A'.

We now examine the same problem from the ensemble point of view.

3.2 A system in the canonical ensemble

We consider an ensemble of $\mathcal{N}$ identical systems (which may be labelled as $1, 2, \ldots, \mathcal{N}$), sharing a total energy $\mathcal{E}$; let $E_r (r = 0, 1, 2, \ldots)$ denote the energy eigenvalues of the systems. If n_r denotes the number of systems which, at any time t, have the energy value E_r, then the set of numbers $\{n_r\}$ must satisfy the obvious conditions

$$\left.\begin{aligned}\sum_r n_r &= \mathcal{N} \\ \sum_r n_r E_r &= \mathcal{E} = \mathcal{N} U,\end{aligned}\right\} \tag{1}$$

where $U (= \mathcal{E}/\mathcal{N})$ denotes the average energy per system in the ensemble. Any set $\{n_r\}$ that satisfies the restrictive conditions (1) represents a possible mode of distribution of the total energy $\mathcal{E}$ among the $\mathcal{N}$ members of the ensemble. Furthermore, any such mode can be realized in a number of ways, for we may effect a reshuffle among those members of the ensemble for which the energy values are different and thereby obtain a state of the ensemble that is distinct from the original one. Denoting the number of different ways of doing so by the symbol $W\{n_r\}$, we have

$$W\{n_r\} = \frac{\mathcal{N}!}{n_0!n_1!n_2!\ldots}. \tag{2}$$

In view of the fact that all possible states of the ensemble, which are compatible with conditions (1), are *equally likely* to occur, the frequency with which the distribution set $\{n_r\}$ may appear will be directly proportional to the number $W\{n_r\}$. Accordingly, the "most probable" mode of distribution will be the one for which the number W is a maximum. We denote the corresponding distribution set by $\{n_r^*\}$; clearly, the set $\{n_r^*\}$ must also satisfy conditions (1). As will be seen in the sequel, the probability of appearance of other modes of distribution, however little they may differ from the most probable mode, is extremely low! Therefore, for all practical purposes, the *most probable distribution set* $\{n_r^*\}$ is the only one we have to contend with.

However, unless this has been mathematically demonstrated, one must take into account *all* possible modes of distribution, as characterized by the various distribution sets $\{n_r\}$, along with their respective weight factors $W\{n_r\}$. Accordingly, the *expectation values*, or *mean values*, $\langle n_r\rangle$ of the numbers n_r would be given by

$$\langle n_r\rangle = \frac{\sum\limits_{\{n_r\}}' n_r W\{n_r\}}{\sum\limits_{\{n_r\}}' W\{n_r\}}, \tag{3}$$

where the primed summations go over all distribution sets that conform to conditions (1). In principle, the mean value $\langle n_r\rangle$, as a fraction of the total number $\mathcal{N}$, should be a natural analog of the probability P_r evaluated in the preceding section. In practice, however, the fraction $n_r^*/\mathcal{N}$ also turns out to be the same.

We now proceed to derive expressions for the numbers n_r^* and $\langle n_r\rangle$, and to show that, in the limit $\mathcal{N}\to\infty$, they are identical.

The method of most probable values

Our aim here is to determine that distribution set which, while satisfying conditions (1), maximizes the weight factor (2). For simplicity, we work with $\ln W$ instead:

$$\ln W = \ln(\mathcal{N}!) - \sum_r \ln(n_r!). \tag{4}$$

Since, in the end, we propose to resort to the limit $\mathcal{N}\to\infty$, the values of n_r (which are going to be of any practical significance) would also, in that limit, tend to infinity. It is, therefore, justified to apply the Stirling formula, $\ln(n!)\approx n\ln n - n$, to (4) and write

$$\ln W = \mathcal{N}\ln\mathcal{N} - \sum_r n_r \ln n_r. \tag{5}$$

If we shift from the set $\{n_r\}$ to a slightly different set $\{n_r + \delta n_r\}$, then expression (5) would change by an amount

$$\delta(\ln W) = -\sum_r (\ln n_r + 1)\delta n_r. \tag{6}$$

Now, if the set $\{n_r\}$ is maximal, the variation $\delta(\ln W)$ should vanish. At the same time, in view of the restrictive conditions (1), the variations δn_r themselves must satisfy the conditions

$$\left.\begin{aligned} \sum_r \delta n_r &= 0 \\ \sum_r E_r \delta n_r &= 0. \end{aligned}\right\} \tag{7}$$

The desired set $\{n_r^*\}$ is then determined by the method of *Lagrange multipliers*,[2] by which the condition determining this set becomes

$$\sum_r \{-(\ln n_r^* + 1) - \alpha - \beta E_r\}\delta n_r = 0, \tag{8}$$

where α and β are the Lagrangian undetermined multipliers that take care of the restrictive conditions (7). In (8), the variations δn_r become completely arbitrary; accordingly, the only way to satisfy this condition is that all its coefficients must vanish identically, that is, for *all r*,

$$\ln n_r^* = -(\alpha + 1) - \beta E_r,$$

which gives

$$n_r^* = C \exp(-\beta E_r), \tag{9}$$

where C is again an undetermined parameter.

To determine C and β, we subject (9) to conditions (1), with the result that

$$\frac{n_r^*}{\mathcal{N}} = \frac{\exp(-\beta E_r)}{\sum_r \exp(-\beta E_r)}, \tag{10}$$

the parameter β being a solution of the equation

$$\frac{\mathcal{E}}{\mathcal{N}} = U = \frac{\sum_r E_r \exp(-\beta E_r)}{\sum_r \exp(-\beta E_r)}. \tag{11}$$

[2]For the method of Lagrange multipliers, see ter Haar and Wergeland (1966, Appendix C.1).

Combining statistical considerations with thermodynamic ones, see Section 3.3, we can show that the parameter β here is exactly the same as the one appearing in Section 3.1, that is, $\beta = 1/kT$.

The method of mean values

Here we attempt to evaluate expression (3) for $\langle n_r \rangle$, taking into account the weight factors (2) and the restrictive conditions (1). To do this, we replace (2) by

$$\tilde{W}\{n_r\} = \frac{\mathcal{N}!\,\omega_0^{n_0}\omega_1^{n_1}\omega_2^{n_2}\ldots}{n_0!\,n_1!\,n_2!\ldots}, \tag{12}$$

with the understanding that in the end all ω_r will be set equal to unity, and introduce a function

$$\Gamma(\mathcal{N}, U) = \sum_{\{n_r\}}{}^{'}\, \tilde{W}\{n_r\}, \tag{13}$$

where the primed summation, as before, goes over all distribution sets that conform to conditions (1). Expression (3) can then be written as

$$\langle n_r \rangle = \omega_r \frac{\partial}{\partial \omega_r}(\ln \Gamma)\bigg|_{\text{all } \omega_r = 1}. \tag{14}$$

Thus, all we need to know here is the dependence of the quantity $\ln \Gamma$ on the parameters ω_r. Now,

$$\Gamma(\mathcal{N}, U) = \mathcal{N}! \sum_{\{n_r\}}{}^{'} \left(\frac{\omega_0^{n_0}}{n_0!} \cdot \frac{\omega_1^{n_1}}{n_1!} \cdot \frac{\omega_2^{n_2}}{n_2!} \cdots \right) \tag{15}$$

but the summation appearing here cannot be evaluated explicitly because it is restricted to those sets only that conform to the pair of conditions (1). If our distribution sets were restricted by the condition $\sum_r n_r = \mathcal{N}$ alone, then the evaluation of (15) would have been trivial; by the multinomial theorem, $\Gamma(\mathcal{N})$ would have been simply $(\omega_0 + \omega_1 + \cdots)^{\mathcal{N}}$. The added restriction $\sum_r n_r E_r = \mathcal{N} U$, however, permits the inclusion of only a "limited" number of terms in the sum — and that constitutes the real difficulty of the problem. Nevertheless, we can still hope to make some progress because, from a physical point of view, we do not require anything more than an *asymptotic* result — one that holds in the limit $\mathcal{N} \to \infty$. The method commonly used for this purpose is the one developed by Darwin and Fowler (1922a,b, 1923), which itself makes use of the *saddle-point method* of integration or the so-called *method of steepest descent.*

We construct a *generating function* $G(\mathcal{N}, z)$ for the quantity $\Gamma(\mathcal{N}, U)$:

$$G(\mathcal{N}, z) = \sum_{U=0}^{\infty} \Gamma(\mathcal{N}, U) z^{\mathcal{N} U} \tag{16}$$

which, in view of equation (15) and the second of the restrictive conditions (1), may be written as

$$G(\mathcal{N},z) = \sum_{U=0}^{\infty}\left[\sum_{\{n_r\}}{}' \frac{\mathcal{N}!}{n_0!\,n_1!\ldots}\left(\omega_0 z^{E_0}\right)^{n_0}\left(\omega_1 z^{E_1}\right)^{n_1}\ldots\right]. \tag{17}$$

It is easy to see that the summation over *doubly* restricted sets $\{n_r\}$, followed by a summation over all possible values of U, is equivalent to a summation over *singly* restricted sets $\{n_r\}$, namely the ones that satisfy only one condition: $\sum_r n_r = \mathcal{N}$. Expression (17) can be evaluated with the help of the multinomial theorem, with the result

$$\begin{aligned} G(\mathcal{N},z) &= \left(\omega_0 z^{E_0} + \omega_1 z^{E_1} + \cdots\right)^{\mathcal{N}} \\ &= [f(z)]^{\mathcal{N}}, \text{ say.} \end{aligned} \tag{18}$$

Now, if we suppose that the E_r (and hence the total energy value $\mathcal{E} = \mathcal{N}U$) are all integers, then, by (16), the quantity $\Gamma(\mathcal{N},U)$ is simply the coefficient of $z^{\mathcal{N}U}$ in the expansion of the function $G(\mathcal{N},z)$ as a power series in z. It can, therefore, be evaluated by the method of residues in the complex z-plane.

To make this plan work, we assume to have chosen, *right at the outset*, a unit of energy so small that, to any desired degree of accuracy, we can regard the energies E_r (and the prescribed total energy $\mathcal{N}U$) as integral multiples of this unit. In terms of this unit, any energy value we come across will be an integer. We further assume, without loss of generality, that the sequence $E_0, E_1, \ldots$ is a *nondecreasing* sequence, *with no common divisor*;[3] also, for the sake of simplicity, we assume that $E_0 = 0$.[4] The solution now is

$$\Gamma(\mathcal{N},U) = \frac{1}{2\pi i}\oint \frac{[f(z)]^{\mathcal{N}}}{z^{\mathcal{N}U+1}}\,dz, \tag{19}$$

where the integration is carried along any closed contour around the origin; of course, we must stay *within* the circle of convergence of the function $f(z)$, so that a need for analytic continuation does not arise.

First of all, we examine the behavior of the integrand as we proceed from the origin along the real positive axis, remembering that all our ω_r are virtually equal to unity and that $0 = E_0 \le E_1 \le E_2 \cdots$. We find that the factor $[f(z)]^{\mathcal{N}}$ starts from the value 1 at $z = 0$, increases monotonically and tends to infinity as z approaches the circle of convergence of $f(z)$, wherever that may be. The factor $z^{-(\mathcal{N}U+1)}$, on the other hand, starts from a positive, infinite value at $z = 0$ and decreases monotonically as z increases. Moreover, the relative rate of increase of the factor $[f(z)]^{\mathcal{N}}$ itself increases monotonically while the relative rate

[3] Actually, this is not a serious restriction at all, for a common divisor, if any, can be removed by selecting the unit of energy correspondingly larger.

[4] This too is not serious, for by doing so we are merely shifting the zero of the energy scale; the mean energy U then becomes $U - E_0$, but we can agree to call it U again.

of decrease of the factor $z^{-(\mathcal{N}U+1)}$ decreases monotonically. Under these circumstances, the integrand must exhibit a minimum (and no other extremum) at some value of z, say x_0, within the circle of convergence. And, in view of the largeness of the numbers $\mathcal{N}$ and $\mathcal{N}U$, this minimum may indeed be very steep!

Thus, at $z = x_0$ the first derivative of the integrand must vanish, while the second derivative must be positive and, hopefully, very large. Accordingly, if we proceed through the point $z = x_0$ in a direction orthogonal to the real axis, the integrand must exhibit an equally steep maximum.[5] Thus, in the complex z-plane, as we move along the real axis our integrand shows a minimum at $z = x_0$, whereas if we move along a path parallel to the imaginary axis but passing through the point $z = x_0$, the integrand shows a maximum there. It is natural to call the point x_0 a *saddle point*; see Figure 3.2. For the contour of integration we choose a circle, with center at $z = 0$ and radius equal to x_0, hoping that on integration along this contour only the immediate neighborhood of the sharp maximum at the point x_0 will make the most dominant contribution to the value of the integral.[6]

To carry out the integration we first locate the point x_0. For this we write our integrand as

$$\frac{[f(z)]^{\mathcal{N}}}{z^{\mathcal{N}U+1}} = \exp[\mathcal{N}g(z)], \tag{20}$$

where

$$g(z) = \ln f(z) - \left(U + \frac{1}{\mathcal{N}}\right)\ln z, \tag{21}$$

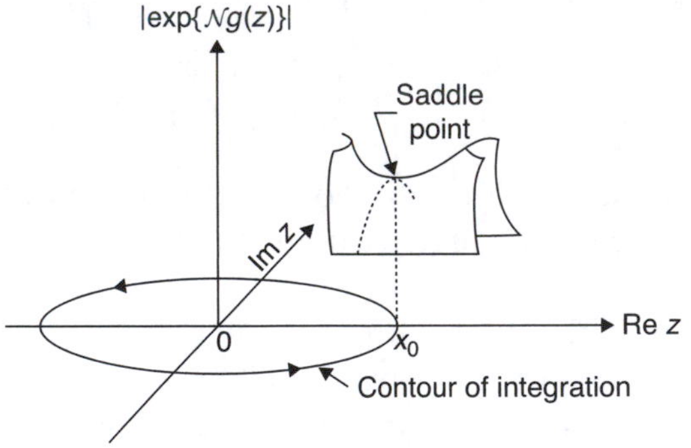

FIGURE 3.2 The saddle point.

[5]This can be seen by noting that (i) an analytic function must possess a *unique* derivative everywhere (so, in our case, it must be zero, irrespective of the direction in which we pass through the point x_0), and (ii) by the Cauchy–Riemann conditions of analyticity, the second derivative of the function with respect to y must be equal and opposite to the second derivative with respect to x.

[6]It is indeed true that, for large $\mathcal{N}$, the contribution from the rest of the circle is negligible. The intuitive reason for this is that the terms $(\omega_r z^{E_r})$, which constitute the function $f(z)$, "reinforce" one another *only* at the point $z = x_0$; elsewhere, there is bound to be disagreement among their phases, so that at *all* other points along the circle, $|f(z)| < f(x_0)$. Now, the factor that actually governs the relative contributions is $[|f(z)|/f(x_0)]^{\mathcal{N}}$; for $\mathcal{N} \gg 1$, this will clearly be negligible. For a rigorous demonstration of this point, see Schrödinger (1960, pp. 31–33).

while

$$f(z) = \sum_r \omega_r z^{E_r}. \tag{22}$$

The number x_0 is then determined by the equation

$$g'(x_0) = \frac{f'(x_0)}{f(x_0)} - \frac{\mathcal{N}U+1}{\mathcal{N}x_0} = 0, \tag{23}$$

which, in view of the fact that $\mathcal{N}U \gg 1$, can be written as

$$U \approx x_0 \frac{f'(x_0)}{f(x_0)} = \frac{\sum_r \omega_r E_r x_0^{E_r}}{\sum_r \omega_r x_0^{E_r}}. \tag{24}$$

We further have

$$\begin{aligned} g''(x_0) &= \left(\frac{f''(x_0)}{f(x_0)} - \frac{[f'(x_0)]^2}{[f(x_0)]^2} \right) + \frac{\mathcal{N}U+1}{\mathcal{N}x_0^2} \\ &\approx \frac{f''(x_0)}{f(x_0)} - \frac{U^2 - U}{x_0^2}. \end{aligned} \tag{25}$$

It will be noted here that, in the limit $\mathcal{N} \to \infty$ and $\mathscr{E} (\equiv \mathcal{N}U) \to \infty$, *with U staying constant*, the number x_0 and the quantity $g''(x_0)$ become independent of $\mathcal{N}$.

Expanding $g(z)$ about the point $z = x_0$, along the direction of integration, that is, along the line $z = x_0 + iy$, we have

$$g(z) = g(x_0) - \frac{1}{2} g''(x_0) y^2 + \cdots;$$

accordingly, the integrand (20) might be approximated as

$$\frac{[f(x_0)]^{\mathcal{N}}}{x_0^{\mathcal{N}U+1}} \exp\left[-\frac{\mathcal{N}}{2} g''(x_0) y^2 \right]. \tag{26}$$

Equation (19) then gives

$$\begin{aligned} \Gamma(\mathcal{N}, U) &\simeq \frac{1}{2\pi i} \frac{[f(x_0)]^{\mathcal{N}}}{x_0^{\mathcal{N}U+1}} \int_{-\infty}^{\infty} \exp\left[-\frac{\mathcal{N}}{2} g''(x_0) y^2 \right] i\, dy \\ &= \frac{[f(x_0)]^{\mathcal{N}}}{x_0^{\mathcal{N}U+1}} \cdot \frac{1}{\{2\pi \mathcal{N} g''(x_0)\}^{1/2}}, \end{aligned} \tag{27}$$

which gives

$$\frac{1}{\mathcal{N}} \ln \Gamma(\mathcal{N}, U) = \{\ln f(x_0) - U \ln x_0\} - \frac{1}{\mathcal{N}} \ln x_0 - \frac{1}{2\mathcal{N}} \ln\{2\pi \mathcal{N} g''(x_0)\}. \tag{28}$$

In the limit $\mathcal{N} \to \infty$ (with U staying constant), the last two terms in this expression tend to zero, with the result

$$\frac{1}{\mathcal{N}} \ln \Gamma(\mathcal{N}, U) = \ln f(x_0) - U \ln x_0. \tag{29}$$

Substituting for $f(x_0)$ and introducing a new variable β, defined by the relationship

$$x_0 \equiv \exp(-\beta), \tag{30}$$

we get

$$\frac{1}{\mathcal{N}} \ln \Gamma(\mathcal{N}, U) = \ln \left\{ \sum_r \omega_r \exp(-\beta E_r) \right\} + \beta U. \tag{31}$$

The expectation value of the number n_r then follows from (14) and (31):

$$\frac{\langle n_r \rangle}{\mathcal{N}} = \left[\frac{\omega_r \exp(-\beta E_r)}{\sum_r \omega_r \exp(-\beta E_r)} + \left\{ -\frac{\sum_r \omega_r E_r \exp(-\beta E_r)}{\sum_r \omega_r \exp(-\beta E_r)} + U \right\} \omega_r \frac{\partial \beta}{\partial \omega_r} \right]_{\text{all } \omega_r = 1}. \tag{32}$$

The term inside the curly brackets vanishes identically because of (24) and (30). It has been included here to emphasize the fact that, for a fixed value of U, the number $\beta (\equiv -\ln x_0)$ in fact depends on the choice of the ω_r; see (24). We will appreciate the importance of this fact when we evaluate the mean square fluctuation in the number n_r; in the calculation of the expectation value of n_r, this does not really matter. We thus obtain

$$\frac{\langle n_r \rangle}{\mathcal{N}} = \frac{\exp(-\beta E_r)}{\sum_r \exp(-\beta E_r)}, \tag{33}$$

which is identical to expression (10) for $n_r^*/\mathcal{N}$. The physical significance of the parameter β is also the same as in that expression, for it is determined by equation (24), with all $\omega_r = 1$, that is, by equation (11) which fits naturally with equation (33) because U is nothing but the ensemble average of the variable E_r:

$$U = \sum_r E_r P_r = \frac{1}{\mathcal{N}} \sum_r E_r \langle n_r \rangle. \tag{34}$$

Finally, we compute fluctuations in the values of the numbers n_r. We have, first of all,

$$\langle n_r^2 \rangle \equiv \frac{\sum'_{\{n_r\}} n_r^2 W\{n_r\}}{\sum'_{\{n_r\}} W\{n_r\}} = \frac{1}{\Gamma} \left(\omega_r \frac{\partial}{\partial \omega_r} \right) \left(\omega_r \frac{\partial}{\partial \omega_r} \right) \Gamma \bigg|_{\text{all } \omega_r = 1}; \tag{35}$$

see equations (12) to (14). It follows that

$$\langle (\Delta n_r)^2 \rangle \equiv \langle \{ n_r - \langle n_r \rangle \}^2 \rangle = \langle n_r^2 \rangle - \langle n_r \rangle^2 = \left(\omega_r \frac{\partial}{\partial \omega_r} \right) \left(\omega_r \frac{\partial}{\partial \omega_r} \right) \ln \Gamma \bigg|_{\text{all } \omega_r = 1}. \tag{36}$$

Substituting from (31) and making use of (32), we get

$$\frac{\langle(\Delta n_r)^2\rangle}{\mathcal{N}} = \omega_r \frac{\partial}{\partial \omega_r}\left[\frac{\omega_r \exp(-\beta E_r)}{\sum\limits_r \omega_r \exp(-\beta E_r)} + \left\{-\frac{\sum\limits_r \omega_r E_r \exp(-\beta E_r)}{\sum\limits_r \omega_r \exp(-\beta E_r)} + U\right\} \omega_r \frac{\partial \beta}{\partial \omega_r}\right]_{\text{all } \omega_r = 1}. \tag{37}$$

We note that the term in the curly brackets would not make any contribution because it is identically zero, *whatever the choice of the* ω_r. However, in the differentiation of the first term, we must not forget to take into account the *implicit* dependence of β on the ω_r, which arises from the fact that unless the ω_r are set equal to unity the relation determining β does contain ω_r; see equations (24) and (30), whereby

$$U = \left.\frac{\sum\limits_r \omega_r E_r \exp(-\beta E_r)}{\sum\limits_r \omega_r \exp(-\beta E_r)}\right|_{\text{all } \omega_r = 1}. \tag{38}$$

A straightforward calculation gives

$$\left.\left(\frac{\partial \beta}{\partial \omega_r}\right)_U\right|_{\text{all } \omega_r = 1} = \frac{E_r - U}{\langle E_r^2\rangle - U^2} \frac{\langle n_r\rangle}{\mathcal{N}}. \tag{39}$$

We can now evaluate (37), with the result

$$\begin{aligned}\frac{\langle(\Delta n_r)^2\rangle}{\mathcal{N}} &= \frac{\langle n_r\rangle}{\mathcal{N}} - \left(\frac{\langle n_r\rangle}{\mathcal{N}}\right)^2 + \frac{\langle n_r\rangle}{\mathcal{N}}(U - E_r)\left.\left(\frac{\partial \beta}{\partial \omega_r}\right)_U\right|_{\text{all } \omega_r = 1} \\ &= \frac{\langle n_r\rangle}{\mathcal{N}}\left[1 - \frac{\langle n_r\rangle}{\mathcal{N}} - \frac{\langle n_r\rangle}{\mathcal{N}} \frac{(E_r - U)^2}{\langle(E_r - U)^2\rangle}\right].\end{aligned} \tag{40}$$

For the relative fluctuation in n_r, we get

$$\left\langle\left(\frac{\Delta n_r}{\langle n_r\rangle}\right)^2\right\rangle = \frac{1}{\langle n_r\rangle} - \frac{1}{\mathcal{N}}\left\{1 + \frac{(E_r - U)^2}{\langle(E_r - U)^2\rangle}\right\}. \tag{41}$$

As $\mathcal{N} \to \infty$, $\langle n_r\rangle$ also $\to \infty$, with the result that the relative fluctuations in n_r tend to zero; accordingly, the canonical distribution becomes infinitely sharp and with it the mean value, the most probable value — in fact, any values of n_r that appear with nonvanishing probability — become essentially the same. And that is the reason why two wildly different methods of obtaining the canonical distribution followed in this section have led to identical results.

3.3 Physical significance of the various statistical quantities in the canonical ensemble

We start with the *canonical distribution*

$$P_r \equiv \frac{\langle n_r \rangle}{\mathcal{N}} = \frac{\exp(-\beta E_r)}{\sum_r \exp(-\beta E_r)}, \tag{1}$$

where β is determined by the equation

$$U = \frac{\sum_r E_r \exp(-\beta E_r)}{\sum_r \exp(-\beta E_r)} = -\frac{\partial}{\partial \beta} \ln\left\{\sum_r \exp(-\beta E_r)\right\}. \tag{2}$$

We now look for a general recipe to extract information about the various macroscopic properties of the given system on the basis of the foregoing statistical results. For this, we recall certain thermodynamic relationships involving the Helmholtz free energy $A(= U - TS)$, namely

$$dA = dU - TdS - SdT = -SdT - PdV + \mu\, dN, \tag{3}$$

$$S = -\left(\frac{\partial A}{\partial T}\right)_{N,V}, \quad P = -\left(\frac{\partial A}{\partial V}\right)_{N,T}, \quad \mu = \left(\frac{\partial A}{\partial N}\right)_{V,T}, \tag{4}$$

and

$$U = A + TS = A - T\left(\frac{\partial A}{\partial T}\right)_{N,V} = -T^2\left[\frac{\partial}{\partial T}\left(\frac{A}{T}\right)\right]_{N,V} = \left[\frac{\partial(A/T)}{\partial(1/T)}\right]_{N,V}, \tag{5}$$

where the various symbols have their usual meanings. Comparing (5) with (2), we infer that there exists a close correspondence between the quantities entering through the statistical treatment and the ones coming from thermodynamics, namely

$$\beta = \frac{1}{kT}, \quad \ln\left\{\sum_r \exp(-\beta E_r)\right\} = -\frac{A}{kT}, \tag{6}$$

where k is a universal constant yet to be determined; soon we shall see that k is indeed the *Boltzmann constant*.

The equations in (6) constitute the most fundamental result of the canonical ensemble theory. Customarily, we write it in the form

$$A(N, V, T) = -kT \ln Q_N(V, T), \tag{7}$$

where

$$Q_N(V, T) = \sum_r \exp(-E_r/kT). \tag{8}$$

The quantity $Q_N(V,T)$ is referred to as the *partition function* of the system; sometimes it is also called the "sum-over-states" (German: *Zustandssumme*). The dependence of Q on T is quite obvious. The dependence on N and V comes through the energy eigenvalues E_r; in fact, any other parameters that might govern the values E_r should also appear in the argument of Q. Moreover, for the quantity $A(N,V,T)$ to be an extensive property of the system, $\ln Q$ must also be an extensive quantity.

Once the Helmholtz free energy is known, the rest of the thermodynamic quantities follow straightforwardly. While the entropy, the pressure and the chemical potential are obtained from formulae (4), the specific heat at constant volume follows from

$$C_V = \left(\frac{\partial U}{\partial T}\right)_{N,V} = -T\left(\frac{\partial^2 A}{\partial T^2}\right)_{N,V} \tag{9}$$

and the Gibbs free energy from

$$G = A + PV = A - V\left(\frac{\partial A}{\partial V}\right)_{N,T} = N\left(\frac{\partial A}{\partial N}\right)_{V,T} = N\mu; \tag{10}$$

see Problem 3.5.

At this stage it appears worthwhile to make a few remarks on the foregoing results. First of all, we note from equations (4) and (6) that the pressure P is given by

$$P = -\frac{\sum_r \frac{\partial E_r}{\partial V}\exp(-\beta E_r)}{\sum_r \exp(-\beta E_r)}, \tag{11}$$

so that

$$PdV = -\sum_r P_r dE_r = -dU. \tag{12}$$

The quantity on the right side of this equation is clearly the change in the average energy of a system (in the ensemble) during a process that alters the energy levels E_r, leaving the probabilities P_r unchanged. The left side then tells us that the volume change dV provides an example of such a process, and the pressure P is the "force" accompanying that process. The quantity P, which was introduced here through the thermodynamic relationship (3), thus acquires a mechanical meaning as well.

The entropy of the system is determined as follows. Since $P_r = Q^{-1}\exp(-\beta E_r)$,

$$\langle \ln P_r\rangle = -\ln Q - \beta\langle E_r\rangle = \beta(A-U) = -S/k,$$

with the result that

$$S = -k\langle \ln P_r\rangle = -k\sum_r P_r \ln P_r. \tag{13}$$

This is an extremely interesting relationship, for it shows that the entropy of a physical system is *solely* and *completely* determined by the probability values P_r (of the system being in different dynamical states accessible to it)!

From the very look of it, equation (13) appears to be of fundamental importance; indeed, it reveals a number of interesting conclusions. One of these relates to a system in its ground state ($T = 0\,\mathrm{K}$). If the ground state is unique, then the system is sure to be found in this particular state and in no other; consequently, P_r is equal to 1 for this state and 0 for all others. Equation (13) then tells us that the entropy of the system is precisely zero, which is essentially the content of the *Nernst heat theorem* or the *third law of thermodynamics*.[7] We also infer that vanishing entropy and perfect statistical order (which implies complete predictability about the system) go together. As the number of accessible states increases, more and more of the P_r become nonzero; the entropy of the system thereby increases. As the number of states becomes exceedingly large, most of the P-values become exceedingly small (and their logarithms assume large, negative values); the net result is that the entropy becomes exceedingly large. Thus, the largeness of entropy and the high degree of statistical disorder (or unpredictability) in the system also go hand in hand.

It is because of this fundamental connection between entropy on one hand and lack of information on the other that equation (13) became the starting point of the pioneering work of Shannon (1948, 1949) in the development of the theory of communication.

It may be pointed out that formula (13) applies in the microcanonical ensemble as well. There, for each member system of the ensemble, we have a group of Ω states, all *equally likely* to occur. The value of P_r is, then, $1/\Omega$ for each of these states and 0 for all others. Consequently,

$$S = -k\sum_{r=1}^{\Omega}\left\{\frac{1}{\Omega}\ln\left(\frac{1}{\Omega}\right)\right\} = k\ln\Omega, \tag{14}$$

which is precisely the central result in the microcanonical ensemble theory; see equation (1.2.6) or (2.3.6).

3.4 Alternative expressions for the partition function

In most physical cases the energy levels accessible to a system are *degenerate*, that is, one has a group of states, g_i in number, all belonging to the same energy value E_i. In such cases it is more useful to write the partition function (3.3.8) as

$$Q_N(V,T) = \sum_i g_i \exp(-\beta E_i); \tag{1}$$

[7] Of course, if the ground state of the system is *degenerate* (with a multiplicity Ω_0), then the ground-state entropy is nonzero and is given by the expression $k\ln\Omega_0$; see equation (14).

the corresponding expression for P_i, the probability that the system be in a state with energy E_i, would be

$$P_i = \frac{g_i \exp(-\beta E_i)}{\sum_i g_i \exp(-\beta E_i)}. \tag{2}$$

Clearly, the g_i states with a common energy E_i are all equally likely to occur. As a result, the probability of a system having energy E_i becomes proportional to the multiplicity g_i of this level; g_i thus plays the role of a "weight factor" for the level E_i. The actual probability is then determined by the weight factor g_i as well as by the Boltzmann factor $\exp(-\beta E_i)$ of the level, as we have in (2). The basic relations established in the preceding section, however, remain unaffected.

Now, in view of the largeness of the number of particles constituting a given system and the largeness of the volume to which these particles are confined, the consecutive energy values E_i of the system are, in general, very close to one another. Accordingly, there lie, within any reasonable interval of energy $(E, E + dE)$, a very large number of energy levels. One may then regard E as a *continuous* variable and write $P(E)dE$ for the probability that the given system, as a member of the canonical ensemble, may have its energy in the range $(E, E + dE)$. Clearly, this probability will be given by the product of the relevant single-state probability and the number of energy states lying in the specified range. Denoting the latter by $g(E)dE$, where $g(E)$ denotes the *density of states* around the energy value E, we have

$$P(E)dE \propto \exp(-\beta E) g(E)dE \tag{3}$$

which, on normalization, becomes

$$P(E)dE = \frac{\exp(-\beta E) g(E)dE}{\int_0^\infty \exp(-\beta E) g(E)dE}. \tag{4}$$

The denominator here is yet another expression for the partition function of the system:

$$Q_N(V, T) = \int_0^\infty e^{-\beta E} g(E)dE. \tag{5}$$

The expression for $\langle f \rangle$, the expectation value of a physical quantity f, may now be written as

$$\langle f \rangle \equiv \sum_i f_i P_i = \frac{\sum_i f(E_i) g_i e^{-\beta E_i}}{\sum_i g_i e^{-\beta E_i}} \to \frac{\int_0^\infty f(E) e^{-\beta E} g(E)dE}{\int_0^\infty e^{-\beta E} g(E)dE}. \tag{6}$$

Before proceeding further, we take a closer look at equation (5) and note that, with $\beta > 0$, the partition function $Q(\beta)$ is just the Laplace transform of the density of states $g(E)$. We may, therefore, write $g(E)$ as the inverse Laplace transform of $Q(\beta)$:

$$g(E) = \frac{1}{2\pi i} \int_{\beta'-i\infty}^{\beta'+i\infty} e^{\beta E} Q(\beta) d\beta \quad (\beta' > 0) \tag{7}$$

$$= \frac{1}{2\pi} \int_{-\infty}^{\infty} e^{(\beta'+i\beta'')E} Q(\beta' + i\beta'') d\beta'', \tag{8}$$

where β is now treated as a complex variable, $\beta' + i\beta''$, while the path of integration runs parallel to, and to the right of, the imaginary axis, that is, along the straight line $\mathrm{Re}\,\beta = \beta' > 0$. Of course, the path may be continuously deformed so long as the integral converges.

3.5 The classical systems

The theory developed in the preceding sections is of very general applicability. It applies to systems in which quantum-mechanical effects are important as well as to those that can be treated classically. In the latter case, our formalism may be written in the language of the phase space; as a result, the summations over quantum states get replaced by integrations over phase space.

We recall the concepts developed in Sections 2.1 and 2.2, especially formula (2.1.3) for the ensemble average $\langle f \rangle$ of a physical quantity $f(q,p)$, namely

$$\langle f \rangle = \frac{\int f(q,p)\rho(q,p) d^{3N}q\, d^{3N}p}{\int \rho(q,p) d^{3N}q\, d^{3N}p}, \tag{1}$$

where $\rho(q,p)$ denotes the density of the representative points (of the systems) in the phase space; we have omitted here the explicit dependence of the function ρ on time t because we are interested in the study of equilibrium situations only. Evidently, the function $\rho(q,p)$ is a measure of the probability of finding a representative point in the vicinity of the phase point (q,p), which in turn depends on the corresponding value $H(q,p)$ of the Hamiltonian of the system. In the canonical ensemble,

$$\rho(q,p) \propto \exp\{-\beta H(q,p)\}; \tag{2}$$

compare to equation (3.1.6). The expression for $\langle f \rangle$ then takes the form

$$\langle f \rangle = \frac{\int f(q,p) \exp(-\beta H) d\omega}{\int \exp(-\beta H) d\omega}, \tag{3}$$

where $d\omega(\equiv d^{3N}q\,d^{3N}p)$ denotes a volume element of the phase space. The denominator of this expression is directly related to the partition function of the system. However, to write the precise expression for the latter, we must take into account the relationship between a volume element in the phase space and the corresponding number of distinct quantum states of the system. This relationship was established in Sections 2.4 and 2.5, whereby an element of volume $d\omega$ in the phase space corresponds to

$$\frac{d\omega}{N!\,h^{3N}} \tag{4}$$

distinct quantum states of the system.[8] The appropriate expression for the partition function would, therefore, be

$$Q_N(V,T) = \frac{1}{N!\,h^{3N}} \int e^{-\beta H(q,p)}\,d\omega; \tag{5}$$

it is understood that the integration in (5) goes over the *whole* of the phase space.

As our first application of this formulation, we consider the example of an ideal gas. Here, we have a system of N identical molecules, assumed to be monatomic (so there are no internal degrees of motion to be considered), confined to a space of volume V and in equilibrium at temperature T. Since there are no intermolecular interactions to be taken into account, the energy of the system is wholly kinetic:

$$H(q,p) = \sum_{i=1}^{N} (p_i^2/2m). \tag{6}$$

The partition function of the system would then be

$$Q_N(V,T) = \frac{1}{N!\,h^{3N}} \int e^{-(\beta/2m)\Sigma_i p_i^2} \prod_{i=1}^{N} (d^3q_i\,d^3p_i). \tag{7}$$

Integrations over the space coordinates are rather trivial; they yield a factor of V^N. Integrations over the momentum coordinates are also quite easy, once we note that integral (7) is simply a product of N identical integrals. Thus, we get

$$Q_N(V,T) = \frac{V^N}{N!\,h^{3N}} \left[\int_0^{\infty} e^{-p^2/2mkT} \left(4\pi p^2\,dp\right) \right]^N \tag{8}$$

$$= \frac{1}{N!} \left[\frac{V}{h^3} (2\pi mkT)^{3/2} \right]^N; \tag{9}$$

[8] Ample justification has already been given for the factor h^{3N}. The factor $N!$ comes from the considerations of Sections 1.5 and 1.6; it arises essentially from the fact that the particles constituting the given system are not only identical but, in fact, *indistinguishable*. For a complete proof of this result, see Section 5.5.

here, use has been made of equation (B.13a). The Helmholtz free energy is then given by, using Stirling's formula (B.29),

$$A(N,V,T) \equiv -kT \ln Q_N(V,T) = NkT\left[\ln\left\{\frac{N}{V}\left(\frac{h^2}{2\pi mkT}\right)^{3/2}\right\} - 1\right]. \tag{10}$$

The foregoing result is identical to equation (1.5.8), which was obtained by following a very different procedure. The simplicity of the present approach is, however, striking. Needless to say, the complete thermodynamics of the ideal gas can be derived from equation (10) in a straightforward manner. For instance,

$$\mu \equiv \left(\frac{\partial A}{\partial N}\right)_{V,T} = kT \ln\left\{\frac{N}{V}\left(\frac{h^2}{2\pi mkT}\right)^{3/2}\right\}, \tag{11}$$

$$P \equiv -\left(\frac{\partial A}{\partial V}\right)_{N,T} = \frac{NkT}{V} \tag{12}$$

and

$$S \equiv -\left(\frac{\partial A}{\partial T}\right)_{N,V} = Nk\left[\ln\left\{\frac{V}{N}\left(\frac{2\pi mkT}{h^2}\right)^{3/2}\right\} + \frac{5}{2}\right]. \tag{13}$$

These results are identical to the ones derived previously, namely (1.5.7), (1.4.2), and (1.5.1a), respectively. In fact, the identification of formula (12) with the ideal-gas law, $PV = nRT$, establishes the identity of the (hitherto undetermined) constant k as the *Boltzmann constant*; see equation (3.3.6). We further obtain

$$U \equiv -\left[\frac{\partial}{\partial \beta}(\ln Q)\right]_{E_r} \equiv -T^2\left[\frac{\partial}{\partial T}\left(\frac{A}{T}\right)\right]_{N,V} \equiv A + TS = \frac{3}{2}NkT, \tag{14}$$

and so on.

At this stage we have an important remark to make. Looking at the form of equation (8) and the manner in which it came about, we may write

$$Q_N(V,T) = \frac{1}{N!}[Q_1(V,T)]^N, \tag{15}$$

where $Q_1(V,T)$ may be regarded as the partition function of a *single* molecule in the system. A little reflection will show that this result obtains essentially from the fact that the basic constituents of our system are noninteracting (and hence the total energy of the system is simply the sum of their individual energies). Clearly, the situation will not be altered even if the molecules in the system had *internal* degrees of motion as well. What is essentially required for equation (15) to be valid is the absence of interactions among the basic constituents of the system (and, of course, the absence of quantum-mechanical correlations).

Going back to the ideal gas, we could as well have started with the density of states $g(E)$. From equation (1.4.17), and in view of the Gibbs correction factor, we have

$$g(E) = \frac{\partial \Sigma}{\partial E} \approx \frac{1}{N!}\left(\frac{V}{h^3}\right)^N \frac{(2\pi m)^{3N/2}}{\{(3N/2)-1\}!} E^{(3N/2)-1}. \tag{16}$$

Substituting this into equation (3.4.5), and noting that the integral involved is equal to $\{(3N/2)-1\}!/\beta^{3N/2}$, we obtain

$$Q_N(\beta) = \frac{1}{N!}\left(\frac{V}{h^3}\right)^N \left(\frac{2\pi m}{\beta}\right)^{3N/2}, \tag{17}$$

which is identical to (9). It may also be noted that if one starts with the single-particle density of states (2.4.7), namely

$$a(\varepsilon) \approx \frac{2\pi V}{h^3}(2m)^{3/2}\varepsilon^{1/2}, \tag{18}$$

computes the single-particle partition function,

$$Q_1(\beta) = \int_0^\infty e^{-\beta\varepsilon} a(\varepsilon) d\varepsilon = \frac{V}{h^3}\left(\frac{2\pi m}{\beta}\right)^{3/2}, \tag{19}$$

and then makes use of formula (15), one would arrive at the same result for $Q_N(V, T)$.

Lastly, we consider the question of determining the density of states, $g(E)$, from the expression for the partition function, $Q(\beta)$ — assuming that the latter is already known; indeed, expression (9) for $Q(\beta)$ was derived without making use of any knowledge regarding the function $g(E)$. According to equation (3.4.7) and (9), we have

$$g(E) = \frac{V^N}{N!}\left(\frac{2\pi m}{h^2}\right)^{3N/2} \frac{1}{2\pi i} \int_{\beta'-i\infty}^{\beta'+i\infty} \frac{e^{\beta E}}{\beta^{3N/2}} d\beta \quad (\beta' > 0). \tag{20}$$

Noting that, for all positive n,

$$\frac{1}{2\pi i} \int_{s'-i\infty}^{s'+i\infty} \frac{e^{sx}}{s^{n+1}} ds = \begin{cases} \frac{x^n}{n!} & \text{for} \quad x \geq 0 \\ 0 & \text{for} \quad x \leq 0, \end{cases} \tag{21}$$ [9]

equation (20) becomes

$$g(E) = \begin{cases} \dfrac{V^N}{N!}\left(\dfrac{2\pi m}{h^2}\right)^{3N/2} \dfrac{E^{(3N/2)-1}}{\{(3N/2)-1\}!} & \text{for} \quad E \geq 0 \\ 0 & \text{for} \quad E \leq 0, \end{cases} \tag{22}$$

[9]For the details of this evaluation, see Kubo (1965, pp. 165–168).

which is indeed the correct result for the density of states of an ideal gas; compare to equation (16). The foregoing derivation may not appear particularly valuable because in the present case we already knew the expression for $g(E)$. However, cases do arise where the evaluation of the partition function of a given system and the consequent evaluation of its density of states turn out to be quite simple, whereas a direct evaluation of the density of states from first principles is rather involved. In such cases, the method given here can indeed be useful; see, for example, Problem 3.15 in comparison with Problems 1.7 and 2.8.

3.6 Energy fluctuations in the canonical ensemble: correspondence with the microcanonical ensemble

In the canonical ensemble, a system can have energy anywhere between zero and infinity. On the other hand, the energy of a system in the microcanonical ensemble is restricted to a very narrow range. How, then, can we assert that the thermodynamic properties of a system derived through the formalism of the canonical ensemble would be the same as the ones derived through the formalism of the microcanonical ensemble? Of course, we do expect that the two formalisms yield identical results, for otherwise our whole scheme would be marred by internal inconsistency. And, indeed, in the case of an ideal classical gas the results obtained by following one approach were precisely the same as the ones obtained by following the other approach. What, then, is the underlying reason for this equivalence?

The answer to this question is obtained by examining the extent of the range over which the energies of the systems in the canonical ensemble have a significant probability to spread; that will tell us the extent to which the canonical ensemble *really* differs from the microcanonical one. To explore this point, we write down the expression for the mean energy

$$U \equiv \langle E \rangle = \frac{\sum_r E_r \exp(-\beta E_r)}{\sum_r \exp(-\beta E_r)} \tag{1}$$

and differentiate it with respect to the parameter β, holding the energy values E_r constant. We obtain

$$\begin{aligned} \frac{\partial U}{\partial \beta} &= -\frac{\sum_r E_r^2 \exp(-\beta E_r)}{\sum_r \exp(-\beta E_r)} + \frac{\left[\sum_r E_r \exp(-\beta E_r)\right]^2}{\left[\sum_r \exp(-\beta E_r)\right]^2} \\ &= -\langle E^2 \rangle + \langle E \rangle^2, \end{aligned} \tag{2}$$

from which it follows that

$$\langle (\Delta E)^2 \rangle \equiv \langle E^2 \rangle - \langle E \rangle^2 = -\left(\frac{\partial U}{\partial \beta}\right) = kT^2 \left(\frac{\partial U}{\partial T}\right) = kT^2 C_V. \tag{3}$$

Note that we have here the specific heat *at constant volume*, because the partial differentiation in (2) was carried out with the E_r kept constant! For the relative root-mean-square fluctuation in E, equation (3) gives

$$\frac{\sqrt{[\langle(\Delta E)^2\rangle]}}{\langle E\rangle} = \frac{\sqrt{(kT^2 C_V)}}{U}, \tag{4}$$

which is $O(N^{-1/2})$, N being the number of particles in the system. Consequently, for large N (which is true for every statistical system) the relative r.m.s. fluctuation in the values of E is quite negligible! Thus, for all practical purposes, a system in the canonical ensemble has an energy equal to, or almost equal to, the mean energy U; the situation in this ensemble is, therefore, practically the same as in the microcanonical ensemble. That explains why the two ensembles lead to practically identical results.

For further understanding of the situation, we consider the manner in which energy is distributed among the various members of the (canonical) ensemble. To do this, we treat E as a continuous variable and start with expression (3.4.3), namely

$$P(E)dE \propto \exp(-\beta E)g(E)dE. \tag{3.4.3}$$

The probability density $P(E)$ is given by the product of two factors: (i) the Boltzmann factor, which monotonically decreases with E, and (ii) the density of states, which monotonically increases with E. The product, therefore, has an extremum at some value of E, say E^*.[10] The value E^* is determined by the condition

$$\frac{\partial}{\partial E}\{e^{-\beta E}g(E)\}\bigg|_{E=E^*} = 0,$$

that is, by

$$\frac{\partial \ln g(E)}{\partial E}\bigg|_{E=E^*} = \beta. \tag{5}$$

Recalling that

$$S = k\ln g \quad \text{and} \quad \left(\frac{\partial S(E)}{\partial E}\right)_{E=U} = \frac{1}{T} = k\beta,$$

the foregoing condition implies that

$$E^* = U. \tag{6}$$

This is a very interesting result, for it shows that, irrespective of the physical nature of the given system, the most probable value of its energy is identical to its mean value. Accordingly, if it is advantageous, we may use one instead of the other.

[10] Subsequently we shall see that this extremum is actually a maximum — and an extremely sharp one at that.

We now expand the logarithm of the probability density $P(E)$ around the value $E^* \approx U$; we get

$$\ln\left[e^{-\beta E}g(E)\right] = \left(-\beta U + \frac{S}{k}\right) + \frac{1}{2}\frac{\partial^2}{\partial E^2}\ln\left\{e^{-\beta E}g(E)\right\}\bigg|_{E=U}(E-U)^2 + \cdots$$

$$= -\beta(U - TS) - \frac{1}{2kT^2C_V}(E-U)^2 + \cdots, \tag{7}$$

from which we obtain

$$P(E) \propto e^{-\beta E}g(E) \simeq e^{-\beta(U-TS)}\exp\left\{-\frac{(E-U)^2}{2kT^2C_V}\right\}. \tag{8}$$

This is a *Gaussian* distribution in E, with mean value U and dispersion $\surd(kT^2C_V)$; compare with equation (3). In terms of the reduced variable E/U, the distribution is again Gaussian, with mean value unity and dispersion $\surd(kT^2C_V)/U$ {which is $O(N^{-1/2})$}; thus, for $N \gg 1$, we have an extremely sharp distribution which, as $N \to \infty$, approaches a delta-function!

It would be instructive here to consider once again the case of a classical ideal gas. Here, $g(E)$ is proportional to $E^{(3N/2-1)}$ and hence increases very fast with E; the factor $e^{-\beta E}$, of course, decreases with E. The product $g(E)\exp(-\beta E)$ exhibits a maximum at $E^* = (3N/2-1)\beta^{-1}$, which is practically the same as the mean value $U = (3N/2)\beta^{-1}$. For values of E significantly different from E^*, the product essentially vanishes (for smaller values of E, due to the relative paucity of the available energy states; for larger values of E, due to the relative depletion caused by the Boltzmann factor). The overall picture is shown in Figure 3.3 where we have displayed the actual behavior of these functions in the special case $N = 10$. The most probable value of E is now $\frac{14}{15}$ of the mean value; so, the distribution is somewhat asymmetrical. The effective width Δ can be readily calculated from (3) and turns out to be $(2/3N)^{1/2}U$, which, for $N = 10$, is about a quarter of U. We can see that, as N becomes large, both E^* and U increase (essentially linearly with N), the ratio E^*/U approaches unity and the distribution tends to become symmetrical about E^*. At the same time, the width Δ increases (but only as $N^{1/2}$); considered in the relative sense, it tends to vanish (as $N^{-1/2}$).

We finally look at the partition function $Q_N(V,T)$, as given by equation (3.4.5), with its integrand replaced by (8). We have

$$Q_N(V,T) \simeq e^{-\beta(U-TS)}\int_0^\infty e^{-(E-U)^2/2kT^2C_V}dE$$

$$\simeq e^{-\beta(U-TS)}\surd(2kT^2C_V)\int_{-\infty}^\infty e^{-x^2}dx$$

$$= e^{-\beta(U-TS)}\surd(2\pi kT^2C_V),$$

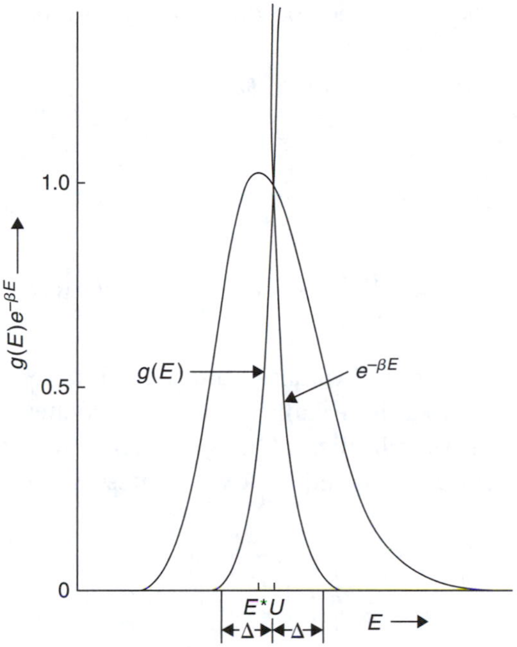

FIGURE 3.3 The actual behavior of the functions $g(E)$, $e^{-\beta E}$, and $g(E)e^{-\beta E}$ for an ideal gas, with $N = 10$. The numerical values of the functions have been expressed as fractions of their respective values at $E = U$.

so that

$$-kT\ln Q_N(V,T) \equiv A \simeq (U - TS) - \frac{1}{2}kT\ln(2\pi kT^2 C_V). \tag{9}$$

The last term, being $O(\ln N)$, is negligible in comparison with the other terms, which are all $O(N)$. Hence,

$$A \approx U - TS. \tag{10}$$

Note that the quantity A in this formula has come through the formalism of the canonical ensemble, while the quantity S has come through a definition belonging to the microcanonical ensemble. The fact that we finally end up with a consistent thermodynamic relationship establishes beyond doubt that these two approaches are, for all practical purposes, identical.

3.7 Two theorems — the "equipartition" and the "virial"

To derive these theorems, we determine the expectation value of the quantity $x_i(\partial H/\partial x_j)$, where $H(q,p)$ is the Hamiltonian of the system (assumed classical) while x_i

and x_j are any two of the 6N generalized coordinates (q, p). In the canonical ensemble,

$$\left\langle x_i \frac{\partial H}{\partial x_j} \right\rangle = \frac{\int \left(x_i \frac{\partial H}{\partial x_j} \right) e^{-\beta H} d\omega}{\int e^{-\beta H} d\omega} \qquad \left(d\omega = d^{3N}q\, d^{3N}p \right). \tag{1}$$

Let us consider the integral in the numerator. Integrating over x_j by parts, it becomes

$$\int \left[-\frac{1}{\beta} x_i e^{-\beta H} \Big|_{(x_j)_1}^{(x_j)_2} + \frac{1}{\beta} \int \left(\frac{\partial x_i}{\partial x_j} \right) e^{-\beta H} dx_j \right] d\omega_{(j)};$$

here, $(x_j)_1$ and $(x_j)_2$ are the "extreme" values of the coordinate x_j, while $d\omega_{(j)}$ denotes "$d\omega$ devoid of dx_j." The integrated part here vanishes because whenever any of the coordinates takes an "extreme" value the Hamiltonian of the system becomes infinite.[11] In the integral that remains, the factor $\partial x_i / \partial x_j$, being equal to δ_{ij}, comes out of the integral sign and we are left with

$$\frac{1}{\beta} \delta_{ij} \int e^{-\beta H} d\omega.$$

Substituting this into (1), we arrive at the remarkable result:

$$\left\langle x_i \frac{\partial H}{\partial x_j} \right\rangle = \delta_{ij} kT, \tag{2}$$

which is independent of the precise form of the function H.

In the special case $x_i = x_j = p_i$, equation (2) takes the form

$$\left\langle p_i \frac{\partial H}{\partial p_i} \right\rangle \equiv \langle p_i \dot{q}_i \rangle = kT, \tag{3}$$

while for $x_i = x_j = q_i$, it becomes

$$\left\langle q_i \frac{\partial H}{\partial q_i} \right\rangle \equiv -\langle q_i \dot{p}_i \rangle = kT. \tag{4}$$

Adding over all i, from $i = 1$ to $3N$, we obtain

$$\left\langle \sum_i p_i \frac{\partial H}{\partial p_i} \right\rangle \equiv \left\langle \sum_i p_i \dot{q}_i \right\rangle = 3NkT \tag{5}$$

[11] For instance, if x_j is a space coordinate, then its extreme values will correspond to "locations at the walls of the container"; accordingly, the potential energy of the system would become infinite. If, on the other hand, x_j is a momentum coordinate, then its extreme values will themselves be $\pm\infty$, in which case the kinetic energy of the system would become infinite.

and

$$\left\langle \sum_i q_i \frac{\partial H}{\partial q_i} \right\rangle \equiv -\left\langle \sum_i q_i \dot{p}_i \right\rangle = 3NkT. \tag{6}$$

Now, in many physical situations the Hamiltonian of the system happens to be a *quadratic* function of its coordinates; so, through a canonical transformation, it can be brought into the form

$$H = \sum_j A_j P_j^2 + \sum_j B_j Q_j^2, \tag{7}$$

where P_j and Q_j are the transformed, canonically conjugate, coordinates while A_j and B_j are certain constants of the problem. For such a system, we clearly have

$$\sum_j \left(P_j \frac{\partial H}{\partial P_j} + Q_j \frac{\partial H}{\partial Q_j} \right) = 2H; \tag{8}$$

accordingly, by equations (3) and (4),

$$\langle H \rangle = \frac{1}{2} fkT, \tag{9}$$

where f is the number of nonvanishing coefficients in expression (7). We, therefore, conclude that each harmonic term in the (transformed) Hamiltonian makes a contribution of $\frac{1}{2}kT$ toward the internal energy of the system and, hence, a contribution of $\frac{1}{2}k$ toward the specific heat C_V. This result embodies the classical theorem of *equipartition of energy* (among the various degrees of freedom of the system). It may be mentioned here that, for the distribution of kinetic energy alone, the equipartition theorem was first stated by Boltzmann (1871).

In our subsequent study we shall find that the equipartition theorem as stated here is not always valid; it applies only when the relevant degrees of freedom can be *freely* excited. At a given temperature T, there may be certain degrees of freedom which, due to the insufficiency of the energy available, are more or less "frozen" due to quantum mechanical effects. Such degrees of freedom do not make a significant contribution toward the internal energy of the system or toward its specific heat; see, for example, Sections 6.5, 7.4, and 8.3. Of course, the higher the temperature of the system the better the validity of this theorem.

We now consider the implications of formula (6). First of all, we note that this formula embodies the so-called *virial theorem* of Clausius (1870) for the quantity $\langle \sum_i q_i \dot{p}_i \rangle$, which is the expectation value of the sum of the products of the coordinates of the various particles in the system and the respective forces acting on them; this quantity is generally referred to as the *virial* of the system and is denoted by the symbol $\mathcal{V}$. The virial theorem then states that

$$\mathcal{V} = -3NkT. \tag{10}$$

The relationship between the virial and other physical quantities of the system is best understood by first looking at a classical gas of noninteracting particles. In this case, the only forces that come into play are the ones arising from the walls of the container; these forces can be designated by an external pressure P that acts on the system by virtue of the fact that it is bounded by the walls of the container. Consequently, we have here a force $-P\,d\boldsymbol{S}$ associated with an element of area $d\boldsymbol{S}$ of the walls; the negative sign appears because the force is directed *inward* while the vector $d\boldsymbol{S}$ is directed outward. The virial of the gas is then given by

$$\mathcal{V}_0 = \left(\sum_i q_i F_i\right)_0 = -P \oint_S \boldsymbol{r} \cdot d\boldsymbol{S}, \qquad (11)^{12}$$

where $\boldsymbol{r}$ is the position vector of a particle that happens to be in the (close) vicinity of the surface element $d\boldsymbol{S}$; accordingly, $\boldsymbol{r}$ may be considered to be the position vector of the surface element itself. By the divergence theorem, equation (11) becomes

$$\mathcal{V}_0 = -P \int_V (\text{div}\,\boldsymbol{r}) dV = -3PV. \tag{12}$$

Comparing (12) with (10), we obtain the well-known result:

$$PV = NkT. \tag{13}$$

The internal energy of the gas, which in this case is wholly kinetic, follows from the equipartition theorem (9) and is equal to $\frac{3}{2}NkT$, $3N$ being the number of degrees of freedom. Comparing this result with (10), we obtain the classical relationship

$$\mathcal{V} = -2K, \tag{14}$$

where K denotes the average kinetic energy of the system.

It is straightforward to apply this theorem to a system of particles interacting through a two-body potential $u(r)$. In the thermodynamic limit, the pressure of a d-dimensional system depends only on the virial terms arising from the forces between pairs of particles:

$$\frac{P}{nkT} = 1 + \frac{1}{NdkT}\left\langle \sum_{i<j} \boldsymbol{F}(\boldsymbol{r}_{ij}) \cdot \boldsymbol{r}_{ij} \right\rangle = 1 - \frac{1}{NdkT}\left\langle \sum_{i<j} \frac{\partial u(r_{ij})}{\partial r_{ij}} r_{ij} \right\rangle. \tag{15}$$

[12] It will be noted that the summation over the various particles of the system, which appears in the definition of the virial, has been replaced by an integration over the surface of the container, for the simple reason that no contribution to the virial arises from the interior of the container.

Equation (15) is called the *virial equation of state*. This equation can also be written in terms of the pair correlation function, equation (10.7.11), and is also used in computer simulations to determine the pressure of the system; see Problem 3.14, Section 10.7, and Section 16.4.

3.8 A system of harmonic oscillators

We shall now examine a system of N, practically independent, harmonic oscillators. This study will not only provide an interesting illustration of the canonical ensemble formulation but will also serve as a basis for some of our subsequent studies in this text. Two important problems in this line are (i) the theory of the black-body radiation (or the "statistical mechanics of photons") and (ii) the theory of lattice vibrations (or the "statistical mechanics of phonons"); see Sections 7.3 and 7.4 for details.

We start with the specialized situation when the oscillators can be treated *classically*. The Hamiltonian of any one of them (assumed to be one-dimensional) is then given by

$$H(q_i, p_i) = \frac{1}{2}m\omega^2 q_i^2 + \frac{1}{2m}p_i^2 \quad (i = 1, \ldots, N). \tag{1}$$

For the single-oscillator partition function, we readily obtain

$$\begin{aligned} Q_1(\beta) &= \int_{-\infty}^{\infty}\int_{-\infty}^{\infty} \exp\left\{-\beta\left(\frac{1}{2}m\omega^2 q^2 + \frac{1}{2m}p^2\right)\right\}\frac{dqdp}{h} \\ &= \frac{1}{h}\left(\frac{2\pi}{\beta m\omega^2}\right)^{1/2}\left(\frac{2\pi m}{\beta}\right)^{1/2} = \frac{1}{\beta\hbar\omega} = \frac{kT}{\hbar\omega}, \end{aligned} \tag{2}$$

where $\hbar = h/2\pi$. This represents a classical counting of the average number of accessible microstates — that is, kT divided by the quantum harmonic oscillator energy spacing. The partition function of the N-oscillator system would then be

$$Q_N(\beta) = [Q_1(\beta)]^N = (\beta\hbar\omega)^{-N} = \left(\frac{kT}{\hbar\omega}\right)^N; \tag{3}$$

note that in writing (3) we have assumed the oscillators to be *distinguishable*. This is so because, as we shall see later, these oscillators are merely a representation of the energy levels available in the system; they are not particles (or even "quasiparticles"). It is actually photons in one case and phonons in the other, which distribute themselves over the various oscillator levels, that are *indistinguishable*!

The Helmholtz free energy of the system is now given by

$$A \equiv -kT\ln Q_N = NkT\ln\left(\frac{\hbar\omega}{kT}\right), \tag{4}$$

whereby

$$\mu = kT \ln\left(\frac{\hbar\omega}{kT}\right), \tag{5}$$

$$P = 0, \tag{6}$$

$$S = Nk\left[\ln\left(\frac{kT}{\hbar\omega}\right) + 1\right], \tag{7}$$

$$U = NkT, \tag{8}$$

and

$$C_P = C_V = Nk. \tag{9}$$

We note that the mean energy per oscillator is in complete agreement with the equipartition theorem, namely $2 \times \frac{1}{2}kT$, for we have here *two* independent quadratic terms in the single-oscillator Hamiltonian.

We may determine the density of states, $g(E)$, of this system from expression (3) for its partition function. We have, in view of (3.4.7),

$$g(E) = \frac{1}{(\hbar\omega)^N} \frac{1}{2\pi i} \int\limits_{\beta'-i\infty}^{\beta'+i\infty} \frac{e^{\beta E}}{\beta^N} d\beta \quad (\beta' > 0),$$

that is,

$$g(E) = \begin{cases} \dfrac{1}{(\hbar\omega)^N} \dfrac{E^{N-1}}{(N-1)!} & \text{for} \quad E \geq 0 \\ 0 & \text{for} \quad E \leq 0. \end{cases} \tag{10}$$

To test the correctness of (10), we may calculate the entropy of the system with the help of this formula. Taking $N \gg 1$ and making use of the Stirling approximation, we get

$$S(N,E) = k \ln g(E) \approx Nk\left[\ln\left(\frac{E}{N\hbar\omega}\right) + 1\right], \tag{11}$$

which gives for the temperature of the system

$$T = \left(\frac{\partial S}{\partial E}\right)_N^{-1} = \frac{E}{Nk}. \tag{12}$$

Eliminating E between these two relations, we obtain precisely our earlier result (7) for the function $S(N,T)$.

We now take up the quantum-mechanical situation, according to which the energy eigenvalues of a one-dimensional harmonic oscillator are given by

$$\varepsilon_n = \left(n + \frac{1}{2}\right)\hbar\omega; \quad n = 0, 1, 2, \ldots \tag{13}$$

Accordingly, we have for the single-oscillator partition function

$$\begin{aligned} Q_1(\beta) &= \sum_{n=0}^{\infty} e^{-\beta(n+1/2)\hbar\omega} = \frac{\exp\left(-\frac{1}{2}\beta\hbar\omega\right)}{1 - \exp(-\beta\hbar\omega)} \\ &= \left\{2\sinh\left(\frac{1}{2}\beta\hbar\omega\right)\right\}^{-1}. \end{aligned} \tag{14}$$

The N-oscillator partition function is then given by

$$\begin{aligned} Q_N(\beta) &= [Q_1(\beta)]^N = \left[2\sinh\left(\frac{1}{2}\beta\hbar\omega\right)\right]^{-N} \\ &= e^{-(N/2)\beta\hbar\omega}\{1 - e^{-\beta\hbar\omega}\}^{-N}. \end{aligned} \tag{15}$$

For the Helmholtz free energy of the system, we get

$$A = NkT\ln\left[2\sinh\left(\frac{1}{2}\beta\hbar\omega\right)\right] = N\left[\frac{1}{2}\hbar\omega + kT\ln\{1 - e^{-\beta\hbar\omega}\}\right], \tag{16}$$

whereby

$$\mu = A/N, \tag{17}$$

$$P = 0, \tag{18}$$

$$\begin{aligned} S &= Nk\left[\frac{1}{2}\beta\hbar\omega\coth\left(\frac{1}{2}\beta\hbar\omega\right) - \ln\left\{2\sinh\left(\frac{1}{2}\beta\hbar\omega\right)\right\}\right] \\ &= Nk\left[\frac{\beta\hbar\omega}{e^{\beta\hbar\omega} - 1} - \ln\{1 - e^{-\beta\hbar\omega}\}\right], \end{aligned} \tag{19}$$

$$U = \frac{1}{2}N\hbar\omega\coth\left(\frac{1}{2}\beta\hbar\omega\right) = N\left[\frac{1}{2}\hbar\omega + \frac{\hbar\omega}{e^{\beta\hbar\omega} - 1}\right], \tag{20}$$

and

$$\begin{aligned} C_P = C_V &= Nk\left(\frac{1}{2}\beta\hbar\omega\right)^2 \operatorname{cosech}^2\left(\frac{1}{2}\beta\hbar\omega\right) \\ &= Nk(\beta\hbar\omega)^2 \frac{e^{\beta\hbar\omega}}{(e^{\beta\hbar\omega} - 1)^2}. \end{aligned} \tag{21}$$

Formula (20) is especially significant, for it shows that the quantum-mechanical oscillators do not obey the equipartition theorem. The mean energy per oscillator is different

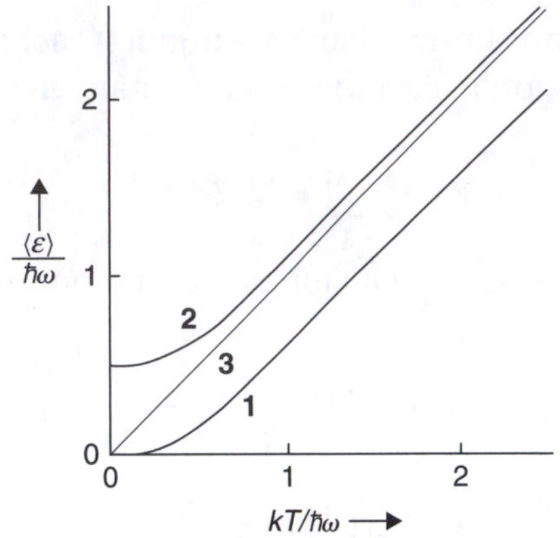

FIGURE 3.4 The mean energy $\langle\varepsilon\rangle$ of a simple harmonic oscillator as a function of temperature. 1, the Planck oscillator; 2, the Schrödinger oscillator; and 3, the classical oscillator.

from the equipartition value kT; actually, it is always greater than kT; see curve 2 in Figure 3.4. Only in the limit of high temperatures, where the thermal energy kT is much larger than the energy quantum $\hbar\omega$, does the mean energy per oscillator tend to the equipartition value. It should be noted here that if the zero-point energy $\frac{1}{2}\hbar\omega$ were not present, the limiting value of the mean energy would be $(kT - \frac{1}{2}\hbar\omega)$, and not kT — we may call such an oscillator the *Planck oscillator*; see curve 1 in Figure 3.4. In passing, we observe that the specific heat (21), which is the same for the Planck oscillator as for the Schrödinger oscillator, is temperature-dependent; moreover, it is always less than, and at high temperatures tends to, the classical value (9).

Indeed, for $kT \gg \hbar\omega$, formulae (14) through (21) go over to their classical counterparts, namely (2) through (9), respectively.

We shall now determine the density of states $g(E)$ of the N-oscillator system from its partition function (15). Carrying out the binomial expansion of this expression, we have

$$Q_N(\beta) = \sum_{R=0}^{\infty} \binom{N+R-1}{R} e^{-\beta(\frac{1}{2}N\hbar\omega + R\hbar\omega)}. \tag{22}$$

Comparing this with the formula

$$Q_N(\beta) = \int_0^{\infty} g(E) e^{-\beta E} dE,$$

we conclude that

$$g(E) = \sum_{R=0}^{\infty} \binom{N+R-1}{R} \delta\left(E - \left\{R + \frac{1}{2}N\right\}\hbar\omega\right), \tag{23}$$

where $\delta(x)$ denotes the Dirac delta function. Equation (23) implies that there are $(N+R-1)!/R!(N-1)!$ microstates available to the system when its energy E has the discrete value $(R+\frac{1}{2}N)\hbar\omega$, where $R=0,1,2,\ldots$, and that no microstate is available for other values of E. This is hardly surprising, but it is instructive to look at this result from a slightly different point of view.

We consider the following problem that arises naturally in the microcanonical ensemble theory. Given an energy E for distribution among a set of N harmonic oscillators, each of which can be in any one of the eigenstates (13), what is the total number of *distinct* ways in which the process of distribution can be carried out? Now, in view of the form of the eigenvalues ε_n, it makes sense to give away, right in the beginning, the zero-point energy $\frac{1}{2}\hbar\omega$ to each of the N oscillators and convert the rest of it into quanta (of energy $\hbar\omega$). Let R be the number of these quanta; then

$$R=\left(E-\frac{1}{2}N\hbar\omega\right)\Big/\hbar\omega. \tag{24}$$

Clearly, R must be an integer; by implication, E must be of the form $(R+\frac{1}{2}N)\hbar\omega$. The problem then reduces to determining the number of *distinct* ways of allotting R quanta to N oscillators, such that an oscillator may have 0 or 1 or 2… quanta; in other words, we have to determine the number of *distinct* ways of putting R *indistinguishable* balls into N *distinguishable* boxes, such that a box may receive 0 or 1 or 2… balls. A little reflection will show that this is precisely the number of permutations that can be realized by shuffling R balls, placed along a row, with $(N-1)$ partitioning lines (that divide the given space into N boxes); see Figure 3.5. The answer clearly is

$$\frac{(R+N-1)!}{R!\,(N-1)!}, \tag{25}$$

which agrees with (23).

We can now determine the entropy of the system from the number (25). Since $N\gg 1$, we have

$$\begin{aligned} S &\approx k\{\ln(R+N)!-\ln R!-\ln N!\} \\ &\approx k\{(R+N)\ln(R+N)-R\ln R-N\ln N\}; \end{aligned} \tag{26}$$

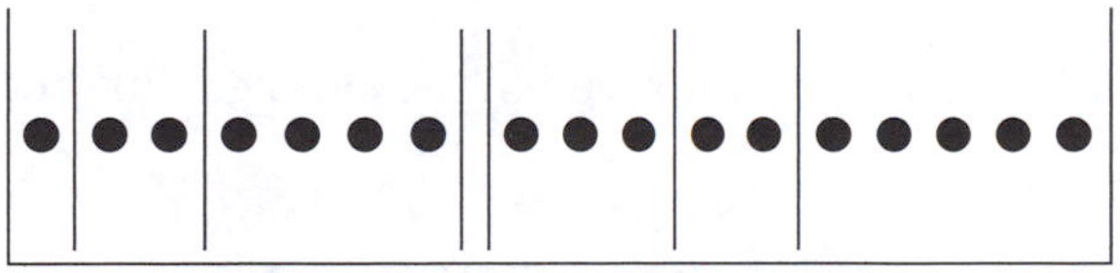

FIGURE 3.5 Distributing 17 *indistinguishable* balls among 7 *distinguishable* boxes. The arrangement shown here represents one of the $23!/17!6!$ distinct ways of carrying out the distribution.

the number R is, of course, a measure of the energy E of the system; see (24). For the temperature of the system, we obtain

$$\frac{1}{T} = \left(\frac{\partial S}{\partial E}\right)_N = \left(\frac{\partial S}{\partial R}\right)_N \frac{1}{\hbar\omega} = \frac{k}{\hbar\omega}\ln\left(\frac{R+N}{R}\right) = \frac{k}{\hbar\omega}\ln\left(\frac{E+\frac{1}{2}N\hbar\omega}{E-\frac{1}{2}N\hbar\omega}\right), \tag{27}$$

so that

$$\frac{E}{N} = \frac{1}{2}\hbar\omega\frac{\exp(\hbar\omega/kT)+1}{\exp(\hbar\omega/kT)-1}, \tag{28}$$

which is identical to (20). It can be further checked that, by eliminating R between (26) and (27), we obtain precisely the formula (19) for $S(N,T)$. Thus, once again, we find that the results obtained by following the microcanonical approach and the canonical approach are the same in the thermodynamic limit.

Finally, we may consider the classical limit when E/N, the mean energy per oscillator, is much larger than the energy quantum $\hbar\omega$, that is, when $R \gg N$. The expression (25) may, in that case, be replaced by

$$\frac{(R+N-1)(R+N-2)\ldots(R+1)}{(N-1)!} \approx \frac{R^{N-1}}{(N-1)!}, \tag{25a}$$

with

$$R \approx E/\hbar\omega.$$

The corresponding expression for the entropy turns out to be

$$S \approx k\{N\ln(R/N)+N\} \approx Nk\left\{\ln\left(\frac{E}{N\hbar\omega}\right)+1\right\}, \tag{26a}$$

which gives

$$\frac{1}{T} = \left(\frac{\partial S}{\partial E}\right)_N \approx \frac{Nk}{E}, \tag{27a}$$

so that

$$\frac{E}{N} \approx kT. \tag{28a}$$

These results are identical to the ones derived in the classical limit earlier in this section.

3.9 The statistics of paramagnetism

Next, we study a system of N magnetic dipoles, each having a magnetic moment $\boldsymbol{\mu}$. In the presence of an external magnetic field $\boldsymbol{H}$, the dipoles will experience a torque tending to align them in the direction of the field. If there were nothing else to check this tendency, the dipoles would align themselves precisely in this direction and we would achieve a complete magnetization of the system. In reality, however, thermal agitation in the system offers resistance to this tendency and, in equilibrium, we obtain only a *partial* magnetization. Clearly, as $T \to 0\,\text{K}$, the thermal agitation becomes ineffective and the system exhibits a complete orientation of the dipole moments, whatever the strength of the applied field; at the other extreme, as $T \to \infty$, we approach a state of complete randomization of the dipole moments, which implies a vanishing magnetization. At intermediate temperatures, the situation is governed by the parameter $(\mu H/kT)$.

The model adopted for this study consists of N identical, localized (and, hence, distinguishable), practically static, mutually noninteracting and freely orientable dipoles. We consider first the case of classical dipoles that can be oriented in any direction relative to the applied magnetic field. It is obvious that the only energy we need to consider here is the potential energy of the dipoles that arises from the presence of the external field $\boldsymbol{H}$ and is determined by the orientations of the dipoles with respect to the direction of the field:

$$E = \sum_{i=1}^{N} E_i = -\sum_{i=1}^{N} \boldsymbol{\mu}_i \cdot \boldsymbol{H} = -\mu H \sum_{i=1}^{N} \cos\theta_i. \tag{1}$$

The partition function of the system is then given by

$$Q_N(\beta) = [Q_1(\beta)]^N, \tag{2}$$

where

$$Q_1(\beta) = \sum_{\theta} \exp(\beta\mu H \cos\theta). \tag{3}$$

The mean magnetic moment $\boldsymbol{M}$ of the system will obviously be in the direction of the field $\boldsymbol{H}$; for its magnitude we shall have

$$\begin{aligned} M_z = N\langle \mu\cos\theta\rangle &= N\frac{\sum_{\theta} \mu\cos\theta\exp(\beta\mu H\cos\theta)}{\sum_{\theta}\exp(\beta\mu H\cos\theta)} \\ &= \frac{N}{\beta}\frac{\partial}{\partial H}\ln Q_1(\beta) = -\left(\frac{\partial A}{\partial H}\right)_T. \end{aligned} \tag{4}$$

Thus, to determine the degree of magnetization in the system all we have to do is to evaluate the single-dipole partition function (3).

First, we proceed classically (after Langevin, 1905a,b). Using $(\sin\theta\, d\theta\, d\phi)$ as the elemental solid angle representing a small range of orientations of the dipole, we get

$$Q_1(\beta) = \int_0^{2\pi}\int_0^{\pi} e^{\beta\mu H\cos\theta} \sin\theta\, d\theta\, d\phi = 4\pi\frac{\sinh(\beta\mu H)}{\beta\mu H}, \tag{5}$$

so that

$$\bar{\mu}_z \equiv \frac{M_z}{N} = \mu \left\{ \coth(\beta\mu H) - \frac{1}{\beta\mu H} \right\} = \mu L(\beta\mu H), \tag{6}$$

where $L(x)$ is the so-called *Langevin function*

$$L(x) = \coth x - \frac{1}{x}; \tag{7}$$

a plot of the Langevin function is shown in Figure 3.6. We note that the parameter $\beta\mu H$ denotes the strength of the (magnetic) potential energy μH compared to the (thermal) kinetic energy kT.

If we have N_0 dipoles per unit volume in the system, then the magnetization of the system, namely the mean magnetic moment per unit volume, is given by

$$M_{z0} = N_0 \bar{\mu}_z = N_0 \mu L(x) \quad (x = \beta\mu H). \tag{8}$$

For magnetic fields so strong (or temperatures so low) that the parameter $x \gg 1$, the function $L(x)$ is almost equal to 1; the system then acquires a state of magnetic saturation:

$$\bar{\mu}_z \simeq \mu \quad \text{and} \quad M_{z0} \simeq N_0 \mu. \tag{9}$$

For temperatures so high (or magnetic fields so weak) that the parameter $x \ll 1$, the function $L(x)$ may be written as

$$\frac{x}{3} - \frac{x^3}{45} + \cdots \tag{10}$$

which, in the lowest approximation, gives

$$M_{z0} \simeq \frac{N_0 \mu^2}{3kT} H. \tag{11}$$

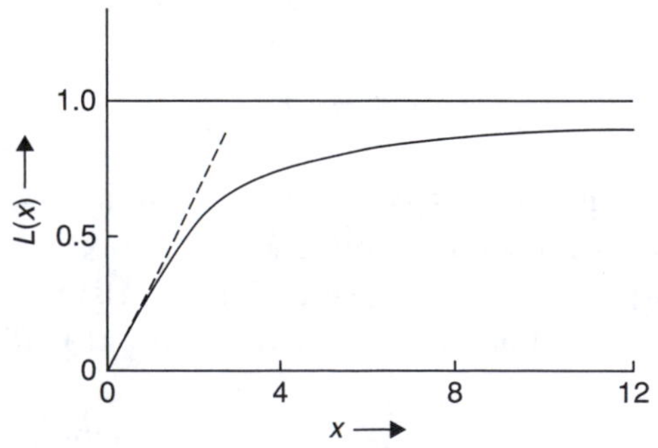

FIGURE 3.6 The Langevin function $L(x)$.

The high-temperature isothermal susceptibility of the system is, therefore, given by

$$\chi_T = \lim_{H \to 0} \left(\frac{\partial M_{z0}}{\partial H} \right)_T \simeq \frac{N_0 \mu^2}{3kT} = \frac{C}{T}, \quad \text{say.} \tag{12}$$

Equation (12) is the *Curie law* of paramagnetism, the parameter C being the *Curie constant* of the system. Figure 3.7 shows a plot of the susceptibility of a powdered sample of copper–potassium sulphate hexahydrate as a function of T^{-1}; the fact that the plot is linear and passes almost through the origin vindicates the Curie law for this particular salt.

We shall now treat the problem of paramagnetism quantum-mechanically. The major modification here arises from the fact that the magnetic dipole moment $\boldsymbol{\mu}$ and its component μ_z in the direction of the applied field cannot have *arbitrary* values. Quite generally, we have a direct relationship between the magnetic moment $\boldsymbol{\mu}$ of a given dipole and its angular momentum $\boldsymbol{l}$:

$$\boldsymbol{\mu} = \left(g \frac{e}{2mc} \right) \boldsymbol{l}, \tag{13}$$

with

$$l^2 = J(J+1)\hbar^2; \quad J = \frac{1}{2}, \frac{3}{2}, \frac{5}{2}, \ldots \quad \text{or} \quad 0, 1, 2, \ldots \tag{14}$$

The quantity $g(e/2mc)$ is the *gyromagnetic ratio* of the dipole while the number g is *Lande's g-factor*. If the net angular momentum of the dipole is due solely to electron spins, then

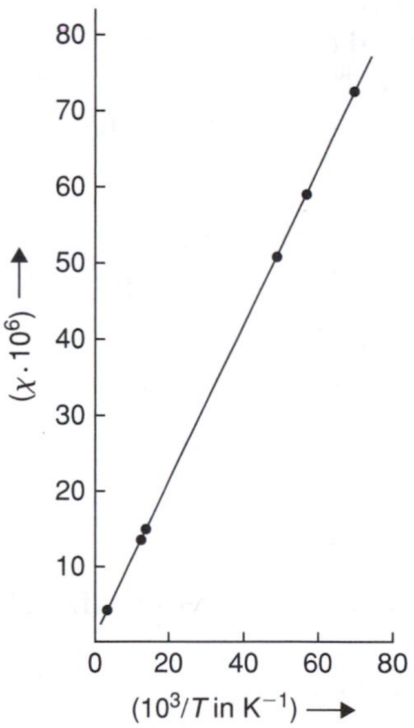

FIGURE 3.7 χ versus $1/T$ plot for a powdered sample of copper–potassium sulphate hexahydrate (after Hupse, 1942).

$g = 2$; on the other hand, if it is due solely to orbital motions, then $g = 1$. In general, however, its origin is mixed; g is then given by the formula

$$g = \frac{3}{2} + \frac{S(S+1) - L(L+1)}{2J(J+1)}, \tag{15}$$

S and L being, respectively, the spin and the orbital quantum numbers of the dipole. Note that there is no upper or lower bound on the values that g can have!

Combining (13) and (14), we can write

$$\mu^2 = g^2 \mu_B^2 J(J+1), \tag{16}$$

where $\mu_B (= e\hbar/2mc)$ is the *Bohr magneton*. The component μ_z of the magnetic moment in the direction of the applied field is, on the other hand, given by

$$\mu_z = g\mu_B m, \quad m = -J, -J+1, \ldots, J-1, J. \tag{17}$$

Thus, a dipole whose magnetic moment $\boldsymbol{\mu}$ conforms to expression (16) can have no other orientations with respect to the applied field except the ones conforming to the values (17) of the component μ_z; obviously, the number of allowed orientations, for a given value of J, is $(2J+1)$. In view of this, the single-dipole partition function $Q_1(\beta)$ is now given by, see (3),

$$Q_1(\beta) = \sum_{m=-J}^{J} \exp(\beta g \mu_B m H). \tag{18}$$

Introducing a parameter x, defined by

$$x = \beta(g\mu_B J)H, \tag{19}$$

equation (18) becomes

$$\begin{aligned} Q_1(\beta) &= \sum_{m=-J}^{J} e^{mx/J} = \frac{e^{-x}\{e^{(2J+1)x/J} - 1\}}{e^{x/J} - 1} \\ &= \frac{e^{(2J+1)x/2J} - e^{-(2J+1)x/2J}}{e^{x/2J} - e^{-x/2J}} \\ &= \sinh\left\{\left(1 + \frac{1}{2J}\right)x\right\} \Big/ \sinh\left\{\frac{1}{2J}x\right\}. \end{aligned} \tag{20}$$

The mean magnetic moment of the system is then given by, see equation (4),

$$\begin{aligned} M_z = N\bar{\mu}_z &= \frac{N}{\beta}\frac{\partial}{\partial H} \ln Q_1(\beta) \\ &= N(g\mu_B J)\left[\left(1 + \frac{1}{2J}\right)\coth\left\{\left(1 + \frac{1}{2J}\right)x\right\} - \frac{1}{2J}\coth\left\{\frac{1}{2J}x\right\}\right]. \end{aligned} \tag{21}$$

Thus

$$\overline{\mu}_z = (g\mu_B J)B_J(x), \tag{22}$$

where $B_J(x)$ is the *Brillouin function* of order J:

$$B_J(x) = \left(1+\frac{1}{2J}\right)\coth\left\{\left(1+\frac{1}{2J}\right)x\right\} - \frac{1}{2J}\coth\left\{\frac{1}{2J}x\right\}. \tag{23}$$

In Figure 3.8 we have plotted the function $B_J(x)$ for some typical values of the quantum number J.

We shall now consider a few special cases. First of all, we note that for strong fields and low temperatures ($x \gg 1$), the function $B_J(x) \simeq 1$ *for all* J, which corresponds to a state of magnetic saturation. On the other hand, for high temperatures and weak fields ($x \ll 1$), the function $B_J(x)$ may be written as

$$\frac{1}{3}(1+1/J)x + \ldots, \tag{24}$$

so that

$$\overline{\mu}_z \simeq \frac{(g\mu_B J)^2}{3kT}\left(1+\frac{1}{J}\right)H = \frac{g^2\mu_B^2 J(J+1)}{3kT}H. \tag{25}$$

The Curie law, $\chi \propto 1/T$, is again obeyed; however, the Curie constant is now given by

$$C_J = \frac{N_0 g^2 \mu_B^2 J(J+1)}{3k} = \frac{N_0 \mu^2}{3k}; \tag{26}$$

see equation (16). It is indeed interesting that the high-temperature results, (25) and (26), directly involve the eigenvalues of the operator μ^2.

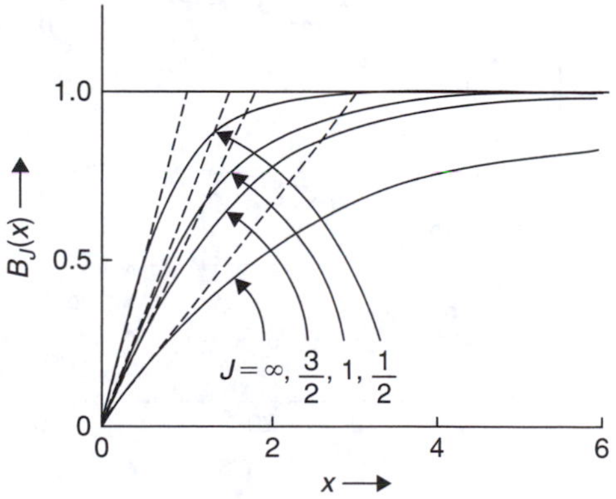

FIGURE 3.8 The Brillouin function $B_J(x)$ for various values of J.

We now look a little more closely at the dependence of the foregoing results on the quantum number J. First of all, we consider the extreme case $J \to \infty$, with the understanding that simultaneously $g \to 0$, such that the value of μ stays constant. From equation (23), we readily observe that, in this limit, the Brillouin function $B_J(x)$ tends to become (i) independent of J and (ii) identical to the Langevin function $L(x)$. This is not surprising because, in this limit, the number of allowed orientations for a magnetic dipole becomes infinitely large, with the result that the problem essentially reduces to its classical counterpart (where one must allow all possible orientations). At the other extreme, we have the case $J = \frac{1}{2}$, which allows only two orientations. The results in this case are very different from the ones for $J \gg 1$. We now have, with $g = 2$,

$$\overline{\mu}_z = \mu_B B_{1/2}(x) = \mu_B \tanh x. \tag{27}$$

For $x \gg 1$, μ_z is very nearly equal to μ_B. For $x \ll 1$, however, $\overline{\mu}_z \simeq \mu_B x$, which corresponds to the Curie constant

$$C_{1/2} = \frac{N_0 \mu_B^2}{k}. \tag{28}$$

In Figure 3.9 we reproduce the experimental values of $\overline{\mu}_z$ (in terms of μ_B) as a function of the quantity H/T, for three paramagnetic salts; the corresponding theoretical plots, namely the curves $gJB_J(x)$, are also included in the figure. The agreement between theory and experiment is indeed good. In passing, we note that, at a temperature of 1.3 K, a field of about 50,000 gauss is sufficient to produce over 99 percent of saturation in these salts.

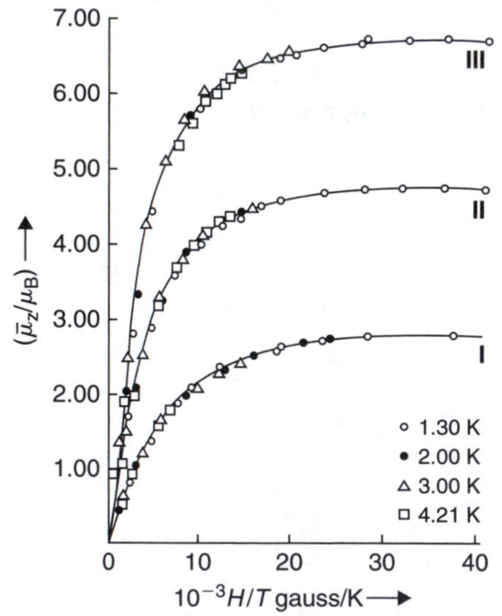

FIGURE 3.9 Plots of $\overline{\mu}_z/\mu_B$ as a function of H/T. The solid curves represent the theoretical results, while the points mark the experimental findings of Henry (1952). Curve I is for potassium chromium alum ($J = \frac{3}{2}, g = 2$), curve II for iron ammonia alum ($J = \frac{5}{2}, g = 2$), and curve III for gadolinium sulphate octahydrate ($J = \frac{7}{2}, g = 2$).

3.10 Thermodynamics of magnetic systems: negative temperatures

For the purpose of this section, it will suffice to consider a system of dipoles with $J = \frac{1}{2}$. Each dipole then has a choice of two orientations, the corresponding energies being $-\mu_B H$ and $+\mu_B H$; let us call these energies $-\varepsilon$ and $+\varepsilon$, respectively. The partition function of the system is then given by

$$Q_N(\beta) = \left(e^{\beta\varepsilon} + e^{-\beta\varepsilon}\right)^N = \{2\cosh(\beta\varepsilon)\}^N; \tag{1}$$

compare to the general expression (3.9.20). Accordingly, the Helmholtz free energy of the system is given by

$$A = -NkT\ln\{2\cosh(\varepsilon/kT)\}, \tag{2}$$

from which

$$S = -\left(\frac{\partial A}{\partial T}\right)_H = Nk\left[\ln\left\{2\cosh\left(\frac{\varepsilon}{kT}\right)\right\} - \frac{\varepsilon}{kT}\tanh\left(\frac{\varepsilon}{kT}\right)\right], \tag{3}$$

$$U = A + TS = -N\varepsilon\tanh\left(\frac{\varepsilon}{kT}\right), \tag{4}$$

$$M = -\left(\frac{\partial A}{\partial H}\right)_T = N\mu_B\tanh\left(\frac{\varepsilon}{kT}\right) \tag{5}$$

and, finally,

$$C = \left(\frac{\partial U}{\partial T}\right)_H = Nk\left(\frac{\varepsilon}{kT}\right)^2\operatorname{sech}^2\left(\frac{\varepsilon}{kT}\right). \tag{6}$$

Equation (5) is essentially the same as (3.9.27); moreover, as expected, $U = -MH$.

The temperature dependence of the quantities S, U, M, and C is shown in Figures 3.10 through 3.13. We note that the entropy of the system is vanishingly small for $kT \ll \varepsilon$; it rises

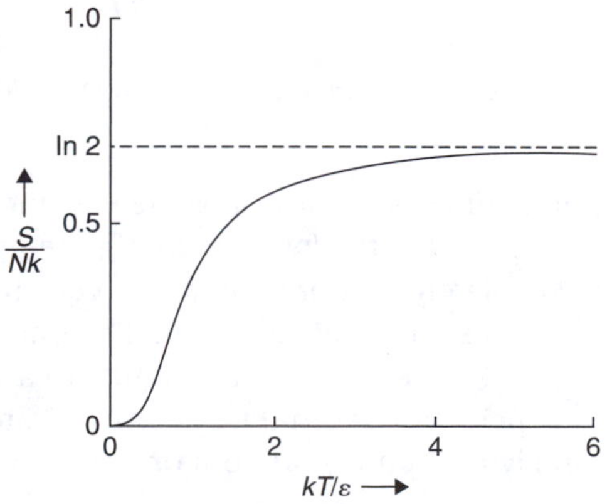

FIGURE 3.10 The entropy of a system of magnetic dipoles (with $J = \frac{1}{2}$) as a function of temperature.

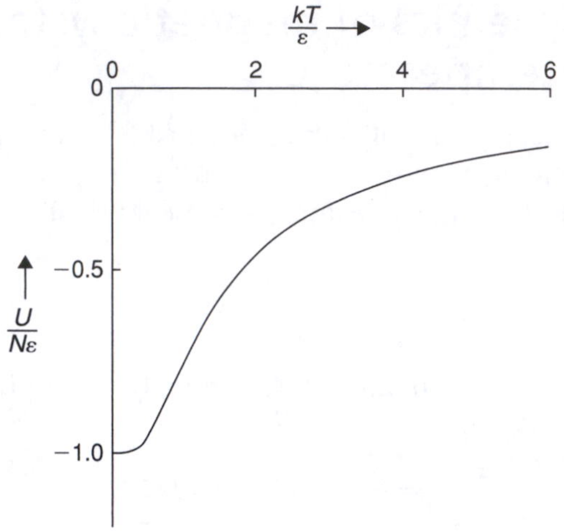

FIGURE 3.11 The energy of a system of magnetic dipoles (with $J = \frac{1}{2}$) as a function of temperature.

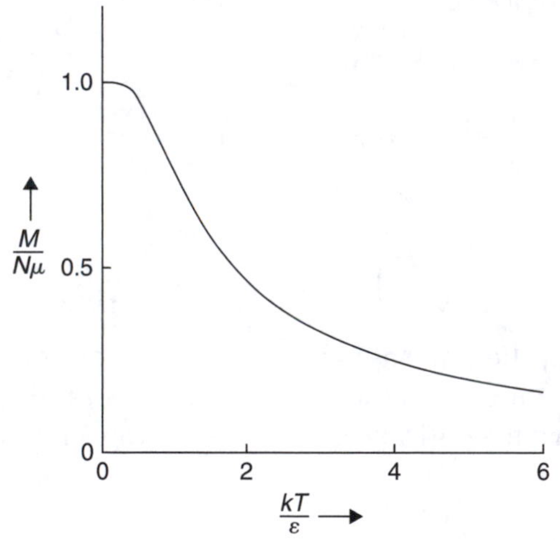

FIGURE 3.12 The magnetization of a system of magnetic dipoles (with $J = \frac{1}{2}$) as a function of temperature.

rapidly when kT is of the order of ε and approaches the limiting value $Nk\ln 2$ for $kT \gg \varepsilon$. This limiting value of S corresponds to the fact that at high temperatures the orientation of the dipoles assumes a completely random character, with the result that the system now has 2^N *equally likely* microstates available to it. The energy of the system attains its lowest value, $-N\varepsilon$, as $T \to 0\,\text{K}$; this clearly corresponds to a state of magnetic saturation and, hence, to a state of perfect order in the system. Toward high temperatures, the energy tends to vanish,[13] implying a purely random orientation of the dipoles and hence

[13]Note that in the present study we are completely disregarding the kinetic energy of the dipoles.

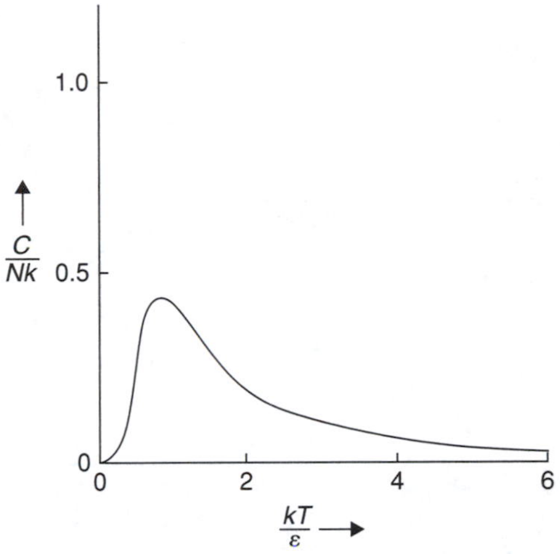

FIGURE 3.13 The specific heat of a system of magnetic dipoles (with $J = \frac{1}{2}$) as a function of temperature.

a complete loss of magnetic order. These features are re-emphasized in Figure 3.12, which depicts the temperature dependence of the magnetization M. The specific heat of the system is vanishingly small at low temperatures but, in view of the fact that the energy of the system tends to a constant value as $T \to \infty$, the specific heat vanishes at high temperatures as well. Somewhere around $T = \varepsilon/k$, it displays a maximum. Writing Δ for the energy difference between the two allowed states of the dipole, the formula for the specific heat can be written as

$$C = Nk\left(\frac{\Delta}{kT}\right)^2 e^{\Delta/kT}(1+e^{\Delta/kT})^{-2}. \tag{7}$$

A specific heat peak of this form is generally known as the *Schottky* anomaly; it is observed in systems that have an excitation gap Δ above the ground state.

Now, throughout our study so far we have considered only those cases for which $T > 0$. For *normal* systems, this is indeed essential, for otherwise we have to contend with canonical distributions that blow up as the energy of the system is indefinitely increased. If, however, the energy of a system is bounded from above, then there is no compelling reason to exclude the possibility of negative temperatures. Such specialized situations do indeed exist, and the system of magnetic dipoles provides a good example thereof. From equation (4), we note that, so long as $U < 0$, $T > 0$ — and that is the only range we covered in Figures 3.10 through 3.13. However, the same equation tells us that if $U > 0$ then $T < 0$, which prompts us to examine the matter a little more closely. For this, we consider the variation of the temperature T and the entropy S with energy U, namely

$$\frac{1}{T} = -\frac{k}{\varepsilon}\tanh^{-1}\left(\frac{U}{N\varepsilon}\right) = \frac{k}{2\varepsilon}\ln\left(\frac{N\varepsilon - U}{N\varepsilon + U}\right) \tag{8}$$

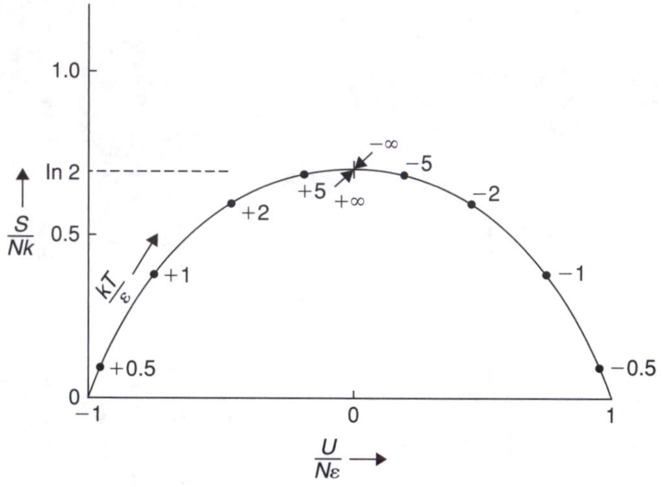

FIGURE 3.14 The entropy of a system of magnetic dipoles (with $J = \frac{1}{2}$) as a function of energy. Some values of the parameter kT/ε are also shown in the figure. The slope at the two endpoints diverges since both ends represent zero temperature but it is difficult to see due to the logarithmic nature of the divergence.

and

$$\frac{S}{Nk} = -\frac{N\varepsilon + U}{2N\varepsilon}\ln\left(\frac{N\varepsilon + U}{2N\varepsilon}\right) - \frac{N\varepsilon - U}{2N\varepsilon}\ln\left(\frac{N\varepsilon - U}{2N\varepsilon}\right); \tag{9}$$

these expressions follow straightforwardly from equations (3) and (4), and are shown graphically in Figures 3.14 and 3.15. We note that for $U = -N\varepsilon$, both S and T vanish. As U increases, they too increase until we reach the special situation where $U = 0$. The entropy is then seen to have attained its maximum value $Nk\ln 2$, while the temperature has reached infinity. Throughout this range, the entropy had been a monotonically increasing function of energy, so T was positive. Now, as U becomes 0_+, (dS/dU) becomes 0_- and T becomes $-\infty$. With a further increase in U, the entropy monotonically decreases; as a result, the temperature continues to be negative, though its magnitude steadily decreases. Finally, we reach the largest value of U, namely $+N\varepsilon$, where the entropy is once again zero and $T = 0_-$.

The region where $U > 0$ (and hence $T < 0$) is indeed abnormal because it corresponds to a magnetization *opposite* in direction to that of the applied field. Nevertheless, it *can* be realized experimentally in the system of nuclear moments of a crystal in which the relaxation time t_1 for mutual interaction among nuclear spins is very small in comparison with the relaxation time t_2 for interaction between the spins and the lattice. Let such a crystal be magnetized in a strong magnetic field and then the field reversed so quickly that the spins are unable to follow the switch-over. This will leave the system in a nonequilibrium state, with energy higher than the new equilibrium value U. During a period of order t_1,

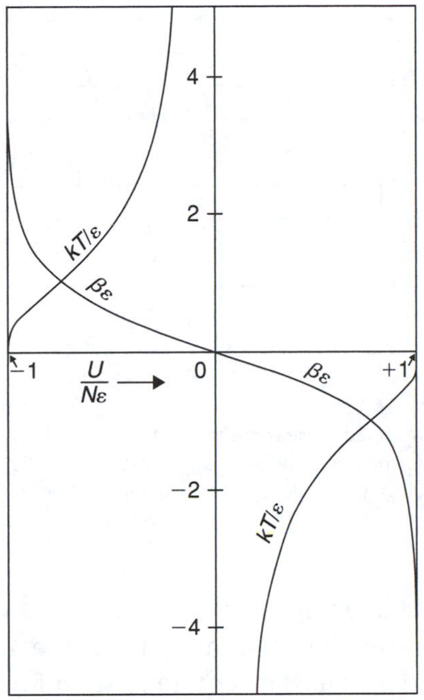

FIGURE 3.15 The temperature parameter kT/ε, and its reciprocal $\beta\varepsilon$, for a system of magnetic dipoles (with $J = \frac{1}{2}$) as a function of energy.

the subsystem of the nuclear spins should be able to attain a state of *internal* equilibrium; this state will have a negative magnetization and will, therefore, correspond to a negative temperature. The subsystem of the lattice, which involves energy parameters that are in principle unbounded, will still be at a positive temperature. During a period of order t_2, the two subsystems would attain a state of mutual equilibrium, which again will have a positive temperature.[14] An experiment of this kind was successfully performed by Purcell and Pound (1951) with a crystal of LiF; in this case, t_1 was of order 10^{-5} sec while t_2 was of order 5 min. A state of negative temperature for the subsystem of spins was indeed attained and was found to persist for a period of several minutes; see Figure 3.16.

Before we close this discussion, a few general remarks seem in order. First of all, we should note that the onset of negative temperatures is possible only if there exists an upper limit on the energy of the given system. In most physical systems this is not the case, simply because most physical systems possess kinetic energy of motion which is obviously unbounded. By the same token, the onset of positive temperatures is related to the

[14]Note that in the latter process, during which the spins realign themselves (now more favorably in the new direction of the field), the energy will flow from the subsystem of the spins to that of the lattice, and not vice versa. This is in perfect agreement with the fact that negative temperatures are *hotter* than positive ones; see the subsequent discussion in the text.

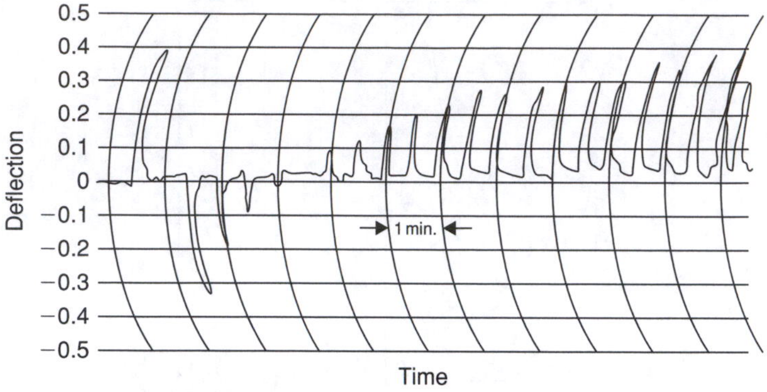

FIGURE 3.16 A typical record of the reversed nuclear magnetization (after Purcell and Pound, 1951). On the left we have a deflection corresponding to normal, equilibrium magnetization ($T \sim 300\,\text{K}$); it is followed by the reversed deflection (corresponding to $T \sim -350\,\text{K}$), which decays through zero deflection (corresponding to a passage from $T = -\infty$ to $T = +\infty$) toward the new equilibrium state that again has a positive T.

existence of a lower limit on the energy of a system; this, however, does not present any problem because, if nothing else, the uncertainty principle alone is sufficient to set such a limit for *every* physical system. Thus, it is quite normal for a system to be at a positive temperature whereas it is very unusual for one to be at a negative temperature.

Now, suppose that we have a system whose energy *cannot* assume unlimited high values. Then, we can surely visualize a temperature T such that the quantity NkT is much larger than any admissible value, E_r, of the energy. At such a high temperature, the mutual interactions of the microscopic entities constituting the system may be regarded as negligible; accordingly, one may write for the partition function of the system

$$Q_N(\beta) \simeq \left[\sum_n e^{-\beta\varepsilon_n}\right]^N. \tag{10}$$

Since, by assumption, all $\beta\varepsilon_n \ll 1$, we have

$$Q_N(\beta) \simeq \left[\sum_n \left\{1 - \beta\varepsilon_n + \frac{1}{2}\beta^2\varepsilon_n^2\right\}\right]^N. \tag{11}$$

Let g denote the number of possible orientations of a microscopic constituent of the system with respect to the direction of the external field; then, the quantities $\sum_n \varepsilon_n^\alpha (\alpha = 0, 1, 2)$ may be replaced by $g\overline{\varepsilon^\alpha}$. We thus get

$$\begin{aligned}\ln Q_N(\beta) &\simeq N\left[\ln g + \ln\left(1 - \beta\bar{\varepsilon} + \frac{1}{2}\beta^2\overline{\varepsilon^2}\right)\right] \\ &\simeq N\left[\ln g - \beta\bar{\varepsilon} + \frac{1}{2}\beta^2\left(\overline{\varepsilon^2} - \bar{\varepsilon}^2\right)\right].\end{aligned} \tag{12}$$

The Helmholtz free energy of the system is then given by

$$A(N,\beta) \simeq -\frac{N}{\beta}\ln g + N\overline{\varepsilon} - \frac{N}{2}\beta\overline{(\varepsilon-\overline{\varepsilon})^2}, \tag{13}$$

from which

$$S(N,\beta) \simeq Nk\ln g - \frac{Nk}{2}\beta^2\overline{(\varepsilon-\overline{\varepsilon})^2}, \tag{14}$$

$$U(N,\beta) \simeq N\overline{\varepsilon} - N\beta\overline{(\varepsilon-\overline{\varepsilon})^2}, \tag{15}$$

and

$$C(N,\beta) \simeq Nk\beta^2\overline{(\varepsilon-\overline{\varepsilon})^2}. \tag{16}$$ [15]

The formulae in equations (12) through (16) determine the thermodynamic properties of the system for $\beta \simeq 0$. The important thing to note here is that they do so not only for $\beta \gtrsim 0$ but also for $\beta \lesssim 0$. In fact, these formulae hold in the vicinity of, and on *both* sides of, the maximum in the $S-U$ curve; see Figure 3.14. Quite expectedly, the maximum value of S is given by $Nk\ln g$, and it occurs at $\beta = \pm 0$; S here decreases both ways, whether U decreases ($\beta > 0$) or increases ($\beta < 0$). It will be noted that the specific heat of the system in either case is positive.

It is not difficult to show that if two systems, characterized by the temperature parameters β_1 and β_2, are brought into thermal contact, then energy will flow from the system with the smaller value of β to the system with the larger value of β; this will continue until the two systems acquire a common value of this parameter. What is more important to note is that this result remains *literally* true even if one or both of the β are negative. Thus, if β_1 is $-$ve while β_2 is $+$ve, then energy will flow from system 1 to system 2, that is, from the system at negative temperature to the one at positive temperature. In this sense, systems at negative temperatures are *hotter* than the ones at positive temperatures; indeed, negative temperatures are above $+\infty$, not below zero!

For further discussion of this topic, reference may be made to a paper by Ramsey (1956).

Problems

3.1. **(a)** Derive formula (3.2.36) from equations (3.2.14) and (3.2.35).
(b) Derive formulae (3.2.39) and (3.2.40) from equations (3.2.37) and (3.2.38).

3.2. Prove that the quantity $g''(x_0)$, see equations (3.2.25), is equal to $\langle (E-U)^2\rangle \exp(2\beta)$. Thus show that equation (3.2.28) is physically equivalent to equation (3.6.9).

3.3. Using the fact that $(1/n!)$ is the coefficient of x^n in the power expansion of the function $\exp(x)$, derive an asymptotic formula for this coefficient by the method of saddle-point integration. Compare your result with the Stirling formula for $n!$.

[15]Compare this result with equation (3.6.3).

3.4. Verify that the quantity $(k/\mathcal{N})\ln\Gamma$, where

$$\Gamma(\mathcal{N}, U) = \sum_{\{n_r\}}{}' W\{n_r\},$$

is equal to the (mean) entropy of the given system. Show that this leads to essentially the same result for $\ln\Gamma$ if we take, in the foregoing summation, *only* the largest term of the sum, namely the term $W\{n_r^*\}$ that corresponds to the *most probable* distribution set.
[Surprised? Well, note the following example:

For all N, the summation over the binomial coefficients ${}^N C_r = N!/[r!(N-r!)]$ gives

$$\sum_{r=0}^{N} {}^N C_r = 2^N;$$

therefore,

$$\ln\left\{\sum_{r=0}^{N} {}^N C_r\right\} = N\ln 2. \tag{a}$$

Now, the largest term in this sum corresponds to $r \simeq N/2$; so, for large N, the logarithm of the largest term is very nearly equal to

$$\ln\{N!\} - 2\ln\{(N/2)!\}$$
$$\approx N\ln N - 2\frac{N}{2}\ln\frac{N}{2} = N\ln 2, \tag{b}$$

which agrees with (a).]

3.5. Making use of the fact that the Helmholtz free energy $A(N, V, T)$ of a thermodynamic system is an *extensive* property of the system, show that

$$N\left(\frac{\partial A}{\partial N}\right)_{V,T} + V\left(\frac{\partial A}{\partial V}\right)_{N,T} = A.$$

[Note that this result implies the well-known relationship: $N\mu = A + PV (\equiv G)$.]

3.6. **(a)** Assuming that the total number of microstates accessible to a given statistical system is Ω, show that the entropy of the system, as given by equation (3.3.13), is maximum when all Ω states are equally likely to occur.

(b) If, on the other hand, we have an ensemble of systems sharing energy (with mean value $\overline{E}$), then show that the entropy, as given by the same formal expression, is maximum when $P_r \propto \exp(-\beta E_r)$, β being a constant to be determined by the given value of $\overline{E}$.

(c) Further, if we have an ensemble of systems sharing energy (with mean value $\overline{E}$) and also sharing particles (with mean value $\overline{N}$), then show that the entropy, given by a similar expression, is maximum when $P_{r,s} \propto \exp(-\alpha N_r - \beta E_s)$, α and β being constants to be determined by the given values of $\overline{N}$ and $\overline{E}$.

3.7. Prove that, quite generally,

$$C_P - C_V = -k\frac{\left[\frac{\partial}{\partial T}\left\{T\left(\frac{\partial \ln Q}{\partial V}\right)_T\right\}\right]_V^2}{\left(\frac{\partial^2 \ln Q}{\partial V^2}\right)_T} > 0.$$

Verify that the value of this quantity for a classical ideal classical gas is Nk.

3.8. Show that, for a classical ideal gas,

$$\frac{S}{Nk} = \ln\left(\frac{Q_1}{N}\right) + T\left(\frac{\partial \ln Q_1}{\partial T}\right)_P.$$

3.9. If an ideal monatomic gas is expanded *adiabatically* to twice its initial volume, what will the ratio of the final pressure to the initial pressure be? If during the process some heat is added to the system, will the final pressure be higher or lower than in the preceding case? Support your answer by deriving the relevant formula for the ratio P_f/P_i.

3.10. **(a)** The volume of a sample of helium gas is increased by withdrawing the piston of the containing cylinder. The final pressure P_f is found to be equal to the initial pressure P_i times $(V_i/V_f)^{1.2}$, V_i and V_f being the initial and final volumes. Assuming that the product PV is always equal to $\frac{2}{3}U$, will (i) the energy and (ii) the entropy of the gas increase, remain constant, or decrease during the process?

(b) If the process were reversible, how much work would be done and how much heat would be added in doubling the volume of the gas? Take $P_i = 1$ atm and $V_i = 1\,\text{m}^3$.

3.11. Determine the work done on a gas and the amount of heat absorbed by it during a compression from volume V_1 to volume V_2, following the law $PV^n = \text{const}$.

3.12. If the "free volume" $\overline{V}$ of a classical system is defined by the equation

$$\overline{V}^N = \int e^{\{\overline{U} - U(\boldsymbol{q}_i)\}/kT} \prod_{i=1}^{N} d^3 q_i,$$

where $\overline{U}$ is the average potential energy of the system and $U(\boldsymbol{q}_i)$ the actual potential energy as a function of the molecular configuration, then show that

$$S = Nk\left[\ln\left\{\frac{\overline{V}}{N}\left(\frac{2\pi mkT}{h^2}\right)^{3/2}\right\} + \frac{5}{2}\right].$$

In what sense is it justified to refer to the quantity $\overline{V}$ as the "free volume" of the system? Substantiate your answer by considering a particular case — for example, the case of a hard sphere gas.

3.13. **(a)** Evaluate the partition function and the major thermodynamic properties of an ideal gas consisting of N_1 molecules of mass m_1 and N_2 molecules of mass m_2, confined to a space of volume V at temperature T. Assume that the molecules of a given kind are mutually indistinguishable, while those of one kind are distinguishable from those of the other kind.

(b) Compare your results with the ones pertaining to an ideal gas consisting of $(N_1 + N_2)$ molecules, *all of one kind*, of mass m, such that $m(N_1 + N_2) = m_1 N_1 + m_2 N_2$.

3.14. Consider a system of N classical particles with mass m moving in a cubic box with volume $V = L^3$. The particles interact via a short-ranged pair potential $u(r_{ij})$ and each particle interacts with each wall with a short-ranged interaction $u_{\text{wall}}(z)$, where z is the perpendicular distance of a particle from the wall. Write down the Lagrangian for this model and use a Legendre transformation to determine the Hamiltonian H.

(a) Show that the quantity $P = -\left(\frac{\partial H}{\partial V}\right) = \frac{-1}{3L^2}\left(\frac{\partial H}{\partial L}\right)$ can clearly be identified as the instantaneous pressure — that is, the force per unit area on the walls.

(b) Reconstruct the Lagrangian in terms of the relative locations of the particles inside the box $\boldsymbol{r}_i = L\boldsymbol{s}_i$, where the variables $\boldsymbol{s}_i$ all lie inside a unit cube. Use a Legendre transformation to determine the Hamiltonian with this set of variables.

(c) Recalculate the pressure using the second version of the Hamiltonian. Show that the pressure now includes three contributions:

(1) a contribution proportional to the kinetic energy,

(2) a contribution related to the forces between pairs of particles, and

(3) a contribution related to the force on the wall.

Show that in the thermodynamic limit the third contribution is negligible compared to the other two. Interpret contributions 1 and 2 and compare to the virial equation of state (3.7.15).

3.15. Show that the partition function $Q_N(V,T)$ of an *extreme* relativistic gas consisting of N monatomic molecules with energy–momentum relationship $\varepsilon = pc$, c being the speed of light, is given by

$$Q_N(V,T) = \frac{1}{N!}\left\{8\pi V\left(\frac{kT}{hc}\right)^3\right\}^N.$$

Study the thermodynamics of this system, checking in particular that

$$PV = \frac{1}{3}U, \quad U/N = 3kT, \quad \text{and} \quad \gamma = \frac{4}{3}.$$

Next, using the inversion formula (3.4.7), derive an expression for the density of states $g(E)$ of this system.

3.16. Consider a system similar to the one in the preceding problem but consisting of $3N$ particles moving in one dimension. Show that the partition function in this case is given by

$$Q_{3N}(L,T) = \frac{1}{(3N)!}\left[2L\left(\frac{kT}{hc}\right)\right]^{3N},$$

L being the "length" of the space available. Compare the thermodynamics and the density of states of this system with the corresponding quantities obtained in the preceding problem.

3.17. If we take the function $f(q,p)$ in equation (3.5.3) to be $U - H(q,p)$, then clearly $\langle f\rangle = 0$; formally, this would mean

$$\int [U - H(q,p)]e^{-\beta H(q,p)}d\omega = 0.$$

Derive, from this equation, expression (3.6.3) for the mean-square fluctuation in the energy of a system embedded in the canonical ensemble.

3.18. Show that for a system in the canonical ensemble

$$\langle(\Delta E)^3\rangle = k^2\left\{T^4\left(\frac{\partial C_V}{\partial T}\right)_V + 2T^3 C_V\right\}.$$

Verify that for an ideal gas

$$\left\langle\left(\frac{\Delta E}{U}\right)^2\right\rangle = \frac{2}{3N} \quad \text{and} \quad \left\langle\left(\frac{\Delta E}{U}\right)^3\right\rangle = \frac{8}{9N^2}.$$

3.19. Consider the long-time averaged behavior of the quantity dG/dt, where

$$G = \sum_i q_i p_i,$$

and show that the validity of equation (3.7.5) implies the validity of equation (3.7.6), and vice versa.

3.20. Show that, for a statistical system in which the interparticle potential energy $u(\boldsymbol{r})$ is a homogeneous function (of degree n) of the particle coordinates, the *virial* $\mathcal{V}$ is given by

$$\mathcal{V} = -3PV - nU$$

and, hence, the *mean kinetic energy* K by

$$K = -\frac{1}{2}\mathcal{V} = \frac{1}{2}(3PV + nU) = \frac{1}{(n+2)}(3PV + nE);$$

here, U denotes the *mean potential energy* of the system while $E = K + U$. Note that this result holds not only for a classical system but for a quantum-mechanical one as well.

3.21. **(a)** Calculate the time-averaged kinetic energy and potential energy of a one-dimensional harmonic oscillator, both classically and quantum-mechanically, and show that the results obtained are consistent with the result established in the preceding problem (with $n = 2$).

(b) Consider, similarly, the case of the hydrogen atom ($n = -1$) on the basis of (i) the Bohr–Sommerfeld model and (ii) the Schrödinger model.

(c) Finally, consider the case of a planet moving in (i) a circular orbit or (ii) an elliptic orbit around the sun.

3.22. The restoring force of an anharmonic oscillator is proportional to the *cube* of the displacement. Show that the mean kinetic energy of the oscillator is *twice* its mean potential energy.

3.23. Derive the virial equation of state equation (3.7.15) from the classical canonical partition function (3.5.5). Show that in the thermodynamic limit the interparticle terms dominate the ones that come from interactions of the particles with the walls of the container.

3.24. Show that in the relativistic case the equipartition theorem takes the form

$$\langle m_0 u^2 (1 - u^2/c^2)^{-1/2} \rangle = 3kT,$$

where m_0 is the rest mass of the particle and u its speed. Check that in the extreme relativistic case the mean thermal energy per particle is twice its value in the nonrelativistic case.

3.25. Develop a *kinetic* argument to show that in a noninteracting system the average value of the quantity $\sum_i p_i \dot{q}_i$ is precisely equal to $3PV$. Hence show that, regardless of relativistic considerations, $PV = NkT$.

3.26. The energy eigenvalues of an s-dimensional harmonic oscillator can be written as

$$\varepsilon_j = (j + s/2)\hbar\omega; \quad j = 0, 1, 2, \ldots$$

Show that the jth energy level has a multiplicity $(j + s - 1)!/j!(s-1)!$. Evaluate the partition function, and the major thermodynamic properties, of a system of N such oscillators, and compare your results with a corresponding system of sN one-dimensional oscillators. Show, in particular, that the chemical potential $\mu_s = s\mu_1$.

3.27. Obtain an asymptotic expression for the quantity $\ln g(E)$ for a system of N quantum-mechanical harmonic oscillators by using the inversion formula (3.4.7) and the partition function (3.8.15). Hence show that

$$\frac{S}{Nk} = \left(\frac{E}{N\hbar\omega} + \frac{1}{2}\right)\ln\left(\frac{E}{N\hbar\omega} + \frac{1}{2}\right) - \left(\frac{E}{N\hbar\omega} - \frac{1}{2}\right)\ln\left(\frac{E}{N\hbar\omega} - \frac{1}{2}\right).$$

[*Hint:* Employ the Darwin–Fowler method.]

3.28. **(a)** When a system of N oscillators with total energy E is in thermal equilibrium, what is the probability p_n that a particular oscillator among them is in the quantum state n?
[*Hint:* Use expression (3.8.25).]
Show that, for $N \gg 1$ and $R \gg n$, $p_n \approx (\overline{n})^n/(\overline{n}+1)^{n+1}$, where $\overline{n} = R/N$.

(b) When an ideal gas of N monatomic molecules with total energy E is in thermal equilibrium, show that the probability of a particular molecule having an energy in the neighborhood of ε is proportional to $\exp(-\beta\varepsilon)$, where $\beta = 3N/2E$.
[*Hint:* Use expression (3.5.16) and assume that $N \gg 1$ and $E \gg \varepsilon$.]

3.29. The potential energy of a one-dimensional, *anharmonic* oscillator may be written as

$$V(q) = cq^2 - gq^3 - fq^4,$$

where c, g, and f are positive constants; quite generally, g and f may be assumed to be very small in value. Show that the leading contribution of anharmonic terms to the heat capacity of the oscillator, assumed classical, is given by

$$\frac{3}{2}k^2\left(\frac{f}{c^2}+\frac{5}{4}\frac{g^2}{c^3}\right)T$$

and, to the same order, the mean value of the position coordinate q is given by

$$\frac{3}{4}\frac{gkT}{c^2}.$$

3.30. The energy levels of a quantum-mechanical, one-dimensional, *anharmonic* oscillator may be approximated as

$$\varepsilon_n=\left(n+\frac{1}{2}\right)\hbar\omega-x\left(n+\frac{1}{2}\right)^2\hbar\omega;\quad n=0,1,2,\ldots$$

The parameter x, usually $\ll 1$, represents the degree of anharmonicity. Show that, to the first order in x and the fourth order in $u(\equiv\hbar\omega/kT)$, the specific heat of a system of N such oscillators is given by

$$C=Nk\left[\left(1-\frac{1}{12}u^2+\frac{1}{240}u^4\right)+4x\left(\frac{1}{u}+\frac{1}{80}u^3\right)\right].$$

Note that the correction term here *increases* with temperature.

3.31. Study, along the lines of Section 3.8, the statistical mechanics of a system of N "Fermi oscillators," which are characterized by only two eigenvalues, namely 0 and ε.

3.32. The quantum states available to a given physical system are (i) a group of g_1 *equally likely* states, with a common energy ε_1 and (ii) a group of g_2 *equally likely* states, with a common energy $\varepsilon_2>\varepsilon_1$. Show that this entropy of the system is given by

$$S=-k[p_1\ln(p_1/g_1)+p_2\ln(p_2/g_2)],$$

where p_1 and p_2 are, respectively, the probabilities of the system being in a state belonging to group 1 or to group 2: $p_1+p_2=1$.

(a) Assuming that the p_i are given by a canonical distribution, show that

$$S=k\left[\ln g_1+\ln\{1+(g_2/g_1)e^{-x}\}+\frac{x}{1+(g_1/g_2)e^{x}}\right],$$

where $x=(\varepsilon_2-\varepsilon_1)/kT$, assumed positive. Compare the special case $g_1=g_2=1$ with that of the Fermi oscillator of the preceding problem.

(b) Verify the foregoing expression for S by deriving it from the partition function of the system.

(c) Check that at $T\to 0$, $S\to k\ln g_1$. Interpret this result physically.

3.33. Gadolinium sulphate obeys Langevin's theory of paramagnetism down to a few degrees Kelvin. Its molecular magnetic moment is 7.2×10^{-23}amp-m^2. Determine the degree of magnetic saturation in this salt at a temperature of 2 K in a field of flux density 2 weber/m^2.

3.34. Oxygen is a paramagnetic gas obeying Langevin's theory of paramagnetism. Its susceptibility per unit volume, at 293 K and at atmospheric pressure, is 1.80×10^{-6} mks units. Determine its molecular magnetic moment and compare it with the Bohr magneton (which is very nearly equal to 9.27×10^{-24}amp-m^2).

3.35. **(a)** Consider a gaseous system of N noninteracting, diatomic molecules, each having an electric dipole moment $\boldsymbol{\mu}$, placed in an external electric field of strength E. The energy of such a molecule will be given by the kinetic energy of rotation as well as translation *plus* the potential energy of orientation in the applied field:

$$\varepsilon=\frac{p^2}{2m}+\left\{\frac{p_\theta^2}{2I}+\frac{p_\phi^2}{2I\sin^2\theta}\right\}-\mu E\cos\theta,$$

where I is the moment of inertia of the molecule. Study the thermodynamics of this system, including the electric polarization and the dielectric constant. Assume that (i) the system is a classical one and (ii) $|\mu E| \ll kT$.[16]

(b) The molecule H_2O has an electric dipole moment of 1.85×10^{-18} e.s.u. Calculate, on the basis of the preceding theory, the dielectric constant of steam at 100°C and at atmospheric pressure.

3.36. Consider a pair of electric dipoles $\boldsymbol{\mu}$ and $\boldsymbol{\mu}'$, oriented in the directions (θ, ϕ) and (θ', ϕ'), respectively; the distance R between their centers is assumed to be fixed. The potential energy in this orientation is given by

$$-\frac{\mu\mu'}{R^3}\{2\cos\theta\cos\theta' - \sin\theta\sin\theta'\cos(\phi - \phi')\}.$$

Now, consider this pair of dipoles to be in thermal equilibrium, their orientations being governed by a canonical distribution. Show that the mean force between these dipoles, at high temperatures, is given by

$$-2\frac{(\mu\mu')^2}{kT}\frac{\hat{\boldsymbol{R}}}{R^7},$$

$\hat{\boldsymbol{R}}$ being the unit vector in the direction of the line of centers.

3.37. Evaluate the high-temperature approximation of the partition function of a system of magnetic dipoles to show that the Curie constant C_J is given by

$$C_J = \frac{N_0 g^2 \mu_B^2}{k}\overline{m^2}.$$

Hence derive the formula (3.9.26).

3.38. Replacing the sum in (3.9.18) by an integral, evaluate $Q_1(\beta)$ of the given magnetic dipole and study the thermodynamics following from it. Compare these results with the ones following from the Langevin theory.

3.39. Atoms of silver vapor, each having a magnetic moment $\mu_B (g = 2, J = \frac{1}{2})$, align themselves either parallel or antiparallel to the direction of an applied magnetic field. Determine the respective fractions of atoms aligned parallel and antiparallel to a field of flux density 0.1 weber/m^2 at a temperature of 1,000 K.

3.40. **(a)** Show that, for any magnetizable material, the heat capacities at constant field H and at constant magnetization M are connected by the relation

$$C_H - C_M = -T\left(\frac{\partial H}{\partial T}\right)_M \left(\frac{\partial M}{\partial T}\right)_H.$$

(b) Show that for a paramagnetic material obeying Curie's law

$$C_H - C_M = CH^2/T^2,$$

where C on the right side of this equation denotes the *Curie constant* of the given sample.

3.41. A system of N spins at a negative temperature $(E > 0)$ is brought into contact with an ideal-gas thermometer consisting of N' molecules. What will the nature of their state of mutual equilibrium be? Will their common temperature be negative or positive, and in what manner will it be affected by the ratio N'/N?

3.42. Consider the system of N magnetic dipoles, studied in Section 3.10, in the microcanonical ensemble. Enumerate the number of microstates, $\Omega(N, E)$, accessible to the system at energy E and evaluate the quantities $S(N, E)$ and $T(N, E)$. Compare your results with equations (3.10.8) and (3.10.9).

[16]The electric dipole moments of molecules are generally of order 10^{-18} e.s.u. (or a *Debye unit*). In a field of 1 e.s.u. (= 300 volts/cm) and at a temperature of 300 K, the parameter $\beta\mu E = O(10^{-4})$.

3.43. Consider a system of charged particles (not dipoles), obeying classical mechanics and classical statistics. Show that the magnetic susceptibility of this system is identically zero (Bohr–van Leeuwen theorem).
[Note that the Hamiltonian of this system in the presence of a magnetic field $\boldsymbol{H}(=\nabla\times\boldsymbol{A})$ will be a function of the quantities $\boldsymbol{p}_j+(e_j/c)\boldsymbol{A}(\boldsymbol{r}_j)$, and not of the $\boldsymbol{p}_j$ as such. One has now to show that the partition function of the system is independent of the applied field.]

3.44. The expression (3.3.13) for the entropy S is equivalent to Shannon's (1949) definition of the information contained in a message $I=-\sum_r P_r\ln(P_r)$, where P_r represents the probability of message r.

(a) Show that information is maximized if the probabilities of all messages are the same. Any other distribution of probabilities reduces the information. In English, "e" is more common than "z", so $P_{\mathrm{e}}>P_{\mathrm{z}}$, so the information per character in an English message is less than the optimal amount possible based on the number of different characters used in an English text.

(b) The information in a text is also affected by correlations between characters in the text. For example, in English, "q" is always followed by "u", so this pair of characters contains the same information as "q" alone. The probability of a character indexed by r followed immediately by character indexed by r' is $P_{r,r'}=P_rP_{r'}G_{r,r'}$, where $G_{r,r'}$ is the character-pair correlation function. If pairs of characters are uncorrelated, then $G_{r,r'}=1$. Show that if characters are uncorrelated then the information in a two-character message is twice the information of a single-character message and that correlations ($G_{r,r'}\neq 1$) reduce the information content. [*Hint:* Use the inequality $\ln x\leq x-1$.]

(c) Write a computer program to determine the information per character in a text file by determining the single-character probabilities P_r and character-pair correlations $G_{r,r'}$. Computers usually use one full byte per character to store information. Since one byte can store 256 different messages, the potential information per byte is $\ln 256=8\ln 2\equiv 8\,\text{bits}$. Show that the information per character in your text file is considerably less than 8 bits and explain why it is possible for file-compression algorithms to reduce the size of a computer file without sacrificing any of the information contained in the file.

4 The Grand Canonical Ensemble

In the preceding chapter we developed the formalism of the canonical ensemble and established a scheme of operations for deriving the various thermodynamic properties of a given physical system. The effectiveness of that approach became clear from the examples discussed there; it will become even more vivid in the subsequent studies carried out in this text. However, for a number of problems, both physical and chemical, the usefulness of the canonical ensemble formalism turns out to be rather limited and it appears that a further generalization of this formalism is called for. The motivation that brings about this generalization is physically of the same nature as the one that led us from the microcanonical to the canonical ensemble — it is just the next natural step from there. It comes from the realization that not only the energy of a system but the number of particles as well is hardly ever measured in a "direct" manner; we only estimate it through an indirect probing into the system. Conceptually, therefore, we may regard both N and E as *variables* and identify their expectation values, $\langle N\rangle$ and $\langle E\rangle$, with the corresponding thermodynamic quantities.

The procedure for studying the statistics of the variables N and E is self-evident. We may *either* (i) consider the given system A as immersed in a large reservoir A' with which it can exchange both energy and particles *or* (ii) regard it as a member of what we may call a *grand canonical ensemble*, which consists of the given system A and a large number of (mental) copies thereof, the members of the ensemble carrying out a mutual exchange of both energy and particles. The end results, in either case, are asymptotically the same.

4.1 Equilibrium between a system and a particle-energy reservoir

We consider the given system A as immersed in a large reservoir A', with which it can exchange both energy and particles; see Figure 4.1. After some time has elapsed, the system and the reservoir are supposed to attain a state of mutual equilibrium. Then, according to Section 1.3, the system and the reservoir will have a common temperature T and a common chemical potential μ. The fraction of the total number of particles $N^{(0)}$ and the fraction of the total energy $E^{(0)}$ that the system A can have at any time t are, however, variables (whose values, in principle, can lie anywhere between zero and unity). If, at a particular instant of time, the system A happens to be in *one* of its states characterized by the number N_r of particles and the amount E_s of energy, then the number of particles in the reservoir would be N_r' and its energy E_s', such that

$$N_r + N_r' = N^{(0)} = \text{const.} \tag{1}$$

© 2011 Elsevier Ltd. All rights reserved.

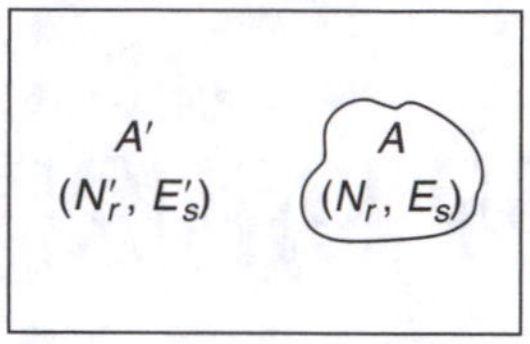

FIGURE 4.1 A statistical system immersed in a particle–energy reservoir.

and

$$E_s + E'_s = E^{(0)} = \text{const.} \tag{2}$$

Again, since the reservoir is supposed to be much larger than the given system, the values of N_r and E_s that are going to be of practical importance will be very small fractions of the total magnitudes $N^{(0)}$ and $E^{(0)}$, respectively; therefore, for all practical purposes,[1]

$$\frac{N_r}{N^{(0)}} = \left(1 - \frac{N'_r}{N^{(0)}}\right) \ll 1 \tag{3}$$

and

$$\frac{E_s}{E^{(0)}} = \left(1 - \frac{E'_s}{E^{(0)}}\right) \ll 1. \tag{4}$$

Now, in the manner of Section 3.1, the probability $P_{r,s}$ that, at any time t, the system A is found to be in an (N_r, E_s)-state would be directly proportional to the number of microstates $\Omega'(N'_r, E'_s)$ that the reservoir can have for the corresponding macrostate (N'_r, E'_s). Thus,

$$P_{r,s} \propto \Omega'(N^{(0)} - N_r, E^{(0)} - E_s). \tag{5}$$

Again, in view of (3) and (4), we can write

$$\begin{aligned} \ln \Omega'(N^{(0)} - N_r, E^{(0)} - E_s) &= \ln \Omega'(N^{(0)}, E^{(0)}) \\ &\quad + \left(\frac{\partial \ln \Omega'}{\partial N'}\right)_{N'=N^{(0)}} (-N_r) + \left(\frac{\partial \ln \Omega'}{\partial E'}\right)_{E'=E^{(0)}} (-E_s) + \cdots \\ &\simeq \ln \Omega'(N^{(0)}, E^{(0)}) + \frac{\mu'}{kT'} N_r - \frac{1}{kT'} E_s; \end{aligned} \tag{6}$$

see equations (1.2.3), (1.2.7), (1.3.3), and (1.3.5). Here, μ' and T' are, respectively, the chemical potential and the temperature of the reservoir (and hence of the given system

[1]Note that A here could well be a relatively small "part" of a given system $A^{(0)}$, while A' represents the "rest" of $A^{(0)}$. That would give a truly practical perspective to the grand canonical formalism.

as well). From (5) and (6), we obtain the desired result:

$$P_{r,s} \propto \exp(-\alpha N_r - \beta E_s), \tag{7}$$

where

$$\alpha = -\mu/kT, \quad \beta = 1/kT. \tag{8}$$

On normalization, it becomes

$$P_{r,s} = \frac{\exp(-\alpha N_r - \beta E_s)}{\sum_{r,s} \exp(-\alpha N_r - \beta E_s)}; \tag{9}$$

the summation in the denominator goes over *all* the (N_r, E_s)-states accessible to the system A. Note that our final expression for $P_{r,s}$ is independent of the choice of the reservoir.

We shall now examine the same problem from the ensemble point of view.

4.2 A system in the grand canonical ensemble

We now visualize an ensemble of $\mathcal{N}$ identical systems (which, of course, can be labeled as $1, 2, \ldots, \mathcal{N}$) mutually sharing a total number of particles[2] $\mathcal{N}\overline{N}$ and a total energy $\mathcal{N}\overline{E}$. Let $n_{r,s}$ denote the number of systems that have, at any time t, the number N_r of particles and the amount E_s of energy $(r, s = 0, 1, 2, \ldots)$; then, obviously,

$$\sum_{r,s} n_{r,s} = \mathcal{N}, \tag{1a}$$

$$\sum_{r,s} n_{r,s} N_r = \mathcal{N}\overline{N}, \tag{1b}$$

and

$$\sum_{r,s} n_{r,s} E_s = \mathcal{N}\overline{E}. \tag{1c}$$

Any set $\{n_{r,s}\}$, of the numbers $n_{r,s}$, which satisfies the restrictive conditions (1), represents one of the possible modes of distribution of particles and energy among the members of our ensemble. Furthermore, any such mode of distribution can be realized in $W\{n_{r,s}\}$ different ways, where

$$W\{n_{r,s}\} = \frac{\mathcal{N}!}{\prod_{r,s} (n_{r,s}!)}. \tag{2}$$

[2]For simplicity, we shall henceforth use the symbols $\overline{N}$ and $\overline{E}$ instead of $\langle N \rangle$ and $\langle E \rangle$.

We may now define the *most probable* mode of distribution, $\{n^*_{r,s}\}$, as the one that maximizes expression (2), satisfying at the same time the restrictive conditions (1). Going through the conventional derivation, see Section 3.2, we obtain for a large ensemble

$$\frac{n^*_{r,s}}{\mathcal{N}} = \frac{\exp(-\alpha N_r - \beta E_s)}{\sum\limits_{r,s} \exp(-\alpha N_r - \beta E_s)}; \tag{3}$$

compare to the corresponding equation (3.2.10) for the canonical ensemble. Alternatively, we may define the *expectation* (or *mean*) values of the numbers $n_{r,s}$, namely

$$\langle n_{r,s} \rangle = \frac{\sum\limits_{\{n_{r,s}\}}' n_{r,s} W\{n_{r,s}\}}{\sum\limits_{\{n_{r,s}\}}' W\{n_{r,s}\}}, \tag{4}$$

where the primed summations go over all distribution sets that conform to conditions (1). An asymptotic expression for $\langle n_{r,s} \rangle$ can be derived by using the method of Darwin and Fowler — the only difference from the corresponding derivation in Section 3.2 being that, in the present case, we will have to work with functions of more than one (complex) variable. The derivation, however, runs along similar lines, with the result

$$\operatorname*{Lim}_{\mathcal{N}\to\infty} \frac{\langle n_{r,s} \rangle}{\mathcal{N}} \simeq \frac{n^*_{r,s}}{\mathcal{N}} = \frac{\exp(-\alpha N_r - \beta E_s)}{\sum\limits_{r,s} \exp(-\alpha N_r - \beta E_s)}, \tag{5}$$

in agreement with equation (4.1.9). The parameters α and β, so far undetermined, are eventually determined by the equations

$$\overline{N} = \frac{\sum\limits_{r,s} N_r \exp(-\alpha N_r - \beta E_s)}{\sum\limits_{r,s} \exp(-\alpha N_r - \beta E_s)} \equiv -\frac{\partial}{\partial \alpha}\left\{\ln \sum_{r,s} \exp(-\alpha N_r - \beta E_s)\right\} \tag{6}$$

and

$$\overline{E} = \frac{\sum\limits_{r,s} E_s \exp(-\alpha N_r - \beta E_s)}{\sum\limits_{r,s} \exp(-\alpha N_r - \beta E_s)} \equiv -\frac{\partial}{\partial \beta}\left\{\ln \sum_{r,s} \exp(-\alpha N_r - \beta E_s)\right\}, \tag{7}$$

where the quantities $\overline{N}$ and $\overline{E}$ here are supposed to be preassigned.

4.3 Physical significance of the various statistical quantities

To establish a connection between the statistics of the grand canonical ensemble and the thermodynamics of the system under study, we introduce a quantity q, defined by

$$q \equiv \ln \left\{ \sum_{r,s} \exp(-\alpha N_r - \beta E_s) \right\}; \tag{1}$$

the quantity q is a function of the parameters α and β, and also of all the E_s.[3] Taking the differential of q and making use of equations (4.2.5), (4.2.6), and (4.2.7), we get

$$dq = -\overline{N} d\alpha - \overline{E} d\beta - \frac{\beta}{\mathcal{N}} \sum_{r,s} \langle n_{r,s} \rangle \, dE_s, \tag{2}$$

so that

$$d(q + \alpha\overline{N} + \beta\overline{E}) = \beta \left(\frac{\alpha}{\beta} d\overline{N} + d\overline{E} - \frac{1}{\mathcal{N}} \sum_{r,s} \langle n_{r,s} \rangle \, dE_s \right). \tag{3}$$

To interpret the terms appearing on the right side of this equation, we compare the expression enclosed within the parentheses with the statement of the first law of thermodynamics, that is,

$$\delta Q = d\overline{E} + \delta W - \mu d\overline{N}, \tag{4}$$

where the various symbols have their usual meanings. The following correspondence now seems inevitable:

$$\delta W = -\frac{1}{\mathcal{N}} \sum_{r,s} \langle n_{r,s} \rangle \, dE_s, \quad \mu = -\alpha/\beta, \tag{5}$$

with the result that

$$d(q + \alpha\overline{N} + \beta\overline{E}) = \beta \delta Q. \tag{6}$$

The parameter β, being the integrating factor for the heat δQ, must be equivalent to the reciprocal of the absolute temperature T, so we may write

$$\beta = 1/kT \tag{7}$$

and, hence,

$$\alpha = -\mu/kT. \tag{8}$$

[3] This quantity was first introduced by Kramers, who called it the *q-potential*.

The quantity $(q+\alpha\overline{N}+\beta\overline{E})$ would then be identified with the thermodynamic variable S/k; accordingly,

$$q = \frac{S}{k} - \alpha\overline{N} - \beta\overline{E} = \frac{TS + \mu\overline{N} - \overline{E}}{kT}. \tag{9}$$

However, $\mu\overline{N}$ is identically equal to G, the Gibbs free energy of the system, and hence to $(\overline{E} - TS + PV)$. So, finally,

$$q \equiv \ln\left\{\sum_{r,s} \exp(-\alpha N_r - \beta E_s)\right\} = \frac{PV}{kT}. \tag{10}$$

Equation (10) provides the essential link between the thermodynamics of the given system and the statistics of the corresponding grand canonical ensemble. It is, therefore, a relationship of central importance in the formalism developed in this chapter.

To derive further results, we prefer to introduce a parameter z, defined by the relation

$$z \equiv e^{-\alpha} = e^{\mu/kT}; \tag{11}$$

the parameter z is generally referred to as the *fugacity* of the system. In terms of z, the q-potential takes the form

$$q \equiv \ln\left\{\sum_{r,s} z^{N_r} e^{-\beta E_s}\right\} \tag{12}$$

$$= \ln\left\{\sum_{N_r=0}^{\infty} z^{N_r} Q_{N_r}(V,T)\right\} \quad \text{(with } Q_0 \equiv 1\text{)}, \tag{13}$$

so we may write

$$q(z,V,T) \equiv \ln \mathcal{Q}(z,V,T), \tag{14}$$

where

$$\mathcal{Q}(z,V,T) \equiv \sum_{N_r=0}^{\infty} z^{N_r} Q_{N_r}(V,T) \quad \text{(with } Q_0 \equiv 1\text{)}. \tag{15}$$

Note that, in going from expression (12) to (13), we have (mentally) carried out a summation over the energy values E_s, with N_r fixed, thus giving rise to the partition function $Q_{N_r}(V,T)$; of course, the dependence of Q_{N_r} on V comes from the dependence of the E_s on V. In going from (13) to (14), we have (again mentally) carried out a summation over all the numbers $N_r = 0, 1, 2, \cdots, \infty$, thus giving rise to the *grand partition function* $\mathcal{Q}(z,V,T)$ of the system. The q-potential, which we have already identified with PV/kT, is, therefore, the logarithm of the grand partition function.

It appears that in order to evaluate the grand partition function $\mathcal{Q}(z, V, T)$ we have to go through the routine of evaluating the partition function $Q(N, V, T)$. In principle, this is indeed true. In practice, however, we find that on many occasions an explicit evaluation of the partition function is extremely hard while considerable progress can be made in the evaluation of the grand partition function. This is particularly true when we deal with systems in which the influence of quantum statistics and/or interparticle interactions is important; see Sections 6.2 and 10.1. The formalism of the grand canonical ensemble then proves to be of considerable value.

We are now in a position to write down the full recipe for deriving the leading thermodynamic quantities of a given system from its q-potential. We have, first of all, for the pressure of the system

$$P(z, V, T) = \frac{kT}{V} q(z, V, T) \equiv \frac{kT}{V} \ln \mathcal{Q}(z, V, T). \tag{16}$$

Next, writing N for $\overline{N}$ and U for $\overline{E}$, we obtain with the help of equations (4.2.6), (4.2.7), and (11)

$$N(z, V, T) = z\left[\frac{\partial}{\partial z} q(z, V, T)\right]_{V,T} = kT\left[\frac{\partial}{\partial \mu} q(\mu, V, T)\right]_{V,T} \tag{17}$$

and

$$U(z, V, T) = -\left[\frac{\partial}{\partial \beta} q(z, V, T)\right]_{z,V} = kT^2\left[\frac{\partial}{\partial T} q(z, V, T)\right]_{z,V}. \tag{18}$$

Eliminating z between equations (16) and (17), one obtains the equation of state, that is, the (P, V, T)-relationship, of the system. On the other hand, eliminating z between equations (17) and (18), one obtains U as a function of N, V, and T, which readily leads to the specific heat at constant volume as $(\partial U/\partial T)_{N,V}$. The Helmholtz free energy is given by the formula

$$\begin{aligned} A &= N\mu - PV = NkT \ln z - kT \ln \mathcal{Q}(z, V, T) \\ &= -kT \ln \frac{\mathcal{Q}(z, V, T)}{z^N}, \end{aligned} \tag{19}$$

which may be compared with the canonical ensemble formula $A = -kT \ln Q(N, V, T)$; see also Problem 4.2. Finally, we have for the entropy of the system

$$S = \frac{U - A}{T} = kT\left(\frac{\partial q}{\partial T}\right)_{z,V} - Nk \ln z + kq. \tag{20}$$

4.4 Examples

We shall now study a couple of simple problems, with the explicit purpose of demonstrating how the method of the q-potential works. This is not intended to be a demonstration of the power of this method, for we shall consider here only those problems that can be solved equally well by the methods of the preceding chapters. The real power of the new method will become apparent only when we study problems involving quantum-statistical effects and effects arising from interparticle interactions; many such problems will appear in the remainder of the text.

The first problem we propose to consider here is that of the classical ideal gas. In Section 3.5 we showed that the partition function $Q_N(V,T)$ of this system could be written as

$$Q_N(V,T) = \frac{[Q_1(V,T)]^N}{N!}, \tag{1}$$

where $Q_1(V,T)$ may be regarded as the partition function of a single particle in the system. First of all, we should note that equation (1) does not imply any restrictions on the particles having *internal* degrees of motion; those degrees of motion, if present, would affect the results only through Q_1. Second, we should recall that the factor $N!$ in the denominator arises from the fact that the particles constituting the gas are, in fact, *indistinguishable*. Closely related to the indistinguishability of the particles is the fact that they are *nonlocalized*, for otherwise we could distinguish them through their very sites; compare, for instance, the system of harmonic oscillators, which was studied in Section 3.8. Now, since our particles are nonlocalized they can be *anywhere* in the space available to them; consequently, the function Q_1 will be directly proportional to V:

$$Q_1(V,T) = Vf(T), \tag{2}$$

where $f(T)$ is a function of temperature alone. We thus obtain for the grand partition function of the gas

$$\begin{aligned}\mathcal{Q}(z,V,T) &= \sum_{N_r=0}^{\infty} z^{N_r} Q_{N_r}(V,T) = \sum_{N_r=0}^{\infty} \frac{\{zVf(T)\}^{N_r}}{N_r!} \\ &= \exp\{zVf(T)\},\end{aligned} \tag{3}$$

which gives

$$q(z,V,T) = zVf(T). \tag{4}$$

Formula (4.3.16) through (4.3.20) then lead to the following results:

$$P = zkTf(T), \tag{5}$$

$$N = zVf(T), \tag{6}$$

$$U = zVkT^2 f'(T), \tag{7}$$

$$A = NkT\ln z - zVkTf(T), \tag{8}$$

and

$$S = -Nk\ln z + zVk\{Tf'(T) + f(T)\}. \tag{9}$$

Eliminating z between (5) and (6), we obtain the equation of state of the system:

$$PV = NkT. \tag{10}$$

We note that equation (10) holds irrespective of the form of the function $f(T)$. Next, eliminating z between (6) and (7), we obtain

$$U = NkT^2 f'(T)/f(T), \tag{11}$$

which gives

$$C_V = Nk\frac{2Tf(T)f'(T) + T^2\{f(T)f''(T) - [f'(T)]^2\}}{[f(T)]^2}. \tag{12}$$

In simple cases, the function $f(T)$ turns out to be directly proportional to a certain power of T. Supposing that $f(T) \propto T^n$, equations (11) and (12) become

$$U = n(NkT) \tag{11a}$$

and

$$C_V = n(Nk). \tag{12a}$$

Accordingly, the pressure in such cases is directly proportional to the energy density of the gas, the constant of proportionality being $1/n$. The reader will recall that the case $n = 3/2$ corresponds to a nonrelativistic gas while $n = 3$ corresponds to an extreme relativistic one.

Finally, eliminating z between equation (6) and equations (8) and (9), we obtain A and S as functions of N, V, and T. This essentially completes our study of the classical ideal gas.

The next problem to be considered here is that of a system of independent, *localized* particles — a model which, in some respects, approximates a solid. Mathematically, the problem is similar to that of a system of harmonic oscillators. In either case, the microscopic entities constituting the system are mutually *distinguishable*. The partition function $Q_N(V,T)$ of such a system can be written as

$$Q_N(V,T) = [Q_1(V,T)]^N. \tag{13}$$

At the same time, in view of the localized nature of the particles, the single-particle partition function $Q_1(V,T)$ is essentially independent of the volume occupied by the system. Consequently, we may write

$$Q_1(V,T) = \phi(T), \tag{14}$$

where $\phi(T)$ is a function of temperature alone. We then obtain for the grand partition function of the system

$$\mathcal{Q}(z,V,T) = \sum_{N_r=0}^{\infty} [z\phi(T)]^{N_r} = [1 - z\phi(T)]^{-1}; \tag{15}$$

clearly, the quantity $z\phi(T)$ must stay below unity, so that the summation over N_r is convergent.

The thermodynamics of the system follows straightforwardly from equation (15). We have, to begin with,

$$P \equiv \frac{kT}{V} q(z,T) = -\frac{kT}{V} \ln\{1 - z\phi(T)\}. \tag{16}$$

Since both z and T are intensive variables, the right side of (16) vanishes as $V \to \infty$. Hence, in the thermodynamic limit, $P = 0$.[4] For other quantities of interest, we obtain, with the help of equations (4.3.17) through (4.3.20),

$$N = \frac{z\phi(T)}{1 - z\phi(T)}, \tag{17}$$

$$U = \frac{zkT^2\phi'(T)}{1 - z\phi(T)}, \tag{18}$$

$$A = NkT\ln z + kT\ln\{1 - z\phi(T)\}, \tag{19}$$

and

$$S = -Nk\ln z - k\ln\{1 - z\phi(T)\} + \frac{zkT\phi'(T)}{1 - z\phi(T)}. \tag{20}$$

From (17), we get

$$z\phi(T) = \frac{N}{N+1} \simeq 1 - \frac{1}{N} \quad (N \gg 1). \tag{21}$$

[4] It will be seen in the sequel that P actually vanishes like $(\ln N)/N$.

It follows that

$$1 - z\phi(T) = \frac{1}{N+1} \simeq \frac{1}{N}. \tag{22}$$

Equations (17) through (20) now give

$$U/N = kT^2\phi'(T)/\phi(T), \tag{18a}$$

$$A/N = -kT\ln\phi(T) + O\left(\frac{\ln N}{N}\right), \tag{19a}$$

and

$$S/Nk = \ln\phi(T) + T\phi'(T)/\phi(T) + O\left(\frac{\ln N}{N}\right). \tag{20a}$$

Substituting

$$\phi(T) = [2\sinh(\hbar\omega/2kT)]^{-1} \tag{23}$$

into these formulae, we obtain results pertaining to a system of *quantum-mechanical*, one-dimensional harmonic oscillators. The substitution

$$\phi(T) = kT/\hbar\omega, \tag{24}$$

on the other hand, leads to results pertaining to a system of *classical*, one-dimensional harmonic oscillators.

As a corollary, we examine here the problem of *solid–vapor equilibrium*. Consider a single-component system, having two phases — solid and vapor — in equilibrium, contained in a closed vessel of volume V at temperature T. Since the phases are free to exchange particles, a state of mutual equilibrium would imply that their chemical potentials are equal; this, in turn, means that they have a common fugacity as well. Now, the fugacity z_g of the gaseous phase is given by, see equation (6),

$$z_g = \frac{N_g}{V_g f(T)}, \tag{25}$$

where N_g is the number of particles in the gaseous phase and V_g the volume occupied by them; in a typical case, $V_g \simeq V$. The fugacity z_s of the solid phase, on the other hand, is given by equation (21):

$$z_s \simeq \frac{1}{\phi(T)}. \tag{26}$$

Equating (25) and (26), we obtain for the *equilibrium particle density* in the vapor phase

$$N_g/V_g = f(T)/\phi(T). \tag{27}$$

Now, if the density in the vapor phase is sufficiently low and the temperature of the system sufficiently high, the vapor pressure P would be given by

$$P_{\text{vapor}} = \frac{N_g}{V_g} kT = kT \frac{f(T)}{\phi(T)}. \tag{28}$$

To be specific, we may assume the vapor to be monatomic; the function $f(T)$ is then of the form

$$f(T) = (2\pi mkT)^{3/2}/h^3. \tag{29}$$

On the other hand, if the solid phase can be approximated by a set of three-dimensional harmonic oscillators characterized by a single frequency ω (the *Einstein* model), the function $\phi(T)$ would be

$$\phi(T) = [2\sinh(h\omega/2kT)]^{-3}. \tag{30}$$

However, there is one important difference here. An atom in a solid is energetically more stabilized than an atom that is free — that is why a certain threshold energy is required to transform a solid into separate atoms. Let ε denote the value of this energy per atom, which in a way implies that the zeros of the energy spectra ε_g and ε_s, which led to the functions (29) and (30), respectively, are displaced with respect to one another by an amount ε. A *true* comparison between the functions $f(T)$ and $\phi(T)$ must take this into account. As a result, we obtain for the vapor pressure

$$P_{\text{vapor}} = kT \left(\frac{2\pi mkT}{h^2}\right)^{3/2} [2\sinh(\hbar\omega/2kT)]^3 e^{-\varepsilon/kT}. \tag{31}$$

In passing, we note that equation (27) also gives us the necessary condition for the formation of the solid phase. The condition clearly is:

$$N > V\frac{f(T)}{\phi(T)}, \tag{32}$$

where N is the total number of particles in the system. Alternatively, this means that

$$T < T_c, \tag{33}$$

where T_c is a *characteristic* temperature determined by the implicit relationship

$$\frac{f(T_c)}{\phi(T_c)} = \frac{N}{V}. \tag{34}$$

Once the two phases appear, the number $N_g(T)$ will have a value determined by equation (27) while the remainder, $N - N_g$, will constitute the solid phase.

4.5 Density and energy fluctuations in the grand canonical ensemble: correspondence with other ensembles

In a grand canonical ensemble, the variables N and E, for any member of the ensemble, can lie anywhere between zero and infinity. Therefore, on the face of it, the grand canonical ensemble appears to be very different from its predecessors — the canonical and the microcanonical ensembles. However, as far as thermodynamics is concerned, the results obtained from this ensemble turn out to be identical to the ones obtained from the other two. Thus, in spite of strong facial differences, the overall behavior of a given physical system is practically the same whether it belongs to one kind of ensemble or another. The basic reason for this is that the "relative fluctuations" in the values of the quantities that vary from member to member in an ensemble are practically negligible. Therefore, in spite of the different surroundings that different ensembles provide to a given physical system, the overall behavior of the system is not significantly affected.

To appreciate this point, we shall evaluate the relative fluctuations in the particle density n and the energy E of a given physical system in the grand canonical ensemble. Recalling that

$$\overline{N} = \frac{\sum\limits_{r,s} N_r e^{-\alpha N_r - \beta E_s}}{\sum\limits_{r,s} e^{-\alpha N_r - \beta E_s}}, \tag{1}$$

it readily follows that

$$\left(\frac{\partial \overline{N}}{\partial \alpha}\right)_{\beta, E_s} = -\overline{N^2} + \overline{N}^2. \tag{2}$$

Thus

$$\overline{(\Delta N)^2} \equiv \overline{N^2} - \overline{N}^2 = -\left(\frac{\partial \overline{N}}{\partial \alpha}\right)_{T,V} = kT\left(\frac{\partial \overline{N}}{\partial \mu}\right)_{T,V}. \tag{3}$$

From (3), we obtain for the relative mean-square fluctuation in the particle density $n\,(=N/V)$

$$\frac{\overline{(\Delta n)^2}}{\overline{n}^2} = \frac{\overline{(\Delta N)^2}}{\overline{N}^2} = \frac{kT}{\overline{N}^2}\left(\frac{\partial \overline{N}}{\partial \mu}\right)_{T,V}. \tag{4}$$

In terms of the variable $v\,(= V/\overline{N})$, we may write

$$\frac{\overline{(\Delta n)^2}}{\overline{n}^2} = \frac{kTv^2}{V^2}\left(\frac{\partial (V/v)}{\partial \mu}\right)_{T,V} = -\frac{kT}{V}\left(\frac{\partial v}{\partial \mu}\right)_T. \tag{5}$$

To put this result into a more practical form, we recall the thermodynamic relation

$$d\mu = v\,dP - s\,dT, \tag{6}$$

according to which $d\mu$ (at constant T) $= v\,dP$. Equation (5) then takes the form

$$\frac{\overline{(\Delta n)^2}}{\overline{n}^2} = -\frac{kT}{V}\frac{1}{v}\left(\frac{\partial v}{\partial P}\right)_T = \frac{kT}{V}\kappa_T, \tag{7}$$

where κ_T is the isothermal compressibility of the system.

Thus, the relative root-mean-square fluctuation in the particle density of the given system is *ordinarily* $O(N^{-1/2})$ and, hence, negligible. However, there are exceptions, like the ones met with in situations accompanying *phase transitions*. In those situations, the compressibility of a given system can become excessively large, as is evidenced by an almost "flattening" of the isotherms. For instance, at a critical point the compressibility diverges, so it is no longer intensive. Finite-size scaling theory described in Chapters 12 and 14 indicates that at the critical point the isothermal compressibility scales with system size as $\kappa_T(T_c) \sim N^{\gamma/d\nu}$ where γ and ν are certain critical exponents and d is the dimension. For the case of experimental liquid–vapor critical points, $\kappa_T(T_c) \sim N^{0.63}$. Accordingly, the root-mean-square density fluctuations grow faster than $N^{1/2}$ — in this case, like $N^{0.82}$. Thus, in the region of phase transitions, especially at the critical points, we encounter unusually large fluctuations in the particle density of the system. Such fluctuations indeed exist and account for phenomena like *critical opalescence*. It is clear that under these circumstances the formalism of the grand canonical ensemble could, in principle, lead to results that are not necessarily identical to the ones following from the corresponding canonical ensemble. In such cases, it is the formalism of the grand canonical ensemble that will have to be preferred because only this one will provide a correct picture of the actual physical situation.

We shall now examine fluctuations in the energy of the system. Following the usual procedure, we obtain

$$\overline{(\Delta E)^2} \equiv \overline{E^2} - \overline{E}^2 = -\left(\frac{\partial \overline{E}}{\partial \beta}\right)_{z,V} = kT^2\left(\frac{\partial U}{\partial T}\right)_{z,V}. \tag{8}$$

To put expression (8) into a more comprehensible form, we write

$$\left(\frac{\partial U}{\partial T}\right)_{z,V} = \left(\frac{\partial U}{\partial T}\right)_{N,V} + \left(\frac{\partial U}{\partial N}\right)_{T,V}\left(\frac{\partial N}{\partial T}\right)_{z,V}, \tag{9}$$

where the symbol N is being used interchangeably for $\overline{N}$. Now, in view of the fact that

$$N = -\left(\frac{\partial}{\partial \alpha}\ln \mathcal{Q}\right)_{\beta,V}, \quad U = -\left(\frac{\partial}{\partial \beta}\ln \mathcal{Q}\right)_{\alpha,V}, \tag{10}$$

we have

$$\left(\frac{\partial N}{\partial \beta}\right)_{\alpha,V} = \left(\frac{\partial U}{\partial \alpha}\right)_{\beta,V} \tag{11}$$

and, hence,

$$\left(\frac{\partial N}{\partial T}\right)_{z,V} = \frac{1}{T}\left(\frac{\partial U}{\partial \mu}\right)_{T,V}. \tag{12}$$

Substituting expressions (9) and (12) into equation (8) and remembering that the quantity $(\partial U/\partial T)_{N,V}$ is the familiar C_V, we get

$$\overline{(\Delta E)^2} = kT^2 C_V + kT\left(\frac{\partial U}{\partial N}\right)_{T,V}\left(\frac{\partial U}{\partial \mu}\right)_{T,V}. \tag{13}$$

Invoking equations (3.6.3) and (3), we finally obtain

$$\overline{(\Delta E)^2} = \langle(\Delta E)^2\rangle_{\text{can}} + \left\{\left(\frac{\partial U}{\partial N}\right)_{T,V}\right\}^2 \overline{(\Delta N)^2}. \tag{14}$$

Formula (14) is highly instructive; it tells us that the mean-square fluctuation in the energy E of a system in the grand canonical ensemble is equal to the value it would have in the canonical ensemble *plus* a contribution arising from the fact that now the particle number N is also fluctuating. Again, under ordinary circumstances, the relative root-mean-square fluctuation in the energy density of the system would be practically negligible. However, in the region of phase transitions, unusually large fluctuations in the value of this variable can arise by virtue of the second term in the formula.

4.6 Thermodynamic phase diagrams

One of the great successes of thermodynamics and statistical mechanics over the last 150 years has been in the study of phase transitions. Statistical mechanics provides the basis for accurate models for a wide variety of thermodynamic phases of materials and has led to a detailed understanding of phase transitions and critical phenomena.

Condensed materials exist in a variety of phases that depend on thermodynamic parameters such as temperature, pressure, magnetic field, and so on. Thermodynamics and statistical mechanics can be used to determine the properties of individual phases, and the locations and characteristics of the phase transitions that occur between those phases. Thermodynamic phases are regions in the phase diagram where the thermodynamics properties are analytic functions of the thermodynamic parameters, while phase transitions are points, lines, or surfaces in the phase diagram where the thermodynamic properties are nonanalytic. Much of the remainder of this text is devoted to using statistical mechanics to explain the properties of material phases and phase transitions.

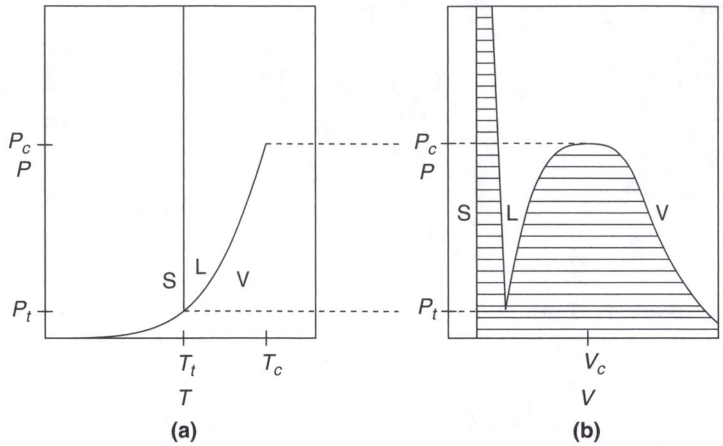

FIGURE 4.2 Sketches (not-to-scale) of the P–T (a) and P–V (b) phase diagrams for argon. This geometry is generic for a wide range of materials. The letters S, L, and V denote solid, liquid, and vapor phases.

It is instructive to examine the structure of phase diagrams. Argon provides a good example because the structure of its phase diagram is similar to that of many other materials (see Figure 4.2). At moderate temperatures and pressures, the stable thermodynamic phases of argon are solid, liquid, and vapor. At high temperatures there is a supercritical fluid phase that smoothly connects the liquid and vapor phases. Most materials, including argon, exhibit multiple solid phases especially at high pressures and low temperatures. Figure 4.2(a) is the phase diagram in the P–T plane and shows the solid–liquid coexistence line, the liquid–vapor coexistence line, and the solid–vapor coexistence line. The three lines meet at the triple point (T_t, P_t) and the liquid–vapor coexistence line ends at the critical point (T_c, P_c). The triple point values and critical point values for argon are $T_t = 83.8$ K, $P_t = 68.9$ kPa, $T_c = 150.7$ K, and $P_c = 4.86$ MPa, respectively.

Figure 4.2(b) is the phase diagram in the P–V plane and shows the pressure versus the specific volume $v (= V/N)$ on the coexistence lines. The dashed lines indicate the triple point pressure and critical pressure in both figures. The horizontal tie lines are the portions of isotherms as they cross coexistence lines and show the discontinuities of v. The tie lines in order from bottom to top are: sublimation tie lines connecting the solid and vapor phases, the triple point tie line that connects all three phases, and a series of solid–liquid and liquid–vapor tie lines. Notice that the liquid and vapor specific volumes continuously approach each other and are both equal to the critical specific volume v_c at the critical point.

The properties of the vapor, liquid, and solid phases are:

- The vapor phase is a low-density gas that is accurately described by the ideal-gas equation of state $P = nkT$ with corrections that are described by the virial expansion; see Chapters 6 and 10.

- The liquid phase is a dense fluid with strong interactions between the atoms. The fluid exhibits characteristic short-range pair correlations and scattering structure, as discussed in Section 10.7. The structure factor and the pair correlation function for argon, as determined from neutron scattering, are shown in Figure 10.8. For temperatures above the critical temperature T_c, one cannot distinguish between liquid and vapor. The density in this supercritical phase is a smooth function of temperature and pressure from the low-density vapor to the high-density liquid. Virial expansions developed in Sections 10.1 through 10.3 aptly describe the supercritical region. Strictly speaking, one can only distinguish between the liquid and vapor phases *on* the liquid–vapor coexistence line since it is possible to evolve smoothly from one phase to the other without crossing a phase boundary.
- The solid phase is a face-centered cubic crystal structure with long-range order, so the scattering structure factor displays Bragg peaks as described in Section 10.7.B. The thermodynamic properties of solid phases are described in Section 7.3.

All equilibrium thermodynamic properties within a single phase are analytic functions of the thermodynamic parameters while phase transitions are defined as places in the phase diagram where equilibrium thermodynamic properties are not analytic. Coexistence lines, or first-order phase transition lines, separate different phases in the P–T phase diagram as shown in Figure 4.2(a). Thermodynamic densities are discontinuous across coexistence lines. This is displayed on the P–V phase diagram in Figure 4.2(b) by horizontal tie lines that connect different values the specific volume takes in the two phases. Generally, all densities such as the specific volume $v = V/N$, entropy per particle $s = S/N$, internal energy density $u = U/V$, and so on, are discontinuous across first-order phase transition lines. The slopes of the coexistence lines in the P–V phase diagram depend on the latent heat of the transition and the specific volumes of the coexisting phases; see Section 4.7. All three phases coexist at the triple point.

The liquid–vapor coexistence line extends from the triple point to the critical point at the end of the first-order phase transition line. The specific volume is discontinuous on the liquid–vapor coexistence line but the size of the discontinuty vanishes at the critical point where the specific volume is v_c; see Figure 4.2(b). All densities are continuous functions of T and P through the critical point. For this reason, critical points are called continuous transitions or, sometimes, second-order phase transitions. Even though thermodynamic densities are continuous, the thermodynamic behavior at the critical point is nonanalytic since, for example, the specific heat and isothermal compressibility both diverge at the critical point. Another characteristic property of critical points is the divergence of the correlation length, which results in a universal behavior of critical points for broad classes of materials. The theory of critical points is developed in Chapters 12, 13, and 14.

Classical statistical mechanics provides a framework for understanding the phase diagrams and thermodynamic properties of a wide variety of materials. However, quantum mechanics and quantum statistics play an important role at low temperatures when the size of the thermal deBroglie wavelength $\lambda = h/\sqrt{2\pi mkT}$ is of the same order as the

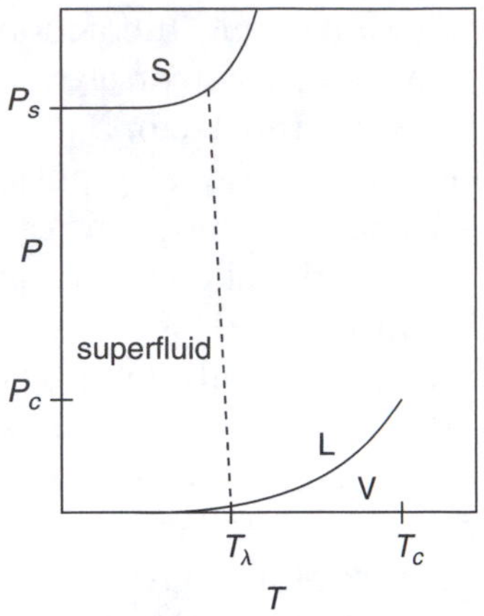

FIGURE 4.3 Sketch of the P–T phase diagram for helium-4. The letters S, L, and V denote solid, liquid, and vapor phases. The critical point is $T_c = 5.19\,\text{K}$ and $P_c = 227\,\text{kPa} = 2.24\,\text{atm}$. The solid–liquid coexistence curve starts at $P_s = 2.5\,\text{MPa} = 25\,\text{atm}$ at $T = 0\,\text{K}$ and does not intersect the liquid–vapor coexistence curve. The λ-line is the continuous phase transition between the normal liquid and the superfluid phase. The superfluid phase transition temperature at the liquid–vapor coexistence line is $T_\lambda = 2.18\,\text{K}$.

average distance between molecules. This is the case with liquid helium at temperatures below a few degrees kelvin. The phase diagram of helium-4 is shown in Figure 4.3. Some aspects of the phase diagram are similar to the phase diagram of argon. Both helium and argon have liquid–vapor coexistence lines that end in critical points and both have crystalline solid phases at low temperatures.

Three differences between the two phase diagrams are most notable: the solid phase for helium only exists for pressures greater than $P_s = 2.5\,\text{GPa} = 25\,\text{atm}$, the liquid phase of helium extends all the way to zero temperature, and helium-4 exhibits a superfluid phase below $T_\lambda = 2.18\text{K}$. The superfluid phase exhibits remarkable properties: zero viscosity, quantized flow, propagating heat modes, and macroscopic quantum coherence. This extraordinary behavior is due to the Bose-Einstein statistics of ^{4}He atoms and a Bose–Einstein condensation into a macroscopic quantum state as discussed in Sections 7.1 and 11.2 through 11.6. Even the solid phase of helium-4 shows evidence of a macroscopic quantum state with the observation of a “supersolid” phase by Kim and Chan (2004).

By contrast, ^{3}He atoms obey Fermi–Dirac statistics and display very different behaviors from ^{4}He atoms at low temperatures. The geometry of the phase diagram of helium-3 is similar to that of helium-4 except that the critical temperature is lower ($T_c = 3.35\,\text{K}$ compared to $5.19\,\text{K}$) and the solid phase forms at 30 atm of pressure rather than 25 atm. The dramatic difference is the lack of a superfluid phase near 1 K in helium-3. Helium-3 remains a normal liquid all the way down to about 10 millikelvin. The properties of the normal liquid phase of helium-3 are described by the theory of degenerate Fermi gases and the Fermi liquid theory developed in Chapter 8 and Sections 11.7 and 11.8. The superfluid state that forms at millikelvin temperatures is the result of Bardeen, Cooper, and Schrieffer (BCS) p-wave pairing between atoms near the Fermi surface; this pairing is discussed in Section 11.9.

4.7 Phase equilibrium and the Clausius–Clapeyron equation

The thermodynamic properties of the phases of a material determine the geometry of the phase diagram. In particular, the Gibbs free energy

$$G(N,P,T) = U - TS + PV = A + PV = \mu(P,T)N \tag{1}$$

determines the locations of the phase boundaries. Note that the chemical potential is the Gibbs free energy per molecule; see Problem 4.6 and Appendix H. Consider a cylinder containing N molecules held at constant pressure P and constant temperature T, that is, in an isothermal, isobaric assembly. Suppose the cylinder initially contains two phases: vapor (A) and liquid (B) so that the total number of molecules is $N = N_A + N_B$ and the Gibbs free energy is $G = G_{\rm A}(N_{\rm A}, P, T) + G_{\rm B}(N_{\rm B}, P, T)$. If the two phases do not coexist at this pressure and temperature, the numbers of molecules in each phase will change as the system approaches equilibrium. As the number of molecules in each phase changes, the Gibbs free energy changes by an amount

$$dG = \left(\frac{\partial G_{\rm A}}{\partial N_{\rm A}}\right)_{T,P} dN_{\rm A} + \left(\frac{\partial G_{\rm B}}{\partial N_{\rm B}}\right)_{T,P} dN_{\rm B} = (\mu_{\rm A} - \mu_{\rm B})dN_{\rm A}, \tag{2}$$

where $dN_{\rm A}$ is the change in the number of molecules in phase A.

The Gibbs free energy is minimized at equilibrium, so $dG \leq 0$. If $\mu_{\rm A} > \mu_{\rm B}$, the number of molecules in phase B will increase and the number in phase A will decrease as the system approaches equilibrium. If $\mu_{\rm A} < \mu_{\rm B}$, the number of molecules in phase A will increase and the number in phase B will decrease. If the chemical potentials are equal, the Gibbs free energy is independent of the number of molecules in the two phases. Therefore, the chemical potentials are equal at coexistence:

$$\mu_{\rm A} = \mu_{\rm B}. \tag{3}$$

Let's consider the familiar example of water. At normal pressures and temperatures, water has three phases: liquid water, solid ice, and water vapor, and its P–T phase diagram is similar to that shown for argon in Figure 4.2(a) — the P–V phase diagram for water is somewhat different because the density of the liquid phase is larger than the density of the solid ice phase; see Problems 4.15 and 4.20. At $P = 1$ atm, water and water vapor coexist at $T = 100\,^\circ$C, the "boiling point" — while boiling is a nonequilibrium process, boiling begins at the temperature at which the equilibrium vapor pressure is equal to the local atmospheric pressure. Consider a two-phase sample of water and water vapor at $T = 99\,^\circ$C. A two-phase sample containing both liquid water and water vapor is easy to create in a constant volume assembly. If there is sufficient volume available, liquid water will evaporate until the water vapor pressure reaches the coexistence pressure at that temperature $P_\sigma(99\,^\circ\text{C}) = 0.965$ atm. If the applied pressure is then increased to, and held constant at, $P = 1$ atm while maintaining a constant temperature of $T = 99\,^\circ$C, the system will be out of equilibrium. At constant pressure, the system will return to equilibrium by decreasing its volume as water vapor condenses into the liquid phase until the the system is completely liquid water. This lowers the Gibbs free energy until it has the equilibrium value determined by the chemical potential of liquid water at this pressure and temperature.

On the other hand, if $T = 100\,^\circ\mathrm{C}$ and $P = 1\,\mathrm{atm}$, the chemical potentials of the liquid and vapor phases are equal, so any combination of water vapor and liquid water has the same Gibbs free energy. The proportion of water and vapor will change as heat is added or removed. The latent heat of vaporization of water $L_v = 540\,\mathrm{cal/g} = 2260\,\mathrm{kJ/kg}$ is the heat needed to convert liquid into vapor.

The coexistence pressure $P_\sigma(T)$ defines the phase boundary between any two phases in the P–T plane, as shown in Figure 4.2(a). From equation (3), the coexistence pressure obeys

$$\mu_{\mathrm{A}}(P_\sigma(T), T) = \mu_{\mathrm{B}}(P_\sigma(T), T). \tag{4}$$

The derivatives of the chemical potentials are related by

$$\left(\frac{\partial \mu_{\mathrm{A}}}{\partial T}\right)_P + \left(\frac{\partial \mu_{\mathrm{A}}}{\partial P}\right)_T \frac{dP_\sigma}{dT} = \left(\frac{\partial \mu_{\mathrm{B}}}{\partial T}\right)_P + \left(\frac{\partial \mu_{\mathrm{B}}}{\partial P}\right)_T \frac{dP_\sigma}{dT}, \tag{5}$$

while the entropy per particle $s = S/N$ and specific volume $v = V/N$ are given by

$$s = -\left(\frac{\partial \mu}{\partial T}\right)_P, \tag{6a}$$

$$v = \left(\frac{\partial \mu}{\partial P}\right)_T; \tag{6b}$$

see equation (4.5.6). Equations (5) and (6) give the *Clausius–Clapeyron equation*

$$\frac{dP_\sigma}{dT} = \frac{s_B - s_A}{v_B - v_A} = \frac{\Delta s}{\Delta v} = \frac{L}{T\Delta v}, \tag{7}$$

where $L = T\Delta s$ is the latent heat per particle. The slope of the coexistence curve depends on the discontinuities of the entropy per particle and the volume per particle. Equation (7) applies very generally to all first-order phase transitions and can be used to determine the coexistence curve as a function of temperature; see Section 4.4, Problems 4.11, and 4.14 through 4.16.

At a triple point, the chemical potentials of three phases are equal:

$$\mu_{\mathrm{A}} = \mu_{\mathrm{B}} = \mu_{\mathrm{C}}. \tag{8}$$

The slopes of the three coexistence lines that define the triple point are related since $\Delta s_{AB} + \Delta s_{BC} + \Delta s_{CA} = 0$ and $\Delta v_{AB} + \Delta v_{BC} + \Delta v_{CA} = 0$. This guarantees that each coexistence line between two phases at the triple point "points into" the third phase; see Problem 4.17.

Problems

4.1. Show that the entropy of a system in the grand canonical ensemble can be written as

$$S = -k \sum_{r,s} P_{r,s} \ln P_{r,s},$$

where $P_{r,s}$ is given by equation (4.1.9).

4.2. In the thermodynamic limit (when the extensive properties of the system become infinitely large, while the intensive ones remain constant), the q-potential of the system may be calculated by taking only the largest term in the sum

$$\sum_{N_r=0}^{\infty} z^{N_r} Q_{N_r}(V,T).$$

Verify this statement and interpret the result physically.

4.3. A vessel of volume $V^{(0)}$ contains $N^{(0)}$ molecules. Assuming that there is no correlation whatsoever between the locations of the various molecules, calculate the probability, $P(N,V)$, that a region of volume V (located anywhere in the vessel) contains exactly N molecules.

(a) Show that $\overline{N} = N^{(0)}p$ and $(\Delta N)_{\text{r.m.s.}} = \{N^{(0)}p(1-p)\}^{1/2}$, where $p = V/V^{(0)}$.

(b) Show that if both $N^{(0)}p$ and $N^{(0)}(1-p)$ are large numbers, the function $P(N,V)$ assumes a Gaussian form.

(c) Further, if $p \ll 1$ and $N \ll N^{(0)}$, show that the function $P(N,V)$ assumes the form of a Poisson distribution:

$$P(N) = e^{-\overline{N}} \frac{(\overline{N})^N}{N!}.$$

4.4. The probability that a system in the grand canonical ensemble has *exactly* N particles is given by

$$p(N) = \frac{z^N Q_N(V,T)}{\mathcal{Q}(z,V,T)}.$$

Verify this statement and show that in the case of a classical, ideal gas the distribution of particles among the members of a grand canonical ensemble is identically a Poisson distribution. Calculate the root-mean-square value of (ΔN) for this system both from the general formula (4.5.3) and from the Poisson distribution, and show that the two results are the same.

4.5. Show that expression (4.3.20) for the entropy of a system in the grand canonical ensemble can also be written as

$$S = k\left[\frac{\partial}{\partial T}(Tq)\right]_{\mu,V}.$$

4.6. Define the isobaric partition function

$$Y_N(P,T) = \frac{1}{\lambda^3}\int_0^{\infty} Q_N(V,T)e^{-\beta PV}\,dV.$$

Show that in the thermodynamic limit the Gibbs free energy (4.7.1) is proportional to $\ln Y_N(P,T)$. Evaluate the isobaric partition function for a classical ideal gas and show that $PV = NkT$. [The factor of the cube of the thermal deBroglie wavelength, λ^3, serves to make the partition function dimensionless and does not contribute to the Gibbs free energy in the thermodynamic limit.]

4.7. Consider a classical system of noninteracting, diatomic molecules enclosed in a box of volume V at temperature T. The Hamiltonian of a single molecule is given by

$$H(\boldsymbol{r}_1,\boldsymbol{r}_2,\boldsymbol{p}_1,\boldsymbol{p}_2) = \frac{1}{2m}(p_1^2+p_2^2) + \frac{1}{2}K|\boldsymbol{r}_1-\boldsymbol{r}_2|^2.$$

Study the thermodynamics of this system, including the dependence of the quantity $\langle r_{12}^2\rangle$ on T.

4.8. Determine the grand partition function of a gaseous system of "magnetic" atoms (with $J = \frac{1}{2}$ and $g = 2$) that can have, in addition to the kinetic energy, a magnetic potential energy equal to $\mu_B H$ or $-\mu_B H$, depending on their orientation with respect to an applied magnetic field H. Derive an expression for the magnetization of the system, and calculate how much heat will be given off by the system when the magnetic field is reduced from H to zero at constant volume and constant temperature.

4.9. Study the problem of solid–vapor equilibrium (Section 4.4) by setting up the grand partition function of the system.

4.10. A surface with N_0 adsorption centers has $N(\le N_0)$ gas molecules adsorbed on it. Show that the chemical potential of the adsorbed molecules is given by

$$\mu = kT \ln \frac{N}{(N_0 - N)a(T)},$$

where $a(T)$ is the partition function of a single adsorbed molecule. Solve the problem by constructing the grand partition function as well as the partition function of the system. [Neglect the intermolecular interaction among the adsorbed molecules.]

4.11. Study the state of equilibrium between a gaseous phase and an adsorbed phase in a single-component system. Show that the pressure in the gaseous phase is given by the Langmuir equation

$$P_g = \frac{\theta}{1-\theta} \times \text{(a certain function of temperature)},$$

where θ is the equilibrium fraction of the adsorption sites that *are* occupied by the adsorbed molecules.

4.12. Show that for a system in the grand canonical ensemble

$$\{\overline{(NE)} - \overline{N}\,\overline{E}\} = \left(\frac{\partial U}{\partial N}\right)_{T,V} \overline{(\Delta N)^2}.$$

4.13. Define a quantity J as

$$J = E - N\mu = TS - PV.$$

Show that for a system in the grand canonical ensemble

$$\overline{(\Delta J)^2} = kT^2 C_V + \left\{\left(\frac{\partial U}{\partial N}\right)_{T,V} - \mu\right\}^2 \overline{(\Delta N)^2}.$$

4.14. Assuming that the latent heat of vaporization of water $L_v = 2260\,\text{kJ/kg}$ is independent of temperature and the specific volume of the liquid phase is negligible compared to the specific volume of the vapor phase, $v_{\text{vapor}} = kT/P_\sigma(T)$, integrate the Clausius–Clapeyron equation (4.7.7) to obtain the coexistence pressure as a function of temperature. Compare your result to the experimental vapor pressure of water from the triple point to 200 °C. The equilibrium vapor pressure at 373 K is 101 kPa = 1 atm.

4.15. Assuming that the latent heat of sublimation of ice $L_s = 2500\,\text{kJ/kg}$ is independent of temperature and the specific volume of the solid phase is negligible compared to the specific volume of the vapor phase, $v_{\text{vapor}} = kT/P_\sigma(T)$, integrate the Clausius–Clapeyron equation (4.7.7) to obtain the coexistence pressure as a function of temperature. Compare your result to the experimental vapor pressure of ice from $T = 0$ to the triple point. The equilibrium vapor pressure at the triple point is 612 Pa.

4.16. Calculate the slope of the solid-liquid transition line for water near the triple point $T = 273.16\,\text{K}$, given that the latent heat of melting is 80 cal/g, the density of the liquid phase is $1.00\,\text{g/cm}^3$, and the density of the ice phase is $0.92\,\text{g/cm}^3$. Estimate the melting temperature at $P = 100\,\text{atm}$.

4.17. Show that the Clausius–Clapeyron equation (4.7.7) guarantees that each of the coexistence curves at the triple point of a material "points into" the third phase; for example, the slope of the solid–vapor coexistence line has a value in-between the slopes of the the the solid–liquid and liquid–vapor coexistence lines.

4.18. Sketch the *P–V* phase diagram for helium-4 using the sketch of the *P–T* phase diagram in Figure 4.3.

4.19. Derive the equivalent of the Clausius–Clapeyron equation (4.7.7) for the slope of the coexistence chemical potential as a function of temperature. Use the fact that the pressures $P(\mu, T)$ in two different phases are equal on the coexistence curve.

4.20. Sketch the *P–T* and *P–V* phase diagrams of water, taking into account the fact that the mass density of the liquid phase is larger than the mass density of the solid phase.

5 Formulation of Quantum Statistics

The scope of the ensemble theory developed in Chapters 2 through 4 is extremely general, though the applications considered so far were confined either to classical systems or to quantum-mechanical systems composed of *distinguishable* entities. When it comes to quantum-mechanical systems composed of *indistinguishable* entities, as most physical systems are, considerations of the preceding chapters have to be applied with care. One finds that in this case it is advisable to rewrite ensemble theory in a language that is more natural to a quantum-mechanical treatment, namely the language of the operators and the wavefunctions. Insofar as statistics are concerned, this rewriting of the theory may not seem to introduce any new physical ideas as such; nonetheless, it provides us with a tool that is highly suited for studying typical quantum systems. And once we set out to study these systems in detail, we encounter a stream of new, and altogether different, physical concepts. In particular, we find that the behavior of even a noninteracting system, such as the ideal gas, departs considerably from the pattern set by the classical treatment. In the presence of interactions, the pattern becomes even more complicated. Of course, in the limit of high temperatures and low densities, the behavior of all physical systems tends *asymptotically* to what we expect on classical grounds. In the process of demonstrating this point, we automatically obtain a criterion that tells us whether a given physical system may or may not be treated classically. At the same time, we obtain rigorous evidence in support of the procedure, employed in the previous chapters, for computing the number, Γ, of microstates (corresponding to a given macrostate) of a given system from the volume, ω, of the relevant region of its phase space, namely $\Gamma \approx \omega/h^f$, where f is the number of "degrees of freedom" in the problem.

5.1 Quantum-mechanical ensemble theory: the density matrix

We consider an ensemble of $\mathcal{N}$ identical systems, where $\mathcal{N} \gg 1$. These systems are characterized by a (common) Hamiltonian, which may be denoted by the operator $\hat{H}$. At time t, the physical states of the various systems in the ensemble will be characterized by the wavefunctions $\psi(\boldsymbol{r}_i, t)$, where $\boldsymbol{r}_i$ denote the position coordinates relevant to the system under study. Let $\psi^k(\boldsymbol{r}_i, t)$ denote the (normalized) wavefunction characterizing the physical state in which the kth system of the ensemble happens to be at time t; naturally, $k = 1, 2, \ldots, \mathcal{N}$. The time variation of the function $\psi^k(t)$ will be determined by the

© 2011 Elsevier Ltd. All rights reserved.

Schrödinger equation[1]

$$\hat{H}\psi^k(t) = i\hbar\dot{\psi}^k(t). \tag{1}$$

Introducing a complete set of orthonormal functions ϕ_n, the wavefunctions $\psi^k(t)$ may be written as

$$\psi^k(t) = \sum_n a_n^k(t)\phi_n, \tag{2}$$

where

$$a_n^k(t) = \int \phi_n^* \psi^k(t) d\tau; \tag{3}$$

here, ϕ_n^* denotes the complex conjugate of ϕ_n while $d\tau$ denotes the volume element of the coordinate space of the given system. Clearly, the physical state of the kth system can be described equally well in terms of the coefficients $a_n^k(t)$. The time variation of these coefficients will be given by

$$\begin{aligned} i\hbar\dot{a}_n^k(t) = i\hbar\int \phi_n^*\dot{\psi}^k(t)d\tau &= \int \phi_n^*\hat{H}\psi^k(t)d\tau \\ &= \int \phi_n^*\hat{H}\left\{\sum_m a_m^k(t)\phi_m\right\}d\tau \\ &= \sum_m H_{nm}a_m^k(t), \end{aligned} \tag{4}$$

where

$$H_{nm} = \int \phi_n^*\hat{H}\phi_m d\tau. \tag{5}$$

The physical significance of the coefficients $a_n^k(t)$ is evident from equation (2). They are the *probability amplitudes* for the various systems of the ensemble to be in the various states ϕ_n; to be practical, the number $|a_n^k(t)|^2$ represents the probability that a measurement at time t finds the kth system of the ensemble to be in the particular state ϕ_n. Clearly, we must have

$$\sum_n |a_n^k(t)|^2 = 1 \qquad \text{(for all } k\text{)}. \tag{6}$$

We now introduce the *density operator* $\hat{\rho}(t)$, as defined by the matrix elements

$$\rho_{mn}(t) = \frac{1}{\mathcal{N}}\sum_{k=1}^{\mathcal{N}}\left\{a_m^k(t)a_n^{k*}(t)\right\}; \tag{7}$$

clearly, the matrix element $\rho_{mn}(t)$ is the ensemble average of the quantity $a_m(t)a_n^*(t)$, which, as a rule, varies from member to member in the ensemble. In particular, the diagonal element $\rho_{nn}(t)$ is the ensemble average of the probability $|a_n(t)|^2$, the latter

[1]For simplicity of notation, we suppress the coordinates $\boldsymbol{r}_i$ in the argument of the wavefunction ψ^k.

itself being a (quantum-mechanical) average. Thus, we encounter here a double-averaging process — once due to the probabilistic aspect of the wavefunctions and again due to the statistical aspect of the ensemble. The quantity $\rho_{nn}(t)$ now represents the probability that a system, chosen *at random* from the ensemble, at time t, is found to be in the state ϕ_n. In view of equations (6) and (7),

$$\sum_n \rho_{nn} = 1. \tag{8}$$

We shall now determine the equation of motion for the density matrix $\rho_{mn}(t)$. We obtain, with the help of the foregoing equations,

$$\begin{aligned} i\hbar\dot{\rho}_{mn}(t) &= \frac{1}{\mathcal{N}}\sum_{k=1}^{\mathcal{N}}\left[i\hbar\left\{\dot{a}_m^k(t)a_n^{k*}(t) + a_m^k(t)\dot{a}_n^{k*}(t)\right\}\right] \\ &= \frac{1}{\mathcal{N}}\sum_{k=1}^{\mathcal{N}}\left[\left\{\sum_l H_{ml}a_l^k(t)\right\}a_n^{k*}(t) - a_m^k(t)\left\{\sum_l H_{nl}^* a_l^{k*}(t)\right\}\right] \\ &= \sum_l \{H_{ml}\rho_{ln}(t) - \rho_{ml}(t)H_{ln}\} \\ &= (\hat{H}\hat{\rho} - \hat{\rho}\hat{H})_{mn}; \end{aligned} \tag{9}$$

here, use has been made of the fact that, in view of the Hermitian character of the operator $\hat{H}, H_{nl}^* = H_{ln}$. Using the commutator notation, equation (9) may be written as

$$i\hbar\dot{\hat{\rho}} = [\hat{H}, \hat{\rho}]_-. \tag{10}$$

Equation (10) is the quantum-mechanical analog of the classical equation (2.2.10) of Liouville. As expected in going from a classical equation of motion to its quantum-mechanical counterpart, the Poisson bracket $[\rho, H]$ has given place to the commutator $(\hat{\rho}\hat{H} - \hat{H}\hat{\rho})/i\hbar$.

If the given system is known to be in a state of equilibrium, the corresponding ensemble must be *stationary*, that is, $\dot{\rho}_{mn} = 0$. Equations (9) and (10) then tell us that, for this to be the case, (i) the density operator $\hat{\rho}$ must be an explicit function of the Hamiltonian operator $\hat{H}$ (for then the two operators will necessarily commute) and (ii) the Hamiltonian must not depend explicitly on time, that is, we must have (i) $\hat{\rho} = \hat{\rho}(\hat{H})$ and (ii) $\dot{\hat{H}} = 0$. Now, if the basis functions ϕ_n were the eigenfunctions of the Hamiltonian itself, then the matrices H and ρ would be diagonal:

$$H_{mn} = E_n\delta_{mn}, \quad \rho_{mn} = \rho_n\delta_{mn}. \tag{11}$$ [2]

[2] It may be noted that in this (so-called energy) representation the density operator $\hat{\rho}$ may be written as

$$\hat{\rho} = \sum_n |\phi_n\rangle\rho_n\langle\phi_n|, \tag{12}$$

for then

$$\rho_{kl} = \sum_n \langle\phi_k|\phi_n\rangle\rho_n\langle\phi_n|\phi_l\rangle = \sum_n \delta_{kn}\rho_n\delta_{nl} = \rho_k\delta_{kl}.$$

The diagonal element ρ_n, being a measure of the probability that a system, chosen *at random* (and at *any* time) from the ensemble, is found to be in the eigenstate ϕ_n, will naturally depend on the corresponding eigenvalue E_n of the Hamiltonian; the precise nature of this dependence is, however, determined by the "kind" of ensemble we wish to construct.

In any representation other than the energy representation, the density matrix may or may not be diagonal. However, quite generally, it will be symmetric:

$$\rho_{mn} = \rho_{nm}. \tag{13}$$

The physical reason for this symmetry is that, in statistical equilibrium, the tendency of a physical system to switch from one state (in the new representation) to another must be counterbalanced by an equally strong tendency to switch between the same states in the reverse direction. This condition of *detailed balancing* is essential for the maintenance of an equilibrium distribution within the ensemble.

Finally, we consider the expectation value of a physical quantity G, which is dynamically represented by an operator $\hat{G}$. This will be given by

$$\langle G \rangle = \frac{1}{\mathcal{N}} \sum_{k=1}^{\mathcal{N}} \int \psi^{k*} \hat{G} \psi^k d\tau. \tag{14}$$

In terms of the coefficients a_n^k,

$$\langle G \rangle = \frac{1}{\mathcal{N}} \sum_{k=1}^{\mathcal{N}} \left[\sum_{m,n} a_n^{k*} a_m^k G_{nm} \right], \tag{15}$$

where

$$G_{nm} = \int \phi_n^* \hat{G} \phi_m d\tau. \tag{16}$$

Introducing the density matrix ρ, equation (15) becomes

$$\langle G \rangle = \sum_{m,n} \rho_{mn} G_{nm} = \sum_m (\hat{\rho}\hat{G})_{mm} = \mathrm{Tr}(\hat{\rho}\hat{G}). \tag{17}$$

Taking $\hat{G} = \hat{1}$, where $\hat{1}$ is the unit operator, we have

$$\mathrm{Tr}(\hat{\rho}) = 1, \tag{18}$$

which is identical to (8). It should be noted here that if the original wavefunctions ψ^k were not normalized then the expectation value $\langle G \rangle$ would be given by the formula

$$\langle G \rangle = \frac{\mathrm{Tr}(\hat{\rho}\hat{G})}{\mathrm{Tr}(\hat{\rho})} \tag{19}$$

instead. In view of the mathematical structure of formulae (17) and (19), the expectation value of any physical quantity G is *manifestly* independent of the choice of the basis $\{\phi_n\}$, as it indeed should be.

5.2 Statistics of the various ensembles

5.2.A The microcanonical ensemble

The construction of the microcanonical ensemble is based on the premise that the systems constituting the ensemble are characterized by a fixed number of particles N, a fixed volume V, and an energy lying within the interval $\left(E - \frac{1}{2}\Delta, E + \frac{1}{2}\Delta\right)$, where $\Delta \ll E$. The total number of distinct microstates accessible to a system is then denoted by the symbol $\Gamma(N, V, E; \Delta)$ and, by assumption, any one of these microstates is as likely to occur as any other. This assumption enters into our theory in the nature of a postulate and is often referred to as the postulate of *equal a priori probabilities* for the various accessible states.

Accordingly, the density matrix ρ_{mn} (which, in the energy representation, must be a diagonal matrix) will be of the form

$$\rho_{mn} = \rho_n \delta_{mn}, \tag{1}$$

with

$$\rho_n = \begin{cases} 1/\Gamma & \text{for each of the accessible states,} \\ 0 & \text{for all other states;} \end{cases} \tag{2}$$

the normalization condition (5.1.18) is clearly satisfied. As we already know, the thermodynamics of the system is completely determined from the expression for its entropy which, in turn, is given by

$$S = k \ln \Gamma. \tag{3}$$

Since Γ, the total number of distinct, accessible states, is supposed to be computed quantum-mechanically, taking due account of the indistinguishability of the particles right from the beginning, no paradox, such as Gibbs', is now expected to arise. Moreover, if the quantum state of the system turns out to be unique ($\Gamma = 1$), the entropy of the system will identically vanish. This provides us with a sound theoretical basis for the hitherto empirical theorem of Nernst (also known as the *third law of thermodynamics*).

The situation corresponding to the case $\Gamma = 1$ is usually referred to as a *pure* case. In such a case, the construction of an ensemble is essentially superfluous, because every system in the ensemble has got to be in one and the same state. Accordingly, there is only one diagonal element ρ_{nn} that is nonzero (actually equal to unity), while all others are zero. The density matrix, therefore, satisfies the relation

$$\rho^2 = \rho. \tag{4}$$

In a different representation, the pure case will correspond to

$$\rho_{mn} = \frac{1}{\mathcal{N}} \sum_{k=1}^{\mathcal{N}} a_m^k a_n^{k*} = a_m a_n^* \tag{5}$$

because all values of k are now literally equivalent. We then have

$$\begin{aligned} \rho_{mn}^2 &= \sum_l \rho_{ml} \rho_{ln} = \sum_l a_m a_l^* a_l a_n^* \\ &= a_m a_n^* \quad \left(\text{because} \sum_l a_l^* a_l = 1\right) \\ &= \rho_{mn}. \end{aligned} \tag{6}$$

Relation (4) thus holds in all representations.

A situation in which $\Gamma > 1$ is usually referred to as a *mixed* case. The density matrix, in the energy representation, is then given by equations (1) and (2). If we now change over to any other representation, the general form of the density matrix should remain the same, namely (i) the off-diagonal elements should continue to be zero, while (ii) the diagonal elements (over the allowed range) should continue to be equal to one another. Now, had we constructed our ensemble on a representation other than the energy representation right from the beginning, how could we have possibly anticipated *ab initio* property (i) of the density matrix, though property (ii) could have been easily invoked through a postulate of *equal a priori probabilities*? To ensure that property (i), as well as property (ii), holds in *every* representation, we must invoke yet another postulate, namely the postulate of *random a priori phases* for the probability amplitudes a_n^k, which in turn implies that the wavefunction ψ^k, for all k, is an *incoherent* superposition of the basis $\{\phi_n\}$. As a consequence of this postulate, coupled with the postulate of equal *a priori* probabilities, we would have in any representation

$$\begin{aligned} \rho_{mn} \equiv \frac{1}{\mathcal{N}} \sum_{k=1}^{\mathcal{N}} a_m^k a_n^{k*} &= \frac{1}{\mathcal{N}} \sum_{k=1}^{\mathcal{N}} |a|^2 e^{i\left(\theta_m^k - \theta_n^k\right)} \\ &= c \left\langle e^{i\left(\theta_m^k - \theta_n^k\right)} \right\rangle \\ &= c\delta_{mn}, \end{aligned} \tag{7}$$

as it should be for a microcanonical ensemble.

Thus, contrary to what might have been expected on customary grounds, to secure the physical situation corresponding to a microcanonical ensemble, we require in general two postulates instead of one! The second postulate arises solely from quantum-mechanics and is intended to ensure noninterference (and hence a complete absence of correlations) among the member systems; this, in turn, enables us to form a mental picture of each system of the ensemble, one at a time, completely disentangled from other systems.

5.2.B The canonical ensemble

In this ensemble the macrostate of a member system is defined through the parameters N, V, and T; the energy E is now a variable quantity. The probability that a system, chosen *at random* from the ensemble, possesses an energy E_r is determined by the Boltzmann factor $\exp(-\beta E_r)$, where $\beta = 1/kT$; see Sections 3.1 and 3.2. The density matrix in the energy representation is, therefore, taken as

$$\rho_{mn} = \rho_n \delta_{mn}, \tag{8}$$

with

$$\rho_n = C \exp(-\beta E_n); \qquad n = 0, 1, 2, \ldots \tag{9}$$

The constant C is determined by the normalization condition (5.1.18), whereby

$$C = \frac{1}{\sum_n \exp(-\beta E_n)} = \frac{1}{Q_N(\beta)}, \tag{10}$$

where $Q_N(\beta)$ is the *partition function* of the system. In view of equations (5.1.12), see footnote 2, the density operator in this ensemble may be written as

$$\begin{aligned} \hat{\rho} &= \sum_n |\phi_n\rangle \frac{1}{Q_N(\beta)} e^{-\beta E_n} \langle \phi_n| \\ &= \frac{1}{Q_N(\beta)} e^{-\beta \hat{H}} \sum_n |\phi_n\rangle\langle\phi_n| \\ &= \frac{1}{Q_N(\beta)} e^{-\beta \hat{H}} = \frac{e^{-\beta \hat{H}}}{\mathrm{Tr}\left(e^{-\beta \hat{H}}\right)}, \end{aligned} \tag{11}$$

for the operator $\sum_n |\phi_n\rangle\langle\phi_n|$ is identically the unit operator. It is understood that the operator $\exp(-\beta\hat{H})$ in equation (11) stands for the sum

$$\sum_{j=0}^{\infty} (-1)^j \frac{(\beta \hat{H})^j}{j!}. \tag{12}$$

The expectation value $\langle G\rangle_N$ of a physical quantity G, which is represented by an operator $\hat{G}$, is now given by

$$\begin{aligned}\langle G\rangle_N = \mathrm{Tr}(\hat{\rho}\hat{G}) &= \frac{1}{Q_N(\beta)}\mathrm{Tr}\left(\hat{G}e^{-\beta\hat{H}}\right)\\ &= \frac{\mathrm{Tr}\left(\hat{G}e^{-\beta\hat{H}}\right)}{\mathrm{Tr}\left(e^{-\beta\hat{H}}\right)};\end{aligned} \tag{13}$$

the suffix N here emphasizes the fact that the averaging is being done over an ensemble *with N fixed.*

5.2.C The grand canonical ensemble

In this ensemble the density operator $\hat{\rho}$ operates on a Hilbert space with an indefinite number of particles. The density operator must therefore commute not only with the Hamiltonian operator $\hat{H}$ but also with a number operator $\hat{n}$ whose eigenvalues are $0, 1, 2, \ldots$. The precise form of the density operator can now be obtained by a straightforward generalization of the preceding case, with the result

$$\hat{\rho} = \frac{1}{\mathcal{Q}(\mu, V, T)}e^{-\beta(\hat{H}-\mu\hat{n})}, \tag{14}$$

where

$$\mathcal{Q}(\mu, V, T) = \sum_{r,s} e^{-\beta(E_r - \mu N_s)} = \mathrm{Tr}\{e^{-\beta(\hat{H}-\mu\hat{n})}\}. \tag{15}$$

The ensemble average $\langle G\rangle$ is now given by

$$\begin{aligned}\langle G\rangle &= \frac{1}{\mathcal{Q}(\mu, V, T)}\mathrm{Tr}\left(\hat{G}e^{-\beta\hat{H}}e^{\beta\mu\hat{n}}\right)\\ &= \frac{\sum_{N=0}^{\infty} z^N \langle G\rangle_N Q_N(\beta)}{\sum_{N=0}^{\infty} z^N Q_N(\beta)},\end{aligned} \tag{16}$$

where $z(\equiv e^{\beta\mu})$ is the *fugacity* of the system while $\langle G\rangle_N$ is the canonical-ensemble average, as given by equation (13). The quantity $\mathcal{Q}(\mu, V, T)$ appearing in these formulae is, clearly, the *grand partition function* of the system.

5.3 Examples

5.3.A An electron in a magnetic field

We consider, for illustration, the case of a single electron that possesses an intrinsic spin $\frac{1}{2}\hbar\hat{\boldsymbol{\sigma}}$ and a magnetic moment μ_B, where $\hat{\boldsymbol{\sigma}}$ is the Pauli spin operator and $\mu_B = e\hbar/2mc$.

The spin of the electron can have two possible orientations, ↑ or ↓, with respect to an applied magnetic field $\boldsymbol{B}$. If the applied field is taken to be in the direction of the z-axis, the configurational Hamiltonian of the spin takes the form

$$\hat{H} = -\mu_B(\hat{\boldsymbol{\sigma}} \cdot \boldsymbol{B}) = -\mu_B B\hat{\sigma}_z. \tag{1}$$

In the representation that makes $\hat{\sigma}_z$ diagonal, namely

$$\hat{\sigma}_x = \begin{pmatrix} 0 & 1 \\ 1 & 0 \end{pmatrix}, \quad \hat{\sigma}_y = \begin{pmatrix} 0 & -i \\ i & 0 \end{pmatrix}, \quad \hat{\sigma}_z = \begin{pmatrix} 1 & 0 \\ 0 & -1 \end{pmatrix}, \tag{2}$$

the density matrix in the canonical ensemble would be

$$\begin{aligned} (\hat{\rho}) &= \frac{(e^{-\beta\hat{H}})}{\mathrm{Tr}(e^{-\beta\hat{H}})} \\ &= \frac{1}{e^{\beta\mu_B B} + e^{-\beta\mu_B B}} \begin{pmatrix} e^{\beta\mu_B B} & 0 \\ 0 & e^{-\beta\mu_B B} \end{pmatrix}. \end{aligned} \tag{3}$$

We thus obtain for the expectation value σ_z

$$\langle\sigma_z\rangle = \mathrm{Tr}(\hat{\rho}\hat{\sigma}_z) = \frac{e^{\beta\mu_B B} - e^{-\beta\mu_B B}}{e^{\beta\mu_B B} + e^{-\beta\mu_B B}} = \tanh(\beta\mu_B B), \tag{4}$$

in perfect agreement with the findings of Sections 3.9 and 3.10.

5.3.B A free particle in a box

We now consider the case of a free particle, of mass m, in a cubical box of side L. The Hamiltonian of the particle is given by

$$\hat{H} = -\frac{\hbar^2}{2m}\nabla^2 = -\frac{\hbar^2}{2m}\left(\frac{\partial^2}{\partial x^2} + \frac{\partial^2}{\partial y^2} + \frac{\partial^2}{\partial z^2}\right), \tag{5}$$

while the eigenfunctions of the Hamiltonian that satisfy periodic boundary conditions,

$$\begin{aligned} \phi(x+L, y, z) &= \phi(x, y+L, z) = \phi(x, y, z+L) \\ &= \phi(x, y, z), \end{aligned} \tag{6}$$

are given by

$$\phi_E(r) = \frac{1}{L^{3/2}} \exp(i\boldsymbol{k} \cdot \boldsymbol{r}), \tag{7}$$

the corresponding eigenvalues E being

$$E = \frac{\hbar^2 k^2}{2m}, \tag{8}$$

and the corresponding wave vector $\boldsymbol{k}$ being

$$\boldsymbol{k} \equiv (k_x, k_y, k_z) = \frac{2\pi}{L}(n_x, n_y, n_z); \tag{9}$$

the quantum numbers n_x, n_y, and n_z must be integers (positive, negative, or zero). Symbolically, we may write

$$\boldsymbol{k} = \frac{2\pi}{L}\boldsymbol{n}, \tag{10}$$

where $\boldsymbol{n}$ is a vector with integral components $0, \pm 1, \pm 2, \ldots$.

We now proceed to evaluate the density matrix $(\hat{\rho})$ of this system in the canonical ensemble; we shall do so in the *coordinate* representation. In view of equation (5.2.11), we have

$$\begin{aligned}\langle \boldsymbol{r}|e^{-\beta \hat{H}}|\boldsymbol{r}'\rangle &= \sum_E \langle \boldsymbol{r}|E\rangle e^{-\beta E}\langle E|\boldsymbol{r}'\rangle \\ &= \sum_E e^{-\beta E}\phi_E(\boldsymbol{r})\phi_E^*(\boldsymbol{r}').\end{aligned} \tag{11}$$

Substituting from equation (7) and making use of relations (8) and (10), we obtain

$$\begin{aligned}\langle \boldsymbol{r}|e^{-\beta \hat{H}}|\boldsymbol{r}'\rangle &= \frac{1}{L^3}\sum_{\boldsymbol{k}} \exp\left[-\frac{\beta\hbar^2}{2m}k^2 + i\boldsymbol{k}\cdot(\boldsymbol{r}-\boldsymbol{r}')\right] \\ &\approx \frac{1}{(2\pi)^3}\int \exp\left[-\frac{\beta\hbar^2}{2m}k^2 + i\boldsymbol{k}\cdot(\boldsymbol{r}-\boldsymbol{r}')\right] d^3k \\ &= \left(\frac{m}{2\pi\beta\hbar^2}\right)^{3/2} \exp\left[-\frac{m}{2\beta\hbar^2}|\boldsymbol{r}-\boldsymbol{r}'|^2\right];\end{aligned} \tag{12}$$

see equations (B.41) and (B.42) in Appendix B. It follows that

$$\begin{aligned}\mathrm{Tr}(e^{-\beta \hat{H}}) &= \int \langle \boldsymbol{r}|e^{-\beta \hat{H}}|\boldsymbol{r}\rangle d^3r \\ &= V\left(\frac{m}{2\pi\beta\hbar^2}\right)^{3/2}.\end{aligned} \tag{13}$$

The expression in equation (13) is indeed the partition function, $Q_1(\beta)$, of a single particle confined to a box of volume V; see equation (3.5.19). Dividing (12) by (13), we obtain for the density matrix in the coordinate representation

$$\langle \boldsymbol{r}|\hat{\rho}|\boldsymbol{r}'\rangle = \frac{1}{V}\exp\left[-\frac{m}{2\beta\hbar^2}|\boldsymbol{r}-\boldsymbol{r}'|^2\right]. \tag{14}$$

As expected, the matrix $\rho_{\boldsymbol{r},\boldsymbol{r}'}$ is symmetric between the states $\boldsymbol{r}$ and $\boldsymbol{r}'$. Moreover, the diagonal element $\langle \boldsymbol{r}|\rho|\boldsymbol{r}\rangle$, which represents the *probability density* for the particle to be in the neighborhood of the point $\boldsymbol{r}$, is independent of $\boldsymbol{r}$; this means that, in the case of a single free particle, *all* positions within the box are equally likely to obtain. A nondiagonal element $\langle \boldsymbol{r}|\rho|\boldsymbol{r}'\rangle$, on the other hand, is a measure of the probability of "spontaneous transition" between the position coordinates $\boldsymbol{r}$ and $\boldsymbol{r}'$ and is therefore a measure of the relative "intensity" of the wave packet (associated with the particle) at a distance $|\boldsymbol{r}-\boldsymbol{r}'|$ from the center of the packet. The spatial extent of the wave packet, which is a measure of the uncertainty involved in locating the position of the particle, is clearly of order $\hbar/(mkT)^{1/2}$; the latter is also a measure of the *mean thermal wavelength* of the particle. The spatial spread found here is a purely quantum-mechanical effect; quite expectedly, it tends to vanish at high temperatures. In fact, as $\beta \to 0$, the behavior of the matrix element (14) approaches that of a delta function, which implies a return to the classical picture of a *point* particle.

Finally, we determine the expectation value of the Hamiltonian itself. From equations (5) and (14), we obtain

$$\begin{aligned}\langle H\rangle = \mathrm{Tr}(\hat{H}\hat{\rho}) &= -\frac{\hbar^2}{2mV}\int\left\{\nabla^2 \exp\left[-\frac{m}{2\beta\hbar^2}|\boldsymbol{r}-\boldsymbol{r}'|^2\right]\right\}_{\boldsymbol{r}=\boldsymbol{r}'} d^3r \\ &= \frac{1}{2\beta V}\int\left\{\left[3-\frac{m}{\beta\hbar^2}|\boldsymbol{r}-\boldsymbol{r}'|^2\right]\exp\left[-\frac{m}{2\beta\hbar^2}|\boldsymbol{r}-\boldsymbol{r}'|^2\right]\right\}_{\boldsymbol{r}=\boldsymbol{r}'} d^3r \\ &= \frac{3}{2\beta} = \frac{3}{2}kT, \end{aligned} \tag{15}$$

which was indeed expected. Otherwise, too,

$$\langle H\rangle = \frac{\mathrm{Tr}\left(\hat{H}e^{-\beta\hat{H}}\right)}{\mathrm{Tr}\left(e^{-\beta\hat{H}}\right)} = -\frac{\partial}{\partial\beta}\ln \mathrm{Tr}\left(e^{-\beta\hat{H}}\right) \tag{16}$$

which, on combination with (13), leads to the same result.

5.3.C A linear harmonic oscillator

Next, we consider the case of a linear harmonic oscillator whose Hamiltonian is given by

$$\hat{H} = -\frac{\hbar^2}{2m}\frac{\partial^2}{\partial q^2} + \frac{1}{2}m\omega^2 q^2, \tag{17}$$

with eigenvalues

$$E_n = \left(n+\frac{1}{2}\right)\hbar\omega; \qquad n = 0, 1, 2, \ldots \tag{18}$$

and eigenfunctions

$$\phi_n(q) = \left(\frac{m\omega}{\pi\hbar}\right)^{1/4}\frac{H_n(\xi)}{(2^n n!)^{1/2}}e^{-(1/2)\xi^2}, \tag{19}$$

where

$$\xi = \left(\frac{m\omega}{\hbar}\right)^{1/2} q \tag{20}$$

and

$$H_n(\xi) = (-1)^n e^{\xi^2} \left(\frac{d}{d\xi}\right)^n e^{-\xi^2}. \tag{21}$$

The matrix elements of the operator $\exp(-\beta\hat{H})$ in the q-representation are given by

$$\begin{aligned}\langle q|e^{-\beta\hat{H}}|q'\rangle &= \sum_{n=0}^{\infty} e^{-\beta E_n}\phi_n(q)\phi_n(q') \\ &= \left(\frac{m\omega}{\pi\hbar}\right)^{1/2} e^{-(1/2)(\xi^2+\xi'^2)} \sum_{n=0}^{\infty} \left\{ e^{-(n+1/2)\beta\hbar\omega} \frac{H_n(\xi)H_n(\xi')}{2^n n!} \right\}.\end{aligned} \tag{22}$$

The summation over n is somewhat difficult to evaluate; nevertheless, the final result is[3]

$$\begin{aligned}\langle q|e^{-\beta\hat{H}}|q'\rangle &= \left[\frac{m\omega}{2\pi\hbar\sinh(\beta\hbar\omega)}\right]^{1/2} \\ &\times \exp\left[-\frac{m\omega}{4\hbar}\left\{(q+q')^2\tanh\left(\frac{\beta\hbar\omega}{2}\right) + (q-q')^2\coth\left(\frac{\beta\hbar\omega}{2}\right)\right\}\right],\end{aligned} \tag{23}$$

which gives

$$\begin{aligned}\mathrm{Tr}\left(e^{-\beta\hat{H}}\right) &= \int_{-\infty}^{\infty} \langle q|e^{-\beta\hat{H}}|q\rangle dq \\ &= \left[\frac{m\omega}{2\pi\hbar\sinh(\beta\hbar\omega)}\right]^{1/2} \int_{-\infty}^{\infty} \exp\left[-\frac{m\omega q^2}{\hbar}\tanh\left(\frac{\beta\hbar\omega}{2}\right)\right] dq \\ &= \frac{1}{2\sinh\left(\frac{1}{2}\beta\hbar\omega\right)} = \frac{e^{-(1/2)\beta\hbar\omega}}{1-e^{-\beta\hbar\omega}}.\end{aligned} \tag{24}$$

Expression (24) is indeed the *partition function* of a linear harmonic oscillator; see equation (3.8.14). At the same time, we find that the *probability density* for the oscillator coordinate to be in the vicinity of the value q is given by

$$\langle q|\hat{\rho}|q\rangle = \left[\frac{m\omega\tanh\left(\frac{1}{2}\beta\hbar\omega\right)}{\pi\hbar}\right]^{1/2} \exp\left[-\frac{m\omega q^2}{\hbar}\tanh\left(\frac{\beta\hbar\omega}{2}\right)\right]; \tag{25}$$

[3]The mathematical details of this derivation can be found in Kubo (1965, pp. 175–177).

we note that this is a Gaussian distribution in q, with mean value zero and root-mean-square deviation

$$q_{\text{r.m.s.}} = \left[\frac{\hbar}{2m\omega\tanh\left(\frac{1}{2}\beta\hbar\omega\right)}\right]^{1/2}. \tag{26}$$

The probability distribution (25) was first derived by Bloch in 1932. In the classical limit ($\beta\hbar\omega \ll 1$), the distribution becomes *purely thermal* — free from quantum effects:

$$\langle q|\hat{\rho}|q\rangle \approx \left(\frac{m\omega^2}{2\pi kT}\right)^{1/2} \exp\left[-\frac{m\omega^2 q^2}{2kT}\right], \tag{27}$$

with dispersion $(kT/m\omega^2)^{1/2}$. At the other extreme ($\beta\hbar\omega \gg 1$), the distribution becomes *purely quantum-mechanical* — free from thermal effects:

$$\langle q|\hat{\rho}|q\rangle \approx \left(\frac{m\omega}{\pi\hbar}\right)^{1/2} \exp\left[-\frac{m\omega q^2}{\hbar}\right], \tag{28}$$

with dispersion $(\hbar/2m\omega)^{1/2}$. Note that the limiting distribution (28) is precisely the one expected for an oscillator in its ground state ($n = 0$), that is one with probability density $\phi_0^2(q)$; see equations (19) through (21).

In view of the fact that the mean energy of the oscillator is given by

$$\langle H\rangle = -\frac{\partial}{\partial\beta}\ln \operatorname{Tr}\left(e^{-\beta\hat{H}}\right) = \frac{1}{2}\hbar\omega\coth\left(\frac{1}{2}\beta\hbar\omega\right), \tag{29}$$

we observe that the temperature dependence of the distribution (25) is *solely* determined by the expectation value $\langle H\rangle$. Actually, we can write

$$\langle q|\hat{\rho}|q\rangle = \left(\frac{m\omega^2}{2\pi\langle H\rangle}\right)^{1/2} \exp\left[-\frac{m\omega^2 q^2}{2\langle H\rangle}\right], \tag{30}$$

with

$$q_{\text{r.m.s.}} = \left(\frac{\langle H\rangle}{m\omega^2}\right)^{1/2}. \tag{31}$$

It is now straightforward to see that the mean value of the potential energy ($\frac{1}{2}m\omega^2 q^2$) of the oscillator is $\frac{1}{2}\langle H\rangle$; accordingly, the mean value of the kinetic energy ($p^2/2m$) will also be the same.

5.4 Systems composed of indistinguishable particles

We shall now formulate the quantum-mechanical description of a system of N identical particles. To fix ideas, we consider a gas of *noninteracting* particles; the findings of this study will be of considerable relevance to other systems as well.

Now, the Hamiltonian of a system of N noninteracting particles is simply a sum of the individual single-particle Hamiltonians:

$$\hat{H}(\boldsymbol{q},\boldsymbol{p}) = \sum_{i=1}^{N} \hat{H}_i(q_i,p_i); \tag{1}$$

here, (q_i,p_i) are the coordinates and momenta of the ith particle while $\hat{H}_i$ is its Hamiltonian.[4] Since the particles are identical, the Hamiltonians $\hat{H}_i (i = 1,2,\ldots,N)$ are *formally* the same; they only differ in the values of their arguments. The time-independent Schrödinger equation for the system is

$$\hat{H}\psi_E(\boldsymbol{q}) = E\psi_E(\boldsymbol{q}), \tag{2}$$

where E is an eigenvalue of the Hamiltonian and $\psi_E(\boldsymbol{q})$ the corresponding eigenfunction. In view of (1), we can write a straightforward solution of the Schrödinger equation, namely

$$\psi_E(\boldsymbol{q}) = \prod_{i=1}^{N} u_{\varepsilon_i}(q_i), \tag{3}$$

with

$$E = \sum_{i=1}^{N} \varepsilon_i; \tag{4}$$

the factor $u_{\varepsilon_i}(q_i)$ in (3) is an eigenfunction of the single-particle Hamiltonian $\hat{H}_i(q_i,p_i)$, with eigenvalue ε_i:

$$\hat{H}_i u_{\varepsilon_i}(q_i) = \varepsilon_i u_{\varepsilon_i}(q_i). \tag{5}$$

Thus, a stationary state of the given system may be described in terms of the single-particle states of the constituent particles. In general, we may do so by specifying the set of numbers $\{n_i\}$ to represent a particular state of the system; this would imply that there are n_i particles in the eigenstate characterized by the energy value ε_i. Clearly, the distribution set

[4] We are studying here a single-component system composed of "spinless" particles. Generalization to a system composed of particles with spin and to a system composed of two or more components is quite straightforward.

$\{n_i\}$ must conform to the conditions

$$\sum_i n_i = N \tag{6}$$

and

$$\sum_i n_i \varepsilon_i = E. \tag{7}$$

Accordingly, the wavefunction of this state may be written as

$$\psi_E(\boldsymbol{q}) = \prod_{m=1}^{n_1} u_1(m) \prod_{m=n_1+1}^{n_1+n_2} u_2(m) \ldots, \tag{8}$$

where the symbol $u_i(m)$ stands for the single-particle wavefunction $u_{\varepsilon_i}(q_m)$.

Now, suppose we effect a permutation among the coordinates appearing on the right side of (8); as a result, the coordinates $(1, 2, \ldots, N)$ get replaced by $(P1, P2, \ldots, PN)$, say. The resulting wavefunction, which we may call $P\psi_E(\boldsymbol{q})$, will be

$$P\psi_E(\boldsymbol{q}) = \prod_{m=1}^{n_1} u_1(Pm) \prod_{m=n_1+1}^{n_1+n_2} u_2(Pm) \ldots. \tag{9}$$

In classical physics, where the particles of a given system, even though identical, are regarded as mutually *distinguishable*, any permutation that brings about an interchange of particles in two *different* single-particle states is recognized to have led to a *new, physically distinct*, microstate of the system. For example, classical physics regards a microstate in which the so-called 5th particle is in the state u_i and the so-called 7th particle in the state $u_j (j \neq i)$ as *distinct* from a microstate in which the 7th particle is in the state u_i and the 5th particle in the state u_j. This leads to

$$\frac{N!}{n_1! n_2! \ldots} \tag{10}$$

(supposedly distinct) microstates of the system, corresponding to a given mode of distribution $\{n_i\}$. The number (10) would then be ascribed as a "statistical weight factor" to the distribution set $\{n_i\}$. Of course, the "correction" applied by Gibbs, which has been discussed in Sections 1.5 and 1.6, reduces this weight factor to

$$W_c\{n_i\} = \frac{1}{n_1! n_2! \ldots}. \tag{11}$$

And the only way one could understand the physical basis of that "correction" was in terms of the inherent *indistinguishability* of the particles.

According to quantum physics, however, the situation remains unsatisfactory even after the Gibbs correction has been incorporated, for, strictly speaking, an interchange among identical particles, *even* if they are in different single-particle states, should not lead to a new microstate of the system! Thus, if we want to take into account the indistinguishability of the particles properly, we must not regard a microstate in which the "5th" particle is in the state u_i and the "7th" in the state u_j as *distinct* from a microstate in which the "7th" particle is in the state u_i and the "5th" in the state u_j (even if $i \neq j$), for the labeling of the particles as No. 1, No. 2, and so on (which one often resorts to) is at most a matter of convenience — it is *not* a matter of reality. In other words, all that matters in the description of a particular state of the given system is the set of numbers n_i that tell us *how many* particles there are in the various single-particle states u_i; the question, "*which* particle is in *which* single-particle state?" has no relevance at all.

Accordingly, the microstates resulting from any permutation P among the N particles (so long as the numbers n_i remain the same) must be regarded as *one and the same* microstate. For the same reason, the weight factor associated with a distribution set $\{n_i\}$, provided that the set is not disallowed on some other physical grounds, should be identically equal to unity, whatever the values of the numbers n_i may be:

$$W_q\{n_i\} \equiv 1. \tag{12}$$[5]

Indeed, if for some physical reason the set $\{n_i\}$ is disallowed, the weight factor W_q for that set should be identically equal to zero; see, for instance, equation (19).

At the same time, a wavefunction of the type (8), which we may call *Boltzmannian* and denote by the symbol $\psi_{\text{Boltz}}(\boldsymbol{q})$, is inappropriate for describing the state of a system composed of indistinguishable particles because an interchange of arguments among the factors u_i and u_j, where $i \neq j$, would lead to a wavefunction that is both mathematically and physically different from the one we started with. Now, since a mere interchange of the particle coordinates must not lead to a new microstate of the system, the wavefunction $\psi_E(\boldsymbol{q})$ must be constructed in such a way that, for all practical purposes, it is insensitive to any interchange among its arguments. The simplest way to do this is to set up a linear combination of all the $N!$ functions of the type (9) that obtain from (8) by all possible permutations among its arguments; of course, the combination must be such that if a permutation of coordinates is carried out in it, then the wavefunctions ψ and $P\psi$ must satisfy the property

$$|P\psi|^2 = |\psi|^2. \tag{13}$$

This leads to the following possibilities:

$$P\psi = \psi \quad \text{for all } P, \tag{14}$$

[5]It may be mentioned here that as early as in 1905 Ehrenfest pointed out that to obtain Planck's formula for the black-body radiation one must assign equal *a priori* probabilities to the various distribution sets $\{n_i\}$.

which means that the wavefunction is *symmetric* in all its arguments, or

$$P\psi = \begin{cases} +\psi & \text{if } P \text{ is an } \textit{even} \text{ permutation,} \\ -\psi & \text{if } P \text{ is an } \textit{odd} \text{ permutation,} \end{cases} \tag{15}$$ [6]

which means that the wavefunction is *antisymmetric* in its arguments. We call these wavefunctions ψ_S and ψ_A, respectively; their mathematical structure is given by

$$\psi_S(\boldsymbol{q}) = \text{const.} \sum_P P\psi_{\text{Boltz}}(\boldsymbol{q}) \tag{16}$$

and

$$\psi_A(\boldsymbol{q}) = \text{const.} \sum_P \delta_P P\psi_{\text{Boltz}}(\boldsymbol{q}), \tag{17}$$

where δ_P in the expression for ψ_A is $+1$ or -1 according to whether the permutation P is even or odd.

We note that the function $\psi_A(\boldsymbol{q})$ can be written in the form of a *Slater determinant*:

$$\psi_A(\boldsymbol{q}) = \text{const.} \begin{vmatrix} u_i(1) & u_i(2) & \cdots & u_i(N) \\ u_j(1) & u_j(2) & \cdots & u_j(N) \\ \cdot & \cdot & \cdots & \cdot \\ \cdot & \cdot & \cdots & \cdot \\ \cdot & \cdot & \cdots & \cdot \\ u_l(1) & u_l(2) & \cdots & u_l(N) \end{vmatrix}, \tag{18}$$

where the leading diagonal is precisely the Boltzmannian wavefunction while the other terms of the expansion are the various permutations thereof; positive and negative signs in the combination (17) appear automatically as we expand the determinant. On interchanging a pair of arguments (which amounts to interchanging the corresponding columns of the determinant), the wavefunction ψ_A merely changes its sign, as it indeed should. However, if two or more particles happen to be in the same single-particle state, then the corresponding rows of the determinant become identical and the wavefunction vanishes.[7] Such a state is physically impossible to realize. We therefore conclude that if a system composed of indistinguishable particles is characterized by an antisymmetric wavefunction,

[6] An even (odd) permutation is one that can be arrived at from the original order by an even (odd) number of "pair interchanges" among the arguments. For example, of the six permutations

$$(1,2,3), \quad (2,3,1), \quad (3,1,2), \quad (1,3,2), \quad (3,2,1), \quad \text{and} \quad (2,1,3),$$

of the arguments 1, 2, and 3, the first three are *even* permutations while the last three are *odd*. A single interchange, among any two arguments, is clearly an *odd* permutation.

[7] This is directly related to the fact that if we effect an interchange among two particles in the *same* single-particle state, then $P\psi_A$ will obviously be identical to ψ_A. At the same time, if we also have $P\psi_A = -\psi_A$, then ψ_A must be identically zero.

then the particles of the system must all be in different single-particle states — a result equivalent to *Pauli's exclusion principle* for electrons.

Conversely, a statistical system composed of particles obeying an exclusion principle must be described by a wavefunction that is antisymmetric in its arguments. The statistics governing the behavior of such particles is called *Fermi–Dirac*, or simply *Fermi*; statistics and the constituent particles themselves are referred to as *fermions*. The statistical weight factor $W_{\text{F.D.}}\{n_i\}$ for such a system is unity so long as the n_i in the distribution set are either 0 or 1; otherwise, it is zero:

$$W_{\text{F.D.}}\{n_i\} = \begin{cases} 1 & \text{if} \quad \sum_i n_i^2 = N, \\ 0 & \text{if} \quad \sum_i n_i^2 > N. \end{cases} \tag{19}$$[8]

No such problems arise for systems characterized by symmetric wavefunctions: in particular, we have no restriction whatsoever on the values of the numbers n_i. The statistics governing the behavior of such systems is called *Bose–Einstein*, or simply *Bose*, statistics and the constituent particles themselves are referred to as *bosons*.[9] The weight factor $W_{\text{B.E.}}\{n_i\}$ for such a system is identically equal to 1, whatever the values of the numbers n_i:

$$W_{\text{B.E.}}\{n_i\} = 1; \qquad n_i = 0, 1, 2, \ldots. \tag{20}$$

It should be pointed out here that there exists an intimate connection between the statistics governing a particular species of particles and the intrinsic spin of the particles. For instance, particles with an integral spin (in units of $\hbar$, of course) obey Bose–Einstein statistics, while particles with a half-odd integral spin obey Fermi–Dirac statistics. Examples in the first category are photons, phonons, π-mesons, gravitons, He^4-atoms, and so on, while those in the second category are electrons, nucleons (protons and neutrons), μ-mesons, neutrinos, He^3-atoms, and so on.

Finally, it must be emphasized that, although we have derived our conclusions here on the basis of a study of noninteracting systems, the basic results hold for interacting systems as well. In general, the desired wavefunction $\psi(\boldsymbol{q})$ will not be expressible in terms of the single-particle wavefunctions $u_i(q_m)$; nonetheless, it will have to be either of the kind $\psi_S(\boldsymbol{q})$, satisfying equation (14), or of the kind $\psi_A(\boldsymbol{q})$, satisfying equation (15).

[8]Note that the condition $\sum_i n_i^2 = N$ would be implies that all n_i are either 0 or 1. On the other hand, if any of the n_i are greater than 1, the sum $\sum_i n_i^2$ would be greater than N.

[9]Possibilities other than Bose–Einstein and Fermi–Dirac statistics can arise in which the wavefunction changes by a complex phase factor $e^{i\theta}$ when particles are interchanged. For topological reasons, this can only happen in two dimensions. Quasiparticle excitations with this property are called *anyons* and, if θ is a rational fraction (other than 1 or 1/2) of 2π, are said to have *fractional statistics* and they play an important role in the theory of the fractional quantum Hall effect; see Wilczek (1990) and Ezawa (2000).

5.5 The density matrix and the partition function of a system of free particles

Suppose that the given system, which is composed of N indistinguishable, noninteracting particles confined to a cubical box of volume V, is a member of a canonical ensemble characterized by the temperature parameter β. The *density matrix* of the system in the coordinate representation will be[10]

$$\langle \boldsymbol{r}_1,\ldots,\boldsymbol{r}_N|\hat{\rho}|\boldsymbol{r}'_1,\ldots,\boldsymbol{r}'_N\rangle = \frac{1}{Q_N(\beta)}\langle \boldsymbol{r}_1,\ldots,\boldsymbol{r}_N|e^{-\beta\hat{H}}|\boldsymbol{r}'_1,\ldots,\boldsymbol{r}'_N\rangle, \tag{1}$$

where $Q_N(\beta)$ is the *partition function* of the system:

$$Q_N(\beta) = \mathrm{Tr}(e^{-\beta\hat{H}}) = \int \langle \boldsymbol{r}_1,\ldots,\boldsymbol{r}_N|e^{-\beta\hat{H}}|\boldsymbol{r}_1,\ldots,\boldsymbol{r}_N\rangle d^{3N}r. \tag{2}$$

For brevity, we denote the vector $\boldsymbol{r}_i$ by the letter i and the primed vector $\boldsymbol{r}'_i$ by i'. Further, let $\psi_E(1,\ldots,N)$ denote the eigenfunctions of the Hamiltonian, the suffix E representing the corresponding eigenvalues. We then have

$$\langle 1,\ldots,N|e^{-\beta\hat{H}}|1',\ldots,N'\rangle = \sum_E e^{-\beta E}\left[\psi_E(1,\ldots,N)\psi_E^*(1',\ldots,N')\right], \tag{3}$$

where the summation goes over all possible values of E; compare to equation (5.3.11).

Since the particles constituting the given system are noninteracting, we may express the eigenfunctions $\psi_E(1,\ldots,N)$ and the eigenvalues E in terms of the single-particle wavefunctions $u_i(m)$ and the single-particle energies ε_i. Moreover, we find it advisable to work with the wave vectors $\boldsymbol{k}_i$ rather than the energies ε_i; so we write

$$E = \frac{\hbar^2K^2}{2m} = \frac{\hbar^2}{2m}(k_1^2 + k_2^2 + \cdots + k_N^2), \tag{4}$$

where the k_i on the right side are the wave vectors of the individual particles. Imposing periodic boundary conditions, the *normalized* single-particle wavefunctions are

$$u_{\boldsymbol{k}}(\boldsymbol{r}) = V^{-1/2}\exp\{i(\boldsymbol{k}\cdot\boldsymbol{r})\}, \tag{5}$$

with

$$\boldsymbol{k} = 2\pi V^{-1/3}\boldsymbol{n}; \tag{6}$$

here, $\boldsymbol{n}$ stands for a three-dimensional vector whose components have values $0, \pm 1, \pm 2, \ldots$. The wavefunction ψ of the total system would then be, see equations (5.4.16)

[10] For a general survey of the density matrix and its applications, see ter Haar (1961).

and (5.4.17),

$$\psi_{\boldsymbol{K}}(1,\ldots,N) = (N!)^{-1/2} \sum_P \delta_P P\{u_{\boldsymbol{k}_1}(1)\ldots u_{\boldsymbol{k}_N}(N)\}, \tag{7}$$

where the magnitudes of the individual k_i are such that

$$(k_1^2 + \cdots + k_N^2) = K^2. \tag{8}$$

The number δ_P in the expression for $\psi_{\boldsymbol{K}}$ is identically equal to $+1$ if the particles are bosons; for fermions, it is $+1$ or -1 according to whether the permutation P is even or odd. Thus, quite generally, we may write

$$\delta_P = (\pm 1)^{[P]}, \tag{9}$$

where $[P]$ denotes the order of the permutation; note that the upper sign in this expression holds for bosons while the lower sign holds for fermions. The factor $(N!)^{-1/2}$ has been introduced here to ensure the normalization of the total wavefunction.

Now, it makes no difference to the wavefunction (7) whether the permutations P are carried out on the coordinates $1,\ldots,N$ or on the wave vectors $\boldsymbol{k}_1,\ldots,\boldsymbol{k}_N$, because after all we are going to sum over *all* the $N!$ permutations. Denoting the permuted coordinates by $P1,\ldots,PN$ and the permuted wave vectors by $P\boldsymbol{k}_1,\ldots,P\boldsymbol{k}_N$, equation (7) may be written as

$$\psi_{\boldsymbol{K}}(1,\ldots,N) = (N!)^{-1/2} \sum_P \delta_P \{u_{\boldsymbol{k}_1}(P1)\ldots u_{\boldsymbol{k}_N}(PN)\} \tag{10a}$$

$$= (N!)^{-1/2} \sum_P \delta_P \{u_{P\boldsymbol{k}_1}(1)\ldots u_{P\boldsymbol{k}_N}(N)\}. \tag{10b}$$

Equations (10a and 10b) may now be substituted into (3), with the result

$$\langle 1,\ldots,N|e^{-\beta\hat{H}}|1',\ldots,N'\rangle = (N!)^{-1} \sum_{\boldsymbol{K}} e^{-\beta\hbar^2K^2/2m}$$
$$\times \left[\sum_{\tilde{P}} \delta_{\tilde{P}} \{u_{\boldsymbol{k}_1}(P1)\ldots u_{\boldsymbol{k}_N}(PN)\} \sum_{\tilde{P}} \delta_{\tilde{P}} \{u^*_{\tilde{P}\boldsymbol{k}_1}(1')\ldots u^*_{\tilde{P}\boldsymbol{k}_N}(N')\} \right], \tag{11}$$

where P and $\tilde{P}$ are any of the $N!$ possible permutations. Now, since a permutation among the $\boldsymbol{k_i}$ changes the wavefunction ψ at most by a sign, the quantity $[\psi\psi^*]$ in (11) is insensitive to such a permutation; the same holds for the exponential factor as well. The summation over $\boldsymbol{K}$ is, therefore, equivalent to $(1/N!)$ times a summation over all the vectors $\boldsymbol{k}_1,\ldots,\boldsymbol{k}_N$ *independently of one another.*

Next, in view of the N-fold summation over the $\boldsymbol{k_i}$, all the permutations $\tilde{P}$ will make equal contributions toward the sum (because they differ from one another only in the ordering of the $\boldsymbol{k}_i$). Therefore, we may consider only one of these permutations, say the one for which $\tilde{P}\boldsymbol{k}_1 = \boldsymbol{k}_1, \ldots, \tilde{P}\boldsymbol{k}_N = \boldsymbol{k}_N$ (and hence $\delta_{\tilde{P}} = 1$ for both kinds of statistics), and include a factor of $(N!)$. The net result is:

$$\langle 1, \ldots, N | e^{-\beta \hat{H}} | 1', \ldots, N' \rangle = (N!)^{-1} \sum_{\boldsymbol{k}_1, \ldots, \boldsymbol{k}_N}$$

$$e^{-\beta\hbar^2(k_1^2 + \cdots + k_N^2)/2m} \left[\sum_P \delta_P \left\{ u_{\boldsymbol{k}_1}(P1) u^*_{\boldsymbol{k}_1}(1') \right\} \ldots \left\{ u_{\boldsymbol{k}_N}(PN) u^*_{\boldsymbol{k}_N}(N') \right\} \right]. \tag{12}$$

Substituting from (5) and noting that, in view of the largeness of V, the summations over the $\boldsymbol{k}_i$ may be replaced by integrations, equation (12) becomes

$$\langle 1, \ldots, N | e^{-\beta \hat{H}} | 1', \ldots, N' \rangle$$

$$= \frac{1}{N!(2\pi)^{3N}} \sum_P \delta_P \left[\int e^{-\beta\hbar^2 k_1^2/2m + i\boldsymbol{k}_1 \cdot (P1 - 1')} d^3k_1 \ldots \right.$$

$$\left. \int e^{-\beta\hbar^2 k_N^2/2m + i\boldsymbol{k}_N \cdot (PN - N')} d^3k_N \right] \tag{13}$$

$$= \frac{1}{N!} \left(\frac{m}{2\pi\beta\hbar^2} \right)^{3N/2} \sum_P \delta_P [f(P1 - 1') \ldots f(PN - N')], \tag{14}$$

where

$$f(\boldsymbol{\xi}) = \exp\left(-\frac{m}{2\beta\hbar^2} \xi^2 \right). \tag{15}$$

Here, use has been made of the mathematical result (5.3.12), which is clearly a special case of the present formula.

Introducing the *mean thermal wavelength*, often referred to as the *thermal deBroglie wavelength*,

$$\lambda = \frac{h}{(2\pi mkT)^{1/2}} = \hbar \left(\frac{2\pi\beta}{m} \right)^{1/2}, \tag{16}$$

and rewriting our coordinates as $\boldsymbol{r}_1, \ldots, \boldsymbol{r}_N$, the diagonal elements among (14) take the form

$$\langle \boldsymbol{r}_1, \ldots, \boldsymbol{r}_N | e^{-\beta \hat{H}} | \boldsymbol{r}_1, \ldots, \boldsymbol{r}_N \rangle = \frac{1}{N! \lambda^{3N}} \sum_P \delta_P [f(P\boldsymbol{r}_1 - \boldsymbol{r}_1) \ldots f(P\boldsymbol{r}_N - \boldsymbol{r}_N)], \tag{17}$$

where

$$f(\boldsymbol{r}) = \exp(-\pi r^2/\lambda^2). \tag{18}$$

To obtain the partition function of the system, we have to integrate (17) over all the coordinates involved. However, before we do that, we would like to make some observations on the summation $\sum_P$. First of all, we note that the leading term in this summation, namely the one for which $P\mathbf{r}_i = \mathbf{r}_i$, is identically equal to unity (because $f(0) = 1$). This is followed by a group of terms in which *only one* pair interchange (among the coordinates) has taken place; a typical term in this group will be $f(\mathbf{r}_j - \mathbf{r}_i)f(\mathbf{r}_i - \mathbf{r}_j)$ where $i \neq j$. This group of terms is followed by other groups of terms in which *more than one* pair interchange has taken place. Thus, we may write

$$\sum_P = 1 \pm \sum_{i<j} f_{ij}f_{ji} + \sum_{i<j<k} f_{ij}f_{jk}f_{ki} \pm \cdots, \tag{19}$$

where $f_{ij} \equiv f(\mathbf{r}_i - \mathbf{r}_j)$; again, note that the upper (lower) signs in this expansion pertain to a system of bosons (fermions). Now, the function f_{ij} vanishes rapidly as the distance r_{ij} becomes much larger than the mean thermal wavelength λ. It then follows that if the mean interparticle distance, $(V/N)^{1/3}$, in the system is much larger than the mean thermal wavelength, that is, if

$$n\lambda^3 = \frac{nh^3}{(2\pi mkT)^{3/2}} \ll 1, \tag{20}$$

where n is the particle density in the system, then the sum $\sum_P$ in (19) may be approximated by unity. Accordingly, the partition function of the system would become, see equation (17),

$$Q_N(V,T) \equiv \mathrm{Tr}\left(e^{-\beta\hat{H}}\right) \approx \frac{1}{N!\lambda^{3N}} \int 1\,(d^{3N}r) = \frac{1}{N!}\left(\frac{V}{\lambda^3}\right)^N. \tag{21}$$

This is precisely the result obtained earlier for the classical ideal gas; see equation (3.5.9). Thus, we have obtained from our quantum-mechanical treatment the precise classical limit for the partition function $Q_N(V,T)$. Incidentally, we have achieved something more. First, we have automatically recovered here the Gibbs correction factor $(1/N!)$, which was introduced into the classical treatment on an *ad hoc*, semi-empirical basis. We, of course, tried to understand its origin in terms of the inherent indistinguishability of the particles. Here, we see it coming in a very natural manner and its source indeed lies in the symmetrization of the wavefunctions of the system (which is ultimately related to the indistinguishability of the particles); compare to Problem 5.4.

Second, we find here a formal justification for computing the number of microstates of a system corresponding to a given region of its phase space by dividing the volume of that region into cells of a "suitable" size and then counting instead the number of these cells. This correspondence becomes all the more transparent by noting that formula (21) is exactly equivalent to the classical expression

$$Q_N(V,T) = \frac{1}{N!}\int e^{-\beta(p_1^2+\cdots+p_N^2)/2m}\left(\frac{d^{3N}q\,d^{3N}p}{\omega_0}\right), \tag{22}$$

with $\omega_0 = h^{3N}$. Thirdly, in deriving the classical limit we have also evolved a criterion that enables us to determine whether a given physical system can be treated classically; mathematically, this criterion is given by condition (20). Now, in statistical mechanical studies, a system that cannot be treated classically is said to be *degenerate*; the quantity $n\lambda^3$ may, therefore, be regarded as a *degeneracy discriminant*. Accordingly, the condition that classical considerations may be applicable to a given physical system is that "the value of the degeneracy discriminant of the system be much less than unity."

Next, we note that, in the classical limit, the diagonal elements of the density matrix are given by

$$\langle \boldsymbol{r}_1,\ldots,\boldsymbol{r}_N|\hat{\rho}|\boldsymbol{r}_1,\ldots,\boldsymbol{r}_N\rangle \approx \left(\frac{1}{V}\right)^N, \tag{23}$$

which is simply a product of N factors, each equal to $(1/V)$. Recalling that, for a single particle in a box of volume V, $\langle \boldsymbol{r}|\hat{\rho}|\boldsymbol{r}\rangle = (1/V)$, see equation (5.3.14), we infer that in the classical limit there is no spatial correlation among the various particles of the system. In general, however, spatial correlations exist even if the particles are supposedly noninteracting; these correlations arise from the symmetrization of the wavefunctions and their magnitude is quite significant if the interparticle distances in the system are comparable with the mean thermal wavelength of the particles. To see this more clearly, we consider the simplest relevant case, namely the one with $N = 2$. The sum $\sum_P$ is now exactly equal to $1 \pm [f(r_{12})]^2$. Accordingly,

$$\langle \boldsymbol{r}_1,\boldsymbol{r}_2|e^{-\beta\hat{H}}|\boldsymbol{r}_1,\boldsymbol{r}_2\rangle = \frac{1}{2\lambda^6}\left[1 \pm \exp\left(-2\pi r_{12}^2/\lambda^2\right)\right] \tag{24}$$

and hence

$$\begin{aligned}
Q_2(V,T) &= \frac{1}{2\lambda^6}\iint\left[1 \pm \exp(-2\pi r_{12}^2/\lambda^2\right]d^3r_1 d^3r_2 \\
&= \frac{1}{2}\left(\frac{V}{\lambda^3}\right)^2\left[1 \pm \frac{1}{V}\int_0^\infty \exp(-2\pi r^2/\lambda^2)\,4\pi r^2\,dr\right] \qquad (25)\\
&= \frac{1}{2}\left(\frac{V}{\lambda^3}\right)^2\left[1 \pm \frac{1}{2^{3/2}}\left(\frac{\lambda^3}{V}\right)\right] \\
&\approx \frac{1}{2}\left(\frac{V}{\lambda^3}\right)^2. \qquad (26)
\end{aligned}$$

Combining (24) and (26), we obtain

$$\langle \boldsymbol{r}_1, \boldsymbol{r}_2 | \hat{\rho} | \boldsymbol{r}_1, \boldsymbol{r}_2 \rangle \approx \frac{1}{V^2}\left[1 \pm \exp\left(-2\pi r_{12}^2/\lambda^2\right)\right]. \tag{27}$$

Thus, if r_{12} is comparable to λ, the probability density (27) may differ considerably from the classical value $(1/V)^2$. In particular, the probability density for a pair of *bosons* to be a distance r apart is larger than the classical, r-independent value by a factor of $[1 + \exp(-2\pi r^2/\lambda^2)]$, which becomes as high as 2 as $r \to 0$. The corresponding result for a pair of *fermions* is smaller than the classical value by a factor of $[1 - \exp(-2\pi r^2/\lambda^2)]$, which becomes as low as 0 as $r \to 0$. Thus, we obtain a *positive* spatial correlation among particles obeying Bose statistics and a *negative* spatial correlation among particles obeying Fermi statistics; see also Section 6.3.

Another way of expressing correlations (among otherwise noninteracting particles) is by introducing a *statistical* interparticle potential $v_s(r)$ and then treating the particles classically (see Uhlenbeck and Gropper, 1932). The potential $v_s(r)$ must be such that the Boltzmann factor $\exp(-\beta v_s)$ is precisely equal to the pair correlation function [...] in (27), that is,

$$v_s(r) = -kT \ln\left[1 \pm \exp\left(-2\pi r^2/\lambda^2\right)\right]. \tag{28}$$

Figure 5.1 shows a plot of the statistical potential $v_s(r)$ for a pair of bosons or fermions. In the Bose case, the potential is throughout attractive, thus giving rise to a "statistical attraction" among bosons; in the Fermi case, it is throughout repulsive, giving rise to a "statistical repulsion" among fermions. In either case, the potential vanishes rapidly as r becomes larger than λ; accordingly, its influence becomes less and less important as the temperature of the system rises.

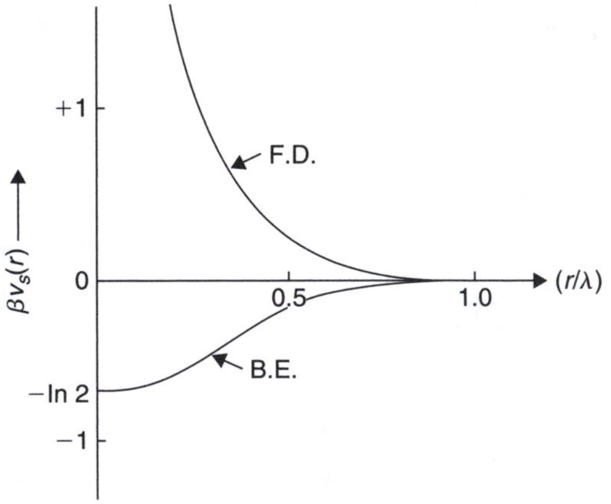

FIGURE 5.1 The statistical potential $v_s(r)$ between a pair of particles obeying Bose–Einstein statistics or Fermi–Dirac statistics.

Problems

5.1. Evaluate the density matrix ρ_{mn} of an electron spin in the representation that makes $\hat{\sigma}_x$ diagonal. Next, show that the value of $\langle\sigma_z\rangle$, resulting from this representation, is precisely the same as the one obtained in Section 5.3.

Hint: The representation needed here follows from the one used in Section 5.3 by carrying out a transformation with the help of the unitary operator

$$\hat{U} = \begin{pmatrix} 1/\sqrt{2} & 1/\sqrt{2} \\ -1/\sqrt{2} & 1/\sqrt{2} \end{pmatrix}.$$

5.2. Prove that

$$\langle q|e^{-\beta\hat{H}}|q'\rangle = \exp\left[-\beta\hat{H}\left(-i\hbar\frac{\partial}{\partial q}, q\right)\right]\delta(q-q'),$$

where $\hat{H}(-i\hbar\,\partial/\partial q, q)$ is the Hamiltonian operator of the system in the q-representation, which formally operates on the Dirac delta function $\delta(q-q')$. Writing the δ-function in a suitable form, apply this result to (i) a free particle and (ii) a linear harmonic oscillator.

5.3. Derive the density matrix ρ for (i) a free particle and (ii) a linear harmonic oscillator in the *momentum* representation and study its main properties along the lines of Section 5.3.

5.4. Study the density matrix and the partition function of a system of free particles, using the *unsymmetrized* wavefunction (5.4.3) instead of the *symmetrized* wavefunction (5.5.7). Show that, following this procedure, one encounters neither the Gibbs' correction factor $(1/N!)$ nor a spatial correlation among the particles.

5.5. Show that in the *first* approximation the partition function of a system of N noninteracting, indistinguishable particles is given by

$$Q_N(V,T) = \frac{1}{N!\lambda^{3N}} Z_N(V,T),$$

where

$$Z_N(V,T) = \int \exp\left\{-\beta\sum_{i<j} v_s(r_{ij})\right\} d^{3N}r,$$

$v_s(r)$ being the statistical potential (5.5.28). Hence evaluate tht *first-order* correction to the equation of state of this system.

5.6. Determine the values of the degeneracy discriminant $(n\lambda^3)$ for hydrogen, helium, and oxygen at NTP. Make an estimate of the respective temperature ranges where the magnitude of this quantity becomes comparable to unity and hence quantum effects become important.

5.7. Show that the quantum-mechanical partition function of a system of N *interacting* particles approaches the classical form

$$Q_N(V,T) = \frac{1}{N!h^{3N}} \int e^{-\beta E(\boldsymbol{q},\boldsymbol{p})} d^{3N}q\, d^{3N}p$$

as the mean thermal wavelength λ becomes much smaller than (i) the mean interparticle distance $(V/N)^{1/3}$ and (ii) a characteristic length r_0 of the interparticle potential.[11]

5.8. Prove the following theorem due to Peierls.[12]

"If $\hat{H}$ is the hermitian Hamiltonian operator of a given physical system and $\{\phi_n\}$ an arbitrary orthonormal set of wavefunctions satisfying the symmetry requirements and the boundary

[11]See Huang (1963, Section 10.2).

[12]See Peierls (1938) and Huang (1963, Section 10.3).

conditions of the problem, then the partition function of the system satisfies the following inequality:

$$Q(\beta) \geq \sum_n \exp\{-\beta\langle\phi_n|\hat{H}|\phi_n\rangle\};$$

the equality holds when $\{\phi_n\}$ constitute a complete orthonormal set of eigenfunctions of the Hamiltonian itself."

6 The Theory of Simple Gases

We are now fully equipped with the formalism required for determining the macroscopic properties of a large variety of physical systems. In most cases, however, derivations run into serious mathematical difficulties, with the result that one is forced to restrict one's analysis either to simpler kinds of systems or to simplified models of actual systems. In practice, even these restricted studies are carried out in a series of stages, the first stage of the process being highly "idealized." The best example of such an idealization is the familiar *ideal* gas, a study of which is not only helpful in acquiring facility with the mathematical procedures but also throws considerable light on the physical behavior of gases actually met with in nature. In fact, it also serves as a base on which the theory of real gases can be founded; see Chapter 10.

In this chapter we propose to derive, and at some length discuss, the most basic properties of simple gaseous systems obeying quantum statistics; the discussion will include some of the essential features of diatomic and polyatomic gases and chemical equilibrium.

6.1 An ideal gas in a quantum-mechanical microcanonical ensemble

We consider a gaseous system of N noninteracting, *indistinguishable* particles confined to a space of volume V and sharing a given energy E. The statistical quantity of interest in this case is $\Omega(N,V,E)$ which, by definition, denotes the number of *distinct* microstates accessible to the system under the macrostate (N,V,E). While determining this number, we must remember that a failure to take into account the indistinguishability of the particles in a proper manner could lead to results which, except in the classical limit, may not be acceptable. With this in mind, we proceed as follows.

Since, for large V, the single-particle energy levels in the system are very close to one another, we may divide the energy spectrum into a large number of "groups of levels," which may be referred to as *energy cells*; see Figure 6.1. Let ε_i denote the average energy of a level, and g_i the (arbitrary) number of levels, in the ith cell; we assume that all $g_i \gg 1$. In a particular situation, we may have n_1 particles in the first cell, n_2 particles in the second cell, and so on. Clearly, the distribution set $\{n_i\}$ must conform to the conditions

$$\sum_i n_i = N \tag{1}$$

© 2011 Elsevier Ltd. All rights reserved.

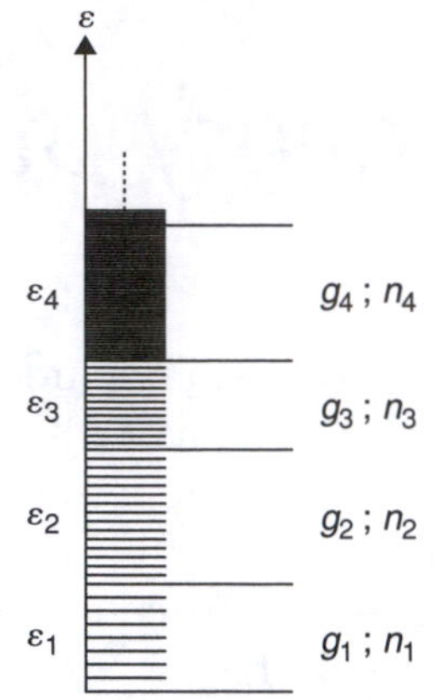

FIGURE 6.1 The grouping of the single-particle energy levels into "cells."

and

$$\sum_i n_i \varepsilon_i = E. \tag{2}$$

Then

$$\Omega(N, V, E) = \sum_{\{n_i\}}{}' W\{n_i\}, \tag{3}$$

where $W\{n_i\}$ is the number of *distinct* microstates associated with the distribution set $\{n_i\}$ while the primed summation goes over all distribution sets that conform to conditions (1) and (2). Next,

$$W\{n_i\} = \prod_i w(i), \tag{4}$$

where $w(i)$ is the number of *distinct* microstates associated with the ith cell of the spectrum (the cell that contains n_i particles, to be accommodated among g_i levels) while the product goes over all the cells in the spectrum. Clearly, $w(i)$ is the number of distinct ways in which the n_i identical, and *indistinguishable*, particles can be distributed among the g_i levels of the ith cell. This number, in the Bose–Einstein case, is given by, see equation (3.8.25),

$$w_{\text{B.E.}}(i) = \frac{(n_i + g_i - 1)!}{n_i!(g_i - 1)!}, \tag{5}$$

so that

$$W_{\text{B.E.}}\{n_i\} = \prod_i \frac{(n_i + g_i - 1)!}{n_i!(g_i - 1)!}. \tag{6}$$

In the Fermi–Dirac case, no single level can accommodate more than one particle; accordingly, the number n_i cannot exceed g_i. The number $w(i)$ is then given by the "number of ways in which the g_i levels can be divided into two subgroups — one consisting of n_i levels (which will have one particle each) and the other consisting of $(g_i - n_i)$ levels (which will be unoccupied)." This number is given by

$$w_{\text{F.D.}}(i) = \frac{g_i!}{n_i!\,(g_i - n_i)!}, \tag{7}$$

so that

$$W_{\text{F.D.}}\{n_i\} = \prod_i \frac{g_i!}{n_i!\,(g_i - n_i)!}. \tag{8}$$

For completeness, we may include the classical — or what is generally known as the Maxwell–Boltzmann — case as well. There, the particles are regarded as *distinguishable*, with the result that any of the n_i particles may be put into any of the g_i levels, independently of one another, and the resulting states may all be regarded as distinct; the number of these states is clearly $(g_i)^{n_i}$. Moreover, the distribution set $\{n_i\}$ in this case is itself regarded as obtainable in

$$\frac{N!}{n_1!\,n_2!\ldots} \tag{9}$$

different ways which, on the introduction of the Gibbs correction factor, lead to a "weight factor" of

$$\frac{1}{n_1!\,n_2!\ldots} = \prod_i \frac{1}{n_i!}; \tag{10}$$

see also Section 1.6, especially equation (1.6.2). Combining these two results, we obtain

$$W_{\text{M.B.}}\{n_i\} = \prod_i \frac{(g_i)^{n_i}}{n_i!}. \tag{11}$$

Now, the entropy of the system would be given by

$$S(N,V,E) = k \ln \Omega(N,V,E) = k \ln \left[{\sum_{\{n_i\}}}' W\{n_i\} \right]. \tag{12}$$

It can be shown that, under the conditions of our analysis, the logarithm of the sum on the right side of (12) can be approximated by the logarithm of the largest term in the sum; see Problem 3.4. We may, therefore, replace (12) by

$$S(N,V,E) \approx k \ln W\{n_i^*\}, \tag{13}$$

where $\{n_i^*\}$ is the distribution set that maximizes the number $W\{n_i\}$; the numbers n_i^* are clearly the *most probable values* of the distribution numbers n_i. The maximization, however, is to be carried out under the restrictions that the quantities N and E remain constant. This can be done by the method of Lagrange's undetermined multipliers; see Section 3.2. Our condition for determining the most probable distribution set $\{n_i^*\}$ now turns out to be, see equations (1), (2), and (13),

$$\delta \ln W\{n_i\} - \left[\alpha \sum_i \delta n_i + \beta \sum_i \varepsilon_i \delta n_i\right] = 0. \tag{14}$$

For $\ln W\{n_i\}$, we obtain from equations (6), (8), and (11), assuming that not only all g_i but also all $n_i \gg 1$ (so that the Stirling approximation $\ln(x!) \approx x \ln x - x$ can be applied to all the factorials that appear in these expressions),

$$\begin{aligned} \ln W\{n_i\} &= \sum_i \ln w(i) \\ &\approx \sum_i \left[n_i \ln\left(\frac{g_i}{n_i} - a\right) - \frac{g_i}{a} \ln\left(1 - a\frac{n_i}{g_i}\right)\right], \end{aligned} \tag{15}$$

where $a = -1$ for the B.E. case, $+1$ for the F.D. case, and 0 for the M.B. case. Equation (14) then becomes

$$\sum_i \left[\ln\left(\frac{g_i}{n_i} - a\right) - \alpha - \beta\varepsilon_i\right]_{n_i = n_i^*} \delta n_i = 0. \tag{16}$$

In view of the arbitrariness of the increments δn_i in (16), we must have (for all i)

$$\ln\left(\frac{g_i}{n_i^*} - a\right) - \alpha - \beta\varepsilon_i = 0, \tag{17}$$

so that[1]

$$n_i^* = \frac{g_i}{e^{\alpha + \beta\varepsilon_i} + a}. \tag{18}$$

The fact that n_i^* turns out to be *directly* proportional to g_i prompts us to interpret the quantity

$$\frac{n_i^*}{g_i} = \frac{1}{e^{\alpha + \beta\varepsilon_i} + a}, \tag{18a}$$

which is actually the most probable number of particles *per* energy level in the ith cell, as the most probable number of particles in a *single* level of energy ε_i. Incidentally, our final result (18a) is totally independent of the manner in which the energy levels of the particles

[1] For a critique of this derivation, see Landsberg (1954a, 1961).

are grouped into cells, so long as the number of levels in each cell is sufficiently large. As shown in Section 6.2, formula (18a) can also be derived without grouping energy levels into cells at all; in fact, it is only then that this result becomes truly acceptable!

Substituting (18) into (15), we obtain for the entropy of the gas

$$\frac{S}{k} \approx \ln W\{n_i^*\} = \sum_i \left[n_i^* \ln\left(\frac{g_i}{n_i^*} - a\right) - \frac{g_i}{a} \ln\left(1 - a\frac{n_i^*}{g_i}\right)\right]$$
$$= \sum_i \left[n_i^* (\alpha + \beta\varepsilon_i) + \frac{g_i}{a} \ln\left\{1 + ae^{-\alpha - \beta\varepsilon_i}\right\}\right]. \tag{19}$$

The first sum on the right side of (19) is identically equal to αN while the second sum is identically equal to βE. For the third sum, therefore, we have

$$\frac{1}{a}\sum_i g_i \ln\left\{1 + ae^{-\alpha - \beta\varepsilon_i}\right\} = \frac{S}{k} - \alpha N - \beta E. \tag{20}$$

Now, the physical interpretation of the parameters α and β here is going to be precisely the same as in Section 4.3, namely

$$\alpha = -\frac{\mu}{kT}, \quad \beta = \frac{1}{kT}; \tag{21}$$

for confirmation see Section 6.2. The right side of equation (20) is, therefore, equal to

$$\frac{S}{k} + \frac{\mu N}{kT} - \frac{E}{kT} = \frac{G - (E - TS)}{kT} = \frac{PV}{kT}. \tag{22}$$

The thermodynamic pressure of the system is, therefore, given by

$$PV = \frac{kT}{a}\sum_i \left[g_i \ln\left\{1 + ae^{-\alpha - \beta\varepsilon_i}\right\}\right]. \tag{23}$$

In the Maxwell–Boltzmann case ($a \to 0$), equation (23) takes the form

$$PV = kT\sum_i g_i e^{-\alpha - \beta\varepsilon_i} = kT\sum_i n_i^* = NkT, \tag{24}$$

which is the familiar equation of state of the classical ideal gas. Note that equation (24) for the Maxwell–Boltzmann case holds irrespective of the details of the energy spectrum ε_i.

It will be recognized that the expression $a^{-1}\sum_i[\,]$ in equation (23), being equal to the thermodynamic quantity (PV/kT), ought to be identical to the q-potential of the ideal gas. One may, therefore, expect to obtain from this expression all the macroscopic properties of this system. However, before demonstrating this, we would like to first develop the formal theory of an ideal gas in the canonical and grand canonical ensembles.

6.2 An ideal gas in other quantum-mechanical ensembles

In the canonical ensemble the thermodynamics of a given system is derived from its partition function

$$Q_N(V,T) = \sum_E e^{-\beta E}, \tag{1}$$

where E denotes the energy eigenvalues of the system while $\beta = 1/kT$. Now, an energy value E can be expressed in terms of the single-particle energies ε; for instance,

$$E = \sum_\varepsilon n_\varepsilon \varepsilon, \tag{2}$$

where n_ε is the number of particles in the single-particle energy state ε. The values of the numbers n_ε must satisfy the condition

$$\sum_\varepsilon n_\varepsilon = N. \tag{3}$$

Equation (1) may then be written as

$$Q_N(V,T) = {\sum_{\{n_\varepsilon\}}}' g\{n_\varepsilon\} e^{-\beta \sum_\varepsilon n_\varepsilon \varepsilon}, \tag{4}$$

where $g\{n_\varepsilon\}$ is the *statistical weight factor* appropriate to the distribution set $\{n_\varepsilon\}$ and the summation $\sum'$ goes over all distribution sets that conform to the restrictive condition (3). The statistical weight factor in different cases is given by

$$g_{\text{B.E.}}\{n_\varepsilon\} = 1, \tag{5}$$

$$g_{\text{F.D.}}\{n_\varepsilon\} = \begin{cases} 1 & \text{if all } n_\varepsilon = 0 \text{ or } 1 \\ 0 & \text{otherwise,} \end{cases} \tag{6}$$

and

$$g_{\text{M.B.}}\{n_\varepsilon\} = \prod_\varepsilon \frac{1}{n_\varepsilon!}. \tag{7}$$

Note that in the present treatment we are dealing with single-particle states as *individual* states, without requiring them to be grouped into cells; indeed, the weight factors (5), (6), and (7) follow straightforwardly from their respective predecessors (6.1.6), (6.1.8), and (6.1.11) by putting all $g_i = 1$.

First of all, we work out the Maxwell–Boltzmann case. Substituting (7) into (4), we get

$$Q_N(V,T) = \sum_{\{n_\varepsilon\}}{}' \left[\left(\prod_\varepsilon \frac{1}{n_\varepsilon!} \right) \prod_\varepsilon \left(e^{-\beta\varepsilon} \right)^{n_\varepsilon} \right]$$

$$= \frac{1}{N!} \sum_{\{n_\varepsilon\}}{}' \left[\frac{N!}{\prod_\varepsilon n_\varepsilon!} \prod_\varepsilon \left(e^{-\beta\varepsilon} \right)^{n_\varepsilon} \right]. \tag{8}$$

Since the summation here is governed by condition (3), it can be evaluated with the help of the multinomial theorem, with the result

$$Q_N(V,T) = \frac{1}{N!} \left[\sum_\varepsilon e^{-\beta\varepsilon} \right]^N$$

$$= \frac{1}{N!} [Q_1(V,T)]^N, \tag{9}$$

in agreement with equation (3.5.15). The evaluation of Q_1 is, of course, straightforward. One obtains, using the asymptotic formula (2.4.7) for the number of single-particle states with energies lying between ε and $\varepsilon + d\varepsilon$,

$$Q_1(V,T) \equiv \sum_\varepsilon e^{-\beta\varepsilon} \approx \frac{2\pi V}{h^3} (2m)^{3/2} \int_0^\infty e^{-\beta\varepsilon} \varepsilon^{1/2} d\varepsilon$$

$$= V/\lambda^3, \tag{10}$$

where $\lambda\, [= h/(2\pi mkT)^{1/2}]$ is the mean thermal wavelength of the particles. Hence

$$Q_N(V,T) = \frac{V^N}{N! \lambda^{3N}}, \tag{11}$$

from which complete thermodynamics of this system can be derived; see, for example, Section 3.5. Further, we obtain for the grand partition function of this system

$$\mathcal{Q}(z,V,T) = \sum_{N=0}^\infty z^N Q_N(V,T) = \exp(zV/\lambda^3); \tag{12}$$

compare to equation (4.4.3). We know that the thermodynamics of the system follows equally well from the expression for $\mathcal{Q}$.

In the Bose–Einstein and Fermi–Dirac cases, we obtain, by substituting (5) and (6) into (4),

$$Q_N(V,T) = \sum_{\{n_\varepsilon\}}{}' \left(e^{-\beta \sum_\varepsilon n_\varepsilon \varepsilon} \right); \tag{13}$$

the difference between the two cases, B.E. and F.D., arises from the values that the numbers n_ε can take. Now, in view of restriction (3) on the summation $\sum'$, an explicit evaluation of the partition function Q_N in these cases is rather cumbersome. The grand partition function $\mathcal{Q}$, on the other hand, turns out to be more easily tractable; we have

$$\mathcal{Q}(z,V,T) = \sum_{N=0}^{\infty}\left[z^N \sum_{\{n_\varepsilon\}}' e^{-\beta\sum_\varepsilon n_\varepsilon \varepsilon}\right] \tag{14a}$$

$$= \sum_{N=0}^{\infty}\left[\sum_{\{n_\varepsilon\}}' \prod_\varepsilon \left(ze^{-\beta\varepsilon}\right)^{n_\varepsilon}\right]. \tag{14b}$$

Now, the double summation in (14b) — first over the numbers n_ε constrained by a *fixed* value of the total number N, and then over all possible values of N — is equivalent to a summation over all possible values of the numbers n_ε, *independently of one another*. Hence, we may write

$$\mathcal{Q}(z,V,T) = \sum_{n_0,n_1,\ldots}\left[\left(ze^{-\beta\varepsilon_0}\right)^{n_0}\left(ze^{-\beta\varepsilon_1}\right)^{n_1}\ldots\right]$$

$$= \left[\sum_{n_0}\left(ze^{-\beta\varepsilon_0}\right)^{n_0}\right]\left[\sum_{n_1}\left(ze^{-\beta\varepsilon_1}\right)^{n_1}\right]\ldots. \tag{15}$$

Now, in the Bose–Einstein case the n_ε can be either 0 or 1 or 2 or ..., while in the Fermi–Dirac case they can be only 0 or 1. Therefore,

$$\mathcal{Q}(z,V,T) = \begin{cases} \prod_\varepsilon \dfrac{1}{(1-ze^{-\beta\varepsilon})} & \text{in the B.E. case, with } ze^{-\beta\varepsilon} < 1 \quad (16a) \\ \prod_\varepsilon (1+ze^{-\beta\varepsilon}) & \text{in the F.D. case.} \quad (16b) \end{cases}$$

The q-potential of the system is thus given by

$$q(z,V,T) \equiv \frac{PV}{kT} \equiv \ln \mathcal{Q}(z,V,T)$$

$$= \mp\sum_\varepsilon \ln(1 \mp ze^{-\beta\varepsilon}); \tag{17}$$

compare to equation (6.1.23), with $g_i = 1$. The identification of the fugacity z with the quantity $e^{-\alpha}$ of equation (6.1.23) is quite natural; accordingly, $\alpha = -\mu/kT$. As usual, the upper (lower) sign in equation (17) corresponds to the Bose (Fermi) case.

In the end, we may write our results for q in a form applicable to all three cases:

$$q(z,V,T) \equiv \frac{PV}{kT} = \frac{1}{a}\sum_\varepsilon \ln(1 + aze^{-\beta\varepsilon}), \tag{18}$$

where $a = -1, +1$, or 0, depending on the statistics governing the system. In particular, the classical case ($a \to 0$) gives

$$q_{\text{M.B.}} = z \sum_{\varepsilon} e^{-\beta\varepsilon} = zQ_1, \tag{19}$$

in agreement with equation (4.4.4). From (18), it follows that

$$\overline{N} \equiv z\left(\frac{\partial q}{\partial z}\right)_{V,T} = \sum_{\varepsilon} \frac{1}{z^{-1}e^{\beta\varepsilon} + a} \tag{20}$$

and

$$\overline{E} \equiv -\left(\frac{\partial q}{\partial \beta}\right)_{z,V} = \sum_{\varepsilon} \frac{\varepsilon}{z^{-1}e^{\beta\varepsilon} + a}. \tag{21}$$

At the same time, the *mean occupation number* $\langle n_\varepsilon \rangle$ of level ε turns out to be, see equations (14a) and (17),

$$\begin{aligned}\langle n_\varepsilon \rangle &= \frac{1}{\mathcal{Q}}\left[-\frac{1}{\beta}\left(\frac{\partial \mathcal{Q}}{\partial \varepsilon}\right)_{z,T,\text{ all other } \varepsilon}\right] \\ &\equiv -\frac{1}{\beta}\left(\frac{\partial q}{\partial \varepsilon}\right)_{z,T,\text{ all other } \varepsilon} \\ &= \frac{1}{z^{-1}e^{\beta\varepsilon} + a},\end{aligned} \tag{22}$$

in keeping with equations (20) and (21). Comparing our final result (22) with its counterpart (6.1.18a), we find that the *mean* value $\langle n \rangle$ and the *most probable* value n^* of the occupation number n of a single-particle state are indeed identical.

6.3 Statistics of the occupation numbers

Equation (6.2.22) gives the *mean occupation number* of a single-particle state with energy ε as an explicit function of the quantity $(\varepsilon - \mu)/kT$:

$$\langle n_\varepsilon \rangle = \frac{1}{e^{(\varepsilon-\mu)/kT} + a}. \tag{1}$$

The functional behavior of this number is shown in Figure 6.2. In the Fermi–Dirac case ($a = +1$), the mean occupation number never exceeds unity, for the variable n_ε itself cannot have a value other than 0 or 1. Moreover, for $\varepsilon < \mu$ and $|\varepsilon - \mu| \gg kT$, the mean occupation number tends to its maximum possible value 1. In the Bose–Einstein case ($a = -1$), we must have $\mu <$ *all* ε; see equation (6.2.16a). In fact, when μ becomes equal to the lowest value of ε, say ε_0, the occupancy of that particular level becomes infinitely high, which leads to the phenomenon of *Bose–Einstein condensation*; see Sections 7.1 and 7.2. For

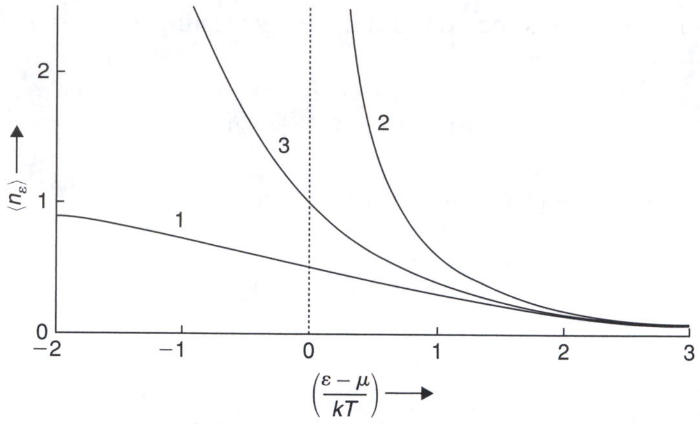

FIGURE 6.2 The mean occupation number $\langle n_\varepsilon \rangle$ of a single-particle energy state ε in a system of noninteracting particles: curve 1 is for fermions, curve 2 for bosons, and curve 3 for the Maxwell–Boltzmann particles.

$\mu < \varepsilon_0$, all values of $(\varepsilon - \mu)$ are positive and the behavior of all $\langle n_\varepsilon \rangle$ is nonsingular. Finally, in the Maxwell–Boltzmann case $(a = 0)$, the mean occupation number takes the familiar form

$$\langle n_\varepsilon \rangle_{\text{M.B.}} = \exp\{(\mu - \varepsilon)/kT\} \propto \exp(-\varepsilon/kT). \tag{2}$$

The important thing to note here is that the distinction between the quantum statistics $(a = \pm 1)$ and the classical statistics $(a = 0)$ becomes imperceptible when, for all values of ε that are of practical interest,

$$\exp\{(\varepsilon - \mu)/kT\} \gg 1. \tag{3}$$

In that event, equation (1) essentially reduces to (2) and we may write, instead of (3),

$$\langle n_\varepsilon \rangle \ll 1. \tag{4}$$

Condition (4) is quite understandable, for it implies that the probability of any of the n_ε being greater than unity is quite negligible, with the result that the classical weight factors $g\{n_\varepsilon\}$, as given by equation (6.2.7), become essentially equal to 1. The distinction between the classical treatment and the quantum-mechanical treatment then becomes rather insignificant. Correspondingly, we find, see Figure 6.2, that for large values of $(\varepsilon - \mu)/kT$ the quantum curves 1 and 2 essentially merge into the classical curve 3. Since we already know that the higher the temperature of the system the better the validity of the classical treatment, condition (3) also implies that μ, the chemical potential of the system, must be negative and large in magnitude. This means that the fugacity $z[\equiv \exp(\mu/kT)]$ of the system must be much smaller than unity; see also equation (6.2.22). One can see, from equations (4.4.6) and (4.4.29), that this is further equivalent to the requirement

$$\frac{N\lambda^3}{V} \ll 1, \tag{5}$$

which agrees with condition (5.5.20).

We shall now examine statistical fluctuations in the variable n_ε. Going a step further from the calculation that led to equation (6.2.22), we have

$$\langle n_\varepsilon^2 \rangle = \frac{1}{\mathcal{Q}} \left[\left(-\frac{1}{\beta} \frac{\partial}{\partial \varepsilon} \right)^2 \mathcal{Q} \right]_{z,T,\text{ all other } \varepsilon}; \tag{6}$$

it follows that

$$\begin{aligned} \langle n_\varepsilon^2 \rangle - \langle n_\varepsilon \rangle^2 &= \left[\left(-\frac{1}{\beta} \frac{\partial}{\partial \varepsilon} \right)^2 \ln \mathcal{Q} \right]_{z,T,\text{ all other } \varepsilon} \\ &= \left[\left(-\frac{1}{\beta} \frac{\partial}{\partial \varepsilon} \right) \langle n_\varepsilon \rangle \right]_{z,T}. \end{aligned} \tag{7}$$

For the relative mean-square fluctuation, we obtain (irrespective of the statistics obeyed by the particles)

$$\frac{\langle n_\varepsilon^2 \rangle - \langle n_\varepsilon \rangle^2}{\langle n_\varepsilon \rangle^2} = \left(\frac{1}{\beta} \frac{\partial}{\partial \varepsilon} \right) \left\{ \frac{1}{\langle n_\varepsilon \rangle} \right\} = z^{-1} e^{\beta\varepsilon}; \tag{8}$$

of course, the actual value of this quantity will depend on the statistics of the particles because, for a given particle density (N/V) and a given temperature T, the value of z will be different for different statistics.

It seems more instructive to write (8) in the form

$$\frac{\langle n_\varepsilon^2 \rangle - \langle n_\varepsilon \rangle^2}{\langle n_\varepsilon \rangle^2} = \frac{1}{\langle n_\varepsilon \rangle} - a. \tag{9}$$

In the classical case $(a = 0)$, the relative fluctuation is *normal*. In the Fermi–Dirac case, it is given by $1/\langle n_\varepsilon \rangle - 1$, which is below normal and tends to vanish as $\langle n_\varepsilon \rangle \to 1$. In the Bose–Einstein case, the fluctuation is clearly *above normal*.[2] Obviously, this result would apply to a gas of photons and, hence, to the oscillator states in the black-body radiation. In the latter context, Einstein derived this result as early as 1909 following Planck's approach and even pointed out that the term 1 in the expression for the fluctuation may be attributed to the wave character of the radiation and the term $1/\langle n_\varepsilon \rangle$ to the particle character of the photons; for details, see Kittel (1958), ter Haar (1968).

Closely related to the subject of fluctuations is the problem of "statistical correlations in photon beams," which have been observed experimentally (see Hanbury Brown and

[2]The special case of fluctuations in the *ground state occupation number*, n_0, of a Bose–Einstein system has been discussed by Wergeland (1969) and by Fujiwara, ter Haar, and Wergeland (1970).

Twiss, 1956, 1957, 1958) and have been explained theoretically in terms of the quantum-statistical nature of these fluctuations (see Purcell, 1956; Kothari and Auluck, 1957). For further details, refer to Mandel, Sudarshan, and Wolf (1964); and Holliday and Sage (1964).

For greater understanding of the statistics of the occupation numbers, we evaluate the quantity $p_\varepsilon(n)$, which is the probability that there are exactly n particles in a state of energy ε. Referring to equation (6.2.14b), we infer that $p_\varepsilon(n) \propto (ze^{-\beta\varepsilon})^n$. On normalization, it becomes in the Bose–Einstein case

$$\begin{aligned} p_\varepsilon(n)|_{\text{B.E.}} &= \left(ze^{-\beta\varepsilon}\right)^n \left[1 - ze^{-\beta\varepsilon}\right] \\ &= \left(\frac{\langle n_\varepsilon \rangle}{\langle n_\varepsilon \rangle + 1}\right)^n \frac{1}{\langle n_\varepsilon \rangle + 1} = \frac{(\langle n_\varepsilon \rangle)^n}{(\langle n_\varepsilon \rangle + 1)^{n+1}}. \end{aligned} \tag{10}$$

In the Fermi–Dirac case, we get

$$\begin{aligned} p_\varepsilon(n)|_{\text{F.D.}} &= \left(ze^{-\beta\varepsilon}\right)^n \left[1 + ze^{-\beta\varepsilon}\right]^{-1} \\ &= \begin{cases} 1 - \langle n_\varepsilon \rangle & \text{for} \quad n = 0 \\ \langle n_\varepsilon \rangle & \text{for} \quad n = 1. \end{cases} \end{aligned} \tag{11}$$

In the Maxwell–Boltzmann case, we have $p_\varepsilon(n) \propto (ze^{-\beta\varepsilon})^n / n!$ instead; see equation (6.2.8). On normalization, we get

$$p_\varepsilon(n)|_{\text{M.B.}} = \frac{\left(ze^{-\beta\varepsilon}\right)^n / n!}{\exp\left(ze^{-\beta\varepsilon}\right)} = \frac{(\langle n_\varepsilon \rangle)^n}{n!} e^{-\langle n_\varepsilon \rangle}. \tag{12}$$

Distribution (12) is clearly a *Poisson distribution*, for which the mean square deviation of the variable in question is equal to the mean value itself; compare to equation (9), with $a = 0$. It also resembles the distribution of the total particle number N in a grand canonical ensemble consisting of ideal, classical systems; see Problem 4.4. We also note that the ratio $p_\varepsilon(n)/p_\varepsilon(n-1)$ in this case varies inversely with n, which is a "normal" statistical behavior of uncorrelated events.

On the other hand, the distribution in the Bose–Einstein case is *geometric*, with a constant common ratio $\langle n_\varepsilon \rangle / (\langle n_\varepsilon \rangle + 1)$. This means that the probability of a state ε acquiring one more particle for itself is independent of the number of particles already occupying that state; thus, in comparison with the "normal" statistical behavior, bosons exhibit a special tendency of "bunching" together, which means a *positive* statistical correlation among them. In contrast, fermions exhibit a *negative* statistical correlation.

6.4 Kinetic considerations

The thermodynamic pressure of an ideal gas is given by equation (6.1.23) or (6.2.18). In view of the largeness of volume V, the single-particle energy states ε would be so close to one another that a summation over them may be replaced by integration. One thereby gets

$$P = \frac{kT}{a}\int_0^\infty \ln\left[1 + aze^{-\beta\varepsilon(p)}\right]\frac{4\pi p^2 dp}{h^3}$$

$$= \frac{4\pi kT}{ah^3}\left[\frac{p^3}{3}\ln\left[1 + aze^{-\beta\varepsilon(p)}\right]\Bigg|_0^\infty + \int_0^\infty \frac{p^3}{3}\frac{aze^{-\beta\varepsilon(p)}}{1 + aze^{-\beta\varepsilon(p)}}\beta\frac{d\varepsilon}{dp}dp\right].$$

The integrated part vanishes at both limits while the rest of the expression reduces to

$$P = \frac{4\pi}{3h^3}\int_0^\infty \frac{1}{z^{-1}e^{\beta\varepsilon(p)} + a}\left(p\frac{d\varepsilon}{dp}\right)p^2 dp. \tag{1}$$

Now, the total number of particles in the system is given by

$$N = \int \langle n_p \rangle \frac{Vd^3p}{h^3} = \frac{4\pi V}{h^3}\int_0^\infty \frac{1}{z^{-1}e^{\beta\varepsilon(p)} + a}p^2 dp. \tag{2}$$

Comparing (1) and (2), we can write

$$P = \frac{1}{3}\frac{N}{V}\left\langle p\frac{d\varepsilon}{dp}\right\rangle = \frac{1}{3}n\langle pu\rangle, \tag{3}$$

where n is the particle density in the gas and u the speed of an individual particle. If the relationship between the energy ε and the momentum p is of the form $\varepsilon \propto p^s$, then

$$P = \frac{s}{3}n\langle\varepsilon\rangle = \frac{s}{3}\frac{E}{V}; \tag{4}$$

the particular cases $s = 1$ and $s = 2$ are pretty easy to recognize. It should be noted here that results (3) and (4) hold independently of the statistics obeyed by the particles.

The structure of formula (3) suggests that the pressure of the gas arises essentially from the physical motion of the particles; it should, therefore, be derivable from kinetic considerations alone. To do this, we consider the bombardment, by the particles of the gas, on the walls of the container. Let us take, for example, an element of area dA on one of the walls normal to the z-axis, see Figure 6.3, and focus our attention on those particles whose velocity lies between $\boldsymbol{u}$ and $\boldsymbol{u} + d\boldsymbol{u}$; the number of such particles per unit volume may be denoted by $nf(\boldsymbol{u})d\boldsymbol{u}$, where

$$\int_{\text{all } \boldsymbol{u}} f(\boldsymbol{u})d\boldsymbol{u} = 1. \tag{5}$$

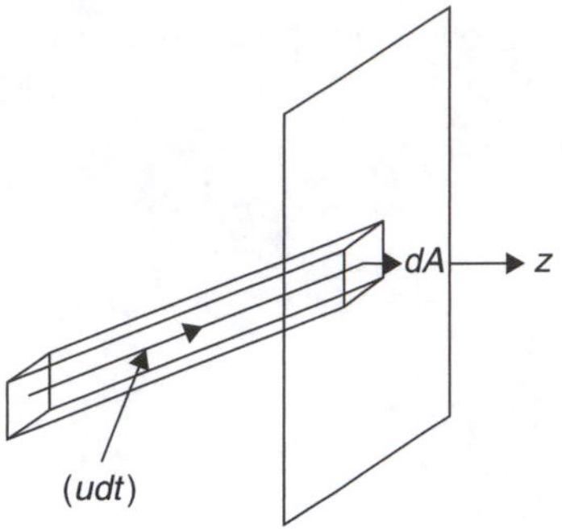

FIGURE 6.3 The molecular bombardment on one of the walls of the container.

Now, the question is: how many of these particles will strike the area $d\boldsymbol{A}$ in time dt? The answer is: all those particles that happen to lie in a cylindrical region of base $d\boldsymbol{A}$ and height $\boldsymbol{u}dt$, as shown in Figure 6.3. Since the volume of this region is $(d\boldsymbol{A}\cdot\boldsymbol{u})dt$, the number of such particles would be $\{(d\boldsymbol{A}\cdot\boldsymbol{u})dt \times nf(\boldsymbol{u})d\boldsymbol{u}\}$. On reflection from the wall, the normal component of the momentum of a particle would undergo a change from p_z to $-p_z$; as a result, the normal momentum imparted by these particles per unit time to a unit area of the wall would be $2\,p_z\{u_z nf(\boldsymbol{u})d\boldsymbol{u}\}$. Integrating this expression over all relevant $\boldsymbol{u}$, we obtain the total normal momentum imparted per unit time to a unit area of the wall by all the particles of the gas which, by definition, is the *kinetic* pressure of the gas:

$$P = 2n \int_{u_x=-\infty}^{\infty} \int_{u_y=-\infty}^{\infty} \int_{u_z=0}^{\infty} p_z u_z f(\boldsymbol{u}) du_x du_y du_z. \qquad (6)^3$$

Since (i) $f(\boldsymbol{u})$ is a function of u alone and (ii) the product $(p_z u_z)$ is an even function of u_z, the foregoing result may be written as

$$P = n \int_{\text{all } \boldsymbol{u}} (p_z u_z) f(\boldsymbol{u}) d\boldsymbol{u}. \qquad (7)$$

Comparing (7) with (5), we obtain

$$P = n\langle p_z u_z\rangle = n\langle pu\cos^2\theta\rangle \qquad (8)$$

$$= \frac{1}{3} n\langle pu\rangle, \qquad (9)$$

which is identical to (3).

[3]Clearly, only those velocities for which $u_z > 0$ are relevant here.

In a similar manner, we can determine the rate of *effusion* of the gas particles through a hole (of unit area) in the wall. This is given by, compared to (6),

$$R = n \int_{u_x=-\infty}^{\infty} \int_{u_y=-\infty}^{\infty} \int_{u_z=0}^{\infty} u_z f(\boldsymbol{u}) du_x du_y du_z \tag{10}$$

$$= n \int_{\phi=0}^{2\pi} \int_{\theta=0}^{\pi/2} \int_{u=0}^{\infty} \{u \cos\theta f(\boldsymbol{u})\}(u^2 \sin\theta \, du \, d\theta \, d\phi); \tag{11}$$

note that the condition $u_z > 0$ restricts the range of the angle θ between the values 0 and $\pi/2$. Carrying out integrations over θ and ϕ, we obtain

$$R = n\pi \int_0^{\infty} f(\boldsymbol{u}) u^3 \, du. \tag{12}$$

In view of the fact that

$$\int_0^{\infty} f(\boldsymbol{u})(4\pi u^2 \, du) = 1, \tag{5a}$$

equation (12) may be written as

$$R = \frac{1}{4} n \langle u \rangle. \tag{13}$$

Again, this result holds independently of the statistics obeyed by the particles.

It is obvious that the velocity distribution among the effused particles is considerably different from the one among the particles inside the container. This is due to the fact that, firstly, the velocity component u_z of the effused particles must be positive (which introduces an element of anisotropy into the distribution) and, secondly, the particles with larger values of u_z appear with an extra weightage, the weightage being directly proportional to the value of u_z; see equation (10). As a result of this, (i) the effused particles carry with them a net forward momentum, thus causing the container to experience a recoil force, and (ii) they carry away a relatively large amount of energy per particle, thus leaving the gas in the container at not only a progressively decreasing pressure and density but also a progressively decreasing temperature; see Problem 6.14.

6.5 Gaseous systems composed of molecules with internal motion

In most of our studies so far we have considered only the translational part of the molecular motion. Though this aspect of motion is invariably present in a gaseous system, other aspects, which are essentially concerned with the *internal* motion of the molecules, also exist. It is only natural that in the calculation of the physical properties of such a system, contributions arising from these motions are also taken into account. In doing so, we shall assume that (i) the effects of the intermolecular interactions are negligible and (ii) the nondegeneracy criterion

$$n\lambda^3 = \frac{nh^3}{(2\pi mkT)^{3/2}} \ll 1 \tag{5.5.20}$$

is fulfilled; this makes our system an *ideal, Boltzmannian* gas. Under these assumptions, which hold sufficiently well in a large number of applications, the partition function of the system is given by

$$Q_N(V,T) = \frac{1}{N!}[Q_1(V,T)]^N, \tag{1}$$

where

$$Q_1(V,T) = \frac{V}{\lambda^3} j(T); \tag{2}$$

the factor within the curly brackets is the familiar translational partition function of a molecule, while $j(T)$ is the partition function corresponding to internal motions. The latter may be written as

$$j(T) = \sum_i g_i e^{-\varepsilon_i/kT}, \tag{3}$$

where ε_i is the energy associated with a state of internal motion (characterized by the quantum numbers i), while g_i is the multiplicity of that state.

The contributions made by the internal motions of the molecules, over and above the translational degrees of freedom, follow straightforwardly from the function $j(T)$. We obtain

$$A_{\text{int}} = -NkT \ln j, \tag{4}$$

$$\mu_{\text{int}} = -kT \ln j, \tag{5}$$

$$S_{\text{int}} = Nk\left(\ln j + T\frac{\partial}{\partial T}\ln j\right), \tag{6}$$

$$U_{\text{int}} = NkT^2 \frac{\partial}{\partial T}\ln j, \tag{7}$$

and

$$(C_V)_{\text{int}} = Nk\frac{\partial}{\partial T}\left\{T^2\frac{\partial}{\partial T}\ln j\right\}. \tag{8}$$

Thus, the central problem in this study is to derive an explicit expression for the function $j(T)$ from a knowledge of the internal states of the molecules. For this, we note that the internal state of a molecule is determined by (i) the electronic state, (ii) the state of the nuclei, (iii) the vibrational state, and (iv) the rotational state. Rigorously speaking, these four modes of excitation mutually interact; in many cases, however, they can be treated independently of one another. We can then write

$$j(T) = j_{\text{elec}}(T) j_{\text{nuc}}(T) j_{\text{vib}}(T) j_{\text{rot}}(T), \tag{3a}$$

with the result that the net contribution made by the internal motions to the various thermodynamic properties of the system is given by a simple sum of the four respective contributions. There is one interaction, however, that plays a special role in the case of homonuclear molecules, such as AA, and which is between the states of the nuclei and the rotational states. In such a case, we better write

$$j(T) = j_{\text{elec}}(T) j_{\text{nuc-rot}}(T) j_{\text{vib}}(T). \tag{3b}$$

We now examine this problem for various systems in the order of increasing complexity.

6.5.A Monatomic molecules

For simplicity, we consider a monatomic gas at temperatures such that the thermal energy kT is small in comparison with the ionization energy ε_{ion}; for different atoms, this amounts to the condition $T \ll \varepsilon_{\text{ion}}/k \sim 10^4 - 10^5$ K. At these temperatures, the number of ionized atoms in the gas would be insignificant. The same would be true of atoms in the excited states, for the separation of any of the excited states from the ground state of the atom is generally of the same order of magnitude as the ionization energy itself. Thus, we may regard all atoms in the gas to be in their (electronic) ground state.

Now, there is a special class of atoms, namely He, Ne, A, ..., which, in their ground state, possess neither orbital angular momentum nor spin ($L = S = 0$). Their (electronic) ground state is clearly a singlet, with $g_e = 1$. The nucleus, however, possesses a degeneracy that arises from the possibility of different orientations of the nuclear spin.[4] If the value of this spin is S_n, the corresponding degeneracy factor $g_n = 2S_n + 1$. Moreover, a monatomic molecule cannot have any vibrational or rotational states. The internal partition function (3a) of such a molecule is, therefore, given by

$$j(T) = (g)_{\text{gr.st.}} = g_e \cdot g_n = 2S_n + 1. \tag{9}$$

[4]As is well known, the presence of the nuclear spin gives rise to the so-called *hyperfine structure* in the electronic states. However, the intervals of this structure are such that, for practically all temperatures of interest, they are small in comparison with kT; for concreteness, these intervals correspond to T-values of the order of 10^{-1} to 10^0 K. Accordingly, in the evaluation of the partition function $j(T)$, the hyperfine splitting of the electronic state may be disregarded while the multiplicity introduced by the nuclear spin may be taken into account through a degeneracy factor.

Equations (4) through (8) then tell us that the internal motions in this case contribute only toward properties such as the chemical potential and the entropy of the gas; they do not contribute toward the internal energy and the specific heat.

If, on the other hand, the ground state does not possess orbital angular momentum but possesses spin ($L = 0$, $S \neq 0$ — as, for example, in the case of alkali atoms), then the ground state will still have no fine structure; it will, however, have a degeneracy $g_e = 2S + 1$. As a result, the internal partition function $j(T)$ will get multiplied by a factor of $(2S + 1)$ and the properties such as the chemical potential and the entropy of the gas will get modified accordingly.

In other cases, the ground state of the atom may possess both orbital angular momentum and spin ($L \neq 0, S \neq 0$); the ground state would then possess a definite fine structure. The intervals of this structure are, in general, comparable to kT; hence, in the evaluation of the partition function, the energies of the various components of the fine structure will have to be taken into account. Since these components differ from one another in the value of the total angular momentum J, the relevant partition function may be written as

$$j_{\text{elec}}(T) = \sum_J (2J + 1)e^{-\varepsilon_J/kT}. \tag{10}$$

The foregoing expression simplifies considerably in the following limiting cases:

(a) $kT \gg$ all ε_J; then

$$j_{\text{elec}}(T) \simeq \sum_J (2J + 1) = (2L + 1)(2S + 1). \tag{10a}$$

(b) $kT \ll$ all ε_J; then

$$j_{\text{elec}}(T) \simeq (2J_0 + 1)e^{-\varepsilon_0/kT}, \tag{10b}$$

where J_0 is the total angular momentum, and ε_0 the energy, of the atom in the lowest state. In either case, the electronic motion makes no contribution toward the specific heat of the gas. Of course, at intermediate temperatures, we do obtain a contribution toward this property. And, in view of the fact that both at high and low temperatures the specific heat tends to be equal to the translational value $\frac{3}{2}Nk$, it must pass through a maximum at a temperature comparable to the separation of the fine structure levels.[5] Needless to say, the multiplicity $(2S_n + 1)$ introduced by the nuclear spin must be taken into account in each case.

6.5.B Diatomic molecules

Now we consider a diatomic gas at temperatures such that kT is small compared to the energy of dissociation; for different molecules, this amounts once again to the condition

[5]It seems worthwhile to note here that the values of $\Delta\varepsilon_J/k$ for the components of the normal triplet term of oxygen are 230 K and 320 K, while those for the normal quintuplet term of iron range from 600 to 1,400 K.

$T \ll \varepsilon_{\text{diss}}/k \sim 10^4 - 10^5$ K. At these temperatures the number of dissociated molecules in the gas would be insignificant. At the same time, in most cases, there would be practically no molecules in the excited states as well, for the separation of any of these states from the ground state of the molecule is in general comparable to the dissociation energy itself.[6] Accordingly, in the evaluation of $j(T)$, we have to take into account only the lowest electronic state of the molecule.

The lowest electronic state, in most cases, is nondegenerate: $g_e = 1$. We then need not consider any further the question of the electronic state making a contribution toward the thermodynamic properties of the gas. However, certain molecules (though not very many) have, in their lowest electronic state, either (i) a nonzero orbital angular momentum ($\Lambda \neq 0$) or (ii) a nonzero spin ($S \neq 0$) or (iii) both. In case (i), the electronic state acquires a twofold degeneracy corresponding to the two possible orientations of the oribital angular momentum relative to the molecular axis;[7] as a result, $g_e = 2$. In case (ii), the state acquires a degeneracy $2S + 1$ corresponding to the space quantization of the spin.[8]

In both these cases the chemical potential and the entropy of the gas are modified by the multiplicity of the electronic state, while the energy and the specific heat remain unaffected. In case (iii), we encounter a fine structure that necessitates a rather detailed study because the intervals of this structure are generally of the same order of magnitude as kT. In particular, for a doublet fine-structure term, such as the one that arises in the molecule NO ($\Pi_{1/2,3/2}$ with a separation of 178 K, the components themselves being Λ-doublets), we have for the electronic partition function

$$j_{\text{elec}}(T) = g_0 + g_1 e^{-\Delta/kT}, \tag{11}$$

where g_0 and g_1 are the degeneracy factors of the two components while Δ is their separation energy. The contribution made by (11) toward the various thermodynamic properties of the gas can be calculated with the help of formulae (4) through (8). In particular, we obtain for the contribution toward the specific heat

$$(C_V)_{\text{elec}} = Nk \frac{(\Delta/kT)^2}{[1 + (g_0/g_1)e^{\Delta/kT}]\,[1 + (g_1/g_0)e^{-\Delta/kT}]}. \tag{12}$$

We note that this contribution vanishes both for $T \ll \Delta/k$ and for $T \gg \Delta/k$ and is maximum for a certain temperature $\sim\Delta/k$; compare to the corresponding situation in the case of monatomic molecules.

[6] An odd case arises with oxygen. The separation between its normal term $^3\Sigma$ and the first excited term $^1\Delta$ is about 11,250 K, whereas the dissociation energy is about 55,000 K. The relevant factor $e^{-\varepsilon_1/kT}$, therefore, can be quite significant even when the factor $e^{-\varepsilon_{\text{diss}}/kT}$ is not, say for $T \sim 2000$ to 6000 K.

[7] Strictly speaking, the term in question splits into two levels — the so-called Λ-doublet. The separation of the levels, however, is such that we can safely neglect it.

[8] The separation of the resulting levels is again negligible from the thermodynamic point of view; as an example, one may cite the very narrow triplet term of O_2.

We now consider the effect of the vibrational states of the molecules on the thermodynamic properties of the gas. To have an idea of the temperature range over which this effect would be significant, we note that the magnitude of the corresponding quantum of energy, namely $\hbar\omega$, for different diatomic gases is of order 10^3 K. Thus, we would obtain full contributions (consistent with the dictates of the equipartition theorem) at temperatures of the order of 10^4 K or more, and practically no contribution at temperatures of the order of 10^2 K or less. Let us assume that the temperature is not high enough to excite vibrational states of large energy; the oscillations of the nuclei then remain small in amplitude and hence harmonic. The energy levels for a mode of frequency ω are then given by the well-known expression $(n+\frac{1}{2})\hbar\omega$.[9]

The evaluation of the vibrational partition function $j_{\text{vib}}(T)$ is quite elementary; see Section 3.8. In view of the rapid convergence of the series involved, the summation may formally be extended to $n=\infty$. The corresponding contributions toward the various thermodynamic properties of the system are then given by equations (3.8.16) through (3.8.21). In particular,

$$(C_V)_{\text{vib}} = Nk\left(\frac{\Theta_v}{T}\right)^2 \frac{e^{\Theta_v/T}}{(e^{\Theta_v/T}-1)^2}; \quad \Theta_v = \frac{\hbar\omega}{k}. \tag{13}$$

We note that for $T \gg \Theta_v$, the vibrational specific heat is very nearly equal to the equipartition value Nk; otherwise, it is always less than Nk. In particular, for $T \ll \Theta_v$, the specific heat tends to zero (see Figure 6.4); the vibrational degrees of freedom are then said to be "frozen."

At sufficiently high temperatures, when vibrations with large n are also excited, the effects of anharmonicity and of interaction between the vibrational and the rotational modes of the molecule can become important.[10] However, since this happens only at large n, the relevant corrections to the various thermodynamic quantities can be determined even classically; see Problems 3.29 and 3.30. One finds that the first-order correction to C_{vib} is directly proportional to the temperature of the gas.

Finally, we consider the effect of (i) the states of the nuclei and (ii) the rotational states of the molecule; wherever necessary, we shall take into account the mutual interaction of these modes. This interaction is of no relevance in the case of *heteronuclear* molecules, such as AB; it is, however, important in the case of *homonuclear* molecules, such as AA. We may, therefore, consider the two cases separately.

The states of the nuclei in the heteronuclear case may be treated separately from the rotational states of the molecule. Proceeding in the same manner as for monatomic molecules, we conclude that the effect of the nuclear states is adequately taken care of

[9]It may be pointed out that the vibrational motion of a molecule is influenced by the centrifugal force arising from the molecular rotation. This leads to an interaction between the rotational and the vibrational modes. However, unless the temperature is too high, this interaction can be neglected and the two modes treated independently of one another.

[10]In principle, these two effects are of the same order of magnitude.

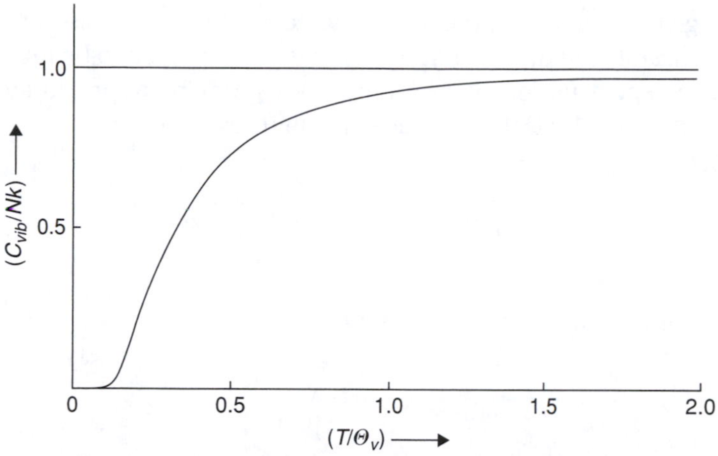

FIGURE 6.4 The vibrational specific heat of a gas of diatomic molecules. At $T = \Theta_v$, the specific heat is already about 93 percent of the equipartition value.

through a degeneracy factor g_n. Denoting the spins of the two nuclei by S_A and S_B,

$$g_n = (2S_A + 1)(2S_B + 1). \tag{14}$$

As before, we obtain a finite contribution toward the chemical potential and the entropy of the gas but none toward the internal energy and the specific heat.

Now, the rotational levels of a linear "rigid" rotator, with two degrees of freedom (for the axis of rotation) and the principal moments of inertia $(I, I, 0)$, are given by

$$\varepsilon_{\text{rot}} = l(l+1)\hbar^2/2I, \quad l = 0, 1, 2, \ldots; \tag{15}$$

here, $I = \mu r_0^2$ where $\mu[= m_1 m_2/(m_1 + m_2)]$ is the *reduced mass* of the nuclei and r_0 the *equilibrium distance* between them. The rotational partition function of the molecule is then given by

$$\begin{aligned} j_{\text{rot}}(T) &= \sum_{l=0}^{\infty}(2l+1)\exp\left\{-l(l+1)\frac{\hbar^2}{2IkT}\right\} \\ &= \sum_{l=0}^{\infty}(2l+1)\exp\left\{-l(l+1)\frac{\Theta_r}{T}\right\}; \quad \Theta_r = \frac{\hbar^2}{2Ik}. \end{aligned} \tag{16}$$

The values of Θ_r, for all gases except the ones involving the isotopes H and D, are much smaller than room temperature. For example, the value of Θ_r for HCl is about 15 K, for N_2, O_2, and NO it lies between 2 K and 3 K, while for Cl_2 it is about one-third of a degree. On the other hand, the values of Θ_r for H_2, D_2, and HD are, respectively, 85 K, 43 K, and 64 K.

These numbers give us an idea of the respective temperature ranges in which the effects arising from the *discreteness* of the rotational states are expected to be important.

For $T \gg \Theta_r$, the spectrum of the rotational states may be approximated by a continuum. The summation in (16) is then replaced by an integration:

$$j_{\text{rot}}(T) \approx \int_0^\infty (2l+1)\exp\left\{-l(l+1)\frac{\Theta_r}{T}\right\} dl = \frac{T}{\Theta_r}. \tag{17}$$

The rotational specific heat is then given by

$$(C_V)_{\text{rot}} = Nk, \tag{18}$$

consistent with the equipartition theorem.

A better evaluation of the sum in (16) can be made with the help of the Euler–Maclaurin formula, namely

$$\sum_{n=0}^{\infty} f(n) = \int_0^\infty f(x)dx + \frac{1}{2}f(0) - \frac{1}{12}f'(0) + \frac{1}{720}f'''(0) - \frac{1}{30,240}f^{\text{v}}(0) + \cdots. \tag{19}$$

Writing

$$f(x) = (2x+1)\exp\{-x(x+1)\Theta_r/T\},$$

one obtains

$$j_{\text{rot}}(T) = \frac{T}{\Theta_r} + \frac{1}{3} + \frac{1}{15}\frac{\Theta_r}{T} + \frac{4}{315}\left(\frac{\Theta_r}{T}\right)^2 + \cdots, \tag{20}$$

which is the so-called *Mulholland's formula*; as expected, the main term of this formula is identical to the classical partition function (17). The corresponding result for the specific heat is

$$(C_V)_{\text{rot}} = Nk\left\{1 + \frac{1}{45}\left(\frac{\Theta_r}{T}\right)^2 + \frac{16}{945}\left(\frac{\Theta_r}{T}\right)^3 + \cdots\right\}, \tag{21}$$

which shows that at high temperatures the rotational specific heat *decreases* with temperature and ultimately tends to the classical value Nk. Thus, at high (but finite) temperatures the rotational specific heat of a diatomic gas is greater than the classical value. On the other hand, it must go to zero as $T \to 0$. We, therefore, conclude that it passes through at least one maximum. Numerical studies show that there is only one maximum that appears at a temperature of about $0.8\Theta_r$ and has a value of about $1.1Nk$; see Figure 6.5.

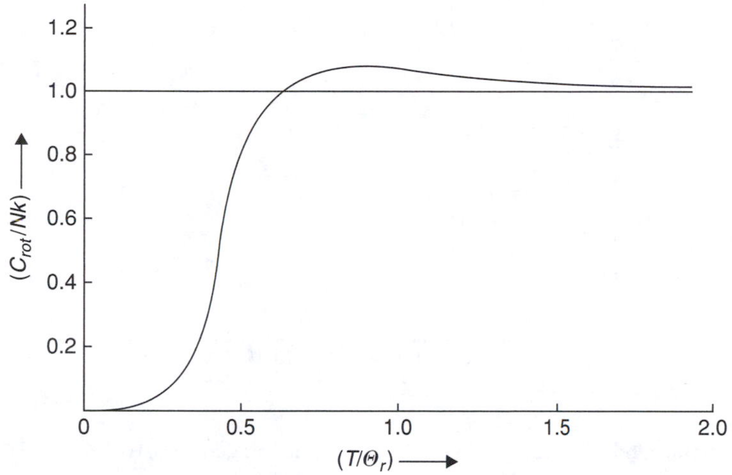

FIGURE 6.5 The rotational specific heat of a gas of heteronuclear diatomic molecules.

In the other limiting case, when $T \ll \Theta_r$, one may retain only the first few terms of the sum in (16); then

$$j_{\text{rot}}(T) = 1 + 3e^{-2\Theta_r/T} + 5e^{-6\Theta_r/T} + \cdots, \tag{22}$$

from which one obtains, in the *lowest* approximation,

$$(C_V)_{\text{rot}} \simeq 12Nk\left(\frac{\Theta_r}{T}\right)^2 e^{-2\Theta_r/T}. \tag{23}$$

Thus, as $T \to 0$, the specific heat drops exponentially to zero; see again Figure 6.5. We, therefore, conclude that at low enough temperatures the rotational degrees of freedom of the molecules are also "frozen."

At this stage it appears worthwhile to remark that, since the internal motions of the molecules do not make any contribution toward the pressure of the gas (A_{int} being independent of V), the quantity $(C_P - C_V)$ is the same for a diatomic gas as for a monatomic one. Moreover, under the assumptions made in the very beginning of this section, the value of this quantity at all temperatures of interest would be equal to the classical value Nk. Thus, at sufficiently low temperatures (when rotational as well as vibrational degrees of freedom of the molecules are "frozen"), we have, by virtue of the translational motion alone,

$$C_V = \frac{3}{2}Nk, \quad C_P = \frac{5}{2}NK; \quad \gamma = \frac{5}{3}. \tag{24}$$

As temperature rises, the rotational degrees of freedom begin to "loosen up" until we reach temperatures that are much larger than Θ_r but much smaller than Θ_v; the rotational degrees of freedom are then fully excited while the vibrational ones are still "frozen."

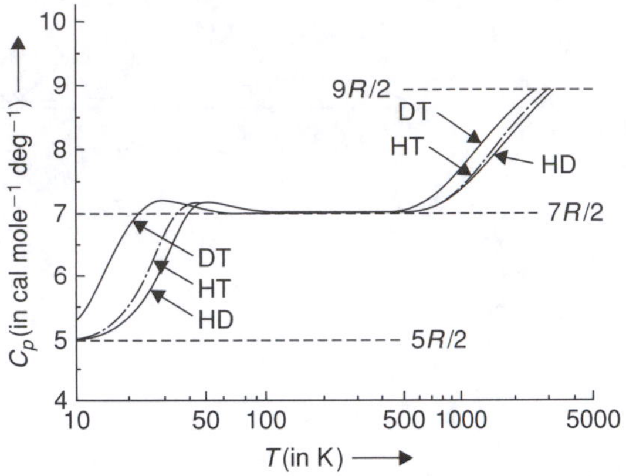

FIGURE 6.6 The rotational-vibrational specific heat, C_P, of the diatomic gases HD, HT, and DT.

Accordingly, for $\Theta_r \ll T \ll \Theta_v$,

$$C_V = \frac{5}{2}Nk, \quad C_P = \frac{7}{2}Nk; \quad \gamma = \frac{7}{5}. \tag{25}$$

As temperature rises further, the vibrational degrees of freedom as well start loosening up, until we reach temperatures that are much larger than Θ_v. Then, the vibrational degrees of freedom are also fully excited and we have

$$C_V = \frac{7}{2}Nk, \quad C_P = \frac{9}{2}Nk; \quad \gamma = \frac{9}{7}. \tag{26}$$

These features are displayed in Figure 6.6 where the experimental results for C_P are plotted for three gases HD, HT, and DT. We note that, in view of the considerable difference between the values of Θ_r and Θ_v, the situation depicted by (25) prevails over a considerably large range of temperatures. In passing, it may be pointed out that, for most diatomic gases, the situation at room temperatures corresponds to the one depicted by (25).

We now study the case of *homonuclear* molecules, such as *AA*. To start with, we consider the limiting case of high temperatures where classical approximation is admissible. The rotational motion of the molecule may then be visualized as a rotation of the molecular axis, that is, the line joining the two nuclei, about an "axis of rotation" that is perpendicular to the molecular axis and passes through the center of mass of the molecule. Then, the two opposing positions of the molecular axis, namely the ones corresponding to the azimuthal angles ϕ and $\phi + \pi$, differ simply by an interchange of the two identical nuclei and, hence, correspond to only one *distinct* state of the molecule. Therefore, in the evaluation of the partition function, the range of the angle ϕ should be taken as $(0, \pi)$ instead of the customary $(0, 2\pi)$. Moreover, since the energy of rotational motion does not depend on angle ϕ, the only effect of this on the partition function of the molecule would be to reduce it by a factor of 2. We thus obtain, *in the classical approximation,*[11]

$$j_{\text{nuc-rot}}(T) = (2S_A + 1)^2 \frac{T}{2\Theta_r}. \tag{27}$$

Obviously, the factor 2 here will not affect the specific heat of the gas; in the classical approximation, therefore, the specific heat of a gas of homonuclear molecules is the same as that of a corresponding gas of heteronuclear molecules.

In contrast, significant changes result at relatively lower temperatures where the states of rotational motion have to be treated as *discrete*. These changes arise from the coupling between the nuclear and the rotational states that in turn arises from the symmetry character of the nuclear-rotational wavefunction. As discussed in Section 5.4, the total wavefunction of a physical state must be either symmetric or antisymmetric (depending on the statistics obeyed by the particles involved) with respect to an interchange of two identical particles. Now, the rotational wavefunction of a diatomic molecule is symmetric or antisymmetric depending on whether the quantum number l is even or odd. The nuclear wavefunction, on the other hand, consists of a linear combination of the spin functions of the two nuclei and its symmetry character depends on the manner in which the combination is formed. It is not difficult to see that, of the $(2S_A + 1)^2$ different combinations that one constructs, exactly $(S_A + 1)(2S_A + 1)$ are symmetric with respect to an interchange of the nuclei and the remaining $S_A(2S_A + 1)$ antisymmetric.[12] In constructing the total wavefunction, as a product of the nuclear and the rotational wavefunctions, we then proceed as follows:

(i) If the nuclei are fermions ($S_A = \frac{1}{2}, \frac{3}{2}, \ldots$), as in the molecule H_2, the total wavefunction must be antisymmetric. To secure this, we may associate any one of the $S_A(2S_A + 1)$ antisymmetric nuclear wavefunctions with any one of the even-l rotational wavefunctions *or* any one of the $(S_A + 1)(2S_A + 1)$ symmetric nuclear wavefunctions with any one of the odd-l rotational wavefunctions. Accordingly, the nuclear-rotational partition function of such a molecule would be

$$j^{\text{(F.D.)}}_{\text{nuc-rot}}(T) = S_A(2S_A + 1) r_{\text{even}} + (S_A + 1)(2S_A + 1) r_{\text{odd}}, \tag{28}$$

[11]It seems instructive to outline here the purely classical derivation of the rotational partition function. Specifying the rotation of the molecule by the angles (θ, ϕ) and the corresponding momenta (p_θ, p_ϕ), the kinetic energy assumes the form

$$\varepsilon_{\text{rot}} = \frac{1}{2I} p_\theta^2 + \frac{1}{2I \sin^2\theta} p_\phi^2,$$

from which

$$j_{\text{rot}}(T) = \frac{1}{h^2} \int e^{-\varepsilon_{\text{rot}}/kT} (dp_\theta dp_\phi d\theta\, d\phi) = \frac{IkT}{\pi \hbar^2} \int_0^{\phi_{\max}} d\phi.$$

For heteronuclear molecules $\phi_{\max} = 2\pi$, while for homonuclear ones $\phi_{\max} = \pi$.

[12]See, for example, Schiff (1968, Section 41).

where

$$r_{\text{even}} = \sum_{l=0,2,\ldots}^{\infty} (2l+1)\exp\{-l(l+1)\Theta_r/T\} \tag{29}$$

and

$$r_{\text{odd}} = \sum_{l=1,3,\ldots}^{\infty} (2l+1)\exp\{-l(l+1)\Theta_r/T\}. \tag{30}$$

(ii) If the nuclei are bosons ($S_A = 0, 1, 2, \ldots$), as in the molecule D_2, the total wavefunction must be symmetric. To secure this, we may associate any one of the $(S_A+1)(2S_A+1)$ symmetric nuclear wavefunctions with any one of the even-l rotational wavefunctions *or* any one of the $S_A(2S_A+1)$ antisymmetric nuclear wavefunctions with any one of the odd-l rotational wavefunctions. We then have

$$j^{(\text{B.E.})}_{\text{nuc-rot}}(T) = (S_A+1)(2S_A+1)r_{\text{even}} + S_A(2S_A+1)r_{\text{odd}}. \tag{31}$$

At high temperatures, it is the larger values of l that contribute most to the sums (29) and (30). The difference between the two sums is then negligibly small, and we have

$$r_{\text{even}} \simeq r_{\text{odd}} \simeq \frac{1}{2} j_{\text{rot}}(T) = T/2\Theta_r; \tag{32}$$

see equations (16) and (17). Consequently,

$$j^{(\text{B.E.})}_{\text{nuc-rot}} \simeq j^{(\text{F.D.})}_{\text{nuc-rot}} = (2S_A+1)^2 T/2\Theta_r, \tag{33}$$

in agreement with our previous result (27). Under these circumstances, the statistics governing the nuclei does not make a significant difference to the thermodynamic behaviour of the gas.

Things change when the temperature of the gas is in a range comparable to the value of Θ_r. It seems most reasonable then to regard the gas as a mixture of two components, generally referred to as *ortho*- and *para*-, whose relative concentrations in equilibrium are determined by the relative magnitudes of the two parts of the partition function (28) or (31), as the case may be. Customarily, the name ortho- is given to that component that carries the larger statistical weight. Thus, in the case of fermions (as in H_2), the ortho- to para-ratio is given by

$$n^{(\text{F.D.})} = \frac{(S_A+1)r_{\text{odd}}}{S_A r_{\text{even}}}, \tag{34}$$

while in the case of bosons (as in D_2), the corresponding ratio is given by

$$n^{(\text{B.E.})} = \frac{(S_A+1)r_{\text{even}}}{S_A r_{\text{odd}}}. \tag{35}$$

As temperature rises, the factor r_{odd}/r_{even} tends to unity and the ratio n, in each case, approaches the temperature-independent value $(S_A+1)/S_A$. In the case of H_2, this limiting value is 3 (since $S_A = \frac{1}{2}$) while in the case of D_2 it is 2 (since $S_A = 1$). At sufficiently low temperatures, one may retain only the main terms of the sums (29) and (30), with the result that

$$\frac{r_{odd}}{r_{even}} \simeq 3\exp\left(-\frac{2\Theta_r}{T}\right) \quad (T \ll \Theta_r), \tag{36}$$

which tends to zero as $T \to 0$. The ratio n then tends to zero in the case of fermions and to infinity in the case of bosons. Hence, as $T \to 0$, the hydrogen gas is wholly para-, while deuterium is wholly ortho-; of course, in each case, the molecules do settle down in the rotational state $l = 0$.

At intermediate temperatures, one has to work with the equilibrium ratio (34), or (35), and with the composite partition function (28), or (31), in order to compute the thermodynamic properties of the gas. One finds, however, that the theoretical results so derived do not *generally* agree with the ones obtained experimentally. This discrepancy was resolved by Dennison (1927) who pointed out that the samples of hydrogen, or deuterium, ordinarily subjected to experiment are not in thermal equilibrium as regards the relative magnitudes of the ortho- and para-components. These samples are ordinarily prepared and kept at room temperatures that are well above Θ_r, with the result that the ortho- to para-ratio in them is very nearly equal to the limiting value $(S_A+1)S_A$.

If now the temperature is lowered, one would expect this ratio to change in accordance with equation (34), or (35). However, it does not do so for the following reason. Since the transition of a molecule from one form of existence to another involves the flipping of the spin of one of its nuclei, the transition probability of the process is quite small. Actually, the periods involved are of the order of a year! Obviously, one cannot expect to attain the true equilibrium ratio n during the short times available. Consequently, even at lower temperatures, what one generally has is a *nonequilibrium* mixture of two independent substances, the relative concentration of which is preassigned. The partition functions (28) and (31) as such are, therefore, inapplicable; we rather have directly for the specific heat

$$C^{(\text{F.D.})} = \frac{S_A}{2S_A+1} C_{even} + \frac{S_A+1}{2S_A+1} C_{odd} \tag{37}$$

and

$$C^{(\text{B.E.})} = \frac{S_A+1}{2S_A+1} C_{even} + \frac{S_A}{2S_A+1} C_{odd}, \tag{38}$$

where

$$C_{even/odd} = Nk\frac{\partial}{\partial T}\left\{T^2(\partial/\partial T)\ln r_{even/odd}\right\}. \tag{39}$$

We have, therefore, to compute C_{even} and C_{odd} separately and then derive the net value of the rotational specific heat with the help of formula (37) or (38), as the case may be.

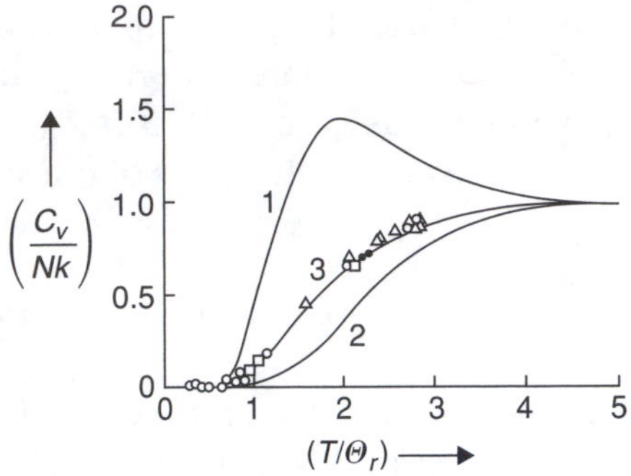

FIGURE 6.7 The theoretical specific heat of a 1:3 mixture of para-hydrogen and ortho-hydrogen. The experimental points originate from various sources listed in Wannier (1966).

Figure 6.7 shows the relevant results for hydrogen. Curves 1 and 2 correspond to the para-hydrogen (C_{even}) and the ortho-hydrogen (C_{odd}), respectively, while curve 3 represents the weighted mean, as given by equation (37). The experimental results are also shown in the figure; the agreement between theory and experiment is clearly good.

Further evidence in favor of Dennison's explanation is obtained by performing experiments with ortho–para mixtures of different relative concentration. This can be done by speeding up the ortho–para conversion by passing hydrogen over activated charcoal. By doing this at various temperatures, and afterwards removing the catalyst, one can fix the ratio n at any desired value. The specific heat then follows a curve obtained by mixing C_{even} and C_{odd} with appropriate weight factors. Further, if one measures the specific heat of the gas in such a way that the ratio n, at every temperature T, has the value that is given by formula (34), it indeed follows the curve obtained from expression (28) for the partition function.

6.5.C Polyatomic molecules

Once again, the translational degrees of freedom of the molecules contribute their usual share, $\frac{3}{2}k$ per molecule, toward the specific heat of the gas. As regards the lowest electronic state, it is, in most cases, far below any of the excited states; nevertheless, it generally possesses a multiplicity (depending on the orbital and spin angular momenta of the state) that can be taken care of by a degeneracy factor g_e. As regards the rotational states, they can be treated classically because the large values of the moments of inertia characteristic of polyatomic molecules make the quantum of rotational energy, $\hbar^2/2I_i$, much smaller than the thermal energy kT at practically all temperatures of interest. Consequently, the interaction between the rotational states and the states of the nuclei can also be treated classically. As a result, the nuclear-rotational partition function is given by the product of the respective partition functions, divided by a symmetry number γ that denotes the number of physically indistinguishable configurations realized during one complete rotation of the molecule:[13]

$$j_{\text{nuc-rot}}(T) = \frac{g_{\text{nuc}} j_{\text{rot}}^C(T)}{\gamma}; \tag{40}$$

compare to equation (27). Here, $j_{\text{rot}}^C(T)$ is the rotational partition function of the molecule evaluated in the classical approximation (without paying regard to the presence of *identical* nuclei, if any); it is given by

$$j_{\text{rot}}^C(T) = \pi^{1/2} \left(\frac{2I_1 kT}{\hbar^2}\right)^{1/2} \left(\frac{2I_2 kT}{\hbar^2}\right)^{1/2} \left(\frac{2I_3 kT}{\hbar^2}\right)^{1/2} \tag{41}$$

where I_1, I_2, and I_3 are the principal moments of inertia of the molecule; see Problem 6.27.[14] The rotational specific heat is then given by

$$C_{\text{rot}} = Nk \frac{\partial}{\partial T}\left\{T^2 \frac{\partial}{\partial T} \ln j_{\text{rot}}^C(T)\right\} = \frac{3}{2}Nk, \tag{42}$$

consistent with the equipartition theorem.

As regards vibrational states, we first note that, unlike a diatomic molecule, a polyatomic molecule has not one but several vibrational degrees of freedom. In particular, a noncollinear molecule consisting of n atoms has $3n - 6$ vibrational degrees of freedom, six degrees of freedom out of the total $3n$ having gone into the translational and rotational motions. On the other hand, a collinear molecule consisting of n atoms would have $3n - 5$ vibrational degrees of freedom, for the rotational motion in this case has only two, not three, degrees of freedom. The vibrational degrees of freedom correspond to a set of normal modes characterized by a set of frequencies ω_i. It might happen that some of these frequencies have identical values; we then speak of *degenerate frequencies*.[15]

In the harmonic approximation, these normal modes may be treated independently of one another. The vibrational partition function of the molecule is then given by the product of the partition functions corresponding to individual normal modes, that is,

$$j_{\text{vib}}(T) = \prod_i \frac{e^{-\Theta_i/2T}}{1 - e^{-\Theta_i/T}}; \quad \Theta_i = \frac{\hbar\omega_i}{k}, \tag{43}$$

[13]For example, the symmetry number γ for H_2O (isosceles triangle) is 2, for NH_3 (regular triangular pyramid) it is 3, while for CH_4 (tetrahedron) and C_6H_6 (regular hexagon) it is 12. For heteronuclear molecules, the symmetry number is unity.

[14]In the case of a collinear molecule, such as N_2O or CO_2, there are only two degrees of freedom for rotation; consequently, $j_{\text{rot}}^C(T)$ is given by $(2IkT/\hbar^2)$, where I is the (common) value of the two moments of inertia of the molecule; see equation (17). Of course, we must also take into account the symmetry number γ. In the examples quoted here, the molecule N_2O, being spatially asymmetric (NNO), has symmetry number 1, while the molecule CO_2, being spatially symmetric (OCO), has symmetry number 2.

[15]For example, of the four frequencies characterizing the normal modes of vibration of the collinear molecule OCO, two that correspond to the (transverse) bending modes, namely $\underset{\downarrow}{\mathrm{O}}\,\overset{\uparrow}{\mathrm{C}}\,\underset{\downarrow}{\mathrm{O}}$, are equal while the others that correspond to (longitudinal) oscillations along the molecular axis, namely $\leftarrow$O C$\rightarrow$ $\leftarrow$O and $\leftarrow$O C O$\rightarrow$, are different; see Problem 6.28.

and the vibrational specific heat is given by the sum of the contributions arising from the individual modes:

$$C_{\text{vib}} = Nk\sum_i \left\{ \left(\frac{\Theta_i}{T}\right)^2 \frac{e^{\Theta_i/T}}{(e^{\Theta_i/T}-1)^2} \right\}. \tag{44}$$

In general, the various Θ_i are of order 10^3 K; for instance, in the case of CO_2, which was cited in footnote 15, $\Theta_1 = \Theta_2 = 960$ K, $\Theta_3 = 1{,}990$ K, and $\Theta_4 = 3{,}510$ K. For temperatures large in comparison with all Θ_i, the specific heat would be given by the equipartition value, namely Nk for each of the normal modes. In practice, however, this limit can hardly be realized because the polyatomic molecules generally break up well before such high temperatures are reached. Secondly, the different frequencies ω_i of a polyatomic molecule are generally spread over a rather wide range of values. Consequently, as temperature rises, different modes of vibration get gradually "included" into the process; in between these "inclusions," the specific heat of the gas may stay constant over considerably large stretches of temperature.

6.6 Chemical equilibrium

The equilibrium amounts of chemicals in a chemical reaction are determined by the chemical potentials of each of the species. Consider the following chemical reaction between chemical species A and B to form species X and Y with stoichiometric coefficients ν_A, ν_B, ν_X, and ν_Y:

$$\nu_A A + \nu_B B \rightleftarrows \nu_X X + \nu_Y Y. \tag{1}$$

Each individual reaction that occurs changes the number of molecules of each species according to the stoichiometric coefficients. If the initial numbers of molecules of the species are N_A^0, N_B^0, N_X^0, and N_Y^0, then the numbers of each species after ΔN chemical reactions have occurred would be $N_A = N_A^0 - \nu_A \Delta N$, $N_B = N_B^0 - \nu_B \Delta N$, $N_X = N_X^0 + \nu_X \Delta N$, and $N_Y = N_Y^0 + \nu_Y \Delta N$. If $\Delta N > 0$, the reaction has proceeded in the positive direction increasing the numbers of X and Y. If $\Delta N < 0$, the reaction has proceeded in the direction of increasing the numbers of A and B. If the reaction takes place in a closed isothermal system with fixed pressure, the Gibbs free energy $G(N_A, N_B, N_X, N_Y, P, T)$ is changed by the amount

$$\Delta G = (-\nu_A\mu_A - \nu_B\mu_B + \nu_X\mu_X + \nu_Y\mu_Y)\Delta N, \tag{2}$$

where $\mu_A = \left(\frac{\partial G}{\partial N_A}\right)_{T,P}$ is the chemical potential of species A, and so on; see Sections 3.3, 4.7, and Appendix H. Since the Gibbs free energy decreases as a system approaches equilibrium, $\Delta G \leq 0$. When the system reaches chemical equilibrium, the Gibbs free energy reaches its minimum value so $\Delta G = 0$. This gives us the general relationship for chemical equilibrium of the reaction in equation (1), namely

$$\nu_A \mu_A + \nu_B \mu_B = \nu_X \mu_X + \nu_Y \mu_Y. \tag{3}$$

Note that if a chemical species that acts as a catalyst is added in equal amounts to both sides of equation (1), the equilibrium relation (3) is unaffected. Therefore, a catalyst may serve to increase the rate of approach toward equilibrium, without affecting the equilibrium condition itself.

If the free energy can be approximated as a sum of the free energies of the individual species such as in an ideal gas or a dilute solution, then we can derive a simple relation between the equilibrium densities of the species. Following from equations (3.5.10) and (6.5.4), the Helmholtz free energy of a classical ideal gas consisting of molecules with internal degrees of freedom can be written as

$$A(N,V,T) = N\varepsilon + NkT \ln\left(\frac{N\lambda^3}{V}\right) - NkT - NkT \ln j(T), \tag{4}$$

where ε is the ground state energy of the molecule, $\lambda = h/\sqrt{2\pi mkT}$ is the thermal deBroglie wavelength, and $j(T)$ is the partition function for the internal degrees of freedom of the molecule. This gives for the chemical potential of species A

$$\mu_A = \left(\frac{\partial A}{\partial N_A}\right)_{T,V} = \varepsilon_A + kT \ln\left(n_A \lambda_A^3\right) - kT \ln j_A(T), \tag{5}$$

where n_A is the number density of species A. The equilibrium condition then (3) gives

$$\frac{[X]^{\nu_X} [Y]^{\nu_Y}}{[A]^{\nu_A} [B]^{\nu_B}} = K(T) = \exp\left(-\beta \Delta\mu^{(0)}\right), \tag{6}$$

where $[A] = n_A/n_0$, and so on,

$$\Delta\mu^{(0)} = \nu_X \mu_X^{(0)} + \nu_Y \mu_Y^{(0)} - \nu_A \mu_A^{(0)} - \nu_B \mu_B^{(0)}, \tag{7a}$$

$$\mu_A^{(0)} = \varepsilon_A + kT \ln\left(n_0 \lambda_A^3\right) - kT \ln j_A(T), \text{ etc.} \tag{7b}$$

and $K(T)$ is the equilibrium constant.

Equation (6) is called the law of mass action. The quantity n_0 is a standard number density and $\mu_A^{(0)}$, and so on, are the chemical potentials of the species at temperature T and standard number density n_0. The quantity $\Delta\mu^{(0)}$ represents the Gibbs free energy change per chemical reaction at standard density. Note that the reaction constant $K(T)$ is a function only of the temperature and determines the densities of the components in equilibrium at temperature T through equation (6). The standard number density for gases is usually chosen to be the number density of an ideal gas at temperature T and standard pressure, that is, $n_0 = (1\,\text{atm})/kT$. The standard density for aqueous solutions is usually chosen to be one mole per liter. The Gibbs free energy in chemical tables is expressed relative to the standard states of the elements.

We now examine a specific example, the combustion of hydrocarbons with oxygen in an internal combustion engine. The reaction used to power clean buses and automobiles using natural gas is

$$CH_4 + 2O_2 \rightleftarrows CO_2 + 2H_2O\,. \tag{8}$$

The primary reaction products are carbon dioxide and water vapor but carbon monoxide is also produced by the reaction

$$2CH_4 + 3O_2 \rightleftarrows 2CO + 4H_2O\,. \tag{9}$$

A primary goal for a clean burning engine is to combust nearly all the hydrocarbon fuel while producing as little carbon monoxide as possible. By combining reactions (8) and (9), we get a direct reaction between carbon monoxide, oxygen, and carbon dioxide:

$$2CO + O_2 \rightleftarrows 2CO_2\,. \tag{10}$$

Equation (6) now gives the equilibrium ratio of CO to CO_2 as

$$\frac{[CO]}{[CO_2]} = \sqrt{\frac{1}{K(T)[O_2]}}\,. \tag{11}$$

At $T \approx 1{,}500\,K$, as combustion occurs inside the cylinder of the engine, the equilibrium constant $K \approx 10^{10}$ so carbon monoxide is present as a combustion product in the few parts per million range — combustion reactions that are not in equilibrium can have CO concentrations well above the equilibrium value. The exhaust gases cool quickly during the power stroke of the engine. As these gases exit the exhaust at $T \approx 600\,K$, the equilibrium constant $K \approx 10^{40}$, which should result in almost no carbon monoxide in the exhaust stream. However, the reaction rate is typically too slow to keep the CO concentration in chemical equilibrium during the rapid cooling, so the amount of CO present in the exhaust stream remains close to the larger value determined at the higher temperature.[16] Fortunately, the leftover carbon monoxide can be converted into carbon dioxide at the exhaust temperature in a catalytic converter that uses platinum and palladium as catalysts to increase the reaction rate. Equation (11) indicates that the carbon monoxide fraction is reduced by increasing the amount of oxygen present in the reaction. This is accomplished by running the engine with a hydrocarbon/air ratio that is a little bit short of the stoichiometric point of equation (8). This reduces the amount of CO left from the combustion itself and also leaves excess O_2 in the exhaust stream for use in the catalytic converter.

[16]Very similar effects happened during the early stages of the universe as the temperature cooled but the cooling rate was too rapid for some constituents to remain in thermal equilibrium; see Chapter 9.

Problems

6.1. Show that the entropy of an ideal gas in thermal equilibrium is given by the formula

$$S = k \sum_{\varepsilon} [\langle n_\varepsilon + 1\rangle \ln\langle n_\varepsilon + 1\rangle - \langle n_\varepsilon\rangle \ln\langle n_\varepsilon\rangle]$$

in the case of *bosons* and by the formula

$$S = k \sum_{\varepsilon} [-\langle 1 - n_\varepsilon\rangle \ln\langle 1 - n_\varepsilon\rangle - \langle n_\varepsilon\rangle \ln\langle n_\varepsilon\rangle]$$

in the case of *fermions*. Verify that these results are consistent with the general formula

$$S = -k \sum_{\varepsilon} \left\{ \sum_{n} p_\varepsilon(n) \ln p_\varepsilon(n) \right\},$$

where $p_\varepsilon(n)$ is the probability that there are exactly n particles in the energy state ε.

6.2. Derive, for all three statistics, the relevant expressions for the quantity $\langle n_\varepsilon^2\rangle - \langle n_\varepsilon\rangle^2$ from the respective probabilities $p_\varepsilon(n)$. Show that, quite generally,

$$\langle n_\varepsilon^2\rangle - \langle n_\varepsilon\rangle^2 = kT \left(\frac{\partial \langle n_\varepsilon\rangle}{\partial \mu} \right)_T;$$

compare with the corresponding result, (4.5.3), for a system embedded in a grand canonical ensemble.

6.3. Refer to Section 6.2 and show that, if the occupation number n_ε of an energy level ε is restricted to the values $0, 1, \ldots, l$, then the mean occupation number of that level is given by

$$\langle n_\varepsilon\rangle = \frac{1}{z^{-1}e^{\beta\varepsilon} - 1} - \frac{l+1}{(z^{-1}e^{\beta\varepsilon})^{l+1} - 1}.$$

Check that while $l = 1$ leads to $\langle n_\varepsilon\rangle_{\text{F.D.}}$, $l \to \infty$ leads to $\langle n_\varepsilon\rangle_{\text{B.E.}}$.

6.4. The potential energy of a system of charged particles, characterized by particle charge e and number density $n(\boldsymbol{r})$, is given by

$$U = \frac{e^2}{2} \iint \frac{n(\boldsymbol{r})n(\boldsymbol{r}')}{|\boldsymbol{r} - \boldsymbol{r}'|} d\boldsymbol{r}d\boldsymbol{r}' + e \int n(\boldsymbol{r})\phi_{\text{ext}}(\boldsymbol{r})d\boldsymbol{r},$$

where $\phi_{\text{ext}}(\boldsymbol{r})$ is the potential of an external electric field. Assume that the entropy of the system, apart from an additive constant, is given by the formula

$$S = -k \int n(\boldsymbol{r}) \ln n(\boldsymbol{r}) d\boldsymbol{r};$$

compare to formula (3.3.13). Using these expressions, derive the equilibrium equations satisfied by the number density $n(\boldsymbol{r})$ and the total potential $\phi(\boldsymbol{r})$, the latter being

$$\phi_{\text{ext}}(\boldsymbol{r}) + e \int \frac{n(\boldsymbol{r}')}{|\boldsymbol{r} - \boldsymbol{r}'|} d\boldsymbol{r}'.$$

6.5. Show that the root-mean-square deviation in the molecular energy ε, in a system obeying Maxwell–Boltzmann distribution, is $\sqrt{(2/3)}$ times the mean molecular energy $\bar{\varepsilon}$. Compare this result with that of Problem 3.18.

6.6. Show that, for any law of distribution of molecular speeds,

$$\left\{ \langle u\rangle \left\langle \frac{1}{u} \right\rangle \right\} \geq 1.$$

Check that the value of this quantity for the Maxwellian distribution is $4/\pi$.

6.7. Through a small window in a furnace, which contains a gas at a high temperature T, the spectral lines emitted by the gas molecules are observed. Because of molecular motions, each spectral line exhibits *Doppler broadening*. Show that the variation of the relative intensity $I(\lambda)$ with wavelength λ in a line is given by

$$I(\lambda) \propto \exp\left\{-\frac{mc^2(\lambda-\lambda_0)^2}{2\lambda_0^2 kT}\right\},$$

where m is the molecular mass, c the speed of light, and λ_0 the mean wavelength of the line.

6.8. An ideal classical gas composed of N particles, each of mass m, is enclosed in a vertical cylinder of height L placed in a uniform gravitational field (of acceleration g) and is in thermal equilibrium; ultimately, both N and $L \to \infty$. Evaluate the partition function of the gas and derive expressions for its major thermodynamic properties. Explain why the specific heat of this system is larger than that of a corresponding system in free space.

6.9. *Centrifuge-based uranium enrichment*: Natural uranium is composed of two isotopes: ^{238}U and ^{235}U, with percentages of 99.27% and 0.72%, respectively. If uranium hexafluoride gas UF_6 is injected into a rapidly spinning hollow metal cylinder with inner radius R, the equilibrium pressure of the gas is largest at the inner radius and isotopic concentration differences between the axis and the inner radius allow enrichment of the concentration of ^{235}U.

(a) Write down the Lagrangian $\mathcal{L}(\{q_k, \dot{q}_k\})$ for particles of mass m moving in a cylindrical coordinate system rotating at angular velocity ω and use a Legendre transformation

$$\mathcal{H}(\{q_k, p_k\}) = \sum_k p_k \dot{q}_k - \mathcal{L},$$

to show that the one-particle Hamiltonian $\mathcal{H}$ in that cylindrical coordinate system is

$$\mathcal{H}(r,\theta,z,p_r,p_\theta,p_z) = \frac{p_r^2}{2m} + \frac{(p_\theta^2 - mr^2\omega)^2}{2mr^2} + \frac{p_z^2}{2m}.$$

Ignore the internal degrees of freedom of the molecules since they will not affect the density as a function of position. Show that the one-particle partition function shown here can be written as

$$Q_1(V,T) = \frac{1}{h^3}\int_{-\infty}^{\infty} dp_r \int_{-\infty}^{\infty} dp_\theta \int_{-\infty}^{\infty} dp_z \int_0^R dr \int_0^{2\pi} d\theta \int_0^H dz \exp(-\beta\mathcal{H}),$$

by constructing the Jacobian of transformation between the cartesian and the cylindrical coordinates for the phase space integral. Evaluate the partition function Q_1 in a closed form and determine the Helmholtz free energy of this system.

(b) Determine the number density $n(r)$ as a function of the distance r from the axis for the N molecules of gas in the rotating cylinder. Show that, in the limit $\omega \to 0$, the density becomes uniform with the value $n = N/\pi R^2 H$. Find an expression for the ratio of the pressure at the inner radius of the cylinder R to the pressure at the axis of the cylinder as a function of ω and R.

(c) Evaluate the pressure ratios for the two isotopically different UF_6 gases at room temperature for the case $\omega R = 500\,\text{m/s}$. Show that the pressure ratio for ^{238}U is approximately 20% larger than the pressure ratio for ^{235}U so that extracting gas near the axis results in an enriched concentration of ^{235}U. A series of centrifuges can be used to raise the concentration of ^{235}U to create a fissionable grade of uranium for use in power-generating reactors or in nuclear weapons. Not surprisingly, this technology is a major concern for possible nuclear proliferation.

6.10. **(a)** Show that, if the temperature is uniform, the pressure of a classical gas in a uniform gravitational field decreases with height according to the *barometric formula*

$$P(z) = P(0) \exp\{-mgz/kT\},$$

where the various symbols have their usual meanings.[17]

(b) Derive the corresponding formula for an *adiabatic* atmosphere, that is, the one in which (PV^γ), rather than (PV), stays constant. Also study the variation, with height, of the temperature T and the density n in such an atmosphere.

6.11. **(a)** Show that the momentum distribution of particles in a *relativistic* Boltzmannian gas, with $\varepsilon = c(p^2 + m_0^2c^2)^{1/2}$, is given by

$$f(\boldsymbol{p})d\boldsymbol{p} = Ce^{-\beta c(p^2+m_0^2c^2)^{1/2}}p^2\,dp,$$

with the normalization constant

$$C = \frac{\beta}{m_0^2 c K_2(\beta m_0 c^2)},$$

$K_\nu(z)$ being a modified Bessel function.

(b) Check that in the nonrelativistic limit ($kT \ll m_0c^2$) we recover the Maxwellian distribution,

$$f(\boldsymbol{p})d\boldsymbol{p} = \left(\frac{\beta}{2\pi m_0}\right)^{3/2} e^{-\beta p^2/2m_0}(4\pi p^2\,dp),$$

while in the extreme relativistic limit ($kT \gg m_0c^2$) we obtain

$$f(\boldsymbol{p})d\boldsymbol{p} = \frac{(\beta c)^3}{8\pi} e^{-\beta pc}(4\pi p^2\,dp).$$

(c) Verify that, quite generally,

$$\langle pu \rangle = 3kT.$$

6.12. **(a)** Considering the loss of translational energy suffered by the molecules of a gas on reflection from a *receding* wall, derive, for a quasistatic adiabatic expansion of an ideal nonrelativistic gas, the well-known relation

$$PV^\gamma = \text{const.},$$

where $\gamma = (3a+2)/3a$, a being the ratio of the total energy to the translational energy of the gas.

(b) Show that, in the case of an extreme relativistic gas, $\gamma = (3a+1)/3a$.

6.13. **(a)** Determine the number of impacts made by gas molecules on a unit area of the wall in a unit time for which the angle of incidence lies between θ and $\theta + d\theta$.

(b) Determine the number of impacts made by gas molecules on a unit area of the wall in a unit time for which the speed of the molecules lies between u and $u + du$.

(c) A molecule AB dissociates if it hits the surface of a solid catalyst with a normal translational energy greater than 10^{-19} J. Show that the rate of the dissociative reaction $AB \to A + B$ is more than doubled by raising the temperature of the gas from 300 K to 310 K.

6.14. Consider the effusion of molecules of a Maxwellian gas through an opening of area a in the walls of a vessel of volume V.

(a) Show that, while the molecules inside the vessel have a mean kinetic energy $\frac{3}{2}kT$, the effused ones have a mean kinetic energy $2kT$, T being the *quasistatic* equilibrium temperature of the gas.

[17] This formula was first given by Boltzmann (1879). For a critical study of its derivation, see Walton (1969).

(b) Assuming that the effusion is so slow that the gas inside is always in a state of quasistatic equilibrium, determine the manner in which the density, the temperature, and the pressure of the gas vary with time.

6.15. A polyethylene balloon at an altitude of 30,000 m is filled with helium gas at a pressure of 10^{-2} atm and a temperature of 300 K. The balloon has a diameter of 10 m, and has numerous pinholes of diameter 10^{-5} m each. How many pinholes per square meter of the surface of the balloon must there be if 1 percent of the gas were to leak out in 1 hour?

6.16. Consider two Boltzmannian gases A and B, at pressures P_A and P_B and temperatures T_A and T_B, respectively, contained in two regions of space that communicate through a very narrow opening in the partitioning wall; see Figure 6.8. Show that the dynamic equilibrium resulting from the mutual effusion of the two kinds of molecules satisfies the condition

$$P_A/P_B = (m_A T_A/m_B T_B)^{1/2},$$

rather than $P_A = P_B$ (which would be the case if the equilibrium had resulted from a hydrodynamic flow).

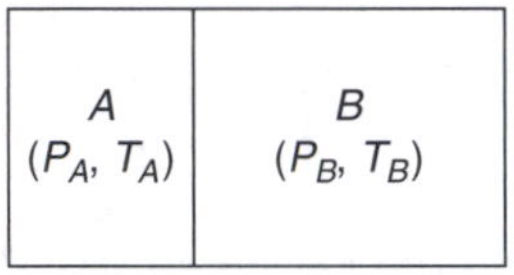

FIGURE 6.8 The molecules of the gases A and B undergoing a two-way effusion.

6.17. A *small* sphere, with initial temperature T, is immersed in an ideal Boltzmannian gas at temperature T_0. Assuming that the molecules incident on the sphere are first absorbed and then reemitted with the temperature of the sphere, determine the variation of the temperature of the sphere with time.
[*Note*: The radius of the sphere may be assumed to be much smaller than the mean free path of the molecules.]

6.18. Show that the mean value of the *relative* speed of two molecules in a Maxwellian gas is $\sqrt{2}$ times the mean speed of a molecule with respect to the walls of the container.
[*Note*: A similar result for the root-mean-square speeds (instead of the mean speeds) holds under much more general conditions.]

6.19. What is the probability that two molecules picked at random from a Maxwellian gas will have a total energy between E and $E + dE$? Verify that $\langle E \rangle = 3kT$.

6.20. The energy difference between the lowest electronic state $^1S_0$ and the first excited state $^3S_1$ of the helium atom is 159,843 cm^{-1}. Evaluate the relative fraction of the excited atoms in a sample of helium gas at a temperature of 6000 K.

6.21. Derive an expression for the equilibrium constant $K(T)$ for the reaction $H_2 + D_2 \leftrightarrow 2HD$ at temperatures high enough to allow classical approximation for the rotational motion of the molecules. Show that $K(\infty) = 4$.

6.22. With the help of the Euler–Maclaurin formula (6.5.19), derive high-temperature expansions for r_{even} and r_{odd}, as defined by equations (6.5.29) and (6.5.30), and obtain corresponding expansions for C_{even} and C_{odd}, as defined by equation (6.5.39). Compare the mathematical trend of these results with the nature of the corresponding curves in Figure 6.7. Also study the low-temperature behavior of the two specific heats and once again compare your results with the relevant parts of the aforementioned curves.

6.23. The potential energy between the atoms of a hydrogen molecule is given by the (semiempirical) *Morse potential*

$$V(r) = V_0\{e^{-2(r-r_0)/a} - 2e^{-(r-r_0)/a}\},$$

where $V_0 = 7 \times 10^{-12}$ erg, $r_0 = 8 \times 10^{-9}$ cm, and $a = 5 \times 10^{-9}$ cm. Evaluate the rotational and vibrational quanta of energy, and estimate the temperatures at which the rotational and vibrational modes of the molecules would begin to contribute toward the specific heat of the hydrogen gas.

6.24. Show that the fractional change in the equilibrium value of the internuclear distance of a diatomic molecule, as a result of rotation, is given by

$$\frac{\Delta r_0}{r_0} \simeq \left(\frac{\hbar}{\mu r_0^2 \omega}\right)^2 J(J+1) = 4\left(\frac{\Theta_r}{\Theta_v}\right)^2 J(J+1);$$

here, ω is the angular frequency of the vibrational state in which the molecule happens to be. Estimate the numerical value of this fraction in a typical case.

6.25. The ground state of an oxygen atom is a triplet, with the following *fine structure*:

$$\varepsilon_{J=2} = \varepsilon_{J=1} - 158.5\ \text{cm}^{-1} = \varepsilon_{J=0} - 226.5\ \text{cm}^{-1}.$$

Calculate the relative fractions of the atoms occupying different J-levels in a sample of atomic oxygen at 300 K.

6.26. Calculate the contribution of the first excited electronic state, namely $^1\Delta$ with $g_e = 2$, of the O_2 molecule toward the Helmholtz free energy and the specific heat of oxygen gas at a temperature of 5000 K; the separation of this state from the ground state, namely $^3\Sigma$ with $g_e = 3$, is 7824 cm^{-1}. How would these results be affected if the parameters Θ_r and Θ_v of the O_2 molecule had different values in the two electronic states?

6.27. The rotational kinetic energy of a rotator with three degrees of freedom can be written as

$$\varepsilon_{\text{rot}} = \frac{M_\xi^2}{2I_1} + \frac{M_\eta^2}{2I_2} + \frac{M_\zeta^2}{2I_3},$$

where (ξ, η, ζ) are coordinates in a rotating frame of reference whose axes coincide with the principal axes of the rotator, while (M_ξ, M_η, M_ζ) are the corresponding angular momenta. Carrying out integrations in the phase space of the rotator, derive expression (6.5.41) for the partition function $j_{\text{rot}}(T)$ in the classical approximation.

6.28. Determine the translational, rotational, and vibrational contributions toward the molar entropy and the molar specific heat of carbon dioxide at NTP. Assume the ideal-gas formulae and use the following data: molecular weight $M = 44.01$; moment of inertia I of a CO_2 molecule $= 71.67 \times 10^{-40}$ g cm^2; wave numbers of the various modes of vibration: $\bar{\nu}_1 = \bar{\nu}_2 = 667.3$ cm^{-1}, $\bar{\nu}_3 = 1383.3$ cm^{-1}, and $\bar{\nu}_4 = 2439.3$ cm^{-1}.

6.29. Determine the molar specific heat of ammonia at a temperature of 300 K. Assume the ideal-gas formula and use the following data: the principal moments of inertia: $I_1 = 4.44 \times 10^{-40}$ g cm^2, $I_2 = I_3 = 2.816 \times 10^{-40}$ g cm^2; wave numbers of the various modes of vibration: $\bar{\nu}_1 = \bar{\nu}_2 = 3336$ cm^{-1}, $\bar{\nu}_3 = \bar{\nu}_4 = 950$ cm^{-1}, $\bar{\nu}_5 = 3414$ cm^{-1}, and $\bar{\nu}_6 = 1627$ cm^{-1}.

6.30. Derive the equilibrium concentration equation (6.6.6) from the equilibrium condition (6.6.3).

6.31. Use the following values to determine the equilibrium constant for the reaction $2CO + O_2 \rightleftarrows 2CO_2$. At a combustion temperature of $T = 1500$ K: $\beta\mu^{(0)}_{CO_2} = -60.95$, $\beta\mu^{(0)}_{CO} = -35.18$, and $\beta\mu^{(0)}_{O_2} = -27.08$. Use this data to compute the fraction $[CO]/[CO_2]$ for the case of $[O_2] = 0.01$. Repeat for a catalytic converter temperature of $T = 600$ K, where $\beta\mu^{(0)}_{CO_2} = -103.45$, $\beta\mu^{(0)}_{CO} = -45.38$, and $\beta\mu^{(0)}_{O_2} = -23.49$.

6.32. Derive an expression for the equilibrium constant $K(T)$ for the reaction $N_2 + O_2 \rightleftarrows 2NO$ in terms of the ground state energy change $\Delta\varepsilon_0 = 2\varepsilon_{NO} - \varepsilon_{N_2} - \varepsilon_{O_2}$ and the vibrational and rotational partition functions of the diatomic molecules, using results from Section 6.5. Give predictions for the ranges of temperatures where the rotational modes are classically excited but the vibration modes are suppressed and for higher temperatures where both the rotational and vibrational models are classically excited.

6.33. Analyze the combustion reaction

$$CH_4 + 2O_2 \rightleftarrows CO_2 + 2H_2O, \tag{6.6.8}$$

assuming that at combustion temperatures the equilibrium constant $K(T) \gg 1$. Show that conducting combustion at the stoichiometric point or just a bit short of the stoichiometric point (so there is enough oxygen to oxidize all of the methane) will lead to low amounts of CH_4 in the exhaust. Determine the equilibrium amount of CH_4 in terms of the initial excess amount of O_2. Determine the equilibrium constant at $T = 1500\,\text{K}$ from the data $\beta\mu^{(0)}_{CO_2} = -60.95$, $\beta\mu^{(0)}_{O_2} = -27.08$, $\beta\mu^{(0)}_{CH_4} = -31.95$, and $\beta\mu^{(0)}_{H_2O} = -44.62$.

6.34. Determine the equilibrium ionization fraction for the reaction

$$\text{Na} \rightleftarrows \text{Na}^+ + e^-$$

in a sodium vapor. Treat all three species as ideal classical monatomic gases. The ionization energy of sodium is 5.139 eV, Na^+ ions are spin-zero, and neutral Na and free e^- are both spin-$\frac{1}{2}$. Derive the Saha equation for the ionized fraction $[\text{Na}^+]/([\text{Na}] + [\text{Na}^+])$ for a neutral plasma as a function of temperature at a fixed total density. Plot the ionized fraction as a function of temperature for some chosen total density.

[Note that, this calculation is very similar to the one concerning ionized hydrogen fraction as a function of temperature during the recombination era in the early universe; see Section 9.8.]

7 Ideal Bose Systems

In continuation of Sections 6.1 through 6.3, we shall now investigate in detail the physical behavior of a class of systems in which, while the intermolecular interactions are still negligible, the effects of quantum statistics (which arise from the indistinguishability of the particles) assume an increasingly important role. This means that the temperature T and the particle density n of the system no longer conform to the criterion

$$n\lambda^3 \equiv \frac{nh^3}{(2\pi mkT)^{3/2}} \ll 1, \tag{5.5.20}$$

where $\lambda\{\equiv h/(2\pi mkT)^{1/2}\}$ is the *mean thermal wavelength* or *thermal deBroglie wavelength* of the particles. In fact, the quantity $n\lambda^3$ turns out to be a very appropriate parameter, in terms of which the various physical properties of the system can be adequately expressed. In the limit $n\lambda^3 \to 0$, all physical properties go over smoothly to their classical counterparts. For small, but not negligible, values of $n\lambda^3$, the various quantities pertaining to the system can be expanded as power series in this parameter; from these expansions one obtains the first glimpse of the manner in which departure from classical behavior sets in. When $n\lambda^3$ becomes of the order of unity, the behavior of the system becomes significantly different from the classical one and is characterized by quantum effects. A study of the system under these circumstances brings us face to face with a set of phenomena unknown in classical statistics.

It is evident that a system is more likely to display quantum behavior when it is at a relatively low temperature and/or has a relatively high density of particles.[1] Moreover, the smaller the particle mass the larger the quantum effects.

Now, when $n\lambda^3$ is of the order of unity, then not only does the behavior of a system exhibit significant departure from typical classical behavior but it is also influenced by whether the particles constituting the system obey Bose–Einstein statistics or Fermi–Dirac statistics. Under these circumstances, the properties of the two kinds of systems are themselves very different. In the present chapter we consider systems belonging to the first category while the succeeding chapter will deal with systems belonging to the second category.

[1] Actually it is the ratio $n/T^{3/2}$, rather than the quantities n and T separately, that determines the degree of degeneracy in a given system. For instance, white dwarf stars, even at temperatures of order 10^7 K, constitute statistically degenerate systems; see Section 8.5.

© 2011 Elsevier Ltd. All rights reserved.

7.1 Thermodynamic behavior of an ideal Bose gas

We obtained, in Sections 6.1 and 6.2, the following formulae for an ideal Bose gas:

$$\frac{PV}{kT} \equiv \ln \mathcal{Q} = -\sum_{\varepsilon} \ln(1 - ze^{-\beta\varepsilon}) \tag{1}$$

and

$$N \equiv \sum_{\varepsilon} \langle n_{\varepsilon} \rangle = \sum_{\varepsilon} \frac{1}{z^{-1}e^{\beta\varepsilon} - 1}, \tag{2}$$

where $\beta = 1/kT$, while z is the fugacity of the gas which is related to the chemical potential μ through the formula

$$z \equiv \exp(\mu/kT); \tag{3}$$

as noted earlier, $ze^{-\beta\varepsilon}$, for all ε, is less than unity. In view of the fact that, for large V, the spectrum of the single-particle states is almost a continuous one, the summations on the right sides of equations (1) and (2) may be replaced by integrations. In doing so, we make use of the asymptotic expression (2.4.7) for the nonrelativistic density of states $a(\varepsilon)$ in the neighborhood of a given energy ε, namely[2]

$$a(\varepsilon)d\varepsilon = (2\pi V/h^3)(2m)^{3/2}\varepsilon^{1/2}d\varepsilon. \tag{4}$$

We, however, note that by substituting this expression into our integrals we are inadvertently giving a weight *zero* to the energy level $\varepsilon = 0$. This is wrong because in a quantum-mechanical treatment we must give a statistical weight unity to each nondegenerate single-particle state in the system. It is, therefore, advisable to take this particular state out of the sum in question before carrying out the integration; for a rigorous justification of this (unusual) step, see Appendix F. We thus obtain

$$\frac{P}{kT} = -\frac{2\pi}{h^3}(2m)^{3/2}\int_0^{\infty} \varepsilon^{1/2} \ln(1 - ze^{-\beta\varepsilon})d\varepsilon - \frac{1}{V}\ln(1 - z) \tag{5}$$

and

$$\frac{N}{V} = \frac{2\pi}{h^3}(2m)^{3/2}\int_0^{\infty} \frac{\varepsilon^{1/2}d\varepsilon}{z^{-1}e^{\beta\varepsilon} - 1} + \frac{1}{V}\frac{z}{1 - z}; \tag{6}$$

of course, the lower limit of these integrals can still be taken as 0, because the state $\varepsilon = 0$ is not going to contribute toward them anyway.

Before proceeding further, a word about the relative importance of the last terms in equations (5) and (6). For $z \ll 1$, which corresponds to situations not far removed from

[2]The theory of this section is restricted to a system of *nonrelativistic* particles. For the more general case, see Kothari and Singh (1941) and Landsberg and Dunning-Davies (1965).

the classical limit, each of these terms is of order $1/N$ and, therefore, negligible. However, as z increases and assumes values close to unity, the term $z/(1-z)V$ in (6), which is identically equal to N_0/V (N_0 being the number of particles in the ground state $\varepsilon = 0$), can well become a significant fraction of the quantity N/V; this accumulation of a macroscopic fraction of the particles into a single state $\varepsilon = 0$ leads to the phenomenon of *Bose–Einstein condensation*. Nevertheless, since $z/(1-z) = N_0$ and hence $z = N_0/(N_0+1)$, the term $\{-V^{-1}\ln(1-z)\}$ in (5) is equal to $\{V^{-1}\ln(N_0+1)\}$, which is at most $O(N^{-1}\ln N)$; this term is, therefore, negligible for *all* values of z and hence may be dropped altogether.

We now obtain from equations (5) and (6), on substituting $\beta\varepsilon = x$,

$$\frac{P}{kT} = -\frac{2\pi(2mkT)^{3/2}}{h^3}\int_0^\infty x^{1/2}\ln(1-ze^{-x})dx = \frac{1}{\lambda^3}g_{5/2}(z) \tag{7}$$

and

$$\frac{N-N_0}{V} = \frac{2\pi(2mkT)^{3/2}}{h^3}\int_0^\infty \frac{x^{1/2}dx}{z^{-1}e^x - 1} = \frac{1}{\lambda^3}g_{3/2}(z), \tag{8}$$

where

$$\lambda = h/(2\pi mkT)^{1/2}, \tag{9}$$

while $g_\nu(z)$ are *Bose–Einstein functions* defined by, see Appendix D,

$$g_\nu(z) = \frac{1}{\Gamma(\nu)}\int_0^\infty \frac{x^{\nu-1}dx}{z^{-1}e^x - 1} = z + \frac{z^2}{2^\nu} + \frac{z^3}{3^\nu} + \cdots; \tag{10}$$

note that to write (7) in terms of the function $g_{5/2}(z)$ we first carried out an integration by parts. Equations (7) and (8) are our basic results; on elimination of z, they would give us the *equation of state* of the system.

The internal energy of this system is given by

$$\begin{aligned} U \equiv -\left(\frac{\partial}{\partial\beta}\ln\mathcal{Q}\right)_{z,V} &= kT^2\left\{\frac{\partial}{\partial T}\left(\frac{PV}{kT}\right)\right\}_{z,V} \\ &= kT^2 V g_{5/2}(z)\left\{\frac{d}{dT}\left(\frac{1}{\lambda^3}\right)\right\} = \frac{3}{2}kT\frac{V}{\lambda^3}g_{5/2}(z); \end{aligned} \tag{11}$$

here, use has been made of equation (7) and of the fact that $\lambda \propto T^{-1/2}$. Thus, quite generally, our system satisfies the relationship

$$P = \frac{2}{3}(U/V). \tag{12}$$

For small values of z, we can make use of expansion (10); at the same time, we can neglect N_0 in comparison with N. An elimination of z between equations (7) and (8) can then be carried out by first inverting the series in (8) to obtain an expansion for z in powers of $n\lambda^3$ and then substituting this expansion into the series appearing in (7). The equation of state thereby takes the form of the *virial expansion*,

$$\frac{PV}{NkT} = \sum_{l=1}^{\infty} a_l \left(\frac{\lambda^3}{v}\right)^{l-1}, \tag{13}$$

where $v(\equiv 1/n)$ is the volume per particle; the coefficients a_l, which are referred to as the *virial coefficients* of the system, turn out to be

$$\left.\begin{aligned} a_1 &= 1, \\ a_2 &= -\frac{1}{4\sqrt{2}} = -0.17678, \\ a_3 &= -\left(\frac{2}{9\sqrt{3}} - \frac{1}{8}\right) = -0.00330, \\ a_4 &= -\left(\frac{3}{32} + \frac{5}{32\sqrt{2}} - \frac{1}{2\sqrt{6}}\right) = -0.00011, \end{aligned}\right\} \tag{14}$$

and so on. For the specific heat of the gas, we obtain

$$\begin{aligned} \frac{C_V}{Nk} &\equiv \frac{1}{Nk}\left(\frac{\partial U}{\partial T}\right)_{N,V} = \frac{3}{2}\left\{\frac{\partial}{\partial T}\left(\frac{PV}{Nk}\right)\right\}_v \\ &= \frac{3}{2}\sum_{l=1}^{\infty} \frac{5-3l}{2} a_l \left(\frac{\lambda^3}{v}\right)^{l-1} \\ &= \frac{3}{2}\left[1 + 0.0884\left(\frac{\lambda^3}{v}\right) + 0.0066\left(\frac{\lambda^3}{v}\right)^2 + 0.0004\left(\frac{\lambda^3}{v}\right)^3 + \cdots\right]. \end{aligned} \tag{15}$$

As $T \to \infty$ (and hence $\lambda \to 0$), both the pressure and the specific heat of the gas approach their classical values, namely nkT and $\frac{3}{2}Nk$, respectively. We also note that at finite, but large, temperatures the specific heat of the gas is larger than its limiting value; in other words, the (C_V, T)-curve has a negative slope at high temperatures. On the other hand, as $T \to 0$, the specific heat must go to zero. Consequently, it must pass through a maximum somewhere. As seen later, this maximum is in the nature of a cusp that appears at a critical temperature T_c; the derivative of the specific heat is found to be discontinuous at this temperature (see Figure 7.4 later in this section).

As the temperature of the system falls (and the value of the parameter λ^3/v grows), expansions such as (13) and (15) do not remain useful. We then have to work with formulae (7), (8), and (11) as such. The precise value of z is now obtained from equation (8), which may be rewritten as

$$N_e = V\frac{(2\pi mkT)^{3/2}}{h^3} g_{3/2}(z), \tag{16}$$

where N_e is the number of particles in the excited states ($\varepsilon \neq 0$); of course, unless z gets extremely close to unity, $N_e \simeq N$.[3] It is obvious that, for $0 \leq z \leq 1$, the function $g_{3/2}(z)$ increases monotonically with z and is *bounded*, its largest value being

$$g_{3/2}(1) = 1 + \frac{1}{2^{3/2}} + \frac{1}{3^{3/2}} + \cdots \equiv \zeta\left(\frac{3}{2}\right) \simeq 2.612; \tag{17}$$

see equation (D.5) in Appendix D. Hence, for all z of interest,

$$g_{3/2}(z) \leq \zeta\left(\frac{3}{2}\right). \tag{18}$$

Consequently, for given V and T, the total (equilibrium) number of particles in all the excited states taken together is also bounded, that is,

$$N_e \leq V \frac{(2\pi mkT)^{3/2}}{h^3} \zeta\left(\frac{3}{2}\right). \tag{19}$$

Now, so long as the actual number of particles in the system is less than this limiting value, everything is well and good; practically all the particles in the system are distributed over the excited states and the precise value of z is determined by equation (16), with $N_e \simeq N$. However, if the actual number of particles exceeds this limiting value, then it is natural that the excited states will receive as many of them as they can hold, namely

$$N_e = V \frac{(2\pi mkT)^{3/2}}{h^3} \zeta\left(\frac{3}{2}\right), \tag{20}$$

while the rest will be pushed *en masse* into the ground state $\varepsilon = 0$ (whose capacity, under all circumstances, is essentially unlimited):

$$N_0 = N - \left\{ V \frac{(2\pi mkT)^{3/2}}{h^3} \zeta\left(\frac{3}{2}\right) \right\}. \tag{21}$$

The precise value of z is now determined by the formula

$$z = \frac{N_0}{N_0 + 1} \simeq 1 - \frac{1}{N_0} \tag{22}$$

which, for all practical purposes, is unity. This curious phenomenon of a macroscopically large number of particles accumulating in a single quantum state ($\varepsilon = 0$) is generally referred to as the phenomenon of *Bose–Einstein condensation*. In a certain sense, this phenomenon is akin to the familiar process of a vapor condensing into the liquid state, which takes place in the ordinary physical space. Conceptually, however, the two processes are very different. Firstly, the phenomenon of Bose–Einstein condensation is purely

[3] Remember that the largest value z can have *in principle* is unity. In fact, as $T \to 0$, $z = N_0/(N_0 + 1) \to N/(N+1)$, which is very nearly unity (but certainly on the right side of it).

of quantum origin (occurring even in the absence of intermolecular forces); secondly, it takes place at best in the momentum space and not in the coordinate space.[4]

The condition for the onset of Bose–Einstein condensation is

$$N > VT^{3/2}\frac{(2\pi mk)^{3/2}}{h^3}\zeta\left(\frac{3}{2}\right) \tag{23}$$

or, if we hold N and V constant and vary T,

$$T < T_c = \frac{h^2}{2\pi mk}\left\{\frac{N}{V\zeta\left(\frac{3}{2}\right)}\right\}^{2/3}; \tag{24^5}$$

here, T_c denotes a characteristic temperature that depends on the particle mass m and the particle density N/V in the system. Accordingly, for $T < T_c$, the system may be looked on as a mixture of two "phases":

(i) a *normal* phase, consisting of $N_e\,\{= N(T/T_c)^{3/2}\}$ particles distributed over the excited states ($\varepsilon \neq 0$), and
(ii) a *condensed* phase, consisting of $N_0\,\{= (N - N_e)\}$ particles accumulated in the ground state ($\varepsilon = 0$).

Figure 7.1 shows the manner in which the complementary fractions (N_e/N) and (N_0/N) vary with T. For $T > T_c$, we have the normal phase alone; the number of particles in the ground state, namely $z/(1-z)$, is $O(1)$, which is completely negligible in comparison with the total number N. Clearly, the situation is *singular* at $T = T_c$. For later reference, we note that, at $T \to T_c$ from below, the condensate fraction vanishes as follows:

$$\frac{N_0}{N} = 1 - \left(\frac{T}{T_c}\right)^{3/2} \approx \frac{3}{2}\frac{T_c - T}{T_c}. \tag{25}$$

A knowledge of the variation of z with T is also of interest here. It is, however, simpler to consider the variation of z with (v/λ^3), the latter being proportional to $T^{3/2}$. For $0 \le (v/\lambda^3) \le (2.612)^{-1}$, which corresponds to $0 \le T \le T_c$, the parameter $z \simeq 1$; see equation (22). For $(v/\lambda^3) > (2.612)^{-1}$, $z < 1$ and is determined by the relationship

$$g_{3/2}(z) = (\lambda^3/v) < 2.612; \tag{26^6}$$

[4]Of course, the repercussions of this phenomenon in the coordinate space are no less curious. It prepares the stage for the onset of *superfluidity*, a quantum manifestation discussed in Section 7.6.

[5]For a rigorous discussion of the onset of Bose–Einstein condensation, see Landsberg (1954b), where an attempt has also been made to coordinate much of the previously published work on this topic. For a more recent study, see Greenspoon and Pathria (1974), Pathria (1983), and Appendix F.

[6]An equivalent relationship is $g_{3/2}(z)/g_{3/2}(1) = (T_c/T)^{3/2} < 1$.

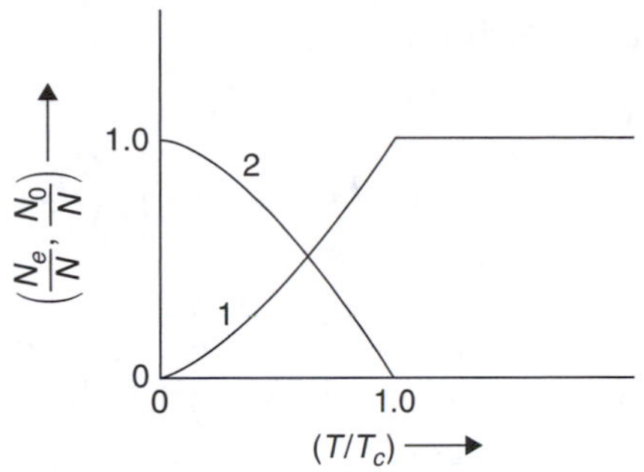

FIGURE 7.1 Fractions of the normal phase and the condensed phase in an ideal Bose gas as a function of the temperature parameter (T/T_c).

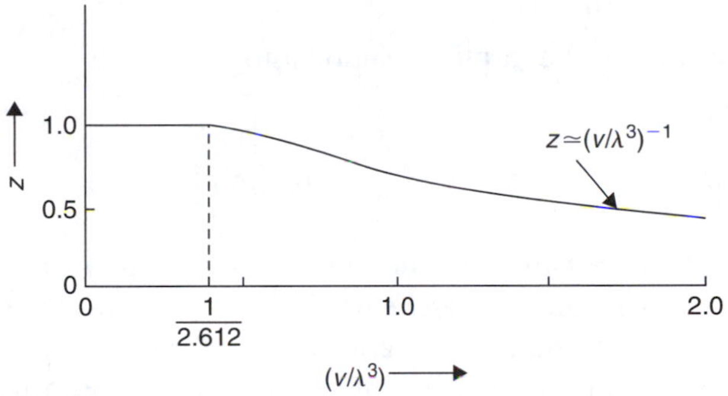

FIGURE 7.2 The fugacity of an ideal Bose gas as a function of (v/λ^3).

see equation (8). For $(v/\lambda^3) \gg 1$, we have $g_{3/2}(z) \ll 1$ and, hence, $z \ll 1$. Under these circumstances, $g_{3/2}(z) \simeq z$; see equation (10). Therefore, in this region, $z \simeq (v/\lambda^3)^{-1}$, in agreement with the classical case.[7] Figure 7.2 shows the variation of z with (v/λ^3).

Next, we examine the (P, T)-diagram of this system, that is, the variation of P with T, *keeping v fixed*. Now, for $T < T_c$, the pressure is given by equation (7), with z replaced by unity:

$$P(T) = \frac{kT}{\lambda^3}\zeta\left(\frac{5}{2}\right), \tag{27}$$

which is proportional to $T^{5/2}$ and is *independent* of v — implying infinite compressibility. At the transition point the value of the pressure is

$$P(T_c) = \left(\frac{2\pi m}{h^2}\right)^{3/2}(kT_c)^{5/2}\zeta\left(\frac{5}{2}\right); \tag{28}$$

[7]Equation (6.2.12) gives, for an ideal classical gas, $\ln \mathcal{Q} = zV/\lambda^3$. Accordingly, $N \equiv z(\partial \ln \mathcal{Q}/\partial z) = z(V/\lambda^3)$, with the result that $z = (\lambda^3/v)$.

with the help of (24), this can be written as

$$P(T_c) = \frac{\zeta\left(\frac{5}{2}\right)}{\zeta\left(\frac{3}{2}\right)}\left(\frac{N}{V}kT_c\right) \simeq 0.5134\left(\frac{N}{V}kT_c\right). \tag{29}$$

Thus, the pressure exerted by the particles of an ideal Bose gas at the transition temperature T_c is about one-half of that exerted by the particles of an equivalent Boltzmannian gas.[8] For $T > T_c$, the pressure is given by

$$P = \frac{N}{V}kT\frac{g_{5/2}(z)}{g_{3/2}(z)}, \tag{30}$$

where $z(T)$ is determined by the implicit relationship

$$g_{3/2}(z) = \frac{\lambda^3}{v} = \frac{N}{V}\frac{h^3}{(2\pi mkT)^{3/2}}. \tag{26a}$$

Unless T is very high, the pressure P cannot be expressed in any simpler terms; of course, for $T \gg T_c$, the virial expansion (13) can be used. As $T \to \infty$, the pressure approaches the classical value NkT/V. All these features are shown in Figure 7.3. The transition line in the figure portrays equation (27). The actual (P, T)-curve follows this line from $T = 0$ up to $T = T_c$ and thereafter departs, tending asymptotically to the classical limit. It may be pointed out that the region to the right of the transition line belongs to the normal phase alone, the line itself belongs to the mixed phase, while the region to the left is inaccessible to the system.

In view of the direct relationship between the internal energy of the gas and its pressure, see equation (12), Figure 7.3 depicts equally well the variation of U with T (of course, with v fixed). Its slope should, therefore, be a measure of the specific heat $C_V(T)$ of the gas. We readily observe that the specific heat is vanishingly small at low temperatures and rises with T until it reaches a maximum at $T = T_c$; thereafter, it decreases, tending asymptotically to the constant classical value. Analytically, for $T \leq T_c$, we obtain [see equations (15) and (27)]

$$\frac{C_V}{Nk} = \frac{3}{2}\frac{V}{N}\zeta\left(\frac{5}{2}\right)\frac{d}{dT}\left(\frac{T}{\lambda^3}\right) = \frac{15}{4}\zeta\left(\frac{5}{2}\right)\frac{v}{\lambda^3}, \tag{31}$$

[8] Actually, for all $T \leq T_c$, we can write

$$P(T) = P(T_c) \cdot (T/T_c)^{5/2} \simeq 0.5134(N_e kT/V).$$

We infer that, while particles in the condensed phase do not exert any pressure at all, particles in the excited states are about half as effective as in the Boltzmannian case.

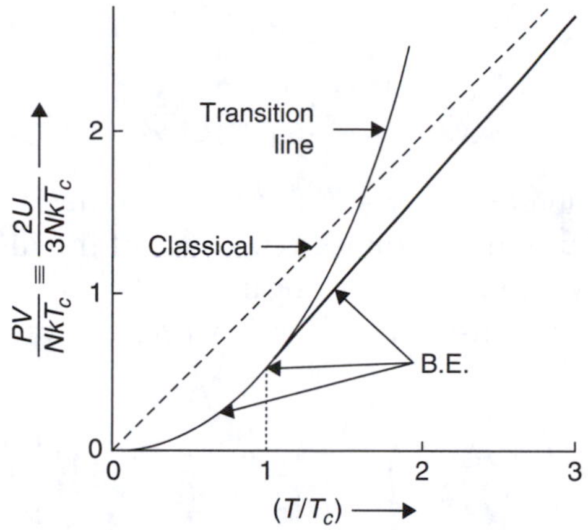

FIGURE 7.3 The pressure and the internal energy of an ideal Bose gas as a function of the temperature parameter (T/T_c).

which is proportional to $T^{3/2}$. At $T = T_c$, we have

$$\frac{C_V(T_c)}{Nk} = \frac{15}{4}\frac{\zeta\left(\frac{5}{2}\right)}{\zeta\left(\frac{3}{2}\right)} \simeq 1.925, \tag{32}$$

which is significantly higher than the classical value 1.5. For $T > T_c$, we obtain an implicit formula. First of all,

$$\frac{C_V}{Nk} = \left[\frac{\partial}{\partial T}\left(\frac{3}{2}T\frac{g_{5/2}(z)}{g_{3/2}(z)}\right)\right]_v; \tag{33}$$

see equations (11) and (26). To carry out the differentiation, we need to know $(\partial z/\partial T)_v$; this can be obtained from equation (26) with the help of the recurrence relation (D.10) in Appendix D. On one hand, since $g_{3/2}(z) \propto T^{-3/2}$,

$$\left[\frac{\partial}{\partial T}g_{3/2}(z)\right]_v = -\frac{3}{2T}g_{3/2}(z); \tag{34}$$

on the other,

$$z\frac{\partial}{\partial z}g_{3/2}(z) = g_{1/2}(z). \tag{35}$$

Combining these two results, we obtain

$$\frac{1}{z}\left(\frac{\partial z}{\partial T}\right)_v = -\frac{3}{2T}\frac{g_{3/2}(z)}{g_{1/2}(z)}. \tag{36}$$

Equation (33) now gives

$$\frac{C_V}{Nk} = \frac{15}{4}\frac{g_{5/2}(z)}{g_{3/2}(z)} - \frac{9}{4}\frac{g_{3/2}(z)}{g_{1/2}(z)}; \tag{37}$$

the value of z, as a function of T, is again to be determined from equation (26). In the limit $z \to 1$, the second term in (37) vanishes because of the divergence of $g_{1/2}(z)$, while the first term gives exactly the result appearing in (32). The specific heat is, therefore, continuous at the transition point. Its derivative is, however, discontinuous, the magnitude of the discontinuity being

$$\left(\frac{\partial C_V}{\partial T}\right)_{T=T_c-0} - \left(\frac{\partial C_V}{\partial T}\right)_{T=T_c+0} = \frac{27Nk}{16\pi T_c}\left\{\zeta\left(\frac{3}{2}\right)\right\}^2 \simeq 3.665\frac{Nk}{T_c}; \tag{38}$$

see Problem 7.6. For $T > T_c$, the specific heat decreases steadily toward the limiting value

$$\left(\frac{C_V}{Nk}\right)_{z\to 0} = \frac{15}{4} - \frac{9}{4} = \frac{3}{2}. \tag{39}$$

Figure 7.4 shows all these features of the (C_V, T)-relationship. It may be noted that it was the similarity of this curve with the experimental one for liquid He^4 (Figure 7.5) that prompted F. London to suggest, in 1938, that the curious phase transition that occurs in liquid He^4 at a temperature of about 2.19 K might be a manifestation of the *Bose–Einstein condensation* taking place in the liquid. Indeed, if we substitute, in (24), data for liquid He^4, namely $m = 6.65 \times 10^{-24}$ g and $V = 27.6\,\text{cm}^3/\text{mole}$, we obtain for T_c a value of about 3.13 K, which is not drastically different from the observed transition temperature of the liquid. Moreover, the interpretation of the phase transition in liquid He^4 as Bose–Einstein condensation provides a theoretical basis for the *two-fluid model* of this liquid, which was empirically put forward by Tisza (1938a,b) to explain the physical behavior of the liquid below the transition temperature.

According to London, the N_0 particles that occupy a single, entropyless state ($\varepsilon = 0$) could be identified with the "superfluid component" of the liquid and the N_e particles that occupy the excited states ($\varepsilon \neq 0$) with the "normal component." As required in the

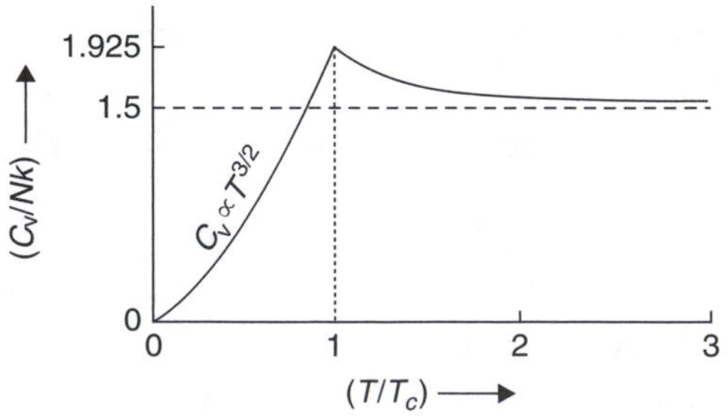

FIGURE 7.4 The specific heat of an ideal Bose gas as a function of the temperature parameter (T/T_c).

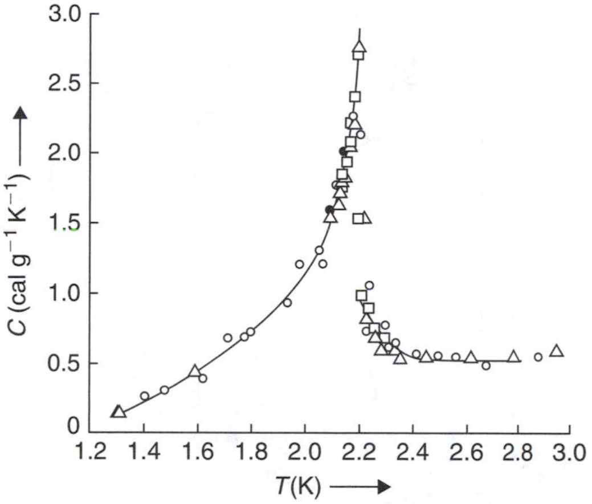

FIGURE 7.5 The specific heat of liquid He^4 under its own vapor pressure (after Keesom and coworkers).

model of Tisza, the superfluid fraction makes its appearance at the transition temperature T_c, and builds up at the cost of the normal fraction until at $T = 0$ the whole fluid becomes superfluid; compare to Figure 7.1. Of course, the actual temperature dependence of these fractions, and of other physical quantities pertaining to liquid He^4, is considerably different from what the simple-minded ideal Bose gas suggests. London had expected that the inclusion of intermolecular interactions would improve the quantitative agreement. Although this expectation has been partially vindicated, there have been other advances in the field that provide alternative ways of looking at the helium problem; see Section 7.6. Nevertheless, many of the features provided by London's interpretation of this phenomenon continue to be of value.

Historically, the experimental measurements of the specific heat of liquid He^4, which led to the discovery of this so-called He I–He II transition, were first made by Keesom in 1927 and 1928. Struck by the shape of the (C_V, T)-curve, Keesom gave this transition the name λ-transition; as a result, the term transition temperature (or transition point) also came to be known as λ-temperature (or λ-point).

We shall now look at the *isotherms* of the ideal Bose gas; that is, the variation of the pressure of the gas with its volume, keeping T fixed. The Bose–Einstein condensation now sets in at a characteristic volume v_c, given by

$$v_c = \lambda^3/\zeta\left(\frac{3}{2}\right); \tag{40}$$

see (23). We note that $v_c \propto T^{-3/2}$. For $v < v_c$, the pressure of the gas is independent of v and is given by

$$P_0 = \frac{kT}{\lambda^3}\zeta\left(\frac{5}{2}\right); \tag{41}$$

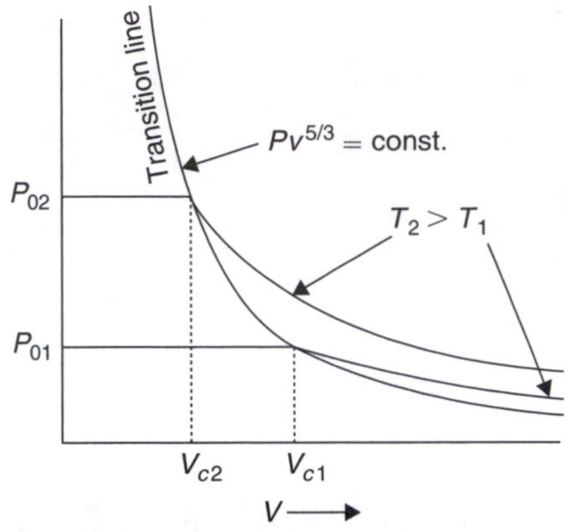

FIGURE 7.6 The isotherms of an ideal Bose gas.

see (27). The region of the mixed phase in the (P, v)-diagram is marked by a boundary line (called the *transition line*) given by the equation

$$P_0 v_c^{5/3} = \frac{h^2}{2\pi m} \frac{\zeta\left(\frac{5}{2}\right)}{\left\{\zeta\left(\frac{3}{2}\right)\right\}^{5/3}} = \text{const.}; \tag{42}$$

see Figure 7.6. Clearly, the region to the left of this line belongs to the mixed phase, while the region to the right belongs to the normal phase alone.

Finally, we examine the *adiabats* of the ideal Bose gas. For this, we need an expression for the entropy of the system. Making use of the thermodynamic formula

$$U - TS + PV \equiv N\mu \tag{43}$$

and the expressions for U and P obtained above, we get

$$\frac{S}{Nk} \equiv \frac{U + PV}{NkT} - \frac{\mu}{kT} = \begin{cases} \dfrac{5}{2}\dfrac{g_{5/2}(z)}{g_{3/2}(z)} - \ln z & \text{for } T > T_c, \quad (44a) \\[2ex] \dfrac{5}{2}\dfrac{v}{\lambda^3}\zeta\left(\dfrac{5}{2}\right) & \text{for } T \le T_c; \quad (44b) \end{cases}$$

again, the value of $z(T)$, for $T > T_c$, is to be obtained from equation (26). Now, a reversible adiabatic process implies the constancy of S and N. For $T > T_c$, this implies the constancy of z as well and in turn, by (26), the constancy of (v/λ^3). For $T \le T_c$, it again implies the same. We thus obtain, quite generally, the following relationship between the volume and the temperature of the system when it undergoes a reversible adiabatic process:

$$vT^{3/2} = \text{const.} \tag{45}$$

The corresponding relationship between the pressure and the temperature is

$$P/T^{5/2} = \text{const.}; \tag{46}$$

see equations (7) and (27). Eliminating T, we obtain

$$Pv^{5/3} = \text{const.} \tag{47}$$

as the equation for an adiabat of the ideal Bose gas.

Incidentally, the foregoing results are exactly the same as for an ideal classical gas. There is, however, a significant difference between the two cases; that is, while the exponent $\frac{5}{3}$ in formula (47) is identically equal to the ratio of the specific heats C_P and C_V in the case of the ideal classical gas, it is not so in the case of the ideal Bose gas. For the latter, this ratio is given by

$$\gamma \equiv \frac{C_P}{C_V} = 1 + \frac{4}{9}\frac{C_V}{Nk}\frac{g_{1/2}(z)}{g_{3/2}(z)} \tag{48a}$$

$$= \frac{5}{3}\frac{g_{5/2}(z)g_{1/2}(z)}{\{g_{3/2}(z)\}^2}; \tag{48b}$$

see Problems 7.4 and 7.5. It is only for $T \gg T_c$ that $\gamma \simeq \frac{5}{3}$. At any finite temperature, $\gamma > \frac{5}{3}$ and as $T \to T_c, \gamma \to \infty$. Equation (47), on the other hand, holds for *all* T.

In the mixed-phase region ($T < T_c$), the entropy of the gas may be written as

$$S = N_e \cdot \frac{5}{2}k\frac{\zeta\left(\frac{5}{2}\right)}{\zeta\left(\frac{3}{2}\right)} \propto N_e; \tag{49}$$

see equations (20) and (44b). As expected, the N_0 particles that constitute the "condensate" do not contribute toward the entropy of the system, while the N_e particles that constitute the normal part contribute an amount of $\frac{5}{2}k\zeta(\frac{5}{2})/\zeta(\frac{3}{2})$ per particle.

7.2 Bose–Einstein condensation in ultracold atomic gases

The first demonstration of Bose–Einstein condensation in ultracold atomic gases came in 1995. Cornell and Wieman Bose-condensed ^{87}Rb (Anderson, Ensher, Matthews, Wieman, and Cornell (1995)) and Ketterle Bose-condensed ^{23}Na (Davis, Mewes, Andrews, van Druten, Durfee, Kurn, and Ketterle (1995)) using magneto-optical traps (MOTs) and magnetic traps to cool vapors of tens of thousands of atoms to temperatures of a few nanokelvin.[9] A survey of the theory and experiments can be found in Pitaevskii and Stringari (2003), Leggett (2006), and Pethick and Smith (2008).

The first step of the cooling of the atomic vapor uses three sets of counter-propagating laser beams oriented along cartesian axes that are tuned just below the resonant frequency

[9]Since 1995, many isotopes have been Bose-condensed including ^{7}Li, ^{23}Na, ^{41}K, ^{52}Cr, ^{84}Sr, ^{85}Rb, ^{87}Rb, ^{133}Cs, and ^{174}Yb. The first molecular Bose–Einstein condensates were created in 2003 by the research groups of Rudolf Grimm at the University of Innsbruck, Deborah S. Jin at the University of Colorado at Boulder, and Wolfgang Ketterle at Massachusetts Institute of Technology.

of the atoms in the trap. Atoms that are stationary are just off resonance and so rarely absorb a photon. Moving atoms are Doppler shifted on resonance to the laser beam that is propagating opposite to the velocity vector of the atom. Those atoms preferentially absorb photons from that direction and then reemit in random directions, resulting in a net momentum kick opposite to the direction of motion. This results in an "optical molasses" that slows the atoms. This cooling method is constrained by the "recoil limit" in which the atoms have a minimum momentum of the order of the momentum of the photons used to cool the gas. This gives a limiting temperature of $(hf)^2/2mc^2k \approx 1\,\mu\text{K}$, where f is the frequency of the spectral line used for cooling and m is the mass of an atom.

In the next step of the cooling process, the lasers are turned off and a spatially varying magnetic field creates an attractive anisotropic harmonic oscillator potential near the center of the magnetic trap

$$V(\mathbf{r}) = \frac{1}{2}m\left(\omega_1^2 x^2 + \omega_2^2 y^2 + \omega_3^2 z^2\right). \tag{1}$$

The frequencies of the trap ω_α are controlled by the applied magnetic field. One can then lower the trap barrier using a resonant transition to remove the highest energy atoms in the trap. If the atoms in the vapor are sufficiently coupled to one other, then the remaining atoms in the trap are cooled by evaporation.

If the interactions between the atoms in the gas can be neglected, the energy of each atom in the harmonic oscillator potential is

$$\varepsilon_{l_1,l_2,l_3} = \hbar\omega_1 l_1 + \hbar\omega_2 l_2 + \hbar\omega_3 l_3 + \frac{1}{2}\hbar(\omega_1 + \omega_2 + \omega_3), \tag{2}$$

where $l_\alpha\,(= 0, 1, 2, \ldots \infty)$ are the quantum numbers of the harmonic oscillator. If the three frequencies are all the same, then the quantum degeneracy of a level with energy $\varepsilon = \hbar\omega(l + 3/2)$ is $(l+1)(l+2)/2$; see Problem 3.26.

For the general anisotropic case, the smoothed density of states as a function of energy (suppressing the zero point energy and assuming $\varepsilon \gg \hbar\omega_\alpha$) is given by

$$a(\varepsilon) = \int_0^\infty\!\int_0^\infty\!\int_0^\infty \delta\left(\varepsilon - \hbar\omega_1 l_1 - \hbar\omega_2 l_2 - \hbar\omega_3 l_3\right) dl_1\, dl_2\, dl_3 = \frac{\varepsilon^2}{2\left(\hbar\omega_0\right)^3}, \tag{3}$$

where $\omega_0 = (\omega_1\omega_2\omega_3)^{1/3}$; this assumes a single spin state per atom. The thermodynamic potential Π, see Appendix H, for bosons in the trap is then given by

$$\Pi(\mu, T) = -\frac{(kT)^4}{2\left(\hbar\omega_0\right)^3}\int_0^\infty x^2 \ln\left(1 - e^{-x}e^{\beta\mu}\right) dx = \frac{(kT)^4}{\left(\hbar\omega_0\right)^3} g_4(z), \tag{4}$$

where $z = \exp(\beta\mu)$ is the fugacity and $g_\nu(z)$ is defined in Appendix D. Volume is not a parameter in the thermodynamic potential since the atoms are confined by the harmonic trap. The average number of atoms in the excited states in the trap is

$$N(\mu, T) = \left(\frac{\partial \Pi}{\partial \mu}\right)_T = \left(\frac{kT}{\hbar\omega_0}\right)^3 g_3(z) \,. \tag{5}$$

For fixed N, the chemical potential monotonically increases as temperature is lowered until Bose–Einstein condensation occurs when $\mu = 0$ ($z = 1$). The critical temperature for N trapped atoms is then given by

$$\frac{kT_c}{\hbar\omega_0} = \left(\frac{N}{\zeta(3)}\right)^{1/3}, \tag{6}$$

where $\zeta(3) = g_3(1) \simeq 1.202$. While the spacing of the energy levels is of order $\hbar\omega_0$, the critical temperature for condensation is much larger than the energy spacing of the lowest levels for $N \gg 1$. A typical magnetic trap oscillation frequency $f \approx 100$Hz. For $N = 2 \times 10^4$, as in Cornell and Wieman's original experiment, $kT_c/\hbar\omega_0 \approx 25.5$. The observed critical temperature was about 170nK (Anderson et al. (1995)).

For $T < T_c$, the number of atoms in the excited states is

$$\frac{N_{\text{excited}}}{N} = \frac{\zeta(3)}{N}\left(\frac{kT}{\hbar\omega_0}\right)^3 = \left(\frac{T}{T_c}\right)^3, \tag{7}$$

so the fraction of atoms that condense into the ground state of the harmonic oscillator is

$$\frac{N_0}{N} = 1 - \left(\frac{T}{T_c}\right)^3; \tag{8}$$

see de Groot, Hooyman, and ten Seldam (1950), and Bagnato, Pritchard, and Kleppner (1987). In the thermodynamic limit, a nonzero fraction of the atoms occupy the ground state for $T < T_c$. By contrast, the occupancy of the first excited state is only of order $N^{1/3}$, so in the thermodynamic limit the occupancy fraction in each excited state is zero. A comparison of the experimentally measured Bose-condensed fraction with equation (8) is shown in Figure 7.7.

7.2.A Detection of the Bose–Einstein condensate

The linear size of the ground state wavefunction in cartesian direction α is

$$a_\alpha = \sqrt{\frac{\hbar}{m\omega_\alpha}}, \tag{9}$$

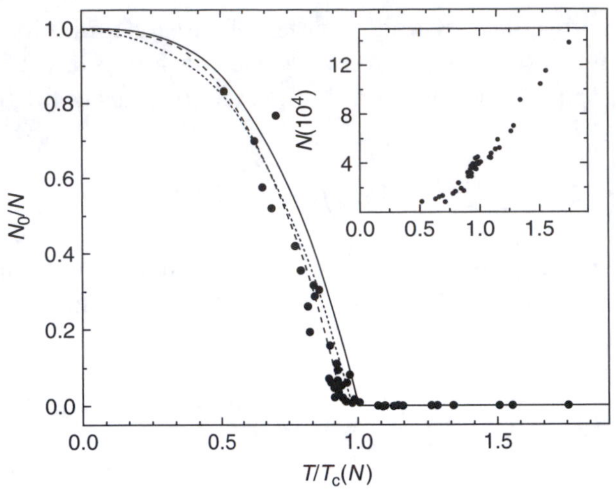

FIGURE 7.7 Experimental measurement of the Bose-condensed fraction vs. temperature, as compared to equation (8). The scaled temperature on the horizontal axis is the temperature divided by the N-dependent critical temperature given in equation (6). The inset shows the total number of atoms in the trap after the evaporative cooling. From Ensher et al. (1996). Reprinted with permission; copyright © 1996, American Physical Society.

while the linear size of the thermal distribution of the noncondensed atoms in that direction is

$$a_{\text{thermal}} = \sqrt{\frac{kT}{m\omega_\alpha^2}} = a_\alpha \sqrt{\frac{kT}{\hbar\omega_\alpha}}. \tag{10}$$

At trap frequency $f = 100\,\text{Hz}$ and temperature $T = 100\,\text{nK}$, these sizes are about $1\,\mu\text{m}$ and $5\,\mu\text{m}$, respectively. Instead of measuring the atoms directly in the trapping potential, experimenters usually measure the momentum distribution of the ultracold gas by a time-of-flight experiment. At time $t = 0$, the magnetic field is turned off suddenly, eliminating the trapping potential. The atomic cloud then expands according to the momentum distribution the atoms had in the harmonic trap. The cloud is allowed to expand for about 100 milliseconds. The speed of the atoms at this temperature is a few millimeters per second, so the cloud expands to a few hundred microns in this period of time. The cloud is then illuminated with a laser pulse on resonance with the atoms, leaving a shadow on a CCD in the image plane of the optics. The size and shape of the light intensity pattern directly measures the momentum distribution the atoms had in the trap at $t = 0$. The expanding cloud can be divided into two components, the N_0 atoms that had been Bose-condensed into the ground state and the remaining $N - N_0$ atoms that were in the excited states of the harmonic oscillator potential. The Bose-condensed atoms have smaller momenta than the atoms that were in the excited states. After time t, the quantum evolution of the ground state has a spatial number density

$$n_0(\boldsymbol{r},t) = N_0\,|\psi_0(\boldsymbol{r},t)|^2 = \frac{N_0}{\pi^{3/2}}\prod_{\alpha=1}^{3}\left[\frac{1}{a_\alpha\sqrt{1+\omega_\alpha^2 t^2}}\exp\left(\frac{-r_\alpha^2}{a_\alpha^2\,(1+\omega_\alpha^2 t^2)}\right)\right]; \tag{11}$$

see Pitaevskii and Stringari (2003), Pethick and Smith (2008), and Problem 7.15.

The atoms that are not condensed into the ground state can be treated semiclassically, that is, the position-momentum distribution function is treated classically while the density follows the Bose–Einstein distribution function:

$$f(\boldsymbol{r},\boldsymbol{p},0) = \frac{1}{\exp\left(\frac{\beta p^2}{2m}+\frac{\beta m}{2}\,(\omega_1^2x^2+\omega_2^2y^2+\omega_3^2z^2)-\beta\mu\right)-1}. \tag{12}$$

After the potential is turned off at $t=0$, the distribution evolves ballistically:

$$f(\boldsymbol{r},\boldsymbol{p},t) = f\left(\boldsymbol{r}+\frac{\boldsymbol{p}t}{m},\boldsymbol{p},0\right). \tag{13}$$

The spatial number density of atoms in the excited states is

$$n_{\text{excited}}(\boldsymbol{r},t) = \frac{1}{h^3}\int f\left(\boldsymbol{r}+\frac{\boldsymbol{p}t}{m},\boldsymbol{p},t\right)\mathrm{d}\boldsymbol{p}, \tag{14}$$

which can be integrated to give

$$n_{\text{excited}}(\boldsymbol{r},t) = \frac{1}{\lambda^3}\sum_{j=1}^{\infty}\frac{e^{\beta\mu j}}{j^{3/2}}\left\{\prod_{\alpha=1}^{3}\left[\frac{1}{\sqrt{1+\omega_\alpha^2 t^2}}\exp\left(\frac{-\beta j m\omega_\alpha^2 r_\alpha^2}{2\,(1+\omega_\alpha^2 t^2)}\right)\right]\right\}, \tag{15}$$

where $\lambda = h/\sqrt{2\pi mkT}$ is the thermal deBroglie wavelength; see Pethick and Smith (2008), and Problem 7.16. The integrals over the condensed state and the excited states correctly count all the atoms:

$$N_0 = \int n_0(\boldsymbol{r},t)d\boldsymbol{r}, \tag{16a}$$

$$N - N_0 = \int n_{\text{excited}}(\boldsymbol{r},t)d\boldsymbol{r} = N_{\text{excited}}; \tag{16b}$$

see Problem 7.18.

Note that at early times ($\omega_\alpha t \ll 1$) both the condensed and the excited distributions are anisotropic due to the anisotropic trapping potential. However, at late times ($\omega_\alpha t \gg 1$), the atoms from the excited states form a spherically symmetric cloud because of the isotropic momentum dependence of the $t=0$ distribution function. By contrast, the atoms that were condensed into the ground state expand anisotropically due to the different spatial extents of the ground state wavefunction at $t=0$. The direction that has the largest

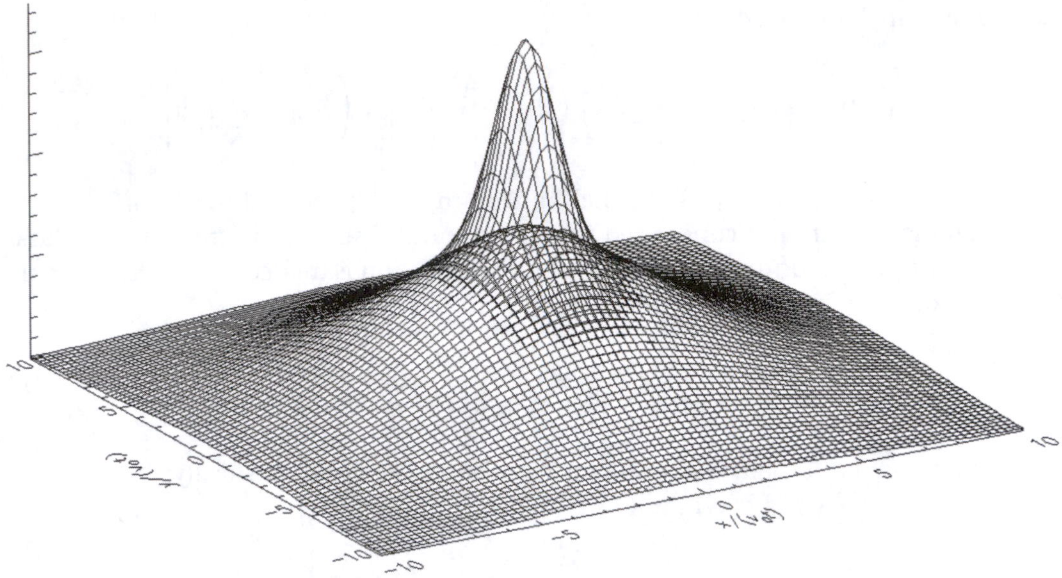

FIGURE 7.8 The two-dimensional time-of-flight number density equations (11) and (15) at late times ($\omega_0 t \gg 1$) for $T/T_c = 0.98$ using the experimental parameters in Anderson et al. (1995): $N = 2 \times 10^4$ atoms in the trap and $\omega_2 = \sqrt{8}\omega_1$. The plot shows the full density and, underneath, the broader isotropic density just due to the excited states. The z-dimension has been integrated out. The Bose-condensed peak is anisotropic: the y-direction spread is $8^{1/4} = 1.68$ times larger than in the x-direction while the broad peak caused by the excited states is isotropic. The distance scale $v_0 t = t\sqrt{\hbar\omega_1/m}$ determines the width of the distribution that results from the Bose-condensed peak in the x-direction at late times; compare to Figure 7.9.

ω_α is quantum mechanically squeezed the most at $t = 0$; so, according to the uncertainty principle, it expands the fastest. This is an important feature of the experimental data that confirms the onset of Bose–Einstein condensation;[10] see Figures 7.8 and 7.9.

7.2.B Thermodynamic properties of the Bose–Einstein condensate

The temperature, condensate fraction, and internal energy can all be observed using time-of-flight measurements. The internal energy can also be written in terms of the function $g_\nu(z)$:

$$U(\mu, T) = \int_0^\infty \frac{\varepsilon^3}{2(\hbar\omega_0)^3} \frac{1}{e^{\beta(\varepsilon-\mu)} - 1} d\varepsilon = 3\frac{(kT)^4}{(\hbar\omega_0)^3} g_4(z). \tag{17}$$

[10]Repulsive interactions between atoms create additional forces that modify the time-of-flight expansion. This is especially important in condensates with a very large numbers of atoms, as many as 10^7 or more in some experiments; see Section 11.2.A.

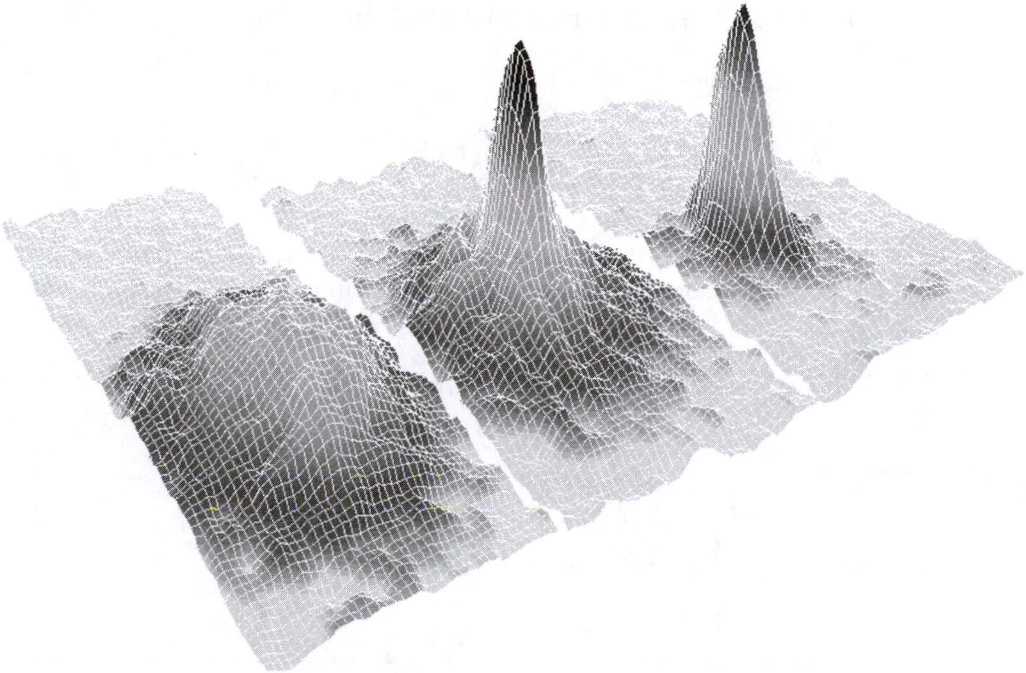

FIGURE 7.9 Time-of-flight images from the first observation of Bose–Einstein condensation in a dilute vapor of ^{87}Rb by Anderson et al. (1995) at temperatures just above and below the phase transition temperature. The anisotropic pattern of the Bose-condensed fraction is evident; compare to Figure 7.8. Courtesy of NIST/JILA/University of Colorado.

The heat capacity at constant number can be written as

$$C_N(T) = \left(\frac{\partial U}{\partial T}\right)_N = \left(\frac{\partial U}{\partial T}\right)_\mu + \left(\frac{\partial U}{\partial \mu}\right)_T \left(\frac{\partial \mu}{\partial T}\right)_N$$
$$= \left(\frac{\partial U}{\partial T}\right)_\mu - \frac{\left(\frac{\partial U}{\partial T}\right)_\mu \left(\frac{\partial N}{\partial T}\right)_\mu}{\left(\frac{\partial N}{\partial \mu}\right)_T}. \tag{18}$$

Equations (5) and (6) can be used to determine the fugacity z numerically, as shown in Figure 7.10(a). The fugacity can then be used in equation (17) to obtain the scaled internal energy

$$\frac{U}{NkT_c} = \begin{cases} 3\left(\frac{T}{T_c}\right)^4 \frac{\zeta(4)}{\zeta(3)} & \text{for } T \le T_c, \\ 3\left(\frac{T}{T_c}\right)^4 \frac{g_4(z)}{\zeta(3)} & \text{for } T \ge T_c; \end{cases} \tag{19}$$

see Figures 7.10(b) and 7.12. The scaled specific heat is given by

$$\frac{C_N}{Nk} = \begin{cases} \dfrac{12\zeta(4)}{\zeta(3)}\left(\dfrac{T}{T_c}\right)^3 & \text{for } T < T_c, \\ \dfrac{1}{\zeta(3)}\left(\dfrac{T}{T_c}\right)^3\left(12g_4(z) - \dfrac{9g_3^2(z)}{g_2(z)}\right) & \text{for } T > T_c, \end{cases} \tag{20}$$

and is shown in Figure 7.11. Unlike the case of Bose–Einstein condensation of free particles in a box (Figure 7.4), the specific heat of a condensate in a harmonic trap displays a

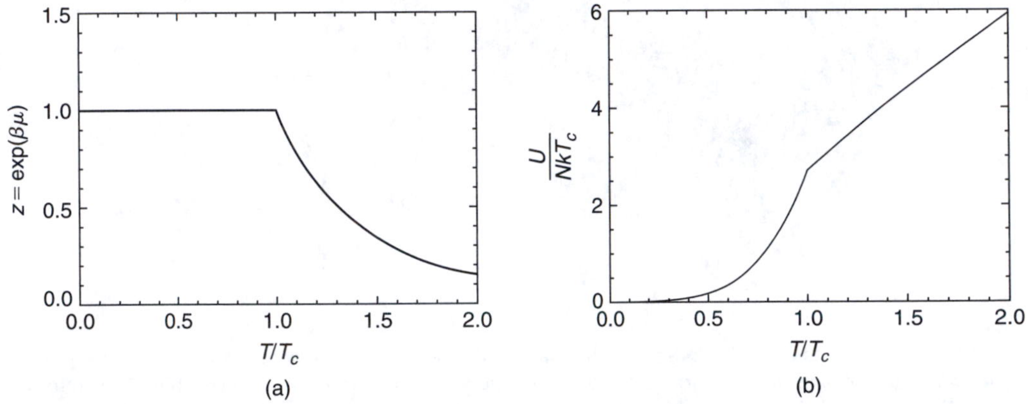

FIGURE 7.10 Fugacity (a) and scaled internal energy (b) vs. scaled temperature (T/T_c) for a Bose–Einstein condensate in a harmonic trap.

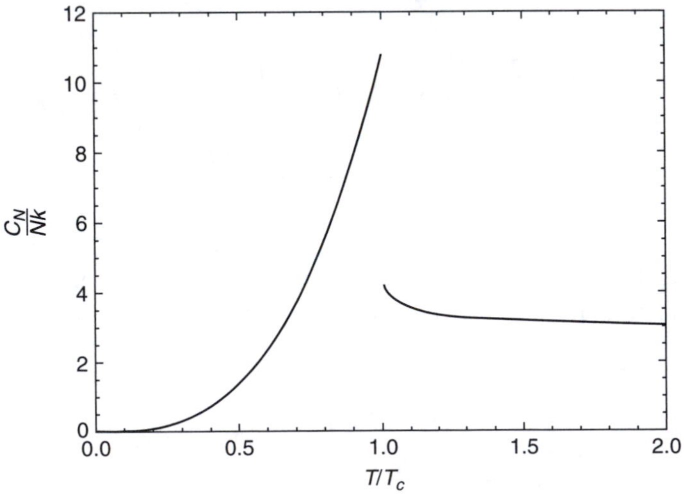

FIGURE 7.11 Scaled specific heat of a Bose–Einstein condensate in a harmonic trap as a function of the scaled temperature (T/T_c); compare with Figure 7.4 for a free-particle Bose gas.

discontinuity at the critical temperature:

$$\frac{C_N}{Nk} \to \begin{cases} \dfrac{12\zeta(4)}{\zeta(3)} \simeq 10.805 & \text{as } T \to T_c^-, \\[2ex] \dfrac{12\zeta(4)}{\zeta(3)} - \dfrac{9\zeta(3)}{\zeta(2)} \simeq 4.228 & \text{as } T \to T_c^+. \end{cases} \tag{21}$$

Figure 7.12 shows experimental data for the internal energy of a Bose–Einstein condensate of ^{87}Rb. The break in slope is an indication of the discontinuous specific heat. Naturally, in a system with a finite number of particles, all nonanalyticities associated with the phase transition are removed. When N is finite, the condensate fraction approaches zero smoothly and the discontinuity in the heat capacity is rounded off. Pathria (1998) has derived N-dependent temperature markers that indicate the onset of Bose–Einstein condensation in terms of the condensate fraction and the specific heat; see also Kirsten and Toms (1996) and Haugerud, Haugest, and Ravndal (1997).

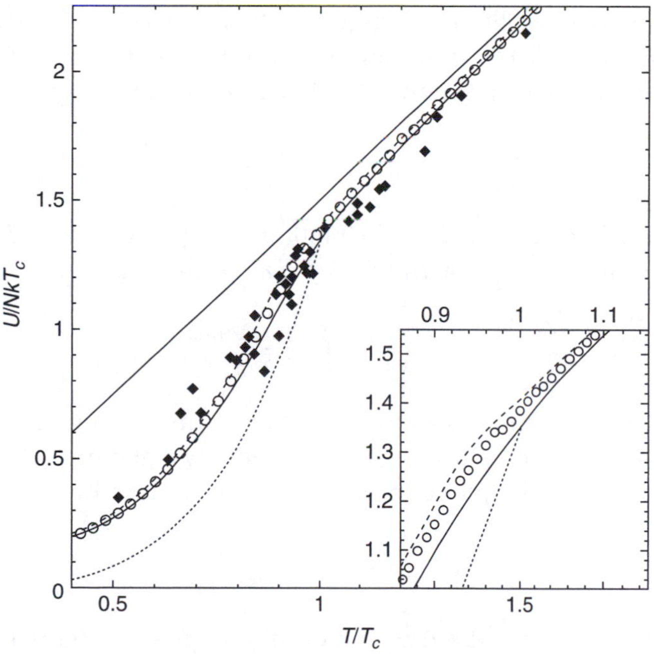

FIGURE 7.12 Comparison of the experimental measurements of Ensher et al. (1996) (*diamonds*) with the noninteracting internal energy result — see equation (19) and Figure 7.10(b) — (*dotted curve*), the zero-order solution including interactions (*full curve*), first-order perturbative treatment (*dashed curve*), and numerical solution (*circles*). The straight line is the classical Maxwell–Boltzmann result. The inset is an enlargement of the region around the critical temperature. The break in slope is an indication of the discontinuity in the thermodynamic limit specific heat shown in Figure 7.11; from Minguzzi, Conti, and Tosi (1997). Reprinted with permission; copyright © 1997, American Institute of Physics.

7.3 Thermodynamics of the blackbody radiation

One of the most important applications of Bose–Einstein statistics is to investigate the equilibrium properties of the *blackbody radiation*. We consider a radiation cavity of volume V at temperature T. Historically, this system has been looked on from two, practically identical but conceptually different, points of view:

(i) as an assembly of *harmonic oscillators* with quantized energies $(n_s + \frac{1}{2})\hbar\omega_s$, where $n_s = 0, 1, 2, \ldots$, and ω_s is the (angular) frequency of an oscillator, or
(ii) as a gas of identical and indistinguishable quanta — the so-called *photons* — the energy of a photon (corresponding to the frequency ω_s of the radiation mode) being $\hbar\omega_s$.

The first point of view is essentially the one adopted by Planck (1900), except that we have also included here the zero-point energy of the oscillator; for the thermodynamics of the radiation, this energy is of no great consequence and may be dropped altogether. The oscillators, being distinguishable from one another (by the very values of ω_s), would obey Maxwell–Boltzmann statistics; however, the expression for the single-oscillator partition function $Q_1(V, T)$ would be different from the classical expression because now the energies accessible to the oscillator are discrete, rather than continuous; compare to equations (3.8.2) and (3.8.14). The expectation value of the energy of a Planck oscillator of frequency ω_s is then given by equation (3.8.20), excluding the zero-point term $\frac{1}{2}\hbar\omega_s$:

$$\langle \varepsilon_s \rangle = \frac{\hbar\omega_s}{e^{\hbar\omega_s/kT} - 1}. \tag{1}$$

Now, the number of normal modes of vibration per unit volume of the cavity in the frequency range $(\omega, \omega + d\omega)$ is given by the *Rayleigh expression*

$$2 \cdot 4\pi \left(\frac{1}{\lambda}\right)^2 d\left(\frac{1}{\lambda}\right) = \frac{\omega^2 d\omega}{\pi^2 c^3}, \tag{2}$$

where the factor 2 has been included to take into account the duplicity of the transverse modes;[11] the symbol c here denotes the speed of light. By equations (1) and (2), the energy density associated with the frequency range $(\omega, \omega + d\omega)$ is given by

$$u(\omega)d\omega = \frac{\hbar}{\pi^2 c^3} \frac{\omega^3 d\omega}{e^{\hbar\omega/kT} - 1}, \tag{3}$$

which is *Planck's formula* for the distribution of energy over the blackbody spectrum. Integrating (3) over all values of ω, we obtain the total energy density in the cavity.

The second point of view originated with Bose (1924) and Einstein (1924, 1925). Bose investigated the problem of the "distribution of *photons* over the various energy levels" in the system; however, instead of worrying about the allocation of the various photons

[11]As is well-known, the longitudinal modes play no role in the case of radiation.

to the various energy levels (as one would have ordinarily done), he concentrated on the statistics of the energy levels themselves! He examined questions such as the "probability of an energy level $\varepsilon_s (= \hbar\omega_s)$ being occupied by n_s photons at a time," "the mean values of n_s and ε_s," and so on. The statistics of the energy levels is indeed Boltzmannian; the mean values of n_s and ε_s, however, turn out to be

$$\langle n_s \rangle = \sum_{n_s=0}^{\infty} n_s e^{-n_s\hbar\omega_s/kT} \Bigg/ \sum_{n_s=0}^{\infty} e^{-n_s\hbar\omega_s/kT} = \frac{1}{e^{\hbar\omega_s/kT} - 1} \tag{4}$$

and hence

$$\langle \varepsilon_s \rangle = \hbar\omega_s \langle n_s \rangle = \frac{\hbar\omega_s}{e^{\hbar\omega_s/kT} - 1}, \tag{5}$$

identical with our earlier result (1). To obtain the number of photon states with momenta lying between $\hbar\omega/c$ and $\hbar(\omega + d\omega)/c$, Bose made use of the connection between this number and the "volume of the relevant region of the phase space," with the result

$$g(\omega)d\omega \approx 2 \cdot \frac{V}{h^3}\left\{4\pi\left(\frac{\hbar\omega}{c}\right)^2\left(\frac{\hbar\, d\omega}{c}\right)\right\} = \frac{V\omega^2 d\omega}{\pi^2 c^3}, \qquad (6)^{12}$$

which is also identical to our earlier result (2). Thus, he finally obtained the distribution formula of Planck. It must be noted here that, *although emphasis lay elsewhere*, the mathematical steps that led Bose to his final result went literally parallel to the ones occurring in the oscillator approach!

Einstein, on the other hand, went deeper into the problem and pondered over the statistics of both the photons and the energy levels, *taken together*. He inferred (from Bose's treatment) that the basic fact to keep in mind during the process of distributing photons over the various energy levels is that the photons are *indistinguishable* — a fact that had been implicitly taken care of in Bose's treatment. Einstein's derivation of the desired distribution was essentially the same as given in Section 6.1, with one important difference, that is, since the total number of photons in any given volume was indefinite, the constraint of a *fixed N* was no longer present. As a result, the Lagrange multiplier α did not enter into the discussion and to that extent the final formula for $\langle n_\varepsilon \rangle$ was simpler:

$$\langle n_\varepsilon \rangle = \frac{1}{e^{\varepsilon/kT} - 1}; \tag{7}$$

compare to equation (6.1.18a) or (6.2.22). The foregoing result is identical to (4), with $\varepsilon = h\omega_s$. The subsequent steps in Einstein's treatment were the same as in Bose's.

[12]The factor 2 in this expression arises essentially from the same cause as in the Rayleigh expression (2). However, in the present context, it would be more appropriate to regard it as representing the two states of polarization of the photon spin.

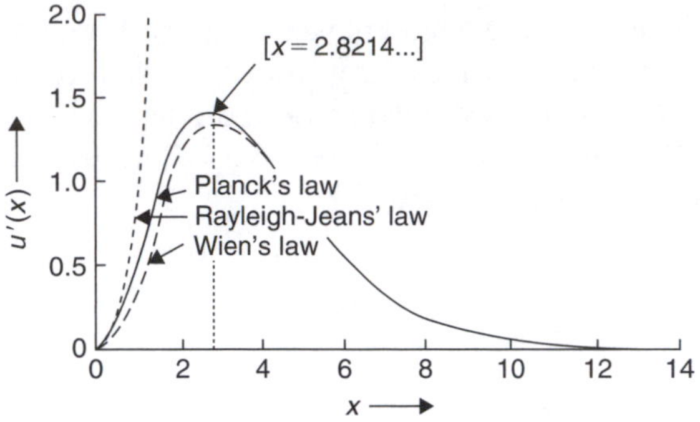

FIGURE 7.13 The spectral distribution of energy in the blackbody radiation. The solid curve represents the quantum-theoretical formula of Planck. The long-wavelength approximation of Rayleigh–Jeans and the short-wavelength approximation of Wien are also shown.

Looking back at the two approaches, we note that there is a complete correspondence between them — "an oscillator in the eigenstate n_s, with energy $(n_s + \frac{1}{2})\hbar\omega_s$" in the first approach corresponds to "the occupation of the energy level $\hbar\omega_s$ by n_s photons" in the second approach, "the average energy $\langle\varepsilon_s\rangle$ of an oscillator" corresponds to "the mean occupation number $\langle n_s\rangle$ of the corresponding energy level," and so on.[13]

Figure 7.13 shows a plot of the distribution function (3), which may be written in the dimensionless form

$$u'(x)dx = \frac{x^3 dx}{e^x - 1}, \tag{8}$$

where

$$u'(x) = \frac{\pi^2 \hbar^3 c^3}{(kT)^4} u(x) \quad \text{and} \quad x = \frac{\hbar\omega}{kT}. \tag{9}$$

For long wavelengths ($x \ll 1$), formula (8) reduces to the classical approximation of Rayleigh (1900) and Jeans (1905), namely[14]

$$u'(x) \approx x^2, \tag{10}$$

while for short wavelengths ($x \gg 1$), it reduces to the rival formula of Wien (1896), namely

$$u'(x) \approx x^3 e^{-x}. \tag{11}$$

[13]Compared to the standard Bose–Einstein result (7.1.2), formula (7) suggests that we are dealing here with a case for which z is precisely equal to unity. It is not difficult to see that this is due to the fact that the total number of particles in the present case is *indefinite*. For then, their equilibrium number $\overline{N}$ has to be determined by the condition that the free energy of the system is at its minimum, that is, $\{(\partial A/\partial N)_{N=\overline{N}}\}_{V,T} = 0$, which, by definition, implies that $\mu = 0$ and hence $z = 1$.

[14]The Rayleigh–Jeans formula follows directly if we use for $\langle\varepsilon_s\rangle$ the equipartition value kT rather than the quantum-theoretical value (1).

For comparison, the limiting forms (10) and (11) are also included in the figure. We note that the areas under the Planck curve and the Wien curve are $\pi^4/15(\simeq 6.49)$ and 6, respectively. The Rayleigh–Jeans curve, however, suffers from a high-frequency catastrophe!

For the total energy density in the cavity, we obtain from equations (8) and (9)

$$\frac{U}{V} = \int_0^\infty u(x)dx = \frac{(kT)^4}{\pi^2\hbar^3c^3}\int_0^\infty \frac{x^3dx}{e^x - 1}$$
$$= \frac{\pi^2k^4}{15\hbar^3c^3}T^4. \qquad (12)^{15}$$

If there is a small opening in the walls of the cavity, the photons will "effuse" through it. The net rate of flow of the radiation, per unit area of the opening, will be given by, see equation (6.4.12),

$$\frac{1}{4}\frac{U}{V}c = \frac{\pi^2k^4}{60\hbar^3c^2}T^4 = \sigma T^4, \qquad (13)$$

where

$$\sigma = \frac{\pi^2k^4}{60\hbar^3c^2} = 5.670\times 10^{-8}\,\mathrm{W\,m^{-2}\,K^{-4}}. \qquad (14)$$

Equation (13) describes the *Stefan–Boltzmann* law of blackbody radiation, σ being the *Stefan constant*. This law was deduced from experimental observations by Stefan in 1879; five years later, Boltzmann derived it from thermodynamic considerations.

For further study of thermodynamics, we evaluate the grand partition function of the photon gas. Using equation (6.2.17) with $z = 1$, we obtain

$$\ln \mathcal{Q}(V,T) \equiv \frac{PV}{kT} = -\sum_\varepsilon \ln(1 - e^{-\varepsilon/kT}). \qquad (15)$$

Replacing summation by integration and making use of the extreme relativistic formula

$$a(\varepsilon)d\varepsilon = 2V\frac{4\pi p^2dp}{h^3} = \frac{8\pi V}{h^3c^3}\varepsilon^2 d\varepsilon, \qquad (16)$$

we obtain, after an integration by parts,

$$\ln \mathcal{Q}(V,T) \equiv \frac{PV}{kT} = \frac{8\pi V}{3h^3c^3}\frac{1}{kT}\int_0^\infty \frac{\varepsilon^3 d\varepsilon}{e^{\varepsilon/kT} - 1}.$$

[15]Here, use has been made of the fact that the value of the definite integral is $6\zeta(4) = \pi^4/15$; see Appendix D.

By a change of variable, this becomes

$$PV = \frac{8\pi V}{3h^3c^3}(kT)^4 \int_0^\infty \frac{x^3 dx}{e^x - 1}$$

$$= \frac{8\pi^5 V}{45h^3c^3}(kT)^4 = \frac{1}{3}U. \tag{17}$$

We thus obtain the well-known result of the radiation theory; that is, the pressure of the radiation is equal to one-third its energy density; see also equations (6.4.3) and (6.4.4). Next, since the chemical potential of the system is zero, the Helmholtz free energy is equal to $-PV$; therefore

$$A = -PV = -\frac{1}{3}U, \tag{18}$$

whereby

$$S \equiv \frac{U - A}{T} = \frac{4}{3}\frac{U}{T} \propto VT^3 \tag{19}$$

and

$$C_V = T\left(\frac{\partial S}{\partial T}\right)_V = 3S. \tag{20}$$

If the radiation undergoes a reversible adiabatic change, the law governing the variation of T with V would be, see (19),

$$VT^3 = \text{const.} \tag{21}$$

Combining (21) with the fact that $P \propto T^4$, we obtain an equation for the *adiabats* of the system, namely

$$PV^{4/3} = \text{const.} \tag{22}$$

It should be noted, however, that the ratio C_P/C_V of the photon gas is *not* 4/3; it is infinite!

Finally, we derive an expression for the equilibrium number $\overline{N}$ of photons in the radiation cavity. We obtain

$$\overline{N} = \frac{V}{\pi^2 c^3} \int_0^\infty \frac{\omega^2 d\omega}{e^{\hbar\omega/kT} - 1}$$

$$= V\frac{2\zeta(3)(kT)^3}{\pi^2\hbar^3c^3} \propto VT^3. \tag{23}$$

Instructive though it may be, formula (23) cannot be taken at its face value because in the present problem the magnitude of the *fluctuations* in the variable N, which is determined by the quantity $(\partial P/\partial V)^{-1}$, is infinitely large; see equation (4.5.7).

One of the most important examples of blackbody radiation is the 2.7 K cosmic microwave background, which is a remnant from the Big Bang. Equations (12) and (23) play an important role in our understanding of the thermodynamics of the early universe; see Problem 7.24 and Chapter 9.

7.4 The field of sound waves

A problem mathematically similar to the one discussed in Section 7.3 arises from the vibrational modes of a macroscopic body, specifically a solid. As in the case of black-body radiation, the problem of the vibrational modes of a solid can be studied equally well by regarding the system as a collection of harmonic oscillators or by regarding it as an enclosed region containing a gas of sound quanta — the so-called *phonons*. To illustrate this point, we consider the Hamiltonian of a *classical* solid composed of N atoms whose positions in space are specified by the coordinates $(x_1, x_2, \ldots, x_{3N})$. In the state of lowest energy, the values of these coordinates may be denoted by $(\overline{x}_1, \overline{x}_2, \ldots, \overline{x}_{3N})$. Denoting the displacements $(x_i - \overline{x}_i)$ of the atoms from their equilibrium positions by the variables $\xi_i (i = 1, 2, \ldots, 3N)$, the kinetic energy of the system in configuration (x_i) is given by

$$K = \frac{1}{2} m \sum_{i=1}^{3N} \dot{x}_i^2 = \frac{1}{2} m \sum_{i=1}^{3N} \dot{\xi}_i^2, \tag{1}$$

and the potential energy by

$$\begin{aligned} \Phi \equiv \Phi(x_i) = \Phi(\overline{x}_i) &+ \sum_i \left(\frac{\partial \Phi}{\partial x_i} \right)_{(x_i)=(\overline{x}_i)} (x_i - \overline{x}_i) \\ &+ \sum_{i,j} \frac{1}{2} \left(\frac{\partial^2 \Phi}{\partial x_i \partial x_j} \right)_{(x_i)=(\overline{x}_i)} (x_i - \overline{x}_i)(x_j - \overline{x}_j) + \cdots . \end{aligned} \tag{2}$$

The main term in this expansion represents the (minimum) energy of the solid when all the atoms are at rest at their mean positions $\overline{x}_i$; this energy may be denoted by the symbol Φ_0. The next set of terms is identically zero because the function $\Phi(x_i)$ has a minimum at $(x_i) = (\overline{x}_i)$ and hence all its first derivatives vanish there. The second-order terms of the expansion represent the *harmonic component* of the vibrations of the atoms about their mean positions. If we assume that the overall amplitude of these vibrations is not large we may retain only the harmonic terms of the expansion and neglect all successive ones; we are then working in the so-called *harmonic approximation*. Thus, we may write

$$H = \Phi_0 + \left\{ \sum_i \frac{1}{2} m \dot{\xi}_i^2 + \sum_{i,j} \alpha_{ij} \xi_i \xi_j \right\}, \tag{3}$$

where

$$\alpha_{ij} = \frac{1}{2}\left(\frac{\partial^2 \Phi}{\partial x_i \partial x_j}\right)_{(x_i)=(\bar{x}_i)}. \tag{4}$$

We now introduce a linear transformation, from the coordinates ξ_i to the so-called *normal coordinates* q_i, and choose the transformation matrix such that the new expression for the Hamiltonian does not contain any cross terms, that is,

$$H = \Phi_0 + \sum_i \frac{1}{2} m\left(\dot{q}_i^2 + \omega_i^2 q_i^2\right), \tag{5}$$

where $\omega_i (i = 1, 2, \ldots, 3N)$ are the *characteristic frequencies* of the *normal modes* of the system and are determined essentially by the quantities α_{ij} or, in turn, by the nature of the potential energy function $\Phi(x_i)$. Equation (5) suggests that the energy of the solid, over and above the (minimum) value Φ_0, may be considered as arising from a set of $3N$ one-dimensional, *noninteracting*, harmonic oscillators whose characteristic frequencies ω_i are determined by the interatomic interactions in the system.

Classically, each of the $3N$ normal modes of vibration corresponds to a wave of distortion of the lattice, that is, a sound wave. Quantum-mechanically, these modes give rise to quanta, called *phonons*, in much the same way as the vibrational modes of the electromagnetic field give rise to photons. There is one important difference, however, that is, while the number of normal modes in the case of an electromagnetic field is infinite, the number of normal modes (or the number of phonon energy levels) in the case of a solid is fixed by the number of lattice sites.[16] This introduces certain differences in the thermodynamic behavior of the sound field in contrast to the thermodynamic behavior of the radiation field; however, at low temperatures, where the high-frequency modes of the solid are not very likely to be excited, these differences become rather insignificant and we obtain a striking similarity between the two sets of results.

The thermodynamics of the solid can now be studied along the lines of Section 3.8. First of all, we note that the eigenvalues of the Hamiltonian (5) are

$$E\{n_i\} = \Phi_0 + \sum_i \left(n_i + \frac{1}{2}\right)\hbar\omega_i, \tag{6}$$

where the numbers n_i denote the "states of excitation" of the various oscillators (or, equally well, the occupation numbers of the various phonon levels in the system). The internal energy of the system is then given by

$$U(T) = \left\{\Phi_0 + \sum_i \frac{1}{2}\hbar\omega_i\right\} + \sum_i \frac{\hbar\omega_i}{e^{\hbar\omega_i/kT} - 1}. \tag{7}$$

[16]Of course, the number of phonons themselves is indefinite. As a result, the chemical potential of the phonon gas is also zero.

The expression within the curly brackets gives the energy of the solid at absolute zero. The term Φ_0 is negative and larger in magnitude than the total zero-point energy, $\sum_i \frac{1}{2}\hbar\omega_i$, of the oscillators: together, they determine the *binding energy* of the lattice. The last term in (7) represents the temperature-dependent part of the energy,[17] which determines the specific heat of the solid:

$$C_V(T) \equiv \left(\frac{\partial U}{\partial T}\right)_V = k \sum_i \frac{(\hbar\omega_i/kT)^2 e^{\hbar\omega_i/kT}}{(e^{\hbar\omega_i/kT} - 1)^2}. \tag{8}$$

To proceed further, we need to know the frequency spectrum of the solid. To obtain this from first principles is not an easy task. Accordingly, one obtains this spectrum either through experiment or by making certain plausible assumptions about it. Einstein, who was the first to apply the quantum concept to the theory of solids (1907), assumed, for simplicity, that the frequencies ω_i are all equal. Denoting this (common) value by ω_E, the specific heat of the solid is given by

$$C_V(T) = 3NkE(x), \tag{9}$$

where $E(x)$ is the so-called *Einstein function*:

$$E(x) = \frac{x^2 e^x}{(e^x - 1)^2}, \tag{10}$$

with

$$x = \hbar\omega_E/kT = \Theta_E/T. \tag{11}$$

The dashed curve in Figure 7.14 depicts the variation of the specific heat with temperature, as given by the Einstein formula (9). At sufficiently high temperatures, where $T \gg \Theta_E$ and hence $x \ll 1$, the Einstein result tends toward the classical one, namely $3Nk$.[18] At sufficiently low temperatures, where $T \ll \Theta_E$ and hence $x \gg 1$, the specific heat falls exponentially fast and tends to zero as $T \to 0$. The theoretical rate of fall, however, turns out to be too fast in comparison with the observed one. Nevertheless, Einstein's approach did at least provide a theoretical basis for understanding the observed departure of the specific heat of solids from the classical law of Dulong and Petit, whereby $C_V = 3R \simeq 5.96$ calories per mole per degree.

Debye (1912), on the other hand, allowed a *continuous spectrum* of frequencies, cut off at an upper limit ω_D such that the total number of normal modes of vibration is $3N$,

[17] The thermal energy of the solid may well be written as $\sum_i \langle n_i \rangle \hbar\omega_i$, where $\langle n_i \rangle \{= (e^{\hbar\omega_i/kT} - 1)^{-1}\}$ is the *mean occupation number* of the phonon level ε_i. Clearly, the phonons, like photons, obey Bose–Einstein statistics, with $\mu = 0$.

[18] Actually, when the temperature is high enough, so that *all* $(\hbar\omega_i/kT) \ll 1$, the general formula (8) itself reduces to the classical one. This corresponds to the situation when each of the $3N$ modes of vibration possesses a thermal energy kT.

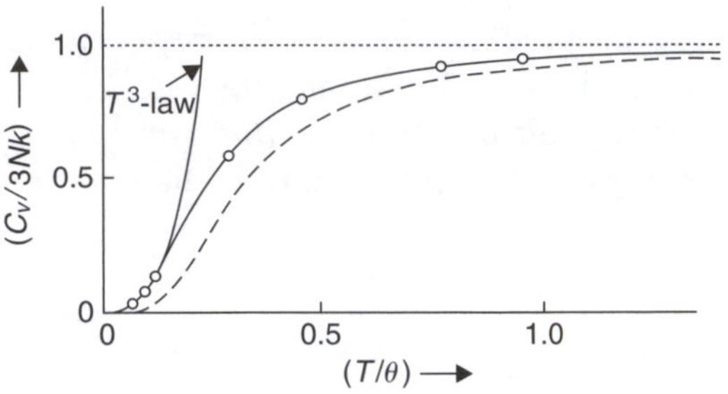

FIGURE 7.14 The specific heat of a solid, according to the Einstein model: - - -, and according to the Debye model: ——. The circles denote the experimental results for copper.

that is

$$\int_0^{\omega_D} g(\omega)d\omega = 3N, \tag{12}$$

where $g(\omega)d\omega$ denotes the number of normal modes of vibration whose frequency lies in the range $(\omega, \omega + d\omega)$. For $g(\omega)$, Debye adopted the Rayleigh expression (7.3.2), modified so as to suit the problem under study. Writing c_L for the velocity of propagation of the longitudinal modes and c_T for that of the transverse modes (and noting that, for any frequency ω, the transverse mode is doubly degenerate), equation (12) becomes

$$\int_0^{\omega_D} V\left(\frac{\omega^2 d\omega}{2\pi^2 c_L^3} + \frac{\omega^2 d\omega}{\pi^2 c_T^3}\right) = 3N, \tag{13}$$

from which one obtains for the cutoff frequency

$$\omega_D^3 = 18\pi^2 \frac{N}{V}\left(\frac{1}{c_L^3} + \frac{2}{c_T^3}\right)^{-1}. \tag{14}$$

Accordingly, the Debye spectrum may be written as

$$g(\omega) = \begin{cases} \dfrac{9N}{\omega_D^3}\omega^2 & \text{for } \omega \le \omega_D, \\ 0 & \text{for } \omega > \omega_D. \end{cases} \tag{15}$$

Before we proceed further to calculate the specific heat of solids on the basis of the Debye spectrum, two remarks seem to be in order. First, the Debye spectrum is only an idealization of the actual situation obtaining in a solid; it may, for instance, be compared with a typical spectrum such as the one shown in Figure 7.15. While for low-frequency modes (the so-called *acoustic* modes) the Debye approximation is reasonable, serious discrepancies are seen in the case of high-frequency modes (the so-called *optical* modes). At

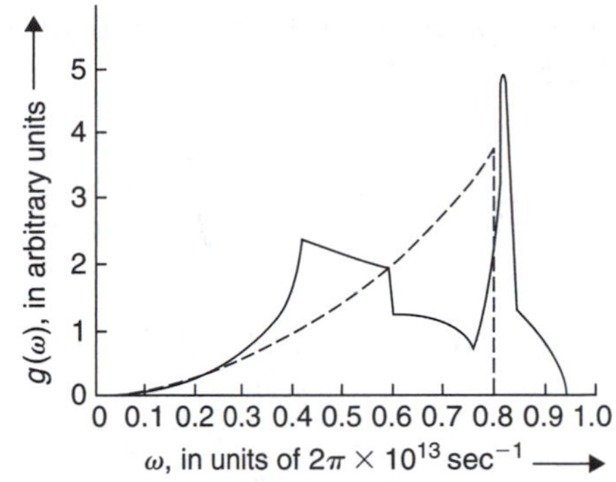

FIGURE 7.15 The normal-mode frequency distribution $g(\omega)$ for aluminum. The solid curve is derived from x-ray scattering measurements [Walker (1956)] while the dashed curve represents the corresponding Debye approximation.

any rate, for "averaged" quantities, such as the specific heat, the finer details of the spectrum are not very important. Second, the longitudinal and the transverse modes of the solid should have their own cutoff frequencies, $\omega_{D,L}$ and $\omega_{D,T}$ say, rather than a common cutoff at ω_D, for the simple reason that, of the $3N$ normal modes of the lattice, exactly N are longitudinal and $2N$ transverse. Accordingly, we should have, instead of (13),

$$\int_0^{\omega_{D,L}} V\frac{\omega^2 d\omega}{2\pi^2 c_L^3} = N \quad \text{and} \quad \int_0^{\omega_{D,T}} V\frac{\omega^2 d\omega}{\pi^2 c_T^3} = 2N. \tag{16}$$

We note that the two cutoffs correspond to a *common wavelength* $\lambda_{\text{min}}\{= (4\pi V/3N)^{1/3}\}$, which is comparable to the *mean interatomic distance* in the solid. This is quite reasonable because, for wavelengths shorter than λ_{min}, it would be rather meaningless to speak of a wave of atomic displacements.

In the Debye approximation, formula (8) gives

$$C_V(T) = 3NkD(x_0), \tag{17}$$

where $D(x_0)$ is the so-called *Debye function*:

$$D(x_0) = \frac{3}{x_0^3}\int_0^{x_0} \frac{x^4 e^x dx}{(e^x - 1)^2}, \tag{18}$$

with

$$x_0 = \frac{\hbar\omega_D}{kT} = \frac{\Theta_D}{T}, \tag{19}$$

Θ_D being the so-called *Debye temperature* of the solid. Integrating by parts, the expression for the Debye function becomes

$$D(x_0) = -\frac{3x_0}{e^{x_0} - 1} + \frac{12}{x_0^3} \int_0^{x_0} \frac{x^3 dx}{e^x - 1}. \tag{20}$$

For $T \gg \Theta_D$, which means $x_0 \ll 1$, the function $D(x_0)$ may be expressed as a power series in x_0:

$$D(x_0) = 1 - \frac{x_0^2}{20} + \cdots. \tag{21}$$

Thus, as $T \to \infty, C_V \to 3Nk$; moreover, according to this theory, the classical result should be applicable to within $\frac{1}{2}$ percent so long as $T > 3\Theta_D$. For $T \ll \Theta_D$, which means $x_0 \gg 1$, the function $D(x_0)$ may be written as

$$\begin{aligned} D(x_0) &= \frac{12}{x_0^3} \int_0^{\infty} \frac{x^3 dx}{e^x - 1} + O(e^{-x_0}), \\ &\approx \frac{4\pi^4}{5x_0^3} = \frac{4\pi^4}{5}\left(\frac{T}{\Theta_D}\right)^3. \end{aligned} \tag{22}$$

Thus, at low temperatures the specific heat of the solid obeys the *Debye T^3-law*:

$$C_V = \frac{12\pi^4}{5} Nk \left(\frac{T}{\Theta_D}\right)^3 = 464.4 \left(\frac{T}{\Theta_D}\right)^3 \text{ cal mole}^{-1}\text{K}^{-1}. \tag{23}$$

It is clear from equation (23) that a measurement of the low-temperature specific heat of a solid should enable us not only to check the validity of the T^3-law but also to obtain an empirical value of the Debye temperature Θ_D.[19] The value of Θ_D can also be obtained by computing the cutoff frequency ω_D from a knowledge of the parameters $N/V, c_L$ and c_T; see equations (14) and (19). The closeness of these estimates is further evidence in favor of Debye's theory. Once Θ_D is known, the whole temperature range can be covered theoretically by making use of the tabulated values of the function $D(x_0)$.[20] A typical case was shown earlier in Figure 7.14. We saw that not only was the T^3-law obeyed at low temperatures, but also the agreement between theory and experiment was good throughout the range of observations.

[19] It can be shown that, according to this theory, deviations from the T^3-law should not exceed 2 percent so long as $T < \Theta_D/10$. However, in the case of *metals*, one cannot expect to reach a true T^3-region because, well before that, the specific heat of the electron gas might become a dominant contribution (see Section 8.3); unless the two contributions are separated out, one is likely to obtain a somewhat suppressed value of Θ_D from these observations.

[20] See, for example, Fowler and Guggenheim (1960, p. 144).

As another illustration of agreement in the low-temperature regime, we include here another plot, Figure 7.16, which is based on data obtained with the KCl crystal at temperatures below 5 K; see Keesom and Pearlman (1953). Here, the observed values of C_V/T are plotted against T^2. It is evident that the data fall quite well on a straight line from whose slope the value of Θ_D can be determined. One thus obtains, for KCl, $\Theta_D = 233 \pm 3$ K, which is in reasonable agreement with the values of 230 to 246 K coming from various estimates of the relevant elastic constants.

In Table 7.1 we list the values of Θ_D for several crystals, as derived from the specific heat measurements and from the values of the elastic constants.

In general, if the specific heat measurements of a given system conform to a T^3-law, one may infer that the thermal excitations in the system are accounted for *solely* by phonons. We expect something similar to happen in liquids as well, with two important differences. First, since liquids cannot withstand shear stress they cannot sustain transverse modes of vibration; a liquid composed of N atoms will, therefore, have only N longitudinal modes of vibration. Second, the normal modes of a liquid cannot be expected to be strictly harmonic; consequently, in addition to phonons, we might have other types of excitation such as *vortex flow* or *turbulence* (or even a modified kind of excitation, such as *rotons* in liquid He^4).

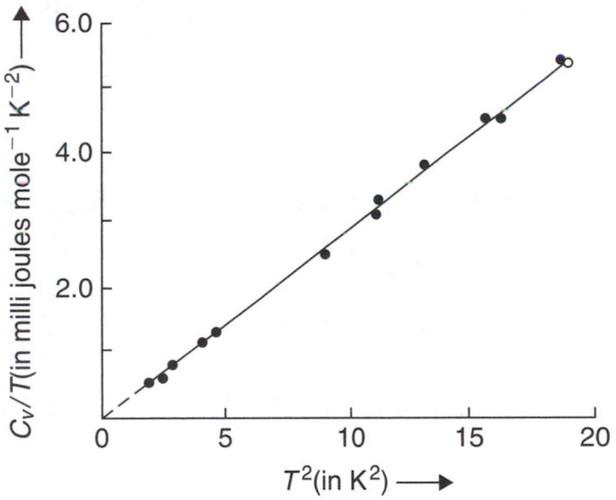

FIGURE 7.16 A plot of (C_V/T) versus T^2 for KCl, showing the validity of the Debye T^3-law. The experimental points are from Keesom and Pearlman (1953).

Table 7.1 The Values of the Debye Temperature Θ_D for Different Crystals

Crystal	Pb	Ag	Zn	Cu	Al	C	NaCl	KCl	MgO
Θ_D from the specific heat measurements	88	215	308	345	398	~1850	308	233	~850
Θ_D from the elastic constants	73	214	305	332	402	–	320	240	~950

Now, helium is the only substance that remains liquid at temperatures low enough to exhibit the T^3-behavior. In the case of the lighter isotope, He^3, the results are strongly influenced by the Fermi–Dirac statistics; as a result, a specific heat proportional to the *first* power of T dominates the scene (see Section 8.1). In the case of the heavier isotope, He^4, the low-temperature situation is completely governed by phonons; accordingly, we expect its specific heat to be given by, see equations (16) and (23),

$$C_V = \frac{4\pi^4}{5} Nk \left(\frac{kT}{\hbar \omega_D} \right)^3, \tag{24}$$

where

$$\omega_D = \left(\frac{6\pi^2 N}{V} \right)^{1/3} c, \tag{25}$$

c being the velocity of sound in the liquid. The specific heat per unit mass of the liquid is then given by

$$c_V = \frac{2\pi^2 k^4}{15 \rho \hbar^3 c^3} T^3, \tag{26}$$

where ρ is the mass density. Substituting $\rho = 0.1455\,\text{g/cm}^3$ and $c = 238\,\text{m/s}$, the foregoing result becomes

$$c_V = 0.0209 T^3 \,\text{joule}\,\text{g}^{-1}\text{K}^{-1}. \tag{27}$$

The experimental measurements of Wiebes et al. (1957), for $0 < T < 0.6\,\text{K}$, conformed to the expression

$$c_V = (0.0204 \pm 0.0004) T^3 \,\text{joule}\,\text{g}^{-1}\text{K}^{-1}. \tag{28}$$

The agreement between the theoretical result and the experimental observations is clearly good.

7.5 Inertial density of the sound field

For further understanding of the low-temperature behavior of liquid He^4, we determine the "inertial mass" associated with a gas of sound quanta in thermal equilibrium. For this, we consider "a phonon gas in mass motion," for then by determining the relationship between the momentum $\boldsymbol{P}$ of the gas and the velocity $\boldsymbol{v}$ of its mass motion we can readily evaluate the property in question. Now, since the total number of phonons in the system is indefinite, the problem is free from the constraint of a fixed N; consequently, the undetermined multiplier α may be taken to be identically zero. However, we now have a new constraint on the system, namely that of a fixed total momentum $\boldsymbol{P}$, additional to the constraint of the fixed total energy E. Under these constraints, the mean occupation number of the phonon level $\varepsilon(\boldsymbol{p})$ would be

$$\langle n(\boldsymbol{p})\rangle = \frac{1}{\exp(\beta\varepsilon + \boldsymbol{\gamma}\cdot\boldsymbol{p}) - 1}. \tag{1}$$

As usual, the parameter β is equal to $1/kT$. To determine $\boldsymbol{\gamma}$, it seems natural to evaluate the drift velocity of the gas. Choosing the z-axis in the direction of the mass motion, the magnitude v of the drift velocity will be given by "the mean value of the component u_z of the individual phonon velocities":

$$v = \langle u\cos\theta\rangle. \tag{2}$$

Now, for phonons

$$\varepsilon = pc \quad \text{and} \quad u \equiv \frac{d\varepsilon}{dp} = c, \tag{3}$$

where c is the *velocity of sound* in the medium. Moreover, by reasons of symmetry, we expect the undetermined vector $\boldsymbol{\gamma}$ to be either parallel or antiparallel to the direction of mass motion; hence, we may write

$$\boldsymbol{\gamma}\cdot\boldsymbol{p} = \gamma_z p_z = \gamma_z p\cos\theta. \tag{4}$$

In view of equations (1), (3), and (4), equation (2) becomes

$$v = \frac{\int_0^\infty\int_0^\pi [\exp\{\beta pc(1 + (\gamma_z/\beta c)\cos\theta)\} - 1]^{-1}(c\cos\theta)(p^2 dp\, 2\pi\sin\theta\, d\theta)}{\int_0^\infty\int_0^\pi [\exp\{\beta pc(1 + (\gamma_z/\beta c)\cos\theta)\} - 1]^{-1}(p^2 dp\, 2\pi\sin\theta\, d\theta)}. \tag{5}$$

Making the substitutions

$$\cos\theta = \eta, \quad p(1 + (\gamma_z/\beta c)\eta) = p'$$

and cancelling away the integrations over p', we obtain

$$v = c\frac{\int_{-1}^{1}(1 + (\gamma_z/\beta c)\eta)^{-3}\eta\, d\eta}{\int_{-1}^{1}(1 + (\gamma_z/\beta c)\eta)^{-3} d\eta} = -\gamma_z/\beta.$$

It follows that

$$\boldsymbol{\gamma} = -\beta\boldsymbol{v}. \tag{6}$$

Accordingly, the expression for the mean occupation number becomes

$$\langle n(\boldsymbol{p})\rangle = \frac{1}{\exp\{\beta(\varepsilon - \boldsymbol{v}\cdot\boldsymbol{p})\} - 1}. \tag{7}$$

A comparison of (7) with the corresponding result in the rest frame of the gas, namely

$$\langle n_0(\boldsymbol{p}_0)\rangle = \frac{1}{\exp(\beta\varepsilon_0) - 1}, \tag{8}$$

shows that the change caused by the imposition of mass motion on the system is nothing but a straightforward manifestation of the *Galilean transformation* between the two frames of reference.

Alternatively, equation (7) may be written as

$$\langle n(\boldsymbol{p})\rangle = \frac{1}{\exp(\beta p' c) - 1} = \frac{1}{\exp\{\beta pc(1-(v/c)\cos\theta)\} - 1}. \tag{9}$$

As such, formula (9) lays down a serious restriction on the drift velocity v, that is, it must not exceed c, the velocity of the phonons, for otherwise some of the occupation numbers would become negative! Actually, as our subsequent analysis will show, the formalism developed in this section breaks down as v approaches c. The velocity c may, therefore, be regarded as the *critical velocity* for the flow of the phonon gas:

$$(v_c)_{\text{ph}} = c. \tag{10}$$

The relevance of this result to the problem of superfluidity in liquid helium II will be seen in the following section.

Next we now calculate the total momentum $\boldsymbol{P}$ of the phonon gas:

$$\boldsymbol{P} = \sum_{\boldsymbol{p}} \langle n(\boldsymbol{p})\rangle \boldsymbol{p}. \tag{11}$$

Indeed, the vector $\boldsymbol{P}$ will be parallel to the vector $\boldsymbol{v}$, the latter being already in the direction of the z-axis. We have, therefore, to calculate only the z-component of the momentum:

$$\begin{aligned} P = P_z &= \sum_{\boldsymbol{p}} \langle n(\boldsymbol{p})\rangle p_z \\ &= \int_0^\infty \int_0^\pi \frac{p\cos\theta}{\exp\{\beta pc(1-(v/c)\cos\theta)\} - 1} \left(\frac{Vp^2 dp\, 2\pi \sin\theta\, d\theta}{h^3}\right) \\ &= \frac{2\pi V}{h^3} \int_0^\infty \frac{p'^3 dp'}{\exp(\beta p' c) - 1} \int_0^\pi \{1-(v/c)\cos\theta\}^{-4} \cos\theta \sin\theta\, d\theta \\ &= V\frac{16\pi^5}{45h^3c^3\beta^4} \cdot \frac{v/c^2}{(1-v^2/c^2)^3}. \end{aligned} \tag{12}$$

The total energy E of the gas is given by

$$\begin{aligned}E &= \sum_{\boldsymbol{p}} \langle n(\boldsymbol{p})\rangle pc \\ &= \frac{2\pi Vc}{h^3}\int_0^\infty \frac{p'^3 dp'}{\exp(\beta p'c)-1}\int_0^\pi \{1-(v/c)\cos\theta\}^{-4}\sin\theta\, d\theta \\ &= V\frac{4\pi^5}{15h^3c^3\beta^4}\frac{1+\frac{1}{3}v^2/c^2}{(1-v^2/c^2)^3}. \end{aligned} \tag{13}$$

It is now natural to regard the ratio P/v as the "inertial mass" of the phonon gas. The corresponding mass density ρ is, therefore, given by

$$\rho = \frac{P}{vV} = \frac{16\pi^5 k^4 T^4}{45h^3c^5}\frac{1}{(1-v^2/c^2)^3}. \tag{14}$$

For $(v/c) \ll 1$, which is generally true, the mass density of the phonon gas is given by

$$(\rho_0)_{\mathrm{ph}} = \frac{16\pi^5 k^4}{45h^3c^5}T^4 = \frac{4}{3c^2}(E_0/V). \tag{15}$$

Substituting the value of c for liquid He^4 at low temperatures, the phonon mass density, as a fraction of the actual density of the liquid, is given by

$$(\rho_0)_{\mathrm{ph}}/\rho_{\mathrm{He}} = 1.22\times 10^{-4}T^4; \tag{16}$$

thus, for example, at $T = 0.3\,\mathrm{K}$ the value of this fraction turns out to be about 9.9×10^{-7}. Now, at a temperature like 0.3 K, phonons are the only excitations in liquid He^4 that need to be considered; the calculated result should, therefore, correspond to the "ratio of the density ρ_n of the *normal fluid* in the liquid to the total density ρ of the liquid." It is practically impossible to make a *direct* determination of a fraction as small as that; however, *indirect* evaluations that make use of other experimentally viable properties of the liquid provide a striking confirmation of the foregoing result; see Figure 7.17.

7.6 Elementary excitations in liquid helium II

Landau (1941, 1947) developed a simple theoretical scheme that explains reasonably well the behavior of liquid helium II at low temperatures not too close to the λ-point. According to this scheme, the liquid is treated as a weakly excited quantum-mechanical system, in which deviations from the ground state ($T = 0\,\mathrm{K}$) are described in terms of "a gas of elementary excitations" hovering over a *quiescent* background. The gas of excitations corresponds to the "normal fluid," while the quiescent background represents the "superfluid."

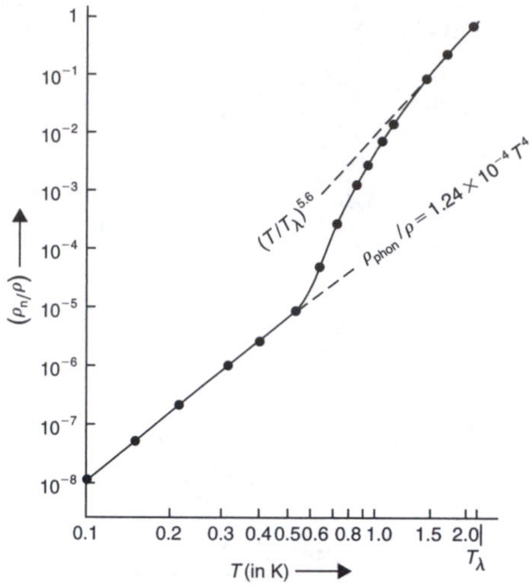

FIGURE 7.17 The normal fraction (ρ_n/ρ), as obtained from experimental data on (i) the velocity of second sound and (ii) the entropy of liquid He II (after de Klerk, Hudson, and Pellam, 1953).

At $T = 0$ K, there are no excitations at all ($\rho_n = 0$) and the whole of the fluid constitutes the superfluid background ($\rho_s = \rho_{He}$). At higher temperatures, we may write

$$\rho_s(T) = \rho_{He}(T) - \rho_n(T), \tag{1}$$

so that at $T = T_\lambda$, $\rho_n = \rho_{He}$ and $\rho_s = 0$. At $T > T_\lambda$, the liquid behaves in all respects as a normal fluid, commonly known as liquid helium I.

Guided by purely empirical considerations, Landau also proposed an energy–momentum relationship $\varepsilon(p)$ for the elementary excitations in liquid helium II. At low momenta, the relationship between ε and p was linear (which is characteristic of phonons), while at higher momenta it exhibited a nonmonotonic character. The excitations were assumed to be bosons and, at low temperatures (when their number is not very large), mutually noninteracting; the macroscopic properties of the liquid could then be calculated by following a straightforward statistical-mechanical approach. It was found that Landau's theory could explain quite successfully the observed properties of liquid helium II over a temperature range of about 0 to 2 K; however, it still remained to be verified that the actual excitations in the liquid did, in fact, conform to the proposed energy spectrum.

Following a suggestion by Cohen and Feynman (1957), a number of experimental workers set out to investigate the spectrum of excitations in liquid helium II by scattering long-wavelength neutrons ($\lambda \gtrsim 4$Å) from the liquid. At temperatures below 2 K, the most important scattering process is the one in which a neutron creates a *single* excitation in the liquid. By measuring the modified wavelength λ_f of the neutrons scattered at an angle ϕ, the energy ε and the momentum p of the excitation created in the scattering process could be determined on the basis of the relevant conservation laws:

$$\varepsilon = h^2(\lambda_i^{-2} - \lambda_f^{-2})/2m, \tag{2}$$

$$p^2 = h^2(\lambda_i^{-2} + \lambda_f^{-2} - 2\lambda_i^{-1}\lambda_f^{-1}\cos\phi), \tag{3}$$

where λ_i is the initial wavelength of the neutrons and m the neutron mass. By varying ϕ, or λ_i, one could map the entire spectrum of the excitations.

The first exhaustive investigation along these lines was carried out by Yarnell et al. (1959); their results, shown in Figure 7.18, possess a striking resemblance to the empirical spectrum proposed by Landau. The more important features of the spectrum, which was obtained at a temperature of 1.1 K, are the following:

(i) If we fit a linear, *phonon-like* spectrum ($\varepsilon = pc$) to points in the vicinity of $p/\hbar = 0.55\text{Å}^{-1}$, we obtain for c a value of (239 ± 5) m/s, which is in excellent agreement with the measured value of the velocity of sound in the liquid, namely about 238 m/s.

(ii) The spectrum passes through a maximum value of $\varepsilon/k = (13.92 \pm 0.10)$ K at $p/\hbar = (1.11 \pm 0.02)\,\text{Å}^{-1}$.

(iii) This is followed by a minimum at $p/\hbar = (1.92 \pm 0.01)\text{Å}^{-1}$, whose neighborhood may be represented by Landau's *roton* spectrum:

$$\varepsilon(p) = \Delta + \frac{(p - p_0)^2}{2\mu}, \tag{4}$$

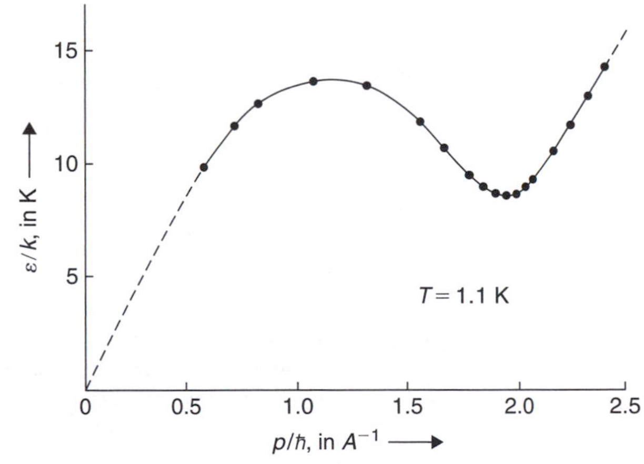

FIGURE 7.18 The energy spectrum of the elementary excitations in liquid He II at 1.1 K [after Yarnell et al. (1959)]; the dashed line emanating from the origin has a slope corresponding to the velocity of sound in the liquid, namely (239 ± 5) m/s.

with

$$\Delta/k = (8.65 \pm 0.04)\text{K},$$
$$p_0/\hbar = (1.92 \pm 0.01)\text{Å}^{-1}, \tag{5}$$[21]

and

$$\mu = (0.16 \pm 0.01)m_{\text{He}}.$$

(iv) Above $p/\hbar \simeq 2.18\text{Å}^{-1}$, the spectrum rises linearly, again with a slope equal to c. Data were also obtained at temperatures 1.6 K and 1.8 K. The spectrum was found to be of the same general shape as at 1.1 K; only the value of Δ was slightly lower.

In a later investigation, Henshaw and Woods (1961) extended the range of observation at both ends of the spectrum; their results are shown in Figure 7.19. On the lower side, they carried out measurements down to $p/\hbar = 0.26\text{Å}^{-1}$ and found that the experimental

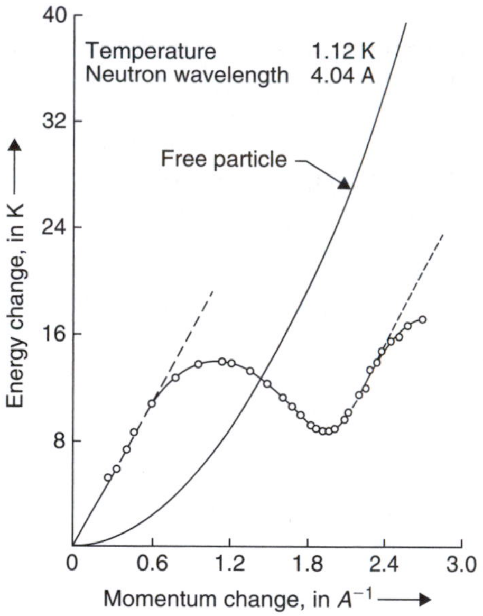

FIGURE 7.19 The energy spectrum of the elementary excitations in liquid He II at 1.12 K (after Henshaw and Woods, 1961); the dashed straight lines have a common slope corresponding to the velocity of sound in the liquid, namely 237 m/s. The parabolic curve rising from the origin represents the energy spectrum, $\varepsilon(p) = p^2/2m$, of *free* helium atoms.

[21]The term "roton" for these excitations was coined by Landau who had originally thought that these excitations might, in some way, represent local disturbances of a *rotational* character in the liquid. However, subsequent theoretical work, especially that of Feynman (1953, 1954) and of Brueckner and Sawada (1957), did not support this contention. Nevertheless, the term "roton" has remained.

points indeed lie on a straight line (of slope 237 m/s). On the upper side, they pushed their measurements up to $p/\hbar = 2.68\,\text{Å}^{-1}$ and found that, after passing through a minimum at $1.91\,\text{Å}^{-1}$, the curve rises with an increasing slope up to about $2.4\,\text{Å}^{-1}$ at which point the second derivative $\partial^2\varepsilon/\partial p^2$ changes sign; the subsequent trend of the curve suggests the possible existence of a second maximum in the spectrum![22]

To evaluate the thermodynamics of liquid helium II, we first of all note that at sufficiently low temperatures we have only low-lying excitations, namely the phonons. The thermodynamic behavior of the liquid is then governed by formulae derived in Sections 7.4 and 7.5. At temperatures higher than about 0.5 K, the second group of excitations, namely the rotons (with momenta in the vicinity of p_0), also shows up. Between 0.5 K and about 1 K, the behavior of the liquid is governed by phonons and rotons together. Above 1 K, however, the phonon contributions to the various thermodynamic properties of the liquid become rather unimportant; then, rotons are the only excitations that need to be considered.

We shall now study the temperature dependence of the roton contributions to the various thermodynamic properties of the liquid. In view of the continuity of the energy spectrum, it is natural to expect that, like phonons, rotons also obey Bose–Einstein statistics. Moreover, their total number N in the system is quite indefinite; consequently, their chemical potential μ is identically zero. We then have for the mean occupation numbers of the rotons

$$\langle n(\boldsymbol{p})\rangle = \frac{1}{\exp\{\beta\varepsilon(p)\} - 1}, \tag{6}$$

where $\varepsilon(p)$ is given by equations (4) and (5). Now, at all temperatures of interest (namely $T \leq 2\,\text{K}$), the minimum value of the term $\exp\{\beta\varepsilon(p)\}$, namely $\exp(\Delta/kT)$, is considerably larger than unity. We may, therefore, write

$$\langle n(\boldsymbol{p})\rangle \simeq \exp\{-\beta\varepsilon(p)\}. \tag{7}$$

The q-potential of the system of rotons is, therefore, given by

$$q(V,T) \equiv \frac{PV}{kT} = -\sum_{\boldsymbol{p}} \ln[1 - \exp\{-\beta\varepsilon(p)\}] \simeq \sum_{\boldsymbol{p}} \exp\{-\beta\varepsilon(p)\} \simeq \overline{N}, \tag{8}$$

where $\overline{N}$ is the "equilibrium" number of rotons in the system. The summation over $\boldsymbol{p}$ may be replaced by integration, with the result

$$\frac{PV}{kT} = \overline{N} = \frac{V}{h^3}\int_0^\infty e^{-\left\{\Delta + \frac{(p-p_0)^2}{2\mu}\right\}\Big/kT}(4\pi p^2\,dp). \tag{9}$$

[22]This seems to confirm a remarkable prediction by Pitaevskii (1959) that an end point in the spectrum might occur at a "critical" value p_c of the excitation momentum where ε_c is equal to 2Δ and $(\partial\varepsilon/\partial p)_c$ is zero.

Substituting $p = p_0 + (2\mu kT)^{1/2}x$, we get

$$\frac{PV}{kT} = \overline{N} = \frac{4\pi p_0^2 V}{h^3} e^{-\Delta/kT}(2\mu kT)^{1/2} \int e^{-x^2} \left\{1 + \frac{(2\mu kT)^{1/2}}{p_0} x\right\}^2 dx. \tag{10}$$

The "relevant" range of the variable x that makes a significant contribution toward this integral is fairly symmetric about the value $x = 0$; consequently, the net effect of the *linear* term in the integrand is vanishingly small. The *quadratic* term too is unimportant because its coefficient $(2\mu kT)/p_0^2 \ll 1$. Thus, all we have to consider is the integral of $\exp(-x^2)$. Now, one can readily verify that the limits of this integral are such that, without seriously affecting the value of the integral, they may be taken as $-\infty$ and $+\infty$; the value of the integral is then simply $\pi^{1/2}$. We thus obtain

$$\frac{PV}{kT} = \overline{N} = \frac{4\pi p_0^2 V}{h^3}(2\pi\mu kT)^{1/2} e^{-\Delta/kT}. \qquad (11)^{23}$$

The free energy of the roton gas is given by (since $\mu = 0$)

$$A = -PV = -\overline{N}kT \propto T^{3/2} e^{-\Delta/kT}, \tag{12}$$

which gives

$$S = -\left(\frac{\partial A}{\partial T}\right)_V = -A\left\{\frac{3}{2T} + \frac{\Delta}{kT^2}\right\} = \overline{N}k\left\{\frac{3}{2} + \frac{\Delta}{kT}\right\}, \tag{13}$$

$$U = A + TS = \overline{N}\left(\Delta + \frac{1}{2}kT\right) \qquad (14)^{24}$$

and

$$C_V = \left(\frac{\partial U}{\partial T}\right)_V = \overline{N}k\left\{\frac{3}{4} + \frac{\Delta}{kT} + \left(\frac{\Delta}{kT}\right)^2\right\}. \tag{15}$$

Clearly, as $T \to 0$, all these results tend to zero (essentially exponentially).

We now determine the inertial mass density of the roton gas. Proceeding as in Section 7.5, we obtain for a gas of excitations with energy spectrum $\varepsilon(p)$

$$\rho_0 = \frac{M_0}{V} = \lim_{v \to 0} \frac{1}{\boldsymbol{v}} \int n(\varepsilon - \boldsymbol{v} \cdot \boldsymbol{p})\boldsymbol{p} \frac{d^3p}{h^3}, \tag{16}$$

[23] Looking back at integral (9), what we have done here amounts to replacing p^2 in the integrand by its mean value p_0^2 and then carrying out integration over the "complete" range of the variable $(p - p_0)$.

[24] This result is highly suggestive of the fact that for rotons there is only *one* true degree of freedom, namely the *magnitude* of the roton momentum, that is thermally effective!

where $n(\varepsilon - \boldsymbol{v}\cdot\boldsymbol{p})$ is the mean occupation number of the state $\varepsilon(p)$, as observed in a frame of reference K with respect to which the gas is in *mass motion* with a drift velocity $\boldsymbol{v}$.[25] For small v, the function $n(\varepsilon - \boldsymbol{v}\cdot\boldsymbol{p})$ may be expanded as a Taylor series in $\boldsymbol{v}$ and only the terms $n(\varepsilon) - (\boldsymbol{v}\cdot\boldsymbol{p})\partial n(\varepsilon)/\partial\varepsilon$ retained. The integral over the first part denotes the momentum density of the system, as observed in the rest frame K_0, and is identically zero. We are thus left with

$$\rho_0 = -\frac{1}{h^3}\int p^2\cos^2\theta\,\frac{\partial n(\varepsilon)}{\partial\varepsilon}(p^2dp\,2\pi\sin\theta d\theta)$$
$$= -\frac{4\pi}{3h^3}\int_0^\infty \frac{\partial n(\varepsilon)}{\partial\varepsilon}p^4dp, \tag{17}$$

which holds for *any* energy spectrum and for *any* statistics.

For phonons, we obtain

$$(\rho_0)_{\text{ph}} = -\frac{4\pi}{3h^3c}\int_0^\infty \frac{dn(p)}{dp}p^4dp$$
$$= -\frac{4\pi}{3h^3c}\left[n(p)\cdot p^4\Big|_0^\infty - \int_0^\infty n(p)\cdot 4p^3dp\right]$$
$$= \frac{4}{3c^2}\int_0^\infty n(p)\cdot pc\left(\frac{4\pi p^2dp}{h^3}\right) = \frac{4}{3c^2}(E_0)_{\text{ph}}/V, \tag{18}$$

which is identical to our earlier result (7.5.15).

For rotons, $n(\varepsilon) \simeq \exp(-\beta\varepsilon)$; hence, $\partial n(\varepsilon)/\partial\varepsilon \simeq -\beta n(\varepsilon)$. Accordingly, by (17),

$$(\rho_0)_{\text{rot}} = \frac{4\pi\beta}{3h^3}\int n(\varepsilon)p^4dp$$
$$= \frac{\beta}{3}\langle p^2\rangle\frac{\overline{N}}{V} \simeq \frac{p_0^2}{3kT}\frac{\overline{N}}{V} \tag{19}$$
$$= \frac{4\pi p_0^4}{3h^3}\left(\frac{2\pi\mu}{kT}\right)^{1/2}e^{-\Delta/kT}; \tag{20}$$

At very low temperatures ($T < 0.3$ K), the roton contribution toward the inertia of the fluid is negligible in comparison with the phonon contribution. At relatively higher temperatures ($T \sim 0.6$ K), the two contributions become comparable. At temperatures above 1 K, the roton contribution is far more dominant than the phonon contribution; at such temperatures, the roton density alone accounts for the density ρ_n of the normal fluid.

[25]The drift velocity $\boldsymbol{v}$ must satisfy the condition $(\boldsymbol{v}\cdot\boldsymbol{p}) \le \varepsilon$, for otherwise some of the occupation numbers will become negative! This leads to the existence of a *critical velocity* v_c for these excitations, such that for v exceeding v_c the formalism developed here would break down. It is not difficult to see that this (critical) velocity is given by the relation $v_c = (\varepsilon/p)_{\min}$, as in equation (24).

It would be instructive to determine the *critical temperature* T_c at which the theoretical value of the density ρ_n became equal to the actual density ρ_{He} of the liquid; this would mean the disappearance of the *superfluid component* of the liquid (and hence a transition from liquid He II to liquid He I). In this manner, we find that $T_c \simeq 2.5$K, as opposed to the experimental value of T_λ, which is $\simeq 2.19$ K. The comparison is not too bad, considering the fact that in the present calculation we have assumed the roton gas to be a *noninteracting* system right up to the transition point; in fact, due to the presence of an exceedingly large number of excitations at higher temperatures, this assumption would not remain plausible.

Equation (19) suggests that a roton excitation possesses an *effective mass* $p_0^2/3kT$. Numerically, this is about 10 to 15 times the mass of a helium atom (and, hence, orders of magnitude larger than the parameter μ of the roton spectrum). However, the more important aspect of the roton effective mass is that it is *inversely* proportional to the temperature of the roton gas! Historically, this aspect was first discovered empirically by Landau (1947) on the basis of the experimental data on the velocity of second sound in liquid He II and its specific heat. Now, since the effective mass of an excitation is generally determined by the quantity $\langle p^2 \rangle/3kT$, Landau concluded that the quantity $\langle p^2 \rangle$ of the relevant excitations in this liquid must be temperature-independent. Thus, as the temperature of the liquid rises, the mean value of p^2 of the excitations must stay constant; this value may be denoted by p_0^2. The mean value of ε, on the other hand, must rise with temperature. The only way to reconcile the two was to invoke a *nonmonotonic* spectrum with a minimum at $p = p_0$.

Finally, we would like to touch on the question of the *critical velocity* of superflow. For this, we consider a mass M of excitation-free superfluid in mass motion; its kinetic energy E and momentum $\boldsymbol{P}$ are given by $\frac{1}{2}Mv^2$ and $M\boldsymbol{v}$, respectively. Any changes in these quantities are related as follows:

$$\delta E = (\boldsymbol{v} \cdot \delta \boldsymbol{P}). \tag{21}$$

Supposing that these changes came about as a result of the creation of an excitation $\varepsilon(\boldsymbol{p})$ in the fluid, we must have, by the principles of conservation,

$$\delta E = -\varepsilon \quad \text{and} \quad \delta \boldsymbol{P} = -\boldsymbol{p}. \tag{22}$$

Equations (21) and (22) lead to the result

$$\varepsilon = (\boldsymbol{v} \cdot \boldsymbol{p}) \leq vp. \tag{23}$$

Thus, it is impossible to create an excitation $\varepsilon(p)$ in the fluid unless the drift velocity v of the fluid is greater than, or at least equal to, the quantity (ε/p). Accordingly, if v is less than even the *lowest* value of ε/p, no excitation at all can be created in the fluid, which will therefore maintain its superfluid character. We thus obtain a condition for the maintenance of superfluidity, namely

$$v < v_c = (\varepsilon/p)_{\min}, \tag{24}$$

which is known as the *Landau criterion* for superflow. The velocity v_c is called the *critical velocity* of superflow; it marks an "upper limit" to the flow velocities at which the fluid exhibits superfluid behavior. The observed magnitude of the critical velocity varies significantly with the geometry of the channel employed; as a rule, the narrower the channel the larger the critical velocity. The observed values of v_c range from about 0.1 cm/s to about 70 cm/s.

The theoretical estimates of v_c are clearly of interest. On one hand, we find that if the excitations obey the ideal-gas relationship, namely $\varepsilon = p^2/2m$, then the critical velocity turns out to be exactly zero. Any velocity v is then greater than the critical velocity; accordingly, no superflow is possible at all. This is a very significant result, for it brings out very clearly the fact that interatomic interactions in the liquid, which give rise to an excitation spectrum different from the one characteristic of the ideal gas, play a fundamental role in bringing about the phenomenon of superfluidity. Thus, while an ideal Bose gas does undergo the phenomenon of Bose–Einstein condensation, it cannot support the phenomenon of superfluidity as such! On the other hand, we find that (i) for phonons, $v_c = c \simeq 2.4 \times 10^4$ cm/s and (ii) for rotons, $v_c = \{(p_0^2 + 2\mu\Delta)^{1/2} - p_0\}/\mu \simeq \Delta/p_0 \simeq 6.3 \times 10^3$ cm/s, which are too high in comparison with the observed values of v_c. In fact, there is another type of collective excitations that can appear in liquid helium II, namely *quantized vortex rings*, with energy–momentum relationship of the form: $\varepsilon \propto p^{1/2}$. The critical velocity for the creation of these rings turns out to be numerically consistent with the experimental findings; not only that, the dependence of v_c on the geometry of the channel can also be understood in terms of the size of the rings created.

For a review of this topic, especially in regard to Feynman's contributions, see Mehra and Pathria (1994); see also Sections 11.4 through 11.6 of this text.

Quantized dissipationless bosonic flow has also been observed in the *solid* phase of helium-4. This "supersolid" behavior was observed by Kim and Chan (2004a, 2004b) using a torsional oscillator containing solid helium infused silica with atomic-sized pores. At $P = 60$ atm, the torsional frequency increases abruptly for temperatures below 175 mK. These authors interpret this result as helium atoms in the solid phase in the pores being free to flow without dissipation.

Problems

7.1. By considering the order of magnitude of the occupation numbers $\langle n_\varepsilon \rangle$, show that it makes no difference to the final results of Section 7.1 if we combine a *finite* number of $(\varepsilon \neq 0)$-terms of the sum (7.1.2) with the $(\varepsilon = 0)$-part of equation (7.1.6) or include them in the integral over ε.

7.2. Deduce the virial expansion (7.1.13) from equations (7.1.7) and (7.1.8), and verify the quoted values of the virial coefficients.

7.3. Combining equations (7.1.24) and (7.1.26), and making use of the first two terms of formula (D.9) in Appendix D, show that, as T approaches T_c from above, the parameter $\alpha(= -\ln z)$ of the ideal Bose gas assumes the form

$$\alpha \approx \frac{1}{\pi}\left(\frac{3\zeta(3/2)}{4}\right)^2\left(\frac{T-T_c}{T_c}\right)^2.$$

7.4. Show that for an ideal Bose gas

$$\frac{1}{z}\left(\frac{\partial z}{\partial T}\right)_P = -\frac{5}{2T}\frac{g_{5/2}(z)}{g_{3/2}(z)};$$

compare this result with equation (7.1.36). Hence show that

$$\gamma \equiv \frac{C_P}{C_V} = \frac{(\partial z/\partial T)_P}{(\partial z/\partial T)_v} = \frac{5}{3}\frac{g_{5/2}(z)g_{1/2}(z)}{\{g_{3/2}(z)\}^2},$$

as in equation (7.1.48b). Check that, as T approaches T_c from above, both γ and C_P diverge as $(T-T_c)^{-1}$.

7.5. **(a)** Show that the isothermal compressibility κ_T and the adiabatic compressibility κ_S of an ideal Bose gas are given by

$$\kappa_T = \frac{1}{nkT}\frac{g_{1/2}(z)}{g_{3/2}(z)}, \quad \kappa_S = \frac{3}{5nkT}\frac{g_{3/2}(z)}{g_{5/2}(z)},$$

where $n(= N/V)$ is the particle density in the gas. Note that, as $z \to 0$, κ_T and κ_S approach their respective classical values, namely $1/P$ and $1/\gamma P$. How do they behave as $z \to 1$?

(b) Making use of the thermodynamic relations

$$C_P - C_V = T\left(\frac{\partial P}{\partial T}\right)_V\left(\frac{\partial V}{\partial T}\right)_P = TV\kappa_T\left(\frac{\partial P}{\partial T}\right)_V^2$$

and

$$C_P/C_V = \kappa_T/\kappa_S,$$

derive equations (7.1.48a) and (7.1.48b).

7.6. Show that for an ideal Bose gas the temperature derivative of the specific heat C_V is given by

$$\frac{1}{Nk}\left(\frac{\partial C_V}{\partial T}\right)_V = \begin{cases} \dfrac{1}{T}\left[\dfrac{45}{8}\dfrac{g_{5/2}(z)}{g_{3/2}(z)} - \dfrac{9}{4}\dfrac{g_{3/2}(z)}{g_{1/2}(z)} - \dfrac{27}{8}\dfrac{\{g_{3/2}(z)\}^2 g_{-1/2}(z)}{\{g_{1/2}(z)\}^3}\right] & \text{for } T > T_c, \\ \dfrac{45}{8}\dfrac{v}{T\lambda^3}\zeta\left(\frac{5}{2}\right) & \text{for } T < T_c. \end{cases}$$

Using these results and the main term of formula (D.9), verify equation (7.1.38).

7.7. Evaluate the quantities $(\partial^2 P/\partial T^2)_v$, $(\partial^2\mu/\partial T^2)_v$, and $(\partial^2\mu/\partial T^2)_P$ for an ideal Bose gas and check that your results satisfy the thermodynamic relationships

$$C_V = VT\left(\frac{\partial^2 P}{\partial T^2}\right)_v - NT\left(\frac{\partial^2\mu}{\partial T^2}\right)_v,$$

and

$$C_P = -NT\left(\frac{\partial^2\mu}{\partial T^2}\right)_P.$$

Examine the behavior of these quantities as $T \to T_c$ from above and from below.

7.8. The velocity of sound in a fluid is given by the formula

$$w = \sqrt{(\partial P/\partial \rho)_s},$$

where ρ is the mass density of the fluid. Show that for an ideal Bose gas

$$w^2 = \frac{5kT}{3m}\frac{g_{5/2}(z)}{g_{3/2}(z)} = \frac{5}{9}\langle u^2 \rangle,$$

where $\langle u^2 \rangle$ is the mean square speed of the particles in the gas.

7.9. Show that for an ideal Bose gas

$$\langle u \rangle \left\langle \frac{1}{u} \right\rangle = \frac{4}{\pi}\frac{g_1(z)g_2(z)}{\{g_{3/2}(z)\}^2},$$

u being the speed of a particle. Examine and interpret the limiting cases $z \to 0$ and $z \to 1$; compare with Problem 6.6.

7.10. Consider an ideal Bose gas in a uniform gravitational field of acceleration g. Show that the phenomenon of Bose–Einstein condensation in this gas sets in at a temperature T_c given by

$$T_c \simeq T_c^0 \left[1 + \frac{8}{9}\frac{1}{\zeta\left(\frac{3}{2}\right)}\left(\frac{\pi mgL}{kT_c^0}\right)^{1/2} \right],$$

where L is the height of the container and $mgL \ll kT_c^0$. Also show that the condensation here is accompanied by a discontinuity in the specific heat C_V of the gas:

$$(\Delta C_V)_{T=T_c} \simeq -\frac{9}{8\pi}\zeta\left(\frac{3}{2}\right)Nk\left(\frac{\pi mgL}{kT_c^0}\right)^{1/2};$$

see Eisenschitz (1958).

7.11. Consider an ideal Bose gas consisting of molecules with internal degrees of freedom. Assuming that, besides the ground state $\varepsilon_0 = 0$, only the first excited state ε_1 of the *internal* spectrum needs to be taken into account, determine the condensation temperature of the gas as a function of ε_1. Show that, for $(\varepsilon_1/kT_c^0) \gg 1$,

$$\frac{T_c}{T_c^0} \simeq 1 - \frac{\frac{2}{3}}{\zeta\left(\frac{3}{2}\right)} e^{-\varepsilon_1/kT_c^0}$$

while, for $(\varepsilon_1/kT_c^0) \ll 1$,

$$\frac{T_c}{T_c^0} \simeq \left(\frac{1}{2}\right)^{2/3} \left[1 + \frac{2^{4/3}}{3\zeta\left(\frac{3}{2}\right)}\left(\frac{\pi \varepsilon_1}{kT_c^0}\right)^{1/2} \right].$$

[*Hint*: To obtain the last result, use the first two terms of formula (D.9) in Appendix D.]

7.12. Consider an ideal Bose gas in the grand canonical ensemble and study fluctuations in the total number of particles N and the total energy E. Discuss, in particular, the situation when the gas becomes highly degenerate.

7.13. Consider an ideal Bose gas confined to a region of area A in *two* dimensions. Express the number of particles in the excited states, N_e, and the number of particles in the ground state, N_0, in terms of z, T, and A, and show that the system does not exhibit Bose–Einstein condensation unless $T \to 0\,\text{K}$.

Refine your argument to show that, if the area A and the total number of particles N are held fixed and we require both N_e and N_0 to be of order N, then we do achieve condensation when

$$T \sim \frac{h^2}{mkl^2}\frac{1}{\ln N},$$

where $l[\sim\sqrt{(A/N)}]$ is the mean interparticle distance in the system. Of course, if both A and $N\to\infty$, keeping l fixed, then the desired T does go to zero.

7.14. Consider an n-dimensional Bose gas whose single-particle energy spectrum is given by $\varepsilon \propto p^s$, where s is some positive number. Discuss the onset of Bose–Einstein condensation in this system, especially its dependence on the numbers n and s. Study the thermodynamic behavior of this system and show that,

$$P=\frac{s}{n}\frac{U}{V},\quad C_V(T\to\infty)=\frac{n}{s}Nk,\quad \text{and}\quad C_P(T\to\infty)=\left(\frac{n}{s}+1\right)Nk.$$

7.15. At time $t=0$, the ground state wavefunction of a one-dimensional quantum harmonic oscillator with potential $V(x)=\frac{1}{2}m\omega_0^2x^2$ is given by

$$\psi(x,0)=\frac{1}{\pi^{1/4}\sqrt{a}}\exp\left(-\frac{x^2}{2a^2}\right),$$

where $a=\sqrt{\frac{\hbar}{m\omega_0}}$. At $t=0$, the harmonic potential is abruptly removed. Use the momentum representation of the wavefunction at $t=0$ and the time-dependent Schrodinger equation to determine the spatial wavefunction and density at time $t>0$; compare to equation (7.2.11).

7.16. At time $t=0$, a collection of *classical* particles is in equilibrium at temperature T in a three-dimensional harmonic oscillator potential $V(\boldsymbol{r})=\frac{1}{2}m\omega_0^2|\boldsymbol{r}|^2$. At $t=0$, the harmonic potential is abruptly removed. Use the momentum distribution at $t=0$ to determine the spatial density at time $t>0$. Show that this is equivalent to the high temperature limit of equation (7.2.15).

7.17. As shown in Section 7.1, $n\lambda^3$ is a measure of the quantum nature of the system. Use equations (7.2.11) and (7.2.15) to determine $n\lambda^3$ at the center of the harmonic trap at $T=T_c/2$ for the condensed and noncondensed fractions.

7.18. Show that the integral of the semiclassical spatial density in equation (7.2.15) gives the correct counting of the atoms that are not condensed into the ground state.

7.19. Construct a theory for N bosons in an isotropic two-dimensional trap. This corresponds to a trap in which the energy level spacing due to excitations in the z direction is much larger than the spacing in the other directions. Determine the density of states $a(\varepsilon)$ of this system. Can a Bose–Einstein condensate form in this trap? If so, find the critical temperature as a function of the trapping frequencies and N. How much larger must the frequency in the third direction be for the system to display two-dimensional behavior?

7.20. The (canonical) partition function of the blackbody radiation may be written as

$$Q(V,T)=\prod_{\omega}Q_1(\omega,T),$$

so that

$$\ln Q(V,T)=\sum_{\omega}\ln Q_1(\omega,T)\approx\int_0^\infty \ln Q_1(\omega,T)g(\omega)d\omega;$$

here, $Q_1(\omega,T)$ is the single-oscillator partition function given by equation (3.8.14) and $g(\omega)$ is the density of states given by equation (7.3.2). Using this information, evaluate the Helmholtz free energy of the system and derive other thermodynamic properties such as the pressure P and the (thermal) energy density U/V. Compare your results with the ones derived in Section 7.3 from the q-potential of the system.

7.21. Show that the mean energy per photon in a blackbody radiation cavity is very nearly $2.7kT$.

7.22. Considering the volume dependence of the frequencies ω of the vibrational modes of the radiation field, establish relation (7.3.17) between the pressure P and the energy density U/V.

7.23. The sun may be regarded as a black body at a temperature of 5800 K. Its diameter is about 1.4×10^9 m while its distance from the earth is about 1.5×10^{11} m.

(a) Calculate the total radiant intensity (in W/m^2) of sunlight at the surface of the earth.
(b) What pressure would it exert on a perfectly absorbing surface placed normal to the rays of the sun?
(c) If a flat surface on a satellite, which faces the sun, were an ideal absorber and emitter, what equilibrium temperature would it ultimately attain?

7.24. Calculate the photon number density, entropy density, and energy density of the 2.725 K cosmic microwave background.

7.25. Figure 7.20 is a plot of $C_V(T)$ against T for a solid, the limiting value $C_V(\infty)$ being the classical result $3Nk$. Show that the shaded area in the figure, namely

$$\int_0^\infty \{C_V(\infty) - C_V(T)\}dT,$$

is exactly equal to the zero-point energy of the solid. Interpret the result physically.

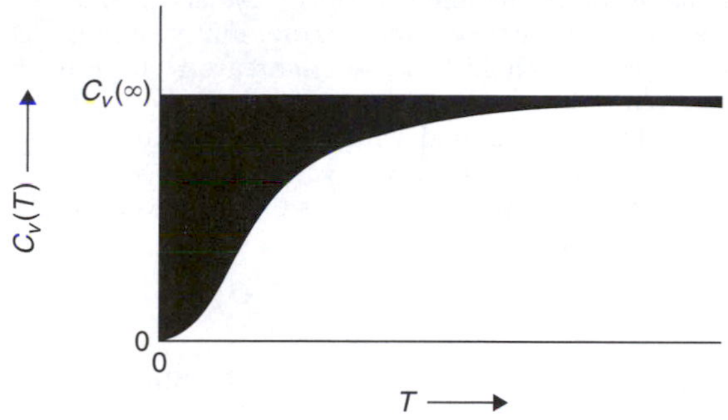

FIGURE 7.20

7.26. Show that the zero-point energy of a Debye solid composed of N atoms is equal to $\frac{9}{8}Nk\Theta_D$.
[Note that this implies, for each vibrational mode of the solid, a mean zero-point energy $\frac{3}{8}k\Theta_D$, that is, $\overline{\omega} = \frac{3}{4}\omega_D$.]

7.27. Show that, for $T \ll \Theta_D$, the quantity $(C_P - C_V)$ of a Debye solid varies as T^7 and hence the ratio $(C_P/C_V) \simeq 1$.

7.28. Determine the temperature T, in terms of the Debye temperature Θ_D, at which one-half of the oscillators in a Debye solid are expected to be in the excited states.

7.29. Determine the value of the parameter Θ_D for liquid He^4 from the empirical result (7.4.28).

7.30. **(a)** Compare the "mean thermal wavelength" λ_T of neutrons at a typical room temperature with the "minimum wavelength" λ_{min} of phonons in a typical crystal.
(b) Show that the frequency ω_D for a sodium chloride crystal is of the same order of magnitude as the frequency of an electromagnetic wave in the infrared.

7.31. Proceeding under conditions (7.4.16) rather than (7.4.13), show that

$$C_V(T) = Nk\{D(x_{0,L}) + 2D(x_{0,T})\},$$

where $x_{0,L} = (\hbar\omega_{D,L}/kT)$ and $x_{0,T} = (\hbar\omega_{D,T}/kT)$. Compare this result with equation (7.4.17), and estimate the nature and the magnitude of the error involved in the latter.

7.32. A mechanical system consisting of n identical masses (each of mass m) connected in a straight line by identical springs (of stiffness K) has natural vibrational frequencies given by

$$\omega_r = 2\sqrt{\left(\frac{K}{m}\right)}\sin\left(\frac{r}{n}\cdot\frac{\pi}{2}\right); r = 1, 2, \ldots (n-1).$$

Correspondingly, a *linear* molecule composed of n identical atoms may be regarded as having a vibrational spectrum given by

$$\nu_r = \nu_c \sin\left(\frac{r}{n}\cdot\frac{\pi}{2}\right); r = 1,2,\ldots(n-1),$$

where ν_c is a characteristic vibrational frequency of the molecule. Show that this model leads to a vibrational specific heat per molecule that varies as T^1 at low temperatures and tends to the limiting value $(n-1)k$ at high temperatures.

7.33. Assuming the dispersion relation $\omega = Ak^s$, where ω is the angular frequency and k the wave number of a vibrational mode existing in a solid, show that the respective contribution toward the specific heat of the solid *at low temperatures* is proportional to $T^{3/s}$.
[Note that while $s = 1$ corresponds to the case of elastic waves in a lattice, $s = 2$ applies to spin waves propagating in a ferromagnetic system.]

7.34. Assuming the excitations to be phonons ($\omega = Ak$), show that their contribution toward the specific heat of an n-dimensional Debye system is proportional to T^n.
[Note that the elements selenium and tellurium form crystals in which atomic chains are arranged in parallel so that in a certain sense they behave as one-dimensional; accordingly, over a certain range of temperatures, the T^1-law holds. For a similar reason, graphite obeys a T^2-law over a certain range of temperatures.]

7.35. The (minimum) potential energy of a solid, when all its atoms are "at rest" at their equilibrium positions, may be denoted by the symbol $\Phi_0(V)$, where V is the volume of the solid. Similarly, the normal frequencies of vibration, ω_i $(i = 1,2,\ldots,3N-6)$, may be denoted by the symbols $\omega_i(V)$. Show that the pressure of this solid is given by

$$P = -\frac{\partial \Phi_0}{\partial V} + \gamma\frac{U'}{V},$$

where U' is the internal energy of the solid arising from the vibrations of the atoms, while γ is the *Grüneisen constant*:

$$\gamma = -\frac{\partial \ln \omega}{\partial \ln V} \approx \frac{1}{3}.$$

Assuming that, for $V \simeq V_0$,

$$\Phi_0(V) = \frac{(V-V_0)^2}{2\kappa_0 V_0},$$

where κ_0 and V_0 are constants and $\kappa_0 C_V T \ll V_0$, show that the coefficient of thermal expansion (at constant pressure $P \simeq 0$) is given by

$$\alpha \equiv \frac{1}{V}\left(\frac{\partial V}{\partial T}\right)_{N,P} = \frac{\gamma\kappa_0 C_V}{V_0}.$$

Also show that

$$C_P - C_V = \frac{\gamma^2 \kappa_0 C_V^2 T}{V_0}.$$

7.36. Apply the general formula (6.4.3) for the kinetic pressure of a gas, namely

$$P = \frac{1}{3} n\langle pu\rangle,$$

to a gas of rotons and verify that the result so obtained agrees with the Boltzmannian relationship $P = nkT$.

7.37. Show that the free energy A and the inertial density ρ of a roton gas in *mass motion* are given by

$$A(v) = A(0)\frac{\sinh x}{x}$$

and

$$\rho(v) = \rho(0)\frac{3(x\cosh x - \sinh x)}{x^3},$$

where $x = vp_0/kT$.

7.38. Integrating (7.6.17) by parts, show that the effective mass of an excitation, whose energy–momentum relationship is denoted by $\varepsilon(p)$, is given by

$$m_{\text{eff}} = \left\langle \frac{1}{3p^2} \left\{ \frac{d}{dp} \left(p^4 \frac{dp}{d\varepsilon} \right) \right\} \right\rangle.$$

Check the validity of this result by considering the examples of (i) an ideal-gas particle, (ii) a phonon, and (iii) a roton.

8

Ideal Fermi Systems

8.1 Thermodynamic behavior of an ideal Fermi gas

According to Sections 6.1 and 6.2, we obtain for an ideal Fermi gas

$$\frac{PV}{kT} \equiv \ln \mathcal{Q} = \sum_{\varepsilon} \ln(1 + ze^{-\beta\varepsilon}) \tag{1}$$

and

$$N \equiv \sum_{\varepsilon} \langle n_{\varepsilon} \rangle = \sum_{\varepsilon} \frac{1}{z^{-1}e^{\beta\varepsilon} + 1}, \tag{2}$$

where $\beta = 1/kT$ and $z = \exp(\mu/kT)$. Unlike the Bose case, the parameter z in the Fermi case can take on *unrestricted* values: $0 \leq z < \infty$. Moreover, in view of the Pauli exclusion principle, the question of a large number of particles occupying a single energy state does not even arise in this case; hence, there is no phenomenon like Bose–Einstein condensation here. Nevertheless, at sufficiently low temperatures, Fermi gas displays its own brand of quantal behavior, a detailed study of which is of great physical interest.

If we replace summations over ε by corresponding integrations, equations (1) and (2) in the case of a nonrelativistic gas become

$$\frac{P}{kT} = \frac{g}{\lambda^3} f_{5/2}(z) \tag{3}$$

and

$$\frac{N}{V} = \frac{g}{\lambda^3} f_{3/2}(z), \tag{4}$$

where g is a weight factor arising from the "internal structure" of the particles (e.g., *spin*), λ is the mean thermal wavelength of the particles

$$\lambda = h/(2\pi mkT)^{1/2}, \tag{5}$$

while $f_{\nu}(z)$ are Fermi–Dirac functions defined by, see Appendix E,

$$f_{\nu}(z) = \frac{1}{\Gamma(\nu)} \int_0^{\infty} \frac{x^{\nu-1}dx}{z^{-1}e^x + 1} = z - \frac{z^2}{2^{\nu}} + \frac{z^3}{3^{\nu}} - \cdots. \tag{6}$$

© 2011 Elsevier Ltd. All rights reserved.

Eliminating z between equations (3) and (4), we obtain the *equation of state* of the Fermi gas.

The internal energy U of the Fermi gas is given by

$$U \equiv -\left(\frac{\partial}{\partial\beta}\ln \mathcal{Q}\right)_{z,V} = kT^2\left(\frac{\partial}{\partial T}\ln \mathcal{Q}\right)_{z,V}$$
$$= \frac{3}{2}kT\frac{gV}{\lambda^3}f_{5/2}(z) = \frac{3}{2}NkT\frac{f_{5/2}(z)}{f_{3/2}(z)}; \tag{7}$$

thus, quite generally, this system satisfies the relationship

$$P = \frac{2}{3}(U/V). \tag{8}$$

The specific heat C_V of the gas can be obtained by differentiating (7) with respect to T, keeping N and V constant, and making use of the relationship

$$\frac{1}{z}\left(\frac{\partial z}{\partial T}\right)_v = -\frac{3}{2T}\frac{f_{3/2}(z)}{f_{1/2}(z)}, \tag{9}$$

which follows from equation (4) and the recurrence formula (E.6) in Appendix E. The final result is

$$\frac{C_V}{Nk} = \frac{15}{4}\frac{f_{5/2}(z)}{f_{3/2}(z)} - \frac{9}{4}\frac{f_{3/2}(z)}{f_{1/2}(z)}. \tag{10}$$

For the Helmholtz free energy of the gas, we get

$$A \equiv N\mu - PV = NkT\left\{\ln z - \frac{f_{5/2}(z)}{f_{3/2}(z)}\right\}, \tag{11}$$

and for the entropy

$$S \equiv \frac{U-A}{T} = Nk\left\{\frac{5}{2}\frac{f_{5/2}(z)}{f_{3/2}(z)} - \ln z\right\}. \tag{12}$$

In order to determine the various properties of the Fermi gas in terms of the particle density $n(=N/V)$ and the temperature T, we need to know the functional dependence of the parameter z on n and T; this information is formally contained in the implicit relationship (4). For detailed studies, one is sometimes obliged to make use of numerical evaluation of the functions $f_\nu(z)$; for physical understanding, however, the various limiting forms of these functions serve the purpose well (see Appendix E).

Now, if the density of the gas is very low and/or its temperature very high, then the situation might correspond to

$$f_{3/2}(z) = \frac{n\lambda^3}{g} = \frac{nh^3}{g(2\pi mkT)^{3/2}} \ll 1; \tag{13}$$

we then speak of the gas as being *nondegenerate* and, therefore, equivalent to a classical ideal gas discussed in Section 3.5. In view of expansion (6), this implies that $z \ll 1$ and hence $f_\nu(z) \simeq z$. Expressions for the various thermodynamic properties of the gas then become

$$P = NkT/V, \quad U = \frac{3}{2}NkT, \quad C_V = \frac{3}{2}Nk, \tag{14}$$

$$A = NkT\left\{\ln\left(\frac{n\lambda^3}{g}\right) - 1\right\}, \tag{15}$$

and

$$S = Nk\left\{\frac{5}{2} - \ln\left(\frac{n\lambda^3}{g}\right)\right\}. \tag{16}$$

If the parameter z is small in comparison with unity but not very small, then we should make a fuller use of series (6) in order to eliminate z between equations (3) and (4). The procedure is just the same as in the corresponding Bose case, that is, we first invert the series appearing in (4) to obtain an expansion for z in powers of $(n\lambda^3/g)$ and then substitute this expansion into the series appearing in (3). The equation of state then takes the form of the *virial expansion*

$$\frac{PV}{NkT} = \sum_{l=1}^{\infty}(-1)^{l-1}a_l\left(\frac{\lambda^3}{gv}\right)^{l-1}, \tag{17}$$

where $v = 1/n$, while the coefficients a_l are the same as quoted in (7.1.14) but alternate in sign compared to the Bose case. For the specific heat, in particular, we obtain

$$\begin{aligned} C_V &= \frac{3}{2}Nk\sum_{l=1}^{\infty}(-1)^{l-1}\frac{5-3l}{2}a_l\left(\frac{\lambda^3}{gv}\right)^{l-1} \\ &= \frac{3}{2}Nk\left[1 - 0.0884\left(\frac{\lambda^3}{gv}\right) + 0.0066\left(\frac{\lambda^3}{gv}\right)^2 - 0.0004\left(\frac{\lambda^3}{gv}\right)^3 + \cdots\right]. \end{aligned} \tag{18}$$

Thus, at finite temperatures, the specific heat of the gas is smaller than its limiting value $\frac{3}{2}Nk$. As will be seen in the sequel, the specific heat of the ideal Fermi gas decreases *monotonically* as the temperature of the gas falls; see Figure 8.2 later in the section and compare it with the corresponding Figure 7.4 for the ideal Bose gas.

If the density n and the temperature T are such that the parameter $(n\lambda^3/g)$ is of order unity, the foregoing expansions cannot be of much use. In that case, one may have to make recourse to numerical calculation. However, if $(n\lambda^3/g) \gg 1$, the functions involved can be expressed as *asymptotic* expansions in powers of $(\ln z)^{-1}$; we then speak of the gas as being *degenerate*. As $(n\lambda^3/g) \to \infty$, our functions assume a closed form, with the result that the expressions for the various thermodynamic quantities pertaining to the system become highly simplified; we then speak of the gas as being *completely degenerate*. For simplicity, we first discuss the main features of the system in a state of complete degeneracy.

In the limit $T \to 0$, which implies $(n\lambda^3/g) \to \infty$, the mean occupation numbers of the single-particle state $\varepsilon(\boldsymbol{p})$ become

$$\langle n_\varepsilon \rangle \equiv \frac{1}{e^{(\varepsilon-\mu)/kT}+1} = \begin{cases} 1 & \text{for} \quad \varepsilon < \mu_0 \\ 0 & \text{for} \quad \varepsilon > \mu_0, \end{cases} \tag{19}$$

where μ_0 is the chemical potential of the system at $T = 0$. The function $\langle n_\varepsilon \rangle$ is thus a *step function* that stays constant at the (highest) value 1 right from $\varepsilon = 0$ to $\varepsilon = \mu_0$ and then suddenly drops to the (lowest) value 0; see the dotted line in Figure 8.1. Thus, at $T = 0$, all single-particle states up to $\varepsilon = \mu_0$ are "completely" filled, with one particle per state (in accordance with the Pauli principle), while all single-particle states with $\varepsilon > \mu_0$ are empty. The limiting energy μ_0 is generally referred to as the *Fermi energy* of the system and is denoted by the symbol ε_F; the corresponding value of the single-particle momentum is referred to as the *Fermi momentum* and is denoted by the symbol p_F. The defining equation for these parameters is

$$\int_0^{\varepsilon_F} a(\varepsilon)d\varepsilon = N, \tag{20}$$

where $a(\varepsilon)$ denotes the *density of states* of the system and is given by the general expression

$$a(\varepsilon) = \frac{gV}{h^3} 4\pi p^2 \frac{dp}{d\varepsilon}. \tag{21}$$

We readily obtain

$$N = \frac{4\pi g V}{3h^3} p_F^3, \tag{22}$$

which gives

$$p_F = \left(\frac{3N}{4\pi g V}\right)^{1/3} h; \tag{23}$$

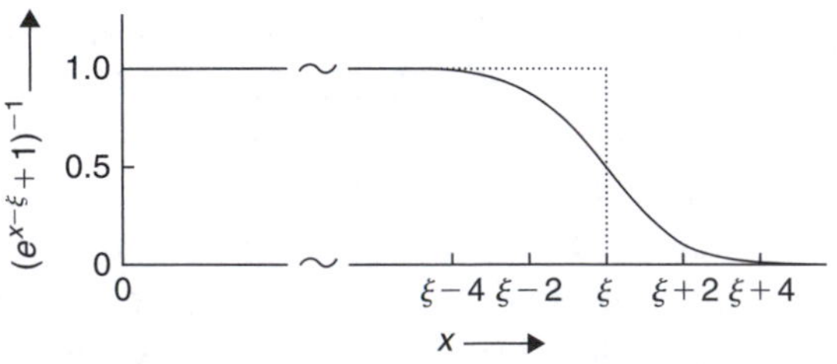

FIGURE 8.1 Fermi distribution at low temperatures, with $x = \varepsilon/kT$ and $\xi = \mu/kT$. The rectangle denotes the limiting distribution as $T \to 0$; in that case, the Fermi function is unity for $\varepsilon < \mu_0$ and zero for $\varepsilon > \mu_0$.

accordingly, in the *nonrelativistic* case,

$$\varepsilon_F = \left(\frac{3N}{4\pi gV}\right)^{2/3}\frac{h^2}{2m} = \left(\frac{6\pi^2 n}{g}\right)^{2/3}\frac{\hbar^2}{2m}. \tag{24}$$

The ground-state, or zero-point, energy of the system is then given by

$$\begin{aligned} E_0 &= \frac{4\pi gV}{h^3}\int_0^{p_F}\left(\frac{p^2}{2m}\right)p^2\,dp \\ &= \frac{2\pi gV}{5mh^3}p_F^5, \end{aligned} \tag{25}$$

which gives

$$\frac{E_0}{N} = \frac{3p_F^2}{10m} = \frac{3}{5}\varepsilon_F. \tag{26}$$

The ground-state pressure of the system is in turn given by

$$P_0 = \frac{2}{3}(E_0/V) = \frac{2}{5}n\varepsilon_F. \tag{27}$$

Substituting for ε_F, the foregoing expression takes the form

$$P_0 = \left(\frac{6\pi^2}{g}\right)^{2/3}\frac{\hbar^2}{5m}n^{5/3} \propto n^{5/3}. \tag{28}$$

The zero-point motion seen here is clearly a quantum effect arising because of the Pauli principle, according to which, even at $T = 0\,\mathrm{K}$, the particles constituting the system cannot settle down into a *single* energy state (as we had in the Bose case) and are therefore spread over a requisite number of lowest available energy states. As a result, the Fermi system, even at absolute zero, is quite live!

For a discussion of properties such as the specific heat and the entropy of the system, we must extend our study to *finite* temperatures. If we decide to restrict ourselves to low temperatures, then deviations from the ground-state results will not be too large; accordingly, an analysis based on the asymptotic expansions of the functions $f_\nu(z)$ would be quite appropriate. However, before we do that it seems useful to carry out a physical assessment of the situation with the help of the expression

$$\langle n_\varepsilon\rangle = \frac{1}{e^{(\varepsilon-\mu)/kT}+1}. \tag{29}$$

The situation corresponding to $T = 0$ is summarized in equation (19) and is shown as a step function in Figure 8.1. Deviations from this, when T is finite (but still much smaller than the *characteristic* temperature μ_0/k), will be significant *only* for those values of ε for which the magnitude of the quantity $(\varepsilon - \mu)/kT$ is of order unity (for otherwise the exponential term in (29) will not be much different from its ground-state value, namely, $e^{\pm\infty}$); see the solid curve in Figure 8.1.

We, therefore, conclude that the thermal excitation of the particles occurs only in a narrow energy range that is located around the energy value $\varepsilon = \mu_0$ and has a width $O(kT)$. The fraction of the particles that are thermally excited is, therefore, $O(kT/\varepsilon_F)$ — the bulk of the system remaining uninfluenced by the rise in temperature.[1] This is the most characteristic feature of a degenerate Fermi system and is essentially responsible for both qualitative and quantitative differences between the physical behavior of this system and that of a corresponding classical system.

To conclude the argument, we observe that since the thermal energy per "excited" particle is $O(kT)$, the thermal energy of the whole system will be $O(Nk^2T^2/\varepsilon_F)$; accordingly, the specific heat of the system will be $O(Nk \cdot kT/\varepsilon_F)$. Thus, the low-temperature specific heat of a Fermi system differs from the classical value $\frac{3}{2}Nk$ by a factor that not only reduces it considerably in magnitude but also makes it temperature-dependent (varying as T^1). It will be seen repeatedly that the first-power dependence of C_V on T is a typical feature of Fermi systems at low temperatures.

For an analytical study of the Fermi gas at finite, but low, temperatures, we observe that the value of z, which was infinitely large at absolute zero, is now finite, though still large in comparison with unity. The functions $f_\nu(z)$ can, therefore, be expressed as asymptotic expansions in powers of $(\ln z)^{-1}$; see Sommerfeld's lemma (E.17) in Appendix E. For the values of ν we are presently interested in, namely $\frac{5}{2}, \frac{3}{2}$, and $\frac{1}{2}$, we have *to the first approximation*

$$f_{5/2}(z) = \frac{8}{15\pi^{1/2}}(\ln z)^{5/2}\left[1 + \frac{5\pi^2}{8}(\ln z)^{-2} + \cdots\right], \tag{30}$$

$$f_{3/2}(z) = \frac{4}{3\pi^{1/2}}(\ln z)^{3/2}\left[1 + \frac{\pi^2}{8}(\ln z)^{-2} + \cdots\right], \tag{31}$$

and

$$f_{1/2}(z) = \frac{2}{\pi^{1/2}}(\ln z)^{1/2}\left[1 - \frac{\pi^2}{24}(\ln z)^{-2} + \cdots\right]. \tag{32}$$

Substituting (31) into (4), we obtain

$$\frac{N}{V} = \frac{4\pi g}{3}\left(\frac{2m}{h^2}\right)^{3/2}(kT\ln z)^{3/2}\left[1 + \frac{\pi^2}{8}(\ln z)^{-2} + \cdots\right]. \tag{33}$$

[1]We, therefore, speak of the totality of the energy levels filled at $T = 0$ as "the Fermi sea" and the small fraction of the particles that are excited near the top, when $T > 0$, as a "mist above the sea." Physically speaking, the origin of this behavior again lies in the Pauli exclusion principle, according to which a fermion of energy ε cannot absorb a quantum of thermal excitation ε_T if the energy level $\varepsilon + \varepsilon_T$ is already filled. Since $\varepsilon_T = O(kT)$, only those fermions that occupy energy levels near the top level ε_F, up to a depth $O(kT)$, can be thermally excited to go over to the unfilled energy levels.

In the zeroth approximation, this gives

$$kT\ln z \equiv \mu \simeq \left(\frac{3N}{4\pi gV}\right)^{2/3}\frac{h^2}{2m}, \tag{34}$$

which is identical to the ground-state result $\mu_0 = \varepsilon_F$; see equation (24). In the next approximation, we obtain

$$kT\ln z \equiv \mu \simeq \varepsilon_F\left[1 - \frac{\pi^2}{12}\left(\frac{kT}{\varepsilon_F}\right)^2\right]. \tag{35}$$

Substituting (30) and (31) into (7), we obtain

$$\frac{U}{N} = \frac{3}{5}(kT\ln z)\left[1 + \frac{\pi^2}{2}(\ln z)^{-2} + \cdots\right]; \tag{36}$$

with the help of (35), this becomes

$$\frac{U}{N} = \frac{3}{5}\varepsilon_F\left[1 + \frac{5\pi^2}{12}\left(\frac{kT}{\varepsilon_F}\right)^2 + \cdots\right]. \tag{37}$$

The pressure of the gas is then given by

$$P = \frac{2}{3}\frac{U}{V} = \frac{2}{5}n\varepsilon_F\left[1 + \frac{5\pi^2}{12}\left(\frac{kT}{\varepsilon_F}\right)^2 + \cdots\right]. \tag{38}$$

As expected, the main terms of equations (37) and (38) are identical to the ground-state results (26) and (27). From the temperature-dependent part of (37), we obtain for the low-temperature specific heat of the gas

$$\frac{C_V}{Nk} = \frac{\pi^2}{2}\frac{kT}{\varepsilon_F} + \cdots. \tag{39}$$

Thus, for $T \ll T_F$, where $T_F\,(= \varepsilon_F/k)$ is the *Fermi temperature* of the system, the specific heat varies as the first power of temperature; moreover, in magnitude, it is considerably smaller than the classical value $\frac{3}{2}Nk$. The overall variation of C_V with T is shown in Figure 8.2.

The Helmholtz free energy of the system follows directly from equations (35) and (38):

$$\begin{aligned}\frac{A}{N} &= \mu - \frac{PV}{N}\\ &= \frac{3}{5}\varepsilon_F\left[1 - \frac{5\pi^2}{12}\left(\frac{kT}{\varepsilon_F}\right)^2 + \cdots\right],\end{aligned} \tag{40}$$

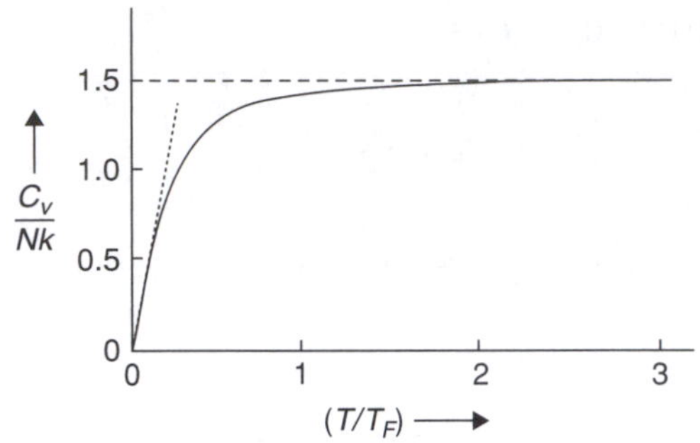

FIGURE 8.2 The specific heat of an ideal Fermi gas; the dotted line depicts the *linear* behavior at low temperatures.

which gives

$$\frac{S}{Nk} = \frac{\pi^2}{2}\frac{kT}{\varepsilon_F} + \cdots. \tag{41}$$

Thus, as $T \to 0, S \to 0$ in accordance with the third law of thermodynamics.

8.2 Magnetic behavior of an ideal Fermi gas

We now turn our attention to studying the equilibrium state of a gas of noninteracting fermions in the presence of an external magnetic field $\boldsymbol{B}$. The main problem here is to determine the net magnetic moment $\boldsymbol{M}$ acquired by the gas (as a function of $\boldsymbol{B}$ and $\boldsymbol{T}$) and then calculate the susceptibility $\chi(T)$. The answer naturally depends on the intrinsic magnetic moment $\boldsymbol{\mu}^*$ of the particles and the corresponding multiplicity factor $(2J+1)$; see, for instance, the treatment given in Section 3.9. According to the Boltzmannian treatment, one obtains a (positive) susceptibility $\chi(T)$ which, at high temperatures, obeys the *Curie law*: $\chi \propto T^{-1}$; at low temperatures, one obtains a state of magnetic saturation. However, if we treat the problem on the basis of Fermi statistics we obtain significantly different results, especially at low temperatures.

In particular, since the Fermi gas is pretty live even at absolute zero, no magnetic saturation ever results; we rather obtain a limiting susceptibility χ_0, which is independent of temperature but is dependent on the density of the gas. Studies along these lines were first made by Pauli, in 1927, when he suggested that the conduction electrons in alkali metals be regarded as a "highly degenerate Fermi gas"; these studies enabled him to explain the physics behind the *feeble* and *temperature-independent* character of the paramagnetism of metals. Accordingly, this phenomenon is referred to as *Pauli paramagnetism*—in contrast to the classical Langevin paramagnetism.

In quantum statistics, we encounter yet another effect which is totally absent in classical statistics. This is diamagnetic in character and arises from the quantization of the orbits of *charged* particles in the presence of an external magnetic field or, one may say, from the quantization of the (kinetic) energy of charged particles associated with their motion perpendicular to the direction of the field. The existence of this effect was first established by Landau (1930); so, we refer to it as *Landau diamagnetism*. This leads to an additional susceptibility $\chi(T)$, which, though negative in sign, is somewhat similar to the paramagnetic susceptibility, in that it obeys Curie's law at high temperatures and tends to a temperature-independent but density-dependent limiting value as $T \to 0$. In general, the magnetic behavior of a Fermi gas is determined jointly by the intrinsic magnetic moment of the particles and the quantization of their orbits. If the spin–orbit interaction is negligible, the resultant behavior is given by a simple addition of the two effects.

8.2.A Pauli paramagnetism

The energy of a particle, in the presence of an external magnetic field $\boldsymbol{B}$, is given by

$$\varepsilon = \frac{p^2}{2m} - \boldsymbol{\mu}^* \cdot \mathbf{B}, \tag{1}$$

where $\boldsymbol{\mu}^*$ is the intrinsic magnetic moment of the particle and m its mass. For simplicity, we assume that the particle spin is $\frac{1}{2}$; the vector $\boldsymbol{\mu}^*$ will then be either parallel to the vector $\boldsymbol{B}$ or antiparallel. We thus have two groups of particles in the gas:

(i) those having $\boldsymbol{\mu}^*$ parallel to $\boldsymbol{B}$, with $\varepsilon = p^2/2m - \mu^* B$, and
(ii) those having $\boldsymbol{\mu}^*$ antiparallel to $\boldsymbol{B}$, with $\varepsilon = p^2/2m + \mu^* B$.

At absolute zero, all energy levels up to the Fermi level ε_F will be filled, while all levels beyond ε_F will be empty. Accordingly, the kinetic energy of the particles in the first group will range between 0 and $(\varepsilon_F + \mu^* B)$, while the kinetic energy of the particles in the second group will range between 0 and $(\varepsilon_F - \mu^* B)$. The respective numbers of occupied energy levels (and hence of particles) in the two groups will, therefore, be

$$N^+ = \frac{4\pi V}{3h^3}\{2m(\varepsilon_F + \mu^* B)\}^{3/2} \tag{2}$$

and

$$N^- = \frac{4\pi V}{3h^3}\{2m(\varepsilon_F - \mu^* B)\}^{3/2}. \tag{3}$$

The net magnetic moment acquired by the gas is then given by

$$M = \mu^*(N^+ - N^-) = \frac{4\pi\mu^* V(2m)^{3/2}}{3h^3}\{(\varepsilon_F + \mu^* B)^{3/2} - (\varepsilon_F - \mu^* B)^{3/2}\}. \tag{4}$$

We thus obtain for the low-field susceptibility (per unit volume) of the gas

$$\chi_0 = \underset{B\to 0}{\mathrm{Lim}}\left(\frac{M}{VB}\right) = \frac{4\pi\mu^{*2}(2m)^{3/2}\varepsilon_F^{1/2}}{h^3}. \tag{5}$$

Making use of formula (8.1.24), with $g = 2$, the foregoing result may be written as

$$\chi_0 = \frac{3}{2} n\mu^{*2}/\varepsilon_F. \tag{6}$$

For comparison, the corresponding high-temperature result is given by equation (3.9.26), with $g = 2$ and $J = \frac{1}{2}$:

$$\chi_\infty = n\mu^{*2}/kT. \tag{7}$$

We note that $\chi_0/\chi_\infty = O(kT/\varepsilon_F)$.

To obtain an expression for χ that holds for all T, we proceed as follows. Denoting the number of particles with momentum $\boldsymbol{p}$ and magnetic moment parallel (or antiparallel) to the field by the symbol $n_{\boldsymbol{p}}^+$ (or $n_{\boldsymbol{p}}^-$), the total energy of the gas can be written as

$$\begin{aligned} E_n &= \sum_{\boldsymbol{p}} \left[\left(\frac{p^2}{2m} - \mu^* B \right) n_{\boldsymbol{p}}^+ + \left(\frac{p^2}{2m} + \mu^* B \right) n_{\boldsymbol{p}}^- \right] \\ &= \sum_{\boldsymbol{p}} (n_{\boldsymbol{p}}^+ + n_{\boldsymbol{p}}^-) \frac{p^2}{2m} - \mu^* B (N^+ - N^-), \end{aligned} \tag{8}$$

where N^+ and N^- denote the total number of particles in the two groups, respectively. The partition function of the system is then given by

$$Q(N) = \sum_{\{n_{\boldsymbol{p}}^+\},\{n_{\boldsymbol{p}}^-\}}{}' \exp(-\beta E_n), \tag{9}$$

where the primed summation is subject to the conditions

$$n_{\boldsymbol{p}}^+, n_{\boldsymbol{p}}^- = 0 \text{ or } 1, \tag{10}$$

and

$$\sum_{\boldsymbol{p}} n_{\boldsymbol{p}}^+ + \sum_{\boldsymbol{p}} n_{\boldsymbol{p}}^- = N^+ + N^- = N. \tag{11}$$

To evaluate the sum in (9), we first fix an arbitrary value of the number N^+ (which automatically fixes the value of N^- as well) and sum over *all* $n_{\boldsymbol{p}}^+$ and $n_{\boldsymbol{p}}^-$ that conform to the fixed values of the numbers N^+ and N^- as well as to condition (10). Next, we sum over all possible values of N^+, namely from $N^+ = 0$ to $N^+ = N$. We thus have

$$Q(N) = \sum_{N^+=0}^{N} \left[e^{\beta\mu^* B(2N^+ - N)} \left\{ \sum_{\{n_{\boldsymbol{p}}^+\}}{}'' \exp\left(-\beta \sum_{\boldsymbol{p}} \frac{p^2}{2m} n_{\boldsymbol{p}}^+ \right) \sum_{\{n_{\boldsymbol{p}}^-\}}{}''' \exp\left(-\beta \sum_{\boldsymbol{p}} \frac{p^2}{2m} n_{\boldsymbol{p}}^- \right) \right\} \right]; \tag{12}$$

here, the summation $\sum''$ is subject to the restriction $\sum_{\boldsymbol{p}} n_{\boldsymbol{p}}^{+} = N^{+}$, while $\sum'''$ is subject to the restriction $\sum_{\boldsymbol{p}} n_{\boldsymbol{p}}^{-} = N - N^{+}$.

Now, let $Q_0(\mathcal{N})$ denote the partition function of an ideal Fermi gas of $\mathcal{N}$ "spinless" particles of mass m; then, obviously,

$$Q_0(\mathcal{N}) = \sum_{\{n_{\boldsymbol{p}}\}}' \exp\left(-\beta \sum_{\boldsymbol{p}} \frac{p^2}{2m} n_{\boldsymbol{p}}\right) \equiv \exp\{-\beta A_0(\mathcal{N})\}, \tag{13}$$

where $A_0(\mathcal{N})$ is the free energy of this fictitious system. Equation (12) can then be written as

$$Q(N) = e^{-\beta\mu^* BN} \sum_{N^+=0}^{N} [e^{2\beta\mu^* BN^+} Q_0(N^+) Q_0(N - N^+)], \tag{14}$$

which gives

$$\frac{1}{N} \ln Q(N) = -\beta\mu^* B + \frac{1}{N} \ln \sum_{N^+=0}^{N} [\exp\{2\beta\mu^* BN^+ - \beta A_0(N^+) - \beta A_0(N - N^+)\}]. \tag{15}$$

As before, the logarithm of the sum $\sum_{N^+}$ may be replaced by the logarithm of the largest term in the sum; the error committed in doing so would be negligible in comparison with the term retained. Now, the value $\overline{N^+}$, of N^+, which corresponds to the largest term in the sum, can be ascertained by setting the differential coefficient of the general term, with respect to N^+, equal to zero; this gives

$$2\mu^* B - \left[\frac{\partial A_0(N^+)}{\partial N^+}\right]_{N^+ = \overline{N^+}} - \left[\frac{\partial A_0(N - N^+)}{\partial N^+}\right]_{N^+ = \overline{N^+}} = 0,$$

that is

$$\mu_0(\overline{N^+}) - \mu_0(N - \overline{N^+}) = 2\mu^* B, \tag{16}$$

where $\mu_0(\mathcal{N})$ is the chemical potential of the fictitious system of $\mathcal{N}$ "spinless" fermions.

The foregoing equation contains the general solution being sought. To obtain an explicit expression for χ, we introduce a dimensionless parameter r, defined by

$$M = \mu^*(\overline{N^+} - \overline{N^-}) = \mu^*(2\overline{N^+} - N) = \mu^* Nr \quad (0 \le r \le 1); \tag{17}$$

equation (16) then becomes

$$\mu_0\left(\frac{1+r}{2}N\right) - \mu_0\left(\frac{1-r}{2}N\right) = 2\mu^* B. \tag{18}$$

If the magnetic field B vanishes so does r, which corresponds to a *completely random* orientation of the elementary moments. For small B, r would also be small; so, we may carry out a Taylor expansion of the left side of (18) about $r = 0$. Retaining only the first term of the expansion, we obtain

$$r \simeq \frac{2\mu^* B}{\left.\frac{\partial \mu_0(xN)}{\partial x}\right|_{x=1/2}}. \tag{19}$$

The low-field susceptibility (per unit volume) of the system is then given by

$$\chi = \frac{M}{VB} = \frac{\mu^* N r}{VB} = \frac{2n\mu^{*2}}{\left.\frac{\partial \mu_0(xN)}{\partial x}\right|_{x=1/2}}, \tag{20}$$

which is the desired result valid for all T.

For $T \to 0$, the chemical potential of the fictitious system can be obtained from equation (8.1.34), with $g = 1$:

$$\mu_0(xN) = \left(\frac{3xN}{4\pi V}\right)^{2/3} \frac{h^2}{2m},$$

which gives

$$\left.\frac{\partial \mu_0(xN)}{\partial x}\right|_{x=1/2} = \frac{2^{4/3}}{3}\left(\frac{3N}{4\pi V}\right)^{2/3} \frac{h^2}{2m}. \tag{21}$$

On the other hand, the Fermi energy of the actual system is given by the same equation (8.1.34), with $g = 2$:

$$\varepsilon_F = \left(\frac{3N}{8\pi V}\right)^{2/3} \frac{h^2}{2m}. \tag{22}$$

Making use of equations (21) and (22), we obtain from (20)

$$\chi_0 = \frac{2n\mu^{*2}}{\frac{4}{3}\varepsilon_F} = \frac{3}{2} n\mu^{*2}/\varepsilon_F, \tag{23}$$

in complete agreement with our earlier result (6). For finite but low temperatures, one has to use equation (8.1.35) instead of (8.1.34). The final result turns out to be

$$\chi \simeq \chi_0 \left[1 - \frac{\pi^2}{12}\left(\frac{kT}{\varepsilon_F}\right)^2\right]. \tag{24}$$

On the other hand, for $T \to \infty$, the chemical potential of the fictitious system follows directly from equation (8.1.4), with $g = 1$ and $f_{3/2}(z) \simeq z$, with the result

$$\mu_0(xN) = kT \ln(xN\lambda^3/V),$$

which gives

$$\left.\frac{\partial \mu_0(xN)}{\partial x}\right|_{x=1/2} = 2kT. \tag{25}$$

Equation (20) then gives

$$\chi_\infty = n\mu^{*2}/kT, \tag{26}$$

in complete agreement with our earlier result (7). For large but finite temperatures, one has to take $f_{3/2}(z) \simeq z - (z^2/2^{3/2})$. The final result then turns out to be

$$\chi \simeq \chi_\infty \left(1 - \frac{n\lambda^3}{2^{5/2}}\right); \tag{27}$$

the correction term here is proportional to $(T_F/T)^{3/2}$ and tends to zero as $T \to \infty$.

8.2.B Landau diamagnetism

We now study the magnetism arising from the quantization of the orbital motion of (charged) particles in the presence of an external magnetic field. In a uniform field of intensity $\boldsymbol{B}$, directed along the z-axis, a charged particle would follow a helical path whose axis is parallel to the z-axis and whose projection on the (x, y)-plane is a circle. Motion along the z-direction has a constant linear velocity u_z, while that in the (x, y)-plane has a constant angular velocity eB/mc; the latter arises from the Lorentz force, $e(\boldsymbol{u} \times \boldsymbol{B})/c$, experienced by the particle. Quantum-mechanically, the energy associated with the circular motion is *quantized* in units of $e\hbar B/mc$. The energy associated with the linear motion along the z-axis is also quantized but, in view of the smallness of the energy intervals, this may be taken as a continuous variable. We thus have for the total energy of the particle[2]

$$\varepsilon = \frac{e\hbar B}{mc}\left(j + \frac{1}{2}\right) + \frac{p_z^2}{2m} \quad (j = 0, 1, 2, \ldots). \tag{28}$$

Now, these quantized energy levels are degenerate because they result from a "coalescing together" of an almost continuous set of zero-field levels. A little reflection shows that all those levels for which the value of the quantity $(p_x^2 + p_y^2)/2m$ lay between $e\hbar Bj/mc$ and $e\hbar B(j+1)/mc$ now "coalesce together" into a single level characterized by the quantum number j. The number of these levels is given by

$$\begin{aligned}\frac{1}{h^2}\int dx\,dy\,dp_x dp_y &= \frac{L_x L_y}{h^2}\pi\left[2m\frac{e\hbar B}{mc}\{(j+1) - j\}\right] \\ &= L_x L_y \frac{eB}{hc},\end{aligned} \tag{29}$$

[2]See, for instance, Goldman et al. (1960); Problem 6.3.

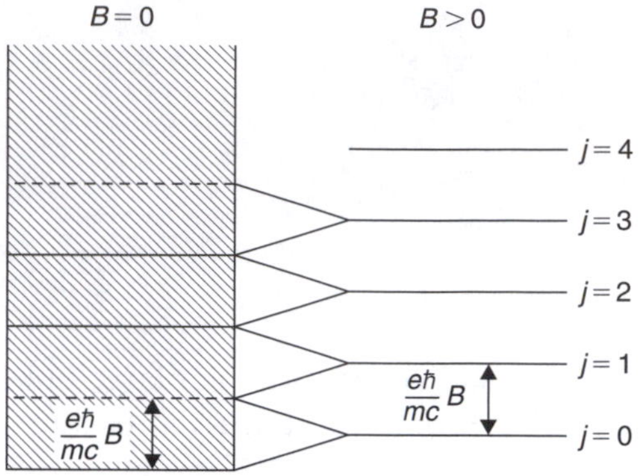

FIGURE 8.3 The single-particle energy levels, for a two-dimensional motion, in the absence of an external magnetic field ($B = 0$) and in the presence of an external magnetic field ($B > 0$).

which is independent of j. The *multiplicity factor* (29) is a quantum-mechanical measure of the freedom available to the particle for the center of its orbit to be "located" anywhere in the total area L_xL_y of the physical space. Figure 8.3 depicts the manner in which the zero-field energy levels of the particle group themselves into a spectrum of oscillator-like levels on the application of the external magnetic field.

The grand partition function of the gas is given by the standard formula

$$\ln \mathcal{Q} = \sum_{\varepsilon} \ln(1 + ze^{-\beta\varepsilon}), \tag{30}$$

where the summation has to be carried over all single-particle states in the system. Substituting (28) for ε, making use of the multiplicity factor (29) and replacing the summation over p_z by an integration, we get

$$\ln \mathcal{Q} = \int_{-\infty}^{\infty} \frac{L_z dp_z}{h} \left[\sum_{j=0}^{\infty} \left(L_xL_y \frac{eB}{hc} \right) \ln\left\{ 1 + ze^{-\beta e\hbar B[j+(1/2)]/mc - \beta p_z^2/2m} \right\} \right]. \tag{31}$$

At high temperatures, $z \ll 1$; so, the system is effectively *Boltzmannian*. The grand partition function then reduces to

$$\begin{aligned} \ln \mathcal{Q} &= \frac{zVeB}{h^2c} \int_{-\infty}^{\infty} e^{-\beta p_z^2/2m} dp_z \sum_{j=0}^{\infty} e^{-\beta e\hbar B[j+(1/2)]/mc} \\ &= \frac{zVeB}{h^2c} \left(\frac{2\pi m}{\beta} \right)^{1/2} \left\{ 2\sinh\left(\frac{\beta e\hbar B}{2mc} \right) \right\}^{-1}. \end{aligned} \tag{32}$$

The equilibrium number of particles $\overline{N}$ and the magnetic moment M of the gas are then given by

$$\overline{N} = \left(z \frac{\partial}{\partial z} \ln \mathcal{Q} \right)_{B,V,T}, \tag{33}$$

and

$$M = \left\langle -\frac{\partial H}{\partial B} \right\rangle = \frac{1}{\beta} \left(\frac{\partial}{\partial B} \ln \mathcal{Q} \right)_{z,V,T}, \tag{34}$$

where H is the Hamiltonian of the system; compare with equation (3.9.4). We thus obtain

$$\overline{N} = \frac{zV}{\lambda^3} \frac{x}{\sinh x}, \tag{35}$$

and

$$M = \frac{zV}{\lambda^3} \mu_{\text{eff}} \left\{ \frac{1}{\sinh x} - \frac{x \cosh x}{\sinh^2 x} \right\}, \tag{36}$$

where $\lambda \{= h/(2\pi mkT)^{1/2}\}$ is the mean thermal wavelength of the particles, while

$$x = \beta \mu_{\text{eff}} B \quad (\mu_{\text{eff}} = eh/4\pi mc). \tag{37}$$

Clearly, if e and m are the electronic charge and the electronic mass, then μ_{eff} is the familiar Bohr magneton μ_B. Combining (35) and (36), we get

$$M = -\overline{N} \mu_{\text{eff}} L(x), \tag{38}$$

where $L(x)$ is the *Langevin function*:

$$L(x) = \coth x - \frac{1}{x}. \tag{39}$$

This result is very similar to the one obtained in the Langevin theory of paramagnetism; see Section 3.9. The presence of the negative sign, however, means that the effect obtained in the present case is *diamagnetic* in nature. We also note that this effect is a direct consequence of quantization; it vanishes if we let $h \to 0$. This is in complete accord with the *Bohr–van Leeuwen theorem*, according to which the phenomenon of diamagnetism does not arise in classical physics; see Problem 3.43.

If the field intensity B and the temperature T are such that $\mu_{\text{eff}} B \ll kT$, then the foregoing results become

$$\overline{N} \simeq \frac{zV}{\lambda^3} \tag{40}$$

and

$$M \simeq -\overline{N} \mu_{\text{eff}}^2 B / 3kT. \tag{41}$$

Equation (40) is in agreement with the zero-field formula $z \simeq n\lambda^3$, while (41) leads to the diamagnetic counterpart of the Curie law:

$$\chi_\infty = \frac{M}{VB} = -\bar{n}\mu_{\text{eff}}^2/3kT; \tag{42}$$

see equation (3.9.12). It should be noted here that the diamagnetic character of this phenomenon is independent of the sign of the electric charge on the particle. For an electron gas, in particular, the net susceptibility at high temperatures is given by the sum of expression (7), with μ^* replaced by μ_B, and expression (42):

$$\chi_\infty = \frac{n\left(\mu_B^2 - \frac{1}{3}\mu_B'^2\right)}{kT}, \tag{43}$$

where $\mu'_B = eh/4\pi m'c$, m' being the *effective mass* of the electron in the given system.

We now look at this problem at *all* temperatures, though we will continue to assume the magnetic field to be weak, so that $\mu_{\text{eff}}B \ll kT$. In view of the latter, the summation in (31) may be handled with the help of the Euler summation formula,

$$\sum_{j=0}^{\infty} f\left(j+\frac{1}{2}\right) \simeq \int_0^{\infty} f(x)dx + \frac{1}{24}f'(0), \tag{44}$$

with the result

$$\begin{aligned}\ln \mathcal{Q} \simeq \frac{VeB}{h^2c}\Bigg[&\int_0^{\infty} dx \int_{-\infty}^{\infty} dp_z \ln\left\{1 + ze^{-\beta(2\mu_{\text{eff}}Bx + p_z^2/2m)}\right\} \\ &-\frac{1}{12}\beta\mu_{\text{eff}}B\int_{-\infty}^{\infty}\frac{dp_z}{z^{-1}e^{\beta(p_z^2/2m)}+1}\Bigg].\end{aligned} \tag{45}$$

The first part here is independent of B, which can be seen by changing the variable from x to $x' = Bx$. The second part, with the substitution $\beta p_z^2/2m = y$, becomes

$$-\frac{\pi V(2m)^{3/2}}{6h^3}(\mu_{\text{eff}}B)^2\beta^{1/2}\int_0^{\infty}\frac{y^{-1/2}dy}{z^{-1}e^y+1}. \tag{46}$$

The low-field susceptibility (per unit volume) of the gas is then given by

$$\begin{aligned}\chi = \frac{M}{VB} &= \frac{1}{\beta VB}\left(\frac{\partial}{\partial B}\ln \mathcal{Q}\right)_{z,V,T} \\ &= -\frac{(2\pi m)^{3/2}\mu_{\text{eff}}^2}{3h^3\beta^{1/2}}f_{1/2}(z),\end{aligned} \tag{47}$$

which is the desired result. Note that, as before, the effect is diamagnetic in character — irrespective of the sign of the charge on the particle.

For $z \ll 1$, $f_{1/2}(z) \simeq z \simeq \overline{n}\lambda^3$; we then recover our previous result (42). For $z \gg 1$ (which corresponds to $T \ll T_F$), $f_{1/2}(z) \approx (2/\pi^{1/2})(\ln z)^{1/2}$; we then get

$$\chi_0 \approx -\frac{2\pi(2m)^{3/2}\mu_{\text{eff}}^2 \varepsilon_F^{1/2}}{3h^3} = -\frac{1}{2} n\mu_{\text{eff}}^2/\varepsilon_F; \tag{48}$$

here, use has also been made of the fact that $(\beta^{-1}\ln z) \simeq \varepsilon_F$. Note that, in magnitude, this result is precisely one-third of the corresponding paramagnetic result (6), provided that we take the μ^* of that expression to be equal to the μ_{eff} of this one.

8.3 The electron gas in metals

One physical system where the application of Fermi–Dirac statistics helped remove a number of inconsistencies and discrepancies is that of conduction electrons in metals. Historically, the electron theory of metals was developed by Drude (1900) and Lorentz (1904–1905), who applied the statistical mechanics of Maxwell and Boltzmann to the electron gas and derived theoretical results for the various properties of metals. The Drude–Lorentz model did provide a reasonable theoretical basis for a partial understanding of the physical behavior of metals; however, it encountered a number of serious problems of a qualitative as well as quantitative nature. For instance, the observed specific heat of metals appeared to be almost completely accountable by the lattice vibrations alone and practically no contribution seemed to be coming from the electron gas. The theory, however, demanded that, on the basis of the equipartition theorem, each electron in the gas should possess a mean thermal energy $\frac{3}{2}kT$ and hence make a contribution of $\frac{3}{2}k$ to the specific heat of the metal. Similarly, one expected the electron gas to exhibit the phenomenon of paramagnetism arising from the intrinsic magnetic moment μ_B of the electrons. According to the classical theory, the paramagnetic susceptibility would be given by (8.2.7), with μ^* replaced by μ_B. Instead, one found that the susceptibility of a normal nonferromagnetic metal was not only independent of temperature but had a magnitude which, at room temperatures, was hardly 1 percent of the expected value.

The Drude–Lorentz theory was also applied to study transport properties of metals, such as the thermal conductivity K and the electrical conductivity σ. While the results for the individual conductivities were not very encouraging, their ratio did conform to the empirical law of Wiedemann and Franz (1853), as formulated by Lorenz (1872), namely that the quantity $K/\sigma T$ was a (universal) constant. The theoretical value of this quantity, which is generally known as the *Lorenz number*, turned out to be $3(k/e)^2 \simeq 2.48 \times 10^{-13}$ e.s.u./deg^2; the corresponding experimental values for most alkali and alkaline–earth metals were, however, found to be scattered around a mean value of 2.72×10^{-13} e.s.u./deg^2. A still more uncomfortable feature of the classical theory was the uncertainty in assigning an appropriate value to the mean free path of the electrons in a given metal and in ascribing to it an appropriate temperature dependence. For these reasons, the problem of the transport properties of metals also remained in a rather unsatisfactory state until the correct lead was provided by Sommerfeld (1928).

The most significant change introduced by Sommerfeld was the replacement of Maxwell–Boltzmann statistics by Fermi–Dirac statistics for describing the electron gas in a metal. With this single stroke of genius, he was able to set most of the things right. To see how it worked, let us first estimate the Fermi energy ε_F of the electron gas in a typical metal, say sodium. Referring to equation (8.1.24), with $g = 2$,

$$\varepsilon_F = \left(\frac{3N}{8\pi V}\right)^{2/3} \frac{h^2}{2m'}, \tag{1}$$

where m' is the *effective mass* of an electron in the gas.[3] The electron density N/V, in the case of a cubic lattice, may be written as

$$\frac{N}{V} = \frac{n_e n_a}{a^3}, \tag{2}$$

where n_e is the number of conduction electrons per atom, n_a the number of atoms per unit cell and a the lattice constant (or the cell length).[4] For sodium, $n_e = 1$, $n_a = 2$, and $a =$ 4.29 Å. Substituting these numbers into (2) and writing $m' = 0.98 m_e$, we obtain from (1)

$$(\varepsilon_F)_{\text{Na}} = 5.03 \times 10^{-12} \text{ erg } = 3.14 \text{ eV}. \tag{3}$$

Accordingly, for the Fermi temperature of the gas is

$$\begin{aligned}(T_F)_{\text{Na}} &= (1.16 \times 10^4) \times \varepsilon_F(\text{ in eV}) \\ &= 3.64 \times 10^4 \text{K},\end{aligned} \tag{4}$$

which is considerably larger than the room temperature $T(\sim 3 \times 10^2 \text{K})$. The ratio T/T_F being of the order of 1 percent, the conduction electrons in sodium constitute a *highly degenerate* Fermi system. This statement, in fact, applies to *all* metals because their Fermi temperatures are generally of order $10^4 - 10^5$ K.

Now, the very fact that the electron gas in metals is a highly degenerate Fermi system is sufficient to explain away some of the basic difficulties of the Drude–Lorentz theory. For instance, the specific heat of this gas would no longer be given by the classical formula,

[3] To justify the assumption that the conduction electrons in a metal may be treated as "free" electrons, it is necessary to ascribe to them an effective mass $m' \neq m$. This is an indirect way of accounting for the fact that the electrons in a metal are not really free; the ratio m'/m accordingly depends on the structural details of the metal and, therefore, varies from metal to metal. In sodium, $m'/m \simeq 0.98$.

[4] Another way of expressing the electron density is to write $N/V = f\rho/M$, where f is the valency of the metal, ρ its mass density, and M the mass of an atom (ρ/M, thus, being the number density of the atoms).

$C_V = \frac{3}{2}Nk$, but rather by equation (8.1.39), namely

$$C_V = \frac{\pi^2}{2}Nk(kT/\varepsilon_F); \tag{5}$$

obviously, the new result is much smaller in value because, at ordinary temperatures, the ratio $(kT/\varepsilon_F) \equiv (T/T_F) = O(10^{-2})$. It is then hardly surprising that, at ordinary temperatures, the specific heat of metals is almost completely determined by the vibrational modes of the lattice and very little contribution comes from the conduction electrons. Of course, as temperature decreases, the specific heat due to lattice vibrations also decreases and finally becomes considerably smaller than the classical value; see Section 7.4, especially Figure 7.14. A stage comes when the two contributions, both nonclassical, become comparable in value. Ultimately, at very low temperatures, the specific heat due to lattice vibrations, being proportional to T^3, becomes *even smaller* than the electronic specific heat, which is proportional to T^1. In general, we may write, for the low-temperature specific heat of a metal,

$$C_V = \gamma T + \delta T^3, \tag{6}$$

where the coefficient γ is given by equation (5) or, more generally, can be shown to be proportional to the density of states at the Fermi energy (see Problem 8.13), while the coefficient δ is given by equation (7.4.23). An experimental determination of the specific heat of metals at low temperatures is, therefore, expected not only to verify the theoretical result based on quantum statistics but also to evaluate some of the parameters of the problem.

Such determinations have been made, among others, by Corak et al. (1955) who worked with copper, silver and, gold in the temperature range 1 to 5 K. Their results for copper are shown in Figure 8.4. The very fact that the (C_V/T) versus T^2 plot is well approximated by a straight line vindicates the theoretical formula (6). Furthermore, the slope of this line gives the value of the coefficient δ, from which one can extract the Debye temperature Θ_D of

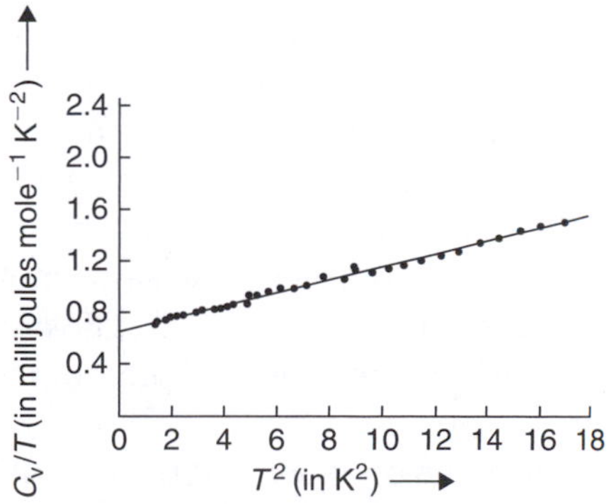

FIGURE 8.4 The observed specific heat of copper at low temperatures (after Corak et al., 1955).

the metal. One thus gets for copper: $\Theta_D = (343.8 \pm 0.5)$ K, which compares favorably with Leighton's theoretical estimate of 345 K (based on the elastic constants of the metal). The intercept on the (C_V/T)-axis yields the value of the coefficient γ, namely (0.688 ± 0.002) millijoule mole^{-1} deg^{-2}, which agrees favorably with Jones' estimate of 0.69 millijoule mole^{-1} deg^{-2} (based on a density-of-states calculation).

The general pattern of the magnetic behavior of the electron gas in nonferromagnetic metals can be understood likewise. In view of the highly degenerate nature of the gas, the magnetic susceptibility χ is given by the Pauli result (8.2.6) plus the Landau result (8.2.48), and *not* by the classical result (8.2.7). In complete agreement with the observation, the new result is (i) independent of temperature and (ii) considerably smaller in magnitude than the classical one.

As regards transport properties K and σ, the new theory again led to the *Wiedemann–Franz law*; the Lorenz number, however, became $(\pi^2/3)(k/e)^2$, instead of the classical $3(k/e)^2$. The resulting theoretical value, namely 2.71×10^{-13} e.s.u./deg^2, turned out to be much closer to the experimental mean value quoted earlier. Of course, the situation regarding individual conductivities and the mean free path of the electrons did not improve until Bloch (1928) developed a theory that took into account interactions among the electron gas and the ion system in the metal. The theory of metals has continued to become more and more sophisticated; the important point to note here is that this development has all along been governed by the new statistics!

Before leaving this topic, we would like to give a brief account of the phenomena of *thermionic* and *photoelectric* emission of electrons from metals. In view of the fact that electronic emission does not take place spontaneously, we infer that the electrons inside a metal find themselves caught in some sort of a "potential well" created by the ions. The detailed features of the potential energy of an electron in this well must depend on the structure of the given metal. For a study of electronic emission, however, we need not worry about these details and may assume instead that the potential energy of an electron stays constant (at a negative value, $-W$, say) throughout the interior of the metal and changes discontinuously to zero at the surface. Thus, while inside the metal, the electrons move about freely and independently of one another; however, as soon as any one of them approaches the surface of the metal and tries to escape, it encounters a potential barrier of height W. Accordingly, only those electrons whose kinetic energy (associated with the motion *perpendicular* to the surface) is greater than W can expect to escape through this barrier. At ordinary temperatures, especially in the absence of any external stimulus, such electrons are too few in any given metal, with the result that there is practically no spontaneous emission from metals. At high temperatures, and more so if there is an external stimulus present, the population of such electrons in a given metal could become large enough to yield a sizeable emission. We then speak of phenomena such as *thermionic effect* and *photoelectric effect*.

Strictly speaking, these phenomena are not equilibrium phenomena because electrons are flowing out *steadily* through the surface of the metal. However, if the number of electrons lost in a given interval of time is small in comparison with their total population in the metal, then the magnitude of the emission current may be calculated on the assumption that the gas inside continues to be in a state of *quasistatic* thermal equilibrium. The mathematical procedure for this calculation is very much the same as the one followed in Section 6.4 (for determining the *rate of effusion* R of the particles of a gas through an opening in the walls of the container). There is one difference, however; whereas in the case of effusion any particle that reached the opening with $u_z > 0$ could escape unquestioned, here we must have $u_z > (2W/m)^{1/2}$, so that the particle in question could successfully cross over the potential barrier at the surface. Moreover, even if this condition is satisfied, there is no guarantee that the particle will really escape because the possibility of an inward reflection cannot be ruled out. In the following discussion, we shall disregard this possibility; however, if one is looking for a numerical comparison of theory with experiment, the results derived here must be multiplied by a factor $(1-r)$, where r is the *reflection coefficient* of the surface.

8.3.A Thermionic emission (the Richardson effect)

The number of electrons emitted per unit area of the metal surface per unit time is given by

$$R = \int\limits_{p_z=(2mW)^{1/2}}^{\infty} \int\limits_{p_x=-\infty}^{\infty} \int\limits_{p_y=-\infty}^{\infty} \left\{ \frac{2dp_x dp_y dp_z}{h^3} \frac{1}{e^{(\varepsilon-\mu)/kT}+1} \right\} u_z; \tag{7}$$

compare with the corresponding expression in Section 6.4. Integration over the variables p_x and p_y may be carried out by changing over to polar coordinates (p', ϕ), with the result

$$\begin{aligned} R &= \frac{2}{h^3} \int\limits_{p_z=(2mW)^{1/2}}^{\infty} \frac{p_z}{m} dp_z \int\limits_{p'=0}^{\infty} \frac{2\pi p' dp'}{\exp\{[(p'^2/2m)+(p_z^2/2m)-\mu]/kT\}+1} \\ &= \frac{4\pi kT}{h^3} \int\limits_{p_z=(2mW)^{1/2}}^{\infty} p_z dp_z \ln[1+\exp\{(\mu - p_z^2/2m)/kT\}] \\ &= \frac{4\pi mkT}{h^3} \int\limits_{\varepsilon_z=W}^{\infty} d\varepsilon_z \ln[1+e^{(\mu-\varepsilon_z)/kT}]. \end{aligned} \tag{8}$$

It so happens that the exponential term inside the logarithm, at all temperatures of interest, is much smaller than unity; see Note 5. We may, therefore, write $\ln(1+x) \simeq x$, with the result

$$R = \frac{4\pi mkT}{h^3} \int_{\varepsilon_z = W}^{\infty} d\varepsilon_z e^{(\mu - \varepsilon_z)/kT}$$
$$= \frac{4\pi mk^2 T^2}{h^3} e^{(\mu - W)/kT}. \tag{9}$$

The thermionic current density is then given by

$$J = eR = \frac{4\pi mek^2}{h^3} T^2 e^{(\mu - W)/kT}. \tag{10}$$

It is only now that the difference between the classical statistics and the Fermi statistics really shows up! In the case of classical statistics, the fugacity of the gas is given by (see equation (8.1.4), with $f_{3/2}(z) \simeq z$)

$$z \equiv e^{\mu/kT} = \frac{n\lambda^3}{g} = \frac{nh^3}{2(2\pi mkT)^{3/2}}; \tag{11}$$

accordingly,

$$J_{\text{class}} = ne \left(\frac{k}{2\pi m} \right)^{1/2} T^{1/2} e^{-\phi/kT} \quad (\phi = W). \tag{12}$$

In the case of Fermi statistics, the chemical potential of the (highly degenerate) electron gas is practically independent of temperature and is very nearly equal to the Fermi energy of the gas ($\mu \simeq \mu_0 \equiv \varepsilon_F$); accordingly,

$$J_{\text{F.D.}} = \frac{4\pi mek^2}{h^3} T^2 e^{-\phi/kT} \quad (\phi = W - \varepsilon_F). \tag{13}$$

The quantity ϕ is generally referred to as the *work function* of the metal. According to (12), ϕ is exactly equal to the height of the surface barrier; according to (13), it is equal to the height of the barrier *over and above* the Fermi level (see Figure 8.5).

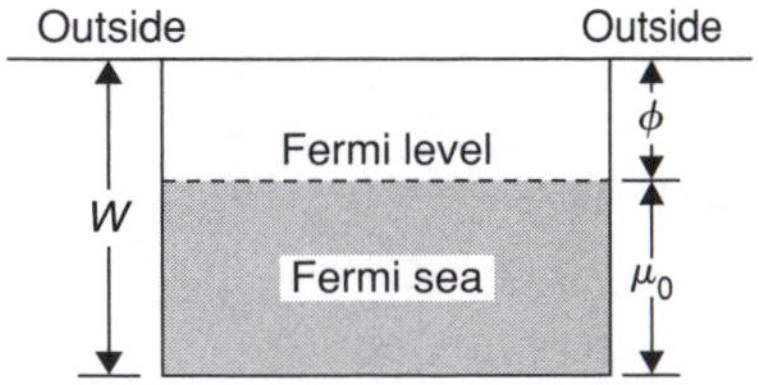

FIGURE 8.5 The work function ϕ of a metal represents the height of the surface barrier *over and above* the Fermi level.

The theoretical results embodied in equations (12) and (13) differ in certain important respects. The most striking difference seems to be in regard to the temperature dependence of the thermionic current density J. However, the major dependence on T comes through the factor $\exp(-\phi/kT)$ — so much so that whether we plot $\ln(J/T^{1/2})$ against $(1/T)$ or $\ln(J/T^2)$ against $(1/T)$ we obtain, in each case, a fairly good straight-line fit. Thus, from the point of view of the temperature dependence of J, a choice between formulae (12) and (13) is rather hard to make. However, the slope of the experimental line should give us *directly* the value of W if formula (12) applies or of $(W - \varepsilon_F)$ if formula (13) applies!

Now, the value of W can be determined independently, for instance, by studying the refractive index of a given metal for de Broglie waves associated with an electron beam impinging on the metal. For a beam of electrons whose initial kinetic energy is E, we have

$$\lambda_{\text{out}} = \frac{h}{\surd(2mE)} \quad \text{and} \quad \lambda_{\text{in}} = \frac{h}{\surd[2m(E+W)]}, \tag{14}$$

so that the refractive index of the metal is given by

$$n = \frac{\lambda_{\text{out}}}{\lambda_{\text{in}}} = \left(\frac{E+W}{E}\right)^{1/2}. \tag{15}$$

By studying electron diffraction for different values of E, one can derive the relevant value of W. In this manner, Davisson and Germer (1927) derived the value of W for a number of metals. For instance, they obtained for tungsten: $W \simeq 13.5\,\text{eV}$. The experimental results on thermionic emission from tungsten are shown in Figure 8.6. The value of ϕ resulting from the slope of the experimental line was about 4.5 eV. The large difference between these

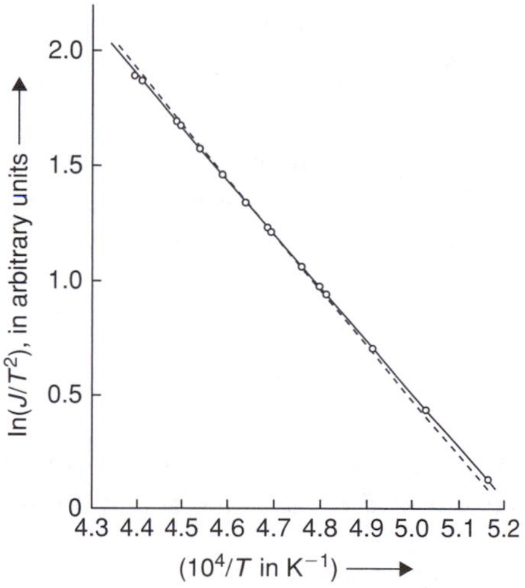

FIGURE 8.6 Thermionic current from tungsten as a function of the temperature of the metal. The continuous line corresponds to $r = \frac{1}{2}$ while the broken line corresponds to $r = 0$, r being the reflection coefficient of the surface.

two values clearly shows that the classical formula (12) does not apply. That the quantum-statistical formula (13) applies is shown by the fact that the Fermi energy of tungsten is very nearly 9 eV; so, the value 4.5 eV for the work function of tungsten is correctly given by the difference between the depth W of the potential well and the Fermi energy ε_F. To quote another example, the experimental value of the work function for nickel was found to be about 5.0 eV, while the theoretical estimate for its Fermi energy turns out to be about 11.8 eV. Accordingly, the depth of the potential well in the case of nickel should be about 16.8 eV. The experimental value of W, obtained by studying electron diffraction in nickel, is indeed (17 ± 1) eV.[5]

The second point of difference between formulae (12) and (13) relates to the actual value of the current obtained. In this respect, the classical formula turns out to be a complete failure while the quantum-statistical formula fares reasonably well. The constant factor in the latter formula is

$$\frac{4\pi mek^2}{h^3} = 120.4 \text{ amp cm}^{-2} \text{deg}^{-2}; \tag{16}$$

of course, this has yet to be multiplied by the *transmission coefficient* $(1 - r)$. The corresponding experimental number, for most metals with clean surfaces, turns out to be in the range 60 to 120 amp cm^{-2} deg^{-2}.

Finally, we examine the influence of a moderately strong electric field on the thermionic emission from a metal — the so-called *Schottky effect*. Denoting the strength of the electric field by F and assuming the field to be uniform and directed perpendicular to the metal surface, the difference Δ between the potential energy of an electron at a distance x above the surface and of one inside the metal is given by

$$\Delta(x) = W - eFx - \frac{e^2}{4x} \quad (x > 0), \tag{17}$$

where the first term arises from the potential well of the metal, the second from the (attractive) field present, and the third from the attraction between the departing electron and the "image" induced in the metal; see Figure 8.7. The largest value of the function $\Delta(x)$ occurs at $x = (e/4F)^{1/2}$, so that

$$\Delta_{\text{max}} = W - e^{3/2} F^{1/2}; \tag{18}$$

thus, the field has effectively reduced the height of the potential barrier by an amount $e^{3/2}F^{1/2}$. A corresponding reduction should take place in the work function as well.

[5] In light of the numbers quoted here, one can readily see that the quantity $e^{(\mu - \varepsilon_z)/kT}$ in equation (8), being *at most* equal to $e^{(\mu_0 - W)/kT} \equiv e^{-\phi/kT}$, is, at all temperatures of interest, much smaller than unity. This means that we are operating here in the (Maxwellian) tail of the Fermi–Dirac distribution and hence the approximation made in going from equation (8) to equation (9) was justified.

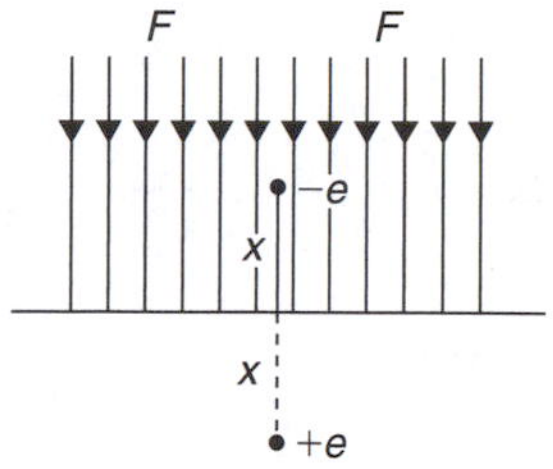

FIGURE 8.7 A schematic diagram to illustrate the Schottky effect.

Accordingly, the thermionic current density in the presence of the field F would be higher than the one in the absence of the field:

$$J_F = J_0 \exp(e^{3/2}F^{1/2}/kT). \tag{19}$$

A plot of $\ln(J_F/J_0)$ against $(F^{1/2}/T)$ should, therefore, be a straight line, with slope $e^{3/2}/k$. Working along these lines, De Bruyne (1928) obtained for the electronic charge a value of 4.84×10^{-10} e.s.u., which is remarkably close to the actual value of e.

The theory of the Schottky effect, as outlined here, holds for field strengths up to about 10^6 volts/cm. For fields stronger than that, one obtains the so-called *cold emission*, which means that the electric field is now strong enough to make the potential barrier practically ineffective; for details, see Fowler and Nordheim (1928).

8.3.B Photoelectric emission (the Hallwachs effect)

The physical situation in the case of photoelectric emission is different from that in the case of thermionic emission, in that there exists now an external agency, the photon in the incoming beam of light, that helps an electron inside the metal in overcoming the potential barrier at the surface. The condition to be satisfied by the momentum component p_z of an electron in order that it could escape from the metal now becomes

$$(p_z^2/2m) + h\nu > W, \tag{20}$$ [6]

where ν is the frequency of the incoming light (assumed monochromatic). Proceeding in the same manner as in the case of thermionic emission, we obtain, instead of (8),

$$R = \frac{4\pi mkT}{h^3} \int_{\varepsilon_z = W - h\nu}^{\infty} d\varepsilon_z \ln[1 + e^{(\mu - \varepsilon_z)/kT}]. \tag{21}$$

We cannot, in general, approximate this integral the way we did there; so the integrand here stays as it is. It is advisable, however, to change over to a new variable x, defined by

$$x = (\varepsilon_z - W + h\nu)/kT, \tag{22}$$

[6] In writing this condition, we have tacitly assumed that the momentum components p_x and p_y of the electron remain unchanged on the absorption of a photon.

whereby equation (21) becomes

$$R = \frac{4\pi m(kT)^2}{h^3} \int_0^\infty dx \ln\left[1 + \exp\left\{\frac{h(\nu - \nu_0)}{kT} - x\right\}\right], \tag{23}$$

where

$$h\nu_0 = W - \mu \simeq W - \varepsilon_F = \phi. \tag{24}$$

The quantity ϕ will be recognized as the (thermionic) work function of the metal; accordingly, the characteristic frequency $\nu_0 (= \phi/h)$ may be referred to as the *threshold frequency* for (photoelectric) emission from the metal concerned.

The current density of photoelectric emission is thus given by

$$J = \frac{4\pi mek^2}{h^3} T^2 \int_0^\infty dx \ln(1 + e^{\delta - x}), \tag{25}$$

where

$$\delta = h(\nu - \nu_0)/kT. \tag{26}$$

Integrating by parts, we find that

$$\int_0^\infty dx \ln(1 + e^{\delta - x}) = \int_0^\infty \frac{x\,dx}{e^{x-\delta} + 1} \equiv f_2(e^\delta); \tag{27}$$

see equation (8.1.6). Accordingly,

$$J = \frac{4\pi mek^2}{h^3} T^2 f_2(e^\delta). \tag{28}$$

For $h(\nu - \nu_0) \gg kT$, $e^\delta \gg 1$ and the function $f_2(e^\delta) \approx \delta^2/2$; see Sommerfeld's lemma (E.17) in Appendix E. Equation (28) then becomes

$$J \approx \frac{2\pi me}{h} (\nu - \nu_0)^2, \tag{29}$$

which is completely independent of T; thus, when the energy of the light quantum is much greater than the work function of the metal, the temperature of the electron gas becomes a "dead" parameter of the problem. At the other extreme, when $\nu < \nu_0$ and $h|\nu - \nu_0| \gg kT$, then $e^\delta \ll 1$ and the function $f_2(e^\delta) \approx e^\delta$. Equation (28) then becomes

$$J \approx \frac{4\pi mek^2}{h^3} T^2 e^{(h\nu - \phi)/kT}, \tag{30}$$

which is just the thermionic current density (13), enhanced by the photon factor $\exp(h\nu/kT)$; in other words, the situation now is very much the same as in the case of thermionic emission, except for a diminished work function $\phi'(=\phi - h\nu)$. At the threshold frequency ($\nu = \nu_0$), $\delta = 0$ and the function $f_2(e^\delta) = f_2(1) = \pi^2/12$; see equation (E.16), with $j = 1$. Equation (28) then gives

$$J_0 = \frac{\pi^3 mek^2}{3h^3} T^2. \tag{31}$$

Figure 8.8 shows a plot of the experimental results for photoelectric emission from palladium ($\phi = 4.97\,\text{eV}$). The agreement with theory is excellent. It will be noted that the plot includes some observations with $\nu < \nu_0$. The fact that we obtain a *finite* photocurrent even for frequencies less than the so-called threshold frequency is fully consistent with the model considered here. The reason for this lies in the fact that, at any *finite* temperature T, there exists in the system a reasonable fraction of electrons whose energy ε exceeds the Fermi energy ε_F by amounts $O(kT)$. Therefore, if the light quantum $h\nu$ gets absorbed by one of these electrons, then condition (20) for photoemission can be satisfied even if $h\nu < (W - \varepsilon_F) = h\nu_0$. Of course, the energy difference $h(\nu_0 - \nu)$ must not be much more than a few times kT, for otherwise the availability of the right kind of electrons will be extremely low. We, therefore, do expect a finite photocurrent for radiation with frequencies less than the threshold frequency ν_0, provided that $h(\nu_0 - \nu) = O(kT)$.

The plot shown in Figure 8.8, namely $\ln(J/T^2)$ versus δ, is generally known as the "Fowler plot." Fitting the observed photoelectric data to this plot, one can obtain the characteristic frequency ν_0 and hence the work function ϕ of the given metal. We have previously seen that the work function of a metal can be derived from thermionic data as well. It is gratifying to note that there is complete agreement between the two sets of results obtained for the work function of the various metals.

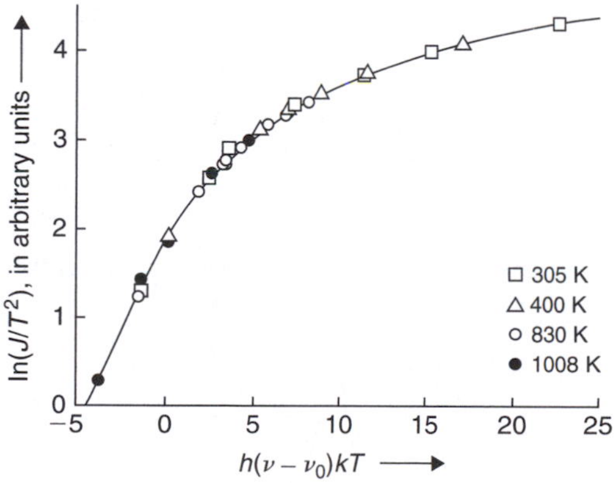

FIGURE 8.8 Photoelectric current from palladium as a function of the quantity $h(\nu - \nu_0)/kT$. The plot includes data taken at several temperatures T for different frequencies ν.

8.4 Ultracold atomic Fermi gases

After the demonstration of Bose–Einstein condensation in ultracold atomic gases in 1995 (Section 7.2), researchers began using laser cooling and magnetic traps to cool gases of fermions to create degenerate Fermi gases of atoms. DeMarco and Jin (1999) created the first degenerate atomic Fermi gas by cooling a dilute vapor of ^{40}K in an atomic trap into the nanokelvin temperature range. The density of states in a harmonic trap is a quadratic function of the energy:

$$a(\varepsilon) = \frac{\varepsilon^2}{2(\hbar\omega_0)^3}, \tag{1}$$

where $\omega_0 = (\omega_1\omega_2\omega_3)^{1/3}$ is the geometric mean of the trap frequencies in the cartesian directions; see equation (7.2.3). The chemical potential and the number of fermions in the trap are related by

$$N(\mu, T) = \frac{1}{2(\hbar\omega_0)^3}\int_0^\infty \frac{\varepsilon^2 d\varepsilon}{e^{\beta(\varepsilon-\mu)}+1}, \tag{2}$$

which gives for the Fermi energy

$$\varepsilon_F = \hbar\omega_0(6N)^{1/3}, \tag{3}$$

a Fermi temperature $T_F = \varepsilon_F/k = 870\,\text{nK}$ for 10^6 atoms in a 100 Hz trap, and a ground-state energy $U_0 = \frac{3}{4}N\varepsilon_F$. The internal energy of the trapped gas can be obtained by time-of-flight measurements as described in Section 7.2 and can be directly compared with the theoretical result

$$\frac{U}{U_0} = 4\left(\frac{T}{T_F}\right)^4 \int_0^\infty \frac{x^3 dx}{e^x e^{-\beta\mu}+1}, \tag{4}$$

where the temperature and the chemical potential are related by

$$3\left(\frac{T}{T_F}\right)^3 \int_0^\infty \frac{x^2 dx}{e^x e^{-\beta\mu}+1} = 1; \tag{5}$$

see Figures 8.9 and 8.10. At low enough temperatures, attractive interactions lead to BEC-BCS condensation, as discussed in Section 11.9.

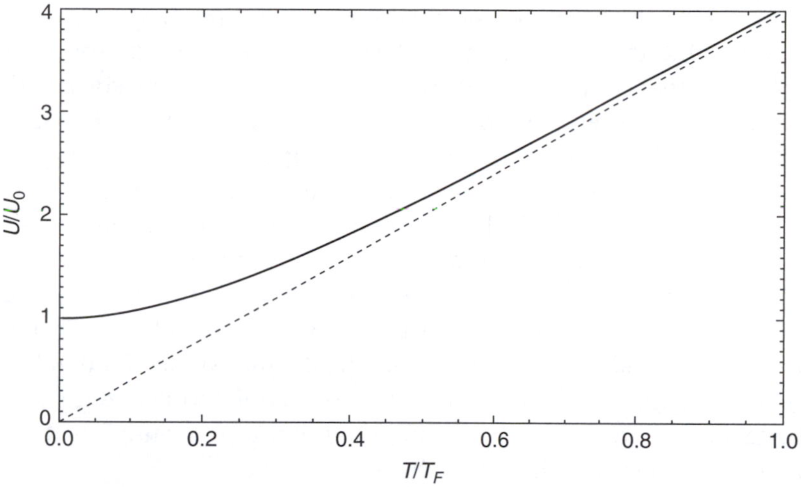

FIGURE 8.9 Scaled internal energy (U/U_0) versus scaled temperature (T/T_F) for an ideal Fermi gas in a harmonic trap from equations (4) and (5). The dotted line is the corresponding classical result $U(T) = 3NkT$.

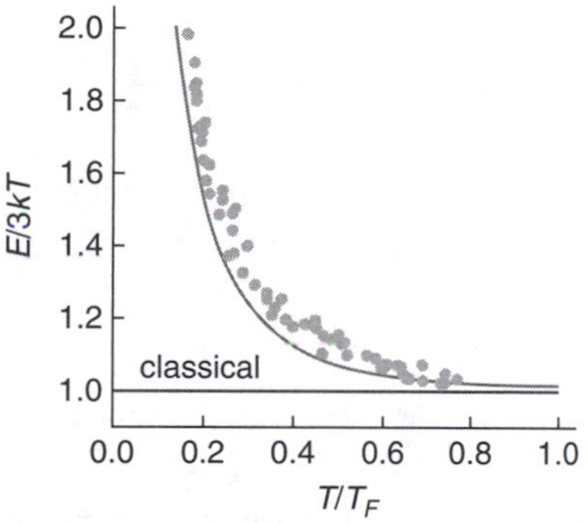

FIGURE 8.10 Experimental results for the mean energy per particle divided by the equipartition value of $3kT$ versus scaled temperature (T/T_F) for ultracold ^{40}K atoms in a harmonic trap compared to the theoretical Fermi gas value from equations (4) and (5). This shows the development of the Fermi degeneracy of the gas at low temperatures; from Jin (2002). Figure courtesy of the IOP. Reprinted with permission; copyright ©2002, American Institute of Physics.

8.5 Statistical equilibrium of white dwarf stars

Historically, the first application of Fermi statistics appeared in the field of astrophysics (Fowler, 1926). It related to the study of thermodynamic equilibrium of *white dwarf stars*—the small-sized stars that are abnormally faint for their (white) color. The general pattern of color–brightness relationship among stars is such that, by and large, a star with red color is expected to be a "dull" star, while one with white color is expected to be a "brilliant" star. However, white dwarf stars constitute an exception to this rule. The reason for this lies in the fact that these stars are relatively old whose hydrogen content is more or less used up, with the result that the thermonuclear reactions in them are now proceeding at a rather low pace, thus making these stars a lot less bright than one would expect on the basis of their color. The material content of white dwarf stars, at the present stage of their career, is mostly helium. And whatever little brightness they presently have derives mostly from the gravitational energy released as a result of a slow contraction of these stars — a mechanism first proposed by Kelvin, in 1861, as a "possible" source of energy for *all* stars!

A typical, though somewhat idealized, model of a white dwarf star consists of a mass $M(\sim 10^{33}\text{g})$ of helium, packed into a ball of mass density $\rho(\sim 10^7 \text{g cm}^{-3})$, at a central temperature $T(\sim 10^7\text{K})$. Now, a temperature of the order of 10^7K corresponds to a mean thermal energy per particle of the order of 10^3 eV, which is much greater than the energy required for ionizing a helium atom. Thus, practically the whole of the helium in the star exists in a state of complete ionization. The microscopic constituents of the star may, therefore, be taken as N electrons (each of mass m) and $\frac{1}{2}N$ helium nuclei (each of mass $\simeq 4m_p$). The mass of the star is then given by

$$M \simeq N(m+2m_p) \simeq 2Nm_p \tag{1}$$

and, hence, the electron density by

$$n = \frac{N}{V} \simeq \frac{M/2m_p}{M/\rho} = \frac{\rho}{2m_p}. \tag{2}$$

A typical value of the electron density in white dwarf stars would, therefore, be $O(10^{30})$ electrons per cm^3. We thus obtain for the Fermi momentum of the electron gas [see equation (8.1.23), with $g = 2$]

$$p_F = \left(\frac{3n}{8\pi}\right)^{1/3} h = O(10^{-17})\,\text{g cm sec}^{-1}, \tag{3}$$

which is rather comparable with the characteristic momentum mc of an electron. The Fermi energy ε_F of the electron gas will, therefore, be comparable with the rest energy mc^2 of an electron, that is, $\varepsilon_F = O(10^6)$ eV and hence the Fermi temperature $T_F = O(10^{10})$ K. In view of these estimates, we conclude that (i) the dynamics of the electrons in this problem is *relativistic*, and (ii) the electron gas, though at a temperature large in comparison with terrestrial standards, is, statistically speaking, in a state of (*almost*) *complete degeneracy*: $(T/T_F) = O(10^{-3})$. The second point was fully appreciated, and duly taken into account, by Fowler himself; the first one was taken care of later, by Anderson (1928) and by Stoner (1929, 1930). The problem, in full generality, was attacked by Chandrasekhar (1931–1935) to whom the final picture of the theory of white dwarf stars is chiefly due; for details, see Chandrasekhar (1939), where a complete bibliography of the subject is given.

Now, the helium nuclei do not contribute as significantly to the dynamics of the problem as do the electrons; in the first approximation, therefore, we may neglect the presence of the nuclei in the system. For a similar reason, we may neglect the effect of the radiation as well. We may thus consider the electron gas alone. Further, for simplicity, we may assume that the electron gas is *uniformly* distributed over the body of the star; we are thus ignoring the spatial variation of the various parameters of the problem — a variation that is physically essential for the very stability of the star! The contention here is that, in spite of neglecting the spatial variation of the parameters involved, we expect that the results obtained here will be correct, *at least* in a qualitative sense.

We study the *ground-state* properties of a degenerate Fermi gas composed of N *relativistic* electrons ($g = 2$). First of all, we have

$$N = \frac{8\pi V}{h^3} \int_0^{p_F} p^2 dp = \frac{8\pi V}{3h^3} p_F^3, \tag{4}$$

which gives

$$p_F = \left(\frac{3n}{8\pi}\right)^{1/3} h. \tag{5}$$

The energy–momentum relation for a relativistic particle is

$$\varepsilon = mc^2[\{1 + (p/mc)^2\}^{1/2} - 1], \tag{6}$$

the speed of the particle being

$$u \equiv \frac{d\varepsilon}{dp} = \frac{(p/m)}{\{1 + (p/mc)^2\}^{1/2}}; \tag{7}$$

here, m denotes the rest mass of the electron. The pressure P_0 of the gas is then given by, see equation (6.4.3),

$$P_0 = \frac{1}{3}\frac{N}{V}\langle pu\rangle_0 = \frac{8\pi}{3h^3} \int_0^{p_F} \frac{(p^2/m)}{\{1 + (p/mc)^2\}^{1/2}} p^2 dp. \tag{8}$$

We now introduce a dimensionless variable θ, defined by

$$p = mc \sinh\theta, \tag{9}$$

which makes

$$u = c \tanh\theta. \tag{10}$$

Equations (4) and (8) then become

$$N = \frac{8\pi V m^3 c^3}{3h^3} \sinh^3\theta_F = \frac{8\pi V m^3 c^3}{3h^3} x^3 \tag{11}$$

and

$$P_0 = \frac{8\pi m^4 c^5}{3h^3} \int_0^{\theta_F} \sinh^4\theta \, d\theta = \frac{\pi m^4 c^5}{3h^3} A(x), \tag{12}$$

where

$$A(x) = x(x^2+1)^{1/2}(2x^2-3) + 3\sinh^{-1} x, \tag{13}$$

with

$$x = \sinh\theta_F = p_F/mc = (3n/8\pi)^{1/3}(h/mc). \tag{14}$$

The function $A(x)$ can be computed for any desired value of x. However, asymptotic results for $x \ll 1$ and $x \gg 1$ are often useful; these are given by (see Kothari and Singh, 1942)

$$\left.\begin{aligned} A(x) &= \tfrac{8}{5}x^5 - \tfrac{4}{7}x^7 + \tfrac{1}{3}x^9 - \tfrac{5}{22}x^{11} + \cdots && \text{for} \quad x \ll 1 \\ &= 2x^4 - 2x^2 + 3(\ln 2x - \tfrac{7}{12}) + \tfrac{5}{4}x^{-2} + \cdots && \text{for} \quad x \gg 1 \end{aligned}\right\}. \tag{15}$$

We shall now consider, somewhat crudely, the equilibrium configuration of this model. In the absence of gravitation, it would be necessary to have "external walls" for keeping the electron gas at a given density n. The gas will exert a pressure $P_0(n)$ on the walls and any compression or expansion (of the gas) will involve an expenditure of work. Assuming the configuration to be spherical, an adiabatic change in V will cause a change in the energy of the gas, as given by

$$dE_0 = -P_0(n)dV = -P_0(R) \cdot 4\pi R^2 dR. \tag{16}$$

In the presence of gravitation, no external walls are needed, but the change in the kinetic energy of the gas, as a result of a change in the size of the sphere, will still be given by formula (16); of course, the expression for P_0, as a function of the "mean" density n, must now take into account the nonuniformity of the system — a fact being disregarded in the present simple-minded treatment. However, equation (16) alone no longer gives us the net change in the energy of the system; if that were the case, the system would expand indefinitely till both n and $P_0(n) \to 0$. Actually, we have now a change in the *potential* energy as well; this is given by

$$dE_g = \left(\frac{dE_g}{dR}\right) dR = \alpha \frac{GM^2}{R^2} dR, \tag{17}$$

where M is the total mass of the gas, G the constant of gravitation, while α is a number (of the order of unity) whose exact value depends on the nature of the spatial variation of n inside the sphere. If the system is in equilibrium, then the net change in its total energy $(E_0 + E_g)$, for an infinitesimal change in its size, should be zero; thus, for equilibrium,

$$P_0(R) = \frac{\alpha}{4\pi} \frac{GM^2}{R^4}. \tag{18}$$

For $P_0(R)$, we may substitute from equation (12), where the parameter x is now given by

$$x = \left(\frac{3n}{8\pi}\right)^{1/3} \frac{h}{mc} = \left(\frac{9N}{32\pi^2}\right)^{1/3} \frac{h/mc}{R}$$

or, in view of (1), by

$$x = \left(\frac{9M}{64\pi^2 m_p}\right)^{1/3} \frac{h/mc}{R} = \left(\frac{9\pi M}{8m_p}\right)^{1/3} \frac{\hbar/mc}{R}. \tag{19}$$

Equation (18) then takes the form

$$A\left(\left\{\frac{9\pi M}{8m_p}\right\}^{1/3} \frac{\hbar/mc}{R}\right) = \frac{3\alpha h^3}{4\pi^2 m^4 c^5} \frac{GM^2}{R^4}$$
$$= 6\pi\alpha \left(\frac{\hbar/mc}{R}\right)^3 \frac{GM^2/R}{mc^2}; \tag{20}$$

the function $A(x)$ is given by equations (13) and (15).

Equation (20) establishes a one-to-one correspondence between the masses M and the radii R of white dwarf stars; it is, therefore, known as the *mass–radius relationship* for these stars. It is rather interesting to see the combinations of parameters that appear in this relationship; we have here (i) the mass of the star in terms of the proton mass, (ii) the radius of the star in terms of the Compton wavelength of the electron, and (iii) the gravitational energy of the star in terms of the rest energy of the electron. This relationship, therefore, exhibits a remarkable blending of quantum mechanics, special relativity, and gravitation.

In view of the implicit character of relationship (20), we cannot express the radius of the star as an explicit function of its mass, except in two extreme cases. For this, we note that, since $M \sim 10^{33}$g, $m_p \sim 10^{-24}$g, and $\hbar/mc \sim 10^{-11}$ cm, the argument of the function $A(x)$ will be of the order of unity when $R \sim 10^8$ cm. We may, therefore, define the two extreme cases as follows:

(i) $R \gg 10^8$ cm, which makes $x \ll 1$ and hence $A(x) \approx \frac{8}{5}x^5$, with the result

$$R \approx \frac{3(9\pi)^{2/3}}{40\alpha} \frac{\hbar^2 M^{-1/3}}{Gmm_p^{5/3}} \propto M^{-1/3}. \tag{21}$$

(ii) $R \ll 10^8$ cm, which makes $x \gg 1$ and hence $A(x) \approx 2x^4 - 2x^2$, with the result

$$R \approx \frac{(9\pi)^{1/3}}{2} \frac{\hbar}{mc} \left(\frac{M}{m_p}\right)^{1/3} \left\{1 - \left(\frac{M}{M_0}\right)^{2/3}\right\}^{1/2}, \tag{22}$$

where

$$M_0 = \frac{9}{64}\left(\frac{3\pi}{\alpha^3}\right)^{1/2} \frac{(\hbar c/G)^{3/2}}{m_p^2}. \tag{23}$$

We thus find that the greater the mass of the white dwarf star, the smaller its size. Not only that, there exists a limiting mass M_0, given by expression (23), that corresponds to a vanishing size of the star. Obviously, for $M > M_0$, our mass–radius relationship does not possess any real solution. We, therefore, conclude that all white dwarf stars in equilibrium must have a mass less than M_0 — a conclusion fully upheld by observation.

The correct limiting mass of a white dwarf star is generally referred to as the *Chandrasekhar limit*. The physical reason for the existence of this limit is that for a mass exceeding this limit the ground-state pressure of the electron gas (that arises from the fact that the electrons obey the Pauli exclusion principle) would not be sufficient to support the star against its "tendency toward a gravitational collapse." The numerical value of the limiting mass, as given by expression (23), turns out to be $\sim 10^{33}$g. Detailed investigations by Chandrasekhar led to the result:

$$M_0 = \frac{5.75}{\mu_e^2}\odot, \tag{24}$$

where $\odot$ denotes the mass of the sun, which is about 2×10^{33}g, while μ_e is a number that represents the degree of ionization of helium in the gas. By definition, $\mu_e = M/Nm_H$; compare to equation (1). Thus, in most cases, $\mu_e \simeq 2$; accordingly $M_0 \simeq 1.44\odot$.

Figure 8.11 shows a plot of the theoretical relationship between the masses and the radii of white dwarf stars. One can see that the behavior in the two extreme regions, namely for $R \gg l$ and $R \ll l$, is described quite well by formulae (21) and (22) of the treatment given here. The Chandrasekhar limit (24) is the mechanism responsible for stellar collapse into neutron stars and black holes. In particular, white dwarf stars whose mass exceeds the Chandrasekhar limit due to influx of matter from a companion binary star are thought to be the primary mechanism for type Ia supernovae; see Hillebrandt and Niemeyer (2000). Such events happen in a typical galaxy on the order of once per hundred years. For a few days after the collapse and subsequent explosion, these supernovae can be comparable in brightness to the remainder of the stars in the galaxy combined. Their well-calibrated light curves provide a bright "standard candle" for determining the distance to remote galaxies used to measure the expansion rate of the universe; see Chapter 9.

8.6 Statistical model of the atom

Another application of the Fermi statistics was made by Thomas (1927) and Fermi (1928) for calculating the charge distribution and the electric field in the extra-nuclear space of a heavy atom. Their approach was based on the observation that the electrons in this system could be regarded as a completely degenerate Fermi gas of *nonuniform* density $n(\mathbf{r})$.

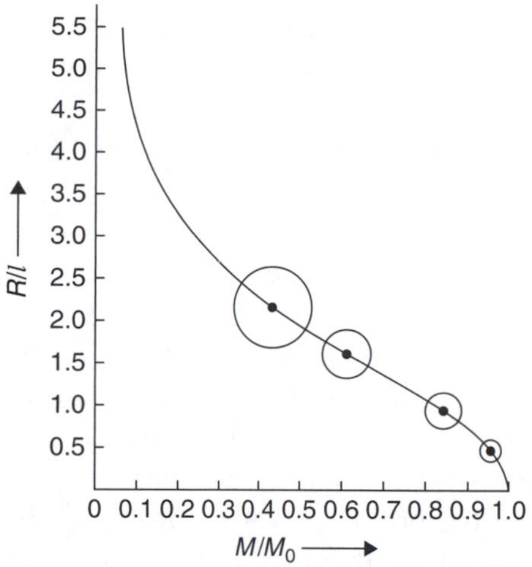

FIGURE 8.11 The mass–radius relationship for white dwarfs (after Chandrasekhar, 1939). The masses are expressed in terms of the limiting mass M_0 and the radii in terms of a characteristic length l, which is given by $7.71\mu_e^{-1} \times 10^8\,\text{cm} \simeq 3.86 \times 10^8\,\text{cm}$.

By considering the equilibrium state of the configuration, one arrives at a differential equation whose solution gives directly the electric potential $\phi(\boldsymbol{r})$ and the electron density $n(\boldsymbol{r})$ at point $\boldsymbol{r}$. By the very nature of the model, which is generally referred to as the *Thomas–Fermi model* of the atom, the resulting function $n(\boldsymbol{r})$ is a smoothly varying function of $\boldsymbol{r}$, devoid of the "peaks" that would remind one of the electron orbits of the Bohr theory. Nevertheless, the model has proved quite useful in deriving composite properties such as the binding energy of the atom. And, after suitable modifications, it has been successfully applied to molecules, solids, and nuclei as well. Here, we propose to outline only the simplest treatment of the model, as applied to an atomic system; for further details and other applications, see Gombás (1949, 1952) and March (1957), where references to other contributions to the subject can also be found.

According to the statistics of a completely degenerate Fermi gas, we have exactly two electrons (with opposite spins) in each elementary cell of the phase space, with $p \leq p_F$; the Fermi momentum p_F of the electron gas is determined by the electron density n, according to the formula

$$p_F = (3\pi^2 n)^{1/3}\hbar. \tag{1}$$

In the system under study, the electron density varies from point to point; so would the value of p_F. We must, therefore, speak of the limiting momentum p_F as a function of $\boldsymbol{r}$, which is clearly a "quasiclassical" description of the situation. Such a description is justifiable if the de Broglie wavelength of the electrons in a given region of space is much smaller than the distance over which the functions $p_F(\boldsymbol{r})$, $n(\boldsymbol{r})$, and $\phi(\boldsymbol{r})$ undergo a significant variation; later on, it will be seen that this requirement is satisfied reasonably well by the heavier atoms.

Now, the total energy ε of an electron at the top of the Fermi sea *at the point* $\boldsymbol{r}$ is given by

$$\varepsilon(\boldsymbol{r}) = \frac{1}{2m} p_F^2(\boldsymbol{r}) - e\phi(\boldsymbol{r}), \tag{2}$$

where e denotes the magnitude of the electronic charge. When the system is in a *stationary* state, the value of $\varepsilon(\boldsymbol{r})$ should be the same throughout, so that electrons anywhere in the system do not have an overall tendency to "flow away" toward other parts of the system. Now, at the boundary of the system, p_F must be zero; by a suitable choice of the zero of energy, we can also have $\phi = 0$ there. Thus, the value of ε at the boundary of the system is zero; so must, then, be the value of ε throughout the system. We thus have, for all $\boldsymbol{r}$,

$$\frac{1}{2m} p_F^2(\boldsymbol{r}) - e\phi(\boldsymbol{r}) = 0. \tag{3}$$

Substituting from (1) and making use of the Poisson equation,

$$\nabla^2 \phi(\boldsymbol{r}) = -4\pi\rho(\boldsymbol{r}) = 4\pi e n(\boldsymbol{r}), \tag{4}$$

we obtain

$$\nabla^2 \phi(\boldsymbol{r}) = \frac{4e(2me)^{3/2}}{3\pi\hbar^3} \{\phi(\boldsymbol{r})\}^{3/2}. \tag{5}$$

Assuming spherical symmetry, equation (5) takes the form

$$\frac{1}{r^2}\frac{d}{dr}\left\{r^2 \frac{d}{dr}\phi(r)\right\} = \frac{4e(2me)^{3/2}}{3\pi\hbar^3} \{\phi(r)\}^{3/2}, \tag{6}$$

which is known as the *Thomas–Fermi equation* of the system. Introducing dimensionless variables x and Φ, defined by

$$x = 2\left(\frac{4}{3\pi}\right)^{2/3} Z^{1/3} \frac{me^2}{\hbar^2} r = \frac{Z^{1/3}}{0.88534 a_B} r \tag{7}$$

and

$$\Phi(x) = \frac{\phi(r)}{Ze/r}, \tag{8}$$

where Z is the atomic number of the system and a_B the first Bohr radius of the hydrogen atom, equation (6) reduces to

$$\frac{d^2\Phi}{dx^2} = \frac{\Phi^{3/2}}{x^{1/2}}. \tag{9}$$

Equation (9) is the *dimensionless Thomas–Fermi equation* of the system. The boundary conditions on the solution to this equation can be obtained as follows. As we approach the nucleus of the system ($r \to 0$), the potential $\phi(r)$ approaches the unscreened value Ze/r; accordingly, we must have: $\Phi(x \to 0) = 1$. On the other hand, as we approach the boundary of the system ($r \to r_0$), $\phi(r)$ in the case of a neutral atom must tend to zero; accordingly, we must have: $\Phi(x \to x_0) = 0$. In principle, these two conditions are sufficient to determine the function $\Phi(x)$ completely. However, it would be helpful if one knew the initial slope of the function as well, which in turn would depend on the precise location of the boundary. Choosing the boundary to be at infinity ($r_0 = \infty$), the appropriate initial slope of the function $\Phi(x)$ turns out to be very nearly -1.5886; in fact, the nature of the solution near the origin is

$$\Phi(x) = 1 - 1.5886x + \frac{4}{3}x^{3/2} + \cdots \tag{10}$$

For $x > 10$, the approximate solution has been determined by Sommerfeld (1932):

$$\Phi(x) \approx \left\{1 + \left(\frac{x^3}{144}\right)^{\lambda}\right\}^{-1/\lambda}, \tag{11}$$

where

$$\lambda = \frac{\sqrt{(73)} - 7}{6} \simeq 0.257. \tag{12}$$

As $x \to \infty$, the solution tends to the simple form: $\Phi(x) \approx 144/x^3$. The complete solution, which is a monotonically decreasing function of x, has been tabulated by Bush and Caldwell (1931). As a check on the numerical results, we note that the solution must satisfy the integral condition

$$\int_0^\infty \Phi^{3/2} x^{1/2} dx = 1, \tag{13}$$

which expresses the fact that the integral of the electron density $n(r)$ over the whole of the space available to the system must be equal to Z, the total number of electrons present.

From the function $\Phi(x)$, one readily obtains the electric potential $\phi(r)$ and the electron density $n(r)$:

$$\phi(r) = \frac{Ze}{r}\Phi\left(\frac{rZ^{1/3}}{0.88534a_B}\right) \propto Z^{4/3} \tag{14}$$

and

$$n(r) = \frac{(2me)^{3/2}}{3\pi^2\hbar^3}\{\phi(r)\}^{3/2} \propto Z^2. \tag{15}$$

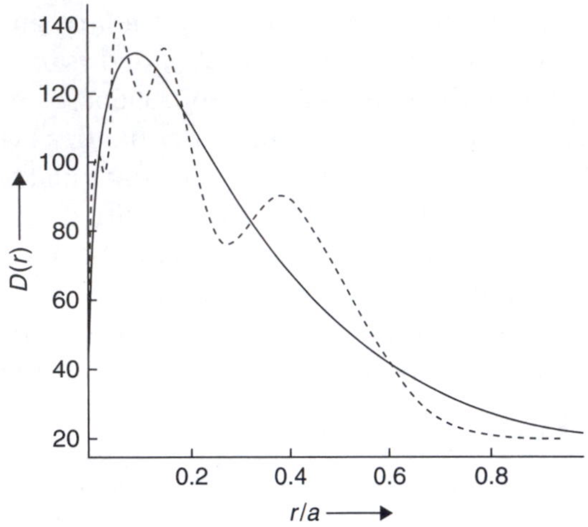

FIGURE 8.12 The electron distribution function $D(r)$ for an atom of mercury. The distance r is expressed in terms of the atomic unit of length $a(=\hbar^2/me^2)$.

A Thomas–Fermi plot of the electron distribution function $D(r)\{= n(r)\cdot 4\pi r^2\}$ for an atom of mercury is shown in Figure 8.12; the actual "peaked" distribution, which conveys unmistakably the preference of the electrons to be in the vicinity of their semiclassical orbits, is also shown in the figure.

To calculate the *binding energy* of the atom, we should determine the total energy of the electron cloud. Now, the mean kinetic energy of an electron at the point $\boldsymbol{r}$ would be $\frac{3}{5}\varepsilon_F(r)$; by equation (3), this is equal to $\frac{3}{5}e\phi(r)$. The total kinetic energy of the electron cloud is, therefore, given by

$$\frac{3}{5}e\int_0^\infty \phi(r)n(r)\cdot 4\pi r^2 dr. \tag{16}$$

For the potential energy of the cloud, we note that a part of the potential $\phi(r)$ at the point $\boldsymbol{r}$ is due to the nucleus of the atom while the rest of it is due to the electron cloud itself; the former is clearly (Ze/r), so the latter must be $\{\phi(r) - Ze/r\}$. The total potential energy of the cloud is, therefore, given by

$$-e\int_0^\infty \left[\frac{Ze}{r} + \frac{1}{2}\left\{\phi(r) - \frac{Ze}{r}\right\}\right] n(r)\cdot 4\pi r^2 dr. \tag{17}$$

We thus obtain for the total energy of the cloud

$$E_0 = \int_0^\infty \left\{\frac{1}{10}e\phi(r) - \frac{1}{2}\frac{Ze^2}{r}\right\} n(r)\cdot 4\pi r^2 dr; \tag{18}$$

of course, the electron density $n(r)$, in terms of the potential function $\phi(r)$, is given by equation (15).

Now, Milne (1927) has shown that the integrals

$$\int_0^\infty \{\phi(r)\}^{5/2} r^2 dr \quad \text{and} \quad \int_0^\infty \{\phi(r)\}^{3/2} r dr, \tag{19}$$

which appear in the expression for E_0, can be expressed directly in terms of the *initial* slope of the function $\Phi(x)$, that is, in terms of the number -1.5886 of equation (10). After a little calculus, one finds that

$$E_0 = \frac{1.5886}{0.88534}\left(\frac{e^2}{2a_B}\right) Z^{7/3}\left(\frac{1}{7}-1\right), \tag{20}$$

from which one obtains for the (Thomas–Fermi) *binding energy* of the atom:

$$E_B = -E_0 = 1.538 Z^{7/3}\chi, \tag{21}$$

where $\chi (= e^2/2a_B \simeq 13.6\,\text{eV})$ is the (actual) binding energy of the hydrogen atom.

It is clear that our statistical result (21) cannot give us anything more than just the first term of an "asymptotic expansion" of the binding energy E_B in powers of the parameter $Z^{-1/3}$. For practical values of Z, other terms of the expansion are also important; however, they cannot be obtained from the simple-minded treatment given here. The interested reader may refer to the review article by March (1957).

In the end we observe that, since the total energy of the electron cloud is proportional to $Z^{7/3}$, the *mean* energy per electron would be proportional to $Z^{4/3}$; accordingly, the mean de Broglie wavelength of the electrons in the cloud would be proportional to $Z^{-2/3}$. At the same time, the overall linear dimensions of the cloud are proportional to $Z^{-1/3}$; see equation (7). We thus find that the quasiclassical description adopted in the Thomas–Fermi model is more appropriate for heavier atoms (so that $Z^{-2/3} \ll Z^{-1/3}$). Otherwise, too, the statistical nature of the approach demands that the number of particles in the system be large.

Problems

8.1 Let the Fermi distribution at low temperatures be represented by a *broken line*, as shown in Figure 8.13, the line being tangential to the actual curve at $\varepsilon = \mu$. Show that this approximate representation yields a "correct" result for the low-temperature specific heat of the Fermi gas, except that the numerical factor turns out to be smaller by a factor of $4/\pi^2$. Discuss, in a qualitative manner, the origin of this numerical discrepancy.

8.2 For a Fermi–Dirac gas, we may define a temperature T_0 at which the chemical potential of the gas is zero ($z = 1$). Express T_0 in terms of the Fermi temperature T_F of the gas.
[*Hint*: Use equation (E.16).]

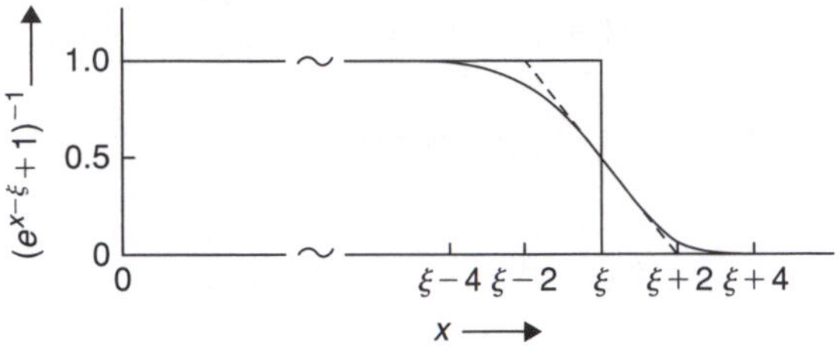

FIGURE 8.13 An *approximate* representation of the Fermi distribution at low temperatures: here, $x = \varepsilon/kT$ and $\xi = \mu/kT$.

8.3 Show that for an ideal Fermi gas

$$\frac{1}{z}\left(\frac{\partial z}{\partial T}\right)_P = -\frac{5}{2T}\frac{f_{5/2}(z)}{f_{3/2}(z)};$$

compare with equation (8.1.9). Hence show that

$$\gamma \equiv \frac{C_P}{C_V} = \frac{(\partial z/\partial T)_P}{(\partial z/\partial T)_v} = \frac{5}{3}\frac{f_{5/2}(z)f_{1/2}(z)}{\{f_{3/2}(z)\}^2}.$$

Check that at low temperatures

$$\gamma \simeq 1 + \frac{\pi^2}{3}\left(\frac{kT}{\varepsilon_F}\right)^2.$$

8.4 **(a)** Show that the isothermal compressibility κ_T and the adiabatic compressibility κ_S of an ideal Fermi gas are given by

$$\kappa_T = \frac{1}{nkT}\frac{f_{1/2}(z)}{f_{3/2}(z)}, \quad \kappa_S = \frac{3}{5nkT}\frac{f_{3/2}(z)}{f_{5/2}(z)},$$

where $n(= N/V)$ is the particle density in the gas. Check that at low temperatures

$$\kappa_T \simeq \frac{3}{2n\varepsilon_F}\left[1 - \frac{\pi^2}{12}\left(\frac{kT}{\varepsilon_F}\right)^2\right], \quad \kappa_S \simeq \frac{3}{2n\varepsilon_F}\left[1 - \frac{5\pi^2}{12}\left(\frac{kT}{\varepsilon_F}\right)^2\right].$$

(b) Making use of the thermodynamic relation

$$C_P - C_V = T\left(\frac{\partial P}{\partial T}\right)_V\left(\frac{\partial V}{\partial T}\right)_P = TV\kappa_T\left(\frac{\partial P}{\partial T}\right)_V^2,$$

show that

$$\frac{C_P - C_V}{C_V} = \frac{4}{9}\frac{C_V}{Nk}\frac{f_{1/2}(z)}{f_{3/2}(z)}$$

$$\simeq \frac{\pi^2}{3}\left(\frac{kT}{\varepsilon_F}\right)^2 \quad (kT \ll \varepsilon_F).$$

(c) Finally, making use of the thermodynamic relation $\gamma = \kappa_T/\kappa_S$, verify the results of Problem 8.3.

8.5 Evaluate $(\partial^2 P/\partial T^2)_v$, $(\partial^2 \mu/\partial T^2)_v$, and $(\partial^2 \mu/\partial T^2)_P$ of an ideal Fermi gas and check that your results satisfy the thermodynamic relations

$$C_V = VT\left(\frac{\partial^2 P}{\partial T^2}\right)_v - NT\left(\frac{\partial^2 \mu}{\partial T^2}\right)_v$$

and

$$C_P = -NT\left(\frac{\partial^2 \mu}{\partial T^2}\right)_P.$$

Examine the low-temperature behavior of these quantities.

8.6 Show that the velocity of sound w in an ideal Fermi gas is given by

$$w^2 = \frac{5kT}{3m}\frac{f_{5/2}(z)}{f_{3/2}(z)} = \frac{5}{9}\langle u^2\rangle,$$

where $\langle u^2\rangle$ is the mean square speed of the particles in the gas. Evaluate w in the limit $z \to \infty$ and compare it with the Fermi velocity u_F.

8.7 Show that for an ideal Fermi gas

$$\langle u\rangle\left\langle\frac{1}{u}\right\rangle = \frac{4}{\pi}\frac{f_1(z)f_2(z)}{\{f_{3/2}(z)\}^2},$$

u being the speed of a particle. Further show that at low temperatures

$$\langle u\rangle\left\langle\frac{1}{u}\right\rangle \simeq \frac{9}{8}\left[1+\frac{\pi^2}{12}\left(\frac{kT}{\varepsilon_F}\right)^2\right];$$

compare with Problem 6.6.

8.8 Obtain numerical estimates of the Fermi energy (in eV) and the Fermi temperature (in K) for the following systems:

(a) conduction electrons in silver, lead, and aluminum;
(b) nucleons in a heavy nucleus, such as $_{80}Hg^{200}$, and
(c) He^3 atoms in liquid helium-3 (atomic volume: 63 Å^3 per atom).

8.9 Making use of another term of the Sommerfeld lemma (E.17), show that *in the second approximation* the chemical potential of a Fermi gas at low temperatures is given by

$$\mu \simeq \varepsilon_F\left[1-\frac{\pi^2}{12}\left(\frac{kT}{\varepsilon_F}\right)^2-\frac{\pi^4}{80}\left(\frac{kT}{\varepsilon_F}\right)^4\right], \qquad (8.1.35a)$$

and the mean energy per particle by

$$\frac{U}{N} \simeq \frac{3}{5}\varepsilon_F\left[1+\frac{5\pi^2}{12}\left(\frac{kT}{\varepsilon_F}\right)^2-\frac{\pi^4}{16}\left(\frac{kT}{\varepsilon_F}\right)^4\right]. \qquad (8.1.37a)$$

Hence determine the T^3-correction to the customary T^1-result for the specific heat of an electron gas. Compare the magnitude of the T^3-term, in a typical metal such as copper, with the low-temperature specific heat arising from the Debye modes of the lattice. For further terms of these expansions, see Kiess (1987).

8.10 Consider an ideal Fermi gas, with energy spectrum $\varepsilon \propto p^s$, contained in a box of "volume" V in a space of n dimensions. Show that, for this system,

(a) $PV = \frac{s}{n}U$;

(b) $\frac{C_V}{Nk} = \frac{n}{s}\left(\frac{n}{s}+1\right)\frac{f_{(n/s)+1}(z)}{f_{n/s}(z)} - \left(\frac{n}{s}\right)^2 \frac{f_{n/s}(z)}{f_{(n/s)-1}(z)}$;

(c) $\frac{C_P - C_V}{Nk} = \left(\frac{sC_V}{nNk}\right)^2 \frac{f_{(n/s)-1}(z)}{f_{(n/s)}(z)}$;

(d) the equation of an *adiabat* is $PV^{1+(s/n)} = \text{const.}$, and

(e) the index $(1+(s/n))$ in the foregoing equation agrees with the ratio (C_P/C_V) of the gas only when $T \gg T_F$. On the other hand, when $T \ll T_F$, the ratio $(C_P/C_V) \simeq 1 + (\pi^2/3)(kT/\varepsilon_F)^2$, irrespective of the values of s and n.

8.11 Examine results (b) and (c) of the preceding problem in the high-temperature limit $(T \gg T_F)$ as well as in the low-temperature limit $(T \ll T_F)$, and compare the resulting expressions with the ones pertaining to a nonrelativistic gas and an extreme relativistic gas in three dimensions.

8.12 Show that, in two dimensions, the specific heat $C_V(N,T)$ of an ideal Fermi gas is identical to the specific heat of an ideal Bose gas, for *all* N and T.
[*Hint*: It will suffice to show that, for given N and T, the thermal energies of the two systems differ at most by a constant. For this, first show that the fugacities, z_F and z_B, of the two systems are mutually related:

$$(1+z_F)(1-z_B) = 1, \quad \text{i.e.,} \quad z_B = z_F/(1+z_F).$$

Next, show that the functions $f_2(z_F)$ and $g_2(z_B)$ are also related:

$$\begin{aligned} f_2(z_F) &= \int_0^{z_F} \frac{\ln(1+z)}{z}dz \\ &= g_2\left(\frac{z_F}{1+z_F}\right) + \frac{1}{2}\ln^2(1+z_F). \end{aligned}$$

It is now straightforward to show that

$$E_F(N,T) = E_B(N,T) + \text{const.},$$

the constant being $E_F(N,0)$.]

8.13 Show that, *quite generally*, the low-temperature behavior of the chemical potential, the specific heat, and the entropy of an ideal Fermi gas is given by

$$\mu \simeq \varepsilon_F\left[1 - \frac{\pi^2}{6}\left(\frac{\partial \ln a(\varepsilon)}{\partial \ln \varepsilon}\right)_{\varepsilon=\varepsilon_F}\left(\frac{kT}{\varepsilon_F}\right)^2\right],$$

and

$$C_V \simeq S \simeq \frac{\pi^2}{3}k^2 T a(\varepsilon_F),$$

where $a(\varepsilon)$ is the *density of (the single-particle) states* in the system. Examine these results for a gas with energy spectrum $\varepsilon \propto p^s$, confined to a space of n dimensions, and discuss the special cases: $s = 1$ and 2, with $n = 2$ and 3.
[*Hint*: Use equation (E.18) from Appendix E.]

8.14 Investigate the Pauli paramagnetism of an ideal gas of fermions with intrinsic magnetic moment $\boldsymbol{\mu}^*$ and spin $J\hbar(J = \frac{1}{2}, \frac{3}{2}, \ldots)$, and derive expressions for the low-temperature and high-temperature susceptibilities of the gas.

8.15 Show that expression (8.2.20) for the paramagnetic susceptibility of an ideal Fermi gas can be written in the form

$$\chi = \frac{n\mu^{*2}}{kT}\frac{f_{1/2}(z)}{f_{3/2}(z)}.$$

Using this result, verify equations (8.2.24) and (8.2.27).

8.16 The observed value of γ, see equation (8.3.6), for sodium is 4.3×10^{-4} cal mole^{-1}K^{-2}. Evaluate the Fermi energy ε_F and the number density n of the conduction electrons in the sodium metal. Compare the latter result with the number density of atoms (given that, for sodium, $\rho = 0.954\,\text{g cm}^{-3}$ and $M = 23$).

8.17 Calculate the fraction of the conduction electrons in tungsten ($\varepsilon_F = 9.0\,\text{eV}$) at 3000 K whose kinetic energy $\varepsilon\,(= \frac{1}{2}mu^2)$ is greater than $W\,(= 13.5\,\text{eV})$. Also calculate the fraction of the electrons whose kinetic energy associated with the z-component of their motion, namely $(\frac{1}{2}mu_z^2)$, is greater than 13.5 eV.

8.18 Show that the ground-state energy E_0 of a *relativistic* gas of electrons is given by

$$E_0 = \frac{\pi V m^4 c^5}{3h^3} B(x),$$

where

$$B(x) = 8x^3\{(x^2+1)^{1/2} - 1\} - A(x),$$

$A(x)$ and x being given by equations (8.5.13) and (8.5.14). Check that the foregoing result for E_0 and equation (8.5.12) for P_0 satisfy the thermodynamic relations

$$E_0 + P_0 V = N\mu_0 \quad \text{and} \quad P_0 = -(\partial E_0/\partial V)_N.$$

8.19 Show that the low-temperature specific heat of the relativisitic Fermi gas, studied in Section 8.5, is given by

$$\frac{C_V}{Nk} = \pi^2 \frac{(x^2+1)^{1/2}}{x^2}\frac{kT}{mc^2} \quad \left(x = \frac{p_F}{mc}\right).$$

Check that this formula gives correct results for the nonrelativistic case as well as for the extreme relativistic one.

8.20 Express the integrals (8.6.19) in terms of the initial slope of the function $\Phi(x)$, and verify equation (8.6.20).

8.21 The total energy E of the electron cloud in an atom can be written as

$$E = K + V_{ne} + V_{ee},$$

where K is the kinetic energy of the electrons, V_{ne} the interaction energy between the electrons and the nucleus, and V_{ee} the mutual interaction energy of the electrons. Show that, according to the Thomas–Fermi model of a neutral atom,

$$K = -E, \quad V_{ne} = +\frac{7}{3}E, \quad \text{and} \quad V_{ee} = -\frac{1}{3}E,$$

so that total $V = V_{ne} + V_{ee} = 2E$. Note that these results are consistent with the *virial theorem*; see Problem 3.20, with $n = -1$.

8.22 Derive equations (8.4.3) through (8.4.5) for a Fermi gas in a harmonic trap. Evaluate equations (8.3.4) and (8.3.5) numerically to reproduce the theoretical curves shown in Figures 8.9 and 8.10.

9 Thermodynamics of the Early Universe

Over the course of the twentieth century, astronomers and astrophysicists gathered a vast body of evidence that indicates the universe began abruptly 13.75 ± 0.11 billion years ago in what became known as the "Big Bang."[1,2] The intense study of the origin and evolution of the universe has led to a convergence of physics and astrophysics. Thermodynamics and statistical mechanics play a crucial role in our understanding of the sequence of transitions that the universe went though shortly after the Big Bang. These transitions left behind mileposts that astrophysicists have exploited to look back into the earliest moments of the universe. The early universe provides particularly good examples for utilizing the properties of ideal classical, Bose, and Fermi gases developed in Chapters 6, 7, and 8, and the theory of chemical equilibrium developed in Section 6.6.

9.1 Observational evidence of the Big Bang

Observational evidence of the Big Bang has grown steadily since Edwin Hubble's discovery in the late 1920s that the universe was expanding. Since that time a coherent standard model for the beginning of the universe has emerged. The following three items describe the key bodies of evidence.

1. Nearly every galaxy in the universe is moving away from every other galaxy and the recessional velocities display an almost linear dependence on the distance between galaxies; see Figure 9.1. Hubble was the first to observe this by measuring both the distances to nearby galaxies and their velocities relative to our own galaxy. The former is based on standard candles, in Hubble's case Cepheid variable stars with known absolute mean luminosity. The latter is based on measurements of the Doppler red shift of spectral lines. Type Ia supernovae are used as the standard candle in the most distant observations made using the Hubble Space Telescope. The data are

[1]For excellent overviews and history of the study of the Big Bang, see *The First Three Minutes: A Modern View of the Origin of the Universe* by Weinberg (1993) and *The Big Bang* by Singh (2005). *Cosmology* by Weinberg (2008) provides an excellent technical survey. The organization of this chapter is based on Weinberg (1993). The 2010 decadal survey of astrophysics *New Worlds, New Horizons in Astronomy and Astrophysics* by the National Academies Press provides an overview of the current state of the field; see *www.nap.edu*.

[2]Steady state cosmology advocate Fred Hoyle coined the term "Big Bang" derisively in a BBC radio broadcast in 1950. To his eternal dismay, the name quickly became popular.

© 2011 Elsevier Ltd. All rights reserved.

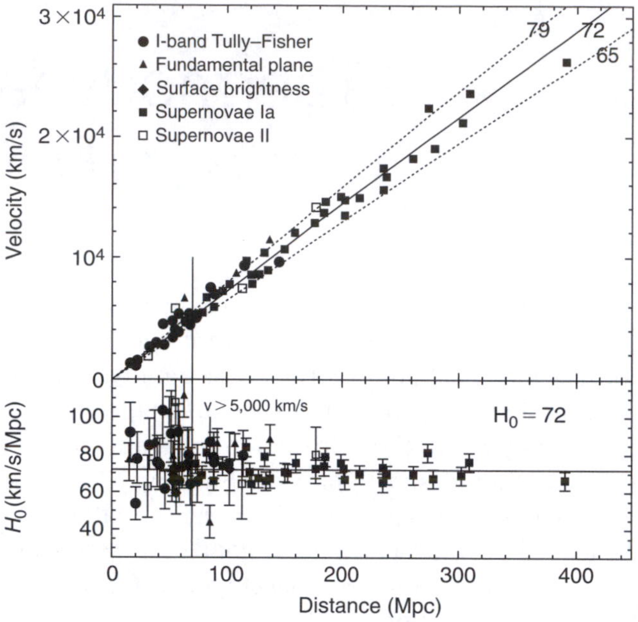

FIGURE 9.1 Hubble diagram of the spectral red-shift velocity of relatively nearby galaxies versus their distance using several astronomical standard candles. The velocity is in km/s and the distance is measured in megaparsecs, where $1\ \text{Mpc} = 3.26 \times 10^6$ light-years. The best fit to the data gives a value of the Hubble parameter as $H_0 = 72 \pm 8\,\text{km}\,\text{s}^{-1}\text{Mpc}^{-1}$. The Hubble parameter has been recently updated by Riess et al. (2009) to give $H_0 = 74.2 \pm 3.6\,\text{km}\,\text{s}^{-1}\,\text{Mpc}^{-1}$. The figure is from Freedman et al. (2001) and is reproduced by permission of the AAS.

encapsulated in the Hubble–Friedmann relation (Friedmann, 1922, 1924),

$$v = \frac{da}{dt} = Ha = \sqrt{\frac{8\pi G u}{3c^2}}\,a. \tag{1}$$

Here a represents the distance between any two points in space that grows with time as the universe expands, v is the recessional velocity, G is the universal constant of gravitation, c is the speed of light, and u is the energy density of the universe.[3] The Hubble parameter, H, is the characteristic expansion rate and is of the order of the inverse of the age of the universe. The particular form of equation (1) assumes, as appears to be the case, that the energy density u is equal to the critical value so that the

[3]Cosmological and general relativistic calculations are usually expressed in terms of the equivalent mass density $\rho = u/c^2 = T^{00}$ where T is the energy-momentum tensor. Astrophysicists usually describe the length scale parameter a in terms of the Doppler shift factor $z = (\lambda/\lambda_{\text{lab}} - 1)$, where λ_{lab} is the laboratory wavelength of a spectral line and λ is the red-shifted value. This gives $z = (T/T_0 - 1)$ where T_0 is the current cosmic microwave background temperature and T is the photon temperature of that era. For example, the Doppler shift from the era of last scattering is

$$z = \frac{3000\,\text{K}}{2.725\,\text{K}} - 1 \simeq 1100,$$

so the universe has expanded by a factor of 1100 since that time.

space-time is flat. This means that the universe is balanced on a knife edge between expanding forever and recollapsing due to gravity. For excellent technical surveys, see Börner (2003) and Weinberg (2008). The measured value of the Hubble parameter is

$$H_0 = 74.2 \pm 3.6\,\mathrm{km\,s^{-1}\,Mpc^{-1}}, \tag{2}$$

where Mpc is a megaparsec, about 3.26×10^6 light-years; see Freedman et al. (2001) and Riess et al. (2009).

2. Penzias and Wilson (1965) observed a nearly uniform and isotropically distributed microwave radiation noise coming from deep space with a blackbody temperature of about 3 K. This cosmic microwave background (CMB) was quickly identified as the remnant blackbody radiation from the era following the Big Bang. Later, balloon experiments and space-based measurements by the Cosmic Background Explorer (COBE) NASA mission showed that the CMB is extremely uniform and isotropic with an average temperature of $T_{\mathrm{CMB}} = 2.725 \pm 0.002\,\mathrm{K}$; see Mather et al. (1994, 1999), Wright et al. (1994), Fixsen et al. (1996), and Figure 9.2. The NASA Wilkinson Microwave Anisotropy Probe (WMAP) mission mapped the angular variation of the CMB temperature. Figure 9.3 shows the $\pm 200\,\mu\mathrm{K}$ CMB temperature variations mapped onto galactic coordinates.

   The CMB represents the photons that were in thermal equilibrium with the high-temperature plasma that existed from the very first moments of the universe until it cooled down to approximately 3000 K about 380,000 years after the Big Bang.

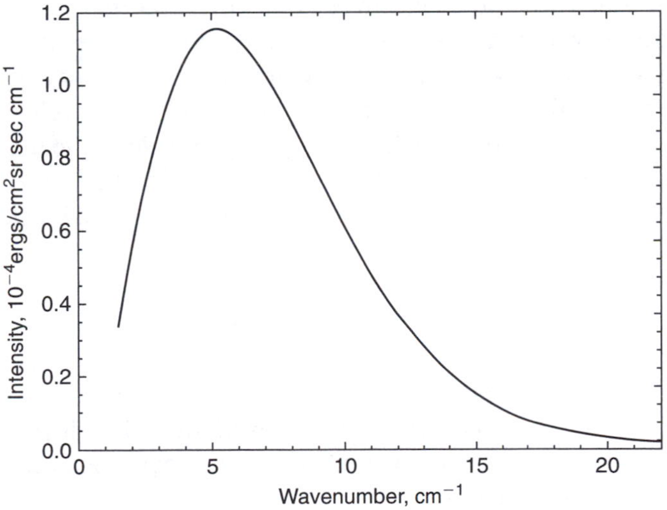

FIGURE 9.2 Cosmic microwave background spectrum from COBE fit numerically to the Planck distribution with an average temperature $T = 2.725 \pm 0.002\,\mathrm{K}$; see equations (7.3.8) and (7.3.9), and Figure 7.13. The error bars at the 43 equally spaced frequencies from the Far Infrared Absolute Spectrophotometer (FIRAS) data are too small to be seen on this scale. Figure courtesy of NASA.

FIGURE 9.3 Measurement of temperature variations in the CMB using 7 years of data from WMAP. This shows the distribution of the CMB blackbody temperature mapped onto galactic coordinates. The variations represent temperature fluctuations of $\pm 200\,\mu$K. Figure courtesy of NASA and the WMAP Science Team.

As the temperature fell below 3000 K, the electrons and protons in the plasma combined for the first time into neutral hydrogen atoms, a period that is known rather oxymoronically as the era of recombination. After this era of "last scattering" of photons by free electrons, the quantum structure of the atoms prevented them from absorbing radiation except at their narrow spectral frequencies, so the universe became transparent and the blackbody radiation quickly fell out of equilibrium with the neutral atoms. As the universe continued to expand, the wavelengths of the blackbody radiation grew linearly with the expansion scale of the universe a. The photon number density fell as a^{-3} and the energy density as a^{-4}, so the Planck distribution, equations (7.3.8) and (7.3.9), was preserved with a blackbody temperature that scaled as

$$T(t)a(t) = \text{const.} \tag{3}$$

Measurements of the Hubble parameter and the COBE and WMAP measurements of the temperature and temperature fluctuations of the CMB allow a determination of the current total energy density of the universe and its composition. The current energy density of the universe is

$$u = \frac{3c^2 H_0^2}{8\pi G} = 8.36 \times 10^{-10}\,\mathrm{J\,m^{-3}}, \tag{4}$$

and is comprised of approximately 72.8 percent dark energy, 22.7 percent dark matter, and 4.56 percent baryonic matter (protons and neutrons).[4] This gives a baryon number density n_B of $0.26\,\mathrm{m}^{-3}$. The number density of photons in a blackbody enclosure as a function of temperature is given by equation (7.3.23):

$$n_\gamma(T) = \frac{2\zeta(3)}{\pi^2}\left(\frac{kT}{\hbar c}\right)^3. \tag{5}$$

At the current temperature of 2.725 K, this gives a CMB photon number density of $n_\gamma = 4.10 \times 10^8\,\mathrm{m}^{-3}$, so the current baryon-to-photon ratio is

$$\eta = \frac{n_B}{n_\gamma} \approx 6 \times 10^{-10}. \tag{6}$$

The ratio η has remained constant as the universe has expanded since both these quantities scale as $a^{-3}(t)$. As we will see later, the numerical value of η played a very important role in the thermal evolution of the early universe.[5]

3. The relative abundances of the light elements ^{1}H, ^{2}H, ^{3}He, ^{4}He, ^{7}Li, and so on created during the first few minutes of the universe are sensitive functions of the baryon-to-photon ratio η; see Figure 9.4. This connection was first explored by George Gamow, Ralph Alpher, and Robert Herman in the late 1940s and early 1950s; see Alpher, Bethe, and Gamow (1948), and Alpher and Herman (1948, 1950).[6]

[4]The energy content of the universe is parameterized in terms of the fraction of the critical density contained in the various constituents. The current values are dark energy: $\Omega_\Lambda = 0.728 \pm 0.016$; baryonic matter: $\Omega_b = 0.0456 \pm 0.0016$; and cold dark matter: $\Omega_c = 0.227 \pm 0.014$. The current age of the universe, or lookback time, is $t_0 = 13.75 \pm 0.11 \times 10^9$ years. The relative contribution from blackbody radiation is about 6×10^{-5}. These concordance values of the parameters are based on WMAP 7-year data and are tabulated in Komatsu et al. (2010). The dark energy is responsible for the accelerating expansion of the universe. The energy density proportions were vastly different in the early universe because they scale differently with the expansion parameter a. At the time of recombination, the proportions were: dark matter 63 percent, baryonic matter 12 percent, relativistic radiation (photons and neutrinos) 25 percent. During the first few moments, relativistic particles provided the dominant contribution to the energy. Using the photon:neutrino:electron ratios of 2 : 21/4 : 7/2 from Table 9.2, the energy content was photons 18.6 percent, neutrinos and antineutrinos 48.8 percent, and electrons and positrons 32.5 percent. While dark energy is currently the dominant contribution to the energy density of the universe, it played only a small role in the early evolution of the universe. Cold dark matter was crucial for the development of the first stars and galaxies at the end of the "dark ages" 100 to 200 million years after the last scattering.

[5]The proper measure here is the ratio of the baryon number density to photon entropy density but, since the CMB photon entropy density and number density both scale as T^3, the ratio is usually quoted in terms of the ratio of the number densities.

[6]George Gamow and Ralph Alpher in 1948, and Alpher and Robert Herman in 1950, proposed a model for nucleosynthesis in a hot, expanding primordial soup of protons, neutrons, and electrons. Alpher and Gamow called this material "ylem." To account for the present abundance of ^{4}He in the universe, Alpher and Herman (1950) proposed a baryon-to-photon ratio of roughly 10^{-9} and predicted a current cosmic microwave background temperature of about 5 K. Gamow added his friend Hans Bethe's name as second author to Alpher, Bethe, and Gamow (1948) as a pun on the Greek alphabet. The paper was published, perhaps not coincidentally, on April 1; see Alpher and Herman (2001), Weinberg (1993), and Singh (2005).

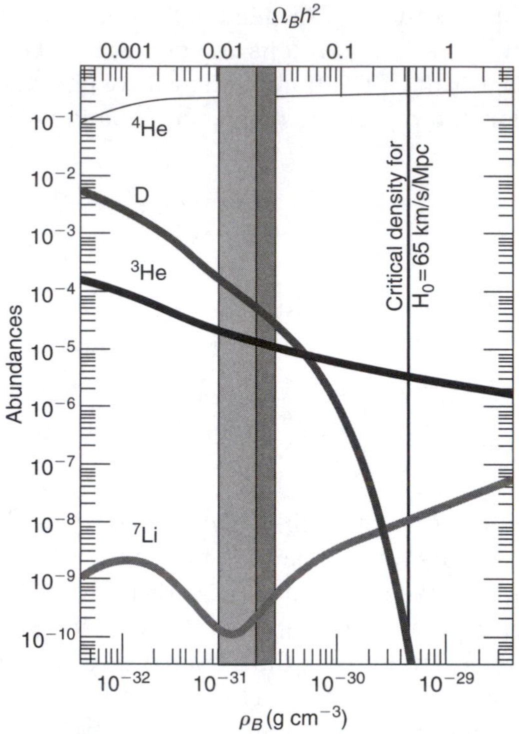

FIGURE 9.4 Calculated primordial abundances of light elements (^{4}He, D=^{2}H, ^{3}He, and ^{7}Li) as functions of the baryon-to-photon ratio. The baryon-to-photon ratio is given by $\eta = 2.7 \times 10^{-8}\Omega_b h^2$, where h is the Hubble parameter in units of 100(km/s)/Mpc and $\Omega_b = 0.046$ is the current baryonic fraction of the mass-energy density of the universe; see Copi, Schramm, and Turner (1997), Schramm and Turner (1998), and Steigman (2006). The experimentally allowed range is in the grey vertical bands. Figure from Schramm and Turner (1998). Reprinted with permission; copyright © 1998, American Physical Society.

9.2 Evolution of the temperature of the universe

As the universe expanded and cooled, the cooling rate was proportional to the Hubble parameter, that is, of the order of the inverse of the age of the universe at that point in its expansion. This led to a sequence of important events when different particles and interactions fell out of equilibrium with the gas of blackbody photons. The neutrinos and neutron-proton conversion reactions fell out of equilibrium at $t \approx 1$ second. Nuclear reactions that formed light nuclei fell out of equilibrium at $t \approx 3$ minutes. Neutral atoms fell out of equilibrium at $t \approx 380,000$ years. All these degrees of freedom froze out when the reaction rates that had kept them in equilibrium with the blackbody photons fell far below the cooling rate of the expanding universe. Each component that fell out of equilibrium left behind a marker of the properties of the universe characteristic of that era. It is these markers that provide evidence of the properties and behavior of the universe during its earliest moments.

From the first moments of the universe up until the recombination era 380,000 years later, the cosmic plasma was in thermal equilibrium with the blackbody radiation through Thomson scattering. Due to the high density of charged particles, the photon scattering mean free time was much shorter than the time scale for temperature changes of the universe as it expanded and cooled, which kept the plasma in thermal equilibrium with the photons. For the first few hundred thousand years of its expansion, the energy density of the universe was dominated by photons and other relativistic particles. This is because the energy density of the blackbody radiation scales as a^{-4} whereas the energy density of nonrelativistic matter scales as a^{-3}. The temperature of the blackbody photons as a function of the age of the universe is shown in Figure 9.5 and Table 9.1.

During the first one-hundredth of a second, the universe expanded and cooled from its singular beginning to a temperature of about 10^{11} K. The physics from this time onward was controlled by the weak and electromagnetic interactions. The strong interactions could be ignored since the baryon-to-photon ratio was so small and the temperature was

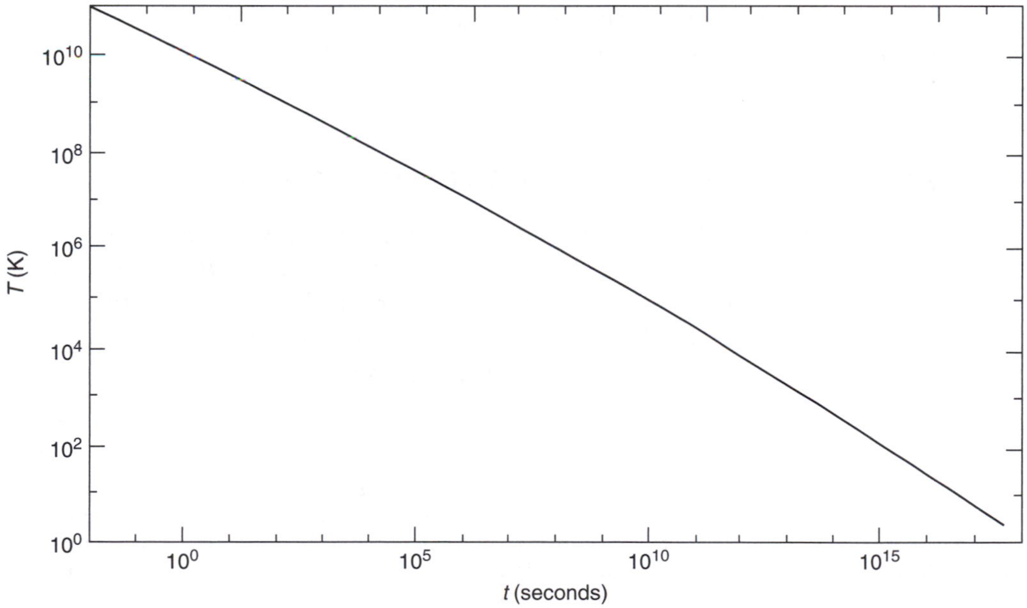

FIGURE 9.5 Sketch of the photon temperature versus the age of the universe. At early times, radiation dominated the energy density, so $T \sim t^{-1/2}$. At later times ($t > 10^{13}$ s) nonrelativistic matter dominated the energy density, so $T \sim t^{-2/3}$. In the current dark energy dominated stage, the universe is beginning to expand exponentially with time, so the photon temperature is beginning to fall exponentially.

Table 9.1 Temperature vs. Age of the Universe

Time (s)	Temperature (K)
0.01	1×10^{11}
0.1	3×10^{10}
1.0	1×10^{10}
12.7	3×10^{9}
168	1×10^{9}
1980	3×10^{8}
1.78×10^{4}	10^{8}
1.20×10^{13}	3000
4.34×10^{17}	2.725

Source: Weinberg (2008).

too low to create additional hadrons.[7,8] We will follow the thermodynamic behavior of the universe from $t = 0.01$ second when the temperature was 10^{11} K to $t = 380,000$ years when the temperature fell below 3000 K. At that point neutral atoms formed, photon scattering ended, and the universe became transparent to radiation. After recombination and last scattering there were no new sources of radiation in the universe since the baryonic matter consisted entirely of neutral atoms. This state of affairs lasted until atoms were first reionized by the gravitational clumping that formed the first stars and galaxies 100 to 200 million years after the Big Bang. This reionization epoch ended the so-called cosmic dark ages.

9.3 Relativistic electrons, positrons, and neutrinos

During the earliest moments of the universe, the temperature was high enough to create several kinds of relativistic particles and antiparticles. If $kT \gg mc^2$, then particle-antiparticle pairs each with mass m can be created from photon-photon interactions. At these temperatures, almost all of the particles that are created will have an energy-momentum relation described by the relativistic limit, namely

$$\varepsilon_{\mathbf{k}} \approx \hbar c k, \tag{1}$$

[7]Before time $t = 0.01$ s, the analysis is more difficult due to the production of strongly interacting particles and antiparticles. At even earlier times, when the temperature was above $kT \approx 300$ MeV ($T = 4 \times 10^{12}$ K), hadrons would have broken apart into a strongly interacting relativistic quark-gluon plasma. The Relativistic Heavy Ion Collider at Brookhaven National Laboratory has succeeded in creating a quark-gluon plasma with the highest temperature matter ever created in the laboratory, $T = 4 \times 10^{12}$ K; see Adare et al. (2010).

[8]The exact mechanism for baryogenesis (i.e., nonzero baryon-to-photon ratio η) is unsettled. It requires, as shown by Sakharov (1967), three things: baryon number nonconservation, C and CP violation, and deviation from equilibrium. All these conditions were satisfied in the earliest moments of the universe (far earlier than the time scales we examine here) but a consensus theory that allows for a baryon asymmetry nearly as large as the observed value of $\eta = 6 \times 10^{-10}$ has not yet emerged.

where $\hbar k$ is the magnitude of the momentum. This relation applies to photons, neutrinos, antineutrinos, electrons, and positrons. The threshold for electron-positron pair formation is $m_e c^2/k = 5.9 \times 10^9$ K. Neutrinos are very light, so we can safely assume that they are relativistic.[9] The relativistic dispersion relation 1 gives essentially the same density of states for all species of relativistic particles:

$$a(\varepsilon) = \frac{g_s}{(2\pi)^3} \int \delta(\varepsilon - \varepsilon_k) d\mathbf{k} = \frac{4\pi g_s}{(2\pi)^3} \int_0^\infty k^2 \delta(\varepsilon - \hbar c k) dk = \frac{g_s \varepsilon^2}{2\pi^2 (\hbar c)^3}, \tag{2}$$

where g_s is the spin degeneracy. Photons have a spin degeneracy $g_s = 2$ (left and right circularly polarized). The other species are all spin-$\frac{1}{2}$ fermions. Electrons and positrons have spin degeneracy $g_s = 2$ while neutrinos and antineutrinos have spin degeneracy $g_s = 1$ since all neutrinos have left-handed helicity.

During this era, because of the charge neutrality of the universe and the small size of the baryon-to-photon ratio η, the number densities of the electrons and positrons were nearly equal, so their chemical potentials were both rather small. Assuming that the net lepton number of the universe is also small, the same applies to the neutrinos and antineutrinos. As explained in Section 7.3, the chemical potential for photons is exactly zero. The pressure, number density, energy density, and entropy density of a relativistic gas of fermions (+) or bosons (−) with zero chemical potential are are given by

$$P(T) = \pm kT \int a(\varepsilon) \ln(1 \pm e^{-\beta\varepsilon}) d\varepsilon = \frac{g_s (kT)^4}{2\pi^2 (\hbar c)^3} \int_0^\infty x^2 \ln\left(1 \pm e^{-x}\right) dx, \tag{3a}$$

$$n(T) = \int a(\varepsilon) \frac{1}{e^{\beta\varepsilon} \pm 1} d\varepsilon = \frac{g_s}{2\pi^2} \left(\frac{kT}{\hbar c}\right)^3 \int_0^\infty \frac{x^2}{e^x \pm 1} dx, \tag{3b}$$

$$u(T) = \int a(\varepsilon) \frac{\varepsilon}{e^{\beta\varepsilon} \pm 1} d\varepsilon = \frac{g_s (kT)^4}{2\pi^2 (\hbar c)^3} \int_0^\infty \frac{x^3}{e^x \pm 1} dx, \tag{3c}$$

$$s(T) = \left(\frac{\partial P}{\partial T}\right)_\mu = \frac{2 g_s k}{\pi^2} \left(\frac{kT}{\hbar c}\right)^3 \int_0^\infty x^2 \ln\left(1 \pm e^{-x}\right) dx. \tag{3d}$$

[9] All three neutrino families are known to have small (but nonzero) mass from neutrino oscillation observations. The electron neutrino is probably the lightest with the experimental limit of $m_{\nu_e} c^2 < 2.2$ eV. The distribution of angular fluctuations of the CMB measured by WMAP puts a limit on the sum of the masses of the neutrinos, $\sum m_\nu c^2 < 0.58$ eV, so we can safely assume that all neutrino species are far lighter than the value of kT during the early universe.

Using the values of the Bose integrals from Appendix D, we arrive at the following expressions for the blackbody photons:

$$P_\gamma(T) = \frac{\pi^2}{45}\frac{(kT)^4}{(\hbar c)^3}, \tag{4a}$$

$$n_\gamma(T) = \frac{2\zeta(3)}{\pi^2}\left(\frac{kT}{\hbar c}\right)^3, \tag{4b}$$

$$u_\gamma(T) = \frac{\pi^2}{15}\frac{(kT)^4}{(\hbar c)^3}, \tag{4c}$$

$$s_\gamma(T) = \frac{4\pi^2 k}{45}\left(\frac{kT}{\hbar c}\right)^3. \tag{4d}$$

All relativistic species with $\mu = 0$ have the same power law temperature dependences for the pressure, energy density, and so on, as the photons, while the Fermi and Bose integrals are the same except for a constant prefactor:

$$\int_0^\infty \frac{x^{n-1}}{e^x+1}dx = \left(1 - \frac{1}{2^{n-1}}\right)\int_0^\infty \frac{x^{n-1}}{e^x-1}dx; \tag{5}$$

see Appendices D and E. The contributions to the pressure, energy density, and entropy density result from counting the spin degeneracies, the number of particles and antiparticles, and accounting for the different Fermi/Bose factors (1 for bosons, 7/8 for fermions). The photons, three generations of neutrinos (electron, muon, and tau neutrinos and their antiparticles), and the electrons and positrons contribute to the total pressure, number density, energy density, and entropy density in the proportions shown in Table 9.2. The counting is presented here, as is usually done in the literature, relative to the contribution per spin state of the photons. The contributions to the number densities are the same except that the Fermi/Bose factor is now 3/4.

Table 9.2 Relativistic Contributions to Pressure, Energy Density, and Entropy Density

Particles	Fermi/Bose Factor	Spin Degeneracy	Number of Species	$2P_{\text{total}}/P_\gamma$
γ	1	2	1	2
ν_e, ν_μ, ν_τ	$\frac{7}{8}$	1	3	$\frac{21}{8}$
$\bar{\nu}_e, \bar{\nu}_\mu, \bar{\nu}_\tau$	$\frac{7}{8}$	1	3	$\frac{21}{8}$
e^-	$\frac{7}{8}$	2	1	$\frac{7}{4}$
e^+	$\frac{7}{8}$	2	1	$\frac{7}{4}$

The totals then are

$$P_{\text{total}}(T) = \left(2 + \frac{21}{4} + \frac{7}{2}\right)\frac{P_\gamma(T)}{2} = \left(\frac{43}{8}\right)P_\gamma(T), \tag{6a}$$

$$u_{\text{total}}(T) = \left(2 + \frac{21}{4} + \frac{7}{2}\right)\frac{u_\gamma(T)}{2} = \left(\frac{43}{8}\right)u_\gamma(T), \tag{6b}$$

$$s_{\text{total}}(T) = \left(2 + \frac{21}{4} + \frac{7}{2}\right)\frac{s_\gamma(T)}{2} = \left(\frac{43}{8}\right)s_\gamma(T), \tag{6c}$$

$$n_{\text{total}}(T) = \left(2 + \frac{9}{2} + 3\right)\frac{n_\gamma(T)}{2} = \left(\frac{19}{4}\right)n_\gamma(T). \tag{6d}$$

The density of the universe was high enough in this era, so the weak and electromagnetic interaction rates kept all these species in thermal equilibrium with one other. Therefore, as the universe expanded adiabatically, the entropy in a comoving volume of linear size a remained constant as the volume expanded from some initial value a_0^3 to a final volume a_1^3:

$$s_{\text{total}}(T_0)a_0^3 = s_{\text{total}}(T_1)a_1^3. \tag{7}$$

Since the entropy density is proportional to T^3, the temperature and length scale at time t are related by

$$T(t)a(t) = \text{const.} \tag{8}$$

This is the same relation that applies for a freely expanding photon gas, see equation (9.1.3), but here it arises from an adiabatic equilibrium process. From equations (9.1.1) and (8), the temperature of the universe as a function of the age of the universe t during this era is

$$T(t) = 10^{10}\,\text{K}\sqrt{\frac{0.992\text{s}}{t}}; \tag{9}$$

see Problem 9.1.

9.4 Neutron fraction

During the first second of the universe, when $T > 10^{10}$ K, and before protons and neutrons combined into nuclei, the weak interaction kept the free neutrons and protons in thermal "beta-equilibrium" with each other and with the photons, neutrinos, electrons, and positrons through the processes

$$n + \nu_e \rightleftarrows p + e^- + \gamma, \tag{1a}$$

$$n + e^+ \rightleftarrows p + \bar{\nu} + \gamma, \tag{1b}$$

$$n \rightleftarrows p + e^- + \bar{\nu} + \gamma. \tag{1c}$$

We can treat this as a chemical equilibrium process, as described in Section 6.6. Since the chemical potentials of the photons, electrons, positrons, neutrinos, and antineutrinos are all zero, the neutron and proton chemical potentials must be equal at equilibrium:

$$\mu_n = \mu_p. \tag{2}$$

At these temperatures ($\approx 10^{11}$ K) and densities ($\approx 10^{32}\, m^{-3}$), the protons and neutrons can be treated as a classical nonrelativistic ideal gas. Following equation (6.6.5), the spin-$\frac{1}{2}$ proton and neutron chemical potentials are

$$\mu_p = m_p c^2 + kT \ln\left(n_p \lambda_p^3\right) - kT \ln 2, \tag{3a}$$

$$\mu_n = m_n c^2 + kT \ln\left(n_n \lambda_n^3\right) - kT \ln 2. \tag{3b}$$

where $\lambda (= h/\sqrt{2\pi mkT})$ is the thermal deBroglie wavelength. The rest energy of the neutron is greater than the rest energy of the proton by

$$m_n c^2 - m_p c^2 = \Delta\varepsilon = 1.293\text{MeV}. \tag{4}$$

Ignoring the small mass difference in the thermal deBroglie wavelength in equations (3a) and (3b) gives

$$n_n = n_p e^{-\beta\Delta\varepsilon}. \tag{5}$$

The baryon number density is the sum of the neutron and proton number densities

$$n_B = n_n + n_p, \tag{6}$$

so the equilibrium neutron fraction is given by

$$q = \frac{n_n}{n_B} = \frac{1}{e^{\beta\Delta\varepsilon} + 1}. \tag{7}$$

The mass difference gives a crossover temperature $T_{np} = \Delta\varepsilon/k \approx 1.50 \times 10^{10}$ K, so the neutron fraction drops from 46 percent when $T = 10^{11}$ K to 16 percent when $T = 9 \times 10^9$ K at $t_1 \approx 1$ second. As the temperature fell below 10^{10} K ($kT = 0.86$ MeV), the weak interaction rate began to fall far below the cooling rate of the universe, so the baryons quickly fell out of equilibrium with the neutrinos. From that time onward the neutrons began to beta-decay with their natural radioactive decay lifetime of $\tau_n = 886$ seconds, so the neutron fraction fell exponentially:

$$q \approx 0.16 \exp\left(\frac{-(t-t_1)}{\tau_n}\right) \quad \text{for } t > t_1 = 1\text{s}. \tag{8}$$

By the time of nucleosynthesis, about 3.7 minutes later, the neutron fraction had dropped to $q \approx 0.12$. At that point, the remaining neutrons bound with protons to form deuterons and other light nuclei. For a discussion of nucleosynthesis, see Section 9.7.

9.5 Annihilation of the positrons and electrons

About one second after the Big Bang, the temperature approached the crossover temperature T_e for creating electron-positron pairs:

$$kT_e = m_e c^2 = 0.511\,\text{MeV}, \tag{1}$$

with $T_e = 5.93 \times 10^9\,\text{K}$. As the temperature fell below T_e, the rate of creating e^+e^- pairs began to fall below the rate at which pairs annihilated. The full relativistic dispersion relation for electrons is

$$\varepsilon_k = \sqrt{(\hbar ck)^2 + (m_e c^2)^2}, \tag{2}$$

which gives for density of states

$$a_e(\varepsilon) = \frac{8\pi}{(2\pi)^3} \int_0^\infty k^2 \delta(\varepsilon - \varepsilon_k) dk = \frac{\varepsilon\sqrt{\varepsilon^2 - (m_e c^2)^2}}{\pi^2(\hbar c)^3} \quad \text{for } \varepsilon \geq m_e c^2. \tag{3}$$

Since the electrons and positrons were in equilibrium with the blackbody photons via the reaction

$$e^+ + e^- \rightleftarrows \gamma + \gamma, \tag{4}$$

the equilibrium equation (6.6.3) implied that the chemical potentials of the species were related by

$$\mu_- + \mu_+ = 2\mu_\gamma = 0. \tag{5}$$

The ratio of the number density of the electrons to that of the photons then was

$$\frac{n_-}{n_\gamma} = \frac{1}{2\zeta(3)} \int_{\beta m_e c^2}^{\infty} \frac{x\sqrt{x^2 - (\beta m_e c^2)^2}}{e^x e^{-\beta\mu_-} + 1} dx, \tag{6}$$

while the positron density ratio was

$$\frac{n_+}{n_\gamma} = \frac{1}{2\zeta(3)} \int_{\beta m_e c^2}^{\infty} \frac{x\sqrt{x^2 - (\beta m_e c^2)^2}}{e^x e^{\beta\mu_-} + 1} dx; \tag{7}$$

see equation (9.1.5). The electron and positron densities became unbalanced as the universe cooled.

Eventually all the positrons got annihilated leaving behind the electrons that currently remain. Charge neutrality of the universe required the difference between the number density of electrons and the number density of positrons to be equal to the number density of protons, $(1-q)n_B$, where q is defined in Section 9.4; hence

$$\frac{(n_- - n_+)}{n_\gamma} = \frac{\sinh(\beta\mu_-)}{2\zeta(3)} \int_{\beta m_e c^2}^{\infty} \frac{x\sqrt{x^2 - (\beta m_e c^2)^2}}{\cosh(x) + \cosh(\beta\mu_-)} dx = (1-q)\eta. \tag{8}$$

We can use equation (8) to determine the electron chemical potential as a function of temperature numerically and then use that value in equations (6) and (7) to determine the electron and positron densities; see Figure 9.6.

Initially, the electron and positron densities both decreased proportional to $\exp(-\beta m_e c^2)$ as the temperature fell below the electron-positron pair threshold, but they

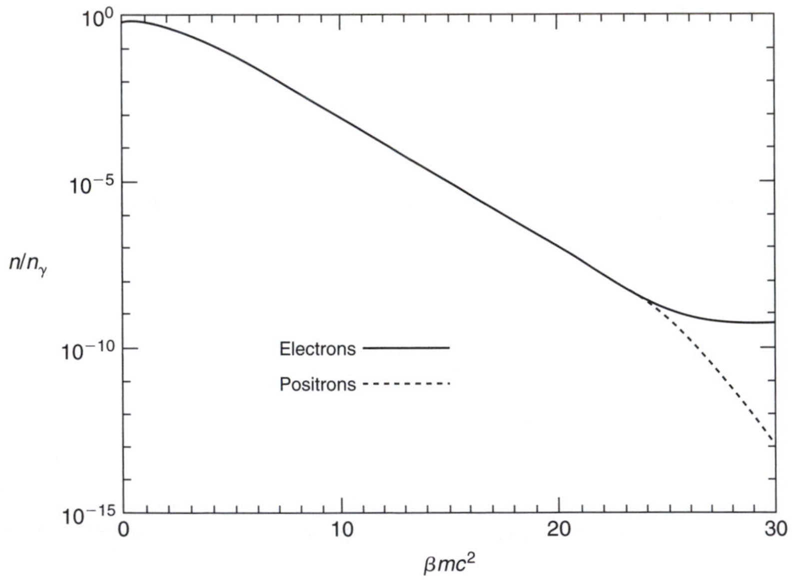

FIGURE 9.6 The ratio of the electron and positron densities to the photon density as a function of $\beta m_e c^2$ during the e^+e^- annihilation for $\eta = 6 \times 10^{-10}$. This era began around temperature 10^{10} K ($\beta m_e c^2 = 1.7$) at time $t = 1$ second and ended when the temperature was about 3×10^8 K ($\beta m_e c^2 = 20$) at time $t = 33$ minutes when the electron number density leveled off at the proton number density.

remained nearly equal to each other until $T \approx m_e c^2/k \ln\left[1/(1-q)\eta\right] \approx 3 \times 10^8\,\mathrm{K}$. At that temperature, the electron density began to level off at the proton density while the positron density continued to fall.

Using the baryon-to-photon ratio $\eta = 6 \times 10^{-10}$, we infer that during the first second of the universe that for every 1.7 billion positrons there must have been *one* extra electron. It is these few extra electrons that will combine with nuclei during the recombination era composing all the atoms now present in the universe. All baryonic matter currently in the universe is the result of this initial asymmetry between matter and antimatter; see footnote 8.

9.6 Neutrino temperature

For temperatures above $T = 10^{10}\,\mathrm{K}$, the rates for the weak interaction reactions (9.4.1) kept the neutrinos in "beta-equilibrium" with the electrons, positrons, and photons. Starting at time $t \approx 1\,\mathrm{s}$, when $T = 10^{10}\,\mathrm{K}$, the weak interaction rates began to fall far below the expansion rate of the universe so the neutrinos quickly fell out of equilibrium. Following the decoupling, the neutrinos expanded freely so the neutrino temperature scaled with the expansion length scale following equation (9.1.3).

The system of electrons, positrons, and photons remained in thermal equilibrium with each other and expanded adiabatically during the electron-positron annihilation era from temperature $T_0 = 10^{10}\,\mathrm{K}$ when the annihilations began, until temperature $T_1 = 3 \times 10^8\,\mathrm{K}$ when nearly all of the positrons had been annihilated. Since this was an adiabatic expansion, we can determine the temperature evolution using entropy conservation. Consider a comoving cubical volume that expanded from an initial linear size a_0 to a final size a_1 during the same time period. The total entropy in the comoving volume at temperature T_0 was due to the photons, electrons, and positrons (refer to Table 9.2):

$$S(T_0) = \frac{11}{4} s_\gamma(T_0) a_0^3, \tag{1}$$

while the entropy at temperature T_1 was due solely to the photons since, by then, nearly all of the electrons and positrons had been annihilated:

$$S(T_1) = s_\gamma(T_1) a_1^3. \tag{2}$$

Entropy conservation during the adiabatic expansion relates the initial and final temperatures as

$$\left(\frac{11}{4}\right)^{1/3} T_0 a_0 = T_1 a_1. \tag{3}$$

In essence, the entropy of the annihilating electrons and positrons was transferred to the photons. Since the neutrino and photon temperatures were equal before the electron-positron annihilation and the neutrinos expanded freely during the annihilation, the neutrino temperature decreased more than the photon temperature during the annihilation era:

$$T_{\nu 1} = (4/11)^{1/3}\, T_1. \tag{4}$$

After the e^+e^- annihilation, both the neutrino and the photon temperatures evolved according to equation (9.1.3) and (9.3.8), so the current temperature of the relic Big Bang neutrinos should be

$$T_\nu = (4/11)^{1/3}\, T_{\text{CMB}} \simeq 1.945\,\text{K}. \tag{5}$$

A measurement of the cosmic neutrino background would provide an excellent additional test of the standard model of the Big Bang but we do not currently have a viable means to measure these very low-energy neutrinos.[10]

9.7 Primordial nucleosynthesis

Light nuclei other than hydrogen first formed between 3 and 4 minutes after the Big Bang when the temperature had cooled to about 10^9 K. Prior to that time, the high-temperature blackbody radiation rapidly photodissociated any deuterium nuclei that happened to form. The first step for the formation of light nuclei from the protons and neutrons is the formation of deuterium because all of the rates for forming nuclei at these densities are dominated by two-body collisions. Once deuterons formed, most of these nuclei would have been quickly converted to helium and other more stable light nuclei in a series of two-body collisions with the remaining protons, neutrons, and with each other. As discussed in Section 9.4, the proton/neutron mixture at this time was about $q = 12$ percent neutrons and $1 - q = 88$ percent protons. By $t \approx 3$ minutes the temperature had fallen to $T \approx 10^9$ K so protons and neutrons could begin to bind themselves into deuterons via the process

$$p + n \rightleftarrows d + \gamma\,. \tag{1}$$

The chemical equilibrium relation for this reaction, see equation (6.6.3), is

$$\mu_p + \mu_n = \mu_d, \tag{2}$$

since the chemical potential of the blackbody photons is zero. At these temperatures and densities the protons, neutrons, and deuterons can be treated as classical ideal gases.

[10]The neutrino elastic scattering cross-section scales like the fourth power of the energy, so the collisions are both very rare and involve very small energy and momentum transfers. This makes direct laboratory detection of the cosmic neutrino background (CνB) infeasible at present; see Gelmini (2005).

The proton and neutron are spin-$\frac{1}{2}$ particles so they have two spin states each while the deuteron is spin-1 and has three spin states:

$$\mu_p = m_p c^2 + kT \ln\left(n_p \lambda_p^3\right) - kT \ln 2, \tag{3a}$$

$$\mu_n = m_n c^2 + kT \ln\left(n_n \lambda_n^3\right) - kT \ln 2, \tag{3b}$$

$$\mu_d = m_d c^2 + kT \ln\left(n_d \lambda_d^3\right) - kT \ln 3. \tag{3c}$$

The binding energy of the deuteron is $\varepsilon_b = m_p c^2 + m_n c^2 - m_d c^2 = 2.20\,\text{MeV}$. Since the deuteron is approximately twice as massive as protons or neutrons, the deuteron number density is given by

$$n_d = \frac{3}{4} n_p n_n \frac{\lambda_p^3 \lambda_n^3}{\lambda_d^3} e^{\beta \varepsilon_b} \approx \frac{3}{\sqrt{2}} n_p n_n \lambda_p^3 e^{\beta \varepsilon_b}. \tag{4}$$

The total number density of baryons is determined by the baryon-to-photon ratio: η: $n_B = \eta n_\gamma = n_p + n_n + 2n_d$. The neutron number density is $q n_B = n_n + n_d$, so the deuteron fraction is given by

$$f_d = \frac{n_d}{n_B} = (1 - q - f_d)(q - f_d)s, \tag{5}$$

where the parameter s is

$$s = \frac{12\zeta(3)}{\sqrt{\pi}} \left(\frac{kT}{m_p c^2}\right)^{3/2} \eta e^{\beta \varepsilon_b}; \tag{6}$$

see also equation (9.1.5). Equation (5) is similar to the Saha equation for the ionization of hydrogen atoms that will be discussed in Section 9.8 and has solution

$$f_d = \frac{1 + s - \sqrt{(1+s)^2 - 4s^2 q(1-q)}}{2s}. \tag{7}$$

For high temperatures, s is small and $f_d \approx q(1-q)s$ while for low temperatures, s is large and $f_d \approx q$, that is, all the neutrons are bound into deuterons. The deuterium fraction as a function of temperature is shown in Figure 9.7. The small values of the baryon-to-photon ratio η and $\varepsilon_b / m_p c^2$ delayed the nucleosynthesis until the temperature had fallen to

$$kT_n \approx \frac{\varepsilon_b}{\ln\left(\frac{1}{\eta}\left(\frac{m_p c^2}{\varepsilon_b}\right)^{3/2}\right)}, \tag{8}$$

providing the time for the neutron fraction to have decayed to $q = 0.12$.

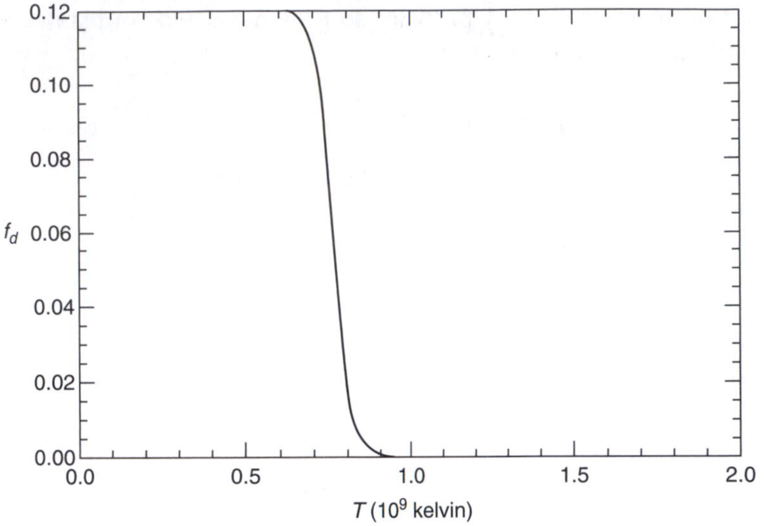

FIGURE 9.7 Plot of the equilibrium deuterium fraction f_d versus temperature T for neutron fraction $q = 0.12$ and baryon-to-photon ratio $\eta = 6 \times 10^{-10}$. As T falls below about 6×10^8 K the neutrons are nearly all bound into deuterons. Further two-body reactions convert most of the deuterium into heavier nuclei, primarily ^{4}He.

The simple equilibrium calculation presented here assumes that no further reactions take place. Including the fast nonequilibrium two-body reactions, namely

$$d + d \rightarrow {}^3\mathrm{H} + p + \gamma, \tag{9a}$$

$$d + d \rightarrow {}^3\mathrm{He} + n + \gamma, \tag{9b}$$

$$d + {}^3\mathrm{H} \rightarrow {}^4\mathrm{He} + n + \gamma, \tag{9c}$$

$$d + {}^3\mathrm{He} \rightarrow {}^4\mathrm{He} + p + \gamma, \tag{9d}$$

results in almost all of the deuterons being cooked into the very stable isotope ^{4}He and small amounts of other light nuclei. Since each ^{4}He nucleus is composed of two protons and two neutrons, this gives a helium mass fraction of $2q = 24$ percent and proton mass fraction of $1 - 2q = 76$ percent. The complete calculation involves nonequilibrium effects modeled with rate equations for each of the nuclear interactions, including those for heavier isotopes, but that only changes the predicted concentration for ^{4}He slightly;[11] see Weinberg (2008). The largest theoretical uncertainty is, remarkably, the uncertainty in the radioactive decay time of the neutron in equation (9.4.8); see Copi, Schramm, and Turner (1997). Calculations of this type were first performed by Gamow, Alpher, and Herman in the late 1940s and early 1950s. Based on current amounts of helium and other light elements

[11]Nuclear reactions continued slowly at a rate that had fallen out of equilibrium and shifted the isotopic ratios until about $t \approx 10$ minutes.

in the universe, Alpher and Herman predicted a 5 K cosmic microwave background over a decade before Penzias and Wilson's discovery; see footnote 6.

9.8 Recombination

After the nucleosynthesis took place in the first few minutes, the universe continued to cool, with the nuclei and electrons remaining as an ordinary plasma in thermal equilibrium with the photons. It took several hundred thousand years for the temperature to drop below the atomic ionization energies of a few electron volts needed for nuclei to capture electrons and form atoms. Hydrogen was the last neutral species to form since it has the smallest ionization energy of a Rydberg ($1\,\mathrm{Ry} = m_e e^4/8\epsilon_0^2 h^2 = 13.6057\,\mathrm{eV}$). At first glance, one would think that atoms form when the temperature falls below $\mathrm{Ry}/k = 158{,}000\,\mathrm{K}$ but, as we will see, the huge number of photons per proton delayed recombination until $T \approx 3000\,\mathrm{K}$. Once all the electrons and protons formed into neutral hydrogen atoms, the universe became transparent due to the last scattering of radiation from free electrons. These CMB blackbody photons were suddenly free to propagate and hence have been traveling unscattered since that time.

The recombination reaction (that is, the inverse of the hydrogen photoionization reaction) is

$$p + e \rightleftarrows \mathrm{H} + \gamma, \tag{1}$$

so the chemical equilibrium relation from Section 6.6 gives

$$\mu_p + \mu_e = \mu_\mathrm{H} \tag{2}$$

since, again, the chemical potential of the blackbody photons is zero.[12] At the temperatures and densities prevailing during this era (a few thousand degrees Kelvin and only about 10^9 atoms per cubic meter), the electrons, protons, and hydrogen atoms can all be treated as classical ideal gases, with the result

$$\mu_p = m_p c^2 + kT \ln(n_p \lambda_p^3) - kT \ln 2, \tag{3a}$$

$$\mu_e = m_e c^2 + kT \ln(n_e \lambda_e^3) - kT \ln 2, \tag{3b}$$

$$\mu_\mathrm{H} = m_\mathrm{H} c^2 + kT \ln(n_\mathrm{H} \lambda_\mathrm{H}^3) - kT \ln 4. \tag{3c}$$

The binding energy of hydrogen is $m_p c^2 + m_e c^2 - m_\mathrm{H} c^2 = 1\,\mathrm{Ry}$. The equilibrium condition (6.6.3) and the ideal gas chemical potential (6.6.5) then give a simple relation between the number densities of the three species:

$$n_\mathrm{H} = n_p n_e \lambda_e^3 e^{\beta \mathrm{Ry}}, \tag{4}$$

[12]The same reaction occured for the deuterons that remained after nucleosynthesis at $t \approx 3$ minutes but the density of the deuterons was 3×10^{-5} times the proton density; refer to Figure 9.4.

where

$$\lambda_e = \frac{h}{\sqrt{2\pi m_e kT}} \tag{5}$$

is the electron thermal deBroglie wavelength. The number densities of free electrons and protons are the same due to charge neutrality:

$$n_e = n_p. \tag{6}$$

The protons remaining after nucleosynthesis are either free or combined into hydrogen atoms, so

$$n_p + n_{\rm H} = (1-2q)n_B = (1-2q)\eta n_\gamma. \tag{7}$$

Putting equations (3), (4), (6), and (7) together and making use of (9.1.5) gives the Saha equation for the neutral hydrogen fraction:

$$f_{\rm H} = \frac{n_H}{n_p + n_{\rm H}} = (1 - f_{\rm H})^2 s, \tag{8}$$

where the parameter s is

$$s = 4\zeta(3)\sqrt{\frac{2}{\pi}}(1-2q)\eta\left(\frac{kT}{m_e c^2}\right)^{3/2} e^{\beta \mathrm{Ry}}. \tag{9}$$

The solution to equation (8), namely

$$f_{\rm H} = \frac{1 + 2s - \sqrt{1+4s}}{2s}, \tag{10}$$

is shown in Figure 9.8. At temperatures above the recombination temperature, s is small so $f_{\rm H}$ is small, making the plasma fully ionized. At low temperatures s is large so $f_{\rm H}$ approaches unity, leaving just neutral atoms. The small values of the baryon-to-photon ratio η and $\mathrm{Ry}/m_e c^2$ make the onset of recombination at temperature

$$kT_r \approx \frac{\mathrm{Ry}}{\ln\left(\frac{1}{\eta}\left(\frac{m_e c^2}{\mathrm{Ry}}\right)^{3/2}\right)}, \tag{11}$$

which delays the last scattering until $T \approx 3000\,\mathrm{K}$; see Figure 9.8.

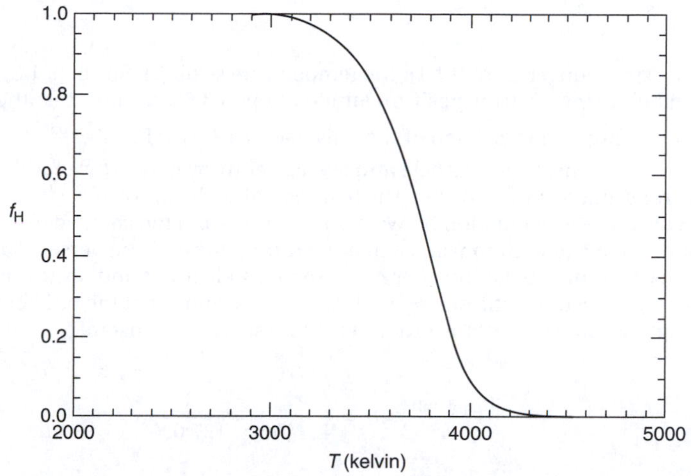

FIGURE 9.8 The equilibrium neutral hydrogen fraction as a function of temperature for baryon-to-phonon ratio $\eta = 6 \times 10^{-10}$ and proton fraction $1 - 2q = 0.76$. By the time temperature $T = 3000\,\mathrm{K}$, 99.5 percent of the free protons and electrons had combined into neutral hydrogen resulting in "last scattering" and the universe became transparent. The age of the universe at that time was about 380,000 years.

9.9 Epilogue

The formation of neutral atoms about 380,000 years after the Big Bang effectively ended the scattering of photons from free charges. The universe became transparent and entered the "dark ages" before the first star formation. The CMB photons were no longer in equilibrium but maintained their Planck distribution as the universe expanded. Small density fluctuations that were present in the electron–proton plasma just before recombination were imprinted on the CMB as temperature fluctuations. The CMB shown in Figure 9.3 earlier displays temperature fluctuations of the order of $\pm 200\,\mu\mathrm{K}$ that represent the density fluctuations in the plasma at the time of recombination. These small mass density fluctuations led to gravitational clumping that resulted in the formation of the first stars and galaxies 100 to 200 million years after the Big Bang. The large fraction of nonbaryonic cold dark matter was crucial in this process. Early stars that exploded as supernovae spewed their heavy elements (carbon, oxygen, silicon, iron, gold, uranium, etc.) into the cosmos. Our own solar system formed from a gas and dust cloud that included heavy elements that had been created in an earlier supernova event. Indeed, "we are stardust."[13]

[13]Joni Mitchell, *Woodstock*; copyright © Siquomb Publishing Company:
"We are stardust
Billion year old carbon
We are golden
Caught in the devil's bargain
And we've got to get ourselves
back to the garden"

Problems

9.1. Use the Hubble expansion relation (9.1.1), the temperature scaling relation (9.1.3), and the energy density relation before the electron-positron annihilation (9.3.6b) to show that the temperature as a function of time during the first second of the universe was $T(t) \approx 10^{10}\,\mathrm{K}\sqrt{\frac{0.992\,\mathrm{s}}{t}}$.

9.2. Determine the average energy per particle and average entropy per particle for the photons, electrons, positrons and neutrinos during the first second of the universe.

9.3. While the electromagnetic interaction between the photons and the charged electrons and positrons kept them in equilibrium with each other during the early universe, show that the direct electromagnetic Coulomb interaction energy between the electrons and positrons was small compared to the relativistic kinetic energy of those species. Show that the ratio between the Coulomb and kinetic energies is of the order of the fine structure constant:

$$\frac{u_{\text{coulomb}}}{u_e} \approx \alpha = \frac{e^2}{4\pi\epsilon_0 \hbar c} = \frac{1}{137.036}.$$

9.4. Show that during the early part of the electron-positron annihilation era, the ratio of the electron number density to the photon number density scaled with temperature as

$$\frac{n_-}{n_\gamma} \approx \frac{n_+}{n_\gamma} \sim \left(\frac{kT}{m_e c^2}\right)^{3/2} \exp\left(-\beta m_e c^2\right).$$

9.5. Show that after nearly all of the positrons were annihilated and the electron number density had nearly leveled off at the proton density, the ratio of the positron number density to the photon number density scaled with temperature as

$$\frac{n_+}{n_\gamma} \sim \left(\frac{kT}{m_e c^2}\right)^{3/2} \exp\left(-2\beta m_e c^2\right).$$

9.6. After the positrons were annihilated, the energy density of the universe was dominated by the photons and the neutrinos. Show that the energy density in that era was: $u_{\text{total}} = (1 + (4/11)^{4/3})u_\gamma$. Next, use the Hubble expansion relation (9.1.1), the temperature scaling relation (9.1.3), and the energy density after the electron-positron annihilation to show that the photon temperature as a function of time was $T(t) \approx 10^{10}\,\mathrm{K}\sqrt{\frac{1.788\,\mathrm{s}}{t}}$. This relation held from $t \approx 100\,\mathrm{s}$ until $t \approx 200{,}000$ years when the energy density due to baryonic and cold dark matter began to dominate.

9.7. How would the primordial helium content of the universe have been affected if the present cosmic background radiation temperature was 27 K instead of 2.7 K? What about 0.27 K?

9.8. Gold-on-gold nuclear collisions at the Relativistic Heavy Ion Collider (RHIC) at the Brookhaven National Laboratory create a quark-gluon plasma with an energy density of about $4\,\mathrm{GeV/fm^3}$; see Adare et al. (2010). Treat nuclear matter as composed of a noninteracting relativistic gas of quarks and gluons. Include the low-mass up and down quarks and their antiparticles (all spin-$\frac{1}{2}$), and spin-1 massless gluons. Like photons, the gluons are bosons, have two spin states each, and are their own antiparticle. There are eight varieties of gluons that change the three color states of the quarks. Only the strongly interacting particles need to be considered due to the tiny size of the plasmas. What is the temperature of the quark-gluon plasma?

9.9. Calculate the energy density versus temperature very early in the universe when the temperatures were above $kT = 300\,\mathrm{MeV}$. At those temperatures, quarks and gluons were released from individual nuclei. Treat the quark-gluon plasma as a noninteracting relativistic gas. At those temperatures, the species that are in equilibrium with one other are: photons, the three neutrino species, electrons and positrons, muons and antimuons, up and down quarks and their antiparticles (all spin-$\frac{1}{2}$), and spin-1 massless gluons. Like photons, the gluons are bosons, have two spin states each, and are their own antiparticle. There are eight varieties of gluons that change the three color states of the quarks. The strange, charm, top, and bottom quarks and tau leptons are heavier than 300 MeV, so they do not contribute substantially at this temperature. Use your result and equation (9.1.1) to determine the temperature evolution as a function of the age of the universe during this era and its age when $kT \approx 300\,\text{MeV}$.

10 Statistical Mechanics of Interacting Systems: The Method of Cluster Expansions

All the systems considered in the previous chapters were composed of, or could be regarded as composed of, *noninteracting* entities. Consequently, the results obtained, though of considerable intrinsic importance, may have limitations when applied to systems that actually exist in nature. For a real contact between the theory and experiment, one must take into account the interparticle interactions operating in the system. This can be done with the help of the formalism developed in Chapters 3 through 5 which, in principle, can be applied to an unlimited variety of physical systems and problems; in practice, however, one encounters in most cases serious difficulties of analysis. These difficulties are less stringent in the case of systems such as low-density gases, for which a corresponding noninteracting system can serve as an approximation. The mathematical expressions for the various physical quantities pertaining to such a system can be written in the form of *series expansions*, whose main terms describe the corresponding ideal-system results while the subsequent terms provide corrections arising from the interparticle interactions in the system. A systematic method of carrying out such expansions, in the case of real gases obeying classical statistics, was developed by Mayer and his collaborators (1937 onward) and is known as the method of *cluster expansions*. Its generalization, which equally well applies to gases obeying quantum statistics, was initiated by Kahn and Uhlenbeck (1938) and was perfected by Lee and Yang (1959a,b; 1960a,b,c).

10.1 Cluster expansion for a classical gas

We start with a relatively simple physical system, namely a single-component, classical, monatomic gas whose potential energy is given by a sum of two-particle interactions u_{ij}. The Hamiltonian of the system is then given by

$$H = \sum_i \left(\frac{1}{2m}p_i^2\right) + \sum_{i<j} u_{ij} \quad (i,j = 1,2,\ldots,N); \tag{1}$$

the summation in the second part goes over all the $N(N-1)/2$ pairs of particles in the system. In general, the potential u_{ij} is a function of the relative position vector $\boldsymbol{r}_{ij}(= \boldsymbol{r}_j - \boldsymbol{r}_i)$;

© 2011 Elsevier Ltd. All rights reserved.

however, if the two-body force is a central one, then the function u_{ij} depends only on the interparticle distance r_{ij}.

With the preceding Hamiltonian, the partition function of the system is given by, see equation (3.5.5),

$$Q_N(V,T) = \frac{1}{N!h^{3N}} \int \exp\left\{-\beta \sum_i \left(\frac{1}{2m}p_i^2\right) - \beta \sum_{i<j} u_{ij}\right\} d^{3N}p\, d^{3N}r. \tag{2}$$

Integration over the momenta of the particles can be carried out straightforwardly, with the result

$$Q_N(V,T) = \frac{1}{N!\lambda^{3N}} \int \exp\left[-\beta \sum_{i<j} u_{ij}\right] d^{3N}r = \frac{1}{N!\lambda^{3N}} Z_N(V,T), \tag{3}$$

where $\lambda\{= h/(2\pi mkT)^{1/2}\}$ is the *mean thermal wavelength* of the particles, while the function $Z_N(V,T)$ stands for the integral over the space coordinates $\mathbf{r}_1, \mathbf{r}_2, \ldots, \mathbf{r}_N$:

$$Z_N(V,T) = \int \exp\left[-\beta \sum_{i<j} u_{ij}\right] d^{3N}r = \int \prod_{i<j} (e^{-\beta u_{ij}})\, d^{3N}r. \tag{4}$$

The function $Z_N(V,T)$ is generally referred to as the *configuration integral* of the system. For a gas of noninteracting particles, the integrand in (4) is unity; we then have

$$Z_N^{(0)}(V,T) = V^N \quad \text{and} \quad Q_N^{(0)}(V,T) = \frac{V^N}{N!\lambda^{3N}}, \tag{5}$$

in agreement with our earlier result (3.5.9).

To treat the nonideal case we introduce, after Mayer, the two-particle function f_{ij}, defined by the relationship

$$f_{ij} = e^{-\beta u_{ij}} - 1. \tag{6}$$

In the absence of interactions, the function f_{ij} is identically zero; in the presence of interactions, it is nonzero but at sufficiently high temperatures it is quite small in comparison with unity. We, therefore, expect that the functions f_{ij} would be quite appropriate for carrying out a high-temperature expansion of the integrand in (4).

A typical plot of the functions u_{ij} and f_{ij} is shown in Figure 10.1; we note that (i) the function f_{ij} is everywhere bounded and (ii) it becomes negligibly small as the interparticle distance r_{ij} becomes large in comparison with the "effective" range, r_0, of the potential.

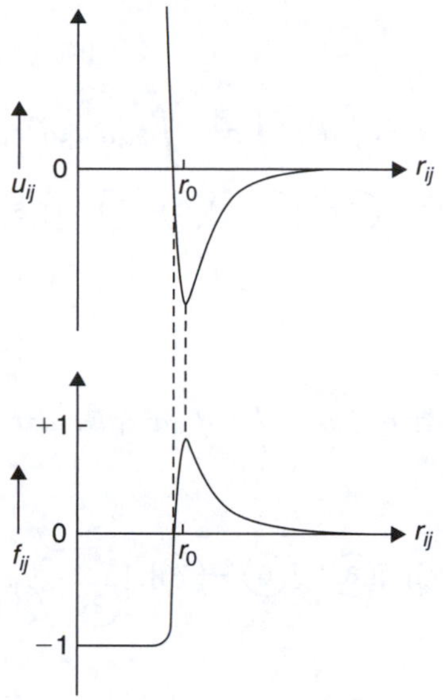

FIGURE 10.1 A typical plot of the two-body potential function u_{ij} and the corresponding Mayer function f_{ij}.

Now, to evaluate the configuration integral (4), we expand its integrand in ascending powers of the functions f_{ij}:

$$Z_N(V,T) = \int \prod_{i<j} (1+f_{ij})\, d^3r_1 \cdots d^3r_N$$

$$= \int \left[1 + \sum f_{ij} + \sum f_{ij} f_{kl} + \cdots \right] d^3r_1 \cdots d^3r_N. \tag{7}$$

A convenient way of enumerating the various terms in (7) is to associate each term with a corresponding N-particle graph. For instance, if N were 8, the terms

$$t_A = \int f_{34} f_{68} d^3r_1 \cdots d^3r_8 \quad \text{and} \quad t_B = \int f_{12} f_{14} f_{67} d^3r_1 \cdots d^3r_8 \tag{8}$$

in the expansion of the configuration integral Z_8 could be associated with the 8-particle graphs

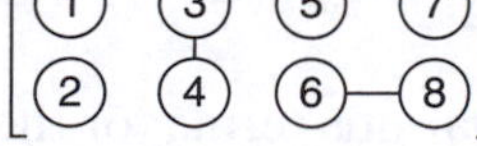

and

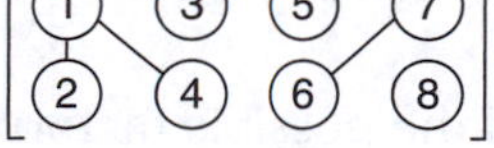

, (9)

respectively. A closer look at the terms t_A and t_B (and at the corresponding graphs) suggests that we better regard these terms as suitably factorized (and the graphs correspondingly decomposed), that is,

$$t_A = \int d^3r_1 \int d^3r_2 \int d^3r_5 \int d^3r_7 \int f_{34} d^3r_3 d^3r_4 \int f_{68} d^3r_6 d^3r_8$$

$$\equiv [\textcircled{1}] \,.\, [\textcircled{2}] \,.\, [\textcircled{5}] \,.\, [\textcircled{7}] \,.\, [\textcircled{3}\!-\!\!-\!\textcircled{4}] \,.\, [\textcircled{6}\!-\!\!-\!\textcircled{8}] \tag{10}$$

and similarly

$$t_B = \int d^3r_3 \int d^3r_5 \int d^3r_8 \int f_{12} f_{14} d^3r_1 d^3r_2 d^3r_4 \int f_{67} d^3r_6 d^3r_7$$

$$\equiv [\textcircled{3}] \,.\, [\textcircled{5}] \,.\, [\textcircled{8}] \,.\, [\textcircled{6}\!-\!\!-\!\textcircled{7}] \,.\, [$$ 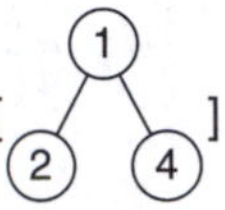 $$]. \tag{11}$$

We may then say that the term t_A in the expansion of the integral Z_8 represents a "configuration" in which there are four "clusters" of one particle each and two "clusters" of two particles each, while the term t_B represents a "configuration" in which there are three "clusters" of one particle each, one "cluster" of two particles and one "cluster" of three particles.

In view of this, we may introduce the notion of an *N-particle graph* which, by definition, is a "collection of N distinct circles, numbered $1, 2, \ldots, N$, with a number of lines linking some (or all) of the circles"; if the distinct pairs (of circles), which are linked through these lines, are denoted by the symbols $\alpha, \beta, \ldots, \lambda$ (each of these symbols denoting a distinct *pair* of indices out of the set $1, 2, \ldots, N$), then the graph represents the term

$$\int (f_\alpha f_\beta \cdots f_\lambda) d^3r_1 \cdots d^3r_N \tag{12}$$

of expansion (7). A graph having the same number of linked pairs as this one but with the set $(\alpha', \beta', \ldots, \lambda')$ *distinct* from the set $(\alpha, \beta, \ldots, \lambda)$ will be counted as a distinct graph, for it represents a different term in the expansion; of course, these terms will belong to one and the same group in the expansion. Now, in view of the one-to-one correspondence between the various terms in the expansion (7) and the various N-particle graphs, we have

$$Z_N(V, T) = \text{sum of all distinct } N\text{-particle graphs.} \tag{13}$$

Further, in view of the possible factorization of the various terms (or the possible decomposition of the various graphs), we may introduce the notion of an *l-cluster* which, by definition, is an "l-particle graph in which each of the l circles, numbered $1, 2, \ldots, l$, is directly or indirectly linked with every other circle." As an example, we write here a

5-particle graph, which is also a 5-cluster:

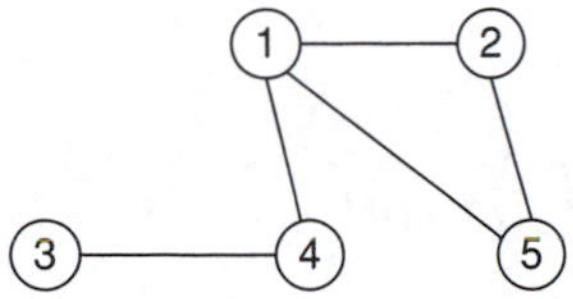

$$\equiv \int f_{12} f_{14} f_{15} f_{25} f_{34} \, d^3 r_1 \cdots d^3 r_5. \tag{14}$$

It is obvious that a cluster as such cannot be decomposed into simpler graphs inasmuch as the corresponding term cannot be factorized into simpler terms. Furthermore, a group of l particles (except when $l = 1$ or 2) can lead to a variety of l-clusters, some of which may be equal in value; for instance, a group of three particles leads to four different 3-clusters, namely

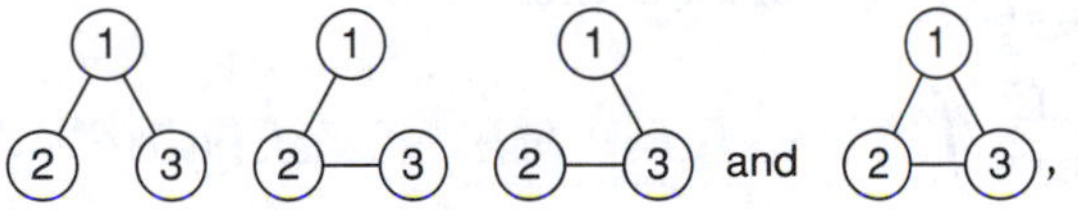

(15)

of which the first three are equal in value. In view of the variety of ways in which an l-cluster can appear, we may introduce the notion of a *cluster integral* b_l, defined by

$$b_l(V, T) = \frac{1}{l!\,\lambda^{3(l-1)} V} \times \text{(the sum of all possible } l\text{-clusters)}. \tag{16}$$

So defined, the cluster integral $b_l(V, T)$ is dimensionless and, in the limit $V \to \infty$, approaches a finite value, $\bar{b}_l(T)$, which is independent of the size and the shape of the container (unless the latter is unduly abnormal). The first property is quite obvious. The second one follows by noting that if we hold one of the l particles fixed, at the point $\boldsymbol{r}_1$ say, and carry out integration over the coordinates of the remaining $(l-1)$ particles, then, because of the fact that the functions f_{ij} extend only over a small finite range of distances, this integration would extend only over a limited region of the space available — a region whose *linear* dimensions are of the order of the range of the functions f_{ij};[1] the result of this integration will be practically independent of the volume of the container.[2] Finally, we integrate over the coordinates $\boldsymbol{r}_1$ of the particle that was held fixed and obtain a straight factor of V; this cancels out the V in the denominator of the defining formula (16). Thus, the dependence of the cluster integral $b_l(V, T)$ on the size of the container is no more than a mere "surface effect" — an effect that disappears as $V \to \infty$, and we end up with a volume-independent number $\bar{b}_l(T)$.

[1]Hence the name "cluster."

[2]Of course, some dependence on the geometry of the container will indeed arise if the fixed particle happened to be close to the walls of the container. This is, however, unimportant when $V \to \infty$.

Some of the simpler cluster integrals are

$$b_1 = \frac{1}{V}[\,①\,] = \frac{1}{V}\int d^3r_1 \equiv 1,$$

$$b_2 = \frac{1}{2\lambda^3 V}[\,①\!\!-\!\!②\,] = \frac{1}{2\lambda^3 V}\iint f_{12}d^3r_1\,d^3r_2$$

$$\approx \frac{1}{2\lambda^3}\int f_{12}d^3r_{12} = \frac{2\pi}{\lambda^3}\int_0^\infty f(r)r^2\,dr \tag{17}$$

$$= \frac{2\pi}{\lambda^3}\int_0^\infty (e^{-u(r)/kT}-1)r^2\,dr, \tag{18}$$

$$b_3 = \frac{1}{6\lambda^6 V} \times \text{ [sum of the clusters (15)]}$$

$$= \frac{1}{6\lambda^6 V}\int (\underbrace{f_{12}f_{13}+f_{12}f_{23}+f_{13}f_{23}}+f_{12}f_{13}f_{23})d^3r_1d^3r_2d^3r_3$$

$$\approx \frac{1}{6\lambda^6 V}\left[3V\iint f_{12}f_{13}d^3r_{12}d^3r_{13} + V\iint f_{12}f_{13}f_{23}d^3r_{12}d^3r_{13}\right]$$

$$= 2b_2^2 + \frac{1}{6\lambda^6}\iint f_{12}f_{13}f_{23}d^3r_{12}d^3r_{13}, \tag{19}$$

and so on.

We now proceed to evaluate the crucial expression in (13). Obviously, an N-particle graph will consist of a number of clusters of which, say, m_1 are 1-clusters, m_2 are 2-clusters, m_3 are 3-clusters, and so on; the numbers $\{m_l\}$ must satisfy the restrictive condition

$$\sum_{l=1}^{N} lm_l = N, \quad m_l = 0,1,2,\ldots,N. \tag{20}$$

However, a given set of numbers $\{m_l\}$ does not specify a unique, single graph; it represents a "collection of graphs" the sum total of which may be denoted by the symbol $S\{m_l\}$. We may then write

$$Z_N(V,T) = \sum_{\{m_l\}}{}' S\{m_l\}, \tag{21}$$

where the primed summation $\sum'$ goes over all sets $\{m_l\}$ that conform to the restrictive condition (20). Equation (21) represents a systematic regrouping of the graphs, as opposed to the simple-minded grouping that first appeared in equation (7).

Our next task consists of evaluating the sum $S\{m_l\}$. To do this, we observe that the "family of graphs" under the distribution set $\{m_l\}$ arises essentially from the following two causes:

(i) there are, in general, many different ways of *assigning* the N particles of the system to the $\sum_l m_l$ clusters, and
(ii) for any given assignment, there are, in general, many different ways of *forming* the various clusters, for even with a given group of l particles an l-cluster (if $l > 2$) can be formed in a number of different ways; see, for example, the four different ways of forming a 3-cluster with a given group of three particles, as listed in (15).

For cause (i), we obtain a straightforward factor of

$$\frac{N!}{(1!)^{m_1}(2!)^{m_2}\cdots} = \frac{N!}{\prod_l (l!)^{m_l}}. \tag{22}$$

Now, if cause (ii) were not there, that is, if all l-clusters were unique in their formation, then the sum $S\{m_l\}$ would be given by the product of the combinatorial factor (22) with the value of any one graph in the setup, namely

$$\prod_l (\text{the value of an } l\text{-cluster})^{m_l}, \tag{23}$$

further corrected for the fact that any two arrangements that differ merely in the exchange of *all* the particles in one cluster with *all* the particles in another cluster of the same size, *must not* be counted as distinct, the corresponding correction factor being

$$\prod_l (1/m_l!). \tag{24}$$

A little reflection now shows that cause (ii) is completely and correctly taken care of if we replace the product of the expressions (23) and (24) by the expression[3]

$$\prod_l [(\text{the } \textit{sum} \text{ of the values of all possible } l\text{-clusters})^{m_l}/m_l!] \tag{25}$$

which, with the help of equation (16), may be written as

$$\prod_l \left[(b_l l! \lambda^{3(l-1)} V)^{m_l}/m_l!\right]. \tag{26}$$

[3]To appreciate the logic behind this replacement, consider expression [ ] in (25) as a multinomial expansion and interpret the various terms of this expansion in terms of the variety of the l-clusters.

The sum $S\{m_l\}$ is now given by the product of factor (22) and expression (26). Substituting this result into (21), we obtain for the configuration integral

$$Z_N(V,T) = N!\lambda^{3N} \sum_{\{m_l\}}{}' \left[\prod_l \left\{ \left(b_l \frac{V}{\lambda^3} \right)^{m_l} \frac{1}{m_l!} \right\} \right]. \tag{27}$$

Here, use has been made of the fact that

$$\prod_l (\lambda^{3l})^{m_l} = \lambda^{3\Sigma_l l m_l} = \lambda^{3N}; \tag{28}$$

see the restrictive condition (20). The partition function of the system now follows from equations (3) and (27), with the result

$$Q_N(V,T) = \sum_{\{m_l\}}{}' \left[\prod_{l=1}^{N} \left\{ \left(b_l \frac{V}{\lambda^3} \right)^{m_l} \frac{1}{m_l!} \right\} \right]. \tag{29}$$

The evaluation of the primed sum in (29) is complicated by the restrictive condition (20), which must be obeyed by every set $\{m_l\}$. We, therefore, move over to the grand partition function of the system:

$$\mathcal{Q}(z,V,T) = \sum_{N=0}^{\infty} z^N Q_N(V,T). \tag{30}$$

Writing

$$z^N = z^{\Sigma_l l m_l} = \prod_l (z^l)^{m_l}, \tag{31}$$

substituting for $Q_N(V,T)$ from (29), and noting that a *restricted* summation over sets $\{m_l\}$, subject to the condition $\sum_l l m_l = N$, followed by a summation over all values of N (from $N = 0$ to $N = \infty$) is equivalent to an *unrestricted* summation over *all possible sets* $\{m_l\}$, we obtain

$$\begin{aligned} \mathcal{Q}(z,V,T) &= \sum_{m_1,m_2,\ldots=0}^{\infty} \left[\prod_{l=1}^{\infty} \left\{ \left(b_l z^l \frac{V}{\lambda^3} \right)^{m_l} \frac{1}{m_l!} \right\} \right] \\ &= \prod_{l=1}^{\infty} \left[\sum_{m_l=0}^{\infty} \left\{ \left(b_l z^l \frac{V}{\lambda^3} \right)^{m_l} \frac{1}{m_l!} \right\} \right] \\ &= \prod_{l=1}^{\infty} \left[\exp\left(b_l z^l \frac{V}{\lambda^3} \right) \right] = \exp\left[\sum_{l=1}^{\infty} b_l z^l \frac{V}{\lambda^3} \right] \end{aligned} \tag{32}$$

and, hence,

$$\frac{1}{V}\ln \mathcal{Q} = \frac{1}{\lambda^3}\sum_{l=1}^{\infty} b_l z^l. \tag{33}$$

In the limit $V \to \infty$,

$$\frac{P}{kT} \equiv \lim_{V\to\infty}\left(\frac{1}{V}\ln \mathcal{Q}\right) = \frac{1}{\lambda^3}\sum_{l=1}^{\infty} \bar{b}_l z^l, \tag{34}$$

and

$$\frac{N}{V} \equiv \lim_{V\to\infty}\left(\frac{z}{V}\frac{\partial \ln \mathcal{Q}}{\partial z}\right) = \frac{1}{\lambda^3}\sum_{l=1}^{\infty} l\bar{b}_l z^l. \tag{35}$$

Equations (34) and (35) constitute the famous *cluster expansions* of the Mayer–Ursell formalism. Eliminating the fugacity z among these equations, we obtain the *equation of state* of the system.

10.2 Virial expansion of the equation of state

The approach developed in the preceding section leads to exact results only if we apply it to the gaseous phase alone. If we attempt to include in our study the phenomena of condensation, the critical point, and the liquid phase, we encounter serious difficulties relating to (i) the limiting procedure involved in equations (10.1.34) and (10.1.35), (ii) the convergence of the summations over l, and (iii) the volume dependence of the cluster integrals b_l. We, therefore, restrict our study to the gaseous phase alone. The equation of state may then be written in the form

$$\frac{Pv}{kT} = \sum_{l=1}^{\infty} a_l(T)\left(\frac{\lambda^3}{v}\right)^{l-1}, \tag{1}$$

where $v(= V/N)$ denotes the volume per particle in the system. Expansion (1), which is supposed to have been obtained by eliminating z between equations (10.1.34) and (10.1.35), is called the *virial expansion* of the system and the numbers $a_l(T)$ the *virial coefficients*.[4] To determine the relationship between the coefficients a_l and the cluster integrals $\bar{b}_l$, we invert equation (10.1.35) to obtain z as a power series in (λ^3/v) and substitute this

[4]For various manipulations of the virial equation of state, see Kilpatrick and Ford (1969).

into (10.1.34). This leads to equation (1), with

$$a_1 = \bar{b}_1 \equiv 1, \tag{2}$$

$$a_2 = -\bar{b}_2 = -\frac{2\pi}{\lambda^3}\int_0^\infty \left(e^{-u(r)/kT} - 1\right) r^2 dr, \tag{3}$$

$$a_3 = 4\bar{b}_2^2 - 2\bar{b}_3 = -\frac{1}{3\lambda^6}\int_0^\infty\int_0^\infty f_{12}f_{13}f_{23}d^3r_{12}d^3r_{13}, \tag{4}$$

$$a_4 = -20\bar{b}_2^3 + 18\bar{b}_2\bar{b}_3 - 3\bar{b}_4 = \cdots, \tag{5}$$

and so on; here, use has also been made of formulae (10.1.17) to (10.1.19). We note that the coefficient a_l is completely determined by the quantities $\bar{b}_1, \bar{b}_2, \ldots, \bar{b}_l$, that is, by the sequence of configuration integrals $Z_1, Z_2, \ldots, Z_l$; see also equations (10.4.5) to (10.4.8).

From equation (4) we observe that the third virial coefficient of the gas is determined solely by the 3-cluster 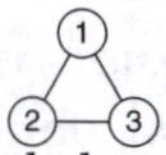

. This suggests that the higher-order virial coefficients may also be determined solely by a special "subgroup" of the various l-clusters. This is indeed true, and the relevant result is that, in the limit of infinite volume,[5]

$$a_l = -\frac{l-1}{l}\beta_{l-1} \quad (l \geq 2), \tag{6}$$

where β_{l-1} is the so-called *irreducible cluster integral,* defined as

$$\beta_{l-1} = \frac{1}{(l-1)!\lambda^{3(l-1)}V} \times \text{(the sum of all irreducible } l\text{-clusters)}; \tag{7}$$

by an *irreducible* l-cluster we mean an "l-particle graph that is multiply-connected (in the sense that there are at least two entirely independent, nonintersecting paths linking each pair of circles in the graph)." For instance, of the four possible 3-clusters, see (10.1.15), only the last one is irreducible. Indeed, if we express equation (4) in terms of this particular cluster and make use of definition (7) for β_2, we do obtain for the third virial coefficient

$$a_3 = -\frac{2}{3}\beta_2, \tag{8}$$

in agreement with the general result (6).[6]

The quantities β_{l-1}, like b_l, are dimensionless and, in the limit $V \to \infty$, approach finite values that are independent of the size and the shape of the container (unless the

[5] For a proof of this result, see Hill (1956, Sections 24 and 25); see also Section 10.4 of the present text.

[6] It may be mentioned here that a 2-cluster is also regarded as an *irreducible* cluster. Accordingly, $\beta_1 = 2b_2$; see equations (10.1.16) and (10.2.7). Equation (3) then gives: $a_2 = -b_2 = -\frac{1}{2}\beta_1$, again in agreement with the general result (6).

latter is unduly abnormal). Moreover, the two sets of quantities are mutually related; see equations (10.4.27) and (10.4.29).

10.3 Evaluation of the virial coefficients

If a given system does not depart much from the ideal-gas behavior, its equation of state is given adequately by the first few virial coefficients. Now, since $a_1 \equiv 1$, the lowest-order virial coefficient that we need to consider here is a_2, which is given by equation (10.2.3):

$$a_2 = -\bar{b}_2 = \frac{2\pi}{\lambda^3} \int_0^\infty \left(1 - e^{-u(r)/kT}\right) r^2\, dr, \tag{1}$$

$u(r)$ being the potential energy of interparticle interaction. A typical plot of the function $u(r)$ was shown earlier in Figure 10.1; a typical semi-empirical formula (Lennard-Jones, 1924) is given by

$$u(r) = 4\varepsilon \left[\left(\frac{\sigma}{r}\right)^{12} - \left(\frac{\sigma}{r}\right)^{6}\right]. \tag{2}$$

The most significant features of an actual interparticle potential are well-simulated by the Lennard-Jones formula (2). For instance, the function $u(r)$ given by (2) exhibits a "minimum," of value $-\varepsilon$, at a distance $r_0 (= 2^{1/6}\sigma)$ and rises to an infinitely large (positive) value for $r < \sigma$ and to a vanishingly small (negative) value for $r \gg \sigma$. The portion to the left of the "minimum" is dominated by *repulsive* interaction that comes into play when two particles come too close to one another, while the portion to the right is dominated by *attractive* interaction that operates between particles when they are separated by a respectable distance. For most practical purposes, the precise form of the repulsive part of the potential is not very important; it may as well be replaced by the crude approximation

$$u(r) = +\infty \quad (\text{for } r < r_0), \tag{3}$$

which amounts to attributing an *impenetrable* core, of diameter r_0, to each particle. The precise form of the attractive part is, however, important; in view of the fact that there exists good theoretical basis for the sixth-power attractive potential (see Problem 3.36), this part may simply be written as

$$u(r) = -u_0 (r_0/r)^6 \quad (r \geq r_0). \tag{4}$$

The potential given by expressions (3) and (4) may, therefore, be used if one is only interested in a qualitative assessment of the situation and not in a quantitative comparison between the theory and experiment.

Substituting (3) and (4) into (1), we obtain for the second virial coefficient

$$a_2 = \frac{2\pi}{\lambda^3}\left[\int_0^{r_0} r^2\,dr + \int_{r_0}^{\infty}\left[1-\exp\left\{\frac{u_0}{kT}\left(\frac{r_0}{r}\right)^6\right\}\right]r^2\,dr\right]. \tag{5}$$

The first integral is straightforward; the second one is considerably simplified if we assume that $(u_0/kT) \ll 1$, which makes the integrand very nearly equal to $-(u_0/kT)(r_0/r)^6$. Equation (5) then gives

$$a_2 \simeq \frac{2\pi r_0^3}{3\lambda^3}\left(1-\frac{u_0}{kT}\right). \tag{6}$$

Substituting (6) into the expansion (10.2.1), we obtain a first-order improvement on the ideal-gas law, namely

$$P \simeq \frac{kT}{v}\left\{1+\frac{2\pi r_0^3}{3v}\left(1-\frac{u_0}{kT}\right)\right\} \tag{7a}$$

$$= \frac{kT}{v}\left\{1+\frac{B_2(T)}{v}\right\}, \text{ say.} \tag{7b}$$

The coefficient B_2, which is also sometimes referred to as the second virial coefficient of the system, is given by

$$B_2 \equiv a_2\lambda^3 \simeq \frac{2\pi r_0^3}{3}\left(1-\frac{u_0}{kT}\right). \tag{8}$$

In our derivation it was explicitly assumed that (i) the potential function $u(r)$ is given by the simplified expressions (3) and (4), and (ii) $(u_0/kT) \ll 1$. We cannot, therefore, expect formula (8) to be a faithful representation of the second virial coefficient of a real gas. Nevertheless, it does correspond, almost exactly, to the *van der Waals approximation* to the equation of state of a real gas. This can be seen by rewriting (7a) in the form

$$\left(P+\frac{2\pi r_0^3 u_0}{3v^2}\right) \simeq \frac{kT}{v}\left(1+\frac{2\pi r_0^3}{3v}\right) \simeq \frac{kT}{v}\left(1-\frac{2\pi r_0^3}{3v}\right)^{-1},$$

which readily leads to the van der Waals equation of state

$$\left(P+\frac{a}{v^2}\right)(v-b) \simeq kT, \tag{9}$$

where

$$a = \frac{2\pi r_0^3 u_0}{3} \quad \text{and} \quad b = \frac{2\pi r_0^3}{3} \equiv 4v_0. \tag{10}$$

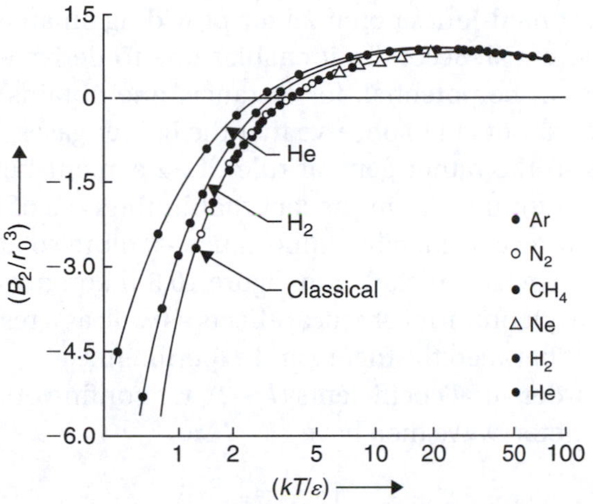

FIGURE 10.2 A dimensionless plot showing the temperature dependence of the second virial coefficient of several gases (after Hirschfelder et al., 1954).

We note that the parameter b in the van der Waals equation of state is exactly four times the actual *molecular volume* v_0, the latter being the "volume of a sphere of diameter r_0"; compare with Problem 1.4. We also note that in this derivation we have assumed that $b \ll v$, which means that the gas is sufficiently dilute for the mean interparticle distance to be much larger than the effective range of the interparticle interaction. Finally, we observe that, according to this simple-minded calculation, the van der Waals parameters a and b are temperature-independent, which in reality is not true.

A realistic study of the second virial coefficient requires the use of a realistic potential, such as the one given by Lennard-Jones, for evaluating the integral in (1). This has indeed been done and the results obtained are shown in Figure 10.2, where the reduced coefficient $B_2'(= B_2/r_0^3)$ is plotted against the reduced temperature $T'(= kT/\varepsilon)$:

$$B_2'(T') = 2\pi \int_0^\infty \left(1 - e^{-u'(r')/T'}\right) r'^2 \, dr', \tag{11}$$

with

$$u'(r') = \left\{ \left(\frac{1}{r'}\right)^{12} - 2\left(\frac{1}{r'}\right)^{6} \right\}, \tag{12}$$

r' being equal to (r/r_0); expressed in this form, the quantity B_2' is a *universal* function of T'. Included in the plot are experimental results for several gases. We note that in most cases the agreement is reasonably good; this is especially satisfying in view of the fact that in each case we had only two adjustable parameters, r_0 and ε, against a much larger number of experimental points available. In the first place, this agreement vindicates the adequacy of the Lennard-Jones potential for providing an analytical description of a typical interparticle potential. Secondly, it enables one to derive empirical values of the respective parameters of the potential; for instance, one obtains for argon: $r_0 = 3.82$ Å and $\varepsilon/k = 120\,\text{K}$.[7] One cannot fail to observe that the lighter gases, hydrogen and helium, constitute exceptions to the rather general rule of agreement between the theory and experiment. The reason for this lies in the fact that in the case of these gases quantum-mechanical effects assume considerable importance — more so at low temperatures. To substantiate this point, we have included in Figure 10.2 theoretical curves for H_2 and He taking into account the quantum-mechanical effects as well; as a result, we find once again a fairly good agreement between the theory and experiment.

As regards higher-order virial coefficients ($l > 2$), we confine our discussion to a gas of hard spheres with diameter D. We then have

$$u(r) = \begin{cases} 0 & \text{if } r > D, \\ \infty & \text{if } r \le D, \end{cases} \tag{13}$$

and, hence,

$$f(r) = \begin{cases} 0 & \text{if } r > D, \\ -1 & \text{if } r \le D. \end{cases} \tag{14}$$

The second virial coefficient of the gas is then given by

$$a_2 = \frac{2\pi D^3}{3\lambda^3} = 4\frac{v_0}{\lambda^3}; \tag{15}$$

compare with equation (6). The third virial coefficient can be determined with the help of equation (10.2.4), namely

$$a_3 = -\frac{1}{3\lambda^6}\int_0^\infty\int_0^\infty f_{12}f_{13}f_{23}\,d^3r_{12}\,d^3r_{13}. \tag{16}$$

To evaluate this integral, we first fix the positions of particles 1 and 2 (such that $r_{12} < D$) and let particle 3 take all possible positions so that we can effect an integration over the variable r_{13}; see Figure 10.3. Since our integrand is equal to -1 when each of the distances r_{13} and r_{23} (like r_{12}) is less than D and 0 otherwise, we have

$$a_3 = \frac{1}{3\lambda^6}\int_{r_{12}=0}^{D}\left\{\int' d^3r_{13}\right\} d^3r_{12}, \tag{17}$$

where the primed integration arises from particle 3 taking all possible positions of interest. In view of the conditions $r_{13} < D$ and $r_{23} < D$, this integral is precisely equal to the "volume

[7] Corresponding values for various other gases have been summarized in Hill (1960, p. 484).

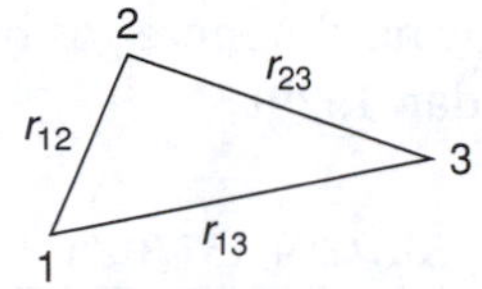

FIGURE 10.3

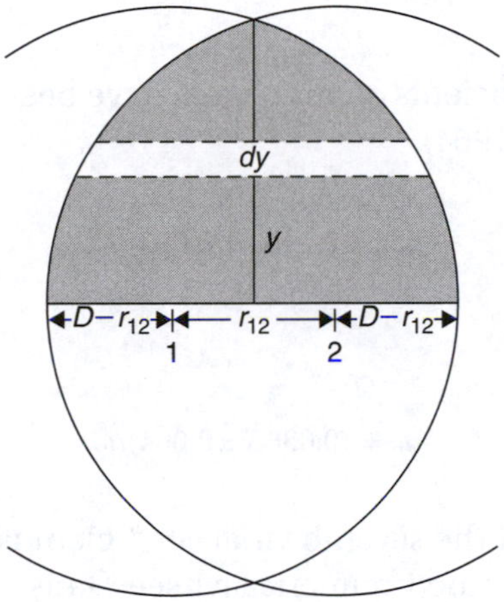

FIGURE 10.4

common to the spheres S_1 and S_2, each of radius D, centered at the *fixed* points 1 and 2"; see Figure 10.4. This in turn can be obtained by calculating the "volume swept by the shaded area in the figure on going through a complete revolution about the line of centers." One gets:

$$\int' d^3r_{13} = \int_0^{\sqrt{[D^2-(r_{12}/2)^2]}} \left\{2(D^2 - y^2)^{1/2} - r_{12}\right\} 2\pi y\, dy. \tag{18}$$

While the quantity within the curly brackets denotes the length of the strip shown in the figure, the element of area $2\pi y\, dy$ arises from the revolution; the limits of integration for y can be checked rather easily. The evaluation of the integral (18) is straightforward; we get

$$\int' d^3r_{13} = \frac{4\pi}{3}\left\{D^3 - \frac{3D^2 r_{12}}{4} + \frac{r_{12}^3}{16}\right\}. \tag{19}$$

Substituting (19) into (17) and carrying out integration over r_{12}, we finally obtain

$$a_3 = \frac{5\pi^2 D^6}{18\lambda^6} = \frac{5}{8}a_2^2. \tag{20}$$

The fourth virial coefficient of the hard-sphere gas has also been evaluated exactly. It is given by (Boltzmann, 1899; Majumdar, 1929) [8]

$$\begin{aligned} a_4 &= \left\{ \frac{1283}{8960} + \frac{3}{2} \cdot \frac{73\surd(2) + 1377\{\tan^{-1}\surd(2) - \pi/4\}}{1120\pi} \right\} a_2^3 \\ &= 0.28695 a_2^3. \end{aligned} \tag{21}$$

The fifth and sixth virial coefficients of this system have been computed numerically, with the results (Ree and Hoover, 1964)

$$a_5 = (0.1103 \pm 0.003) a_2^4, \tag{22}$$

and

$$a_6 = (0.0386 \pm 0.004) a_2^5. \tag{23}$$

Ree and Hoover's estimate of the seventh virial coefficient is $0.0127 a_2^6$. Terms up through 10th order have been determined numerically; see Hansen and McDonald (1986) and Malijevsky and Kolafa (2008). If the virial equation of state for hard spheres is written in terms of the volume packing fraction $\eta = \pi n D^3/6$, the first ten terms are

$$\begin{aligned} \frac{P}{nkT} &= 1 + 4\eta + 10\eta^2 + 18.364768\eta^3 + 28.22445\eta^4 + 39.81545\eta^5 \\ &\quad + 53.3418\eta^6 + 68.534\eta^7 + 85.805\eta^8 + 105.8\eta^9 + \cdots. \end{aligned} \tag{24}$$

Carnahan and Starling (1969) proposed a simple form for the equation of state that closely approximates all of the known virial coefficients:

$$\begin{aligned} \frac{P}{nkT} &\approx \frac{1 + \eta + \eta^2 - \eta^3}{(1-\eta)^3} \\ &= 1 + 4\eta + 10\eta^2 + 18\eta^3 + 28\eta^4 + 40\eta^5 \\ &\quad + 54\eta^6 + 70\eta^7 + 88\eta^8 + 108\eta^9 + 130\eta^{10} + \cdots \end{aligned} \tag{25}$$

This gives an excellent fit to the hard sphere equation of state for the entire fluid phase as determined in computer simulations. The fluid phase is the equilibrium phase for $0 < \eta \lesssim 0.491$. The high-density equilibrium phase of hard spheres is a face-centered cubic solid; see Chapter 16. Many other approximate analytical forms have also been proposed to closely reproduce the virial series; see for instance Mulero et al. (2008).

[8] See also Katsura (1959).

10.4 General remarks on cluster expansions

Shortly after the pioneering work of Mayer and his collaborators, Kahn and Uhlenbeck (1938) initiated the development of a similar treatment for quantum-mechanical systems. Of course, their treatment applied to the limiting case of classical systems as well but it faced certain inherent difficulties of analysis, some of which were later removed by the formal methods developed by Lee and Yang (1959a,b; 1960a,b,c). We propose to examine these developments in the next three sections of this chapter. First, however, we would like to make a few general observations on the problem of cluster expansions. These observations, due primarily to Ono (1951) and Kilpatrick (1953), are of considerable interest insofar as they hold for a very large class of physical systems. For instance, the system may be quantum-mechanical or classical, it may be a multicomponent one or single-component, its molecules may be polyatomic or monatomic, and so on. All we have to assume is that (i) the system is gaseous in state and (ii) its partition functions $Q_N(V,T)$, for some low values of N, can somehow be obtained. We can then calculate the "cluster integrals" $\mathfrak{b}_l$, and the virial coefficients a_l, of the system in the following straightforward manner.

Quite generally, the *grand partition function* of the system can be written as

$$\mathcal{Q}(z,V,T) \equiv \sum_{N=0}^{\infty} Q_N(V,T)z^N = \sum_{N=0}^{\infty} \frac{Z_N(V,T)}{N!}\left(\frac{z}{\lambda^3}\right)^N, \tag{1}$$

where we have introduced the "configuration integrals" $Z_N(V,T)$, defined in analogy with equation (10.1.3) of the classical treatment:

$$Z_N(V,T) \equiv N!\lambda^{3N} Q_N(V,T). \tag{2}$$

Dimensionally, the quantity Z_N is like (a volume)^N; moreover, the quantity Z_0 (like Q_0) is supposed to be identically equal to 1, while $Z_1 (\equiv \lambda^3 Q_1)$ is identically equal to V. We then have, in the limit $V \to \infty$,

$$\frac{P}{kT} \equiv \frac{1}{V}\ln\mathcal{Q} = \frac{1}{V}\ln\left\{1 + \frac{Z_1}{1!}\left(\frac{z}{\lambda^3}\right)^1 + \frac{Z_2}{2!}\left(\frac{z}{\lambda^3}\right)^2 + \cdots\right\} \tag{3}$$

$$= \frac{1}{\lambda^3}\sum_{l=1}^{\infty} \mathfrak{b}_l z^l, \text{ say.} \tag{4}$$

Again, the last expression has been written in analogy with the classical expansion (10.1.34); the coefficients $\mathfrak{b}_l$ may, therefore, be looked upon as the *cluster integrals* of the given system. Expanding (3) as a power series in z and equating respective coefficients with the $\mathfrak{b}_l$ of (4), we obtain

$$\mathfrak{b}_1 = \frac{1}{V} Z_1 \equiv 1, \tag{5}$$

$$\mathfrak{b}_2 = \frac{1}{2!\lambda^3 V}(Z_2 - Z_1^2), \tag{6}$$

$$\mathfrak{b}_3 = \frac{1}{3!\lambda^6 V}(Z_3 - 3Z_2Z_1 + 2Z_1^3), \tag{7}$$

$$\mathfrak{b}_4 = \frac{1}{4!\lambda^9 V}(Z_4 - 4Z_3Z_1 - 3Z_2^2 + 12Z_2Z_1^2 - 6Z_1^4), \tag{8}$$

and so on. We note that, for all $l > 1$, the sum of the coefficients appearing within the parentheses is identically equal to zero. Consequently, in the case of an ideal classical gas, for which $Z_i \equiv V^i$, see equation (10.1.4), all cluster integrals with $l > 1$ vanish. This, in turn, implies the vanishing of all the virial coefficients of the gas (except, of course, a_1, which is identically equal to unity).

Comparing equations (6) through (8) with equation (10.1.16), we find that the expressions involving the products of the various Z_i that appear within the parentheses play the same role here as "the sum of all possible l-clusters" does in the classical case. We therefore expect that, in the limit $V \to \infty$, the $\mathfrak{b}_l$ here would also be independent of the size and the shape of the container (unless the latter is unduly abnormal). This, in turn, requires that the various combinations of the Z_i appearing within the parentheses here must all be proportional to the *first* power of V. This observation leads to the very interesting result, first noticed by Rushbrooke, namely

$$\mathfrak{b}_l = \frac{1}{l!\lambda^{3(l-1)}} \times (\text{the coefficient of } V^l \text{ in the volume expansion of } Z_l). \tag{9}$$

At this stage, it seems worthwhile to point out that the expressions appearing within the parentheses of equations (6) through (8) are well-known in mathematical statistics as the *semi-invariants* of Thiele. The general formula for these expressions is

$$\begin{aligned}(\ldots)_l &\equiv \mathfrak{b}_l\{l!\lambda^{3(l-1)}V\} \\ &= l! \sum_{\{m_i\}}{}' (-1)^{\Sigma_i m_i - 1}\left[\left(\sum_i m_i - 1\right)! \prod_i \left\{\frac{(Z_i/i!)^{m_i}}{m_i!}\right\}\right],\end{aligned} \tag{10}$$

where the primed summation goes over all sets $\{m_i\}$ that conform to the condition

$$\sum_{i=1}^{l} i m_i = l; \quad m_i = 0, 1, 2, \ldots. \tag{11}$$

Relations inverse to (10) can be written down by referring to equation (10.1.29) of the classical treatment; thus

$$Z_M \equiv M!\lambda^{3M} Q_M = M!\lambda^{3M} {\sum_{\{m_l\}}}' \prod_{l=1}^{M} \left\{ \frac{(V \mathfrak{b}_l/\lambda^3)^{m_l}}{m_l!} \right\}, \tag{12}$$

where the primed summation goes over all sets $\{m_l\}$ that conform to the condition

$$\sum_{l=1}^{M} l m_l = M; \quad m_l = 0, 1, 2 \ldots. \tag{13}$$

The calculation of the virial coefficients a_l now consists of a straightforward step that involves a use of formulae (5) through (8) in conjunction with formulae (10.2.2) through (10.2.5). It appears, however, of interest to demonstrate here the manner in which the general relationship (10.2.6) between the virial coefficients a_l and the "irreducible cluster integrals" β_{l-1} arises mathematically. As a bonus, we will acquire yet another interpretation of the β_k.

Now, in view of the relations

$$\frac{P}{kT} \equiv \operatorname*{Lim}_{V\to\infty} \left(\frac{1}{V} \ln \mathcal{Q} \right) = \frac{1}{\lambda^3} \sum_{l=1}^{\infty} \mathfrak{b}_l z^l \tag{14}$$

and

$$\frac{1}{v} \equiv \operatorname*{Lim}_{V\to\infty} \left(\frac{z}{V} \frac{\partial \ln \mathcal{Q}}{\partial z} \right) = \frac{1}{\lambda^3} \sum_{l=1}^{\infty} l \mathfrak{b}_l z^l, \tag{15}$$

we can write

$$\frac{P(z)}{kT} = \int_0^z \frac{1}{v(z)} \frac{dz}{z}. \tag{16}$$

We introduce a new variable x, defined by

$$x = n\lambda^3 = \lambda^3/v. \tag{17}$$

In terms of this variable, equation (15) becomes

$$x(z) = \sum_{l=1}^{\infty} l \mathfrak{b}_l z^l, \tag{18}$$

the inverse of which may be written (see Mayer and Harrison, 1938; Harrison and Mayer, 1938; also Kahn, 1938) as

$$z(x) = x \exp\{-\phi(x)\}. \tag{19}$$

In view of the fact that, for $z \ll 1$, the variables z and x are practically the same, the function $\phi(x)$ must tend to zero as $x \to 0$; it may, therefore, be expressed as a power series in x:

$$\phi(x) = \sum_{k=1}^{\infty} \beta_k x^k. \tag{20}$$

It may be mentioned beforehand that the coefficients β_k of this expansion are ultimately going to be identified with the "irreducible cluster integrals" β_{l-1}. Substituting from equations (17), (19), and (20) into equation (16), we get

$$\begin{aligned}\frac{P(x)}{kT} &= \int_0^x \frac{x}{\lambda^3}\left\{\frac{1}{x} - \phi'(x)\right\} dx = \frac{1}{\lambda^3}\left[x - \int_0^x \left\{\sum_{k=1}^{\infty} k\beta_k x^k\right\} dx\right] \\ &= \frac{x}{\lambda^3}\left[1 - \sum_{k=1}^{\infty}\left(\frac{k}{k+1}\beta_k x^k\right)\right].\end{aligned} \tag{21}$$

Combining (17) and (21), we obtain

$$\frac{Pv}{kT} = 1 - \sum_{k=1}^{\infty}\left(\frac{k}{k+1}\beta_k x^k\right). \tag{22}$$

Comparing this result with the virial expansion (10.2.1), we arrive at the desired relationship:

$$a_l = -\frac{l-1}{l}\beta_{l-1} \qquad (l > 1). \tag{23}$$

For obvious reasons, the β_k appearing here may be regarded as a *generalization* of the irreducible cluster integrals of Mayer.

Finally, we would like to derive a relationship between the β_k and the $\mathfrak{b}_l$. For this, we make use of a theorem due to Lagrange which, for the present purpose, states that "the solution $x(z)$ to the equation

$$z(x) = x/f(x) \tag{24}$$

is given by the series

$$x(z) = \sum_{j=1}^{\infty} \frac{z^j}{j!}\left[\frac{d^{j-1}}{d\xi^{j-1}}\{f(\xi)\}^j\right]_{\xi=0}\text{''}; \tag{25}$$

it is obvious that the expression within the square brackets is $(j-1)!$ times "the coefficient of ξ^{j-1} in the Taylor expansion of the function $\{f(\xi)\}^j$ about the point $\xi = 0$." Applying this theorem to the function

$$f(x) = \exp\{\phi(x)\} = \exp\left\{\sum_{k=1}^{\infty} \beta_k x^k\right\} = \prod_{k=1}^{\infty} \exp(\beta_k x^k), \tag{26}$$

we obtain

$$x(z) = \sum_{j=1}^{\infty} \frac{z^j}{j!}(j-1)! \times \left\{\text{the coefficient of} \xi^{j-1} \text{ in the Taylor expansion} \right.$$

$$\left. \text{of} \prod_{k=1}^{\infty} \exp(j\beta_k \xi^k) \text{ about } \xi = 0\right\}.$$

Comparing this with equation (18), we get

$$\bar{b}_j = \frac{1}{j^2} \times \left\{\text{the coefficient of } \xi^{j-1} \text{ in } \prod_{k=1}^{\infty}\left[\sum_{m_k \geq 0} \frac{(j\beta_k)^{m_k}}{m_k!} \xi^{km_k}\right]\right\}$$

$$= \frac{1}{j^2} \sum_{\{m_k\}}{}' \prod_{k=1}^{j-1} \frac{(j\beta_k)^{m_k}}{m_k!}, \tag{27}$$

where the primed summation goes over all sets $\{m_k\}$ that conform to the condition

$$\sum_{k=1}^{j-1} k m_k = j-1; \quad m_k = 0, 1, 2, \ldots. \tag{28}$$

Formula (27) was first obtained by Maria Goeppert-Mayer in 1937. Its inverse, however, was established much later (Mayer et al., 1942; Kilpatrick, 1953):

$$\beta_{l-1} = \sum_{\{m_i\}}{}' (-1)^{\Sigma_i m_i - 1} \frac{(l-2+\Sigma_i m_i)!}{(l-1)!} \prod_i \frac{(i\bar{b}_i)^{m_i}}{m_i!}, \tag{29}$$

where the primed summation goes over all sets $\{m_i\}$ that conform to the condition

$$\sum_{i=2}^{l} (i-1) m_i = l-1; \quad m_i = 0, 1, 2, \ldots. \tag{30}$$

It is not difficult to see that the highest value of the index i in the set $\{m_i\}$ would be l (the corresponding set having all its m_i equal to 0, except m_l which would be equal to 1); accordingly, the highest order to which the quantities $\bar{b}_i$ would appear in the expression for β_{l-1} is that of $\bar{b}_l$. We thus see, once again, that the virial coefficient a_l is completely determined by the quantities $\bar{b}_1, \bar{b}_2, \ldots, \bar{b}_l$.

10.5 Exact treatment of the second virial coefficient

We now present a formulation, originally from Uhlenbeck and Beth (1936) and Beth and Uhlenbeck (1937), that enables us to make an exact calculation of the second virial coefficient of a quantum-mechanical system from a knowledge of the two-body potential $u(r)$.[9] In view of equation (10.4.6),

$$\bar{b}_2 = -a_2 = \frac{1}{2\lambda^3 V}\left(Z_2 - Z_1^2\right). \tag{1}$$

For the corresponding noninteracting system, one would have

$$\bar{b}_2^{(0)} = -a_2^{(0)} = \frac{1}{2\lambda^3 V}\left(Z_2^{(0)} - Z_1^{(0)2}\right); \tag{2}$$

the superscript (0) on the various symbols here implies that they pertain to the noninteracting system. Combining (1) and (2), and remembering that $Z_1 = Z_1^{(0)} = V$, we obtain

$$\bar{b}_2 - \bar{b}_2^{(0)} = \frac{1}{2\lambda^3 V}\left(Z_2 - Z_2^{(0)}\right) \tag{3}$$

which, by virtue of relation (10.4.2), becomes

$$\bar{b}_2 - \bar{b}_2^{(0)} = \frac{\lambda^3}{V}\left(Q_2 - Q_2^{(0)}\right) = \frac{\lambda^3}{V}\,\mathrm{Tr}\left(e^{-\beta \hat{H}_2} - e^{-\beta \hat{H}_2^{(0)}}\right). \tag{4}$$

For evaluating the trace in (4), we need to know the eigenvalues of the two-body Hamiltonian which, in turn, requires solving the Schrödinger equation[10]

$$\hat{H}_2 \Psi_\alpha(\boldsymbol{r}_1, \boldsymbol{r}_2) = E_\alpha \Psi_\alpha(\boldsymbol{r}_1, \boldsymbol{r}_2), \tag{5}$$

where

$$\hat{H}_2 = -\frac{\hbar^2}{2m}\left(\nabla_1^2 + \nabla_2^2\right) + u(r_{12}). \tag{6}$$

Transforming to the *center-of-mass* coordinates $\boldsymbol{R}\left\{= \frac{1}{2}(\boldsymbol{r}_1 + \boldsymbol{r}_2)\right\}$ and the *relative* coordinates $\boldsymbol{r}\{= (\boldsymbol{r}_2 - \boldsymbol{r}_1)\}$, we have

$$\Psi_\alpha(\boldsymbol{R}, \boldsymbol{r}) = \psi_j(\boldsymbol{R})\psi_n(\boldsymbol{r}) = \left\{\frac{1}{V^{1/2}} e^{i(\mathbf{P}_j \cdot \boldsymbol{R})/\hbar}\right\} \psi_n(\boldsymbol{r}), \tag{7}$$

with

$$E_\alpha = \frac{P_j^2}{2(2m)} + \varepsilon_n. \tag{8}$$

[9] For a discussion of the third virial coefficient, see Pais and Uhlenbeck (1959).

[10] For simplicity, we assume the particles to be "spinless." For the influence of spin, see Problems 10.11 and 10.12.

Here, $\boldsymbol{P}$ denotes the total momentum of the two particles and $2m$ their total mass, while ε denotes the energy associated with the relative motion of the particles; the symbol α refers to the set of quantum numbers j and n that determine the actual values of the variables $\boldsymbol{P}$ and ε. The wave equation for the relative motion will be

$$\left\{-\frac{\hbar^2}{2\left(\frac{1}{2}m\right)}\nabla_r^2 + u(r)\right\}\psi_n(\boldsymbol{r}) = \varepsilon_n\psi_n(\boldsymbol{r}), \tag{9}$$

$\frac{1}{2}m$ being the *reduced* mass of the particles; the normalization condition for the relative wavefunction will be

$$\int |\psi_n(\boldsymbol{r})|^2 d^3r = 1. \tag{10}$$

Equation (4) thus becomes

$$\begin{aligned} b_2 - b_2^{(0)} &= \frac{\lambda^3}{V}\sum_\alpha\left\{e^{-\beta E_\alpha} - e^{-\beta E_\alpha^{(0)}}\right\} \\ &= \frac{\lambda^3}{V}\sum_j e^{-\beta P_j^2/4m}\sum_n\left\{e^{-\beta\varepsilon_n} - e^{-\beta\varepsilon_n^{(0)}}\right\}. \end{aligned} \tag{11}$$

For the first sum, we obtain

$$\sum_j e^{-\beta P_j^2/4m} \approx \frac{4\pi V}{h^3}\int_0^\infty e^{-\beta P^2/4m}P^2 dP = \frac{8^{1/2}V}{\lambda^3}, \tag{12}$$

so that equation (11) becomes

$$b_2 - b_2^{(0)} = 8^{1/2}\sum_n\left\{e^{-\beta\varepsilon_n} - e^{-\beta\varepsilon_n^{(0)}}\right\}. \tag{13}$$

The next step consists of examining the energy spectra, ε_n and $\varepsilon_n^{(0)}$, of the two systems. In the case of a noninteracting system, all we have is a "continuum"

$$\varepsilon_n^{(0)} = \frac{p^2}{2\left(\frac{1}{2}m\right)} = \frac{\hbar^2k^2}{m} \quad (k = p/\hbar), \tag{14}$$

with the standard density of states $g^{(0)}(k)$. In the case of an interacting system, we may have a set of *discrete eigenvalues* ε_B (that correspond to "bound" states), along with a "continuum"

$$\varepsilon_n = \frac{\hbar^2k^2}{m} \quad (k = p/\hbar), \tag{15}$$

with a *characteristic* density of states $g(k)$. Consequently, equation (13) can be written as

$$b_2 - b_2^{(0)} = 8^{1/2}\sum_B e^{-\beta\varepsilon_B} + 8^{1/2}\int_0^\infty e^{-\beta\hbar^2k^2/m}\{g(k) - g^{(0)}(k)\}\,dk, \tag{16}$$

where the summation in the first part goes over all bound states made possible by the two-body interaction.

The next thing to consider here is the density of states $g(k)$. For this, we note that, since the two-body potential is assumed to be central, the wavefunction $\psi_n(\boldsymbol{r})$ for the relative motion may be written as a product of a radial function $\chi(r)$ and a spherical harmonic $Y(\theta,\varphi)$:

$$\psi_{klm}(\boldsymbol{r}) = A_{klm}\frac{\chi_{kl}(r)}{r}Y_{l,m}(\theta,\varphi). \tag{17}$$

Moreover, the requirement of symmetry, namely $\psi(-\boldsymbol{r}) = \psi(\boldsymbol{r})$ for bosons and $\psi(-\boldsymbol{r}) = -\psi(\boldsymbol{r})$ for fermions, imposes the restriction that the quantum number l be even for bosons and odd for fermions. The (outer) boundary condition on the wavefunction may be written as

$$\chi_{kl}(R_0) = 0, \tag{18}$$

where R_0 is a fairly large value (of the variable r) that ultimately goes to infinity. Now, the asymptotic form of the function $\chi_{kl}(r)$ is well-known:

$$\chi_{kl}(r) \propto \sin\left\{kr - \frac{l\pi}{2} + \eta_l(k)\right\}; \tag{19}$$

accordingly, we must have

$$kR_0 - \frac{l\pi}{2} + \eta_l(k) = n\pi, \quad n = 0,1,2,\ldots. \tag{20}$$

The symbol $\eta_l(k)$ here stands for the *scattering phase shift* due to the two-body potential $u(r)$ for the lth partial wave of wave number k.

Equation (20) determines the full spectrum of the partial waves. To obtain from it an expression for the density of states $g_l(k)$, we observe that the wave number difference Δk between two consecutive states n and $n+1$ is given by the formula

$$\left\{R_0 + \frac{d\eta_l(k)}{dk}\right\}\Delta k = \pi, \tag{21}$$

with the result that

$$g_l(k) = \frac{2l+1}{\Delta k} = \frac{2l+1}{\pi}\left\{R_0 + \frac{\partial\eta_l(k)}{\partial k}\right\}; \tag{22}$$

the factor $(2l+1)$ has been included here to take account of the fact that each eigenvalue k pertaining to an lth partial wave is $(2l+1)$-fold degenerate (because the magnetic quantum number m can take *any* of the values $l, (l-1), \ldots, -l$, without affecting the eigenvalue). The total density of states, $g(k)$, of *all* partial waves of wave numbers around the value k is then given by

$$g(k) = \sum_l{}' g_l(k) = \frac{1}{\pi}\sum_l{}' (2l+1)\left\{R_0 + \frac{\partial \eta_l(k)}{\partial k}\right\}; \tag{23}$$

note that the primed summation $\sum'$ goes over $l = 0, 2, 4, \ldots$ in the case of bosons and over $l = 1, 3, 5, \ldots$ in the case of fermions. For the corresponding noninteracting case, we have (since all $\eta_l(k) = 0$)

$$g^{(0)}(k) = \frac{R_0}{\pi}\sum_l{}' (2l+1). \tag{24}$$

Combining (23) and (24), we obtain

$$g(k) - g^{(0)}(k) = \frac{1}{\pi}\sum_l{}' (2l+1)\frac{\partial \eta_l(k)}{\partial k}. \tag{25}$$

Substituting (25) into (16), we obtain the desired result

$$b_2 - b_2^{(0)} = 8^{1/2}\sum_B e^{-\beta\varepsilon_B} + \frac{8^{1/2}}{\pi}\sum_l{}' (2l+1)\int_0^\infty e^{-\beta\hbar^2k^2/m}\frac{\partial \eta_l(k)}{\partial k}dk \tag{26}$$

which, in principle, is calculable for any given potential $u(r)$ through the respective phase shifts $\eta_l(k)$.

Equation (26) can be used for determining the quantity $b_2 - b_2^{(0)}$. To determine b_2 itself, we must know the value of $b_2^{(0)}$. This has already been obtained in Section 7.1 for bosons and in Section 8.1 for fermions; see equations (7.1.13) and (8.1.17). Thus

$$b_2^{(0)} = -a_2^{(0)} = \pm\frac{1}{2^{5/2}}, \tag{27}$$

where the upper sign holds for bosons and the lower sign for fermions. It is worthwhile to note that the foregoing result can be obtained directly from the relationship

$$b_2^{(0)} = \frac{1}{2\lambda^3 V}\left(Z_2^{(0)} - Z_1^{(0)2}\right) = \frac{\lambda^3}{V}\left(Q_2^{(0)} - \frac{1}{2}Q_1^{(0)2}\right)$$

by substituting for $Q_2^{(0)}$ the exact expression (5.5.25):

$$\mathfrak{b}_2^{(0)} = \frac{\lambda^3}{V}\left[\left\{\frac{1}{2}\left(\frac{V}{\lambda^3}\right)^2 \pm \frac{1}{2^{5/2}}\left(\frac{V}{\lambda^3}\right)^1\right\} - \frac{1}{2}\left(\frac{V}{\lambda^3}\right)^2\right] = \pm\frac{1}{2^{5/2}}. \tag{28}$$[11]

It is of interest to note that this result can also be obtained by using the classical formula (10.1.18) and substituting for the two-body potential $u(r)$ the "statistical potential" (5.5.28); thus

$$\mathfrak{b}_2^{(0)} = \frac{2\pi}{\lambda^3}\int_0^\infty \left(e^{-u_s(r)/kT} - 1\right) r^2 dr$$
$$= \pm\frac{2\pi}{\lambda^3}\int_0^\infty e^{-2\pi r^2/\lambda^2} r^2 dr = \pm\frac{1}{2^{5/2}}. \tag{29}$$

As an illustration of this method, we now calculate the second virial coefficient of a gas of hard spheres. The two-body potential in this case may be written as

$$u(r) = \begin{cases} +\infty & \text{for} \quad r < D \\ 0 & \text{for} \quad r > D. \end{cases} \tag{30}$$

The scattering phase shifts $\eta_l(k)$ can now be determined by making use of the (inner) boundary condition, namely $\chi(r) = 0$ for all $r < D$ and hence it vanishes as $r \to D$ from the above. We thus obtain (see, for example, Schiff, 1968)

$$\eta_l(k) = \tan^{-1}\frac{j_l(kD)}{n_l(kD)}, \tag{31}$$

where $j_l(x)$ and $n_l(x)$ are, respectively, the "spherical Bessel functions" and the "spherical Neumann functions":

$$j_0(x) = \frac{\sin x}{x}, \quad j_1(x) = \frac{\sin x - x\cos x}{x^2},$$
$$j_2(x) = \frac{(3 - x^2)\sin x - 3x\cos x}{x^3}, \ldots$$

and

$$n_0(x) = -\frac{\cos x}{x}, \quad n_1(x) = -\frac{\cos x + x\sin x}{x^2},$$
$$n_2(x) = -\frac{(3 - x^2)\cos x + 3x\sin x}{x^3}, \ldots.$$

[11] This calculation incidentally verifies the general formula (10.4.9) for the case $l = 2$. By that formula, the "cluster integral" $\mathfrak{b}_2$ of a given system would be equal to $1/(2\lambda^3)$ times the coefficient of V^1 in the volume expansion of the "configuration integral" Z_2 of the system. In the case under study, this coefficient is $\pm\lambda^3/2^{3/2}$; hence the result.

Accordingly,

$$\eta_0(k) = \tan^{-1}\{-\tan(kD)\} = -kD, \tag{32}$$

$$\eta_1(k) = \tan^{-1}\left\{-\frac{\tan(kD) - kD}{1 + kD\tan(kD)}\right\} = -\{kD - \tan^{-1}(kD)\}$$

$$= -\frac{(kD)^3}{3} + \frac{(kD)^5}{5} - \cdots, \tag{33}$$

$$\eta_2(k) = \tan^{-1}\left\{-\frac{\tan(kD) - 3(kD)/[3-(kD)^2]}{1 + 3(kD)\tan(kD)/[3-(kD)^2]}\right\}$$

$$= -\left\{kD - \tan^{-1}\frac{3(kD)}{3-(kD)^2}\right\} = -\frac{(kD)^5}{45} + \cdots, \tag{34}$$

and so on. We now have to substitute these results into formula (26). However, before doing that we should point out that, in the case of hard-sphere interaction, (i) we cannot have bound states at all and (ii) since, for all $l, \eta_l(0) = 0$, the integral in (26) can be simplified by a prior integration by parts. Thus, we have

$$b_2 - b_2^{(0)} = \frac{8^{1/2}\lambda^2}{\pi^2}\sum_l{}'(2l+1)\int_0^\infty e^{-\beta\hbar^2k^2/m}\eta_l(k)k\,dk. \tag{35}$$

Substituting for $l = 0$ and 2 in the case of bosons and for $l = 1$ in the case of fermions, we obtain (to fifth power in D/λ)

$$b_2 - b_2^{(0)} = -2\left(\frac{D}{\lambda}\right)^1 - \frac{10\pi^2}{3}\left(\frac{D}{\lambda}\right)^5 - \cdots \quad \text{(Bose)} \tag{36}$$

$$= -6\pi\left(\frac{D}{\lambda}\right)^3 + 18\pi^2\left(\frac{D}{\lambda}\right)^5 - \cdots \quad \text{(Fermi)}, \tag{37}$$

which may be compared with the corresponding classical result $-(2\pi/3)(D/\lambda)^3$.

10.6 Cluster expansion for a quantum-mechanical system

When it comes to calculating b_l for $l > 2$ we have no formula comparable in simplicity to formula (10.5.26) for b_2. This is due to the fact that we have no treatment of the l-body problem (for $l > 2$) that is as neat as the phase-shift analysis of the two-body problem. Nevertheless, a formal theory for the calculation of higher-order "cluster integrals" has been developed by Kahn and Uhlenbeck (1938); an elaboration by Lee and Yang (1959a,b; 1960a,b,c) has made this theory almost as good for treating a quantum-mechanical system as Mayer's theory has been for a classical gas. The basic approach in this theory is to evolve a scheme for expressing the grand partition function of the given system in essentially the same way as Mayer's cluster expansion does for a classical gas. However, because of the interplay of quantum-statistical effects and the effects arising from interparticle interactions, the mathematical structure of this theory is considerably involved.

We consider here a quantum-mechanical system of N identical particles enclosed in a box of volume V. The Hamiltonian of the system is assumed to be of the form

$$\hat{H}_N = -\frac{\hbar^2}{2m}\sum_{i=1}^{N}\nabla_i^2 + \sum_{i<j} u(r_{ij}). \tag{1}$$

Now, the partition function of the system is given by

$$\begin{aligned} Q_N(V,T) \equiv \text{Tr}(e^{-\beta \hat{H}_N}) &= \sum_\alpha e^{-\beta E_\alpha} \\ &= \sum_\alpha \int_V \{\Psi_\alpha^*(1,\ldots,N) e^{-\beta \hat{H}_N} \Psi_\alpha(1,\ldots,N)\} d^{3N}r, \end{aligned} \tag{2}$$

where the functions Ψ_α are supposed to form a complete set of (properly symmetrized) orthonormal wavefunctions of the system, while the numbers $1,\ldots,N$ denote the position coordinates $\boldsymbol{r}_1,\ldots,\boldsymbol{r}_N$, respectively. We may as well introduce the *probability density operator* $\hat{W}_N$ of the system through the matrix elements

$$\begin{aligned} \langle 1',\ldots,N'|\hat{W}_N|1,\ldots,N\rangle &\equiv N!\lambda^{3N}\sum_\alpha \{\Psi_\alpha(1',\ldots,N') e^{-\beta \hat{H}_N} \Psi_\alpha^*(1,\ldots,N)\} \\ &= N!\lambda^{3N}\sum_\alpha \{\Psi_\alpha(1',\ldots,N')\Psi_\alpha^*(1,\ldots,N)\} e^{-\beta E_\alpha}. \end{aligned} \tag{3}$$

We denote the diagonal elements of the operator $\hat{W}_N$ by the symbols $W_N(1,\ldots,N)$; thus

$$W_N(1,\ldots,N) = N!\lambda^{3N}\sum_\alpha \{\Psi_\alpha(1,\ldots,N)\Psi_\alpha^*(1,\ldots,N)\} e^{-\beta E_\alpha}, \tag{4}$$

whereby equation (2) takes the form

$$Q_N(V,T) = \frac{1}{N!\lambda^{3N}} \int_V W_N(1,\ldots,N) d^{3N}r = \frac{1}{N!\lambda^{3N}}\, \text{Tr}\,(\hat{W}_N). \tag{5}$$

A comparison of equation (5) with equations (10.1.3) and (10.4.2) shows that the "trace of the probability density operator $\hat{W}_N$" is the analogue of the "configuration integral" Z_N, and the quantity $W_N(1,\ldots,N)d^{3N}r$ is a measure of the probability that the "configuration" of the given system is found to be within the interval $[(\boldsymbol{r}_1,\ldots,\boldsymbol{r}_N),(\boldsymbol{r}_1+d\boldsymbol{r}_1,\ldots,\boldsymbol{r}_N+d\boldsymbol{r}_N)]$.

Before we proceed further, let us acquaint ourselves with some of the basic properties of the matrix elements (3):

(i)
$$\begin{aligned}\langle 1'|\hat{W}_1|1\rangle &= \lambda^3 \sum_{\boldsymbol{p}} \left\{ \frac{1}{\sqrt{V}} e^{i(\boldsymbol{p}\cdot\boldsymbol{r}_1')/\hbar} \frac{1}{\sqrt{V}} e^{-i(\boldsymbol{p}\cdot\boldsymbol{r}_1)/\hbar} \right\} e^{-\beta p^2/2m} \\ &\simeq \frac{\lambda^3}{V} \iiint_{-\infty}^{+\infty} \frac{V d^3p}{h^3} e^{\{i\boldsymbol{p}\cdot(\boldsymbol{r}_1'-\boldsymbol{r}_1)/\hbar - \beta p^2/2m\}} \\ &= e^{-\pi|\boldsymbol{r}_1'-\boldsymbol{r}_1|^2/\lambda^2}; \end{aligned} \tag{6}$$

compare with equation (5.3.14) for the density matrix of a single particle. The foregoing result is a manifestation of the quantum-mechanical, *not quantum-statistical*, correlation between the positions $\boldsymbol{r}$ and $\boldsymbol{r}'$ of a given particle (or, for that matter, any particle in the system). This correlation extends over distances of the order of λ which is, therefore, a measure of the linear dimensions of the wave packet representing the particle. As $T \to \infty$, and hence $\lambda \to 0$, the matrix element (6) tends to zero for all finite values of $|\boldsymbol{r}_1' - \boldsymbol{r}_1|$.

(ii)
$$\langle 1|\hat{W}_1|1\rangle = 1; \tag{7}$$

consequently, by equation (5),

$$Q_1(V,T) = \frac{1}{\lambda^3} \int_V 1 d^3r = \frac{V}{\lambda^3}. \tag{8}$$

(iii) Whatever the symmetry character of the wavefunctions Ψ, the diagonal elements $W_N(1,\ldots,N)$ of the probability density operator $\hat{W}_N$ are *symmetric* in respect of a permutation among the arguments $(1,\ldots,N)$.

(iv) The elements $W_N(1,\ldots,N)$ are *invariant* under a unitary transformation of the set $\{\Psi_\alpha\}$.

(v) Suppose that the coordinates $\boldsymbol{r}_1,\ldots,\boldsymbol{r}_N$ are such that they can be divided into two groups, A and B, with the property that *any* two coordinates, say $\boldsymbol{r}_i$ and $\boldsymbol{r}_j$, of which one belongs to group A and the other to group B, satisfy the conditions that
   (a) the separation r_{ij} is much larger than the mean thermal wavelength λ of the particles, and
   (b) it is also much larger than the effective range r_0 of the two-body potential, then

$$W_N(\boldsymbol{r}_1,\ldots,\boldsymbol{r}_N) \simeq W_A(\boldsymbol{r}_A) W_B(\boldsymbol{r}_B), \tag{9}$$

where $\boldsymbol{r}_A$ and $\boldsymbol{r}_B$ denote *collectively* the coordinates in group A and group B, respectively. It is not easy to furnish here a rigorous mathematical proof of this property, though physically it is quite understandable. One can see this by noting that, in view of conditions (a) and (b), there does not exist any spatial correlation between the particles of group A on one hand and the particles of group B on the other (either by virtue of statistics or by virtue of interparticle interactions). The two groups, therefore, behave toward each other like two *independent* entities. It is then natural that, to a very good approximation, the probability density W_N of the composite configuration be equal to the product of the probability densities W_A and W_B.

We now proceed with the formulation. First of all, to fix ideas about the approach to be followed, we may consider the case with $N = 2$. In that case, as $r_{12} \to \infty$, we expect, in view of property (v), that

$$W_2(1,2) \to W_1(1)W_1(2) = 1. \tag{10}$$

In general, however, $W_2(1,2)$ will be different from $W_1(1)W_1(2)$. Now, if we denote the difference between $W_2(1,2)$ and $W_1(1)W_1(2)$ by the symbol $U_2(1,2)$, then, as $r_{12} \to \infty$,

$$U_2(1,2) \to 0. \tag{11}$$

It is not difficult to see that the quantity $U_2(1,2)$ is the quantum-mechanical analogue of the Mayer function f_{ij}. With this in mind, we introduce a sequence of *cluster functions* $\hat{U}_l$ defined by the hierarchy[12]

$$\langle 1'|\hat{W}_1|1\rangle = \langle 1'|\hat{U}_1|1\rangle, \tag{12}$$

$$\langle 1',2'|\hat{W}_2|1,2\rangle = \langle 1'|\hat{U}_1|1\rangle\langle 2'|\hat{U}_1|2\rangle + \langle 1',2'|\hat{U}_2|1,2\rangle, \tag{13}$$

$$\begin{aligned}\langle 1',2',3'|\hat{W}_3|1,2,3\rangle = {} & \langle 1'|\hat{U}_1|1\rangle\langle 2'|\hat{U}_1|2\rangle\langle 3'|\hat{U}_1|3\rangle \\ & + \langle 1'|\hat{U}_1|1\rangle\langle 2',3'|\hat{U}_2|2,3\rangle \\ & + \langle 2'|\hat{U}_1|2\rangle\langle 1',3'|\hat{U}_2|1,3\rangle \\ & + \langle 3'|\hat{U}_1|3\rangle\langle 1',2'|\hat{U}_2|1,2\rangle \\ & + \langle 1',2',3'|\hat{U}_3|1,2,3\rangle,\end{aligned} \tag{14}$$

and so on. A particular $\hat{U}_l$ is thus defined with the help of the first l equations of the hierarchy. The last equation in this hierarchy will be (writing only the diagonal elements)

$$W_N(1,\ldots,N) = \sum_{\{m_l\}}{}' \left\{ \sum_P \underbrace{[U_1()\cdots U_1()]}_{m_1 \text{ factors}} \underbrace{[U_2()\cdots U_2()]}_{m_2 \text{ factors}} \cdots \right\}, \tag{15}$$

where the primed summation goes over all sets $\{m_l\}$ that conform to the condition

$$\sum_{l=1}^{N} l m_l = N; \quad m_l = 0, 1, 2, \ldots. \tag{16}$$

[12]The functions U_l were first introduced by Ursell, in 1927, in order to simplify the classical configuration integral. Their introduction into the quantum-mechanical treatment is due to Kahn and Uhlenbeck (1938).

Moreover, in selecting the arguments of the various U_l appearing in (15), out of the numbers $1,\ldots,N$, one has to remember that a permutation of the arguments within the same bracket is *not* regarded as leading to anything distinctly different from what one had before the permutation; the symbol $\sum_P$ then denotes a summation over all *distinct* ways of selecting the arguments under the set $\{m_l\}$.

Relations inverse to the preceding ones are easy to obtain. One gets

$$\langle 1'|\hat{U}_1|1\rangle = \langle 1'|\hat{W}_1|1\rangle, \tag{17}$$

$$\langle 1',2'|\hat{U}_2|1,2\rangle = \langle 1',2'|\hat{W}_2|1,2\rangle - \langle 1'|\hat{W}_1|1\rangle\langle 2'|\hat{W}_1|2\rangle, \tag{18}$$

$$\begin{aligned}\langle 1',2',3'|\hat{U}_3|1,2,3\rangle = {} & \langle 1',2',3'|\hat{W}_3|1,2,3\rangle \\ & - \langle 1'|\hat{W}_1|1\rangle\langle 2',3'|\hat{W}_2|2,3\rangle \\ & - \langle 2'|\hat{W}_1|2\rangle\langle 1',3'|\hat{W}_2|1,3\rangle \\ & - \langle 3'|\hat{W}_1|3\rangle\langle 1',2'|\hat{W}_2|1,2\rangle \\ & + 2\langle 1'|\hat{W}_1|1\rangle\langle 2'|\hat{W}_1|2\rangle\langle 3'|\hat{W}_1|3\rangle,\end{aligned} \tag{19}$$

and so on; compare the right sides of these equations with the expressions appearing within the parentheses in equations (10.4.5) through (10.4.7). We note that (i) the coefficient of a general term here is

$$(-1)^{\sum_l m_l - 1}\left(\sum_l m_l - 1\right)!, \tag{20}$$

where $\sum_l m_l$ is the number of the W_n in the term, and (ii) the sum of the coefficients of all the terms on the right side of equations (18), (19), ... is identically zero. Moreover, the diagonal elements $U_l(1,\ldots,l)$, just like the diagonal elements of the operators $\hat{W}_n$, are symmetric in respect of permutations among the arguments $(1,\ldots,l)$, and are determined by the sequence of the diagonal elements $W_1, W_2, \ldots, W_l$. Finally, in view of property (v) of the W_n, as embodied in formula (9), the U_l possess the following property:

$$U_l(1,\ldots,l) \simeq 0 \quad \text{if} \quad r_{ij} \gg \lambda, r_0; \tag{21}$$

here, r_{ij} is the separation between any two of the coordinates $(1,\ldots,l)$.[13]

We now define the "cluster integral" b_l by the formula

$$b_l(V,T) = \frac{1}{l!\,\lambda^{3(l-1)}V}\int U_l(1,\ldots,l)\,d^{3l}r; \tag{22}$$

compare with equation (10.1.16). Clearly, the quantity $b_l(V,T)$ is dimensionless and, by virtue of property (21) of the diagonal elements $U_l(1,\ldots,l)$, is practically independent of V (so long as V is large). In the limit $V \to \infty$, $b_l(V,T)$ tends to a finite volume-independent

[13]This can be seen by examining the break-up of the structure on the right side of any equation in the hierarchy (18, 19, ...) when one or more of the l coordinates in the "cluster" get sufficiently separated from the rest.

value, which may be denoted by $\mathfrak{b}_l(T)$. We then obtain for the partition function of the system, see equations (5) and (15),

$$Q_N(V,T) = \frac{1}{N!\lambda^{3N}} \int d^{3N}r \left\{ \sum_{\{m_l\}}' \left[\sum_P [U_1 \cdots U_1][U_2 \cdots U_2] \cdots \right] \right\} \tag{23}$$

$$= \frac{1}{N!\lambda^{3N}} \sum_{\{m_l\}}' \frac{N!}{(1!)^{m_1}(2!)^{m_2} \cdots m_1! m_2! \cdots} \times \int d^{3N}r \{[U_1 \cdots U_1][U_2 \cdots U_2] \cdots\}. \tag{24}$$

In writing the last result we have made use of the fact that, since a permutation among the arguments of the functions U_l does not affect the value of the integral concerned, the summation over P may be replaced by any one term of the summation, multiplied by the number of *distinct* permutations allowed by the set $\{m_l\}$; compare with the corresponding product of the numbers (10.1.22) and (10.1.24). Making use of the definition (22), equation (24) can be written as

$$Q_N(V,T) = \frac{1}{\lambda^{3N}} \sum_{\{m_l\}}' \left[\prod_{l=1}^{N} \{ (b_l \lambda^{3(l-1)} V)^{m_l} / m_l! \} \right]$$

$$= \sum_{\{m_l\}}' \left[\prod_{l=1}^{N} \left\{ \left(b_l \frac{V}{\lambda^3} \right)^{m_l} \frac{1}{m_l!} \right\} \right]; \tag{25}$$

again, use has been made of the fact that

$$\prod_l (\lambda^{3l})^{m_l} = \lambda^{3\sum_l l m_l} = \lambda^{3N}. \tag{26}$$

Equation (25) is formally identical to equation (10.1.29) of Mayer's theory. The subsequent development of the formalism, leading to the equation of state of the system, is formally identical to that theory. Thus, we finally obtain

$$\frac{P}{kT} = \frac{1}{\lambda^3} \sum_{l=1}^{\infty} \mathfrak{b}_l z^l \quad \text{and} \quad \frac{1}{v} = \frac{1}{\lambda^3} \sum_{l=1}^{\infty} l \mathfrak{b}_l z^l. \tag{27}$$

There are, however, important physical differences. We may recall that the calculation of the cluster integrals $\mathfrak{b}_l$ in the classical case involved the evaluation of a number of finite, $3l$-dimensional integrals. The corresponding calculation in the quantum-mechanical case requires a knowledge of the functions U_l and hence of all W_n, with $n \leq l$; this in turn requires solutions of the n-body Schrödinger equation for all $n \leq l$. The case $l = 2$ can be handled neatly, as was done in Section 10.5. For $l > 2$, the mathematical procedure is rather cumbersome. Nevertheless, Lee and Yang (1959a,b; 1960a,b,c) have evolved a scheme that enables us to calculate the higher $\mathfrak{b}_l$ in successive approximations. According to that scheme, the functions U_l of a given system can be evaluated by "separating out" the effects of statistics from those of interparticle interactions, that is, we first take care of the statistical aspect of the problem and then tackle the dynamical aspect of it. Thus, the whole feat is accomplished in two steps.

First, the U-functions pertaining to the given system are expressed in terms of U-functions pertaining to a corresponding *quantum-mechanical system obeying Boltzmann statistics*, that is, a (fictitious) system described by *unsymmetrized wavefunctions*. This step takes care of the statistics of the given system, that is, of the symmetry properties of the wavefunctions describing the system. Next, the U-functions of the (fictitious) Boltzmannian system are expanded, loosely speaking, in powers of a *binary kernel B* which is obtainable from a solution of the two-body problem with the given interaction. A commendable feature of this method is that it can be applied even if the given interaction contains a singular, repulsive core, that is, even if the potential energy for certain configurations of the system becomes infinitely large. Though the method is admirably systematic and fairly straightforward in principle, its application to real systems is quite complicated. We will, therefore, turn to a more practical method — the method of quantized fields (see Chapter 11) — which has been extremely useful in the study of quantum-mechanical systems composed of interacting particles. For a detailed exposition of the (binary collision) method of Lee and Yang, see Sections 9.7 and 9.8 of the first edition of this book.

In passing, we note yet another important difference between the quantum-mechanical case and the classical one. In the latter case, if interparticle interactions are absent, then all b_l, with $l \geq 2$, vanish. This is not true in the quantum-mechanical case; here, see Sections 7.1 and 8.1,

$$b_l^{(0)} = (\pm 1)^{l-1} l^{-5/2}, \tag{28}$$

of which equation (10.5.27) was a special case.

10.7 Correlations and scattering

Correlations and scattering play an extremely important role in modern statistical mechanics. Different phases most are easily distinguished by different spatial orderings they display. Molecules in a low-density vapor are nearly uncorrelated whereas molecules in a dense liquid can be strongly correlated and display short-range order due to their strong steric repulsions but the correlations decay away rapidly at large distances. In crystalline solids, the location of every particle is highly correlated with the location of all the others, and these correlations do not decay away to zero at large distances between the particles; this is called long-range order. At a critical point, systems display order that lies between short-range and long-range, with so-called quasi-long-range order characterized by a power-law decay of correlations. Crystals and liquid-crystal phases display molecular orientational correlations that can be short-range, long-range, or quasi-long-range in addition to the various spatial orderings of the molecules. Different phases of magnets are distinguished by the spatial orderings of the magnetic dipoles: short-range ordering in paramagnets, long-range ordering in ferromagnets and antiferromagnets, and power-law decay of correlations at magnetic critical points.

Spatial correlation functions are based on n-particle densities. The one-body number density is defined by the average quantity

$$n_1(\boldsymbol{r}) = \left\langle \sum_i \delta(\boldsymbol{r} - \boldsymbol{r}_i) \right\rangle. \tag{1}$$

This defines the local number density in which $n_1(\boldsymbol{r})d\boldsymbol{r}$ is a measure of the probability of finding a particle inside an infinitesimal volume $d\boldsymbol{r}$ located at position $\boldsymbol{r}$. If the system is translationally invariant, the one-body density is the usual number density $n_1(\boldsymbol{r}) = n = \langle N \rangle / V$. The spatial integral of the one-body density over volume V gives the average number of particles in that volume:

$$\int n_1(\boldsymbol{r})d\boldsymbol{r} = \langle N \rangle. \tag{2}$$

The two-body number density is defined as

$$n_2(\boldsymbol{r}, \boldsymbol{r}') = \left\langle \sum_{i \neq j} \delta(\boldsymbol{r} - \boldsymbol{r}_i)\delta(\boldsymbol{r}' - \boldsymbol{r}_j) \right\rangle. \tag{3}$$

The quantity $n_2(\boldsymbol{r}, \boldsymbol{r}')d\boldsymbol{r}d\boldsymbol{r}'$ is a measure of the probability of finding one particle inside the infinitesimal volume $d\boldsymbol{r}$ located at position $\boldsymbol{r}$ and *another* particle inside the infinitesimal volume $d\boldsymbol{r}'$ located at position $\boldsymbol{r}'$. In a dilute classical gas, the particles interact only when they are close to one another, so the probability of finding two different particles at two different locations many atomic diameters apart is simply the product of finding either particle individually, that is, $n_2(\boldsymbol{r}, \boldsymbol{r}') \to n_1(\boldsymbol{r})n_1(\boldsymbol{r}')$ as $|\boldsymbol{r} - \boldsymbol{r}'| \to \infty$. It is the deviation from this uncorrelated behavior that is both interesting and important. The integral of the two-body density over volume V gives

$$\int n_2(\boldsymbol{r}, \boldsymbol{r}')d\boldsymbol{r}d\boldsymbol{r}' = \left\langle N^2 \right\rangle - \langle N \rangle. \tag{4}$$

If the system is translationally and rotationally invariant, the one-body number density is independent of position and the two-body number density depends only on the magnitude of the distance between $\boldsymbol{r}$ and $\boldsymbol{r}'$. This allows us to define the *pair correlation function* $g(r)$:

$$n_2(\boldsymbol{r}, \boldsymbol{r}') = n^2 g\left(|\boldsymbol{r} - \boldsymbol{r}'|\right). \tag{5}$$

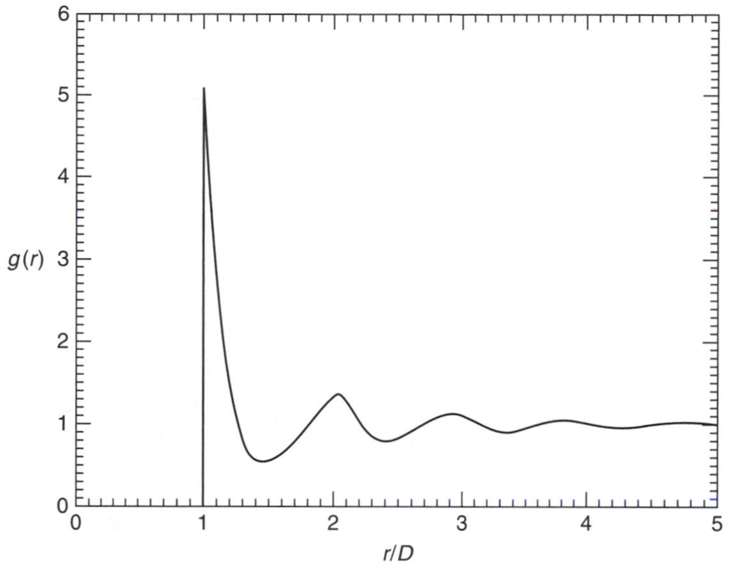

FIGURE 10.5 An approximate pair correlation function for hard spheres with diameter D in three dimensions. The volume fraction $\eta = \pi n D^3/6 \simeq 0.49$ is the fraction of the volume occupied by the particles and is close to the liquid side of the solid-liquid phase transition in the model. The correlation function is calculated using the exact solution of the Percus–Yevick approximation; see Percus and Yevick (1958), Wertheim (1963), and Hansen and McDonald (1986). The correlation length for this case is $\xi \approx 2D$.

In three dimensions, $4\pi n r^2 g(r)\,dr$ is the probability of finding a particle in a spherical shell of radius r and thickness dr, given that another particle is simultaneously located at the origin. The pair correlation function of a classical ideal gas is equal to unity; see the footnote to Problem 10.17.

Figure 10.5 displays the pair correlation function $g(r)$ for a system of hard spheres interacting via pair potential

$$u(r) = \begin{cases} 0 & \text{if } r > D, \\ \infty & \text{if } r \leq D. \end{cases} \tag{6}$$

Clearly, the pair correlation function vanishes for $r < D$ since no two particles in the system can be closer to each other than D due to the infinite repulsion. These steric repulsions result in an oscillatory decay of $g(r)$. The pair correlation function is greater than unity at separations slightly greater than D since the local geometry of the fluid enhances the probability of finding two particles a distance slightly more than D apart; for illustration, see Figure 10.6. The pair correlation function is less than unity at slightly larger distances due to the repulsion of the cluster of particles just outside the hard repulsion distance. The oscillating correlations decay rapidly with distance, so that $g(r)$ approaches unity at large separations. This behavior of the pair correlation function is typical of all dense fluids

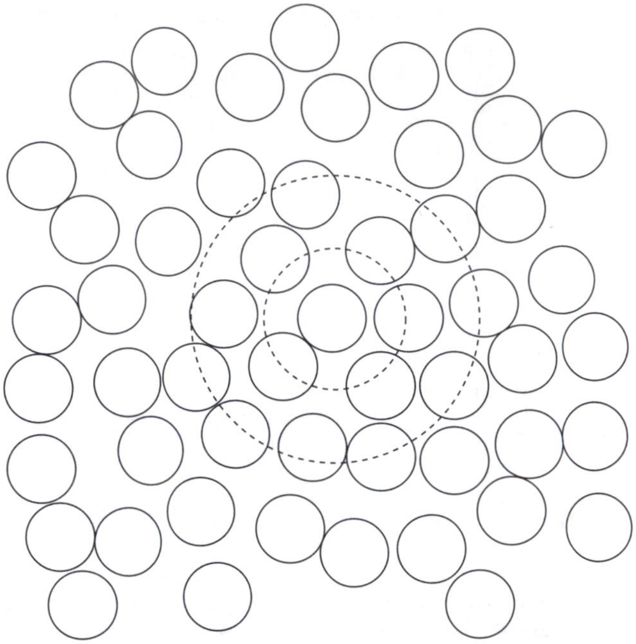

FIGURE 10.6 An equilibrium configuration of hard disks that displays steric effects leading to oscillations in the pair correlation function. The inner dashed circle with radius D is the closest approach distance to the central disk. In this case, the centers of five disks are close to the distance D which contributes to the enhancement in $g(r)$ near $r = D$. The outer dashed circle shows the next shell of particles that contribute to the second peak in $g(r)$. In-between these distances, we have a reduced probability of finding the center of a particle, leading to $g(r) < 1$.

and is called short-range order since the correlations decay exponentially with distance: $g(r) - 1 \sim \exp(-r/\xi)$, where ξ is called the correlation length.

The pair correlation function can be used to directly calculate the pressure in a fluid. For a classical fluid whose potential energy can be written as a sum of pair potentials,

$$U_N(\boldsymbol{r}_1, \boldsymbol{r}_2, \ldots, \boldsymbol{r}_N) = \sum_{i<j} u(r_{ij}), \tag{7}$$

the pressure is determined by the average of the quantity $r(\partial u/\partial r)$ between pairs of particles, as discussed in Section 3.7. In the canonical ensemble, the pressure P is given by

$$P \equiv -\left(\frac{\partial A}{\partial V}\right)_{T,N} = \frac{kT}{Z_N}\left(\frac{\partial Z_N}{\partial V}\right)_{T,N}, \tag{8}$$

where Z_N is the configurational partition function

$$Z_N = \frac{1}{N!}\int d^N\boldsymbol{r}\, \exp\left(-\beta \sum_{i<j} u(r_{ij})\right). \tag{9}$$

The d-dimensional integrals over the volume V can be rewritten in terms of a set of scaled variables $\{\boldsymbol{s}_i\}$ defined by $\boldsymbol{r}_i = V^{1/d}\boldsymbol{s}_i$, so the scaled integrals are over regions with unit volume:

$$Z_N = \frac{V^N}{N!} \int d^N \boldsymbol{s} \exp\left(-\beta \sum_{i<j} u(V^{1/d} s_{ij})\right). \tag{10}$$

Equations (8) and (10) then give

$$P = nkT \left(1 - \frac{n}{2dkT} \int \frac{du}{dr} r g(r) d\boldsymbol{r}\right). \tag{11}$$

This is called the *virial equation of state* and is useful for determining pressure from approximate expressions for the pair correlation function. Compare equation (11) with the form of the virial equation of state in equation (3.7.15).

For the particular case of hard spheres, the discontinuous potential results in the pressure being determined by the pair correlation function at contact. In one, two, and three dimensions, the hard sphere pressure is given by

$$\frac{p^{\mathrm{HS}}}{nkT} = \begin{cases} 1 + \eta g(D^+) & \eta = nD \quad d = 1, \\ 1 + 2\eta g(D^+) & \eta = \frac{\pi}{4} nD^2 \quad d = 2, \\ 1 + 4\eta g(D^+) & \eta = \frac{\pi}{6} nD^3 \quad d = 3, \end{cases} \tag{12}$$

where $g(D^+)$ is the correlation function at contact and η is the volume fraction, that is, the fraction of the d-dimensional volume of the sample occupied by the spheres; see Problem 10.14. Likewise, the internal energy of the fluid can be written as an integral over the pair correlation function and the pair potential:

$$U(N, V, T) = \langle H \rangle = \frac{dNkT}{2} + \frac{nN}{2} \int u(r) g(r) d\boldsymbol{r}. \tag{13}$$

The pair correlation function itself contains all the statistical information needed to construct the full thermodynamic behavior of the system. For example, equation (4) can be used to show that the isothermal compressibility, which is proportional to the number density fluctuations, is also proportional to an integral over the pair correlation function:

$$nkT\kappa_T = \frac{\kappa_T}{\kappa_T^{\mathrm{ideal}}} = 1 + n \int (g(r) - 1) d\boldsymbol{r} = \frac{\langle N^2 \rangle - \langle N \rangle^2}{\langle N \rangle}; \tag{14}$$

this is known as the *compressibility equation of state.* Since $\kappa_T^{-1} = n\left(\frac{\partial P}{\partial n}\right)_T$, one can use equation (14) to determine the pressure and free energy of the system by performing thermodynamic integrations with respect to the particle density.

10.7.A Static structure factor

The pair correlation function $g(r)$ can be measured experimentally using quasielastic scattering. If a sample is illuminated with a monochromatic beam of x-rays, neutrons, visible light, and so on, the scattered intensity as a function of the angle from the incident beam direction is proportional to the Fourier transform of $g(r)$. The quasielastic scattering amplitude from a single particle at location $\boldsymbol{r}_i$ illuminated by a plane wave with amplitude ϕ_0 and wavevector $\boldsymbol{k}_0$ into a detector at location $\boldsymbol{R}$ is

$$\Phi_1(\boldsymbol{k}) = \phi_0 f(\boldsymbol{k}) \frac{e^{i\boldsymbol{k}_0\cdot\boldsymbol{r}_i} e^{i\boldsymbol{k}_1\cdot(\boldsymbol{R}-\boldsymbol{r}_i)}}{|\boldsymbol{R}-\boldsymbol{r}_i|}, \tag{15}$$

where $\boldsymbol{k} = \boldsymbol{k}_1 - \boldsymbol{k}_0$ is the wavevector transfer and $f(\boldsymbol{k})$ is the single-particle scattering form factor; see Figure 10.7. The total scattering amplitude from the N particles in the sample is

$$\Phi_N(\boldsymbol{k}) \approx \frac{\phi_0 f(\boldsymbol{k})}{|\boldsymbol{R}|} e^{i\boldsymbol{k}_1\cdot\boldsymbol{R}} \sum_i e^{-i\boldsymbol{k}\cdot\boldsymbol{r}_i}, \tag{16}$$

where we have assumed that the detector is far from the sample. The scattered intensity from the N-particle sample is

$$I_N(\boldsymbol{k}) = \left|\Phi_N(\boldsymbol{k})\right|^2 \approx \frac{\left|\phi_0 f(\boldsymbol{k})\right|^2}{|\boldsymbol{R}|^2} \left\langle \sum_{i,j} e^{-i\boldsymbol{k}\cdot(\boldsymbol{r}_i-\boldsymbol{r}_j)} \right\rangle = N I_1(\boldsymbol{k}) S(\boldsymbol{k}), \tag{17}$$

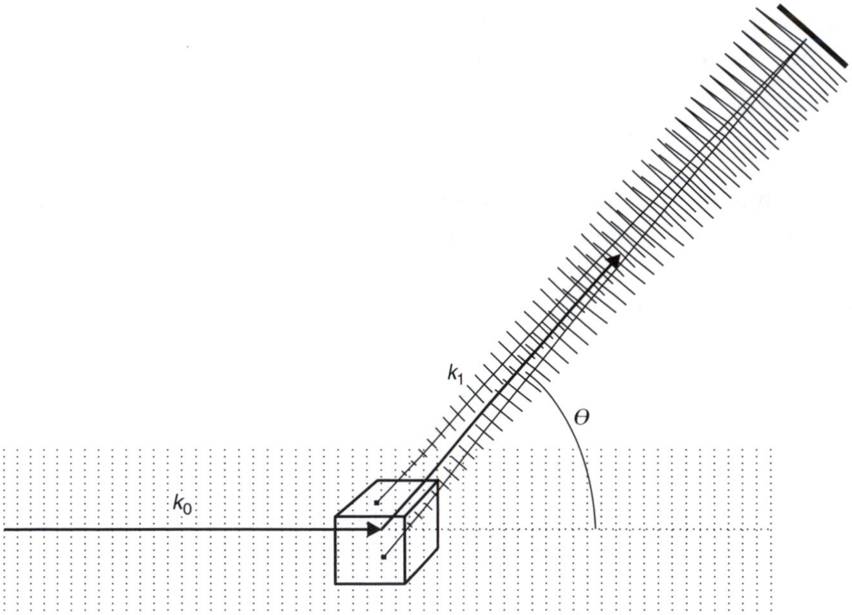

FIGURE 10.7 Scattering from two particles. The incident wavevector is $\boldsymbol{k}_0$, the scattered wavevector toward the detector is $\boldsymbol{k}_1$, and the wavevector transfer is $\boldsymbol{k} = \boldsymbol{k}_1 - \boldsymbol{k}_0$. Since $|\boldsymbol{k}_1| = |\boldsymbol{k}_0|$ for quasielastic scattering, the magnitude of the wavevector transfer is $k = 2k_0 \sin(\theta/2)$, where θ is the angle between $\boldsymbol{k}_0$ and $\boldsymbol{k}_1$.

where $I_1(\boldsymbol{k})$ is the scattering intensity from a single particle and

$$S(\boldsymbol{k}) = \frac{1}{N}\left\langle \sum_{i,j} \exp\left(-i\boldsymbol{k}\cdot(\boldsymbol{r}_i - \boldsymbol{r}_j)\right)\right\rangle \tag{18}$$

is the *static structure factor*. It represents the actual scattering intensity divided by the scattering intensity from an imaginary randomly distributed and, therefore, uncorrelated sample of atoms at the same particle density n.

If the sample is translationally invariant and isotropic, as in a uniform fluid, the static structure factor depends only on the magnitude of the wavevector transfer, that is $S(\boldsymbol{k}) = S(k)$. For that case, $S(k)$ can be written as the Fourier transform of the pair correlation function:

$$S(k) = 1 + \frac{N}{V}\int (g(r)-1)e^{i\boldsymbol{k}\cdot\boldsymbol{r}}d\boldsymbol{r} + \frac{N}{V^2}\left|\int e^{i\boldsymbol{k}\cdot\boldsymbol{r}}d\boldsymbol{r}\right|^2. \tag{19}$$

The final term in equation (19) represents the forward shape scattering of the sample volume. The shape scattering term is negligible for $k \gg 1/L$, so in the thermodynamic limit it can be ignored for $k \neq 0$. The structure factor for isotropic fluids in one, two, and three dimensions is then given by

$$S(k) = 1 + 2n\int_0^\infty (g(r)-1)\cos(kr)dr \qquad d=1, \tag{20a}$$

$$S(k) = 1 + 2\pi n\int_0^\infty r(g(r)-1)J_0(kr)dr \qquad d=2, \tag{20b}$$

$$S(k) = 1 + \frac{4\pi n}{k}\int_0^\infty r(g(r)-1)\sin(kr)dr \quad d=3. \tag{20c}$$

The pair correlation function $g(r)$ can be determined using the inverse Fourier transform of the measured structure factor, as shown in Figure 10.8. For liquids and other short-range ordered materials, the structure factor tends to unity as $k \to \infty$. The value of $S(k)$ as $k \to 0$ is a measure of the number density fluctuations in the sample:

$$\lim_{k\to 0} S(k) = 1 + n\int (g(r)-1)d\boldsymbol{r} = \frac{\kappa_T}{\kappa_T^{\text{ideal}}} = \frac{\langle N^2\rangle - \langle N\rangle^2}{\langle N\rangle}. \tag{21}$$

Equation (21) is called the *fluctuation-compressibility relation* and is the equilibrium limit of the *fluctuation-dissipation theorem* we will discuss in Section 15.6.

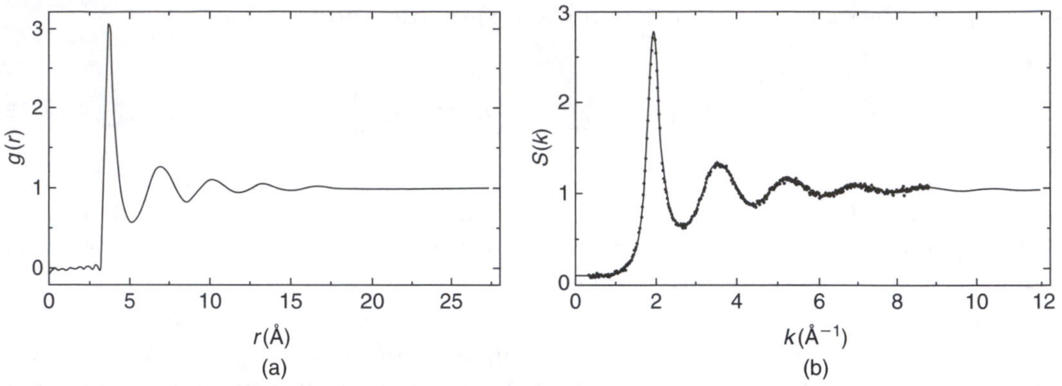

FIGURE 10.8 Experimentally measured pair correlation function $g(r)$ and structure factor $S(k)$ for liquid argon at 85 K. The structure factor (b) is determined from neutron scattering and the pair correlation function (a) is determined from the inverse Fourier transform of the structure factor. The small oscillations in $g(r)$ near $r=0$ are an experimental artifact of the Fourier transformation of the scattering data. This figure displays the typical features of correlations in fluids: nearly zero $g(r)$ at short distances, large $g(r)$ for particles separated by approximately a molecular diameter, oscillatory decay of correlations to unity at large separations, small $S(k)$ at small wavevector due to the small compressibility of dense fluids, and $S(k)$ approaching unity at large wavevectors. Figures from Yarnell, Katz, Wenzel, and Koenig (1973). Reprinted with permission; copyright ©1973, American Physical Society.

10.7.B Scattering from crystalline solids

In an ideal crystalline solid, the atoms in the crystal are located at the sites of a periodic structure. For a simple crystal, identical atoms are sited on a Bravais lattice $\{\boldsymbol{R}\}$. For example, a simple cubic lattice has lattice vectors $\boldsymbol{R} \in \{(n_1\hat{x}+n_2\hat{y}+n_3\hat{z})a\}$, where n_1, n_2, and n_3 are integers and a is the lattice constant. The reciprocal lattice $\{\boldsymbol{G}\}$ is defined by the set of reciprocal lattice vectors $\boldsymbol{G}$ — such that $\boldsymbol{G}\cdot\boldsymbol{R}=2\pi m$, where m is an integer for all $\{\boldsymbol{G}\}$ and $\{\boldsymbol{R}\}$. The reciprocal lattice of the simple cubic lattice is also a simple cubic lattice: $\boldsymbol{G} \in \{(m_1\hat{x}+m_2\hat{y}+m_3\hat{z})\frac{2\pi}{a}\}$, where m_1, m_2, and m_3, are integers. For a perfect Bravais lattice, the structure factor $S(\boldsymbol{k})$ is of the form

$$S(\boldsymbol{k}) = \frac{1}{N}\left\langle \sum_{\boldsymbol{R},\boldsymbol{R}'} e^{i\boldsymbol{k}\cdot(\boldsymbol{R}-\boldsymbol{R}')}\right\rangle = N\sum_{\boldsymbol{G}} \delta_{\boldsymbol{k},\boldsymbol{G}}, \tag{22}$$

where $\delta_{\boldsymbol{k},\boldsymbol{G}}$ is the Kronecker delta. The structure factor is enhanced by a factor of N on each reciprocal lattice vector due to the coherent constructive interference of scattering from the long-range ordered array of atoms. One can determine the crystal structure of the solid from the experimental pattern of these sharp Bragg peaks; see Ashcroft and Mermin (1976).

Thermal excitations cause atoms to deviate from their equilibrium positions. The displaced position of an atom whose equilibrium position is $\boldsymbol{R}$ can be written $\boldsymbol{R}+\boldsymbol{u}(\boldsymbol{R})$, where $\boldsymbol{u}(\boldsymbol{R})$ denotes the displacement from equilibrium. As long as the atoms remain close to their lattice sites, the sharp Bragg peaks in the structure factor will also remain but the intensity of each peak will be reduced by an amount dependent on the average of the squares of the deviations $\langle|\boldsymbol{u}(\boldsymbol{R})|^2\rangle$. This turns out to be the case for normal three-dimensional solids. The structure factor then takes the form

$$S(\boldsymbol{k}) = \frac{1}{N}\sum_{\boldsymbol{R},\boldsymbol{R}'} e^{i\boldsymbol{k}\cdot(\boldsymbol{R}-\boldsymbol{R}')}\left\langle e^{i\boldsymbol{k}\cdot(\boldsymbol{u}(\boldsymbol{R})-\boldsymbol{u}(\boldsymbol{R}'))}\right\rangle. \tag{23}$$

If the excitations about the equilibrium positions are Gaussian (i.e., the terms in the Hamiltonian higher than second order in $\boldsymbol{u}(\boldsymbol{R})$ can be ignored), then the average of the deviations in the exponential can be simplified to give

$$\left\langle e^{i\boldsymbol{k}\cdot(\boldsymbol{u}(\boldsymbol{R})-\boldsymbol{u}(\boldsymbol{R}'))}\right\rangle = e^{-\frac{1}{2}\left\langle|\boldsymbol{k}\cdot(\boldsymbol{u}(\boldsymbol{R})-\boldsymbol{u}(\boldsymbol{R}'))|^2\right\rangle}. \tag{24}$$

If the displacements of the atoms far from each other on the lattice are uncorrelated, as they are in three-dimensional crystals,

$$\frac{1}{2}\left\langle|\boldsymbol{k}\cdot(\boldsymbol{u}(\boldsymbol{R})-\boldsymbol{u}(\boldsymbol{R}'))|^2\right\rangle \approx \frac{k^2\langle\boldsymbol{u}^2\rangle}{3} \quad \text{for } |\boldsymbol{R}-\boldsymbol{R}'|\to\infty, \tag{25}$$

then the structure factor takes the form

$$S(\boldsymbol{k}) = N\sum_{\boldsymbol{G}} W_{\boldsymbol{G}}\delta_{\boldsymbol{k},\boldsymbol{G}}, \tag{26}$$

where

$$W_{\boldsymbol{G}} = \exp\left(-\frac{G^2\langle\boldsymbol{u}^2\rangle}{3}\right) \tag{27}$$

is called the Debye–Waller factor. The random atomic deviations from lattice sites reduces the intensity in the Bragg peaks but the sharp scattering indicative of long-range crystalline order remains intact; see Ashcroft and Mermin (1976).

An interesting variant of this calculation occurs in two-dimensional solids. Peierls (1935) and Landau (1937) showed that harmonic thermal fluctuations in two dimensions destroy crystalline long-range order. This was generalized by Mermin (1968) to show that long-range crystalline order was not possible for any two-dimensional system of particles with short-range interactions. Two-dimensional solids exhibit power-law decay of translational correlations while maintaining long-range order in the lattice orientational correlations. This leads to power-law singularities rather than delta-functions in the static structure factor. It is possible for the solid to melt via two Kosterlitz–Thouless-like continuous transitions rather than a single first-order transition. The intervening "hexatic" phase exhibits short-range translational correlations and quasi-long-range orientational correlations; see Section 13.7, Kosterlitz and Thouless (1972, 1973), Halperin and Nelson (1978), and Young (1979).

Problems

10.1. For imperfect-gas calculations, one sometimes employs the *Sutherland potential*

$$u(r) = \begin{cases} \infty & \text{for } r < D \\ -\varepsilon(D/r)^6 & \text{for } r > D. \end{cases}$$

Using this potential, determine the second virial coefficient of a classical gas. Also determine first-order corrections to the ideal-gas law and to the various thermodynamic properties of the system.

10.2. According to Lennard-Jones, the physical behavior of most real gases can be well understood if the intermolecular potential is assumed to be of the form

$$u(r) = \frac{A}{r^m} - \frac{B}{r^n},$$

where n is very nearly equal to 6 while m ranges between 11 and 13. Determine the second virial coefficient of a Lennard-Jones gas and compare your result with that for a van der Waals gas; see equation (10.3.8).

10.3. (a) Show that for a gas obeying van der Waals equation of state (10.3.9),

$$C_P - C_V = Nk\left\{1 - \frac{2a}{kTv^3}(v-b)^2\right\}^{-1}.$$

(b) Also show that, for a van der Waals gas with *constant* specific heat C_V, an adiabatic process conforms to the equation

$$(v-b)T^{C_V/Nk} = \text{const};$$

compare with equation (1.4.30).

(c) Further show that the temperature change resulting from an expansion of the gas (into vacuum) from volume V_1 to volume V_2 is given by

$$T_2 - T_1 = \frac{N^2 a}{C_V}\left(\frac{1}{V_2} - \frac{1}{V_1}\right).$$

10.4. The coefficient of volume expansion α and the isothermal bulk modulus B of a gas are given by the empirical expressions

$$\alpha = \frac{1}{T}\left(1 + \frac{3a'}{vT^2}\right) \quad \text{and} \quad B = P\left(1 + \frac{a'}{vT^2}\right)^{-1},$$

where a' is a constant parameter. Show that these expressions are mutually compatible. Also derive the equation of state of this gas.

10.5. Show that the first-order Joule–Thomson coefficient of a gas is given by the formula

$$\left(\frac{\partial T}{\partial P}\right)_H = \frac{N}{C_P}\left(T\frac{\partial(a_2\lambda^3)}{\partial T} - a_2\lambda^3\right),$$

where $a_2(T)$ is the second virial coefficient of the gas and H its enthalpy; see equation (10.2.1). Derive an explicit expression for the Joule–Thomson coefficient in the case of a gas with interparticle interaction

$$u(r) = \begin{cases} +\infty & \text{for } 0 < r < D, \\ -u_0 & \text{for } D < r < r_1, \\ 0 & \text{for } r_1 < r < \infty, \end{cases}$$

and discuss the temperature dependence of this coefficient.

10.6. Assume that the molecules of the nitrogen gas interact through the potential of the previous problem. Making use of the experimental data given next, determine the "best" empirical values for the parameters D, r_1, and u_0/k:

T (in K)	100	200	300	400	500
$a_2\lambda^3$ (in K per atm)	-1.80	-4.26×10^{-1}	-5.49×10^{-2}	$+1.12 \times 10^{-1}$	$+2.05 \times 10^{-1}$.

10.7. Determine the lowest-order corrections to the ideal-gas values of the Helmholtz free energy, the Gibbs free energy, the entropy, the internal energy, the enthalpy, and the (constant-volume and constant-pressure) specific heats of a real gas. Discuss the temperature dependence of these corrections in the case of a gas whose molecules interact through the potential of Problem 10.5.

10.8. The molecules of a solid attract one another with a force $F(r) = \alpha(l/r)^5$. Two semi-infinite solids composed of n molecules per unit volume are separated by a distance d, that is, the solids fill the whole of the space with $x \le 0$ and $x \ge d$. Calculate the force of attraction, per unit area of the surface, between the two solids.

10.9. Referring to equation (10.5.31) for the phase shifts $\eta_l(k)$ of a hard-sphere gas, show that for $kD \ll 1$

$$\eta_l(k) \simeq -\frac{(kD)^{2l+1}}{(2l+1)\{1 \cdot 3 \cdots (2l-1)\}^2}.$$

10.10. Using the wavefunctions

$$u_p(\boldsymbol{r}) = \frac{1}{\sqrt{V}} e^{i(\boldsymbol{p}\cdot\boldsymbol{r})\hbar}$$

to describe the motion of a free particle, write down the *symmetrized* wavefunctions for a pair of noninteracting bosons/fermions, and show that

$$\langle 1', 2' | \hat{U}_2^{S/A} | 1, 2 \rangle = \pm \langle 2' | \hat{W}_1 | 1 \rangle \langle 1' | \hat{W}_1 | 2 \rangle.$$

10.11. Show that for a gas composed of particles with spin J

$$\mathfrak{b}_2^S(J) = (J+1)(2J+1)\mathfrak{b}_2^S(0) + J(2J+1)\mathfrak{b}_2^A(0)$$

and

$$\mathfrak{b}_2^A(J) = J(2J+1)\mathfrak{b}_2^S(0) + (J+1)(2J+1)\mathfrak{b}_2^A(0).$$

10.12. Show that the coefficient $\mathfrak{b}_2$ for a quantum-mechanical *Boltzmannian* gas composed of "spinless" particles satisfies the following relations:

$$\mathfrak{b}_2 = \operatorname*{Lim}_{J\to\infty}\left\{\frac{1}{(2J+1)^2}\mathfrak{b}_2^S(J)\right\} = \operatorname*{Lim}_{J\to\infty}\left\{\frac{1}{(2J+1)^2}\mathfrak{b}_2^A(J)\right\}$$
$$= \frac{1}{2}\{\mathfrak{b}_2^S(0) + \mathfrak{b}_2^A(0)\}.$$

Obtain the value of $\mathfrak{b}_2$, to fifth order in (D/λ), by using the Beth–Uhlenbeck expressions in equations (10.5.36) and (10.5.37), and compare your result with the classical value of $\mathfrak{b}_2$, namely $-(2\pi/3)(D/\lambda)^3$.

10.13. Use a virial expansion approach to determine the first few nontrivial order contributions to the pair correlation function $g(r)$ in d dimensions. Show that the pair correlation function is of the form $g(r) = e^{-\beta u(r)}y(r)$, where $u(r)$ is the pair potential and $y(r)$ is a smooth function of r. Show that even for the case of hard sphere interaction, $y(r)$ and its first few derivatives are continuous.

10.14. For the particular case of hard spheres, the pressure in the virial equation of state is determined by evaluating the pair correlation function at contact. Write the pair correlation function as $g(r) = e^{-\beta u(r)}y(r)$ and derive equations (10.7.12) for hard spheres in one, two, and three dimensions.

[*Hint*: For hard spheres, the Boltzmann factor $e^{-\beta u(r)}$ is a Heaviside step function].

10.15. Derive the probability distribution $w(r)$ for the distance to the *closest* neighboring particle using the pair correlation function $g(r)$ and the number density n. Show that in three dimensions

$$w(r) = 4\pi n r^2 g(r) \exp\left(-\int_0^r 4\pi n s^2 g(s) ds\right),$$

and the average closest-neighbor distance for an ideal gas is

$$r_1 = \int_0^\infty r w(r) dr = \Gamma\left(\frac{4}{3}\right)\left(\frac{4\pi n}{3}\right)^{-1/3}.$$

10.16. Consider a gas, of infinite extent, divided into regions A and B by an imaginary sheet running through the system. The molecules of the gas interact through a potential energy function $u(r)$. Show that the average net force $\boldsymbol{F}$ experienced by all the molecules on the A-side of the sheet caused by all the molecules on the B-side are *perpendicular* to the plane of the sheet, and that its magnitude (per unit area) is given by

$$\frac{F}{A} = -\frac{2\pi n^2}{3}\int_0^\infty \left(\frac{du}{dr}\right) g(r) r^3 dr.$$

10.17. Show that for a gas of noninteracting bosons, or fermions, the pair correlation function $g(r)$ is given by the expression

$$g(r) = 1 \pm \frac{g_s}{n^2 h^6}\left|\int_{-\infty}^{\infty} \frac{e^{i(\boldsymbol{p}\cdot\boldsymbol{r})/\hbar}\, d^3p}{e^{(p^2/2m-\mu)/kT} \mp 1}\right|^2,$$

where $g_s\,(=2s+1)$ is the spin multiplicity factor. Note that the upper sign here applies to bosons, the lower one to fermions.[14]

[*Hint*: To solve this problem, one may use the method of second quantization, as developed in Chapter 11. The *particle density operator* $\hat{n}$ is then given by the sum,

$$\sum_{\alpha,\beta} a_\alpha^\dagger a_\beta u_\alpha^*(\boldsymbol{r}) u_\beta(\boldsymbol{r}),$$

whose diagonal terms are directly related to the mean particle density n in the system. The nondiagonal terms give the *density fluctuation operator* $(\hat{n}-n)$, and so on; see equation (11.1.25).]

10.18. Show that, in the case of a degenerate gas of fermions ($T \ll T_F$), the correlation function $g(r)$, for $r \gg \hbar/p_F$, reduces to the expression

$$g(r) - 1 = -\frac{3(mkT)^2}{4p_F^3\hbar r^2}\left\{\sinh\left(\frac{\pi mkTr}{p_F\hbar}\right)\right\}^{-2}.$$

Note that, as $T \to 0$, this expression tends to the limiting form

$$g(r) - 1 = -\frac{3\hbar}{4\pi^2 p_F r^4} \propto \frac{1}{r^4}.$$

[14] Note that, in the classical limit ($\hbar \to 0$), the infinitely rapid oscillations of the factor $\exp\{i(\boldsymbol{p}\cdot\boldsymbol{r})/\hbar\}$ make the integral vanish. Consequently, for an ideal classical gas, the function $g(r)$ is identically equal to 1. Quantum-mechanical systems of identical particles exhibit spatial correlations due to Bose and Fermi statistics even in the absence of interactions. It is not difficult to see that, for $n\lambda^3 \ll 1$ where $\lambda = h/\sqrt{(2\pi mkT)}$,

$$g(r) \simeq 1 \pm \tfrac{1}{g_s}\exp(-2\pi r^2/\lambda^2);$$

compare with equation (5.5.27).

10.19. (a) For a dilute gas, the pair correlation function $g(r)$ may be approximated as

$$g(r) \simeq \exp\{-u(r)/kT\}.$$

Show that, under this approximation, the virial equation of state (10.7.11) takes the form

$$\frac{PV}{NkT} \simeq 1 - 2\pi n \int_0^\infty f(r) r^2 dr,$$

where $f(r)$ [$= \exp\{-u(r)/kT\} - 1$] is the Mayer function, equation (10.1.6).

(b) What form will this result take for a gas of hard spheres? Compare your result with that of Problem 1.4.

10.20. Show that the pressure and Helmholtz free energy of a fluid at temperature T can be determined by performing a thermodynamic integration of the inverse of the isothermal compressibility from the chosen density to the ideal gas reference state.

10.21. Show that, for a general Gaussian distribution of variables u_j, the average of the exponential of a linear combination of the variables obeys the relation

$$\left\langle \exp\left(\sum_j a_j u_j\right)\right\rangle = \exp\left[\frac{1}{2}\left\langle\left(\sum_j a_j u_j\right)^2\right\rangle\right].$$

10.22. Calculate the isothermal compressibility and Helmholtz free energy for the Carnahan–Starling equation of state (10.3.25) and show that the Helmholtz free energy is given by

$$\frac{\beta A}{N} = \frac{\beta A^{\text{ideal}}}{N} + \frac{\eta(4-3\eta)}{(1-\eta)^2},$$

where A^{ideal} is the Helmholtz free energy of a classical monatomic ideal gas at the same density.

10.23. The virial expansion for a two-dimensional system of hard disks gives the following series when expressed in terms of the two-dimensional packing fraction $\eta = \pi n D^2/4$:

$$\frac{P}{nkT} = 1 + 2\eta + 3.128018\eta^2 + 4.257854\eta^3 + 5.33689664\eta^4 + 6.363026\eta^5 + 7.352080\eta^6 + 8.318668\eta^7 + 9.27236\eta^8 + 10.2161\eta^9 + \cdots;$$

see Malijevsky and Kolafa (2008). Propose some simple analytical functions $f(\eta)$ that closely approximate this series.

11

Statistical Mechanics of Interacting Systems: The Method of Quantized Fields

In this chapter we present another method of dealing with systems composed of interacting particles. This method is based on the concept of a *quantized field* that is characterized by the field operators $\psi(\boldsymbol{r})$, and their hermitian conjugates $\psi^{\dagger}(\boldsymbol{r})$, which satisfy a set of well-defined commutation rules. In terms of these operators, one defines a number operator $\hat{N}$ and a Hamiltonian operator $\hat{H}$ that provide a suitable representation for a system composed of any finite number of particles and possessing any finite amount of energy. In view of its formal similarity with the Schrödinger formulation, the formulation in terms of a quantized field is generally referred to as the *second quantization* of the system. For convenience of calculation, the field operators $\psi(\boldsymbol{r})$ and $\psi^{\dagger}(\boldsymbol{r})$ are often expressed as superpositions of a set of single-particle wavefunctions $\{u_{\alpha}(\boldsymbol{r})\}$, with coefficients a_{α} and $a_{\alpha}^{\dagger}$; the latter turn out to be the *annihilation* and *creation* operators, which again satisfy a set of well-defined commutation rules. The operators $\hat{N}$ and $\hat{H}$ then find a convenient expression in terms of the operators a_{α} and $a_{\alpha}^{\dagger}$, and the final formulation is well-suited for a treatment based on operator algebra; as a result, many calculations, which would otherwise be tedious, can be carried out in a more or less straightforward manner.

11.1 The formalism of second quantization

To represent a system of particles by a *quantized field*, we invoke the field operators $\psi(\boldsymbol{r})$ and $\psi^{\dagger}(\boldsymbol{r})$, which are defined for all values of the position coordinate $\boldsymbol{r}$ and which operate on a *Hilbert space*; a vector in this space corresponds to a particular state of the quantized field. The values of the quantities ψ and $\psi^{\dagger}$, at all $\boldsymbol{r}$, represent the *degrees of freedom* of the field; since $\boldsymbol{r}$ is a continuous variable, the number of these degrees of freedom is innumerably infinite. Now, if the given system is composed of bosons, the field operators $\psi(\boldsymbol{r})$ and $\psi^{\dagger}(\boldsymbol{r})$ satisfy the commutation rules

$$[\psi(\boldsymbol{r}), \psi^{\dagger}(\boldsymbol{r}')] = \delta(\boldsymbol{r} - \boldsymbol{r}') \tag{1a}$$

$$[\psi(\boldsymbol{r}), \psi(\boldsymbol{r}')] = [\psi^{\dagger}(\boldsymbol{r}), \psi^{\dagger}(\boldsymbol{r}')] = 0, \tag{1b}$$

© 2011 Elsevier Ltd. All rights reserved.

where the symbol $[A, B]$ stands for the commutator $(AB - BA)$ of the given operators A and B. If, on the other hand, the given system is composed of fermions, then the field operators satisfy the rules

$$\{\psi(\boldsymbol{r}), \psi^\dagger(\boldsymbol{r}')\} = \delta(\boldsymbol{r} - \boldsymbol{r}') \tag{2a}$$

$$\{\psi(\boldsymbol{r}), \psi(\boldsymbol{r}')\} = \{\psi^\dagger(\boldsymbol{r}), \psi^\dagger(\boldsymbol{r}')\} = 0, \tag{2b}$$

where the symbol $\{A, B\}$ stands for the anticommutator $(AB + BA)$ of the given operators A and B. In the case of fermions, the operators $\psi(\boldsymbol{r})$ and $\psi^\dagger(\boldsymbol{r})$ possess certain explicit properties that follow directly from (2b), namely

$$\psi(\boldsymbol{r})\psi(\boldsymbol{r}') = -\psi(\boldsymbol{r}')\psi(\boldsymbol{r}), \quad \therefore \psi(\boldsymbol{r})\psi(\boldsymbol{r}) = 0 \quad \text{for all } \boldsymbol{r}; \tag{2c}$$

similarly,

$$\psi^\dagger(\boldsymbol{r})\psi^\dagger(\boldsymbol{r}') = -\psi^\dagger(\boldsymbol{r}')\psi^\dagger(\boldsymbol{r}), \quad \therefore \psi^\dagger(\boldsymbol{r})\psi^\dagger(\boldsymbol{r}) = 0 \quad \text{for all } \boldsymbol{r}. \tag{2d}$$

Clearly, no such property holds for the field operators pertaining to bosons. In the sequel we shall see that the mathematical difference between the commutation rules (1) for the boson field operators and rules (2) for the fermion field operators is intimately related to the fundamental difference in the symmetry properties of the respective wavefunctions in the Schrödinger formulation. Of course, in their own place, both sets of rules, (1) and (2), are essentially axiomatic.

We now introduce two hermitian operators, the *particle-number operator* $\hat{N}$ and the *Hamiltonian operator* $\hat{H}$, through definitions that hold for bosons as well as fermions:

$$\hat{N} \equiv \int d^3r \psi^\dagger(\boldsymbol{r})\psi(\boldsymbol{r}) \tag{3}$$

and

$$\hat{H} \equiv -\frac{\hbar^2}{2m}\int d^3r \psi^\dagger(\boldsymbol{r})\nabla^2\psi(\boldsymbol{r}) + \frac{1}{2}\iint d^3r_1 d^3r_2 \psi^\dagger(\boldsymbol{r}_1)\psi^\dagger(\boldsymbol{r}_2)u(\boldsymbol{r}_1, \boldsymbol{r}_2)\psi(\boldsymbol{r}_2)\psi(\boldsymbol{r}_1), \tag{4}$$

where $u(\boldsymbol{r}_1, \boldsymbol{r}_2)$ denotes the two-body interaction potential in the given system. It is quite natural to interpret the product $\psi^\dagger(\boldsymbol{r})\psi(\boldsymbol{r})$ as the *number density operator* of the field. The similarity between the foregoing definitions and the expressions for the expectation values of the corresponding physical quantities in the Schrödinger formulation is fairly obvious. However, the similarity is only "formal" because, while there we are concerned with the wavefunctions of the given system (which are *c-numbers*), here we are concerned with the operators of the corresponding matter field. We can easily verify that, irrespective of the commutation rules obeyed by the operators $\psi(\boldsymbol{r})$ and $\psi^{\dagger}(\boldsymbol{r})$, the operators $\hat{N}$ and $\hat{H}$ do commute:

$$[\hat{N}, \hat{H}] = 0; \tag{5}$$

accordingly, the operators $\hat{N}$ and $\hat{H}$ can be diagonalized simultaneously.

We now choose a *complete orthonormal basis* of the Hilbert space, such that any vector $|\Phi_n\rangle$ among the basis is a simultaneous eigenstate of the operators $\hat{N}$ and $\hat{H}$. We may, therefore, denote any particular member of the basis by the symbol $|\Psi_{NE}\rangle$, with the properties

$$\hat{N}|\Psi_{NE}\rangle = N|\Psi_{NE}\rangle, \quad \hat{H}|\Psi_{NE}\rangle = E|\Psi_{NE}\rangle \tag{6}$$

and

$$\langle\Psi_{NE}|\Psi_{NE}\rangle = 1. \tag{7}$$

The vector $|\Psi_{00}\rangle$, which represents the *vacuum state* of the field and is generally denoted by the symbol $|0\rangle$, is assumed to be unique; it possesses the obvious properties

$$\hat{N}|0\rangle = \hat{H}|0\rangle = 0 \quad \text{and} \quad \langle 0|0\rangle = 1. \tag{8}$$

Next we observe that, regardless of whether we employ the boson commutation rules (1) or the fermion rules (2), the operator $\hat{N}$ and the operators $\psi(\boldsymbol{r})$ and $\psi^{\dagger}(\boldsymbol{r})$ satisfy the commutation properties

$$[\psi(\boldsymbol{r}), \hat{N}] = \psi(\boldsymbol{r}) \quad \text{and} \quad [\psi^{\dagger}(\boldsymbol{r}), \hat{N}] = -\psi^{\dagger}(\boldsymbol{r}), \tag{9}$$

from which it follows that

$$\hat{N}\psi(\boldsymbol{r})|\Psi_{NE}\rangle = \left(\psi(\boldsymbol{r})\hat{N} - \psi(\boldsymbol{r})\right)|\Psi_{NE}\rangle = (N-1)\psi(\boldsymbol{r})|\Psi_{NE}\rangle \tag{10}$$

and

$$\hat{N}\psi^{\dagger}(\boldsymbol{r})|\Psi_{NE}\rangle = \left(\psi^{\dagger}(\boldsymbol{r})\hat{N} + \psi^{\dagger}(\boldsymbol{r})\right)|\Psi_{NE}\rangle = (N+1)\psi^{\dagger}(\boldsymbol{r})|\Psi_{NE}\rangle. \tag{11}$$

Clearly, the state $\psi(\boldsymbol{r})|\Psi_{NE}\rangle$ is also an eigenstate of the operator $\hat{N}$, but with eigenvalue $(N-1)$; thus, the application of the operator $\psi(\boldsymbol{r})$ onto the state $|\Psi_{NE}\rangle$ of the field *annihilates* one particle from the field. Similarly, the state $\psi^{\dagger}(\boldsymbol{r})|\Psi_{NE}\rangle$ is an eigenstate of the operator $\hat{N}$, with eigenvalue $(N+1)$; thus, the application of the operator $\psi^{\dagger}(\boldsymbol{r})$ onto the state $|\Psi_{NE}\rangle$ of the field *creates* a particle in the field. In each case, the process (of annihilation or creation) is tied down to the point $\boldsymbol{r}$ of the field; however, the energy associated with the process, which also means the change in the energy of the field, remains undetermined; see equations (18) and (19). By a repeated application of the operator $\psi^{\dagger}$ onto the vacuum state $|0\rangle$, we find that the eigenvalues of the operator $\hat{N}$ are $0, 1, 2, \ldots$.

On the other hand, the application of the operator ψ onto the vacuum state $|0\rangle$ gives nothing but zero because, for obvious reasons, we cannot admit negative eigenvalues for the operator $\hat{N}$. Of course, if we apply the operator ψ onto the state $|\Psi_{NE}\rangle$ repeatedly N times, we end up with the vacuum state; we then have, by virtue of the orthonormality of the basis chosen,

$$\langle \Phi_n | \psi(\boldsymbol{r}_1)\psi(\boldsymbol{r}_2)\ldots\psi(\boldsymbol{r}_N)|\Psi_{NE}\rangle = 0 \tag{12}$$

unless the state $|\Phi_n\rangle$ is itself the vacuum state, in which case we would obtain a nonzero result instead. In terms of this latter result, we may define a function of the N coordinates $\boldsymbol{r}_1, \boldsymbol{r}_2, \ldots, \boldsymbol{r}_N$, namely

$$\Psi_{NE}(\boldsymbol{r}_1,\ldots,\boldsymbol{r}_N) = (N!)^{-1/2}\langle 0|\psi(\boldsymbol{r}_1)\ldots\psi(\boldsymbol{r}_N)|\Psi_{NE}\rangle. \tag{13}$$

Obviously, the function $\Psi_{NE}(\boldsymbol{r}_1,\ldots,\boldsymbol{r}_N)$ has something to do with an assemblage of N particles located at the points $\boldsymbol{r}_1,\ldots,\boldsymbol{r}_N$ of the field because their annihilation from those very points of the field has led us to the vacuum state of the field. To obtain the precise meaning of this function, we first note that in the case of bosons (fermions) this function is symmetric (antisymmetric) with respect to an interchange of any two of the N coordinates; see equations (1b) and (2b), respectively. Secondly, its norm is equal to unity, which can be seen as follows.

By the very definition of $\Psi_{NE}(\boldsymbol{r}_1,\ldots,\boldsymbol{r}_N)$,

$$\begin{aligned}
&\int d^{3N}r\,\Psi^*_{NE}(\boldsymbol{r}_1,\ldots,\boldsymbol{r}_N)\Psi_{NE}(\boldsymbol{r}_1,\ldots,\boldsymbol{r}_N)\\
&= (N!)^{-1}\int d^{3N}r\langle\Psi_{NE}|\psi^\dagger(\boldsymbol{r}_N)\ldots\psi^\dagger(\boldsymbol{r}_1)|0\rangle\langle 0|\psi(\boldsymbol{r}_1)\ldots\psi(\boldsymbol{r}_N)|\Psi_{NE}\rangle\\
&= (N!)^{-1}\int d^{3N}r\sum_n\langle\Psi_{NE}|\psi^\dagger(\boldsymbol{r}_N)\ldots\psi^\dagger(\boldsymbol{r}_1)|\Phi_n\rangle\langle\Phi_n|\psi(\boldsymbol{r}_1)\ldots\psi(\boldsymbol{r}_N)|\Psi_{NE}\rangle\\
&= (N!)^{-1}\int d^{3N}r\langle\Psi_{NE}|\psi^\dagger(\boldsymbol{r}_N)\ldots\psi^\dagger(\boldsymbol{r}_2)\psi^\dagger(\boldsymbol{r}_1)\psi(\boldsymbol{r}_1)\psi(\boldsymbol{r}_2)\ldots\psi(\boldsymbol{r}_N)|\Psi_{NE}\rangle;
\end{aligned}$$

here, use has been made of equation (12), which holds for all $|\Phi_n\rangle$ except for the vacuum state, and of the fact that the summation of $|\Phi_n\rangle\langle\Phi_n|$ over the complete orthonormal set of the basis chosen is equivalent to a unit operator. We now carry out integration over $\boldsymbol{r}_1$, yielding the factor

$$\int d^3r_1\psi^\dagger(\boldsymbol{r}_1)\psi(\boldsymbol{r}_1) = \hat{N}.$$

Next, we carry out integration over $\boldsymbol{r}_2$, yielding the factor

$$\int d^3\boldsymbol{r}_2\psi^\dagger(\boldsymbol{r}_2)\hat{N}\psi(\boldsymbol{r}_2) = \int d^3\boldsymbol{r}_2\psi^\dagger(\boldsymbol{r}_2)\psi(\boldsymbol{r}_2)(\hat{N}-1) = \hat{N}(\hat{N}-1);$$

see equation (10). By iteration, we obtain

$$\begin{aligned}\int d^{3N} r \Psi^*_{NE}(\boldsymbol{r}_1,\ldots,\boldsymbol{r}_N)\Psi_{NE}(\boldsymbol{r}_1,\ldots,\boldsymbol{r}_N) \\ = (N!)^{-1}\langle\Psi_{NE}|\hat{N}(\hat{N}-1)(\hat{N}-2)\ldots \text{ up to } N \text{ factors}|\Psi_{NE}\rangle \\ = (N!)^{-1}N!\,\langle\Psi_{NE}|\Psi_{NE}\rangle = 1.\end{aligned} \tag{14}$$

Finally, we can show that, for bosons as well as fermions, the function $\Psi_{NE}(\boldsymbol{r}_1,\ldots,\boldsymbol{r}_N)$ satisfies the differential equation, see Problem 11.1,

$$\left(-\frac{\hbar^2}{2m}\sum_{i=1}^{N}\nabla_i^2 + \sum_{i<j} u_{ij}\right)\Psi_{NE}(\boldsymbol{r}_1,\ldots,\boldsymbol{r}_N) = E\Psi_{NE}(\boldsymbol{r}_1,\ldots,\boldsymbol{r}_N), \tag{15}$$

which is simply the *Schrödinger equation* of an N-particle system. The function $\Psi_{NE}(\boldsymbol{r}_1,\ldots,\boldsymbol{r}_N)$ is, therefore, the Schrödinger wavefunction of the system, with energy eigenvalue E; accordingly, the product $\Psi^*_{NE}\Psi_{NE}$ is the probability density for the particles of the system to be in the vicinity of the coordinates $(\boldsymbol{r}_1,\ldots,\boldsymbol{r}_N)$, when the system happens to be in an eigenstate with energy E. This establishes the desired correspondence between the quantized field formulation and the Schrödinger formulation. In passing, we place on record the quantized-field expression for the function $\Psi^*_{NE}(\boldsymbol{r}_1,\ldots,\boldsymbol{r}_N)$, which is the complex conjugate of the wavefunction $\Psi_{NE}(\boldsymbol{r}_1,\ldots,\boldsymbol{r}_N)$, namely

$$\Psi^*_{NE}(\boldsymbol{r}_1,\ldots,\boldsymbol{r}_N) = (N!)^{-1/2}\langle\Psi_{NE}|\psi^\dagger(\boldsymbol{r}_N)\ldots\psi^\dagger(\boldsymbol{r}_1)|0\rangle. \tag{16}$$

We now introduce a complete orthonormal set of single-particle wavefunctions $u_\alpha(\boldsymbol{r})$, where the suffix α provides a label for identifying the various single-particle states; it could, for instance, be the energy eigenvalue of the state (or the momentum $\boldsymbol{p}$, along with the spin component σ pertaining to the state). In view of the orthonormality of these wavefunctions,

$$\int d^3 r\, u^*_\alpha(\boldsymbol{r})u_\beta(\boldsymbol{r}) = \delta_{\alpha\beta}. \tag{17}$$

The field operators $\psi(\boldsymbol{r})$ and $\psi^\dagger(\boldsymbol{r})$ may now be expanded in terms of the functions $u_\alpha(\boldsymbol{r})$:

$$\psi(\boldsymbol{r}) = \sum_\alpha a_\alpha u_\alpha(\boldsymbol{r}) \tag{18}$$

and

$$\psi^\dagger(\boldsymbol{r}) = \sum_\alpha a^\dagger_\alpha u^*_\alpha(\boldsymbol{r}). \tag{19}$$

Relations inverse to equation (18) and (19) are

$$a_\alpha = \int d^3r\psi(\mathbf{r})u_\alpha^*(\mathbf{r}) \tag{20}$$

and

$$a_\alpha^\dagger = \int d^3r\psi^\dagger(\mathbf{r})u_\alpha(\mathbf{r}). \tag{21}$$

The coefficients a_α and $a_\alpha^\dagger$, like the field variables $\psi(\mathbf{r})$ and $\psi^\dagger(\mathbf{r})$, are operators that operate on the elements of the relevant Hilbert space. Indeed, the operators a_α and $a_\alpha^\dagger$ now take over the role of the *degrees of freedom* of the field.

Substituting (18) and (19) into the set of rules (1) or (2), and making use of the closure property of the u_α, namely

$$\sum_\alpha u_\alpha(\mathbf{r})u_\alpha^*(\mathbf{r}') = \delta(\mathbf{r}-\mathbf{r}'), \tag{22}$$

we obtain[1] for the operators a_α and $a_\alpha^\dagger$ the commutation relations

$$[a_\alpha, a_\beta^\dagger] = \delta_{\alpha\beta} \tag{23a}$$

$$[a_\alpha, a_\beta] = [a_\alpha^\dagger, a_\beta^\dagger] = 0 \tag{23b}$$

in the case of *bosons*, and

$$\{a_\alpha, a_\beta^\dagger\} = \delta_{\alpha\beta} \tag{24a}$$

$$\{a_\alpha, a_\beta\} = \{a_\alpha^\dagger, a_\beta^\dagger\} = 0 \tag{24b}$$

in the case of *fermions*. In the latter case, the operators a_α and $a_\alpha^\dagger$ possess certain explicit properties that follow directly from (24b), namely

$$a_\alpha a_\beta = -a_\beta a_\alpha, \quad \therefore a_\alpha a_\alpha = 0 \quad \text{for all } \alpha; \tag{24c}$$

similarly

$$a_\alpha^\dagger a_\beta^\dagger = -a_\beta^\dagger a_\alpha^\dagger, \quad \therefore a_\alpha^\dagger a_\alpha^\dagger = 0 \quad \text{for all } \alpha. \tag{24d}$$

No such property holds for operators pertaining to bosons. We will see very shortly that this vital difference between the commutation rules for the boson operators and those for the fermion operators is closely linked with the fact that while fermions have to conform to the restrictions imposed by the Pauli exclusion principle, there are no such restrictions for bosons.

[1] Alternatively, one may employ equations (20) and (21), and make use of rules (1) or (2) along with equation (17).

We now proceed to express operators $\hat{N}$ and $\hat{H}$ in terms of a_α and $a_\alpha^\dagger$. Substituting (18) and (19) into (3), we obtain

$$\begin{aligned}\hat{N} &= \int d^3r \sum_{\alpha,\beta} a_\alpha^\dagger a_\beta u_\alpha^*(\mathbf{r}) u_\beta(\mathbf{r}) = \sum_{\alpha,\beta} a_\alpha^\dagger a_\beta \delta_{\alpha\beta} \\ &= \sum_\alpha a_\alpha^\dagger a_\alpha. \end{aligned} \tag{25}$$

It seems natural to speak of the operator $a_\alpha^\dagger a_\alpha$ as the *particle-number operator* pertaining to the single-particle state α. We denote this operator by the symbol $\hat{N}_\alpha$:

$$\hat{N}_\alpha = a_\alpha^\dagger a_\alpha. \tag{26}$$

It is easy to verify that, for bosons as well as fermions, the operators $\hat{N}_\alpha$ commute with one another; hence, they can be simultaneously diagonalized. Accordingly, we may choose a complete orthonormal basis of the Hilbert space in such a way that any vector belonging to the basis is a simultaneous eigenstate of all the operators $\hat{N}_\alpha$.[2] Let a particular member of the basis be denoted by the vector $|n_0, n_1, \ldots, n_\alpha, \ldots\rangle$, or by the shorter symbol $|\Phi_n\rangle$, with the properties

$$\hat{N}_\alpha |\Phi_n\rangle = n_\alpha |\Phi_n\rangle \tag{27}$$

and

$$\langle \Phi_n | \Phi_n \rangle = 1; \tag{28}$$

the number n_α, being the eigenvalue of the operator $\hat{N}_\alpha$ in the state $|\Phi_n\rangle$ of the field, denotes the number of particles in the single-particle state α of the given system. One particular member of the basis, for which $n_\alpha = 0$ for all α, will represent the *vacuum state* of the field; denoting the vacuum state by the symbol $|\Phi_0\rangle$, we have

$$\hat{N}_\alpha |\Phi_0\rangle = 0 \quad \text{for all } \alpha, \quad \text{and} \quad \langle \Phi_0 | \Phi_0 \rangle = 1. \tag{29}$$

Next we observe that, regardless of whether we employ the boson commutation rules (23) or the fermion rules (24), the operator $\hat{N}_\alpha$ and the operators a_α and $a_\alpha^\dagger$ satisfy the commutation properties

$$[a_\alpha, \hat{N}_\alpha] = a_\alpha \quad \text{and} \quad [a_\alpha^\dagger, \hat{N}_\alpha] = -a_\alpha^\dagger, \tag{30}$$

from which it follows that

$$\hat{N}_\alpha a_\alpha |\Phi_n\rangle = (a_\alpha \hat{N}_\alpha - a_\alpha)|\Phi_n\rangle = (n_\alpha - 1) a_\alpha |\Phi_n\rangle \tag{31}$$

[2] This representation of the field is generally referred to as the *particle-number representation*.

and

$$\hat{N}_\alpha a_\alpha^\dagger |\Phi_n\rangle = (a_\alpha^\dagger \hat{N}_\alpha + a_\alpha^\dagger)|\Phi_n\rangle = (n_\alpha + 1)a_\alpha^\dagger |\Phi_n\rangle. \tag{32}$$

Clearly, the state $a_\alpha|\Phi_n\rangle$ is also an eigenstate of the operator $\hat{N}_\alpha$, but with eigenvalue $(n_\alpha - 1)$; thus, the application of the operator a_α onto the state $|\Phi_n\rangle$ of the field *annihilates* one particle from the field. Similarly, the state $a_\alpha^\dagger|\Phi_n\rangle$ is an eigenstate of the operator $\hat{N}_\alpha$, with eigenvalue $(n_\alpha + 1)$; thus, the application of the operator $a_\alpha^\dagger$ onto the state $|\Phi_n\rangle$ *creates* a particle in the field. The operators a_α and $a_\alpha^\dagger$ are, therefore, referred to as the *annihilation* and *creation operators*. Of course, in each case the process (of annihilation or creation) is tied down to the single-particle state α; however, the precise location of the event (in the coordinate space) remains undetermined; see equations (20) and (21). Now, since the application of the operator a_α or $a_\alpha^\dagger$ onto the state $|\Phi_n\rangle$ of the field does not affect the eigenvalues of the particle-number operators other than $\hat{N}_\alpha$, we may write

$$a_\alpha|n_0, n_1, \ldots, n_\alpha, \ldots\rangle = A(n_\alpha)|n_0, n_1, \ldots, n_\alpha - 1, \ldots\rangle \tag{33}$$

and

$$a_\alpha^\dagger|n_0, n_1, \ldots, n_\alpha, \ldots\rangle = B(n_\alpha)|n_0, n_1, \ldots, n_\alpha + 1, \ldots\rangle, \tag{34}$$

where the factors $A(n_\alpha)$ and $B(n_\alpha)$ can be determined with the help of the commutation rules governing the operators a_α and $a_\alpha^\dagger$. For bosons,

$$A(n_\alpha) = \surd n_\alpha, \quad B(n_\alpha) = \surd(n_\alpha + 1); \tag{35}$$

consequently, if we regard the state $|\Phi_n\rangle$ to have arisen from the vacuum state $|\Phi_0\rangle$ by a repeated application of the creation operators, we can write

$$|\Phi_n\rangle = \frac{1}{\surd(n_0!n_1!\ldots n_\alpha!\ldots)}(a_0^\dagger)^{n_0}(a_1^\dagger)^{n_1}\cdots(a_\alpha^\dagger)^{n_\alpha}\cdots|\Phi_0\rangle. \tag{36}$$

In the case of fermions, the operators $a_\alpha^\dagger$ anticommute, with the result that $a_\alpha^\dagger a_\beta^\dagger = -a_\beta^\dagger a_\alpha^\dagger$; consequently, there would remain an uncertainty of a phase factor ± 1 *unless* the order in which the $a_\alpha^\dagger$ operate on the vacuum state is specified. To be definite, let us agree that, as indicated in equation (36), the $a_\alpha^\dagger$ are arranged in the order of increasing subscripts and the phase factor is then $+1$. Second, since the product $a_\alpha^\dagger a_\alpha^\dagger$ now vanishes, none of the n_α in (36) can exceed unity; the eigenvalues of the fermion operators $\hat{N}_\alpha$ are, therefore, restricted to 0 and 1, which is precisely the requirement of the Pauli exclusion principle.[3] Accordingly, the factor $[\Pi_\alpha(n_\alpha!)]^{-1/2}$ in (36) would be identically equal to unity.

[3]This can also be seen by noting that the fermion operators $\hat{N}_\alpha$ satisfy the identity

$$\hat{N}_\alpha^2 = a_\alpha^\dagger a_\alpha a_\alpha^\dagger a_\alpha = a_\alpha^\dagger(1 - a_\alpha^\dagger a_\alpha)a_\alpha = a_\alpha^\dagger a_\alpha = \hat{N}_\alpha \quad (\text{since } a_\alpha^\dagger a_\alpha^\dagger a_\alpha a_\alpha \equiv 0).$$

The same would be true of the eigenvalues n_α. Hence, $n_\alpha^2 = n_\alpha$, which means that $n_\alpha = 0$ or 1.

In passing, we note that in the case of fermions, operation (33) has meaning only if $n_\alpha = 1$ and operation (34) has meaning only if $n_\alpha = 0$.

Finally, the substitution of expressions (18) and (19) into (4) gives for the Hamiltonian operator of the field

$$\hat{H} = -\frac{\hbar^2}{2m}\sum_{\alpha,\beta}\langle\alpha|\nabla^2|\beta\rangle a_\alpha^\dagger a_\beta + \frac{1}{2}\sum_{\alpha,\beta,\gamma,\lambda}\langle\alpha\beta|u|\gamma\lambda\rangle a_\alpha^\dagger a_\beta^\dagger a_\gamma a_\lambda, \tag{37}$$

where

$$\langle\alpha|\nabla^2|\beta\rangle = \int d^3r u_\alpha^*(\boldsymbol{r})\nabla^2 u_\beta(\boldsymbol{r}) \tag{38}$$

and

$$\langle\alpha\beta|u|\gamma\lambda\rangle = \iint d^3r_1 d^3r_2 u_\alpha^*(\boldsymbol{r}_1)u_\beta^*(\boldsymbol{r}_2)u_{12}u_\gamma(\boldsymbol{r}_2)u_\lambda(\boldsymbol{r}_1). \tag{39}$$

Now, if the single-particle wavefunctions are chosen to be

$$u_\alpha(\boldsymbol{r}) = \frac{1}{\sqrt{V}}e^{i\boldsymbol{p}_\alpha\cdot\boldsymbol{r}/\hbar}, \tag{40}$$

where $\boldsymbol{p}_\alpha$ denotes the momentum of the particle (assumed "spinless"), then the matrix elements (38) and (39) become

$$\langle\alpha|\nabla^2|\beta\rangle = \frac{1}{V}\int d^3r e^{-i\boldsymbol{p}_\alpha\cdot\boldsymbol{r}/\hbar}\left(-\frac{p_\beta^2}{\hbar^2}\right)e^{i\boldsymbol{p}_\beta\cdot\boldsymbol{r}/\hbar} = -\frac{p_\beta^2}{\hbar^2}\delta_{\alpha\beta} \tag{41}$$

and

$$\langle\alpha\beta|u|\gamma\lambda\rangle = \frac{1}{V^2}\iint d^3r_1 d^3r_2 e^{-i(\boldsymbol{p}_\alpha-\boldsymbol{p}_\lambda)\cdot\boldsymbol{r}_1/\hbar}u(\boldsymbol{r}_2-\boldsymbol{r}_1)e^{-i(\boldsymbol{p}_\beta-\boldsymbol{p}_\gamma)\cdot\boldsymbol{r}_2/\hbar}. \tag{42}$$

In view of the fact that the total momentum is conserved in each collision,

$$\boldsymbol{p}_\alpha + \boldsymbol{p}_\beta = \boldsymbol{p}_\gamma + \boldsymbol{p}_\lambda, \tag{43}$$

the matrix element (42) takes the form

$$\begin{aligned}\langle\alpha\beta|u|\gamma\lambda\rangle &= \frac{1}{V^2}\iint d^3r_1 d^3r_2 e^{i(\boldsymbol{p}_\gamma-\boldsymbol{p}_\beta)\cdot(\boldsymbol{r}_2-\boldsymbol{r}_1)/\hbar}u(\boldsymbol{r}_2-\boldsymbol{r}_1)\\ &= \frac{1}{V}\int d^3r e^{i\boldsymbol{p}\cdot\boldsymbol{r}/\hbar}u(\boldsymbol{r}),\end{aligned} \tag{44}$$

where $\boldsymbol{p}$ denotes the *momentum transfer* during the collision:

$$\boldsymbol{p} = (\boldsymbol{p}_\gamma - \boldsymbol{p}_\beta) = -(\boldsymbol{p}_\lambda - \boldsymbol{p}_\alpha). \tag{45}$$

Substituting (41) and (44) into (37), we finally obtain

$$\hat{H} = \sum_{\boldsymbol{p}} \frac{p^2}{2m} a_{\boldsymbol{p}}^\dagger a_{\boldsymbol{p}} + \frac{1}{2} \sum{}' u_{\boldsymbol{p}_1,\boldsymbol{p}_2}^{\boldsymbol{p}_1',\boldsymbol{p}_2'} a_{\boldsymbol{p}_1'}^\dagger a_{\boldsymbol{p}_2'}^\dagger a_{\boldsymbol{p}_2} a_{\boldsymbol{p}_1}, \tag{46}$$

where $u_{\boldsymbol{p}_1,\boldsymbol{p}_2}^{\boldsymbol{p}_1',\boldsymbol{p}_2'}$ denotes the matrix element (44), with

$$\boldsymbol{p} = (\boldsymbol{p}_2 - \boldsymbol{p}_2') = -(\boldsymbol{p}_1 - \boldsymbol{p}_1'); \tag{47}$$

note that the primed summation in the second term of (46) goes only over those values of $\boldsymbol{p}_1, \boldsymbol{p}_2, \boldsymbol{p}_1'$, and $\boldsymbol{p}_2'$ that conserve the total momentum of the particles: $\boldsymbol{p}_1' + \boldsymbol{p}_2' = \boldsymbol{p}_1 + \boldsymbol{p}_2$. It is obvious that the main term in (46) represents the *kinetic energy* of the field ($a_{\boldsymbol{p}}^\dagger a_{\boldsymbol{p}}$ being the particle-number operator pertaining to the single-particle state $\boldsymbol{p}$), while the second term represents the *potential energy*.

In the case of spin-half fermions, the single-particle states have to be characterized not only by the value $\boldsymbol{p}$ of the particle momentum but also by the value σ of the z-component of its spin; accordingly, the creation and annihilation operators would carry double indices. The operator $\hat{H}$ then takes the form

$$\hat{H} = \sum_{\boldsymbol{p},\sigma} \frac{p^2}{2m} a_{\boldsymbol{p}\sigma}^\dagger a_{\boldsymbol{p}\sigma} + \frac{1}{2} \sum{}' u_{\boldsymbol{p}_1\sigma_1,\boldsymbol{p}_2\sigma_2}^{\boldsymbol{p}_1'\sigma_1',\boldsymbol{p}_2'\sigma_2'} a_{\boldsymbol{p}_1'\sigma_1'}^\dagger a_{\boldsymbol{p}_2'\sigma_2'}^\dagger a_{\boldsymbol{p}_2\sigma_2} a_{\boldsymbol{p}_1\sigma_1}; \tag{48}$$

the summation in the second term now goes only over those states (of the two particles) that conform to the conditions of both momentum conservation and spin conservation.

In the following sections we shall apply the formalism of second quantization to investigate low-temperature properties of systems composed of interacting particles. In most cases we shall study these systems under the approximating conditions $a/\lambda \ll 1$ and $na^3 \ll 1$, where a is the *scattering length* of the two-body interaction, λ the *mean thermal wavelength* of the particles, and n the *particle density* in the system. Now, the effective scattering cross-section for the collision of two particles, each of mass m, is primarily determined by the "scattering amplitude" $a(\boldsymbol{p})$, where

$$a(\boldsymbol{p}) = \frac{m}{4\pi\hbar^2} \int u(\boldsymbol{r}) e^{i\boldsymbol{p}\cdot\boldsymbol{r}/\hbar} d^3r, \tag{49}$$

$\boldsymbol{p}$ being the momentum transfer during the collision; if the potential is central, equation (49) takes the form

$$a(p) = \frac{m}{4\pi\hbar^2} \int_0^\infty u(r) \frac{\sin(kr)}{kr} 4\pi r^2 dr \quad \left(k = \frac{p}{\hbar}\right). \tag{50}$$

For low-energy scattering (which implies "slow" collisions), we have the limiting result

$$a = \frac{mu_0}{4\pi\hbar^2}, \quad u_0 = \int u(\boldsymbol{r}) d^3r, \tag{51}$$

the quantity a being the *scattering length* of the given potential.[4] Alternatively, one may employ the S-wave scattering phase shift $\eta_0(k)$, see Section 10.5, and write on one hand

$$\tan\eta_0(k) \simeq -\frac{mk}{4\pi\hbar^2}\int_0^\infty u(r)\frac{\sin^2(kr)}{(kr)^2}4\pi r^2 dr \tag{52}$$

and on the other

$$\cot\eta_0(k) = -\frac{1}{ka} + \frac{1}{2}kr^* + \cdots, \tag{53}$$

where a is the "scattering length" and r^* the "effective range" of the potential. For low-energy scattering, equations (52) and (53) once again lead to (51). In passing, we note that a is positive or negative according as the potential in question is predominantly repulsive or predominantly attractive; unless a statement is made to the contrary, we shall assume a to be positive.

11.2 Low-temperature behavior of an imperfect Bose gas

The Hamiltonian of the quantized field for spinless bosons is given by the expression (11.1.46), where the matrix element $u_{\boldsymbol{p}_1,\boldsymbol{p}_2}^{\boldsymbol{p}_1',\boldsymbol{p}_2'}$ is a function of the momentum $\boldsymbol{p}$ transferred during the collision and is given by formula (11.1.44). At low temperatures the particle momenta are small, so we may insert for the matrix elements $u(\boldsymbol{p})$ their value at $\boldsymbol{p} = 0$, namely u_0/V, where u_0 is given by equation (11.1.51). At the same time, we may retain only those terms in the sum $\sum'$ that pertain to a vanishing momentum transfer. We then have

$$\hat{H} = \sum_{\boldsymbol{p}} \frac{p^2}{2m} a_{\boldsymbol{p}}^\dagger a_{\boldsymbol{p}} + \frac{2\pi a\hbar^2}{mV}\left[\sum_{\boldsymbol{p}} a_{\boldsymbol{p}}^\dagger a_{\boldsymbol{p}}^\dagger a_{\boldsymbol{p}} a_{\boldsymbol{p}} + \sum_{\boldsymbol{p}_1 \neq \boldsymbol{p}_2}\left(a_{\boldsymbol{p}_1}^\dagger a_{\boldsymbol{p}_2}^\dagger a_{\boldsymbol{p}_2} a_{\boldsymbol{p}_1} + a_{\boldsymbol{p}_2}^\dagger a_{\boldsymbol{p}_1}^\dagger a_{\boldsymbol{p}_2} a_{\boldsymbol{p}_1}\right)\right]. \tag{1}$$

Now

$$\sum_{\boldsymbol{p}} a_{\boldsymbol{p}}^\dagger a_{\boldsymbol{p}}^\dagger a_{\boldsymbol{p}} a_{\boldsymbol{p}} = \sum_{\boldsymbol{p}} a_{\boldsymbol{p}}^\dagger (a_{\boldsymbol{p}} a_{\boldsymbol{p}}^\dagger - 1) a_{\boldsymbol{p}} = \sum_{\boldsymbol{p}} (n_{\boldsymbol{p}}^2 - n_{\boldsymbol{p}}) = \sum_{\boldsymbol{p}} n_{\boldsymbol{p}}^2 - N, \tag{2}$$

[4]This result is consistent with the *pseudopotential approach* of Huang and Yang (1957) in which $u(\boldsymbol{r})$ is replaced by the singular potential $(4\pi a\hbar^2/m)\delta(\boldsymbol{r})$, so the integral u_0 becomes $4\pi a\hbar^2/m$. For an exposition of the pseudopotential approach, see Chapter 10 of the first edition of this book.

whereas

$$\sum_{p_1 \neq p_2} a^\dagger_{p_1} a^\dagger_{p_2} a_{p_2} a_{p_1} = \sum_{p_1 \neq p_2} n_{p_1} n_{p_2} = \sum_{p_1} n_{p_1}(N - n_{p_1}) = N^2 - \sum_{p} n^2_{p}, \tag{3}$$

the same being true of the sum over the exchange terms $a^\dagger_{p_2} a^\dagger_{p_1} a_{p_2} a_{p_1}$. Collecting these results, the energy eigenvalues of the system turn out to be

$$E\{n_p\} = \sum_p n_p \frac{p^2}{2m} + \frac{2\pi a \hbar^2}{mV}\left[2N^2 - N - \sum_p n^2_p\right]$$

$$\simeq \sum_p n_p \frac{p^2}{2m} + \frac{2\pi a \hbar^2}{mV}(2N^2 - n_0^2). \qquad (4)^5$$

We first examine the *ground state* of the given system, which corresponds to the distribution set

$$n_p \simeq \begin{cases} N & \text{for } p = 0 \\ 0 & \text{for } p \neq 0, \end{cases} \tag{5}$$

with the result that

$$E_0 \simeq \frac{2\pi a \hbar^2 N^2}{mV}. \tag{6}$$

The ground-state pressure is then given by

$$P_0 = -\left(\frac{\partial E_0}{\partial V}\right)_N = \frac{2\pi a \hbar^2 N^2}{mV^2} = \frac{2\pi a \hbar^2 n^2}{m}, \tag{7}$$

where $n(= N/V)$ is the particle density in the system. This leads to the velocity of sound, c_0, given by

$$c_0^2 = \frac{1}{m}\frac{dP_0}{dn} = \frac{4\pi a \hbar^2 n}{m^2}. \tag{8}$$

Inserting numbers relevant to liquid He^4, namely $a \simeq 2.2$Å, $n = 1/v$ where $v \simeq 45\text{Å}^3$ and $m \simeq 6.65 \times 10^{-24}$ g, we obtain: $c_0 \simeq 125$ m/s. A comparison with the actual velocity of sound in the liquid, which is about 240 m/s, should not be too disheartening, for the theory

[5] In the last step we have replaced the sum $\sum_p n_p^2$ by the single term n_0^2, thus neglecting the partial sum $\sum_{p\neq 0} n_p^2$ in comparison with the number $(2N^2 - n_0^2)$. Justification for this step lies in the fact that, by the theory of fluctuations, the neglected part here will be $O(N)$, and not $O(N^2)$.

developed here was never intended to be applicable to a liquid. Finally, the chemical potential of the system at $T = 0$ K turns out to be

$$\mu_0 = \left(\frac{\partial E_0}{\partial N}\right)_V = \frac{4\pi a\hbar^2 N}{mV} = \frac{4\pi a\hbar^2 n}{m}. \tag{9}$$

At finite but low temperatures, the physical behavior of the system may be studied through its partition function

$$\begin{aligned} Q(N,V,T) &= \sum_{\{n_{\mathbf{p}}\}} \exp(-\beta E\{n_{\mathbf{p}}\}) \\ &= \sum_{\{n_{\mathbf{p}}\}} \exp\left[-\beta\left\{\sum_{\mathbf{p}} n_{\mathbf{p}} \frac{p^2}{2m} + \frac{2\pi a\hbar^2 N^2}{mV}\left(2 - \frac{n_0^2}{N^2}\right)\right\}\right]. \end{aligned} \tag{10}$$

In the lowest approximation, the quantity (n_0/N) appearing here may be replaced by its ideal-gas value, as given in Section 7.1, namely

$$\frac{n_0}{N} = 1 - \frac{\lambda_c^3}{\lambda^3} \quad \left[\lambda = \frac{h}{(2\pi mkT)^{1/2}}, \quad \lambda_c = \{v\zeta(3/2)\}^{1/3}\right] \tag{11a}$$

$$= 1 - \frac{v}{v_c} \quad \left[v = \frac{V}{N}, \quad v_c = \frac{\lambda^3}{\zeta(3/2)}\right]. \tag{11b}$$

We thus obtain, to *first* order in a,

$$\ln Q(N,V,T) \simeq \ln Q_{id}(N,V,T) - \beta\frac{2\pi a\hbar^2 N^2}{mV}\left(1 + \frac{2v}{v_c} - \frac{v^2}{v_c^2}\right). \tag{12}$$

The Helmholtz free energy, per particle, is then given by[6]

$$\frac{1}{N}A(N,V,T) = -\frac{kT}{N}\ln Q(N,V,T) \simeq \frac{1}{N}A_{id}(N,V,T) + \frac{2\pi a\hbar^2}{m}\left(\frac{1}{v} + \frac{2}{v_c} - \frac{v}{v_c^2}\right). \tag{13}$$

The pressure P and the chemical potential μ now follow straightforwardly:

$$P = -\left(\frac{\partial A}{\partial V}\right)_{N,T} = -\left(\frac{\partial(A/N)}{\partial v}\right)_T = P_{id} + \frac{2\pi a\hbar^2}{m}\left(\frac{1}{v^2} + \frac{1}{v_c^2}\right), \tag{14}$$

and

$$\mu = \frac{A}{N} + Pv = \mu_{id} + \frac{4\pi a\hbar^2}{m}\left(\frac{1}{v} + \frac{1}{v_c}\right), \tag{15}$$

which may be compared with the ground-state results (7) and (9) that pertain to $v_c = \infty$.

[6]This and the subsequent results were first derived by Lee and Yang (1958, 1960c) using the binary collision method and by Huang (1959, 1960) using the pseudopotential method.

At the transition point (where $v = v_c$ and $\lambda = \lambda_c$), the pressure P_c and the chemical potential μ_c turn out to be

$$P_c = P_{id} + \frac{4\pi a\hbar^2}{m\lambda_c^6}\left\{\zeta\left(\frac{3}{2}\right)\right\}^2 = \frac{kT_c}{\lambda_c^3}\left[\zeta\left(\frac{5}{2}\right) + 2\left\{\zeta\left(\frac{3}{2}\right)\right\}^2 \frac{a}{\lambda_c}\right] \tag{16}$$

and

$$\mu_c = \mu_{id} + \frac{8\pi a\hbar^2}{m\lambda_c^3}\zeta\left(\frac{3}{2}\right) = 4\zeta\left(\frac{3}{2}\right)kT_c\frac{a}{\lambda_c}; \tag{17}$$

the corresponding value of the fugacity, z_c, is given by

$$z_c = \exp(\mu_c/kT_c) \simeq 1 + 4\zeta(3/2)(a/\lambda_c). \tag{18}$$

For a slightly different approach to this result, see Problem 11.2.

11.2.A Effects of interactions on ultracold atomic Bose–Einstein condensates

In Section 7.2 we discussed Bose–Einstein condensation of noninteracting bosons confined in magnetic traps. The low-energy interactions between atoms are described by the scattering length a, see equation (11.1.51), and the effect of the scattering length on the Bose-condensed ground state of a uniform gas is described by equations (6) through (9). We can include the effect of atomic interactions on the spatially nonuniform ground state using the Gross–Pitaevskii equation; see Pitaevskii (1961), Gross (1961, 1963), Pitaevskii and Stringari (2003), and Leggett (2006). The magnetic trap potential can be approximated by an anisotropic harmonic oscillator potential

$$V(\boldsymbol{r}) = \frac{1}{2}m(\omega_1^2 x^2 + \omega_2^2 y^2 + \omega_3^2 z^2), \tag{19}$$

which leads to the unperturbed single-particle ground-state wavefunction

$$\phi(\boldsymbol{r}) = \frac{1}{\pi^{3/4}\sqrt{a_1 a_2 a_3}}\exp\left[-\frac{1}{2}\left(\frac{x^2}{a_1^2} + \frac{y^2}{a_2^2} + \frac{z^2}{a_3^2}\right)\right], \tag{20}$$

where $a_\alpha = \sqrt{\hbar/m\omega_\alpha}$ is the linear size of the unperturbed harmonic oscillator ground state in Cartesian direction α.

For the noninteracting case at $T = 0$, all the N atoms in the trap occupy this same single-particle state to form a macroscopic quantum state $\Psi(\boldsymbol{r}) = \sqrt{N}\phi(\boldsymbol{r})$. At low energies, the interactions can be approximated by an effective contact potential $u_0\delta(\boldsymbol{r} - \boldsymbol{r}')$ with scattering length a and coupling $u_0 = 4\pi a\hbar^2/m$. This provides a fairly accurate description of interactions in ultracold gases since the kinetic energies and densities of the particles are so small; see equation (11.1.51). If the scattering length a is positive, the interaction is repulsive while if a is negative the interaction is attractive. For atoms, the scattering length is normally of the order of a few Bohr radii but in some atomic isotopes the scattering length can be tuned over a large range, including a change of sign, with only small changes in the magnetic field via a Feshbach resonance. When interactions are included, the mean field energy can be written in terms of the macroscopic quantum-state wavefunction $\Psi(\boldsymbol{r})$, where the macroscopic ground-state number density is given by $n(\boldsymbol{r}) = |\Psi(\boldsymbol{r})|^2$. The Gross–Pitevskii energy functional then is

$$E[\Psi] = \int d\boldsymbol{r}\left[\frac{\hbar^2}{2m}|\nabla\Psi(\boldsymbol{r})|^2 + V(\boldsymbol{r})\,|\Psi(\boldsymbol{r})|^2 + \frac{1}{2}u_0\,|\Psi(\boldsymbol{r})|^4\right]; \tag{21}$$

see Bahm and Pethick (1996), Pitaevskii and Stringari (2003), Leggett (2006), and Pethick and Smith (2008). The energy $E[\Psi]$ can be minimized with respect to Ψ^*, with the constraint

$$N = \int n(\boldsymbol{r})d\boldsymbol{r} = \int |\Psi(\boldsymbol{r})|^2\,d\boldsymbol{r}, \tag{22}$$

using a Lagrange multiplier μ. Setting $\delta E - \mu\delta N = 0$ gives the Gross–Pitaevskii equation

$$-\frac{\hbar^2}{2m}\nabla^2\Psi(\boldsymbol{r}) + V(\boldsymbol{r})\Psi(\boldsymbol{r}) + u_0\,|\Psi(\boldsymbol{r})|^2\,\Psi(\boldsymbol{r}) = \mu\Psi(\boldsymbol{r}). \tag{23}$$

Equations (21) and (23) are quite appropriate for describing the zero-temperature nonuniform Bose gas when the scattering length a is much smaller than the average spacing between the particles. Equation (23) is in the form of a single-particle Schrödinger equation with the addition of a nonlinear term proportional to $u_0\,|\Psi(\boldsymbol{r})|^2$ that gives a mean field coupling of one particle to all the remaining particles in the condensate. To determine the mean field ground-state energy, one solves equation (23) for $\Psi(\boldsymbol{r})$ and uses that solution to evaluate (21). When $a = 0$, the solution is the noninteracting ground-state wavefunction $\Psi(\boldsymbol{r}) = \sqrt{N}\phi(\boldsymbol{r})$. The dimensionless parameter that controls the size of the interaction term is Na/a_{osc} where $a_{\text{osc}} = (a_1a_2a_3)^{1/3}$, so the effects of interactions are largest in systems with large numbers of atoms in the condensate.

Equation (23) can be analyzed analytically in several limiting cases; it can also be studied numerically. In particular, solving equations (21) and (23) for a uniform system with $V = 0$ reproduces the results given in equations (6) through (9). For $a > 0$, the condensate wavefunction expands in every direction relative to the noninteracting Bose–Einstein condensate. For $Na/a_{\text{osc}} \gg 1$, the kinetic energy term can be neglected in the style of a Thomas-Fermi analysis, so the wavefunction becomes approximately

$$\Psi(\boldsymbol{r}) \approx \sqrt{(\mu - V(\boldsymbol{r}))/u_0}\,, \tag{24}$$

where the Thomas–Fermi wavefunction vanishes for $V(\boldsymbol{r}) > \mu$. The chemical potential μ and the number of bosons N are related by

$$N = \frac{8\pi}{15}\left(\frac{2\mu}{m\omega_0^2}\right)^{3/2}\frac{\mu}{u_0}, \tag{25}$$

where $\omega_0 = (\omega_1\omega_2\omega_3)^{1/3}$, so

$$\mu = \frac{1}{2}\left(\frac{15Na}{a_{\rm osc}}\right)^{2/5}\hbar\omega_0. \tag{26}$$

The total energy of the condensate in this limit is

$$E = \frac{5}{7}\mu N. \tag{27}$$

The linear extents of the condensate in the three directions of the trap are given by

$$R_\alpha = \sqrt{\frac{2\mu}{m\omega_\alpha^2}} = a_{\rm osc}\left(\frac{15Na}{a_{\rm osc}}\right)^{1/5}\frac{\omega_0}{\omega_\alpha}, \tag{28}$$

so the repulsive interactions expand the size of the condensate, making the aniostropy of the system larger than that of the noninteracting Bose–Einstein condensate; see Pitaevskii and Stringari (2003); Leggett (2006), and Pethick and Smith (2008). In time-of-flight measurements, the repulsive interactions result in higher velocities in the directions that were most confined in the trap, so the time-of-flight distributions are also more anisotropic than in the noninteracting case; see Holland and Cooper (1996) and Holland et al. (1997).

For attractive interactions with negative scattering lengths, the condensate is ultimately unstable because of the formation of pairs, but a long-lived atomic condensate exists for small negative $Na/a_{\rm osc}$. For the isotropic case, the atomic condensate is metastable in the mean field theory for $-0.575 \lesssim Na/a_{\rm osc} < 0$. Even in the anisotropic case, the condensate is

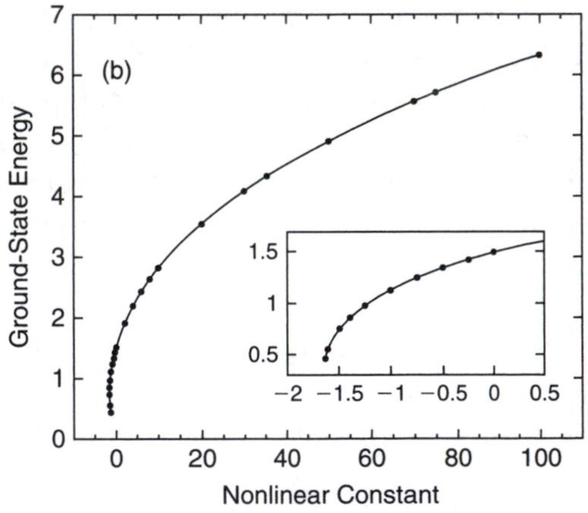

FIGURE 11.1 Ground-state energy of a Bose–Einstein condensate in an isotropic harmonic trap as a function of the scattering length a. The energy is plotted in units of $N\hbar\omega_0$, where ω_0 is the trap frequency. The "nonlinear constant" is proportional to $Na/a_{\rm osc}$, where N is the number of atoms and $a_{\rm osc} = \sqrt{\hbar/m\omega_0}$ is the width of the harmonic oscillator ground state wavefunction. Figure from Ruprecht et al. (1995). Reprinted with permission; copyright © 1995, American Physical Society.

nearly spherical since the solution of equation (23) is dominated by the kinetic and interaction terms; see Ruprecht et al. (1995) and Figure 11.1. Roberts et al. (2001) have used a Feshbach resonance to tune the scattering length of ^{85}Rb to find that the condensate becomes unstable at $N|a|/a_{\text{osc}} \simeq 0.46$.

11.3 Low-lying states of an imperfect Bose gas

In the preceding section we examined first-order corrections to the low-temperature behavior of an imperfect Bose gas arising from interparticle interactions in the system. One important result emerging in that study was a *nonzero* velocity of sound, as given by equation (11.2.8). This raises the possibility that phonons, the quanta of sound field, might play an important role in determining the low-temperature behavior of this system — a role not seen in Section 11.2. To look into this question, we explore the nature of the low-lying states of an imperfect Bose gas, in the hope that we thus discover an energy-momentum relation $\varepsilon(\boldsymbol{p})$ obeyed by the elementary excitations of the system, of which phonons may be an integral part. For this, we have to go a step beyond the approximation adopted in Section 11.2 which, in turn, requires several significant improvements. To keep matters simple, we confine ourselves to situations in which the fraction of particles occupying the state with $\boldsymbol{p} = 0$ is fairly close to 1 while the fraction of particles occupying states with $\boldsymbol{p} \neq 0$ is much less than 1.

Going back to equations (11.2.1) through (11.2.4), we first write

$$2N^2 - n_0^2 = N^2 + (N^2 - n_0^2) \simeq N^2 + 2N(N - n_0) = N^2 + 2N \sum_{\boldsymbol{p}\neq 0} a_{\boldsymbol{p}}^{\dagger} a_{\boldsymbol{p}}. \tag{1}$$

Next, we retain another set of terms from the sum $\sum'$ in equation (11.1.46) — terms that involve a nonzero momentum transfer, namely

$$\sum_{\boldsymbol{p}\neq 0} u(\boldsymbol{p})[a_{\boldsymbol{p}}^{\dagger} a_{-\boldsymbol{p}}^{\dagger} a_0 a_0 + a_0^{\dagger} a_0^{\dagger} a_{\boldsymbol{p}} a_{-\boldsymbol{p}}]. \tag{2}$$

Now, since $a_0^{\dagger} a_0 = n_0 = O(N)$ and $(a_0 a_0^{\dagger} - a_0^{\dagger} a_0) = 1 \ll N$, it follows that $a_0 a_0^{\dagger} = (n_0 + 1) \simeq a_0^{\dagger} a_0$. The operators a_0 and $a_0^{\dagger}$ may, therefore, be treated as c-numbers, each equal to $n_0^{1/2} \simeq N^{1/2}$. At the same time, the amplitude $u(\boldsymbol{p})$ in the case of low-lying states may be replaced by u_0/V, as before. Expression (2) then becomes

$$\frac{u_0 N}{V} \sum_{\boldsymbol{p}\neq 0} (a_{\boldsymbol{p}}^{\dagger} a_{-\boldsymbol{p}}^{\dagger} + a_{\boldsymbol{p}} a_{-\boldsymbol{p}}). \tag{3}$$

In view of these results, the Hamiltonian of the system assumes the form

$$\hat{H} = \sum_{\boldsymbol{p}} \frac{p^2}{2m} a_{\boldsymbol{p}}^\dagger a_{\boldsymbol{p}} + \frac{u_0}{2V} \left[N^2 + N \sum_{\boldsymbol{p} \neq 0} (2a_{\boldsymbol{p}}^\dagger a_{\boldsymbol{p}} + a_{\boldsymbol{p}}^\dagger a_{-\boldsymbol{p}}^\dagger + a_{\boldsymbol{p}} a_{-\boldsymbol{p}}) \right]. \tag{4}$$

Our next task consists of determining an improved relationship between the quantity u_0 and the scattering length a. While the (approximate) result stated in equation (11.1.51) is good enough for evaluating the term involving $N\sum_{\boldsymbol{p}\neq 0}$, it is not so for evaluating the term involving N^2. For this, we note that "if the probability of a particular quantum transition in a given system under the influence of a constant perturbation $\hat{V}$ is, in the *first* approximation, determined by the matrix element V_0^0, then in the *second* approximation we have instead

$$V_0^0 + \sum_{n \neq 0} \frac{V_n^0 V_0^n}{E_0 - E_n},$$

the summation going over the various states of the unperturbed system."

In the present case, we are dealing with a collision process in the two-particle system (with reduced mass $\frac{1}{2}m$), and the role of V_0^0 is played by the quantity

$$u_{00}^{00} = \frac{1}{V} \int u(\boldsymbol{r}) d^3 r = \frac{u_0}{V};$$

see equation (11.1.44) for the matrix element $u_{\boldsymbol{p}_1, \boldsymbol{p}_2}^{\boldsymbol{p}'_1, \boldsymbol{p}'_2}$. Making use of the other matrix elements, we find that in going from the first to second approximation, we have to replace u_0/V by

$$\frac{u_0}{V} + \frac{1}{V^2} \sum_{\boldsymbol{p} \neq 0} \frac{\left| \int d^3 r e^{i\boldsymbol{p}\cdot\boldsymbol{r}/\hbar} u(\boldsymbol{r}) \right|^2}{-p^2/m} \simeq \frac{u_0}{V} - \frac{u_0^2 m}{V^2} \sum_{\boldsymbol{p} \neq 0} \frac{1}{p^2}. \tag{5}$$

Equating (5) with the standard expression $4\pi a\hbar^2/mV$, we obtain, instead of (11.1.51),

$$u_0 \simeq \frac{4\pi a\hbar^2}{m} \left(1 + \frac{4\pi a\hbar^2}{V} \sum_{\boldsymbol{p} \neq 0} \frac{1}{p^2} \right). \tag{6}$$

Substituting (6) into (4), we get

$$\begin{aligned} \hat{H} = & \frac{2\pi a\hbar^2}{m} \frac{N^2}{V} \left(1 + \frac{4\pi a\hbar^2}{V} \sum_{\boldsymbol{p} \neq 0} \frac{1}{p^2} \right) \\ & + \frac{2\pi a\hbar^2}{m} \frac{N}{V} \sum_{\boldsymbol{p} \neq 0} (2a_{\boldsymbol{p}}^\dagger a_{\boldsymbol{p}} + a_{\boldsymbol{p}}^\dagger a_{-\boldsymbol{p}}^\dagger + a_{\boldsymbol{p}} a_{-\boldsymbol{p}}) + \sum_{\boldsymbol{p} \neq 0} \frac{p^2}{2m} a_{\boldsymbol{p}}^\dagger a_{\boldsymbol{p}}. \end{aligned} \tag{7}$$

To evaluate the energy levels of the system one would have to diagonalize the Hamiltonian (7), which can be done with the help of a *linear transformation* of the operators $a_{\boldsymbol{p}}$ and $a^\dagger_{\boldsymbol{p}}$, first employed by Bogoliubov (1947):

$$b_{\boldsymbol{p}} = \frac{a_{\boldsymbol{p}} + \alpha_{\boldsymbol{p}} a^\dagger_{-\boldsymbol{p}}}{\surd(1-\alpha^2_{\boldsymbol{p}})}, \quad b^\dagger_{\boldsymbol{p}} = \frac{a^\dagger_{\boldsymbol{p}} + \alpha_{\boldsymbol{p}} a_{-\boldsymbol{p}}}{\surd(1-\alpha^2_{\boldsymbol{p}})}, \tag{8}$$

where

$$\alpha_{\boldsymbol{p}} = \frac{mV}{4\pi a\hbar^2 N}\left\{\frac{4\pi a\hbar^2 N}{mV} + \frac{p^2}{2m} - \varepsilon(\boldsymbol{p})\right\}, \tag{9}$$

with

$$\varepsilon(\boldsymbol{p}) = \left\{\frac{4\pi a\hbar^2 N}{mV}\frac{p^2}{m} + \left(\frac{p^2}{2m}\right)^2\right\}^{1/2}; \tag{10}$$

clearly, each $\alpha_{\boldsymbol{p}} < 1$. Relations inverse to (8) are

$$a_{\boldsymbol{p}} = \frac{b_{\boldsymbol{p}} - \alpha_{\boldsymbol{p}} b^\dagger_{-\boldsymbol{p}}}{\surd(1-\alpha^2_{\boldsymbol{p}})}, \quad a^\dagger_{\boldsymbol{p}} = \frac{b^\dagger_{\boldsymbol{p}} - \alpha_{\boldsymbol{p}} b_{-\boldsymbol{p}}}{\surd(1-\alpha^2_{\boldsymbol{p}})}. \tag{11}$$

It is straightforward to check that the new operators $b_{\boldsymbol{p}}$ and $b^\dagger_{\boldsymbol{p}}$ satisfy the same commutation rules as the old operators $a_{\boldsymbol{p}}$ and $a^\dagger_{\boldsymbol{p}}$ did, namely

$$[b_{\boldsymbol{p}}, b^\dagger_{\boldsymbol{p}'}] = \delta_{\boldsymbol{p}\boldsymbol{p}'} \tag{12a}$$

$$[b_{\boldsymbol{p}}, b_{\boldsymbol{p}'}] = [b^\dagger_{\boldsymbol{p}}, b^\dagger_{\boldsymbol{p}'}] = 0. \tag{12b}$$

Substituting (11) into (7), we obtain our Hamiltonian in the diagonalized form:

$$\hat{H} = E_0 + \sum_{\boldsymbol{p}\neq 0} \varepsilon(\boldsymbol{p}) b^\dagger_{\boldsymbol{p}} b_{\boldsymbol{p}}, \tag{13}$$

where

$$E_0 = \frac{2\pi a\hbar^2 N^2}{mV} + \frac{1}{2}\sum_{\boldsymbol{p}\neq 0}\left\{\varepsilon(\boldsymbol{p}) - \frac{p^2}{2m} - \frac{4\pi a\hbar^2 N}{mV} + \left(\frac{4\pi a\hbar^2 N}{mV}\right)^2 \frac{m}{p^2}\right\}. \tag{14}$$

In view of the commutation rules (12) and expression (13) for the Hamiltonian operator $\hat{H}$, it seems natural to infer that the operators $b_{\boldsymbol{p}}$ and $b^\dagger_{\boldsymbol{p}}$ are the annihilation and creation operators of certain "quasiparticles" — which represent *elementary excitations* of the system — with the energy–momentum relation given by (10); it is also clear that these quasiparticles obey Bose–Einstein statistics. The quantity $b^\dagger_{\boldsymbol{p}} b_{\boldsymbol{p}}$ is then the particle-number operator for the quasiparticles (or elementary excitations) of momentum $\boldsymbol{p}$, whereby the second part of the Hamiltonian (13) becomes the energy operator corresponding to the elementary excitations present in the system. The first part of the Hamiltonian, given explicitly by equation (14), is therefore the *ground-state* energy of the system. Replacing the summation over $\boldsymbol{p}$ by integration and introducing a dimensionless variable x, defined by

$$x = p\left(\frac{V}{8\pi a\hbar^2 N}\right)^{1/2},$$

we obtain for the ground-state energy of the system

$$E_0 = \frac{2\pi a\hbar^2 N^2}{mV}\left[1 + \left(\frac{128Na^3}{\pi V}\right)^{1/2} \times \int_0^\infty dx \left[x^2\left(x\sqrt{(x^2+2)} - x^2 - 1 + \frac{1}{2x^2}\right)\right]\right]. \tag{15}$$

The value of the integral turns out to be $(128)^{1/2}/15$, with the result

$$\frac{E_0}{N} = \frac{2\pi a\hbar^2 n}{m}\left[1 + \frac{128}{15\pi^{1/2}}(na^3)^{1/2}\right], \tag{16}$$

where n denotes the particle density in the system. Equation (16) represents the first two terms of the expansion of the quantity E_0/N in terms of the low-density parameter $(na^3)^{1/2}$; the first term was already obtained in Section 11.2.[7]

The foregoing result was first derived by Lee and Yang (1957) using the binary collision method; the details of this calculation, however, appeared somewhat later (see Lee and Yang, 1960a; see also Problem 11.6). Using the pseudopotential method, this result was rederived by Lee, Huang, and Yang (1957).

The ground-state pressure of the system is now given by

$$P_0 = -\left(\frac{\partial E_0}{\partial V}\right)_N = n^2\frac{\partial(E_0/N)}{\partial n}$$
$$= \frac{2\pi a\hbar^2 n^2}{m}\left[1 + \frac{64}{5\pi^{1/2}}(na^3)^{1/2}\right], \tag{17}$$

[7]The evaluation of higher-order terms of this expansion necessitates consideration of three-body collisions as well; hence, in general, they cannot be expressed in terms of the scattering length alone. The exceptional case of a hard-sphere gas has been studied by Wu (1959), who obtained (using the pseudopotential method)

$$\frac{E_0}{N} = \frac{2\pi a\hbar^2 n}{m}\left[1 + \frac{128}{15\pi^{1/2}}(na^3)^{1/2} + 8\left(\frac{4\pi}{3} - \sqrt{3}\right)(na^3)\ln(12\pi na^3) + O(na^3)\right],$$

which shows that the expansion does not proceed in simple powers of $(na^3)^{1/2}$.

from which one obtains for the velocity of sound

$$c_0^2 = \frac{1}{m}\frac{dP_0}{dn} = \frac{4\pi a\hbar^2 n}{m^2}\left[1 + \frac{16}{\pi^{1/2}}(na^3)^{1/2}\right]. \tag{18}$$

Equations (17) and (18) are an improved version of the results obtained in Section 11.2.

The ground state of the system is characterized by a total absence of excitations; accordingly, the eigenvalue of the (number) operator $b_{\boldsymbol{p}}^\dagger b_{\boldsymbol{p}}$ of the quasiparticles must be zero for all $\boldsymbol{p} \neq 0$. As for the real particles, there must be some that possess nonzero energies even at absolute zero, for otherwise the system cannot have a finite amount of energy in the ground state. The momentum distribution of the real particles can be determined by evaluating the ground-state expectation values of the number operators $a_{\boldsymbol{p}}^\dagger a_{\boldsymbol{p}}$. Now, in the ground state of the system,

$$a_{\boldsymbol{p}}|\Psi_0\rangle = \frac{1}{\sqrt{(1-\alpha_{\boldsymbol{p}}^2)}}(b_{\boldsymbol{p}} - \alpha_{\boldsymbol{p}} b_{-\boldsymbol{p}}^\dagger)|\Psi_0\rangle = \frac{-\alpha_{\boldsymbol{p}}}{\sqrt{(1-\alpha_{\boldsymbol{p}}^2)}} b_{-\boldsymbol{p}}^\dagger|\Psi_0\rangle \tag{19}$$

because $b_{\boldsymbol{p}}|\Psi_0\rangle \equiv 0$. Constructing the hermitian conjugate of (19) and remembering that $\alpha_{\boldsymbol{p}}$ is real, we have

$$\langle\Psi_0|a_{\boldsymbol{p}}^\dagger = \frac{-\alpha_{\boldsymbol{p}}}{\sqrt{(1-\alpha_{\boldsymbol{p}}^2)}}\langle\Psi_0|b_{-\boldsymbol{p}}. \tag{20}$$

The scalar product of expressions (19) and (20) gives

$$\langle\Psi_0|a_{\boldsymbol{p}}^\dagger a_{\boldsymbol{p}}|\Psi_0\rangle = \frac{\alpha_{\boldsymbol{p}}^2}{1-\alpha_{\boldsymbol{p}}^2}\langle\Psi_0|b_{-\boldsymbol{p}} b_{-\boldsymbol{p}}^\dagger|\Psi_0\rangle = \frac{\alpha_{\boldsymbol{p}}^2}{1-\alpha_{\boldsymbol{p}}^2}; \tag{21}$$

here, use has been made of the facts that (i) $b_{\boldsymbol{p}} b_{\boldsymbol{p}}^\dagger - b_{\boldsymbol{p}}^\dagger b_{\boldsymbol{p}} = 1$ and (ii) in the ground state, for all $\boldsymbol{p} \neq 0, b_{\boldsymbol{p}}^\dagger b_{\boldsymbol{p}} = 0$ (and hence $b_{\boldsymbol{p}} b_{\boldsymbol{p}}^\dagger = 1$). Thus, for $\boldsymbol{p} \neq 0$,

$$\bar{n}_{\boldsymbol{p}} = \frac{\alpha_{\boldsymbol{p}}^2}{1-\alpha_{\boldsymbol{p}}^2} = \frac{x^2+1}{2x\sqrt{(x^2+2)}} - \frac{1}{2}, \tag{22}$$

where $x = p(8\pi a\hbar^2 n)^{-1/2}$. The total number of "excited" particles in the ground state of the system is, therefore, given by

$$\begin{aligned}\sum_{\boldsymbol{p}\neq 0}\bar{n}_{\boldsymbol{p}} &= \sum_{\boldsymbol{p}\neq 0}\frac{\alpha_{\boldsymbol{p}}^2}{1-\alpha_{\boldsymbol{p}}^2} = \sum_{x>0}\frac{1}{2}\left(\frac{x^2+1}{x\sqrt{(x^2+2)}} - 1\right) \\ &\simeq N\left\{\frac{32}{\pi}(na^3)\right\}^{1/2}\int_0^\infty dx\left[x^2\left(\frac{x^2+1}{x\sqrt{(x^2+2)}} - 1\right)\right].\end{aligned} \tag{23}$$

The value of the integral turns out to be $(2)^{1/2}/3$, with the result

$$\sum_{p\neq 0}\overline{n}_{\boldsymbol{p}} \simeq N\frac{8}{3\pi^{1/2}}(na^3)^{1/2}. \tag{24}$$

Accordingly,

$$\overline{n}_0 = N - \sum_{p\neq 0}\overline{n}_{\boldsymbol{p}} \simeq N\left[1 - \frac{8}{3\pi^{1/2}}(na^3)^{1/2}\right]. \tag{25}$$

The foregoing result was first obtained by Lee, Huang, and Yang (1957), using the pseudopotential method. It may be noted here that the importance of the *real-particle* occupation numbers $n_{\boldsymbol{p}}$ in the study of the ground state of an interacting Bose system had been emphasized earlier by Penrose and Onsager (1956).

11.4 Energy spectrum of a Bose liquid

In this section we propose to study the most essential features of the energy spectrum of a Bose liquid and to examine the relevance of this study to the problem of liquid He^4. In this context we have seen that the low-lying states of a *low-density gaseous* system composed of weakly interacting bosons are characterized by the presence of the so-called *elementary excitations* (or "quasiparticles"), which are themselves bosons and whose energy spectrum is given by

$$\varepsilon(p) = \{p^2u^2 + (p^2/2m)^2\}^{1/2}, \tag{1}$$

where

$$u = (4\pi an)^{1/2}(\hbar/m); \tag{2}$$

see equations (11.3.10), (11.3.12), and (11.3.13).[8] For $p \ll mu$, that is, $p \ll \hbar(an)^{1/2}$, the spectrum is essentially linear: $\varepsilon \simeq pu$. The initial slope of the (ε, p)-curve is, therefore, given by the parameter u, which is identical to the limiting value of the velocity of sound in the system; compare (2) with (11.3.18). It is then natural that these low-momentum excitations be identified as *phonons* — the quanta of the sound field. For $p \gg mu$, the spectrum approaches essentially the classical limit: $\varepsilon \simeq p^2/2m + \Delta^*$, where $\Delta^* = mu^2 = 4\pi an\hbar^2/m$. It is important to note that, all along, this energy–momentum relationship is strictly *monotonic* and does not display any "dip" of the kind propounded by Landau (1941, 1947) (for

[8]Spectrum (1) was first obtained by Bogoliubov (1947) by the method outlined in the preceding sections. Using the pseudopotential method, it was rederived by Lee, Huang, and Yang (1957).

liquid He^4) and observed experimentally by Yarnell et al. (1959), and by Henshaw and Woods (1961); see Section 7.6. Thus, the spectrum provided by the theory of the preceding sections simulates the Landau spectrum only to the extent of phonons; it does not account for rotons. This should not be surprising, for the theory in question was intended only for a low-density Bose gas ($na^3 \ll 1$) and not for liquid He^4($na^3 \simeq 0.2$).

The problem of elementary excitations in liquid He^4 was tackled successfully by Feynman who, in 1953 to 1954, developed an atomic theory of a *Bose liquid* at low temperatures. In a series of three fundamental papers starting from first principles, Feynman established the following important results.[9]

(i) In spite of the presence of interatomic forces, a Bose liquid undergoes a phase transition analogous to the momentum-space condensation occurring in the ideal Bose gas; in other words, the original suggestion of London (1938a,b) regarding liquid He^4, see Section 7.1, is essentially correct.

(ii) At sufficiently low temperatures, the only excited states possible in the liquid are the ones related to compressional waves, namely *phonons*. Long-range motions, which leave the density of the liquid unaltered (and consequently imply nothing more than a simple "stirring" of the liquid), do not constitute excited states because they differ from the ground state only in the "permutation" of certain atoms. Motions on an atomic scale are indeed possible, but they require a *minimum* energy Δ for their excitation; clearly, these excitations would show up only at comparatively higher temperatures ($T \sim \Delta/k$) and might well turn out to be Landau's *rotons*.

(iii) The wavefunction of the liquid, in the presence of *an* excitation, should be approximately of the form

$$\Psi = \Phi \sum_i f(\boldsymbol{r}_i), \tag{3}$$

where Φ denotes the ground-state wavefunction of the system while the summation of $f(\boldsymbol{r}_i)$ goes over all the N coordinates $\boldsymbol{r}_1, \ldots, \boldsymbol{r}_N$; the wavefunction Ψ is, clearly, symmetric in its arguments. The exact character of the function $f(\boldsymbol{r})$ can be determined by making use of a variational principle that requires the energy of the state Ψ (and hence the energy associated with the excitation in question) to be a minimum.

The optimal choice for $f(\boldsymbol{r})$ turns out to be, see Problem 11.8,

$$f(\boldsymbol{r}) = \exp i(\boldsymbol{k} \cdot \boldsymbol{r}), \tag{4}$$

[9]The reader interested in pursuing Feynman's line of argument should refer to Feynman's original papers or to a review of Feynman's work on superfluidity by Mehra and Pathria (1994).

with the (minimized) energy value

$$\varepsilon(\boldsymbol{k}) = \frac{\hbar^2 k^2}{2mS(\boldsymbol{k})}, \tag{5}$$

where $S(\boldsymbol{k})$ is the *structure factor* of the liquid, that is, the Fourier transform of the *pair correlation function* $g(\boldsymbol{r})$:

$$S(\boldsymbol{k}) = 1 + n\int (g(\boldsymbol{r}) - 1)e^{i\boldsymbol{k}\cdot\boldsymbol{r}}d\boldsymbol{r}; \tag{6}$$

it may be recalled here that the function $ng(\boldsymbol{r}_2 - \boldsymbol{r}_1)$ is the probability density for finding an atom in the neighborhood of the point $\boldsymbol{r}_2$ when another one is known to be at the point $\boldsymbol{r}_1$; see Section 10.7. The optimal wavefunction is, therefore, given by

$$\Psi = \Phi \sum_i e^{i\boldsymbol{k}\cdot\boldsymbol{r}_i}. \tag{7}$$

Now the momentum associated with this excited state is $\hbar\boldsymbol{k}$ because

$$\boldsymbol{P}\Psi = \left(-i\hbar \sum_i \nabla_i\right)\Psi = \hbar\boldsymbol{k}\Psi, \tag{8}$$

$\boldsymbol{P}\Phi$ being identically equal to zero. Naturally, this would be interpreted as the momentum $\boldsymbol{p}$ associated with the excitation. One thus obtains, from first principles, the energy–momentum relationship for the elementary excitations in a Bose liquid.

On physical grounds one can show that, for small k, the structure factor $S(k)$ rises linearly as $\hbar k/2mc$, reaches a maximum near $k = 2\pi/r_0$ (corresponding to a maximum in the pair correlation function at the nearest-neighbor spacing r_0, which for liquid He^4 is about 3.6 Å) and thereafter decreases to approach, with minor oscillations (corresponding to the subsidiary maxima in the pair correlation function at the spacings of the next nearest neighbors), the limiting value 1 for large k; the limiting value 1 arises from the presence of a delta function in the expression for $g(r)$ (because, as $r_2 \to r_1$, one is sure to find an atom there).[10] Accordingly, the energy $\varepsilon(k)$ of an elementary excitation in liquid He^4 would start linearly as $\hbar kc$, show a "dip" at $k_0 \simeq 2\text{Å}^{-1}$ and rise again to approach the eventual limit of $\hbar^2k^2/2m$.[11] These features are shown in Figure 11.2. Clearly, Feynman's approach merges both phonons and rotons into a single, unified scheme in which they represent

[10]For a microscopic study of the structure factor $S(k)$, see Huang and Klein (1964); also Jackson and Feenberg (1962).

[11]It is natural that at some value of $k < k_0$, the (ε, k)-curve passes through a *maximum*; this happens when $dS/dk = 2S/k$.

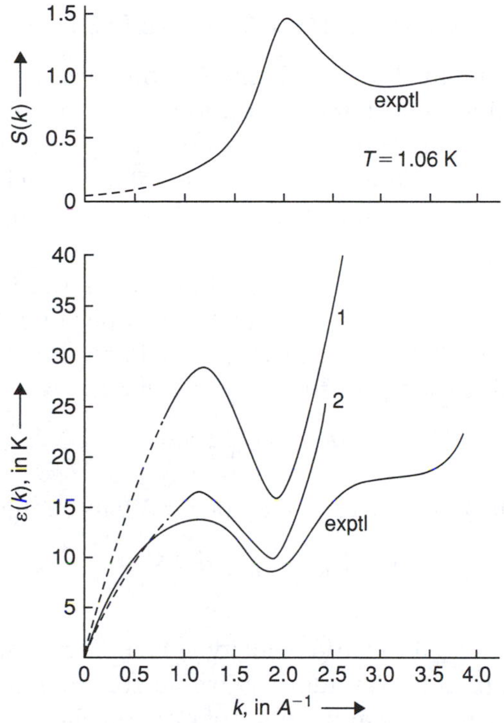

FIGURE 11.2 The energy spectrum of the elementary excitations in liquid He^4. The upper portion shows the structure factor of the liquid, as derived by Henshaw (1960) from experimental data on neutron diffraction. Curve 1 in the lower portion shows the energy–momentum relationship based on the Feynman formula (5) while curve 2 is based on an improved formula by Feynman and Cohen (1956). For comparison, the experimental results of Woods (1966) obtained directly from neutron scattering are also included.

different parts of a common (and continuous) energy spectrum $\varepsilon(k)$, as determined by the structure of the liquid through the function $S(k)$. Since no motion of a rotational character is involved here, the name "roton" is clearly a misnomer.

It seems appropriate to mention here that, soon after the work of London, which advocated a connection between the phase transition in liquid He^4 and the phenomenon of Bose–Einstein condensation, Bijl (1940) investigated the mathematical structure of the wavefunctions appropriate to an interacting Bose gas and the excitation energy associated with those wavefunctions. His picture corresponded very closely to Feynman's and indeed led to the wavefunction (7). Bijl also derived an expression for $\varepsilon(k)$ that was exactly the same as (5). Unfortunately, he could not make much use of his results — primarily because he leaned too heavily on the expansion

$$S(k) = S(0) + C_2 k^2 + C_4 k^4 + \cdots, \tag{9}$$

which, as we now know, represents neither phonons nor rotons.

11.5 States with quantized circulation

We now proceed to examine the possibility of "organized motion" in the ground state of a Bose fluid. In this context, the most important concept is embodied in the *circulation theorem* of Feynman (1955), which establishes a physical basis for the existence of "quantized vortex motion" in the fluid. In the case of liquid helium II, this concept has successfully resolved some of the vital questions that baffled superfluid physicists for a long time.

The ground-state wavefunction of a superfluid, composed of N bosons, may be denoted by a symmetric function $\Phi(\boldsymbol{r}_1, \ldots, \boldsymbol{r}_N)$; if the superfluid does not partake in any organized motion, then Φ will be a pure real number. If, on the other hand, it possesses a *uniform* mass-motion with velocity $\boldsymbol{v}_s$, then its wavefunction would be

$$\Psi = \Phi e^{i(\boldsymbol{P}_s \cdot \boldsymbol{R})/\hbar} = \Phi e^{im(\boldsymbol{v}_s \cdot \Sigma_i \boldsymbol{r}_i)/\hbar}, \tag{1}$$

where $\boldsymbol{P}_s$ denotes the total momentum of the fluid and $\boldsymbol{R}$ its center of mass:

$$\boldsymbol{P}_s = Nm\boldsymbol{v}_s; \quad \boldsymbol{R} = N^{-1} \sum_i \boldsymbol{r}_i. \tag{2}$$

The wavefunction (1) is *exact* if the drift velocity $\boldsymbol{v}_s$ is uniform throughout the fluid. If $\boldsymbol{v}_s$ is nonuniform, then the present wavefunction would still be good locally — in the sense that the phase change $\Delta\phi$ resulting from a "set of local displacements" of the atoms (over distances too small for velocity variations to be appreciable) would be practically the same as the one following from expression (1). Thus, for a given set of displacements $\Delta\boldsymbol{r}_i$ of the atoms constituting the fluid, the change in the phase of the wavefunction would very nearly be

$$\Delta\phi = \frac{m}{\hbar} \sum_i (\boldsymbol{v}_{si} \cdot \Delta\boldsymbol{r}_i), \tag{3}$$

where $\boldsymbol{v}_s$ is now a function of $\boldsymbol{r}$.

The foregoing result may be used for calculating the net phase change resulting from a displacement of atoms along a ring, from their original positions in the ring to the neighboring ones, so that after displacement we obtain a configuration that is physically identical to the one we started with; see Figure 11.3. In view of the symmetry of the wavefunction, the net phase change resulting from such a displacement must be an integral multiple of 2π (so that the wavefunction after the displacement is identical to the one before the displacement):

$$\frac{m}{\hbar} \sum\nolimits_i' (\boldsymbol{v}_{si} \cdot \Delta\boldsymbol{r}_i) = 2\pi n, \quad n = 0, \pm 1, \pm 2, \ldots; \tag{4}$$

the summation $\sum'$ here goes over all the atoms constituting the ring. We note that, for the foregoing result to be valid, it is *only* the individual $\Delta\boldsymbol{r}_i$ that have to be small, *not* the whole perimeter of the ring. Now, for a ring of a *macroscopic* size, one may regard the fluid as a

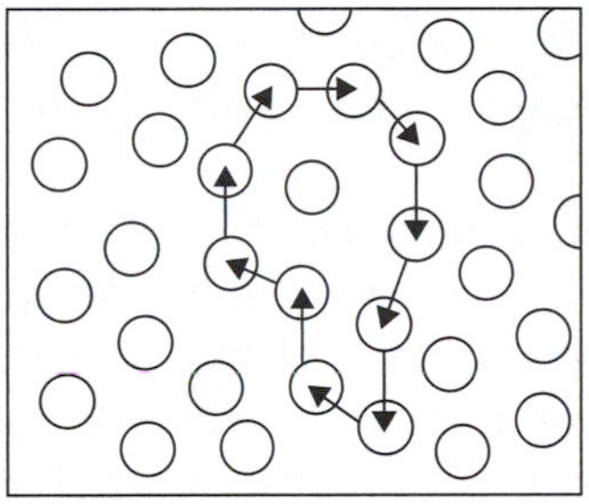

FIGURE 11.3 The wavefunction of the fluid must not change as a result of a permutation of the atoms. If all the atoms are displaced around a ring, as shown, the phase change must be a multiple of 2π.

continuum; equation (4) then becomes

$$\oint \boldsymbol{v}_s \cdot d\boldsymbol{r} = n\frac{h}{m}, \quad n = 0, \pm 1, \pm 2, \ldots. \tag{5}$$

The quantity on the left side of this equation is, by definition, the *circulation* (of the flow) associated with the circuit of integration and is clearly quantized, the "quantum of circulation" being h/m. Equation (5) constitutes the *circulation theorem* of Feynman; it bears a striking resemblance to the quantum condition of Bohr, namely

$$\oint p\,dq = nh, \tag{6}$$

though the region of application here is macroscopic rather than microscopic.[12]

By Stokes' theorem, equation (5) may be written as

$$\int_S (\text{curl}\, \boldsymbol{v}_s) \cdot d\boldsymbol{S} = n\frac{h}{m}, \quad n = 0, \pm 1, \pm 2, \ldots, \tag{7}$$

where $\boldsymbol{S}$ denotes the area enclosed by the circuit of integration. If this area is "simply-connected" and the velocity $\boldsymbol{v}_s$ is continuous throughout the area, then the domain of integration can be shrunk in a *continuous* manner without limit. The integral on the left side is then expected to decrease *continuously* and finally tend to zero. The right side, however, *cannot* vary continuously. We infer that in this particular case the quantum number n must be zero, that is, our integral must be identically vanishing. Thus, "in a simply-connected region, in which the velocity field is continuous throughout, the condition

$$\text{curl}\, \boldsymbol{v}_s = 0 \tag{8}$$

[12]That the vortices in a superfluid may be quantized, the quantum of circulation being h/m, was first suggested by Onsager (1949) in a footnote to a paper dealing with the classical vortex theory and the theory of turbulence!

holds everywhere." This is precisely the condition postulated by Landau (1941), which has been the cornerstone of our understanding of the hydrodynamic behavior of superfluid helium.[13]

Clearly, the Landau condition is only a special case of the Feynman theorem. It is quite possible that in a "multiply-connected" domain, which cannot be shrunk *continuously* to zero (without encountering singularities in the velocity field), the Landau condition may not hold everywhere. A typical example of such a domain is provided by the flow of a *vortex*, which is a planar flow with cylindrical symmetry, such that

$$v_\rho = 0, \quad v_\phi = \frac{K}{2\pi\rho}, \quad v_z = 0, \tag{9}$$

where ρ is the distance measured perpendicular to the axis of symmetry while K is the circulation of the flow:

$$\oint \boldsymbol{v} \cdot d\boldsymbol{r} = \oint v_\phi(\rho\, d\phi) = K; \tag{10}$$

note that the circuit of integration in (10) must enclose the axis of the vortex.

Another version of the foregoing result is

$$\int_S (\text{curl}\, \boldsymbol{v}) \cdot d\boldsymbol{S} = \int_S \left\{ \frac{1}{\rho} \frac{d}{d\rho}(\rho v_\phi) \right\} (2\pi\rho\, d\rho) = K. \tag{11}$$

Now, at all $\rho \neq 0$, curl $\boldsymbol{v} = 0$ but at $\rho = 0$, where v_ϕ is singular, curl $\boldsymbol{v}$ appears to be indeterminate; it is not difficult to see that, at $\rho = 0$, curl $\boldsymbol{v}$ diverges (in such a way that the integral in (11) turns out to be finite). In this context, it seems worthwhile to point out that if we carry out the integration in (10) along a circuit that *does not* enclose the axis of the vortex, or in (11) over a region that does not include the point $\rho = 0$, the result would be identically zero.

At this stage we note that the energy associated with a unit length of a *classical* vortex is given by

$$\frac{\mathcal{E}}{L} = \int_a^b \frac{1}{2} \{2\pi\rho\, d\rho (mn_0)\} \left(\frac{K}{2\pi\rho} \right)^2 = \frac{mn_0 K^2}{4\pi} \ln(b/a). \tag{12}$$

[13]Drawing on the well-known analogy between the phenomena of superfluidity and superconductivity, and the resulting correspondence between the mechanical momentum $m\boldsymbol{v}_s$ of a superfluid particle and the electromagnetic momentum $2e\boldsymbol{A}/c$ of a Cooper pair of electrons, we observe that the relevant counterpart of the Landau condition, in superconductors, would be

$$\text{curl}\, \boldsymbol{A} \equiv \boldsymbol{B} = 0, \tag{8a}$$

which is precisely the *Meissner effect*; furthermore, the appropriate counterpart of the Feynman theorem would be

$$\int_S \boldsymbol{B} \cdot d\boldsymbol{S} = n \frac{hc}{2e}, \tag{7a}$$

which leads to the "quantization of the magnetic flux," the quantum of flux being $hc/2e$.

Here, (mn_0) is the mass density of the fluid (which is assumed to be uniform), the upper limit b is related to the size of the container while the lower limit a depends on the structure of the vortex; in our study, a would be comparable to the interatomic separation.

In the *quantum-mechanical* case we may describe our vortex through a self-consistent wavefunction $\psi(\boldsymbol{r})$, which, in the case of cylindrical symmetry, see equation (9), may be written as

$$\psi(\boldsymbol{r}) = n^{*1/2} e^{is\phi} f_s(\rho), \tag{13}$$

so that

$$n(\boldsymbol{r}) \equiv |\psi(\boldsymbol{r})|^2 = n^* f_s^2(\rho). \tag{14}$$

As $\rho \to \infty$, $f_s(\rho) \to 1$, so that n^* becomes the limiting particle density in the fluid in regions far away from the axis of the vortex. The velocity field associated with this wavefunction will be

$$\boldsymbol{v}(\boldsymbol{r}) = \frac{\hbar}{2im(\psi^*\psi)}(\psi^*\nabla\psi - \psi\nabla\psi^*)$$

$$= \frac{\hbar}{m}\nabla(s\phi) = \left(0, s\frac{\hbar}{m\rho}, 0\right). \qquad (15)^{14}$$

Comparing (15) with (9), we conclude that the circulation K in the present case is sh/m; by the circulation theorem, s must be an integer:

$$s = 0, \pm 1, \pm 2, \ldots. \tag{16}$$

Clearly, the value 0 is of no interest to us. Furthermore, the negative values of s differ from the positive ones only in the "sense of rotation" of the fluid. It is, therefore, sufficient to consider the positive values alone, namely

$$s = 1, 2, 3, \ldots. \tag{17}$$

The function $f_s(\rho)$ appearing in equation (13) may be determined with the help of a Schrödinger equation in which the potential term is itself ψ-dependent, namely

$$\left(-\frac{\hbar^2}{2m}\nabla^2 + u_0|\psi|^2\right)\psi = \varepsilon\psi, \tag{18}$$

[14]It is of interest to see that the angular momentum per particle in the fluid is given by

$$\frac{1}{\psi}\left(\frac{\hbar}{i}\frac{\partial}{\partial\phi}\psi\right) = s\hbar (= mv_\phi \rho);$$

this is again reminiscent of the quantum condition of Bohr.

where u_0 is given by equation (11.1.51):

$$u_0 = 4\pi a\hbar^2/m, \tag{19}$$

a being the scattering length of the interparticle interaction operating in the fluid. The characteristic energy ε follows from the observation that, at large distances from the axis of the vortex, the fluid is essentially uniform in density, with $n(\boldsymbol{r}) \to n^*$; equation (18) then gives

$$\varepsilon = u_0 n^* = 4\pi a\hbar^2 n^*/m, \tag{20}$$

which may be compared with equation (11.2.9). Substituting (20) into (18) and remembering that the flow is cylindrically symmetrical, we get

$$-\left[\frac{1}{\rho}\frac{d}{d\rho}\left\{\rho\frac{d}{d\rho}f_s(\rho)\right\} - \frac{s^2}{\rho^2}f_s(\rho)\right] + 8\pi an^* f_s^3(\rho) = 8\pi an^* f_s(\rho). \tag{21}$$

Expressing ρ in terms of a characteristic length l,

$$\rho = l\rho' \quad \{l = (8\pi an^*)^{-1/2}\}, \tag{22}$$

we obtain

$$\frac{d^2 f_s}{d\rho'^2} + \frac{1}{\rho'}\frac{df_s}{d\rho'} + \left(1 - \frac{s^2}{\rho'^2}\right)f_s - f_s^3 = 0. \tag{23}$$

Toward the axis of the vortex, where $\rho \to 0$, the very high velocity of the fluid (and the very large centrifugal force accompanying it) will push the particles outward, thus causing an enormous decrease in the density of the fluid. Consequently, the function f_s should tend to zero as $\rho \to 0$. This will make the last term in equation (23) negligible and thereby reduce it to the familiar Bessel's equation. Accordingly, for small ρ,

$$f_s(\rho') \sim J_s(\rho') \sim \rho^s, \tag{24}$$

J_s being the *ordinary Bessel function* of order s. For $\rho' \gg 1, f_s \simeq 1$; then, the first two terms of equation (23) become negligible, with the result

$$f_s(\rho') \simeq 1 - \frac{s^2}{2\rho'^2}. \tag{25}$$

The full solution is obtained by integrating the equation numerically; the results so obtained are shown in Figure 11.4 where solutions for $s = 1, 2$, and 3 are displayed.

We thus find that our model of an imperfect Bose gas does allow for the presence of quantized vortices in the system. Not only that, we do not have to invoke here any special assumptions regarding the nature of the "core" of the vortex (as one has to do in the classical theory); our treatment naturally leads to a continual diminution of the particle density

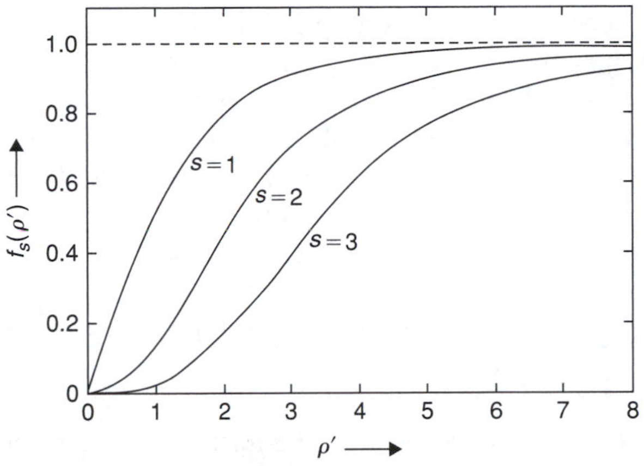

FIGURE 11.4 Solutions of equation (23) for various values of s (after Kawatra and Pathria, 1966).

n as the axial line is approached, so that there does not exist any specific distribution of vorticity around this line. The distance scale, which governs the spatial variation of n, is provided by the parameter l of equation (22); for liquid He^4, $l \simeq 1\,\text{Å}$.

Pitaevskii (1961), who was among the first to demonstrate the possibility of obtaining solutions whose natural interpretation lay in quantized vortex motion (see also Gross, 1961; Weller, 1963), also evaluated the energy per unit length of the vortex. Employing wavefunction (13), with known values of the functions $f_s(\rho)$, Pitaevskii obtained the following results for the energy per unit length of the vortex, with $s = 1, 2$, or 3,

$$\frac{n^* h^2}{4\pi m}\{1\ln(1.46R/l),\quad 4\ln(0.59R/l),\quad 9\ln(0.38R/l)\}, \tag{26}$$

where R denotes the outer radius of the domain involved. The above results may be compared with the "semiclassical" ones, namely

$$\frac{n_0 h^2}{4\pi m}\{1\ln(R/a),\quad 4\ln(R/a),\quad 9\ln(R/a)\}, \tag{27}$$

which follow from formula (12), with K replaced by sh/m and b by R. It is obvious that vortices with $s > 1$ would be relatively unstable because energetically it would be cheaper for a system to have s vortices of unit circulation rather than a single vortex of circulation s.

The existence of quantized vortex lines in liquid helium II has been demonstrated convincingly by the ingenious experiments of Vinen (1958–1961) in which the circulation K around a fine wire immersed in the liquid was measured through the influence it exerts on the transverse vibrations of the wire. Vinen found that while vortices with unit circulation were exceptionally stable those with higher circulation too made an appearance. Repeating Vinen's experiment with thicker wires, Whitmore and Zimmermann (1965) were able to observe stable vortices with circulation up to three quantum units. For a survey of this and other aspects of the superfluid behavior, see Vinen (1968) and Betts et al. (1969). Kim and Chan (2004) have even observed a "supersolid" phase of helium-4 at low temperatures that has the crystalline structure of a solid while also exhibiting superfuid-like flow.

For the relevance of quantized vortex lines to the problem of "rotation" of the superfluid, see Section 10.7 of the first edition of this book.

11.6 Quantized vortex rings and the breakdown of superfluidity

Feynman (1955) was the first to suggest that the formation of vortices in liquid helium II might provide the mechanism responsible for the breakdown of superfluidity in the liquid. He considered the flow of liquid helium II from an orifice of diameter D and, by tentative arguments, found that the velocity v_0 at which the flow energy available would just be sufficient to create quantized vortices in the liquid is given by

$$v_0 = \frac{\hbar}{mD}\ln(D/l). \qquad (1)$$

Thus, for an orifice of diameter 10^{-5} cm, v_0 would be of the order of 1 m/s.[15] It is tempting to identify v_0 with v_c, the *critical velocity* of superflow through the given capillary, despite the fact that this theoretical estimate for v_0 is an order of magnitude higher than the corresponding experimental values of v_c; the latter, for instance, are 13 cm/s, 8 cm/s, and 4 cm/s for capillary diameters 1.2×10^{-5} cm, 7.9×10^{-5} cm, and 3.9×10^{-4} cm, respectively. Nevertheless, the present estimate is far more desirable than the prohibitively large ones obtained earlier on the basis of a possible creation of phonons or rotons in the liquid; see Section 7.6. Moreover, one obtains here a definitive dependence of the critical velocity of superflow on the width of the capillary employed which, at least qualitatively, agrees with the trend seen in the experimental findings. In what follows, we propose to develop Feynman's idea further along the lines suggested by the preceding section.

So far we have been dealing with the so-called *linear vortices* whose velocity field possesses cylindrical symmetry. More generally, however, a vortex line need not be straight—it may be curved and, if it does not terminate on the walls of the container or on the free surface of the liquid, may close on itself. We then speak of a *vortex ring*, which is very much like a smoke ring. Of course, the quantization condition (11.5.5) is as valid for a vortex ring as for a vortex line. However, the dynamical properties of a ring are quite different from those of a line; see, for instance, Figure 11.5, which shows schematically a vortex ring in cross-section, the radius r of the ring being much larger than the core dimension l. The flow velocity v_s at any point in the field is determined by a superposition of the flow velocities due to the various elements of the ring. It is not difficult to see that the velocity field

[15] We have taken here: $l \simeq 1$ Å, so that $\ln(D/l) \simeq 7$.

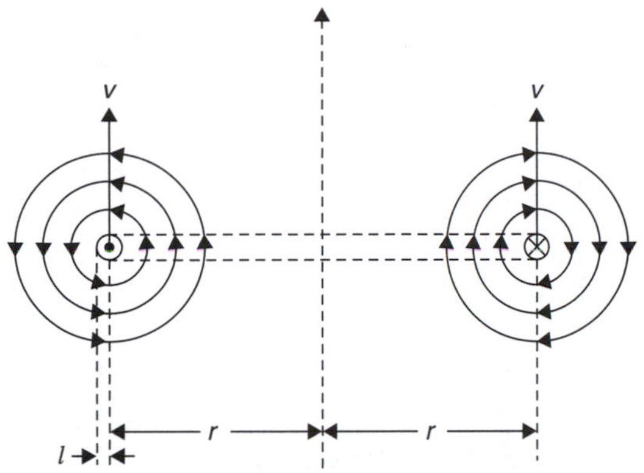

FIGURE 11.5 Schematic illustration of a quantized vortex ring in cross-section.

of the ring, including the ring itself, moves in a direction perpendicular to the plane of the ring, with a velocity[16]

$$v \sim \hbar/2mr; \tag{2}$$

see equation (11.5.15), with $s = 1$ and $\rho \sim 2r$. An estimate of the energy associated with the flow may be obtained from expression (11.5.12), with $L = 2\pi r, K = h/m$, and $b \sim r$; thus

$$\varepsilon \sim 2\pi^2\hbar^2 n_0 m^{-1} r \ln(r/l). \tag{3}$$

Clearly, the dependence of ε on r arises mainly from the factor r and only slightly from the factor $\ln(r/l)$. Therefore, with good approximation, $v \propto \varepsilon^{-1}$, that is, a ring with larger energy moves slower! The reason behind this somewhat startling result is that firstly the larger the ring the larger the distances between the various circulation-carrying elements of the ring (thus reducing the velocity imparted by one element to another) and secondly a larger ring carries with it a larger amount of fluid ($M \propto r^3$), so the total energy associated with the ring is also larger (essentially proportional to Mv^2, i.e., $\propto r$). The product $v\varepsilon$, apart from the slowly varying factor $\ln(r/l)$, is thus a constant, which is equal to $\pi^2\hbar^3 n_0/m^2$.

It is gratifying that vortex rings such as the ones discussed here have been observed and the circulation around them is found to be as close to the quantum h/m as one could expect under the conditions of the experiment. Figure 11.6 shows the experimental results of Rayfield and Reif (1964) for the velocity–energy relationship of free-moving, charge-carrying vortex rings created in liquid helium II by suitably accelerated helium ions. Vortex rings carrying positive as well as negative charge were observed; dynamically, however, they behaved *alike*, as one indeed expects because both the velocity and the energy associated with a vortex ring are determined by the properties of a large amount of fluid carried

[16]This result would be exact if we had a pair of oppositely directed *linear* vortices, with the same cross-section as shown in Figure 11.5. In the case of a ring, the velocity would be somewhat larger.

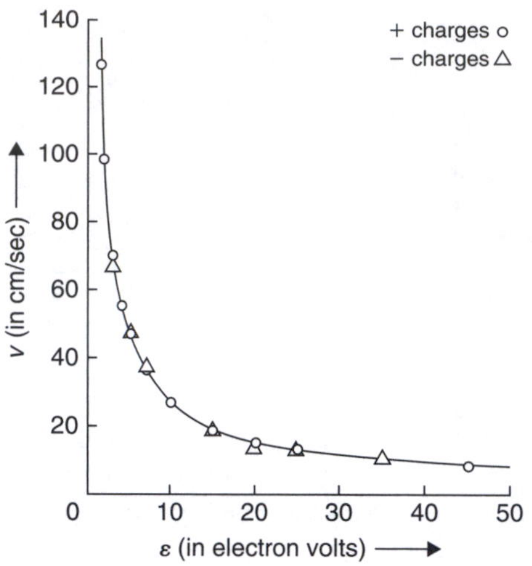

FIGURE 11.6 The velocity–energy relationship of the vortex rings formed in liquid helium II (after Rayfield and Reif, 1964). The points indicate the experimental data, while the curve represents the theoretical relationship based on the "quantum of circulation" h/m.

along with the ring rather than by the small charge coupled to it. Fitting experimental results with the notion of the vortex rings, Rayfield and Reif concluded that their rings carried a circulation of $(1.00 \pm 0.03) \times 10^{-3}\,\text{cm}^2/\text{s}$, which is close to the Onsager–Feynman unit $h/m (= 0.997 \times 10^{-3}\,\text{cm}^2/\text{s})$; moreover, these rings seemed to have a core radius of about 1.2Å, which is comparable with the characteristic parameter l of the fluid.

We shall now show that the dynamics of the quantized vortex rings is such that their creation in liquid helium II does provide a mechanism for the *breakdown of superfluidity*. To see this, it is simplest to consider the case of a superfluid flowing through a capillary of radius R. As the velocity of flow increases and approaches the critical value v_c, quantized vortex rings begin to form and energy dissipation sets in, which in turn brings about the rupture of the superflow. By symmetry, the rings will be so formed that their central plane will be perpendicular to the axis of the capillary and they will be moving in the direction of the main flow. Now, by the Landau criterion (7.6.24), the critical velocity of superflow is directly determined by the energy spectrum of the excitations created:

$$v_c = (\varepsilon/p)_{\text{min}}. \tag{4}$$

We, therefore, require an expression for the momentum p of the vortex ring. In analogy with the classical vortex ring, we may take

$$p = 2\pi^2 \hbar n_0 r^2, \tag{5}$$

which seems satisfactory because (i) it conforms to the general result: $v = (\partial\varepsilon/\partial p)$, though only to a first approximation, and (ii) it leads to the (approximate) dispersion relation:

$\varepsilon \propto p^{1/2}$, which has been separately verified by Rayfield and Reif by subjecting their rings to a transverse electric field. Substituting (3) and (5) into (4), we obtain

$$v_c \sim \left\{ \frac{\hbar}{mr} \ln(r/l) \right\}_{\min}. \tag{6}$$

Now, since the r-dependence of the quantity ε/p arises mainly from the factor $1/r$, the minimum in (6) will be obtained when r has its largest value, namely R, the radius of the capillary. We thus obtain

$$v_c \sim \frac{\hbar}{mR} \ln(R/l), \tag{7}$$

which is very much the same as the original estimate of Feynman — with D replaced by R. Naturally, then, the numerical values of v_c obtained from the new expression (7) continue to be significantly larger than the corresponding experimental values; however, the theory is now much better founded.

Fetter (1963) was the first to account for the fact that, as the radius r of the ring approaches the radius R of the capillary, the influence of the "image vortex" becomes important. The energy of the flow now falls below the asymptotic value given by (3) by a factor of 10 or so which, in turn, reduces the critical velocity by a similar factor. The actual result obtained by Fetter was

$$v_c \simeq \frac{11}{24} \frac{\hbar}{mR} = 0.46 \frac{\hbar}{mR}. \tag{8}$$

Kawatra and Pathria (1966) extended Fetter's calculation by taking into account the boundary effects arising explicitly from the walls of the capillary as well as the ones arising implicitly from the "image vortex"; moreover, in the computation of ε, they employed actual wavefunctions, obtained by solving equation (11.5.23), rather than the analytical approximation employed by Fetter. They obtained

$$v_c \simeq 0.59 \frac{\hbar}{mR}, \tag{9}$$

which is about 30 percent higher than Fetter's value; for comments regarding the "most favorable" location for the formation of the vortex ring in the capillary, see the original reference of Kowatra and Pathria (1966).

11.7 Low-lying states of an imperfect Fermi gas

The Hamiltonian of the quantized field for spin-half fermions ($\sigma = +\frac{1}{2}$ or $-\frac{1}{2}$) is given by equation (11.1.48), namely

$$\hat{H} = \sum_{\boldsymbol{p},\sigma} \frac{p^2}{2m} a^{\dagger}_{\boldsymbol{p}\sigma} a_{\boldsymbol{p}\sigma} + \frac{1}{2} \sum{}' u^{\boldsymbol{p}'_1\sigma'_1, \boldsymbol{p}'_2\sigma'_2}_{\boldsymbol{p}_1\sigma_1, \boldsymbol{p}_2\sigma_2} a^{\dagger}_{\boldsymbol{p}'_1\sigma'_1} a^{\dagger}_{\boldsymbol{p}'_2\sigma'_2} a_{\boldsymbol{p}_2\sigma_2} a_{\boldsymbol{p}_1\sigma_1}, \tag{1}$$

where the matrix elements u are related to the scattering length a of the two-body interaction; the summation in the second part of this expression goes only over those states (of the two particles) that conform to the principles of momentum and spin conservation. As in the Bose case, the matrix elements u in the second sum may be approximated by their values at $\boldsymbol{p}=0$, that is,

$$u_{\boldsymbol{p}_1\sigma_1,\boldsymbol{p}_2\sigma_2}^{\boldsymbol{p}'_1\sigma'_1,\boldsymbol{p}'_2\sigma'_2} \simeq u_{0\sigma_1,0\sigma_2}^{0\sigma'_1,0\sigma'_2}. \tag{2}$$

Then, in view of the antisymmetric character of the product $a_{\boldsymbol{p}_1\sigma_1}a_{\boldsymbol{p}_2\sigma_2}$, see equation (11.1.24c), all those terms of the second sum in (1) that contain identical indices σ_1 and σ_2 vanish on summation over $\boldsymbol{p}_1$ and $\boldsymbol{p}_2$. Similarly, all those terms that contain identical indices σ'_1 and σ'_2 vanish on summation over $\boldsymbol{p}'_1$ and $\boldsymbol{p}'_2$.[17] Thus, for a given set of values of the particle momenta, the only choices for the spin components remaining in the sum are

(i) $\sigma_1=+\frac{1}{2},\quad \sigma_2=-\frac{1}{2};\quad \sigma'_1=+\frac{1}{2},\quad \sigma'_2=-\frac{1}{2}$

(ii) $\sigma_1=+\frac{1}{2},\quad \sigma_2=-\frac{1}{2};\quad \sigma'_1=-\frac{1}{2},\quad \sigma'_2=+\frac{1}{2}$

(iii) $\sigma_1=-\frac{1}{2},\quad \sigma_2=+\frac{1}{2};\quad \sigma'_1=-\frac{1}{2},\quad \sigma'_2=+\frac{1}{2}$

(iv) $\sigma_1=-\frac{1}{2},\quad \sigma_2=+\frac{1}{2};\quad \sigma'_1=+\frac{1}{2},\quad \sigma'_2=-\frac{1}{2}.$

It is not difficult to see that the contribution arising from choice (i) will be identically equal to the one arising from choice (iii), while the contribution arising from choice (ii) will be identically equal to the one arising from choice (iv). We may, therefore, write

$$\hat{H}=\sum_{\boldsymbol{p},\sigma}\frac{p^2}{2m}a^{\dagger}_{\boldsymbol{p}\sigma}a_{\boldsymbol{p}\sigma}+\frac{u_0}{V}\sum{}' a^{\dagger}_{\boldsymbol{p}'_1+}a^{\dagger}_{\boldsymbol{p}'_2-}a_{\boldsymbol{p}_2-}a_{\boldsymbol{p}_1+}, \tag{3}$$

where

$$\frac{u_0}{V}=\left(u_{0+,0-}^{0+,0-}-u_{0+,0-}^{0-,0+}\right), \tag{4}$$

while the indices + and − denote the spin states $\sigma=+\frac{1}{2}$ and $\sigma=-\frac{1}{2}$, respectively; the summation in the second part of (3) now goes over all momenta that conform to the conservation law

$$\boldsymbol{p}'_1+\boldsymbol{p}'_2=\boldsymbol{p}_1+\boldsymbol{p}_2. \tag{5}$$

To evaluate the eigenvalues of Hamiltonian (3), we shall employ the techniques of the *perturbation theory.*

[17] Physically, this means that in the limiting case of *slow collisions* only particles with opposite spins interact with one another.

First of all, we note that the main term in the expression for $\hat{H}$ is already diagonal, and its eigenvalues are

$$E^{(0)} = \sum_{\boldsymbol{p},\sigma} \frac{p^2}{2m} n_{\boldsymbol{p}\sigma}, \tag{6}$$

where $n_{\boldsymbol{p}\sigma}$ is the occupation number of the single-particle state $(\boldsymbol{p},\sigma)$; its mean value, in equilibrium, is given by the Fermi distribution function

$$\bar{n}_{\boldsymbol{p}\sigma} = \frac{1}{z_0^{-1}\exp(p^2/2mkT)+1}. \tag{7}$$

The sum in (6) may be replaced by an integral, with the result (see Section 8.1, with $g = 2$)

$$E^{(0)} = V\frac{3kT}{\lambda^3} f_{5/2}(z_0), \tag{8}$$

where λ is the mean thermal wavelength of the particles,

$$\lambda = h/(2\pi mkT)^{1/2}, \tag{9}$$

while $f_\nu(z_0)$ is the Fermi–Dirac function

$$f_\nu(z_0) = \frac{1}{\Gamma(\nu)}\int_0^\infty \frac{x^{\nu-1}dx}{z_0^{-1}e^x+1} = \sum_{l=1}^\infty (-1)^{l-1}\frac{z_0^l}{l^\nu}; \tag{10}$$

the ideal-gas fugacity z_0 is determined by the total number of particles in the system:

$$N = \sum_{\boldsymbol{p},\sigma} n_{\boldsymbol{p}\sigma} = V\frac{2}{\lambda^3} f_{3/2}(z_0). \tag{11}$$

The *first-order* correction to the energy of the system is given by the diagonal elements of the interaction term, namely the ones for which $\boldsymbol{p}'_1 = \boldsymbol{p}_1$ and $\boldsymbol{p}'_2 = \boldsymbol{p}_2$; thus

$$E^{(1)} = \frac{u_0}{V}\sum_{\boldsymbol{p}_1,\boldsymbol{p}_2} n_{\boldsymbol{p}_1+} n_{\boldsymbol{p}_2-} = \frac{u_0}{V}N^+N^-, \tag{12}$$

where $N^+ (N^-)$ denotes the total number of particles with spin up (down). Substituting the *equilibrium* values $\overline{N^+} = \overline{N^-} = \frac{1}{2}N$, we obtain

$$E^{(1)} = \frac{u_0}{4V}N^2 = V\frac{u_0}{\lambda^6}\{f_{3/2}(z_0)\}^2. \tag{13}$$

Substituting $u_0 \simeq 4\pi a\hbar^2/m$, see equation (11.1.51), we obtain to *first order* in a

$$E_1^{(1)} = \frac{\pi a\hbar^2}{m}\frac{N}{V}N = V\frac{2kT}{\lambda^3}\left(\frac{a}{\lambda}\right)\{f_{3/2}(z_0)\}^2. \tag{14}$$

The *second-order* correction to the energy of the system can be obtained with the help of the formula

$$E_n^{(2)} = \sum_{m \neq n} \frac{|V_{nm}|^2}{E_n - E_m}, \tag{15}$$

where the indices n and m pertain to the unperturbed states of the system. A simple calculation yields:

$$E^{(2)} = 2\frac{u_0^2}{V^2} \sum_{\boldsymbol{p}_1, \boldsymbol{p}_2, \boldsymbol{p}_1'} \frac{n_{\boldsymbol{p}_1+} n_{\boldsymbol{p}_2-} \left(1 - n_{\boldsymbol{p}_1'+}\right)\left(1 - n_{\boldsymbol{p}_2'-}\right)}{\left(p_1^2 + p_2^2 - p_1'^2 - p_2'^2\right)/2m}, \tag{16}$$

where the summation goes over *all* $\boldsymbol{p}_1, \boldsymbol{p}_2$, and $\boldsymbol{p}_1'$ (the value of $\boldsymbol{p}_2'$ being fixed by the requirement of momentum conservation); it is understood that we do not include in the sum (16) any terms for which $p_1^2 + p_2^2 = p_1'^2 + p_2'^2$. It will be noted that the numerator of the summand in (16) is closely related to the fact that the squared matrix element for the transition $(\boldsymbol{p}_1, \boldsymbol{p}_2) \to (\boldsymbol{p}_1', \boldsymbol{p}_2')$ is directly proportional to the probability that "the states $\boldsymbol{p}_1$ and $\boldsymbol{p}_2$ are occupied and at the same time the states $\boldsymbol{p}_1'$ and $\boldsymbol{p}_2'$ are unoccupied."

Now, expression (16) does not in itself exhaust terms of *second order* in a. A contribution of the same order of magnitude arises from expression (12) if for u_0 we employ an expression more accurate than the one just employed. The desired expression can be obtained in the same manner as in the Bose case; check the steps leading to equations (11.2.5) and (11.2.6). In the present case, we obtain

$$\frac{4\pi a\hbar^2}{mV} \simeq \frac{u_0}{V} + 2\frac{u_0^2}{V^2} \sum_{\boldsymbol{p}_1, \boldsymbol{p}_2, \boldsymbol{p}_1'} \frac{1}{\left(p_1^2 + p_2^2 - p_1'^2 - p_2'^2\right)/2m},$$

from which it follows that

$$u_0 \simeq \frac{4\pi a\hbar^2}{m}\left[1 - \frac{8\pi a\hbar^2}{mV} \sum_{\boldsymbol{p}_1, \boldsymbol{p}_2, \boldsymbol{p}_1'} \frac{1}{\left(p_1^2 + p_2^2 - p_1'^2 - p_2'^2\right)/2m}\right]. \tag{17}$$

Substituting (17) into (12), we obtain, apart from the first-order term already given in (14), a second order term, namely

$$E_2^{(1)} = -2\left(\frac{4\pi a\hbar^2}{mV}\right)^2 \sum_{\boldsymbol{p}_1, \boldsymbol{p}_2, \boldsymbol{p}_1'} \frac{n_{\boldsymbol{p}_1+} n_{\boldsymbol{p}_2-}}{\left(p_1^2 + p_2^2 - p_1'^2 - p_2'^2\right)/2m}. \tag{18}$$

For a comparable term given in (16), the approximation $u_0 \simeq 4\pi a\hbar^2/m$ is sufficient, with the result

$$E_2^{(2)} = 2\left(\frac{4\pi a\hbar^2}{mV}\right)^2 \sum_{\boldsymbol{p}_1,\boldsymbol{p}_2,\boldsymbol{p}'_1} \frac{n_{\boldsymbol{p}_1+}n_{\boldsymbol{p}_2-}\left(1-n_{\boldsymbol{p}'_1+}\right)\left(1-n_{\boldsymbol{p}'_2-}\right)}{\left(p_1^2+p_2^2-p_1'^2-p_2'^2\right)/2m}. \tag{19}$$

Combining (18) and (19), we obtain[18]

$$E_2 = E_2^{(1)} + E_2^{(2)} = -2\left(\frac{4\pi a\hbar^2}{mV}\right)^2 \sum_{\boldsymbol{p}_1,\boldsymbol{p}_2,\boldsymbol{p}'_1} \frac{n_{\boldsymbol{p}_1+}n_{\boldsymbol{p}_2-}\left(n_{\boldsymbol{p}'_1+}+n_{\boldsymbol{p}'_2-}\right)}{\left(p_1^2+p_2^2-p_1'^2-p_2'^2\right)/2m}. \tag{20}$$

To evaluate the sum in (20), we prefer to write it as a *symmetrical* summation over the four momenta $\boldsymbol{p}_1, \boldsymbol{p}_2, \boldsymbol{p}'_1$, and $\boldsymbol{p}'_2$ by introducing a Kronecker delta to take care of the momentum conservation; thus

$$E_2 = -2\left(\frac{4\pi a\hbar^2}{mV}\right)^2 \sum_{\boldsymbol{p}_1,\boldsymbol{p}_2,\boldsymbol{p}'_1,\boldsymbol{p}'_2} \frac{n_{\boldsymbol{p}_1+}n_{\boldsymbol{p}_2-}\left(n_{\boldsymbol{p}'_1+}+n_{\boldsymbol{p}'_2-}\right)\delta_{\boldsymbol{p}_1+\boldsymbol{p}_2,\boldsymbol{p}'_1+\boldsymbol{p}'_2}}{\left(p_1^2+p_2^2-p_1'^2-p_2'^2\right)/2m}. \tag{21}$$

It is obvious that the two parts of the sum (21), one arising from the factor $n_{\boldsymbol{p}'_1+}$ and the other from the factor $n_{\boldsymbol{p}'_2-}$, would give identical results on summation. We may, therefore, write

$$E_2 = -4\left(\frac{4\pi a\hbar^2}{mV}\right)^2 \sum_{\boldsymbol{p}_1,\boldsymbol{p}_2,\boldsymbol{p}'_1,\boldsymbol{p}'_2} \frac{n_{\boldsymbol{p}_1+}n_{\boldsymbol{p}_2-}n_{\boldsymbol{p}'_1+}\delta_{\boldsymbol{p}_1+\boldsymbol{p}_2,\boldsymbol{p}'_1+\boldsymbol{p}'_2}}{\left(p_1^2+p_2^2-p_1'^2-p_2'^2\right)/2m}. \tag{22}$$

The sum in (22) can be evaluated by following a procedure due to Huang, Yang, and Luttinger (1957), with the result[19]

$$E_2 = V\frac{8kT}{\lambda^3}\left(\frac{a^2}{\lambda^2}\right)F(z_0), \tag{23}$$

where

$$F(z_0) = -\sum_{r,s,t=1}^{\infty} \frac{(-z_0)^{r+s+t}}{\sqrt{(rst)}(r+s)(r+t)}. \tag{24}$$

[18]We have omitted here terms containing a "product of *four* n's" for the following reason: in view of the fact that the numerator of such terms would be symmetric and the denominator antisymmetric with respect to the exchange operation $(\boldsymbol{p}_1, \boldsymbol{p}_2) \leftrightarrow (\boldsymbol{p}'_1, \boldsymbol{p}'_2)$, their sum over the variables $\boldsymbol{p}_1, \boldsymbol{p}_2, \boldsymbol{p}'_1$ (and $\boldsymbol{p}'_2$) would vanish identically.

[19]For a direct evaluation of the sum (22), in the limit $T \to 0$, see Abrikosov and Khalatnikov (1957). See also Problem 11.12.

Combining (8), (14), and (23), we obtain to *second order* in the scattering length a

$$E = V\frac{kT}{\lambda^3}\left[3f_{5/2}(z_0) + \frac{2a}{\lambda}\{f_{3/2}(z_0)\}^2 + \frac{8a^2}{\lambda^2}F(z_0)\right], \tag{25}$$

where z_0 is determined by (11).

It is now straightforward to obtain the *ground-state energy* of the imperfect Fermi gas ($z_0 \to \infty$); we have simply to know the asymptotic behavior of the functions involved. For the functions $f_\nu(z_0)$, we have from the Sommerfeld lemma (see Appendix E)

$$f_\nu(z_0) \approx (\ln z_0)^\nu / \Gamma(\nu+1), \tag{26}$$

so that

$$f_{5/2}(z_0) \approx \frac{8}{15\pi^{1/2}}(\ln z_0)^{5/2}; \quad f_{3/2}(z_0) \approx \frac{4}{3\pi^{1/2}}(\ln z_0)^{3/2}. \tag{27}$$

Equation (11) then gives

$$n = \frac{N}{V} \approx \frac{8}{3\pi^{1/2}\lambda^3}(\ln z_0)^{3/2}, \tag{28}$$

so that

$$\ln z_0 \approx \lambda^2\left(\frac{3\pi^{1/2}n}{8}\right)^{2/3}. \tag{29}$$

The asymptotic behaviour of $F(z_0)$ is given by

$$F(z_0) \approx \frac{16(11-2\ln 2)}{105\pi^{3/2}}(\ln z_0)^{7/2}; \tag{30}$$

see Problem 11.12. Substituting (27) and (30) into (25), and making use of relation (29), we finally obtain

$$\frac{E_0}{N} = \frac{3}{10}\frac{\hbar^2}{m}(3\pi^2 n)^{2/3} + \frac{\pi a\hbar^2}{m}n\left\{1 + \frac{6}{35}(11-2\ln 2)\left(\frac{3}{\pi}\right)^{1/3}n^{1/3}a\right\}. \tag{31}$$

The ground-state pressure of the gas is then given by

$$\begin{aligned} P_0 &= n^2\frac{\partial(E_0/N)}{\partial n} \\ &= \frac{1}{5}\frac{\hbar^2}{m}(3\pi^2 n)^{2/3}n + \frac{\pi a\hbar^2}{m}n^2\left\{1 + \frac{8}{35}(11-2\ln 2)\left(\frac{3}{\pi}\right)^{1/3}n^{1/3}a\right\}. \end{aligned} \tag{32}$$

We may also calculate the velocity of sound, which directly involves the compressibility of the system, with the result

$$\begin{aligned} c_0^2 &= \frac{\partial P_0}{\partial(mn)} \\ &= \frac{1}{3}\frac{\hbar^2}{m^2}(3\pi^2 n)^{2/3} + \frac{2\pi a\hbar^2}{m^2} n \left\{1 + \frac{4}{15}(11 - 2\ln 2)\left(\frac{3}{\pi}\right)^{1/3} n^{1/3} a\right\}. \end{aligned} \quad (33)$$

The leading terms of the foregoing expressions represent the ground-state results for an ideal Fermi gas, while the remaining terms represent corrections arising from the interparticle interactions.

The result embodied in equation (31) was first obtained by Huang and Yang (1957) by the method of pseudopotentials; Martin and De Dominicis (1957) were the first to attempt an estimate of the third-order correction.[20] Lee and Yang (1957) obtained (31) on the basis of the binary collision method; for the details of their calculation, see Lee and Yang (1959b, 1960a). The same result was derived somewhat later by Galitskii (1958) who employed the method of Green's functions.

11.8 Energy spectrum of a Fermi liquid: Landau's phenomenological theory[21]

In Section 11.4 we discussed the main features of the energy spectrum of a Bose liquid; such a spectrum is generally referred to as a *Bose type* spectrum. A liquid consisting of spin-half fermions, such as liquid He^3, is expected to have a different kind of spectrum which, by contrast, may be called a *Fermi type* spectrum.

Right away we should emphasize that a liquid consisting of fermions may not necessarily possess a spectrum of the Fermi type; the spectrum actually possessed by such a liquid depends crucially on the nature of the interparticle interactions operating in the liquid. The discussion here assumes that the interactions are strictly repulsive so that the fermions have no opportunity to form bosonic pairs. In the present section, we propose to discuss the main features of a spectrum which is characteristically of the Fermi type. The effects of attractive interactions will be discussed in Section 11.9

According to Landau (1956), whose work provides the basic framework for our discussion, the Fermi type spectrum of a quantum liquid is constructed in analogy with the spectrum of an ideal Fermi gas. As is well-known, the ground state of the ideal Fermi gas corresponds to a "complete filling up of the single-particle states with $p \leq p_F$ and a complete absence of particles in the states with $p > p_F$"; the excitation of the system corresponds to a transition of one or more particles from the occupied states to the unoccupied states. The limiting momentum p_F is related to the particle density in the system and, for

[20] The third-order correction has also been discussed by Mohling (1961).
[21] For a *microscopic* theory of a Fermi liquid, see Nozières (1964); see also Tuttle and Mohling (1966).

spin-half particles, is given by

$$p_F = \hbar(3\pi^2 N/V)^{1/3}. \tag{1}$$

In a liquid, we cannot speak of quantum states of *individual* particles. However, as a basis for constructing the desired spectrum, we may assume that, as interparticle interactions are gradually "switched on" and a transition made from the gaseous to the liquid state, the ordering of the energy levels (in the momentum space) remains unchanged. Of course, in this ordering, the role of the gas particles is passed on to the "elementary excitations" of the liquid (also referred to as "quasiparticles"), whose number coincides with the number of particles in the liquid and which also obey Fermi statistics. Each "quasiparticle" possesses a definite momentum $\boldsymbol{p}$, so we can speak of a *distribution function* $n(\boldsymbol{p})$ such that

$$\int n(\boldsymbol{p}) d\tau = N/V, \tag{2}$$

where $d\tau = 2d^3p/h^3$. We then expect that the specification of the function $n(\boldsymbol{p})$ uniquely determines the total energy E of the liquid. Of course, E will not be given by a simple sum of the energies $\varepsilon(\boldsymbol{p})$ of the quasiparticles; it will rather be a *functional* of the distribution function $n(\boldsymbol{p})$. In other words, the energy E will not reduce to the simple integral $\int \varepsilon(\boldsymbol{p}) n(\boldsymbol{p}) V d\tau$, though in the first approximation a *variation* in its value may be written as

$$\delta E = V \int \varepsilon(\boldsymbol{p}) \delta n(\boldsymbol{p}) d\tau, \tag{3}$$

where $\delta n(\boldsymbol{p})$ is an *assumed* variation in the distribution function of the "quasiparticles." The reason E does not reduce to an integral of the quantity $\varepsilon(\boldsymbol{p}) n(\boldsymbol{p})$ is related to the fact that the quantity $\varepsilon(\boldsymbol{p})$ is itself a functional of the distribution function. If the initial distribution function is a step function (which corresponds to the ground state of the system), then the variation in $\varepsilon(\boldsymbol{p})$ due to a *small* deviation of the distribution function from the step function (which implies only *low-lying* excited states of the system) would be given by a *linear* functional relationship:

$$\delta\varepsilon(\boldsymbol{p}) = \int f(\boldsymbol{p}, \boldsymbol{p}') \delta n(\boldsymbol{p}') d\tau'. \tag{4}$$

Thus, the quantities $\varepsilon(\boldsymbol{p})$ and $f(\boldsymbol{p}, \boldsymbol{p}')$ are the first and second *functional derivatives* of E with respect to $n(\boldsymbol{p})$. Inserting spin dependence, we may now write

$$\delta E = \sum_{\boldsymbol{p},\sigma} \varepsilon(\boldsymbol{p},\sigma) \delta n(\boldsymbol{p},\sigma) + \frac{1}{2V} \sum_{\boldsymbol{p},\sigma;\boldsymbol{p}',\sigma'} f(\boldsymbol{p},\sigma;\boldsymbol{p}',\sigma') \delta n(\boldsymbol{p},\sigma) \delta n(\boldsymbol{p}',\sigma'), \tag{5}$$

where δn are *small* variations in the distribution function $n(\boldsymbol{p})$ from the step function (that characterizes the ground state of the system); it is obvious that these variations will be significant only in the vicinity of the limiting momentum p_F, which continues to be given by equation (1). It is thus understood that the quantity $\varepsilon(\boldsymbol{p},\sigma)$ in (5) corresponds to the distribution function $n(\boldsymbol{p},\sigma)$ being infinitesimally close to the step function (of the ground state). One may also note that the function $f(\boldsymbol{p},\sigma;\boldsymbol{p}',\sigma')$, being a second functional derivative of E, must be symmetric in its arguments; often, it is of the form $a+b\hat{\boldsymbol{s}}_1\cdot\hat{\boldsymbol{s}}_2$, where the coefficients a and b depend only on the angle between the momenta $\boldsymbol{p}$ and $\boldsymbol{p}'$.[22] The function f plays a central role in the theory of the Fermi liquid; for an ideal gas, f vanishes.

To discover the formal dependence of the distribution function $n(\boldsymbol{p})$ on the energy $\varepsilon(\boldsymbol{p})$, we note that, in view of the one-to-one correspondence between the energy levels of the liquid and of the ideal gas, the number of microstates (and hence the entropy) of the liquid is given by the same expression as for the ideal gas; see equation (6.1.15), with all $g_i=1$ and $a=+1$, or Problem 6.1:

$$\frac{S}{k}=-\sum_{\boldsymbol{p}}\{n\ln n+(1-n)\ln(1-n)\}\approx -V\int\{n\ln n+(1-n)\ln(1-n)\}d\tau. \tag{6}$$

Maximizing this expression, under the constraints $\delta E=0$ and $\delta N=0$, we obtain for the *equilibrium distribution function*

$$\bar{n}=\frac{1}{\exp\{(\varepsilon-\mu)/kT\}+1}. \tag{7}$$

It should be noted here that, despite its formal similarity with the standard expression for the Fermi–Dirac distribution function, formula (7) is different insofar as the quantity ε appearing here is itself a function of $\bar{n}$; consequently, this formula gives only an *implicit*, and probably a very complicated, expression for the function $\bar{n}$.

A word may now be said about the quantity ε appearing in equation (5). Since this ε corresponds to the *limiting case* of n being a step function, it is expected to be a well-defined function of $\boldsymbol{p}$. Equation (7) then reduces to the usual Fermi–Dirac distribution function, which is indeed an *explicit* function of ε. It is not difficult to see that this reduction remains valid so long as expression (5) is valid, that is, so long as the variations δn are small, which in turn means that $T\ll T_F$. As mentioned earlier, the variation δn will be significant only in the vicinity of the Fermi momentum p_F; accordingly, we will not have much to do with the function $\varepsilon(\boldsymbol{p})$ except when $p\simeq p_F$. We may, therefore, write

$$\varepsilon(p\simeq p_F)=\varepsilon_F+\left(\frac{\partial\varepsilon}{\partial p}\right)_{p=p_F}(p-p_F)+\cdots\simeq\varepsilon_F+u_F(p-p_F), \tag{8}$$

where u_F denotes the "velocity" of the quasiparticles at the Fermi surface. In the case of an ideal gas ($\varepsilon=p^2/2m$), $u_F=p_F/m$. By analogy, we define a parameter m^* such that

$$m^*\equiv\frac{p_F}{u_F}=\frac{p_F}{(\partial\varepsilon/\partial p)_{p=p_F}} \tag{9}$$

[22]Of course, if the functions involved here are spin-dependent, then the factor 2 in the element $d\tau$ (as well as in $d\tau'$) must be replaced by a summation over the spin variable(s).

and call it the *effective mass* of the quasiparticle with momentum p_F (or with $p \simeq p_F$).

Another way of looking at the parameter m^* is due to Brueckner and Gammel (1958), who wrote

$$\varepsilon(p \simeq p_F) = \frac{p^2}{2m} + V(p) = \frac{p^2}{2m^*} + \text{const.}; \tag{10}$$

the philosophy behind this expression is that "for quasiparticles with $p \simeq p_F$, the modification, $V(p)$, brought into the quantity $\varepsilon(p)$ by the presence of inter-particle interactions in the liquid may be represented by a constant term while the kinetic energy, $p^2/2m$, is modified so as to replace the particle mass m by an effective, quasiparticle mass m^*"; in other words, we adopt a *mean field* point of view. Differentiating (10) with respect to p and setting $p = p_F$, we obtain

$$\frac{1}{m^*} = \frac{1}{m} + \frac{1}{p_F}\left(\frac{dV(p)}{dp}\right)_{p=p_F}. \tag{11}$$

The quantity m^*, in particular, determines the low-temperature specific heat of the Fermi liquid. We can readily see that, for $T \ll T_F$, the ratio of the specific heat of a Fermi liquid to that of an ideal Fermi gas is precisely equal to the ratio m^*/m:

$$\frac{(C_V)_{\text{real}}}{(C_V)_{\text{ideal}}} = \frac{m^*}{m}. \tag{12}$$

This follows from the fact that (i) expression (6) for the entropy S, in terms of the distribution function n, is the same for the liquid as for the gas, (ii) the same is true of relation (7) between $\bar{n}$ and ε, and (iii) for the evaluation of the integral in (6) *at low temperatures* only momenta close to p_F are important. Consequently, the result stated in Problem 8.13, namely

$$C_V \simeq S \simeq \frac{\pi^2}{3} k^2 T a(\varepsilon_F), \tag{13}$$

continues to hold — with the sole difference that in the expression for the density of states $a(\varepsilon_F)$, in the vicinity of the Fermi surface, the particle mass m gets replaced by the effective mass m^*; see equation (8.1.21).

We now proceed to establish a relationship between the parameters m and m^* in terms of the characteristic function f. In doing so, we neglect the spin-dependence of f, if any; the necessary modification can be introduced without any difficulty. The guiding principle here is that, in the absence of external forces, the momentum density of the liquid must be equal to the density of mass transfer. The former is given by $\int \boldsymbol{p} n \, d\tau$, while the latter is given by $m \int (\partial\varepsilon/\partial\boldsymbol{p}) n \, d\tau$, $(\partial\varepsilon/\partial\boldsymbol{p})$ being the "velocity" of the quasiparticle with momentum

$\boldsymbol{p}$ and energy ε.[23] Thus

$$\int \boldsymbol{p} n\, d\tau = m \int \frac{\partial \varepsilon}{\partial \boldsymbol{p}} n\, d\tau. \tag{14}$$

Varying the distribution function by δn and making use of equation (4), we obtain

$$\begin{aligned}\int \boldsymbol{p}\delta n\, d\tau &= m \int \frac{\partial \varepsilon}{\partial \boldsymbol{p}} \delta n\, d\tau + m \iint \left\{ \frac{\partial f(\boldsymbol{p}, \boldsymbol{p}')}{\partial \boldsymbol{p}} \delta n' d\tau' \right\} n\, d\tau \\ &= m \int \frac{\partial \varepsilon}{\partial \boldsymbol{p}} \delta n\, d\tau - m \iint f(\boldsymbol{p}, \boldsymbol{p}') \frac{\partial n'}{\partial \boldsymbol{p}'} \delta n\, d\tau\, d\tau';\end{aligned} \tag{15}$$

in obtaining the last expression, we have interchanged the variables $\boldsymbol{p}$ and $\boldsymbol{p}'$ and have also carried out an integration *by parts*. In view of the arbitrariness of the variation δn, equation (15) requires that

$$\frac{\boldsymbol{p}}{m} = \frac{\partial \varepsilon}{\partial \boldsymbol{p}} - \int f(\boldsymbol{p}, \boldsymbol{p}') \frac{\partial n'}{\partial \boldsymbol{p}'} d\tau'. \tag{16}$$

We apply this result to quasiparticles with momenta close to p_F; at the same time, we replace the distribution function n' by a "step" function, whereby

$$\frac{\partial n'}{\partial \boldsymbol{p}'} = -\frac{\boldsymbol{p}'}{p'} \delta(p' - p_F).$$

This enables us to carry out integration over the magnitude p' of the momentum, so that

$$\int f(\boldsymbol{p}, \boldsymbol{p}') \frac{\partial n'}{\partial \boldsymbol{p}'} \frac{2p'^2 dp' d\omega'}{h^3} = -\frac{2p_F}{h^3} \int f(\theta) \boldsymbol{p}'_F\, d\omega', \tag{17}$$

$d\omega'$ being the element of a solid angle; note that we have contracted the arguments of the function f because in simple situations it depends only on the angle between the two momenta. Inserting (17) into (16), with $\boldsymbol{p} = \boldsymbol{p}_F$, making a scalar product with $\boldsymbol{p}_F$ and dividing by p_F^2, we obtain the desired result

$$\frac{1}{m} = \frac{1}{m^*} + \frac{p_F}{2h^3} \cdot 4 \int f(\theta) \cos\theta\, d\omega'. \tag{18}$$

If the function f depends on the spins s_1 and s_2 of the particles involved, then the factor 4 in front of the integral will have to be replaced by a summation over the spin variables.

[23] Since the total number of quasiparticles in the liquid is the same as the total number of real particles, to obtain the net transport of mass by the quasiparticles one has to multiply their number by the mass m of the *real* particle.

We now derive a formula for the velocity of sound at absolute zero. From first principles, we have[24]

$$c_0^2 = \frac{\partial P_0}{\partial(mN/V)} = -\frac{V^2}{mN}\left(\frac{\partial P_0}{\partial V}\right)_N.$$

In the present context, it is preferable to have an expression in terms of the chemical potential of the liquid. This can be obtained by making use of the formula $Nd\mu_0 = VdP_0$, see Problem 1.16, from which it follows that[25]

$$\left(\frac{\partial \mu_0}{\partial N}\right)_V = -\frac{V}{N}\left(\frac{\partial \mu_0}{\partial V}\right)_N = -\frac{V^2}{N^2}\left(\frac{\partial P_0}{\partial V}\right)_N$$

and hence

$$c_0^2 = \frac{N}{m}\left(\frac{\partial \mu_0}{\partial N}\right)_V. \tag{19}$$

Now, $\mu_0 = \varepsilon(p_F) = \varepsilon_F$; therefore, the change $\delta\mu_0$ arising from a change δN in the total number of particles in the system is given by

$$\delta\mu_0 = \frac{\partial \varepsilon_F}{\partial p_F}\delta p_F + \int f(\mathbf{p}_F, \mathbf{p}')\delta n'\, d\tau'. \tag{20}$$

The first part in (20) arises from the fact that a change in the total number of particles in the system inevitably alters the value of the limiting momentum p_F; see equation (1), from which (for constant V)

$$\delta p_F/p_F = \frac{1}{3}\delta N/N$$

and hence

$$\frac{\partial \varepsilon_F}{\partial p_F}\delta p_F = \frac{p_F^2}{3m^*}\frac{\delta N}{N}. \tag{21}$$

[24]At $T = 0, S = 0$; so there is no need to distinguish between the isothermal and adiabatic compressibilities of the liquid.

[25]Since μ_0 is an intensive quantity and, therefore, it depends on N and V only through the ratio N/V, we can write: $\mu_0 = \mu_0(N/V)$. Consequently,

$$\left(\frac{\partial \mu_0}{\partial N}\right)_V = \mu_0'\left(\frac{\partial (N/V)}{\partial N}\right)_V = \mu_0'\frac{1}{V}$$

and

$$\left(\frac{\partial \mu_0}{\partial V}\right)_N = \mu_0'\left(\frac{\partial (N/V)}{\partial V}\right)_N = -\mu_0'\frac{N}{V^2}.$$

Hence

$$\left(\frac{\partial \mu_0}{\partial N}\right)_V = -\frac{V}{N}\left(\frac{\partial \mu_0}{\partial V}\right)_N.$$

The second part arises from equation (4). It will be noted that the variation $\delta n'$ appearing in the integral of equation (20) is significant only for $p' \simeq p_F$; we may, therefore, write

$$\int f(\boldsymbol{p}_F, \boldsymbol{p}')\delta n' d\tau' \simeq \frac{\delta N}{4\pi V} \int f(\theta) d\omega'. \tag{22}$$

Substituting (21) and (22) into (20), we obtain

$$\left(\frac{\partial \mu_0}{\partial N}\right)_V = \frac{p_F^2}{3m^* N} + \frac{1}{4\pi V} \int f(\theta) d\omega'. \tag{23}$$

Making use of equations (18) and (1), we finally obtain

$$c_0^2 = \frac{N}{m}\left(\frac{\partial \mu_0}{\partial N}\right)_V = \frac{p_F^2}{3m^2} + \frac{p_F^3}{6mh^3} \cdot 4 \int f(\theta)(1 - \cos\theta)\, d\omega'. \tag{24}$$

Once again, if the function f depends on the spins of the particles, then the factor 4 in front of the integral will have to be replaced by a summation over the spin variables.

For illustration, we shall apply this theory to the imperfect Fermi gas studied in Section 11.7. To calculate $f(\boldsymbol{p}, \sigma; \boldsymbol{p}', \sigma')$, we have to differentiate twice the sum of expression (11.7.12), with $u_0 = 4\pi a\hbar^2/m$, and expression (11.7.22) with respect to the distribution function $n(\boldsymbol{p}, \sigma)$ and then substitute $p = p' = p_F$. Performing the desired calculation, then changing summations into integrations and carrying out integrations by simple means, we find that the function f is spin-dependent — the spin-dependent term being in the nature of an *exchange term*, proportional to $\hat{\boldsymbol{s}}_1 \cdot \hat{\boldsymbol{s}}_2$. The complete result, according to Abrikosov and Khalatnikov (1957), is

$$f(\boldsymbol{p}, \sigma; \boldsymbol{p}', \sigma') = A(\theta) + B(\theta)\hat{\boldsymbol{s}}_1 \cdot \hat{\boldsymbol{s}}_2, \tag{25}$$

where

$$A(\theta) = \frac{2\pi a\hbar^2}{m}\left[1 + 2a\left(\frac{3N}{\pi V}\right)^{1/3}\left\{2 + \frac{\cos\theta}{2\sin(\theta/2)} \ln\frac{1 + \sin(\theta/2)}{1 - \sin(\theta/2)}\right\}\right]$$

and

$$B(\theta) = -\frac{8\pi a\hbar^2}{m}\left[1 + 2a\left(\frac{3N}{\pi V}\right)^{1/3}\left\{1 - \frac{1}{2}\sin\left(\frac{\theta}{2}\right) \ln\frac{1 + \sin(\theta/2)}{1 - \sin(\theta/2)}\right\}\right],$$

a being the scattering length of the two-body potential and θ the angle between the momentum vectors $\boldsymbol{p}_F$ and $\boldsymbol{p}'_F$. Substituting (25) into formulae (18) and (24), in which the factor 4 is now supposed to be replaced by a summation over the spin variables, we find that while the spin-dependent term $B(\theta)\hat{\boldsymbol{s}}_1 \cdot \hat{\boldsymbol{s}}_2$ does not make any contribution toward the final results, the spin-independent term $A(\theta)$ leads to[26]

$$\frac{1}{m^*} = \frac{1}{m} - \frac{8}{15m}(7\ln 2 - 1)\left(\frac{3N}{\pi V}\right)^{2/3} a^2 \tag{26}$$

and

$$c_0^2 = \frac{p_F^2}{3m^2} + \frac{2\pi a\hbar^2}{m^2}\frac{N}{V}\left[1 + \frac{4}{15}(11 - 2\ln 2)\left(\frac{3N}{\pi V}\right)^{1/3} a\right]; \tag{27}$$

the latter result is identical to expression (11.7.33) derived in the preceding section. Proceeding backward, one can obtain from equation (27) corresponding expressions for the ground-state pressure P_0 and the ground-state energy E_0, namely equations (11.7.32) and (11.7.31), as well as the ground-state chemical potential μ_0, as quoted in Problem 11.15.

11.9 Condensation in Fermi systems

The discussion of the $T = 0$ Fermi liquid in Sections 11.7 and 11.8 applies when the interactions between the fermions are strictly repulsive. The resulting Fermi liquid has a ground state and quasiparticle excitations that are qualitatively similar to the ideal Fermi gas. However, for fermions with attractive interactions, *no matter how weak*, the degenerate Fermi gas is unstable due to the formation of bosonic pairs. This leads to a number of important phenomena including superconductivity in metals, superfluidity in ^{3}He, and condensation in ultracold Fermi gases. In low-temperature superconductors, screening and the electron-phonon interaction result in a retarded attraction between quasiparticles on opposite sides of the Fermi surface. The formation of these so-called Cooper pairs leads to the creation of a superconducting state with critical temperature

$$kT_c \approx \hbar\omega_D \exp\left(-\frac{1}{N(\epsilon_F)|u_0|}\right), \tag{1}$$

where $N(\epsilon_F)$ is the density of states *per spin configuration* at the Fermi surface, u_0 is the weak attractive coupling between electrons, and $\hbar\omega_D$ is the Debye energy discussed in Section 7.4 since the coupling is due to the acoustic phonons. As can be seen from equation (1), the phase transition temperature is nonperturbative in u_0. A complete treatment of superconductivity is far beyond the scope of this section, so we refer the reader to the original papers by Cooper (1956) and Bardeen, Cooper, and Schrieffer (1957) and the texts on superconductivity by Tilley and Tilley (1990) and Tinkham (1996). The case of superfluidity in ^{3}He is surveyed by Vollhardt and Wölfle (1990).

Bosonic condensation has also recently been observed in trapped ultracold atomic Fermi gases. The sign and size of the atomic interactions in ultracold gases can be tuned

[26]In a dense system, such as liquid He^3, the ratio m^*/m would be significantly larger than unity. The experimental work of Roberts and Sydoriak (1955), on the specific heat of liquid He^3, and the theoretical work of Brueckner and Gammel (1958), on the thermodynamics of a dense Fermi gas, suggest that the ratio $(m^*/m)_{\text{He}^3} \simeq 1.85$.

with a magnetic field near Feshbach resonance allowing unprecedented experimental control of interactions. In particular, experimenters can create a low-lying molecular bound state or a weakly attractive interaction without allowing a molecular bound state to form. If interaction between pairs of fermions allows the formation of bound bosonic molecules, the ground state of a degenerate Fermi gas will be destablized since molecules will form and, if the density of the bosonic molecules is large enough, they will Bose-condense — see Greiner, Regal, and Jin (2003); Jochim et al. (2003); and Zwierlein et al. (2003).

For weakly attractive interactions, the fermionic system condenses into a BCS-like state and provides an excellent experimental environment for testing theoretical predictions due to the well-understood nature and experimental control of the atomic interactions. Theory predicts a smooth crossover from BCS to Bose–Einstein condensation (BEC) behavior as the magnitude of the attractive interaction parameter u_0 is varied from values small to large. BCS theory describes the behavior for weak coupling. For broad Feshbach resonances of trapped fermions, the most common experimental situation, the BCS critical temperature is given by

$$\frac{kT_c}{\hbar\omega_0} \approx \frac{\epsilon_F}{\hbar\omega_0} \exp\left(-\frac{\pi}{2k_F\,|a|}\right) \sim N^{1/3} \exp\left(-\frac{a_{\rm osc}}{1.214 N^{1/6}\,|a|}\right), \tag{2}$$

where $\omega_0 = (\omega_1\omega_2\omega_3)^{1/3}$ is the average oscillation frequency of atoms in the trap, $a_{\rm osc} = \sqrt{\hbar/m\omega_0}$, and $k_F = \sqrt{2m\epsilon_F}/\hbar$ is the Fermi wavevector; see Pitaevskii and Stringari (2003), Leggett (2006), and Pethick and Smith (2008). For large negative scattering lengths, the transition temperature smoothly crosses over to the BEC limit with noninteracting Bose-condensation temperature, see equation (7.2.6),

$$\frac{kT_c}{\hbar\omega_0} \approx \left(\frac{N}{2\zeta(3)}\right)^{1/3}, \tag{3}$$

since the number of Cooper pairs is $N/2$. The ratio of the transition temperature in the BEC limit and the Fermi temperature from equation (8.4.3) is

$$\frac{kT_c}{\epsilon_F} \approx \left(\frac{1}{12\zeta(3)}\right)^{1/3} \simeq 0.41. \tag{4}$$

Mean-field analysis of the broad resonance limit (Leggett, 2006) and analytical analysis of the narrow resonance limit (Gurarie and Radzihovsky, 2007) both indicate that the phase transition temperature has a maximum at intermediate coupling. Figure 11.7 is a sketch of the critical temperature as a function of the coupling parameter u_0.

Experimental observations of condensation in a degenerate Fermi gas in the BEC–BCS crossover region by Regal, Greiner, and Jin (2004) are shown in Figure 11.8. They used a Feshbach resonance to tune the scattering length of ^{40}K into an attractive range ($a < 0$) that

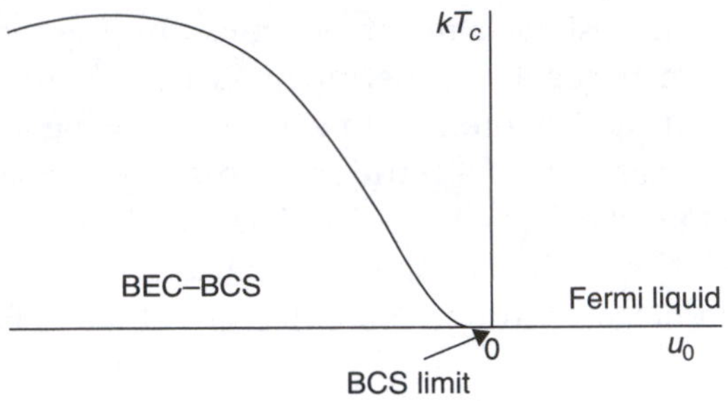

FIGURE 11.7 Sketch of the BEC–BCS phase diagram on the BCS side of the Feshbach resonance for ultracold fermions in an atomic trap. The scattering length a and coupling $u_0 = 4\pi\hbar^2 a/m$ can be tuned from positive values to negative with the help of a magnetic field. Positive (repulsive) couplings result in a Fermi liquid. Negative (attractive) couplings result in a BCS condensation at low temperatures. The nature of the condensed phase varies smoothly from BCS behavior for small negative coupling to Bose–Einstein behavior for large negative coupling. The phase transition temperature has a maximum at intermediate coupling.

FIGURE 11.8 Time-of-flight images showing condensation of fermions in an ultracold atomic gas. The images show the quantum mechanical projection of the fermionic system onto a molecular gas and are shown for three values of the magnetic field on the BCS side of the Feshbach resonance for an ultracold trapped gas of ^{40}K. The temperature of the Fermi gas is $(kT/\epsilon_F) \approx 0.07$. The condensed fraction varies from about 1 to 10 percent of the original cold fermions in the trap; see Regal, Greiner, and Jin (2004). Figure courtesy of NIST/JILA/University of Colorado.

does not allow a two-particle molecular bound state and observed the fermions condensing into a BCS-like macroscopic quantum state. They explored the BEC–BCS crossover behavior by tuning $|a|$ from small values to large.

Problems

11.1. (a) Show that, for bosons as well as fermions,

$$[\psi(\mathbf{r}_j), \hat{H}] = \left(-\frac{\hbar^2}{2m}\nabla_j^2 + \int d^3r\psi^\dagger(\mathbf{r})u(\mathbf{r},\mathbf{r}_j)\psi(\mathbf{r})\right)\psi(\mathbf{r}_j),$$

where $\hat{H}$ is the Hamiltonian operator defined by equation (11.1.4).

(b) Making use of the foregoing result, show that the equation

$$\frac{1}{\sqrt{N!}}\langle 0|\psi(\boldsymbol{r}_1)\ldots\psi(\boldsymbol{r}_N)\hat{H}|\Psi_{NE}\rangle = E\frac{1}{\sqrt{N!}}\langle 0|\psi(\boldsymbol{r}_1)\ldots\psi(\boldsymbol{r}_N)|\Psi_{NE}\rangle$$
$$= E\Psi_{NE}(\boldsymbol{r}_1,\ldots\boldsymbol{r}_N)$$

is equivalent to the Schrödinger equation (11.1.15).

11.2. The grand partition function of a gaseous system composed of mutually interacting bosons is given by

$$\ln \mathcal{Q} \equiv \frac{PV}{kT} = \frac{V}{\lambda^3}\left[g_{5/2}(z) - 2\{g_{3/2}(z)\}^2\frac{a}{\lambda} + O\left(\frac{a^2}{\lambda^2}\right)\right].$$

Study the analytic behavior of this expression near $z = 1$ and show that the system exhibits the phenomenon of Bose–Einstein condensation when its fugacity assumes the critical value

$$z_c = 1 + 4\zeta\left(\frac{3}{2}\right)\frac{a}{\lambda_c} + O\left(\frac{a^2}{\lambda_c^2}\right).$$

Further show that the pressure of the gas at the critical point is given by (Lee and Yang 1958, 1960b)

$$\frac{P_c}{kT_c} = \frac{1}{\lambda_c^3}\left[\zeta\left(\frac{5}{2}\right) + 2\left\{\zeta\left(\frac{3}{2}\right)\right\}^2\frac{a}{\lambda_c} + O\left(\frac{a^2}{\lambda_c^2}\right)\right];$$

compare these results to equations (11.2.16) through (11.2.18).

11.3. For the imperfect Bose gas studied in Section 11.2, calculate the specific heat C_V near absolute zero and show that, as $T \to 0$, the specific heat vanishes in a manner characteristic of a system with an "energy gap" $\Delta = 4\pi a\hbar^2 n/m$.

11.4. **(a)** Show that, to *first* order in the scattering length a, the discontinuity in the specific heat C_V of an imperfect Bose gas at the transition temperature T_c is given by

$$(C_V)_{T=T_{c-}} - (C_V)_{T=T_{c+}} = Nk\frac{9a}{2\lambda_c}\zeta(3/2),$$

while the discontinuity in the bulk modulus K is given by

$$(K)_{T=T_{c-}} - (K)_{T=T_{c+}} = -\frac{4\pi a\hbar^2}{mv_c^2}.$$

(b) Examine the discontinuities in the quantities $(\partial^2 P/\partial T^2)_v$ and $(\partial^2\mu/\partial T^2)_v$ as well, and show that your results are consistent with the thermodynamic relationship

$$C_V = VT\left(\frac{\partial^2 P}{\partial T^2}\right)_v - NT\left(\frac{\partial^2\mu}{\partial T^2}\right)_v.$$

11.5. **(a)** Complete the mathematical steps leading to equations (11.3.15) and (11.3.16).
(b) Complete the mathematical steps leading to equations (11.3.23) and (11.3.24).

11.6. The ground-state pressure of an interacting Bose gas (see Lee and Yang, 1960a) turns out to be

$$P_0 = \frac{\mu_0^2 m}{8\pi a\hbar^2}\left[1 - \frac{64}{15\pi}\frac{\mu_0^{1/2} m^{1/2} a}{\hbar} + \cdots\right],$$

where μ_0 is the ground-state chemical potential of the gas. It follows that

$$n \equiv \left(\frac{dP_0}{d\mu_0}\right) = \frac{\mu_0 m}{4\pi a\hbar^2}\left[1 - \frac{16}{3\pi}\frac{\mu_0^{1/2} m^{1/2} a}{\hbar} + \cdots\right]$$

and

$$\frac{E_0}{V} \equiv (n\mu_0 - P_0) = \frac{\mu_0^2 m}{8\pi a\hbar^2}\left[1 - \frac{32}{5\pi}\frac{\mu_0^{1/2} m^{1/2} a}{\hbar} + \cdots\right].$$

Eliminating μ_0 from these results, derive equations (11.3.16) and (11.3.17).

11.7. Show that in an interacting Bose gas the mean occupation number $\overline{n}_{\boldsymbol{p}}$ of the *real* particles and the mean occupation number $\overline{N}_{\boldsymbol{p}}$ of the *quasiparticles* are connected by the relationship

$$\overline{n}_{\boldsymbol{p}} = \frac{\overline{N}_{\boldsymbol{p}} + \alpha_{\boldsymbol{p}}^2(\overline{N}_{\boldsymbol{p}} + 1)}{1 - \alpha_{\boldsymbol{p}}^2} \qquad (\boldsymbol{p} \neq 0),$$

where $\alpha_{\boldsymbol{p}}$ is given by equations (11.3.9) and (11.3.10). Note that equation (11.3.22) corresponds to the special case $\overline{N}_{\boldsymbol{p}} = 0$.

11.8. The excitation energy of liquid He^4, carrying a *single* excitation above the ground state, is determined by the *minimum* value of the quantity

$$\varepsilon = \int \Psi^* \left\{-\frac{\hbar^2}{2m}\sum_i \nabla_i^2 + V - E_0\right\} \Psi d^{3N} r \Big/ \int \Psi^* \Psi d^{3N} r,$$

where E_0 denotes the ground-state energy of the liquid while Ψ, according to Feynman, is given by equation (11.4.3). Show that the process of minimization of this expression leads to equation (11.4.5) for the energy of the excitation.

[*Hint*: First express ε in the form

$$\varepsilon = \frac{\hbar^2}{2m}\int |\nabla f(\boldsymbol{r})|^2 d^3 r \Big/ \int\int f^*(\boldsymbol{r}_1) f(\boldsymbol{r}_2) g(\boldsymbol{r}_2 - \boldsymbol{r}_1) d^3 r_1 d^3 r_2.$$

Then show that ε is *minimum* when $f(\boldsymbol{r})$ is of the form (11.4.4).]

11.9. Show that, for a sufficiently large momentum $\hbar k$ (in fact, such that the slope $d\varepsilon/dk$ of the energy spectrum is greater than the initial slope $\hbar c$), a state of *double excitation* in liquid He^4 is energetically more favorable than a state of *single excitation*, that is, there exist wavevectors $\boldsymbol{k}_1$ and $\boldsymbol{k}_2$ such that, while $\boldsymbol{k}_1 + \boldsymbol{k}_2 = \boldsymbol{k}, \varepsilon(k_1) + \varepsilon(k_2) < \varepsilon(k)$.

11.10. Using Fetter's analytical approximation,

$$f_1(\rho') = \frac{\rho'}{\sqrt{(1 + \rho'^2)}},$$

for the solution of equation (11.5.23) with $s = 1$, calculate the energy (per unit length) associated with a quantized vortex line of unit circulation. Compare your result with the one quoted in (11.5.26).

11.11. **(a)** Study the nature of the velocity field arising from a pair of parallel vortex lines, with $s_1 = +1$ and $s_2 = -1$, separated by a distance d. Derive and discuss the general equation of the stream lines.

(b) Again, using Fetter's analytical approximation for the functions $f(\rho_1')$ and $f(\rho_2')$, calculate the energy (per unit length) of the system and show that its limiting value, as $d \to 0$, is $11\pi\hbar^2 n_0/12m$. Making use of this result, derive expression (11.6.8) for the critical velocity of superflow.

11.12. Establish the asymptotic formula (11.7.30) for the function $F(z_0)$.

[*Hint*: Write the coefficient that appears in the sum (11.7.24) in the form

$$\frac{1}{\sqrt{(rst)}(r+s)(r+t)}$$
$$= \left(\frac{2}{\sqrt{\pi}}\right)^3 \int_0^\infty e^{-X^2 r - Y^2 s - Z^2 t - \xi(r+s) - \eta(r+t)} dX\, dY\, dZ\, d\xi\, d\eta.$$

Insert this expression into (11.7.24) and carry out summations over r, s, and t, with the result

$$F(z_0) = \frac{8}{\pi^{3/2}} \int_0^\infty \frac{1}{z_0^{-1} e^{X^2+\xi+\eta}+1} \frac{1}{z_0^{-1} e^{Y^2+\xi}+1} \frac{1}{z_0^{-1} e^{Z^2+\eta}+1} dX\, dY\, dZ\, d\xi\, d\eta.$$

In the limit $z_0 \to \infty$, the integrand is essentially equal to 1 in the region R defined by

$$X^2 + \xi + \eta < \ln z_0, \quad Y^2 + \xi < \ln z_0, \quad \text{and} \quad Z^2 + \eta < \ln z_0;$$

outside this region, it is essentially 0. Hence, the dominant term of the asymptotic expansion is

$$\frac{8}{\pi^{3/2}} \int_R 1 \cdot dX\, dY\, dZ\, d\xi\, d\eta,$$

which, in turn, reduces to the double integral

$$\frac{8}{\pi^{3/2}} \iint (\ln z_0 - \xi - \eta)^{1/2} (\ln z_0 - \xi)^{1/2} (\ln z_0 - \eta)^{1/2} d\xi\, d\eta;$$

the limits of integration here are such that not only $\xi < (\ln z_0)$ and $\eta < (\ln z_0)$, but also $(\xi + \eta) < (\ln z_0)$. The rest of the calculation is straightforward.]

11.13. The grand partition function of a gaseous system composed of mutually interacting, spin-half fermions has been evaluated by Lee and Yang (1957), with the result[27]

$$\ln \mathcal{Q} \equiv \frac{PV}{kT} = \frac{V}{\lambda^3} \left[2f_{5/2}(z) - \frac{2a}{\lambda}\{f_{3/2}(z)\}^2 \right.$$
$$\left. + \frac{4a^2}{\lambda^2} f_{1/2}(z)\{f_{3/2}(z)\}^2 - \frac{8a^2}{\lambda^2} F(z) + \cdots \right],$$

[27] For the details of this calculation, see Lee and Yang (1959b) where the case of bosons, as well as of fermions, with spin J has been treated using the *binary collision* method. The second-order result for the case of *spinless* bosons was first obtained by Huang, Yang, and Luttinger (1957) using the *method of pseudopotentials*.

where z is the fugacity of the *actual* system (not of the corresponding noninteracting system, which was denoted by the symbol z_0 in the text); the functions $f_\nu(z)$ and $F(z)$ are defined in a manner similar to equations (11.7.10) and (11.7.24). From this result, one can derive expressions for the quantities $E(z, V, T)$ and $N(z, V, T)$ by using the formulae

$$E(z,V,T) \equiv kT^2 \frac{\partial(\ln \mathcal{Q})}{\partial T} \quad \text{and} \quad N(z,V,T) \equiv \frac{\partial(\ln \mathcal{Q})}{\partial(\ln z)} \left\{ = \frac{2V}{\lambda^3} f_{3/2}(z_0) \right\}.$$

(a) Eliminating z between these two results, derive equation (11.7.25) for E.

(b) Obtain the zero-point value of the chemical potential μ, correct to *second* order in (a/λ), and verify, with the help of equations (11.7.31) and (11.7.32), that

$$(E + PV)_{T=0} = N(\mu)_{T=0}.$$

[*Hint*: At $T = 0\,\mathrm{K}, \mu = (\partial E/\partial N)_V$.]

(c) Show that the low-temperature specific heat and the low-temperature entropy of this gas are given by (see Pathria and Kawatra, 1962)

$$\frac{C_V}{Nk} \simeq \frac{S}{Nk} \simeq \frac{\pi^2}{2}\left(\frac{kT}{\varepsilon_F}\right)\left[1 + \frac{8}{15\pi^2}(7\ln 2 - 1)(k_F a)^2 + \cdots\right],$$

where $k_F = (3\pi^2 n)^{1/3}$. Clearly, the factor within square brackets is to be identified with the ratio m^*/m; see equations (11.8.12) and (11.8.26).

[*Hint*: To determine C_V to the *first* power in T, we must know E to the *second* power in T. For this, we require higher-order terms of the asymptotic expansions of the functions $f_\nu(z)$ and $F(z)$; these are given by

$$f_{5/2}(z) = \frac{8}{15\sqrt{\pi}}(\ln z)^{5/2} + \frac{\pi^{3/2}}{3}(\ln z)^{1/2} + O(1),$$

$$f_{3/2}(z) = \frac{4}{3\sqrt{\pi}}(\ln z)^{3/2} + \frac{\pi^{3/2}}{6}(\ln z)^{-1/2} + O(\ln z)^{-5/2},$$

$$f_{1/2}(z) = \frac{2}{\sqrt{\pi}}(\ln z)^{1/2} - \frac{\pi^{3/2}}{12}(\ln z)^{-3/2} + O(\ln z)^{-7/2},$$

and

$$F(z) = \frac{16(11 - 2\ln 2)}{105\pi^{3/2}}(\ln z)^{7/2} - \frac{2(2\ln 2 - 1)}{3}\pi^{1/2}(\ln z)^{3/2} + O(\ln z)^{5/4}.$$

The first three results here follow from the Sommerfeld lemma (E.17); for the last one, see Yang (1962).]

11.14. The energy spectrum $\varepsilon(p)$ of a gas composed of mutually interacting, spin-half fermions is given by (Galitskii, 1958; Mohling, 1961)

$$\frac{\varepsilon(p)}{p_F^2/2m} \simeq x^2 + \frac{4}{3\pi}(k_F a) + \frac{4}{15\pi^2}(k_F a)^2 \times \left[11 + 2x^4 \ln\frac{x^2}{|x^2-1|} - 10\left(x - \frac{1}{x}\right)\ln\left|\frac{x+1}{x-1}\right| - \frac{(2-x^2)^{5/2}}{x}\ln\left(\frac{1+x\sqrt{(2-x^2)}}{1-x\sqrt{(2-x^2)}}\right)\right],$$

where $x = p/p_F \leq \sqrt{2}$ and $k = p/\hbar$. Show that, for k close to k_F, this spectrum reduces to

$$\frac{\varepsilon(p)}{p_F^2/2m} \simeq x^2 + \frac{4}{3\pi}(k_F a) + \frac{4}{15\pi^2}(k_F a)^2\left[(11 - 2\ln 2) - 4(7\ln 2 - 1)\left(\frac{k}{k_F} - 1\right)\right].$$

Using equations (11.8.10) and (11.8.11), check that this expression leads to the result

$$\frac{m^*}{m} \simeq 1 + \frac{8}{15\pi^2}(7\ln 2 - 1)(k_F a)^2.$$

11.15. In the ground state of a Fermi system, the chemical potential is identical to the Fermi energy: $(\mu)_{T=0} = \varepsilon(p_F)$. Making use of the energy spectrum $\varepsilon(p)$ of the previous problem, we obtain

$$(\mu)_{T=0} \simeq \frac{p_F^2}{2m}\left[1 + \frac{4}{3\pi}(k_F a) + \frac{4}{15\pi^2}(11 - 2\ln 2)(k_F a)^2\right].$$

Integrating this result, rederive equation (11.7.31) for the ground-state energy of the system.

11.16. The energy levels of an imperfect Fermi gas in the presence of an external magnetic field $\boldsymbol{B}$, to *first* order in a, may be written as

$$E_n = \sum_{\boldsymbol{p}}(n_{\boldsymbol{p}}^+ + n_{\boldsymbol{p}}^-)\frac{p^2}{2m} + \frac{4\pi a\hbar^2}{mV}N^+N^- - \mu^* B(N^+ - N^-);$$

see equations (8.2.8) and (11.7.12). Using this expression for E_n and following the procedure adopted in Section 8.2.A, study the magnetic behavior of this gas — in particular, the zero-field susceptibility $\chi(T)$. Also examine the possibility of spontaneous magnetization arising from the interaction term with $a > 0$.

11.17. Rewrite the Gross–Pitaevskii equation and the mean field energy, see equations (11.2.21) and (11.2.23), for an isotropic harmonic oscillator trap with frequency ω_0 in a dimensionless form by defining a dimensionless wavefunction $\psi = a_{\text{osc}}^{3/2}/\Psi N$, a dimensionless length $\boldsymbol{s} = \boldsymbol{r}/a_{\text{osc}}$, and a dimensionless energy $E/N\hbar\omega_0$. Show that the dimensionless parameter that controls the mean field energy is Na/a_{osc}, where N is the number of particles in the condensate, a is the scattering length, and $a_{\text{osc}} = \sqrt{\hbar/m\omega_0}$. Next, show that the dimensionless versions of the Gross–Pitaevskii equation and the mean field energy are

$$-\frac{1}{2}\tilde{\nabla}^2\psi + \frac{1}{2}s^2\psi + \frac{4\pi Na}{a_{\text{osc}}}|\psi|^2\psi = \tilde{\mu}\psi,$$

and

$$\frac{E[\psi]}{N\hbar\omega_0} = \int \left(\frac{1}{2} \left| \tilde{\nabla}\psi \right|^2 + \frac{1}{2} s^2 |\psi|^2 + \frac{2\pi Na}{a_{\text{osc}}} |\psi|^4 \right) ds \, .$$

11.18. Solve the Gross–Pitaevskii equation and evaluate the mean field energy, see equations (11.2.21) and (11.2.23), for a *uniform* Bose gas to show that this method yields precisely equation (11.2.6).

11.19. Solve the Gross–Pitaevskii equation (11.2.23) in a harmonic trap for the case when the scattering length a is zero. Show that this reproduces the properties of the ground state of the noninteracting Bose gas.

11.20. Solve the Gross–Pitaevskii equation and evaluate the mean field energy, see equations (11.2.21) and (11.2.23), for an isotropic harmonic oscillator trap with frequency ω_0 for the case $Na/a_{\text{osc}} \gg 1$ by ignoring the kinetic energy term. Reproduce the results (11.2.25) through (11.2.28).

12 Phase Transitions: Criticality, Universality, and Scaling

Various physical phenomena to which the formalism of statistical mechanics has been applied may, in general, be divided into two categories. In the first category, the microscopic constituents of the given system are, or can be regarded as, practically noninteracting; as a result, the thermodynamic functions of the system follow straightforwardly from a knowledge of the energy levels of the individual constituents. Notable examples of phenomena belonging to this category are the specific heats of gases (Sections 1.4 and 6.5), the specific heats of solids (Section 7.4), chemical reactions and equilibrium constants (Section 6.6), the condensation of an ideal Bose gas (Sections 7.1 and 7.2), the spectral distribution of the blackbody radiation (Section 7.3), the elementary electron theory of metals (Section 8.3), the phenomenon of paramagnetism (Sections 3.9 and 8.2), and so on. In the case of solids, the interatomic interaction does, in fact, play an important physical role; however, since the actual positions of the atoms, over a substantial range of temperatures, do not depart significantly from their mean values, we can rewrite our problem in terms of the so-called *normal* coordinates and treat the given solid as an "assembly of practically noninteracting harmonic oscillators." We note that the most significant feature of the phenomena falling in the first category is that, with the *sole* exception of Bose–Einstein condensation, the thermodynamic functions of the systems involved are smooth and continuous!

Phenomena belonging to the second category, however, present a very different situation. In most cases, one encounters analytic discontinuities or singularities in the thermodynamic functions of the given system which, in turn, correspond to the occurrence of various kinds of *phase transitions*. Notable examples of phenomena belonging to this category are the condensation of gases, the melting of solids, phenomena associated with the coexistence of phases (especially in the neighborhood of a critical point), the behavior of mixtures and solutions (including the onset of phase separation), phenomena of ferromagnetism and antiferromagnetism, the order–disorder transitions in alloys, the superfluid transition from liquid He I to liquid He II, the transition from a normal to a superconducting material, and so on. The characteristic feature of the interparticle interactions in these systems is that they *cannot* be "removed" by means of a transformation of the coordinates of the problem; accordingly, the energy levels of the total system cannot, in any simple manner, be related to the energy levels of the individual constituents. One finds instead that, under favorable circumstances, a large number of microscopic

© 2011 Elsevier Ltd. All rights reserved.

constituents of the system may exhibit a tendency of interacting with one another in a rather *strong, cooperative* fashion. This cooperative behavior assumes macroscopic significance at a particular temperature T_c, known as the *critical temperature* of the system, and gives rise to the kind of phenomena listed previously.

Mathematical problems associated with the study of cooperative phenomena are quite formidable.[1] To facilitate calculations, one is forced to introduce models in which the interparticle interactions are considerably simplified, yet retain characteristics that are essential to the cooperative aspect of the problem. One then hopes that a theoretical study of these simplified models, which still involves serious difficulties of analysis, will reproduce the most basic features of the phenomena exhibited by actual physical systems. For instance, in the case of a magnetic transition, one may consider a lattice structure in which all interactions other than the ones among nearest-neighbor spins are neglected. It turns out that a model as simplified as that captures practically all the essential features of the phenomenon — especially in the close neighborhood of the critical point. The inclusion of interactions among spins farther out than the nearest neighbors does not change these features in any significant manner, nor are they affected by the replacement of one lattice structure by another so long as the dimensionality of the lattice is the same. Not only this, these features may also be shared, with little modification, by many other physical systems undergoing very different kinds of phase transitions, for example, gas–liquid instead of paramagnetic–ferromagnetic. This "unity in diversity" turns out to be a hallmark of the phenomena associated with phase transitions — a subject we propose to explore in considerable detail in this and the following two chapters, but first a few preliminaries.

12.1 General remarks on the problem of condensation

We consider an N-particle system, obeying classical or quantum statistics, with the *proviso* that the total potential energy of the system can be written as a sum of two-particle terms $u(r_{ij})$, with $i < j$. The function $u(r)$ is supposed to satisfy the conditions

$$\left.\begin{aligned} &u(r) = +\infty && \text{for} \quad r \le \sigma, \\ &0 > u(r) > -\varepsilon && \text{for} \quad \sigma < r < r^* \\ &u(r) = 0 && \text{for} \quad r \ge r^* \end{aligned}\right\}; \tag{1}$$

see Figure 12.1. Thus, each particle may be looked upon as a hard sphere of diameter σ, surrounded by an attractive potential of range r^* and of (maximum) depth ε. From a practical point of view, conditions (1) do not entail any "serious restriction" on the two-body potential, for the interparticle potentials ordinarily met with in nature are not materially

[1] In this connection, one should note that the mathematical schemes developed in Chapters 10 and 11 give reliable results only if the interactions among the microscopic constituents of the given system are sufficiently weak — in fact, too weak to bring about *cooperative transitions*.

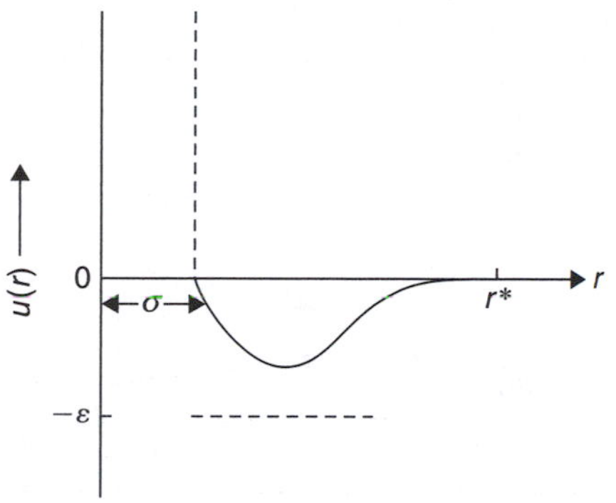

FIGURE 12.1 A sketch of the interparticle potential $u(r)$, as given by equation (1).

different from the one satisfying these conditions. We, therefore, expect that the conclusions drawn from the use of this potential will not be very far from the realities of the actual physical phenomena.

Suppose that we are able to evaluate the *exact* partition function, $Q_N(V,T)$, of the given system. This function will possess certain properties that have been recognized and accepted for quite some time, though a rigorous proof of these was first attempted by van Hove as late as in 1949.[2] These properties can be expressed as follows:

(i) In the thermodynamic limit (i.e., when N and $V \to \infty$ while the ratio N/V stays constant), the quantity $N^{-1} \ln Q$ tends to be a function only of the specific volume $v\ (= V/N)$ and the temperature T; this limiting form may be denoted by the symbol $f(v,T)$. It is natural to identify $f(v,T)$ with the *intensive* variable $-A/NkT$, where A is the Helmholtz free energy of the system. The thermodynamic pressure P is then given by

$$P(v,T) = -\left(\frac{\partial A}{\partial V}\right)_{N,T} = kT\left(\frac{\partial f}{\partial v}\right)_T, \tag{2}$$

which turns out to be a strictly *nonnegative* quantity.

(ii) The function $f(v,T)$ is everywhere *concave*, so the slope $(\partial P/\partial v)_T$ of the (P,v)-curve is *never* positive. While at high temperatures the slope is negative for all v, at lower temperatures there *can* exist a region (or regions) in which the slope is zero and, consequently, the system is infinitely compressible! The existence of such regions, in the (P,v)-diagram, corresponds to the coexistence of two or more phases of different density in the given system; in other words, it constitutes direct evidence of the onset of a phase transition in the system. In this connection it is important to note that, so long as one uses the *exact* partition function of the system, isotherms of the van der Waals type, which possess unphysical regions of positive slope, *never* appear. On the

[2]For historical details, see Griffiths (1972, p. 12).

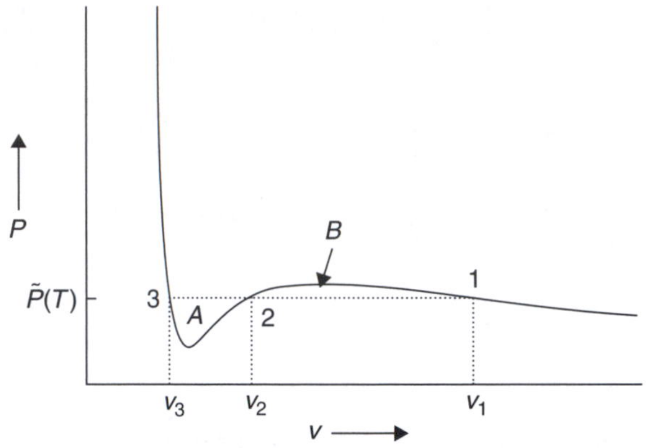

FIGURE 12.2 An unphysical isotherm corrected with the help of the Maxwell construction; the horizontal line is such that the areas A and B are equal. The "corrected" isotherm corresponds to a phase transition, taking place at pressure $\tilde{P}(T)$, with densities v_1^{-1} and v_3^{-1} of the respective phases.

other hand, if the partition function is evaluated under approximations, as we did in the derivation of the van der Waals equation of state in Section 10.3, isotherms with unphysical regions may indeed appear. In that case the isotherms in question have got to be "corrected," by introducing a region of "flatness" ($\partial P/\partial v = 0$), with the help of the *Maxwell construction of equal areas*; see Figure 12.2.[3] The real reason for the appearance of unphysical regions in the isotherms is that the approximate evaluations of the partition function introduce, almost invariably (though implicitly), the restraint of a *uniform* density throughout the system. This restraint eliminates the very possibility of the system passing through states in which there exist, side by side, two phases of different densities; in other words, the existence of a region of "flatness" in the (P, v)-diagram is automatically ruled out. On the other hand, an exact evaluation of the partition function must allow for *all* possible configurations of the system, including the ones characterized by a simultaneous existence of two or more phases of different densities. Under suitable conditions (for instance, when the temperature is sufficiently low), such a configuration might turn out to be the *equilibrium configuration* of the system, with the result that the system shows up in a multiphase, rather than a single-phase, state. We should, in this context, mention that if in the evaluation of the partition function one introduces no other approximation except the assumption of a uniform density in the system, then the resulting isotherms, corrected with the help of the Maxwell construction, would be the *exact* isotherms of the problem.

[3]The physical basis of the Maxwell construction can be seen with the help of the Gibbs free energy density $g(T, P)$. Since $dg = -s dT + v dP$ and along the "corrected" isotherm $dP = dT = 0$, it follows that $g_1 = g_3$; see Figure 12.2. To achieve the same result from the theoretical isotherm (along which $dT = 0$ but $dP \neq 0$), we clearly require that the quantity $v dP$, integrated along the isotherm from state 1 to state 3, must vanish; this leads to the "theorem of equal areas."

(iii) The presence of an absolutely flat portion in an isotherm, with mathematical singularities at its ends, is, strictly speaking, a consequence of the limiting process $N \to \infty$. If N were *finite*, and if the exact partition function were used, then the quantity P', defined by the relation

$$P' = kT\left(\frac{\partial \ln Q}{\partial V}\right)_{N,T}, \tag{3}$$

would be free from mathematical singularities. The ordinarily sharp corners in an isotherm would be rounded off; at the same time, the ordinarily flat portion of the isotherm would not be strictly flat — it would have for large N a small, negative slope. In fact, the quantity P' in this case would not be a function of v and T alone; it would depend on the number N as well, though in a thermodynamically negligible manner.

If we employ the grand partition function $\mathcal{Q}$, as obtained from the exact partition functions Q_N, namely

$$\mathcal{Q}(z, V, T) = \sum_{N \geq 0} Q_N(V, T) z^N, \tag{4}$$

a similar picture results. To see this, we note that for real molecules, with a given V, the variable N will be bounded by an upper limit, say N_m, which is the number of molecules that fill the volume V "tight-packed"; obviously, $N_m \sim V/\sigma^3$. For $N > N_m$, the potential energy of the system will be infinite; accordingly,

$$Q_N(N > N_m) \equiv 0. \tag{5}$$

Hence, for all practical purposes, our power series in (4) is a polynomial in z (which is ≥ 0) and is of degree N_m. Since the coefficients Q_N are all positive and $Q_0 \equiv 1$, the sum $\mathcal{Q} \geq 1$. The thermodynamic potential $\ln \mathcal{Q}$ is, therefore, a *well-behaved* function of the parameters z, V, and T. Consequently, so long as V (and hence N_m) remains finite, we do not expect any singularities or discontinuities in any of the functions derived from this potential. A nonanalytic behavior could appear only in the limit $(V, N_m) \to \infty$.

We now define P' by the relation

$$P' = \frac{kT}{V} \ln \mathcal{Q} \quad (V \text{ finite}); \tag{6}$$

since $\mathcal{Q} \geq 1, P' \geq 0$. The mean number of particles and the mean square deviation in this number are given by the formulae

$$\overline{N} = \left(\frac{\partial \ln \mathcal{Q}}{\partial \ln z}\right)_{V,T} \tag{7}$$

and

$$\overline{N^2} - \overline{N}^2 \equiv \overline{(N - \overline{N})^2} = \left(\frac{\partial \overline{N}}{\partial \ln z}\right)_{V,T}, \tag{8}$$

respectively; see Sections 4.2 and 4.5. Accordingly,

$$\left(\frac{\partial \ln \mathcal{Q}}{\partial \overline{N}}\right)_{V,T} = \left(\frac{\partial \ln \mathcal{Q}}{\partial \ln z}\right)_{V,T} \bigg/ \left(\frac{\partial \overline{N}}{\partial \ln z}\right)_{V,T} = \frac{\overline{N}}{\overline{N^2} - \overline{N}^2}. \tag{9}$$

At the same time, writing $\overline{v}$ for $V/\overline{N}$ and using (6), we have

$$\left(\frac{\partial \ln \mathcal{Q}}{\partial \overline{N}}\right)_{V,T} = \frac{V}{kT}\left(\frac{\partial P'}{\partial \overline{N}}\right)_{V,T} = -\frac{\overline{v}^2}{kT}\left(\frac{\partial P'}{\partial \overline{v}}\right)_{V,T}. \tag{10}$$

Comparing (9) and (10), we obtain

$$\left(\frac{\partial P'}{\partial \overline{v}}\right)_{V,T} = -\frac{kT}{V^2}\frac{\overline{N}^3}{\overline{N^2} - \overline{N}^2}, \tag{11}$$

which is clearly nonpositive.[4] For finite V, expression (11) will never vanish; accordingly, P' will never be strictly constant. Nevertheless, the slope $(\partial P'/\partial \overline{v})$ can, in a certain region, be extremely small — in fact, as small as $O(1/\overline{N})$; such a region would hardly be distinguishable from a phase transition because, on a macroscopic scale, the value of P' in such a region would be as good as a constant.[5]

If we now define the pressure of the system by the limiting relationship

$$P(\overline{v}, T) = \lim_{V\to\infty} P'(\overline{v}, T; V) = kT \lim_{V\to\infty}\left(\frac{1}{V}\ln \mathcal{Q}(z, V, T)\right), \tag{12}$$

then we *can* expect, in a set of isotherms, an absolutely flat portion $(\partial P/\partial \overline{v} \equiv 0)$, with sharp corners implying mathematical singularities. The mean particle density $\overline{n}$ would now be given by

$$\overline{n} = \lim_{V\to\infty}\left[\frac{1}{V}\frac{\partial \ln \mathcal{Q}(z, V, T)}{\partial \ln z}\right]; \tag{13}$$

it seems important to mention here that the operation $V \to \infty$ and the operation $\partial/\partial \ln z$ cannot be interchanged freely.

In passing, we note that the picture emerging from the grand partition function $\mathcal{Q}$, which has been obtained from the exact partition functions Q_N, remains practically

[4]Compare equation (11), which has been derived here *nonthermodynamically*, with equation (4.5.7) derived earlier.

[5]The presence of such a region entails that $(\overline{N^2} - \overline{N}^2)$ be $O(\overline{N}^2)$. This implies that the fluctuations in the variable N be *macroscopically* large, which in turn implies equally large fluctuations in the variable v within the system and hence the coexistence of two or more phases with different values of $\overline{v}$. In a single-phase state, $(\overline{N^2} - \overline{N}^2)$ is $O(\overline{N})$; the slope $(\partial P'/\partial \overline{v})$ is then $O(\overline{N}^0)$, as an intensive quantity should be.

unchanged even if one had employed a set of approximate Q_N. This is so because the argument developed in the preceding paragraphs makes no use whatsoever of the actual form of the functions Q_N. Thus, if an approximate Q_N leads to the van der Waals type of loop in the canonical ensemble, as shown in Figure 12.2, the corresponding set of Q_N, when employed in a grand canonical ensemble, would lead to isotherms free from such loops (Hill, 1953).

Subsequent to van Hove, Yang and Lee (1952) suggested an altemative approach that enables one to carry out a rigorous mathematical discussion of the phenomenon of condensation and of other similar transitions. In their approach, one is primarily concerned with the analytic behavior of the quantities P and $\overline{n}$, of equations (12) and (13), as functions of z at different values of T. The problem is examined in terms of the "zeros of the grand partition function $\mathcal{Q}$ in the *complex* z-plane," with attention focused on the way these zeros are distributed in the plane and the manner in which they evolve as the volume of the system is increased. For real, positive z, $\mathcal{Q} \geq 1$, therefore none of the zeros will lie on the real, positive axis in the z-plane. However, as $V \to \infty$ (and hence the degree of the polynomial (4) and, with it, the number of zeros itself grows to infinity), the distribution of zeros is expected to become *continuous* and, depending on T, may in fact *converge* on the real, positive axis at one or more points z_c. If so, our functions $P(z)$ and $\overline{n}(z)$, even with z varied along the real axis only, may, by virtue of their relationship to the function $\ln \mathcal{Q}$, turn out to be singular at the points $z = z_c$. The presence of such a singularity would imply the onset of a phase transition in the system. For further details of this approach, see Sections 12.3 and 12.4 of the first edition of this book; see also Griffiths (1972, pp. 50–58).

12.2 Condensation of a van der Waals gas

We start with the simplest, and historically the first, theoretical model that undergoes a gas–liquid phase transition. This model is generally referred to as the van der Waals gas and obeys the equation of state, see equation (10.3.9),

$$P = \frac{RT}{v - b} - \frac{a}{v^2}, \tag{1}$$

v being the *molar* volume of the gas; the parameters a and b then also pertain to one mole of the gas. We recall that, while a is a measure of the attractive forces among the molecules of the system, b is a measure of the repulsive forces that come into play when two molecules come too close to one another; accordingly, b is also a measure of the "effective space" occupied by the molecules (by virtue of a finite volume that may be associated with each one of them). In Section 10.3, the equation of state (1) was derived under the express assumption that $v \gg b$; here, we shall pretend, with van der Waals, that this equation holds even when v is comparable to b.

The isotherms following from equation (1) are shown in Figure 12.3. We note that, for temperatures above a *critical temperature* T_c, P decreases monotonically with v. For

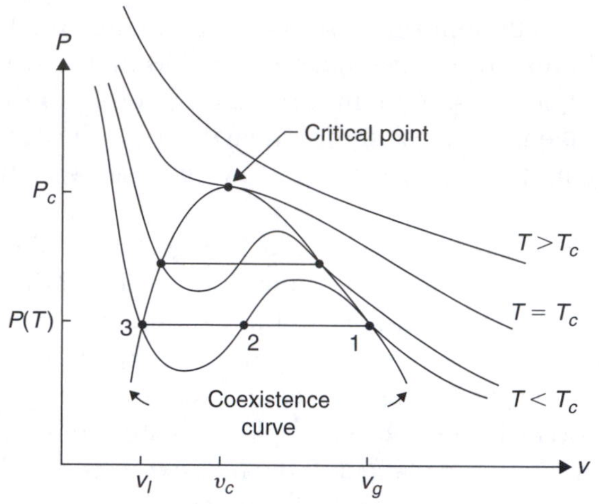

FIGURE 12.3 The isotherms of a van der Waals system; those for $T < T_c$ are "corrected" with the help of the Maxwell construction, thus leading to the coexistence curve at the top of which sits the critical point.

$T < T_c$, however, the relationship is nonmonotonic, so that over a certain range of v we encounter a region where $(\partial P/\partial v) > 0$; such a region is unphysical and must be "corrected" with the help of the Maxwell construction,[6] leading to an isotherm with a flat portion signaling transition from the gaseous state with molar volume $v_g(T)$ to the liquid state with molar volume $v_l(T)$ at a constant pressure $P(T)$. For $v_l < v < v_g$, the system resides in a state of mixed phases — partly liquid, partly gaseous — and, since the passage from one end of the flat portion to the other takes place with $\Delta v \neq 0$ but $\Delta P = 0$, the system is all along in a state of *infinite* compressibility; clearly, we are encountering here a brand of behavior that is patently *singular*. As T increases toward T_c, the transition takes place at a comparatively higher value of P, with v_g less than and v_l more than before — so that, as $T \to T_c$, both v_g and v_l approach a common value v_c that may be referred to as the *critical volume*; the corresponding value of P may then be designated by P_c, and we find ourselves located at the *critical point* of the system. The locus of all such points as $v_l(T)$ and $v_g(T)$ is generally referred to as the *coexistence curve*, for the simple reason that in the region enclosed by this curve the gaseous and the liquid phases mutually coexist; the top of this curve, where $v_l = v_g$, coincides with the critical point itself. Finally, the isotherm

[6]A more precise formulation of the van der Waals theory, as the limit of a theory with an infinite range potential, has been formulated by Kac, Uhlenbeck and Hemmer (1963). They considered the potential

$$u(r) = \begin{cases} +\infty & \text{for} \quad r \leq \sigma \\ -\kappa e^{-\kappa r} & \text{for} \quad r > \sigma, \end{cases}$$

so that the integral $\int_\sigma^\infty u(r)dr$ is simply $-\exp(-\kappa\sigma)$; when $\kappa \to 0$ the potential becomes infinite in range but infinitesimally weak. Kac et al. showed that, in this limit, the model becomes essentially the same as van der Waals' — with one noteworthy improvement, that is, no unphysical regions in the (P, v)–diagram appear and hence no need for the Maxwell construction arises.

pertaining to $T = T_c$, which, of course, passes through the critical point is referred to as the *critical isotherm* of the system; it is straightforward to see that the critical point is a point of inflection of this isotherm, so that both $(\partial P/\partial v)_T$ and $(\partial^2 P/\partial v^2)_T$ vanish at this point. Using (1), we obtain for the coordinates of the critical point

$$P_c = \frac{a}{27b^2}, \quad v_c = 3b, \quad T_c = \frac{8a}{27bR}, \tag{2}$$

so that the number

$$\mathcal{K} \equiv RT_c/P_c v_c = 8/3 = 2.666\ldots. \tag{3}$$

We thus find that, while P_c, v_c, and T_c vary from system to system (through the interaction parameters a and b), the quantity $\mathcal{K}$ has a common, *universal* value for all of them — so long as they all obey the same (i.e., van der Waals) equation of state. The experimental results for $\mathcal{K}$ indeed show that it is nearly the same over a large group of substances; for instance, its value for carbon tetrachloride, ethyl ether, and ethyl formate is 3.677, 3.814, and 3.895, respectively — close, though not exactly the same, and also a long way from the van der Waals value. The concept of universality is, nonetheless, there (even though the van der Waals equation of state may not truly apply).

It is now tempting to see if the equation of state itself can be written in a universal form. We find that this indeed can be done by introducing *reduced variables*

$$P_r = \frac{P}{P_c}, \quad v_r = \frac{v}{v_c}, \quad T_r = \frac{T}{T_c}. \tag{4}$$

Using (1) and (2), we readily obtain the reduced equation of state

$$\left(P_r + \frac{3}{v_r^2}\right)(3v_r - 1) = 8T_r, \tag{5}$$

which is clearly universal for all systems obeying van der Waals' original equation of state (1); all we have done here is to rescale the observable quantities P, v, and T in terms of their critical values and thereby "push the interaction parameters a and b into the background." Now, if two different systems happen to be in states characterized by the same values of v_r and T_r, then their P_r would also be the same; the systems are then said to be in "corresponding states" and, for that reason, the statement just made is referred to as the "law of corresponding states." Clearly, the passage from equation (1) to equation (5) takes us from an expression of diversity to a statement of unity!

We shall now examine the behavior of the van der Waals system in the *close* neighborhood of the critical point. For this, we write

$$P_r = 1 + \pi, \quad v_r = 1 + \psi, \quad T_r = 1 + t. \tag{6}$$

Equation (5) then takes the form

$$\pi\left(2+7\psi+8\psi^2+3\psi^3\right)+3\psi^3=8t\left(1+2\psi+\psi^2\right). \tag{7}$$

First of all, along the critical isotherm ($t=0$) and in the close vicinity of the critical point ($|\pi|,|\psi|\ll 1$), we obtain the simple, asymptotic result

$$\pi\approx-\frac{3}{2}\psi^3, \tag{8}$$

which is indicative of the "degree of flatness" of the critical isotherm at the critical point. Next, we examine the dependence of ψ on t as we approach the critical point from below. For this, we write (7) in the form

$$3\psi^3+8(\pi-t)\psi^2+(7\pi-16t)\psi+2(\pi-4t)\simeq 0. \tag{9}$$

Now, a close look at the (symmetric) shape of the coexistence curve near its top (where $|t|\ll 1$) shows that the three roots ψ_1,ψ_2, and ψ_3 of equation (9), which arise from the limiting behavior of the roots v_1,v_2, and v_3 of the original equation of state (1) as $T\to T_c-$, are such that $|\psi_2|\ll|\psi_{1,3}|$ and $|\psi_1|\simeq|\psi_3|$. This means that, in the region of interest,

$$\pi\approx 4t, \tag{10}$$

so that one of the roots, ψ_2, of equation (9) essentially vanishes while the other two are given by

$$\psi^2+8t\psi+4t\simeq 0. \tag{9a}$$

We expect the middle term here to be negligible (as will be confirmed by the end result), yielding

$$\psi_{1,3}\approx\pm 2|t|^{1/2}; \tag{11}$$

note that the upper sign here pertains to the gaseous phase and the lower sign to the liquid phase.

Finally, we consider the *isothermal compressibility* of the system which, in terms of reduced variables, is determined essentially by the quantity $-(\partial\psi/\partial\pi)_t$. Retaining only the dominant terms, we obtain from (7)

$$-\left(\frac{\partial\psi}{\partial\pi}\right)_t\approx\frac{2}{7\pi+9\psi^2-16t}. \tag{12}$$

For $t>0$, we approach the critical point along the critical isochore ($\psi=0$); equation (12), with the help of equation (10), then gives

$$-\left(\frac{\partial\psi}{\partial\pi}\right)_{t\to 0+}\approx\frac{1}{6t}. \tag{13}$$

For $t < 0$, we approach the critical point along the coexistence curve (on which $\psi^2 \approx -4t$); we now obtain

$$-\left(\frac{\partial \psi}{\partial \pi}\right)_{t \to 0-} \approx \frac{1}{12|t|}. \tag{14}$$

For the record, we quote here results for the specific heat, C_V, of the van der Waals gas (Uhlenbeck, 1966; Thompson, 1988)

$$C_V \approx \begin{cases} (C_V)_{\text{ideal}} + \frac{9}{2}Nk\left(1 + \frac{28}{25}t\right) & (t \leq 0) \quad \text{(15a)} \\ (C_V)_{\text{ideal}} & (t > 0), \quad \text{(15b)} \end{cases}$$

which imply a finite jump at the critical point.

Equations (8), (11), (13), (14), and (15) illustrate the nature of the *critical behavior* displayed by a van der Waals system undergoing the gas–liquid transition. While it differs in several important respects from the critical behavior of real physical systems, it shows up again and again in studies pertaining to other critical phenomena that have apparently nothing to do with the gas–liquid phase transition. In fact, this particular brand of behavior turns out to be a benchmark against which the results of more sophisticated theories are automatically compared.

12.3 A dynamical model of phase transitions

A number of physico-chemical systems that undergo phase transitions can be represented, to varying degrees of accuracy, by an "array of lattice sites, with only nearest-neighbor interaction that depends on the manner of occupation of the neighboring sites." This simple-minded model turns out to be good enough to provide a unified, theoretical basis for understanding a variety of phenomena such as ferromagnetism and antiferromagnetism, gas–liquid and liquid–solid transitions, order–disorder transitions in alloys, phase separation in binary solutions, and so on. There is no doubt that this model considerably oversimplifies the actual physical systems it is supposed to represent; nevertheless, it does retain the essential physical features of the problem — features that account for the propagation of *long-range order* in the system. Accordingly, it does lead to the onset of a phase transition in the given system, which arises in the nature of a *cooperative* phenomenon.

We find it convenient to formulate our problem in the language of ferromagnetism; later on, we shall establish correspondence between this language and the languages appropriate to other physical phenomena. We thus regard each of the N lattice sites to be occupied by an atom possessing a magnetic moment $\boldsymbol{\mu}$, of magnitude $g\mu_B\sqrt{[J(J+1)]}$, which is capable of $(2J+1)$ discrete orientations in space. These orientations define "different possible manners of occupation" of a given lattice site; accordingly, the whole lattice is capable of $(2J+1)^N$ different configurations. Associated with each configuration is an energy E that arises from mutual interactions among the neighboring atoms of the lattice and from the interaction of the whole lattice with an external field $\boldsymbol{B}$. A statistical analysis in the canonical ensemble should then enable us to determine the expectation value, $\overline{M}(B,T)$, of the net magnetization M. The presence of a *spontaneous magnetization* $\overline{M}(0,T)$ at temperatures below a certain (critical) temperature T_c and its absence above that temperature will then be interpreted as a ferromagnetic phase transition in the system at $T = T_c$.

Detailed studies, both theoretical and experimental, have shown that, for all ferromagnetic materials, data on the temperature dependence of the spontaneous magnetization, $\overline{M}(0,T)$, fit best with the value $J = \frac{1}{2}$; see Figure 12.4. One is, therefore, tempted to infer that the phenomenon of ferromagnetism is associated only with the spins of the electrons and not with their orbital motions. This is further confirmed by gyromagnetic experiments (Barnett, 1944; Scott, 1951, 1952), in which one either reverses the magnetization of a freely suspended specimen and observes the resulting rotation or imparts a rotation to the specimen and observes the resulting magnetization; the former is known as the *Einstein–de Haas method*, the latter the *Barnett method*. From these experiments one can derive the relevant g-value of the specimen which, in each case, turns out to be very close to 2; this, as we know, pertains to the electron spin. Therefore, in discussing the problem of ferromagnetism, we may specifically take: $\mu = 2\mu_B\sqrt{[s(s+1)]}$, where s is the quantum number associated with the electron spin. With $s = \frac{1}{2}$, only two orientations are possible for each lattice site, namely $s_z = +\frac{1}{2}$ (with $\mu_z = +\mu_B$) and $s_z = -\frac{1}{2}$ (with $\mu_z = -\mu_B$). The whole lattice is then capable of 2^N configurations; one such configuration is shown in Figure 12.5.

We now consider the nature of the interaction energy between two neighboring spins s_i and s_j. According to quantum mechanics, this energy is of the form $K_{ij} \pm J_{ij}$, where the upper sign applies to "antiparallel" spins ($S = 0$) and the lower sign to "parallel" spins ($S = 1$). Here, K_{ij} is the direct or Coulomb energy between the two spins, while J_{ij} is the

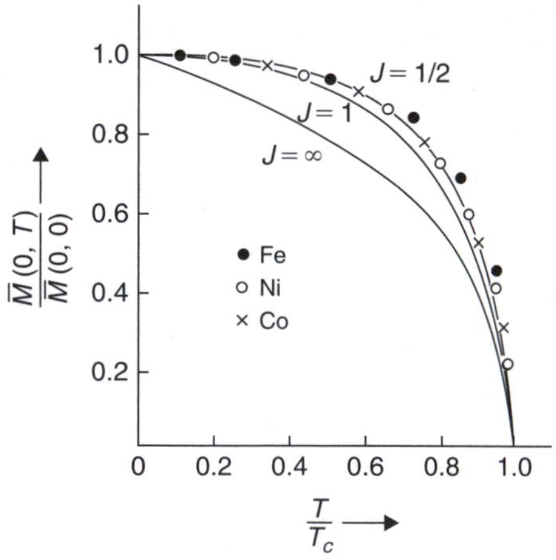

FIGURE 12.4 Spontaneous magnetization of iron, nickel, and cobalt as a function of temperature. Theoretical curves are based on the Weiss theory of ferromagnetism.

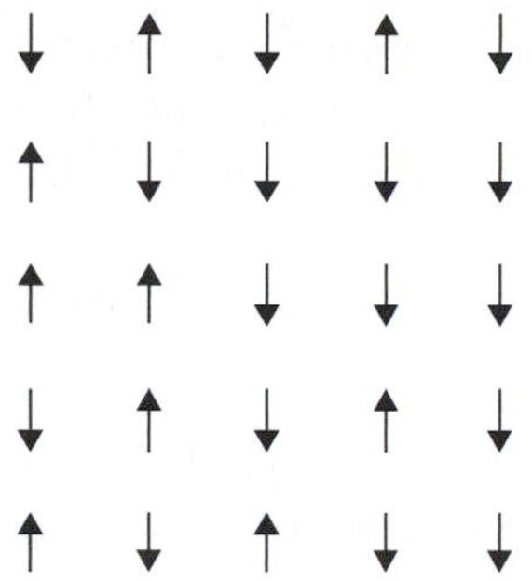

FIGURE 12.5 One of the 2^N possible configurations of a system composed of N spins; here, $N = 25$.

exchange energy between them:

$$K_{ij} = \int \psi_i^*(1)\psi_j^*(2)u_{ij}\psi_j(2)\psi_i(1)d\tau_1 d\tau_2, \tag{1}$$

while

$$J_{ij} = \int \psi_j^*(1)\psi_i^*(2)u_{ij}\psi_j(2)\psi_i(1)d\tau_1 d\tau_2, \tag{2}$$

u_{ij} being the relevant interaction potential. The energy difference between a state of "parallel" spins and one of "antiparallel" spins is given by

$$\varepsilon_{\uparrow\uparrow} - \varepsilon_{\uparrow\downarrow} = -2J_{ij}. \tag{3}$$

If $J_{ij} > 0$, the state $\uparrow\uparrow$ is energetically favored against the state $\uparrow\downarrow$; we then look for the possibility of *ferromagnetism*. If, on the other hand, $J_{ij} < 0$, the situation is reversed and we see the possibility of *antiferromagnetism*.

It seems useful to express the interaction energy of the two states, $\uparrow\uparrow$ and $\downarrow\downarrow$, by a single expression; for this, we consider the eigenvalues of the scalar product

$$\begin{aligned} \mathbf{s}_i \cdot \mathbf{s}_j &= \frac{1}{2}\left\{(\mathbf{s}_i + \mathbf{s}_j)^2 - \mathbf{s}_i^2 - \mathbf{s}_j^2\right\} \\ &= \frac{1}{2}S(S+1) - s(s+1), \end{aligned} \tag{4}$$

which equals $+\frac{1}{4}$ if $S = 1$ and $-\frac{3}{4}$ if $S = 0$. We may, therefore, write for the interaction energy of the spins i and j

$$\varepsilon_{ij} = \text{const.} - 2J_{ij}(\mathbf{s}_i \cdot \mathbf{s}_j), \tag{5}$$

which is consistent with the energy difference (3). The precise value of the constant here is immaterial because the potential energy is arbitrary to the extent of an additive constant anyway. Typically, the exchange interaction J_{ij} falls off rapidly as the separation of the two spins is increased. To a first approximation, therefore, we may regard J_{ij} as negligible for all but nearest-neighbor pairs (for which its value may be denoted by a common symbol J). The interaction energy of the whole lattice is then given by

$$E = \text{const.} - 2J \sum_{\text{n.n.}} (\boldsymbol{s}_i \cdot \boldsymbol{s}_j), \tag{6}$$

where the summation goes over all nearest-neighbor pairs in the lattice. The model based on expression (6) for the interaction energy of the lattice is known as the *Heisenberg model* (1928).

A simpler model results if we use, instead of (6), a *truncated* expression in which the product $(\boldsymbol{s}_i \cdot \boldsymbol{s}_j)$, which is equal to the sum $(s_{ix}s_{jx} + s_{iy}s_{jy} + s_{iz}s_{jz})$, is replaced by a single term $s_{iz}s_{jz}$; one reason for adopting this simpler model is that it does not necessarily require a quantum-mechanical treatment (because all the variables in the truncated expression for E commute). Expression (6) may now be written as

$$E = \text{const.} - J \sum_{\text{n.n.}} \sigma_i \sigma_j, \tag{7}$$

where the new symbol σ_i (or σ_j) $= +1$ for an "up" spin and -1 for a "down" spin; note that, with the introduction of the new symbol, we still have: $\varepsilon_{\uparrow\uparrow} - \varepsilon_{\uparrow\downarrow} = -2J$. The model based on expression (7) is known as the *Ising model*; it originated with Lenz (1920) and was subsequently investigated by his student Ising (1925).[7]

A different model results if we suppress the z-components of the spins and retain the x- and y-components instead. This model was originally introduced by Matsubara and Matsuda (1956) as a model of a quantum lattice gas, with possible relevance to the superfluid transition in liquid He^4. The critical behavior of this so-called *XY* model has been investigated in detail by Betts and coworkers, who have also emphasized the relevance of this model to the study of insulating ferromagnets (see Betts et al., 1968 – 1974).

It seems appropriate to regard the Ising and the *XY* models as special cases of an *anisotropic* Heisenberg model with interaction parameters J_x, J_y, and J_z; while the Ising model represents the situation $J_x, J_y \ll J_z$, the *XY* model represents just the opposite. Introducing a parameter n, which denotes the number of spin components entering into the Hamiltonian of the system, we may regard the Ising, the *XY*, and the Heisenberg models as pertaining to the n-values 1, 2, and 3, respectively. As will be seen later, the parameter n, along with the dimensionality d of the lattice, constitutes the basic set of elements that determine the qualitative nature of the critical behavior of a given system. For the time being, though, we confine our attention to the Ising model, which is not only the simplest one to analyze but also unifies the study of phase transitions in systems as diverse as ferromagnets, gas–liquids, liquid mixtures, binary alloys, and so on.

[7]For an historical account of the origin and development of the Lenz–Ising model, see the review article by Brush (1967). This review gives a large number of other references as well.

To study the statistical mechanics of the Ising model, we disregard the kinetic energy of the atoms occupying the various lattice sites, for the phenomenon of phase transitions is essentially a consequence of the interaction energy among the atoms; in the interaction energy again, we include only the nearest-neighbor contributions, in the hope that the farther-neighbor contributions would not affect the results qualitatively. To fix the z-direction, and to be able to study properties such as magnetic susceptibility, we subject the lattice to an external magnetic field $\boldsymbol{B}$, directed "upward"; the spin σ_i then possesses an additional potential energy $-\mu B\sigma_i$.[8] The Hamiltonian of the system in configuration $\{\sigma_1, \sigma_2, \ldots, \sigma_N\}$ is then given by

$$H\{\sigma_i\} = -J\sum_{\text{n.n.}} \sigma_i\sigma_j - \mu B\sum_i \sigma_i, \tag{8}$$

and the partition function by

$$\begin{aligned} Q_N(B,T) &= \sum_{\sigma_1}\sum_{\sigma_2}\cdots\sum_{\sigma_N} \exp[-\beta H\{\sigma_i\}] \\ &= \sum_{\sigma_1}\sum_{\sigma_2}\cdots\sum_{\sigma_N} \exp\left[\beta J\sum_{\text{n.n.}} \sigma_i\sigma_j + \beta\mu B\sum_i \sigma_i\right]. \end{aligned} \tag{9}$$

The Helmholtz free energy, the internal energy, the specific heat, and the net magnetization of the system then follow from the formulae

$$A(B,T) = -kT\ln Q_N(B,T), \tag{10}$$

$$U(B,T) = -T^2\frac{\partial}{\partial T}\left(\frac{A}{T}\right) = kT^2\frac{\partial}{\partial T}\ln Q_N, \tag{11}$$

$$C(B,T) = \frac{\partial U}{\partial T} = -T\frac{\partial^2 A}{\partial T^2}, \tag{12}$$

and

$$\overline{M}(B,T) = \mu\overline{\left(\sum_i \sigma_i\right)} = \overline{\left(-\frac{\partial H}{\partial B}\right)} = \frac{1}{\beta}\left(\frac{\partial \ln Q_N}{\partial B}\right)_T = -\left(\frac{\partial A}{\partial B}\right)_T. \tag{13}$$

Obviously, the quantity $\overline{M}(0,T)$ gives the *spontaneous magnetization* of the system; if it is nonzero at temperatures below a certain critical temperature T_c, the system would be ferromagnetic for $T < T_c$ and paramagnetic for $T > T_c$. At the transition temperature itself, the system is expected to show some sort of a singular behavior.

It is obvious that the energy levels of the system as a whole will be *degenerate*, in the sense that the various configurations $\{\sigma_i\}$ will not all possess *distinct* energy values. In fact, the energy of a given configuration does not depend on the *detailed* values of all

[8] Henceforth, we use the symbol μ instead of μ_B.

the variables σ_i; it depends only on a few numbers such as the total number N_+ of "up" spins, the total number N_{++} of "up–up" nearest-neighbor pairs, and so on. To see this, we define certain other numbers as well: N_- as the total number of "down" spins, N_{--} as the total number of "down–down" nearest-neighbor pairs, and N_{+-} as the total number of nearest-neighbor pairs with opposite spins. The numbers N_+ and N_- must satisfy the relation

$$N_+ + N_- = N. \tag{14}$$

And if q denotes the *coordination number* of the lattice, that is, the number of nearest neighbors for each lattice site,[9] then we also have the relations

$$qN_+ = 2N_{++} + N_{+-}, \tag{15}$$

$$qN_- = 2N_{--} + N_{+-}. \tag{16}$$

With the help of these relations, we can express all our numbers in terms of any two of them, say N_+ and N_{++}. Thus

$$N_- = N - N_+, \quad N_{+-} = qN_+ - 2N_{++}, \quad N_{--} = \frac{1}{2}qN - qN_+ + N_{++}; \tag{17}$$

it will be noted that the total number of nearest-neighbor pairs of all types is given, quite expectedly, by the expression

$$N_{++} + N_{--} + N_{+-} = \frac{1}{2}qN. \tag{18}$$

Naturally, the Hamiltonian of the system can also be expressed in terms of N_+ and N_{++}; we have from (8), with the help of the relations established above,

$$\begin{aligned} H_N(N_+, N_{++}) &= -J(N_{++} + N_{--} - N_{+-}) - \mu B(N_+ - N_-) \\ &= -J\left(\frac{1}{2}qN - 2qN_+ + 4N_{++}\right) - \mu B(2N_+ - N). \end{aligned} \tag{19}$$

Now, let $g_N(N_+, N_{++})$ be "the number of *distinct* ways in which the N spins of the lattice can be so arranged as to yield certain preassigned values of the numbers N_+ and N_{++}." The partition function of the system can then be written as

$$Q_N(B,T) = \sum_{N_+, N_{++}}{}' g_N(N_+, N_{++}) \exp\{-\beta H_N(N_+, N_{++})\}, \tag{20}$$

[9]The coordination number q for a linear chain is obviously 2; for two-dimensional lattices, namely honeycomb, square, and triangular, it is 3, 4, and 6, respectively; for three-dimensional lattices, namely simple cubic, body-centered cubic, and face-centered cubic, it is 6, 8, and 12, respectively.

that is,

$$e^{-\beta A} = e^{\beta N\left(\frac{1}{2}qJ-\mu B\right)} \sum_{N_+=0}^{N} e^{-2\beta(qJ-\mu B)N_+} \sum_{N_{++}}{}' g_N(N_+,N_{++}) e^{4\beta J N_{++}}, \tag{21}$$

where the primed summation in (21) goes over all values of N_{++} that are consistent with a *fixed* value of N_+ and is followed by a summation over all possible values of N_+, that is, from $N_+ = 0$ to $N_+ = N$. The central problem thus consists in determining the *combinatorial function* $g_N(N_+,N_{++})$ for the various lattices of interest.

12.4 The lattice gas and the binary alloy

Apart from ferromagnets, the Ising model can be readily adapted to simulate the behavior of certain other systems as well. More common among these are the lattice gas and the binary alloy.

The lattice gas

Although it had already been recognized that the results derived for the Ising model would apply equally well to a system of "occupied" and "unoccupied" lattice sites (i.e., to a system of "atoms" and "holes" in a lattice), it was Yang and Lee (1952) who first used the term "lattice gas" to describe such a system. By definition, a lattice gas is a collection of atoms, N_a in number, that can occupy only discrete positions in space — positions that constitute a lattice structure with coordination number q.

Each lattice site can be occupied by *at most* one atom, and the interaction energy between two occupied sites is nonzero, say $-\varepsilon_0$, only if the sites involved constitute a *nearest-neighbor* pair. The configurational energy of the gas is then given by

$$E = -\varepsilon_0 N_{aa}, \tag{1}$$

where N_{aa} is the total number of nearest-neighbor pairs (of occupied sites) in a given configuration of the system. Let $g_N(N_a, N_{aa})$ denote "the number of *distinct* ways in which the N_a atoms of the gas, assumed indistinguishable, can be distributed among the N sites of the lattice so as to yield a certain preassigned value of the number N_{aa}." The partition function of the system, neglecting the kinetic energy of the atoms, is then given by

$$Q_{N_a}(N,T) = \sum_{N_{aa}}{}' g_N(N_a, N_{aa})\, e^{\beta\varepsilon_0 N_{aa}}, \tag{2}$$

where the primed summation goes over all values of N_{aa} that are consistent with the given values of N_a and N; clearly, the number N here plays the role of the "total volume" available to the gas.

Going over to the grand canonical ensemble, we write for the grand partition function of the system

$$\mathcal{Q}(z, N, T) = \sum_{N_a=0}^{N} z^{N_a} Q_{N_a}(N, T). \tag{3}$$

The pressure P and the mean number $\overline{N}_a$ of the atoms in the gas are then given by

$$e^{\beta PN} = \sum_{N_a=0}^{N} z^{N_a} \sum_{N_{aa}}{}' g_N(N_a, N_{aa}) e^{\beta \varepsilon_0 N_{aa}} \tag{4}$$

and

$$\frac{\overline{N}_a}{N} = \frac{1}{v} = \frac{z}{kT}\left(\frac{\partial P}{\partial z}\right)_T; \tag{5}$$

here, v denotes the average volume per particle of the gas (measured in terms of the "volume of a primitive cell of the lattice").

To establish a formal correspondence between the lattice gas and the ferromagnet, we compare the present formulae with the ones established in the preceding section — in particular, formula (4) with formula (12.3.21). The first thing to note here is that the canonical ensemble of the ferromagnet corresponds to the grand canonical ensemble of the lattice gas! The rest of the correspondence is summarized in the following chart:

The lattice gas		*The ferromagnet*
$N_a, N - N_a$	$\leftrightarrow$	$N_+, N - N_+ (= N_-)$
ε_0	$\leftrightarrow$	$4J$
z	$\leftrightarrow$	$\exp\{-2\beta(qJ - \mu B)\}$
P	$\leftrightarrow$	$-\left(\frac{A}{N} + \frac{1}{2}qJ - \mu B\right)$
$\frac{\overline{N}_a}{N}\left(= \frac{1}{v}\right)$	$\leftrightarrow$	$\frac{\overline{N}_+}{N}\left(= \frac{1}{2}\left\{\frac{\overline{M}}{N\mu} + 1\right\}\right)$,

where

$$\overline{M} = \mu(\overline{N}_+ - \overline{N}_-) = \mu(2\overline{N}_+ - N). \tag{6}$$

We also note that the ferromagnetic analogue of formula (5) would be

$$\frac{\overline{N}_+}{N} = \frac{1}{kT}\left\{\frac{\partial\left(A/N + \frac{1}{2}qJ - \mu B\right)}{2\beta\partial(qJ - \mu B)}\right\}_T = \frac{1}{2}\left[-\frac{1}{N\mu}\left(\frac{\partial A}{\partial B}\right)_T + 1\right] \tag{7}$$

which, by equation (12.3.13), assumes the expected form

$$\frac{\overline{N}_+}{N} = \frac{1}{2}\left(\frac{\overline{M}}{N\mu} + 1\right). \tag{8}$$

It is quite natural to ask: does lattice gas correspond to any real physical system in nature? The immediate answer is that if we let the lattice constant tend to zero (thus going from a discrete structure to a continuous one) and also add, to the lattice-gas formulae, terms corresponding to an ideal gas (namely, the kinetic energy terms), then the model might simulate the behavior of a gas of real atoms interacting through a delta function potential. A study of the possibility of a phase transition in such a system may, therefore, be of some value in understanding phase transitions in real gases. The case $\varepsilon_0 > 0$, which implies an *attractive* interaction among the nearest neighbors, has been frequently cited as a possible model for a gas–liquid transition.

On the other hand, if the interaction is *repulsive* ($\varepsilon_0 < 0$), so that configurations with alternating sites being "occupied" and "unoccupied" are the more favored ones, then we obtain a model that arouses interest in connection with the theory of solidification; in such a study, however, the lattice constant has to stay finite. Thus, several authors have pursued the study of the antiferromagnetic version of this model, hoping that this might throw some light on the liquid–solid transition. For a bibliography of these pursuits, see the review article by Brush (1967).

The binary alloy

Much of the early activity in the theoretical analysis of the Ising model was related to the study of order–disorder transitions in alloys. In an alloy — to be specific, a *binary* alloy — we have a lattice structure consisting of two types of atoms, say 1 and 2, numbering N_1 and N_2, respectively. In a configuration characterized by the numbers N_{11}, N_{22}, and N_{12} of the three types of nearest-neighbor pairs, the configurational energy of the alloy may be written as

$$E = \varepsilon_{11}N_{11} + \varepsilon_{22}N_{22} + \varepsilon_{12}N_{12}, \tag{9}$$

where $\varepsilon_{11}, \varepsilon_{22}$, and ε_{12} have obvious meanings. As in the case of a ferromagnet, the various numbers appearing in the expression for E may be expressed in terms of the numbers N, N_1, and N_{11} (of which only N_{11} is variable here). Equation (9) then takes the form

$$\begin{aligned} E &= \varepsilon_{11}N_{11} + \varepsilon_{22}\left(\frac{1}{2}qN - qN_1 + N_{11}\right) + \varepsilon_{12}\left(qN_1 - 2N_{11}\right) \\ &= \frac{1}{2}q\varepsilon_{22}N + q(\varepsilon_{12} - \varepsilon_{22})N_1 + (\varepsilon_{11} + \varepsilon_{22} - 2\varepsilon_{12})N_{11}. \end{aligned} \tag{10}$$

The correspondence between this system and the lattice gas is now straightforward:

The lattice gas		*The binary alloy*
$N_a, N - N_a$	$\leftrightarrow$	$N_1, N - N_1 (= N_2)$
$-\varepsilon_0$	$\leftrightarrow$	$(\varepsilon_{11} + \varepsilon_{22} - 2\varepsilon_{12})$
A	$\leftrightarrow$	$A - \frac{1}{2} q\varepsilon_{22} N - q(\varepsilon_{12} - \varepsilon_{22}) N_1$

The correspondence with a ferromagnet can be established likewise; in particular, this requires that $\varepsilon_{11} = \varepsilon_{22} = -J$ and $\varepsilon_{12} = +J$.

At absolute zero, our alloy will be in the state of minimum configurational energy, which would also be the state of maximum configurational order. We expect that the two types of atoms would then occupy *mutually exclusive* sites, so that one might speak of atoms 1 being only at sites a and atoms 2 being only at sites b. As temperature rises, an exchange of sites results and, in the face of thermal agitation, the order in the system starts giving way. Ultimately, the two types of atoms get so "mixed up" that the very notion of the sites a being the "right" ones for atoms 1 and the sites b being the "right" ones for atoms 2 break down; the system then behaves, from the *configurational* point of view, as an assembly of $N_1 + N_2$ atoms of essentially the same species.

12.5 Ising model in the zeroth approximation

In 1928 Gorsky attempted a statistical study of order–disorder transitions in binary alloys on the basis of the assumption that the work expended in transferring an atom from an "ordered" position to a "disordered" one (or, in other words, from a "right" site to a "wrong" one) is directly proportional to the *degree of order* prevailing in the system. This idea was further developed by Bragg and Williams (1934, 1935) who, for the first time, introduced the concept of *long-range order* in the sense we understand it now and, with relatively simple mathematics, obtained results that could explain the main qualitative features of the relevant experimental data. The basic assumption in the Bragg–Williams approximation was that the energy of an individual atom in the given system is determined by the (*average*) degree of order prevailing in the entire system rather than by the (*fluctuating*) configurations of the neighboring atoms. In this sense, the approximation is equivalent to the *mean molecular field* (or the *internal field*) theory of Weiss, which was put forward in 1907 to explain the magnetic behavior of ferromagnetic materials.

It seems natural to call this approximation the *zeroth* approximation, for its features are totally insensitive to the detailed structure, or even to the dimensionality, of the lattice. We expect that the results following from this approximation will become more reliable as the number of neighbors interacting with a given atom increases (i.e., as $q \to \infty$), thus diminishing the importance of local, fluctuating influences.[10]

We now define a long-range order parameter L in a given configuration by the very suggestive relationship

$$L = \frac{1}{N}\sum_i \sigma_i = \frac{N_+ - N_-}{N} = 2\frac{N_+}{N} - 1 \quad (-1 \le L \le +1), \tag{1}$$

which gives

$$N_+ = \frac{N}{2}(1 + L) \quad \text{and} \quad N_- = \frac{N}{2}(1 - L). \tag{2}$$

The magnetization M is then given by

$$M = (N_+ - N_-)\mu = N\mu L \quad (-N\mu \le M \le +N\mu); \tag{3}$$

the parameter L is, therefore, a direct measure of the net magnetization in the system. For a completely random configuration, $\overline{N}_+ = \overline{N}_- = \frac{1}{2}N$; the expectation values of both L and M are then identically zero.

Now, in the spirit of the present approximation, we replace the first part of the Hamiltonian (12.3.8) by the expression $-J(\frac{1}{2}q\overline{\sigma})\sum_i \sigma_i$, that is, for a given σ_i, we replace each of the $q\sigma_j$ by $\overline{\sigma}$ while the factor $\frac{1}{2}$ is included to avoid duplication in the counting of the nearest-neighbor pairs. Making use of equation (1), and noting that $\overline{\sigma} \equiv \overline{L}$, we obtain for the total configurational energy of the system

$$E = -\frac{1}{2}(qJ\overline{L})NL - (\mu B)NL. \tag{4}$$

The expectation value of E is then given by

$$U = -\frac{1}{2}qJN\overline{L}^2 - \mu BN\overline{L}. \tag{5}$$

In the same approximation, the difference $\Delta\varepsilon$ between the overall configurational energy of an "up" spin and the overall configurational energy of a "down" spin — specifically, the energy expended in changing an "up" spin into a "down" spin — is given by, see equation (12.3.8),

$$\begin{aligned}\Delta\varepsilon &= -J(q\overline{\sigma})\Delta\sigma - \mu B\Delta\sigma \\ &= 2\mu\left(\frac{qJ}{\mu}\overline{\sigma} + B\right),\end{aligned} \tag{6}$$

[10] In connection with the present approximation, we may as well mention that early attempts to construct a theory of binary solutions were based on the assumption that the atoms in the solution mix randomly. One finds that the results following from this assumption of *random mixing* are mathematically equivalent to the ones following from the *mean field approximation*; see Problems 12.12 and 12.13.

for here $\Delta\sigma = -2$. The quantity $qJ\overline{\sigma}/\mu$ thus plays the role of the *internal* (*molecular*) *field* of Weiss; it is determined by (i) the mean value of the long-range order prevailing in the system and by (ii) the strength of the coupling, qJ, between a given spin i and its q nearest neighbors. The relative values of the equilibrium numbers $\overline{N}_+$ and $\overline{N}_-$ then follow from the *Boltzmann principle*, namely

$$\overline{N}_-/\overline{N}_+ = \exp(-\Delta\varepsilon/kT) = \exp\{-2\mu(B' + B)/kT\}, \tag{7}$$

where B' denotes the internal molecular field:

$$B' = qJ\overline{\sigma}/\mu = qJ\left(\overline{M}/N\mu^2\right). \tag{8}$$

Substituting (2) into (7), and keeping in mind equation (8), we obtain for $\overline{L}$

$$\frac{1-\overline{L}}{1+\overline{L}} = \exp\{-2(qJ\overline{L} + \mu B)/kT\} \tag{9}$$

or, equivalently,

$$\frac{qJ\overline{L} + \mu B}{kT} = \frac{1}{2}\ln\frac{1+\overline{L}}{1-\overline{L}} = \tanh^{-1}\overline{L}. \tag{10}$$

To investigate the possibility of spontaneous magnetization, we let $B \to 0$, which leads to the relationship

$$\overline{L}_0 = \tanh\left(\frac{qJ\overline{L}_0}{kT}\right). \tag{11}$$

Equation (11) may be solved graphically; see Figure 12.6. For any temperature T, the appropriate value of $\overline{L}_0(T)$ is determined by the point of intersection of (i) the straight line $y = L_0$ and (ii) the curve $y = \tanh(qJL_0/kT)$. Clearly, the solution $\overline{L}_0 = 0$ is always there; however, we are interested in *nonzero* solutions, if any. For those, we note that, since the slope of the curve (ii) varies from the initial value qJ/kT to the final value zero while the slope of the line (i) is unity throughout, an intersection other than the one at the origin is possible if, and only if,

$$qJ/kT > 1, \tag{12}$$

that is,

$$T < qJ/k = T_c, \quad \text{say}. \tag{13}$$

We thus obtain a *critical temperature* T_c, below which the system *can* acquire a nonzero spontaneous magnetization and above which it *cannot*. It is natural to identify T_c with the *Curie temperature* of the system — the temperature that marks a transition from the ferromagnetic to the paramagnetic behavior of the system or vice versa.

It is clear from Figure 12.6, as well as from equation (11), that if $\overline{L}_0$ is a solution of the problem, then $-\overline{L}_0$ is also a solution. The reason for this duplicity of solutions is that, in

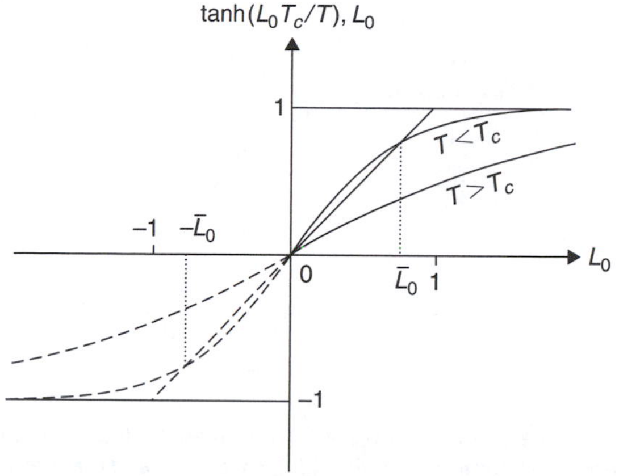

FIGURE 12.6 The graphical solution of equation (11), with $T_c = qJ/k$.

the absence of an external field, there is no way of assigning a "positive," as opposed to a "negative," direction to the alignment of spins. In fact, if B were zero right from the beginning, then the positive solution of equation (11) would be as likely to occur as the negative one—with the result that the *true* expectation value of $L_0(T)$ would be zero. If, on the other hand, B were nonzero to begin with (to be definite, say $B > 0$), then equation (10) for $\overline{L}(B, T)$ would admit only positive solutions and, in the limit $B \to 0+$, we would obtain a positive $\overline{L}_0(T)$. The "up–down symmetry" will then be broken and we will see a *net* alignment of spins in the "up" direction.[11]

The precise variation of $\overline{L}_0(T)$ with T can be obtained by solving equation (11) numerically; the general trend, however, can be seen from Figure 12.6. We note that, at $T = qJ/k \, (= T_c)$, the straight line $y = L_0$ is tangential to the curve $y = \tanh(qJL_0/kT)$ at the origin; the relevant solution then is $\overline{L}_0(T_c) = 0$. As T decreases, the initial slope of the curve becomes larger and the relevant point of intersection moves rapidly away from the origin; accordingly, $\overline{L}_0(T)$ rises rapidly as T decreases below T_c. To obtain an approximate dependence of $\overline{L}_0(T)$ on T near $T = T_c$, we write (11) in the form $\overline{L}_0 = \tanh(\overline{L}_0 T_c/T)$ and use the approximation $\tanh x \simeq x - x^3/3$, to obtain

$$\overline{L}_0(T) \approx \{3(1 - T/T_c)\}^{1/2} \quad (T \lesssim T_c, B \to 0). \tag{14}$$

On the other hand, as $T \to 0, \overline{L}_0 \to 1$, in accordance with the asymptotic relationship

$$\overline{L}_0(T) \approx 1 - 2\exp(-2T_c/T) \quad \{(T/T_c) \ll 1\}. \tag{15}$$

Figure 12.7 shows a plot of $\overline{L}_0(T)$ versus T, along with the relevant experimental results for iron, nickel, cobalt, and magnetite; we find the agreement not too bad.

[11]The concept of "broken symmetry" plays a vital role in this and many other phenomena in physics; for details, see Fisher (1972) and Anderson (1984).

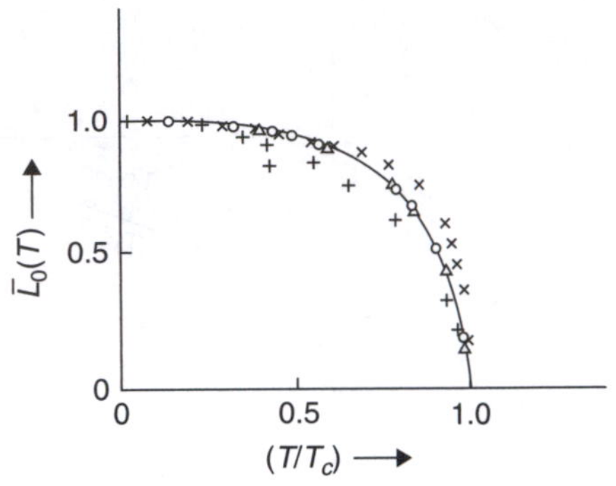

FIGURE 12.7 The spontaneous magnetization of a Weiss ferromagnet as a function of temperature. The experimental points (after Becker) are for iron (×), nickel (o), cobalt (Δ), and magnetite (+).

The field-free configurational energy and the field-free specific heat of the system are given by, see equation (5),

$$U_0(T) = -\frac{1}{2}qJN\bar{L}_0^2 \tag{16}$$

and

$$C_0(T) = -qJN\bar{L}_0\frac{d\bar{L}_0}{dT} = \frac{Nk\bar{L}_0^2}{(T/T_c)^2/(1-\bar{L}_0^2) - T/T_c}, \tag{17}$$

where the last step has been carried out with the help of equation (11). Thus, for all $T > T_c$, both $U_0(T)$ and $C_0(T)$ are identically zero. However, the value of the specific heat at the transition temperature T_c, as approached from below, turns out to be, see equations (14) and (17),

$$C_0(T_c-) = \operatorname*{Lim}_{x\to 0}\left\{\frac{Nk\cdot 3x}{\frac{(1-x)^2}{1-3x} - (1-x)}\right\} = \frac{3}{2}Nk. \tag{18}$$

The specific heat, therefore, displays a discontinuity at the transition point. On the other hand, as $T \to 0$, the specific heat vanishes, in accordance with the formula, see equations (15) and (17),

$$C_0(T) \approx 4Nk\left(\frac{T_c}{T}\right)^2 \exp(-2T_c/T). \tag{19}$$

The full trend of the function $C_0(T)$ is shown in Figure 12.8.

It is important to note that the vanishing of the configurational energy and the specific heat of the system at temperatures above T_c is directly related to the fact that, in the present approximation, the configurational order prevailing in the system at lower temperatures is completely wiped out as $T \to T_c$. Consequently, the configurational entropy and

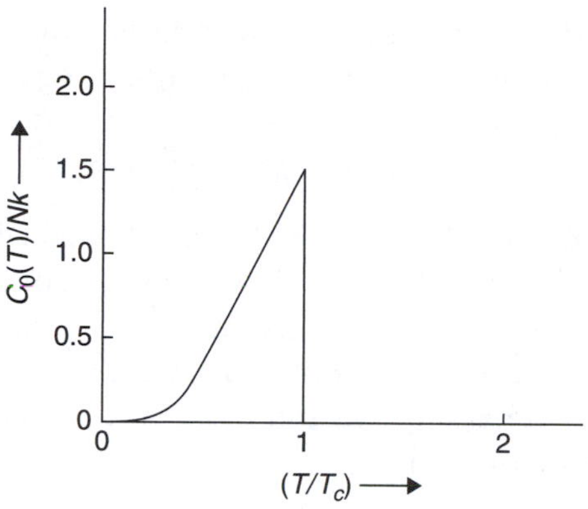

FIGURE 12.8 The field-free specific heat of a Weiss ferromagnet as a function of temperature.

the configurational energy of the system attain their maximum values at $T = T_c$; beyond that, the system remains thermodynamically "inert." As a check, we evaluate the configurational entropy of the system at $T = T_c$; with the help of equations (11) and (17), we get

$$S_0(T_c) = \int_0^{T_c} \frac{C_0(T)dT}{T} = -qJN \int_1^0 \frac{\bar{L}_0}{T} d\bar{L}_0$$

$$= Nk \int_0^1 (\tanh^{-1}\bar{L}_0) d\bar{L}_0 = Nk \ln 2, \tag{20}$$

precisely the result we expect for a system capable of 2^N *equally likely* microstates.[12] The fact that all these microstates are *equally likely* to occur is again related to the fact that for $T \geq T_c$ there is no (configurational) order in the system.

We now proceed to study the magnetic susceptibility of the system. Using equation (10), we get

$$\chi(B,T) = \left(\frac{\partial \overline{M}}{\partial B}\right)_T = N\mu \left(\frac{\partial \bar{L}}{\partial B}\right)_T = \frac{N\mu^2}{k} \frac{1 - \bar{L}^2(B,T)}{T - T_c\{1 - \bar{L}^2(B,T)\}}. \tag{21}$$

For $\bar{L} \ll 1$ (which is true at high temperatures for a wide range of B but is also true near T_c if B is small), we obtain the *Curie–Weiss law*

$$\chi_0(T) \approx (N\mu^2/k)(T - T_c)^{-1} \quad (T \gtrsim T_c, B \to 0), \tag{22a}$$

[12]Recall equation (3.3.14), whereby $S = k \ln \Omega$.

which may be compared with the *Curie law* derived earlier for a paramagnetic system; see equation (3.9.12). For T less than, but close to, T_c we utilize equation (14) as well and get

$$\chi_0(T) \approx (N\mu^2/2k)(T_c - T)^{-1} \quad (T \lesssim T_c, B \to 0). \tag{22b}$$

Experimentally, one finds that the Curie–Weiss law is satisfied with considerable accuracy, except that the empirical value of T_c thus obtained is always somewhat larger than the true transition temperature of the material; for instance, in the case of nickel, the empirical value of T_c obtained in this manner turns out to be about 650 K while the actual transition takes place at about 631 K. In passing, we add that, as $T \to 0$, the low-field susceptibility vanishes, in accordance with the formula

$$\chi_0(T) \approx \frac{4N\mu^2}{kT} \exp(-2T_c/T). \tag{23}$$

Finally, we examine the relationship between $\overline{L}$ and B at $T = T_c$. Using, once again, equation (10) and employing the approximation $\tanh^{-1} x \simeq x + x^3/3$, we get

$$\overline{L} \approx (3\mu B/kT_c)^{1/3} \quad (T = T_c, B \to 0). \tag{24}$$

At this point we wish to emphasize the remarkable similarity that exists between the critical behavior of a gas–liquid system obeying van der Waals equation of state and that of a magnetic system treated in the Bragg–Williams approximation. Even though the two systems are physically very different, the level of approximation is such that the exponents governing power-law behavior of the various physical quantities in the critical region turn out to be the same; compare, for instance, equation (14) with (12.2.11), equations (22a) and (22b) with (12.2.13) and (12.2.14), equation (24) with (12.2.8) — along with the behavior of the specific heat as well. This sort of similarity will be seen again and again whenever we employ an approach similar in spirit to the mean field approach of this section.

Before we close our discussion of the so-called zeroth approximation, we would like to demonstrate that it corresponds exactly to the *random mixing approximation* (which was employed originally in the theory of binary solutions). According to equation (12.3.19), the mean configurational energy in the absence of the external field is given by

$$U_0 = -J\left(\frac{1}{2}qN - 2q\overline{N}_+ + 4\overline{N}_{++}\right). \tag{25}$$

At the same time, equations (2) and (16) of the present approach give

$$\overline{N}_+ = \frac{1}{2}N(1 + \overline{L}_0) \quad \text{and} \quad U_0 = -\frac{1}{2}qJN\overline{L}_0^2. \tag{26}$$

Combining (25) and (26), we obtain

$$\overline{N}_{++} = \frac{1}{8}qN(1 + \overline{L}_0)^2, \tag{27}$$

so that

$$\frac{\overline{N}_{++}}{\frac{1}{2}qN} = \left(\frac{\overline{N}_{+}}{N}\right)^2. \tag{28}$$

Thus, the probability of having an "up–up" nearest-neighbor pair of spins in the lattice is precisely equal to the square of the probability of having an "up" spin; in other words, there does not exist, in spite of the presence of a nearest-neighbor interaction (characterized by the coupling constant J), any specific correlation between the neighboring spins of the lattice. Put differently, there does not exist any *short-range order* in the system, apart from what follows statistically from the long-range order (characterized by the parameter $\overline{L}$). It follows that, in the present approximation, our system consists of a specific number of "up" spins, namely $N(1+\overline{L})/2$, and a corresponding number of "down" spins, namely $N(1-\overline{L})/2$, distributed over the N lattice sites *in a completely random manner* — similar to the mixing of $N(1+\overline{L})/2$ atoms of one kind with $N(1-\overline{L})/2$ atoms of another kind in a completely random manner to obtain a binary solution of N atoms; see also Problem 12.4. For this sort of mixing, we obviously have

$$\overline{N}_{++} = \frac{1}{2}qN\left(\frac{1+\overline{L}}{2}\right)^2, \quad \overline{N}_{--} = \frac{1}{2}qN\left(\frac{1-\overline{L}}{2}\right)^2, \tag{29a}$$

$$\overline{N}_{+-} = 2\cdot\frac{1}{2}qN\left(\frac{1+\overline{L}}{2}\right)\left(\frac{1-\overline{L}}{2}\right), \tag{29b}$$

with the result that

$$\frac{\overline{N}_{++}\overline{N}_{--}}{(\overline{N}_{+-})^2} = \frac{1}{4}. \tag{30}$$

12.6 Ising model in the first approximation

The approaches considered in the preceding section have a natural generalization toward an improved approximation. The mean field approach leads naturally to the *Bethe approximation* (Bethe, 1935; Rushbrooke, 1938), which treats the interaction of a given spin with its nearest neighbors somewhat more accurately. The random mixing approach, on the other hand, leads to the *quasichemical* approximation (Guggenheim, 1935; Fowler and Guggenheim, 1940), which takes into account the *specific* short-range order of the lattice — over and above the one that follows statistically from the long-range order. As shown by Guggenheim (1938) and by Chang (1939), the two methods yield identical results for the Ising model. It seems worthwhile to mention here that the extension of these approximations to higher orders, or their application to the Heisenberg model, does not produce identical results.

In the Bethe approximation, a given spin σ_0 is regarded as the central member of a group, which consists of this spin and its q nearest neighbors, and in writing down the Hamiltonian of this group the interaction between the central spin and its q neighbors is taken into account *exactly* while the interaction of these neighbors with other spins in the lattice is taken into account through a mean molecular field B'. Thus

$$H_{q+1} = -\mu B\sigma_0 - \mu(B+B')\sum_{j=1}^{q}\sigma_j - J\sum_{j=1}^{q}\sigma_0\sigma_j, \tag{1}$$

B being the external magnetic field acting on the lattice. The internal field B' is determined by the *condition of self-consistency*, which requires that the mean value, $\overline{\sigma}_0$, of the central spin be the same as the mean value, $\overline{\sigma}_j$, of any of the q neighbors. The partition function Z of this group of spins as a whole is given by

$$\begin{aligned} Z &= \sum_{\sigma_0,\sigma_j=\pm 1} \exp\left[\frac{1}{kT}\left\{\mu B\sigma_0 + \mu\left(B+B'\right)\sum_{j=1}^{q}\sigma_j + J\sum_{j=1}^{q}\sigma_0\sigma_j\right\}\right] \\ &= \sum_{\sigma_0,\sigma_j=\pm 1} \exp\left[\alpha\sigma_0 + (\alpha+\alpha')\sum_{j=1}^{q}\sigma_j + \gamma\sum_{j=1}^{q}\sigma_0\sigma_j\right], \end{aligned} \tag{2}$$

where

$$\alpha = \frac{\mu B}{kT}, \quad \alpha' = \frac{\mu B'}{kT} \quad \text{and} \quad \gamma = \frac{J}{kT}. \tag{3}$$

Now, the right side of (2) can be written as a sum of two terms, one pertaining to $\sigma_0 = +1$ and the other to $\sigma_0 = -1$, that is,

$$Z = Z_+ + Z_-,$$

where

$$\begin{aligned} Z_\pm &= \sum_{\sigma_j=\pm 1} \exp\left[\pm\alpha + (\alpha+\alpha'\pm\gamma)\sum_{j=1}^{q}\sigma_j\right] \\ &= e^{\pm\alpha}\left[2\cosh\left(\alpha+\alpha'\pm\gamma\right)\right]^q. \end{aligned} \tag{4}$$

The mean value of the central spin is then given by

$$\overline{\sigma}_0 = \frac{Z_+ - Z_-}{Z}, \tag{5}$$

while the mean value of any one of its q neighbors is given by, see (2) and (4),

$$\bar{\sigma}_j = \frac{1}{q}\overline{\left(\sum_{j=1}^{q}\sigma_j\right)} = \frac{1}{q}\left(\frac{1}{Z}\frac{\partial Z}{\partial\alpha'}\right)$$
$$= \frac{1}{Z}\left\{Z_+\tanh\left(\alpha+\alpha'+\gamma\right)+Z_-\tanh\left(\alpha+\alpha'-\gamma\right)\right\}. \tag{6}$$

Equating (5) and (6), we get

$$Z_+\left\{1-\tanh\left(\alpha+\alpha'+\gamma\right)\right\} = Z_-\left\{1+\tanh\left(\alpha+\alpha'-\gamma\right)\right\}. \tag{7}$$

Substituting for Z_+ and Z_- from (4), we finally obtain

$$e^{2\alpha'} = \left\{\frac{\cosh\left(\alpha+\alpha'+\gamma\right)}{\cosh\left(\alpha+\alpha'-\gamma\right)}\right\}^{q-1}. \tag{8}$$

Equation (8) determines α' which, in turn, determines the magnetic behavior of the lattice.

To study the possibility of spontaneous magnetization, we set $\alpha(=\mu B/kT)=0$. Equation (8) then reduces to

$$\alpha' = \frac{q-1}{2}\ln\left\{\frac{\cosh\left(\alpha'+\gamma\right)}{\cosh\left(\alpha'-\gamma\right)}\right\}. \tag{9}$$

In the absence of interactions $(\gamma=0)$, α' is clearly zero. In the presence of interactions $(\gamma\neq 0)$, α' may still be zero unless γ exceeds a certain critical value, γ_c say. To determine this value, we expand the right side of (9) as a Taylor series around $\alpha'=0$, with the result

$$\alpha' = (q-1)\tanh\gamma\left\{\alpha' - \operatorname{sech}^2\gamma\,\frac{\alpha'^3}{3}+\cdots\right\}. \tag{10}$$

We note that, for all γ, $\alpha'=0$ is one possible solution of the problem; this, however, does not interest us. A nonzero solution requires that

$$(q-1)\tanh\gamma > 1,$$

that is,

$$\gamma > \gamma_c = \tanh^{-1}\left(\frac{1}{q-1}\right) = \frac{1}{2}\ln\left(\frac{q}{q-2}\right). \tag{11}$$

In terms of temperature, this means that

$$T < T_c = \frac{2J}{k}\bigg/\ln\left(\frac{q}{q-2}\right), \tag{12}$$

which determines the *Curie temperature* of the lattice. From (10), we also infer that for temperatures less than, but close to, the Curie temperature

$$\alpha' \simeq \left\{3\cosh^2\gamma_c\left[(q-1)\tanh\gamma - 1\right]\right\}^{1/2} \simeq \{3(q-1)(\gamma-\gamma_c)\}^{1/2}$$
$$\simeq \left\{3(q-1)\frac{J}{kT_c}\left(1-\frac{T}{T_c}\right)\right\}^{1/2}. \tag{13}$$

The parameter $\overline{L}$, which is a measure of the long-range order in the system, is, by definition, equal to $\overline{\sigma}$. From equations (5) and (7), we get

$$\overline{L} = \frac{(Z_+/Z_-)-1}{(Z_+/Z_-)+1} = \frac{\sinh(2\alpha+2\alpha')}{\cosh(2\alpha+2\alpha')+\exp(-2\gamma)}. \tag{14}$$

In the limit $B \to 0$ (which means $\alpha \to 0$) and at temperatures less than, but close to, the Curie temperature ($\gamma \gtrsim \gamma_c; \alpha' \simeq 0$), we obtain

$$\overline{L}_0 = \frac{\sinh(2\alpha')}{\cosh(2\alpha')+\exp(-2\gamma)} \simeq \frac{2\alpha'}{1+(q-2)/q} = \frac{q}{q-1}\alpha'. \tag{15}$$

Substituting from (12) and (13), we get

$$\overline{L}_0 \simeq \left[3\frac{q}{q-1}\left\{\frac{q}{2}\ln\left(\frac{q}{q-2}\right)\right\}\left(1-\frac{T}{T_c}\right)\right]^{1/2}. \tag{16}$$

We note that, for $q \gg 1$, equations (12) and (16) reduce to their zeroth-order counter parts (12.5.13) and (12.5.14), respectively; in either case, as $T \to T_c$ from below, $\overline{L}_0$ vanishes as $(T_c - T)^{1/2}$. We also note that the spontaneous magnetization curve in the present approximation has the same general shape as in the zeroth approximation; see Figure 12.7. Of course, in the present case the curve depends explicitly on the coordination number q, being steepest for small q and becoming less steep as q increases — tending ultimately to the limiting form given by the zeroth approximation.

We shall now study *correlations* that might exist among neighboring spins in the lattice. For this, we evaluate the numbers $\overline{N}_{++}, \overline{N}_{--}$, and $\overline{N}_{+-}$ in terms of the parameters α, α', and γ, and compare the resulting expressions with the ones obtained under the mean field approximation. Carrying out summations in (2) over all the spins (of the group) except σ_0 and σ_1, we obtain

$$Z = \sum_{\sigma_0,\sigma_1=\pm1}\left[\exp\{\alpha\sigma_0+(\alpha+\alpha')\sigma_1+\gamma\sigma_0\sigma_1\}\{2\cosh(\alpha+\alpha'+\gamma\sigma_0)\}^{q-1}\right]. \tag{17}$$

Writing this as a sum of three parts pertaining, respectively, to the cases (i) $\sigma_0 = \sigma_1 = +1$, (ii) $\sigma_0 = \sigma_1 = -1$, and (iii) $\sigma_0 = -\sigma_1 = \pm1$, we have

$$Z = Z_{++} + Z_{--} + Z_{+-}, \tag{18}$$

where, naturally enough,

$$\overline{N}_{++} : \overline{N}_{--} : \overline{N}_{+-} :: Z_{++} : Z_{--} : Z_{+-}. \tag{19}$$

We thus obtain, using (8) as well,

$$\begin{aligned}
\overline{N}_{++} &\propto e^{(2\alpha+\alpha'+\gamma)} \left\{2\cosh(\alpha+\alpha'+\gamma)\right\}^{q-1}, \\
\overline{N}_{--} &\propto e^{(-2\alpha-\alpha'+\gamma)} \left\{2\cosh(\alpha+\alpha'-\gamma)\right\}^{q-1} \\
&= e^{(-2\alpha-3\alpha'+\gamma)}\{2\cosh(\alpha+\alpha'+\gamma)\}^{q-1},
\end{aligned}$$

and

$$\begin{aligned}
\overline{N}_{+-} &\propto e^{(-\alpha'-\gamma)} \left\{2\cosh(\alpha+\alpha'+\gamma)\right\}^{q-1} + e^{(\alpha'-\gamma)} \left\{2\cosh(\alpha+\alpha'-\gamma)\right\}^{q-1} \\
&= 2e^{(-\alpha'-\gamma)}\{2\cosh(\alpha+\alpha'+\gamma)\}^{q-1}.
\end{aligned}$$

Normalizing these expressions with the help of the relationship

$$\overline{N}_{++} + \overline{N}_{--} + \overline{N}_{+-} = \frac{1}{2}qN, \tag{20}$$

we obtain the desired results

$$(\overline{N}_{++}, \overline{N}_{--}, \overline{N}_{+-}) = \frac{1}{2}qN \frac{\left(e^{2\alpha+2\alpha'+\gamma}, e^{-2\alpha-2\alpha'+\gamma}, 2e^{-\gamma}\right)}{2\left\{e^{\gamma}\cosh(2\alpha+2\alpha') + e^{-\gamma}\right\}}, \tag{21}$$

whereby

$$\frac{\overline{N}_{++}\overline{N}_{--}}{\left(\overline{N}_{+-}\right)^2} = \frac{1}{4}e^{4\gamma} = \frac{1}{4}e^{4J/kT}. \tag{22}$$

The last result differs significantly from the one that followed from the random mixing approximation, namely (12.5.30). The difference lies in the extra factor $\exp(4J/kT)$ which, for $J > 0$, favors the formation of parallel-spin pairs ↑↑ and ↓↓, as opposed to antiparallel-spin pairs ↑↓ and ↓↑. In fact, one may regard the elementary process

$$\uparrow\uparrow + \downarrow\downarrow \Leftrightarrow 2\uparrow\downarrow, \tag{23}$$

which leaves the total numbers of "up" spins and "down" spins unaltered, as a kind of a "chemical reaction" which, proceeding from left to right, is endothermic (requiring an amount of energy $4J$ to get through) and, proceeding from right to left, is exothermic (releasing an amount of energy $4J$). Equation (22) then constitutes the *law of mass action* for this reaction, the expression on the right side being the *equilibrium constant* of the reaction. Historically, equation (22) was adopted by Guggenheim as the starting point of his "quasichemical" treatment of the Ising model; only later on did he show that his treatment was equivalent to the Bethe approximation expounded here.

Equation (22) tells us that, for $J > 0$, there exists among *like* neighbors (↑ and ↑ or ↓ and ↓) a *positive* correlation and among *unlike* neighbors (↑ and ↓) a *negative* correlation, and that these correlations are a direct consequence of the nearest-neighbor interaction. Accordingly, there must exist a *specific* short-range order in the system, over and above the one that follows statistically from the long-range order. To see this explicitly, we note that even when long-range order disappears ($\alpha + \alpha' = 0$), some short-range order still persists. For instance, from equation (21) we obtain

$$(\overline{N}_{++}, \overline{N}_{--}, \overline{N}_{+-})_{\overline{L}=0} = \frac{1}{2} qN \frac{(e^{\gamma}, e^{\gamma}, 2e^{-\gamma})}{4 \cosh \gamma} \tag{24}$$

which, only in the limit $\gamma \to 0$, goes over to the random-mixing result, see equation (12.5.29) with $\overline{L} = 0$,

$$(\overline{N}_{++}, \overline{N}_{--}, \overline{N}_{+-})_{\overline{L}=0} = \frac{1}{2} qN \frac{(1,1,2)}{4}. \tag{25}$$

In the zeroth approximation, equation (25) is supposed to hold at *all* temperatures above T_c; we now find that a better approximation at these temperatures is provided by (24).

Next, we evaluate the configurational energy U_0 and the specific heat C_0 of the lattice in the absence of the external field ($\alpha = 0$). In view of equation (12.5.25),

$$U_0 = -J\left(\frac{1}{2} qN - 2q\overline{N}_+ + 4\overline{N}_{++}\right)_{\alpha=0}. \tag{26}$$

The expression for $\overline{N}_{++}$ is given by equation (21) while that for $\overline{N}_+$ can be obtained from (14):

$$(\overline{N}_+)_{\alpha=0} = \frac{1}{2} N(1 + \overline{L}_0) = \frac{1}{2} N \frac{\exp(2\alpha') + \exp(-2\gamma)}{\cosh(2\alpha') + \exp(-2\gamma)}. \tag{27}$$

Equation (26) then gives

$$U_0 = -\frac{1}{2} qJN \frac{\cosh(2\alpha') - \exp(-2\gamma)}{\cosh(2\alpha') + \exp(-2\gamma)}, \tag{28}$$

where α' is determined by equation (9). For $T > T_c$, $\alpha' = 0$, so

$$U_0 = -\frac{1}{2} qJN \frac{1 - \exp(-2\gamma)}{1 + \exp(-2\gamma)} = -\frac{1}{2} qJN \tanh \gamma. \tag{29}$$

Obviously, this result arises solely from the *short-range order* that persists in the system even above T_c. As for the specific heat, we get

$$C_0/Nk = \frac{1}{2} q\gamma^2 \operatorname{sech}^2 \gamma \quad (T > T_c). \tag{30}$$

As $T \to \infty$, C_0 vanishes like T^{-2}. We note that a nonzero specific heat above the transition temperature is a welcome feature of the present approximation, for it brings our model

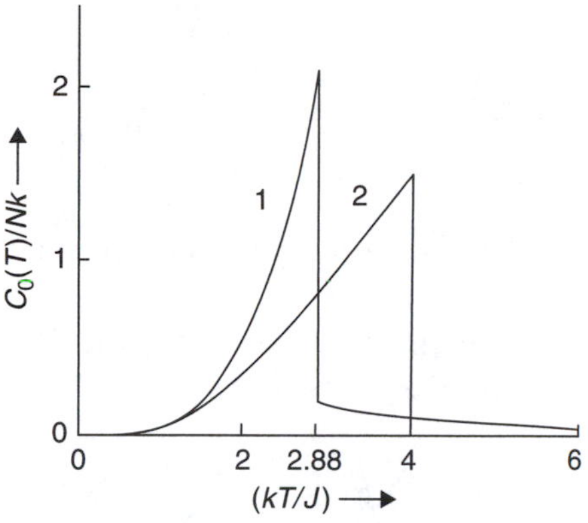

FIGURE 12.9 The field-free specific heat of an Ising lattice with coordination number 4. Curve 1 obtains in the Bethe approximation, curve 2 in the Bragg–Williams approximation.

somewhat closer to real physical systems. In this connection, we recall that in the previous approximation the specific heat was zero for *all* $T > T_c$. Figure 12.9 shows the specific heat of an Ising lattice, with coordination number 4, as given by the Bethe approximation; for comparison, the result of the previous approximation is also included.

We are now in a position to study the specific heat discontinuity at $T = T_c$. The limiting value of C_0, as T approaches T_c from above, can be obtained from equation (30) by letting $\gamma \to \gamma_c$. One obtains, with the help of equation (11),

$$\frac{1}{Nk} C_0(T_c+) = \frac{1}{2} q\gamma_c^2 \operatorname{sech}^2 \gamma_c = \frac{1}{8} \frac{q^2(q-2)}{(q-1)^2} \left\{ \ln\left(\frac{q}{q-2}\right) \right\}^2 . \tag{31}$$

To obtain the corresponding result as T approaches T_c from below, we must use the general expression (28) for U_0, with $\alpha' \to 0$ as $\gamma \to \gamma_c$. Expanding (28) in powers of the quantities $(\gamma - \gamma_c)$ and α', and making use of equation (13), we obtain for $(1 - T/T_c) \ll 1$

$$\begin{aligned} U_0 &= -\frac{1}{2} qJN \left[\frac{1}{(q-1)} + \frac{q(q-2)}{(q-1)^2} (\gamma - \gamma_c) + \frac{q(q-2)}{(q-1)^2} \alpha'^2 + \cdots \right] \\ &= -\frac{1}{2} qJN \left[\frac{1}{(q-1)} + \frac{q(q-2)(3q-2)}{(q-1)^2} \frac{J}{kT_c} \left(1 - \frac{T}{T_c}\right) + \cdots \right]. \end{aligned} \tag{32}$$

Differentiating with respect to T and substituting for T_c, we obtain

$$\frac{1}{Nk} C_0(T_c-) = \frac{1}{8} \frac{q^2(q-2)(3q-2)}{(q-1)^2} \left\{ \ln\left(\frac{q}{q-2}\right) \right\}^2 , \tag{33}$$

which is $(3q-2)$ times larger than the corresponding result for $T = T_c+$; compare with equation (31). The specific-heat discontinuity at the transition point is, therefore, given by

$$\frac{1}{Nk}\Delta C_0 = \frac{3}{8}\frac{q^2(q-2)}{(q-1)}\left\{\ln\left(\frac{q}{q-2}\right)\right\}^2. \tag{34}$$

One may check that, for $q \gg 1$, the foregoing results go over to the ones following from the zeroth approximation.

Finally, we examine the relationship between $\overline{L}$ and B at $T = T_c$. Using equations (8) and (14), with both α and $\alpha' \ll 1$ while $\gamma = \gamma_c$, we get

$$\overline{L} \approx \left\{3q^2\mu B/(q-1)(q-2)kT_c\right\}^{1/3} \quad (T = T_c, B \to 0); \tag{35}$$

compare with equation (12.5.24). For the behavior of χ_0, see Problem 12.16.

In passing, we note that, according to equation (12), the transition temperature for a lattice with $q = 2$ is zero, which essentially means that a one-dimensional Ising chain *does not* undergo a phase transition. This result is in complete agreement with the one following from an exact treatment of the one-dimensional lattice; see Section 13.2. In fact, for a lattice with $q = 2$, any results following from the Bethe approximation are completely identical with the corresponding exact results (see Problem 13.3); on the other hand, the Bragg–Williams approximation is least reliable when $q = 2$.

That T_c for $q = 2$ is zero (rather than $2J/k$) is in line with the fact that, for all q, the first approximation yields a transition temperature closer to the correct value of T_c than does the zeroth approximation. The same is true of the amplitudes that determine the quantitative behavior of the various physical quantities near $T = T_c$, though the exponents in the various power laws governing this behavior remain the same; compare, for instance, equation (16) with (12.5.14), equation (35) with (24) as well as the behavior of the specific heat. In fact, one finds that successive approximations of the mean field approach, while continuing to improve the theoretical value of T_c *and* the quantitative behavior of the various physical quantities (as given by the amplitudes), *do not* modify their qualitative behavior (as determined by the exponents). For an account of the higher-order approximations, see Domb (1960).

One important virtue of the Bethe approximation is that it brings out the role of the dimensionality of the lattice in bringing about a phase transition in the system. The fact that $T_c = 0$ for $q = 2$ and thereon it increases steadily with q leads one to infer that, while a linear Ising chain does not undergo phase transition at any finite T, higher dimensionality does promote the phenomenon. One may, in fact, argue that the absence of a phase transition in a one-dimensional chain is essentially due to the fact that, the interactions being severely short-ranged, "communication" between any two parts of the chain can be completely disrupted by a single defect in-between. The situation remains virtually unaltered even if the range of interactions is allowed to increase — so long as it remains finite. Only when interactions become truly long-ranged, with $J_{ij} \sim |i-j|^{-(1+\sigma)}(\sigma > 0)$, does a phase transition at a finite T become possible — but only if $\sigma < 1$; for $\sigma > 1$, we are back to the case of no phase transition, while the borderline case $\sigma = 1$ remains in doubt. For more details, see Griffiths (1972, pp. 89–94).

Peierls (1936) was the first to demonstrate that at sufficiently low temperatures the Ising model in two or three dimensions *must* exhibit a phase transition. He considered the lattice as made up of two kinds of domains, one consisting of "up" spins and the other of "down" spins, separated by a set of boundaries between the neighboring domains, and argued on energy considerations that in a two- or three-dimensional lattice the long-range order that exists at 0 K would persist at finite temperatures. Again, for details, see Griffiths (1972, pp. 59–66).

12.7 The critical exponents

A basic problem in the theory of phase transitions is to study the behavior of a given system in the neighborhood of its critical point. We know that this behavior is marked by the fact that the various physical quantities pertaining to the system possess singularities at the critical point. It is customary to express these singularities in terms of power laws characterized by a set of *critical exponents* that determine the qualitative nature of the critical behavior of the given system. To begin with, we identify an *order parameter m*, and the corresponding *ordering field h*, such that, in the limit $h \to 0$, m tends to a limiting value m_0, with the property that $m_0 = 0$ for $T \geq T_c$ and $\neq 0$ for $T < T_c$. For a magnetic system, the natural candidate for m is the parameter $\overline{L}(=\overline{\sigma})$ of Sections 12.5 and 12.6, while h is identified with the quantity $\mu B/kT_c$; for a gas–liquid system, one may adopt the density differential $(\rho_l - \rho_c)$ or $|\rho_g - \rho_c|$ for m and the pressure differential $(P - P_c)$ for h. The various critical exponents are then defined as follows.

The manner in which $m_0 \to 0$, as $T \to T_c$ from below, defines the exponent β:

$$m_0 \sim (T_c - T)^{\beta} \quad (h \to 0, T \lesssim T_c). \tag{1}$$

The manner in which the low-field susceptibility χ_0 diverges, as $T \to T_c$ from above (or from below), defines the exponent γ (or γ'):

$$\chi_0 \sim \left(\frac{\partial m}{\partial h}\right)_{T,h\to 0} \sim \begin{cases} (T - T_c)^{-\gamma} & (h \to 0, T \gtrsim T_c) \quad \text{(2a)} \\ (T_c - T)^{-\gamma'} & (h \to 0, T \lesssim T_c); \quad \text{(2b)} \end{cases}$$

in the gas–liquid transition, the role of χ_0 is played by the isothermal compressibility, $\kappa_T = \rho^{-1}(\partial\rho/\partial P)_T$, of the system. Next, we define an exponent δ by the relation

$$m|_{T=T_c} \sim h^{1/\delta} \quad (T = T_c, h \to 0); \tag{3}$$

in the case of a gas–liquid system, δ is a measure of the "degree of flatness" of the critical isotherm at the critical point, for then

$$|P - P_c|\Big|_{T=T_c} \sim |\rho - \rho_c|^{\delta} \quad (T = T_c, P \to P_c). \tag{4}$$

Finally, we define exponents α and α' on the basis of the specific heat, C_V, of the gas–liquid system:

$$C_V \sim \begin{cases} (T - T_c)^{-\alpha} & (T \gtrsim T_c) & (5a) \\ (T_c - T)^{-\alpha'} & (T \lesssim T_c). & (5b) \end{cases}$$

In connection with the foregoing relations, especially equations (5), we wish to emphasize that in certain cases the exponent in question is rather small in value; it is then more appropriate to write

$$f(t) \sim \frac{|t|^{-\lambda} - 1}{\lambda} \quad (|t| \ll 1). \tag{6}$$

Now, if $\lambda > 0$, the function $f(t)$ would have a power-law divergence at $t = 0$; in case $\lambda \to 0$, the function $f(t)$ would have a logarithmic divergence instead:

$$f(t) \sim \ln(1/|t|) \quad (|t| \ll 1). \tag{7}$$

In either case, the derivative $f'(t) \sim |t|^{-(1+\lambda)}$.

A survey of the results derived in Sections 12.2 through 12.6 shows that for a gas–liquid system obeying van der Waals equation of state or for a magnetic system treated in the mean field approximation (it does not matter what order of approximation one is talking about), the various critical exponents are the same:

$$\beta = \frac{1}{2}, \quad \gamma = \gamma' = 1, \quad \delta = 3, \quad \alpha = \alpha' = 0. \tag{8}$$

In Table 12.1 we have compiled experimental data on critical exponents pertaining to a variety of systems including the ones mentioned above; for completeness, we have included here data on another two exponents, ν and η, which will be defined in Section 12.12. We find that, while the observed values of an exponent, say β, differ very little as one goes from system to system within a given category (or even from category to category), these values are considerably different from the ones following from the mean field approximation. Clearly, we need a theory of phase transitions that is basically different from the mean field theory.

To begin with, some questions arise:

(i) Are these exponents completely independent of one another *or* are they mutually related? In the latter case, how many of them are truly independent?
(ii) On what characteristics of the given system do they depend? This includes the question why, for systems differing so much from one another, they differ so little.
(iii) How can they be evaluated from first principles?

The answer to question (i) is simple: yes, the various exponents do obey certain relations and hence are *not* completely independent. These relations appear in the form of inequalities, dictated by the principles of thermodynamics, which will be explored

Table 12.1 Experimental Data on Critical Exponents

Critical Exponents	Magnetic Systems[a]	Gas–liquid Systems[b]	Binary Fluid Mixtures[c]	Binary Alloys[d]	Ferroelectric Systems[e]	Superfluid He^{4}[f]	Mean Field Results
α, α'	0.0–0.2	0.1–0.2	0.05–0.15	---	---	−0.026	0
β	0.30–0.36	0.32–0.35	0.30–0.34	0.305 ± 0.005	0.33–0.34	---	1/2
γ	1.2–1.4	1.2–1.3	1.2–1.4	1.24 ± 0.015	1.0 ± 0.2	inaccessible	1
γ'	1.0–1.2	1.1–1.2	---	1.23 ± 0.025	1.23 ± 0.02	inaccessible	1
δ	4.2–4.8	4.6–5.0	4.0–5.0	---	---	inaccessible	3
ν	0.62–0.68	---	---	0.65 ± 0.02	0.5–0.8	0.675	1/2
η	0.03–0.15	---	---	0.03–0.06	---	---	0

[a] Stierstadt et al. (1990).
[b] Voronel (1976); Rowlinson and Swinton (1982).
[c] Rowlinson and Swinton (1982).
[d] Als-Nielsen (1976); data pertain to beta-brass only.
[e] Kadanoff et al. (1967); Lines and Glass (1977).
[f] Ahlers (1980).

in Section 12.8; in the modern theory of phase transitions, see Sections 12.10 through 12.12 and Chapter 14, the same relations turn up as equalities, and the number of these (restrictive) relations is such that, in most cases only *two* of the exponents are truly independent.

As regards question (ii), it turns out that our exponents depend on a *very small* number of characteristics, or parameters, of the problem, which explains why they differ so little from one system to another in a given category of systems (and also from one category to another, even though systems in those categories are so different from one another). The characteristics that seem to matter are (a) the dimensionality, d, of the space in which the system is embedded, (b) the number of components, n, of the order parameter of the problem, and (c) the range of microscopic interactions in the system.

Insofar as interactions are concerned, all that matters is whether they are *short-ranged* (which includes the special case of nearest-neighbor interactions) or *long-ranged*. In the former case, the values of the critical exponents resulting from nearest-neighbor interactions remain unaltered — regardless of whether further-neighbor interactions are included or not; in the latter case, assuming $J_{ij} \sim |i-j|^{-(d+\sigma)}$ with $\sigma > 0$, the critical exponents depend on σ. Unless a statement is made to the contrary, the microscopic interactions operating in the given system will be assumed to be short-ranged; the critical exponents then depend only on d and n — both of which, for instructional purposes, may be treated as *continuous* variables.

Insofar as d is concerned, we recall the Bethe approximation that highlighted the special role played by the dimensionality of the lattice through, and only through, the coordination number q. We also recall that, while the theoretical value of T_c and the various amplitudes of the problem were influenced by q, the critical exponents were not. In more accurate theories we find that the critical exponents depend more directly on d and only indirectly on q; however, for a given d, they do *not* depend on the structural details of the lattice (including the number q).

Insofar as n is concerned, the major difference lies between the Ising model ($n = 1$) with *discrete* symmetry ($\sigma_i = +1$ or -1) and other models ($n \geq 2$) with *continuous* symmetry ($-1 \leq \sigma_{i\alpha} \leq +1$ for $\alpha = 1, \ldots, n$, with $|\sigma_i| = 1$). In the former case, T_c is zero for $d \leq 1$ and nonzero for $d > 1$; in the latter, T_c is zero for $d \leq 2$ and nonzero for $d > 2$.[13] In either case, the critical exponents depend on both d and n, except that for $d > 4$ they become independent of d and n, and assume values identical to the ones given by the mean field theory; the physical reason behind this overwhelming generality is examined in Section 12.13. In passing, we note that, for given d and n, the critical exponents do not depend on whether the spins constituting the system are treated classically or quantum-mechanically.

As regards question (iii), the obvious procedure for evaluating the critical exponents is to carry out exact (or almost exact) analysis of the various models — a task to which the whole of Chapter 13 is devoted. An alternative approach is provided by the renormalization group theory, which is discussed in Chapter 14. A modest attempt to evaluate the critical exponents is made in Section 12.9, which yields results that are inconsistent with the experiment but teaches us quite a few lessons about the shortcomings of the so-called classical approaches.

12.8 Thermodynamic inequalities

The first rigorous relation linking critical exponents was derived by Rushbrooke (1963) who, on thermodynamic grounds, showed that for any physical system undergoing a phase transition

$$(\alpha' + 2\beta + \gamma') \geq 2. \tag{1}$$

The proof of inequality (1) is straightforward if one adopts a magnetic system as an example. We start with the thermodynamic formula for the difference between the specific heat at constant field C_H and the specific heat at constant magnetization C_M (see Problem 3.40)

$$C_H - C_M = -T\left(\frac{\partial H}{\partial T}\right)_M \left(\frac{\partial M}{\partial T}\right)_H = T\chi^{-1}\left\{\left(\frac{\partial M}{\partial T}\right)_H\right\}^2. \tag{2}$$

Since $C_M \geq 0$, it follows that

$$C_H \geq T\chi^{-1}\left\{\left(\frac{\partial M}{\partial T}\right)_H\right\}^2. \tag{3}$$

Now, letting $H \to 0$ and $T \to T_c$ from below, we get

$$D_1(T_c - T)^{-\alpha'} \geq D_2 T_c (T_c - T)^{\gamma' + 2(\beta - 1)}, \tag{4}$$

[13]The special case $n = 2$ with $d = 2$ is qualitatively different from others; for details, see Section 13.7.

where D_1 and D_2 are positive constants; here, use has been made of power laws (12.7.1, 2b, and 5b).[14] Inequality (4) may as well be written as

$$(T_c - T)^{2-(\alpha'+2\beta+\gamma')} \geq D_2 T_c / D_1. \tag{5}$$

Since $(T_c - T)$ can be made as small as we like, (5) will not hold if $(\alpha' + 2\beta + \gamma') < 2$. The *Rushbrooke inequality* (1) is thus established.

To establish further inequalities, one utilizes the convexity properties of the Helmholtz free energy $A(T,M)$. Since $dA = -SdT + HdM$,

$$\left(\frac{\partial A}{\partial T}\right)_M = -S, \quad \left(\frac{\partial^2 A}{\partial T^2}\right)_M = -\left(\frac{\partial S}{\partial T}\right)_M = -\frac{C_M}{T} \leq 0 \tag{6a, b}$$

and

$$\left(\frac{\partial A}{\partial M}\right)_T = H, \quad \left(\frac{\partial^2 A}{\partial M^2}\right)_T = \left(\frac{\partial H}{\partial M}\right)_T = \frac{1}{\chi} \geq 0. \tag{7a, b}$$

It follows that $A(T,M)$ is concave in T and convex in M. We now proceed to establish the *Griffiths inequality* (1965a, b)

$$\alpha' + \beta(\delta + 1) \geq 2. \tag{8}$$

Consider a magnetic system in zero field and at a temperature $T_1 < T_c$. Then, by (7a), $A(T,M)$ is a function of T only, so we can write

$$A(T_1, M) = A(T_1, 0) \quad (-M_1 \leq M \leq M_1), \tag{9}$$

where M_1 is the spontaneous magnetization at temperature T_1; see Figure 12.10. Applying (6a) to (9), we get

$$S(T_1, M) = S(T_1, 0) \quad (-M_1 \leq M \leq M_1). \tag{10}$$

We now define two new functions

$$A^*(T,M) = \{A(T,M) - A_c\} + (T - T_c)S_c \tag{11}$$

and

$$S^*(T,M) = S(T,M) - S_c, \tag{12}$$

[14]Recalling the correspondence between a gas–liquid system and a magnet, one might wonder why we have employed C_H, rather than C_M, in place of C_V. The reason is that, since we are letting $H \to 0$ and $T \to T_c-, M \to 0$ as well. So, as argued by Fisher (1967), in the limit considered here, C_H and C_M display the *same* singular behavior. In fact, it can be shown that if the ratio $C_M/C_H \to 1$ as $T \to T_c$, then $(\alpha' + 2\beta + \gamma')$ must be greater than 2; on the other hand, if this ratio tends to a value less than 1, then $(\alpha' + 2\beta + \gamma') = 2$. For details, see Stanley (1971), Section 4.1.

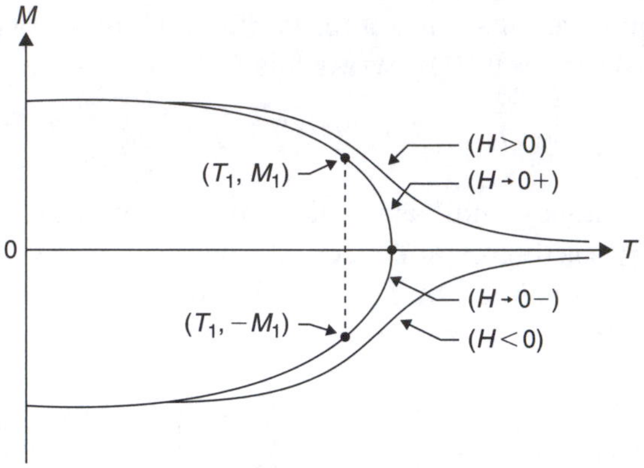

FIGURE 12.10 Magnetization, $M(T,H)$, of a magnetic system for $H > 0$, $H < 0$, and $H \to 0$. Here, M_1 denotes the spontaneous magnetization of the system at a temperature $T_1 < T_c$.

where $A_c = A(T_c, 0)$ and $S_c = S(T_c, 0)$. It follows that

$$\left(\frac{\partial A^*}{\partial T}\right)_M = -S^*, \quad \left(\frac{\partial^2 A^*}{\partial T^2}\right)_M = -\left(\frac{\partial S^*}{\partial T}\right)_M = -\frac{C_M}{T} \le 0. \qquad \text{(13a, b)}$$

Thus, A^* is also concave in T. Geometrically, this means that, for any choice of T_1, the curve $A^*(T)$, with M fixed at M_1, lies below the tangent line at $T = T_1$, that is,

$$A^*(T, M_1) \le A^*(T_1, M_1) + \left(\frac{\partial A^*}{\partial T}\right)_{M_1, T=T_1}(T - T_1); \qquad (14)$$

see Figure 12.11. Letting $T = T_c$ in (14), we get

$$A^*(T_c, M_1) \le A^*(T_1, M_1) - S^*(T_1, M_1)(T_c - T_1) \qquad (15)$$

which, in view of equations (9) through (12), may be written as

$$A^*(T_c, M_1) \le A^*(T_1, 0) - S^*(T_1, 0)(T_c - T_1). \qquad (16)$$

Utilizing, once again, the concavity of the function $A^*(T)$ but this time at $T = T_c$ (with M fixed at zero and the slope $(\partial A^*/\partial T)$ vanishing), we get, see (14),

$$A^*(T, 0) \le A^*(T_c, 0). \qquad (17)$$

Now, letting $T = T_1$ in (17) and noting that $A^*(T_c, 0) = 0$ by definition, we get

$$A^*(T_1, 0) \le 0. \qquad (18)$$

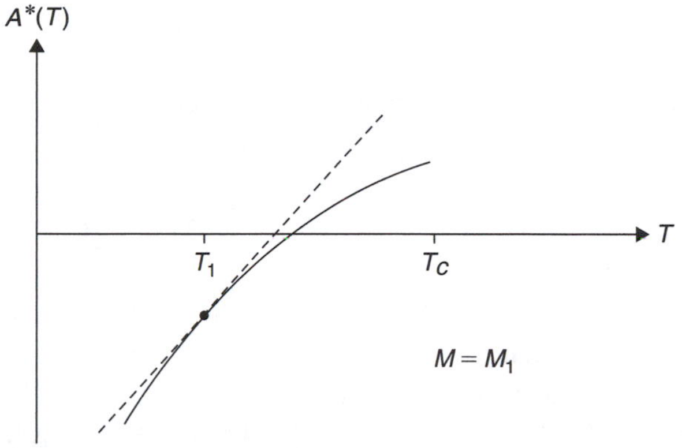

FIGURE 12.11 The function $A^*(T,M)$ of a magnetic system, with magnetization M fixed at M_1. The slope of this curve is $S(T_c,0) - S(T,M_1)$, which is positive for all $T \le T_c$.

Combining (16) and (18), we finally get

$$A^*(T_c,M_1) \le -(T_c - T_1)S^*(T_1,0), \tag{19}$$

valid for all $T_1 < T_c$.

The next step is straightforward. We let $T_1 \to T_c-$, so that $M_1 \to 0$ and along with it

$$A^*(T_c,M_1) = \left[\int_0^{M_1} HdM\right]_{T=T_c} \approx DM_1^{\delta+1} \approx D'(T_c - T_1)^{\beta(\delta+1)}, \tag{20}$$

while

$$S^*(T_1,0) = \int_{T_c}^{T_1} \frac{C(T,0)}{T} dT \approx -\frac{D''}{T_c}(T_c - T_1)^{1-\alpha'}, \tag{21}$$

where D, D', and D'' are positive constants; here, use has been made of power laws (12.7.1, 3, and 5b). Substituting (20) and (21) into (19), we get

$$(T_c - T_1)^{2-\alpha'-\beta(\delta+1)} \ge D'T_c/D''. \tag{22}$$

Again, since $(T_c - T_1)$ can be made as small as we like, (22) will not hold if $\alpha' + \beta(\delta+1) < 2$. The Griffiths inequality (8) is thus established. It will be noted that unlike the Rushbrooke inequality, which related critical exponents pertaining only to $T < T_c$, the present inequality relates two such exponents, α' and β, with one, namely δ, that pertains to the critical isotherm ($T = T_c$).

While inequalities (1) and (8) are thermodynamically exact, Griffiths has derived several others that require certain plausible assumptions on the system in question. We quote two of them here, without proof:

$$\gamma' \geq \beta(\delta - 1) \tag{23}$$

$$\gamma \geq (2 - \alpha)(\delta - 1)/(\delta + 1). \tag{24}$$

For a complete list of such inequalities, see Griffiths (1972), p. 102, where references to original papers are also given.

Before proceeding further, the reader may like to verify that the experimental data on critical exponents, as given earlier in Table 12.1, do indeed conform to the inequalities proved or quoted in this section. It is important in this connection to note that the mean field exponents ($\alpha = \alpha' = 0,\ \beta = 1/2,\ \gamma = \gamma' = 1$, and $\delta = 3$) satisfy all these relations as *equalities*.

12.9 Landau's phenomenological theory

As early as 1937 Landau attempted a unified description of all *second-order* phase transitions — second-order in the sense that the second derivatives of the free energy, namely the specific heat and the magnetic susceptibility (or isothermal compressibility, in the case of fluids), show a divergence while the first derivatives, namely the entropy and the magnetization (or density, in the case of fluids), are continuous at the critical point. He emphasized the importance of an *order parameter* m_0 (which would be zero on the high-temperature side of the transition and nonzero on the low-temperature side) and suggested that the basic features of the critical behavior of a given system may be determined by expanding its free energy in powers of m_0 (for we know that, in the close vicinity of the critical point, $m_0 \ll 1$). He also argued that in the absence of the *ordering field* ($h = 0$) the up–down symmetry of the system would require that the proposed expansion contain only *even* powers of m_0. Thus, the zero-field free energy $\psi_0 (= A_0/NkT)$ of the system may be written as

$$\psi_0(t, m_0) = q(t) + r(t)m_0^2 + s(t)m_0^4 + \cdots \quad \left(t = \frac{T - T_c}{T_c}, |t| \ll 1\right); \tag{1}$$

at the same time, the coefficients $q(t),\ r(t),\ s(t) \ldots$ may be written as

$$q(t) = \sum_{k \geq 0} q_k t^k, \quad r(t) = \sum_{k \geq 0} r_k t^k, \quad s(t) = \sum_{k \geq 0} s_k t^k, \ldots. \tag{2}$$

The equilibrium value of the order parameter is then determined by *minimizing* ψ_0 with respect to m_0; retaining terms only up to the order displayed in (1), which for thermodynamic stability requires that $s(t) > 0$, we obtain

$$r(t)m_0 + 2s(t)m_0^3 = 0. \tag{3}$$

The equilibrium value of m_0 is thus either 0 or $\pm\sqrt{[-r(t)/2s(t)]}$. The first solution is of lesser interest, though this is the only one we will have for $t > 0$; it is the other solutions that lead to the possibility of spontaneous magnetization in the system. To obtain physically sensible results, see equations (9) through (11), we must have in equation (2): $r_0 = 0$, $r_1 > 0$, and $s_0 > 0$, with the result

$$|m_0| \approx [(r_1/2s_0)|t|]^{1/2} \quad (t \lesssim 0), \tag{4}$$

giving $\beta = 1/2$.

The asymptotic expression for the free energy, namely

$$\psi_0(t, m_0) \approx q_0 + r_1 t m_0^2 + s_0 m_0^4 \quad (r_1, s_0 > 0), \tag{5}$$

is plotted in Figure 12.12. We see that, for $t \geq 0$, there is only *one* minimum, which is located at $m_0 = 0$; for $t = 0$, the minimum is rather flat. For $t < 0$, on the other hand, we have *two* minima, located at $m_0 = \pm m_s$, as given by expression (4), with a maximum at $m_0 = 0$. Now, since ψ_0 has to be convex in m_0, so that the susceptibility of the system be nonnegative, see equation (12.8.7b), we must replace the nonconvex portion of the curve, which lies between the points $m_0 = -m_s$ and $m_0 = +m_s$, by a straight line (along which the susceptibility would be infinite). This replacement is reminiscent of the *Maxwell construction* employed in Sections 12.1 and 12.2.

We now subject the system to an *ordering field h*, assumed positive. If the field is weak, the only change in the expression for the free energy would be the addition of a term $-hm$. Disregarding the appearance of any higher powers of (hm) as well as any modifications of

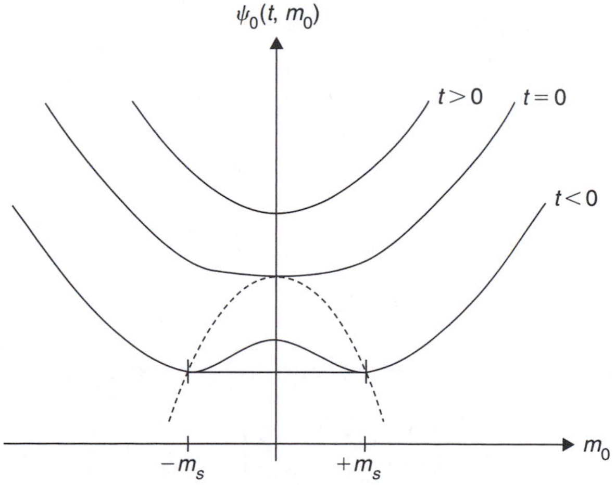

FIGURE 12.12 The free energy $\psi_0(t, m_0)$ of the Landau theory, shown as a function of m_0, for three different values of t. The dashed curve depicts spontaneous magnetization $m_s(t)$, while the horizontal line for $t < 0$ provides the Maxwell construction.

the coefficients already present, we may now write

$$\psi_h(t,m) = -hm + q(t) + r(t)m^2 + s(t)m^4. \tag{6}$$

The equilibrium value of m is now given by[15]

$$-h + 2r(t)m + 4s(t)m^3 = 0. \tag{7}$$

The low-field susceptibility of the system, in units of $N\mu^2/kT$, thus turns out to be

$$\chi = \left(\frac{\partial h}{\partial m}\right)_t^{-1} = \frac{1}{2r(t) + 12s(t)m^2}, \tag{8}$$

valid in the limit $h \to 0$. Now, for $t > 0$, $m \to 0$ and we get

$$\chi \approx 1/2r_1 t \quad (t \gtrsim 0), \tag{9}$$

giving $\gamma = 1$. On the other hand, for $t < 0$, $m \to \surd[(r_1/2s_0)|t|]$, see (4); we then get

$$\chi \approx 1/4r_1|t| \quad (t \lesssim 0), \tag{10}$$

giving $\gamma' = 1$. Finally, if we set $t = 0$ in (7), we obtain the following relation between h and m:

$$h \approx 4s_0 m^3 \quad (h \to 0), \tag{11}$$

giving $\delta = 3$.

We shall now look at the specific heats C_h and C_m. If $t > 0$, then $h \to 0$ implies $m \to 0$, so in this limit there is no difference between C_h and C_m. Equation (1) then gives, in units of Nk,

$$C_h = C_m = -\left(\frac{\partial^2 \psi_0}{\partial t^2}\right)_{m \to 0} = -(2q_2 + 6q_3 t + \cdots) \quad (t \gtrsim 0). \tag{12}$$

For $t < 0$, we have

$$\begin{aligned} C_m &= -\left[(2q_2 + 6q_3 t + \cdots) + (2r_2 + \cdots)m_s^2 + \cdots\right] \\ &= -\left[2q_2 + \{6q_3 - (r_1 r_2/s_0)\}t + \ldots\right] \quad (t \lesssim 0). \end{aligned} \tag{13}$$

Next, using equation (12.8.2) along with (4) and (10), we have

$$C_h - C_m = \left(\frac{\partial h}{\partial m}\right)_t \left\{\left(\frac{\partial m}{\partial t}\right)_h\right\}^2 \approx \frac{r_1^2}{2s_0} \quad (t \lesssim 0). \tag{14}$$

[15]It may be mentioned here that the passage from equation (1) to (6) is equivalent to effecting a Legendre transformation from the Helmholtz free energy A to the Gibbs free energy $G\ (= A - HM)$, and equation (7) is analogous to the relation $(\partial A/\partial M)_T = H$.

It follows that, while C_m possesses a cusp-like singularity at $t = 0$, C_h undergoes a jump discontinuity of magnitude

$$(C_h)_{t\to 0-} - (C_h)_{t\to 0+} \approx r_1^2/2s_0. \tag{15}$$

It follows that $\alpha = \alpha' = 0$.

The most striking feature of the Landau theory is that it gives exactly the same critical exponents as the mean field theory of Sections 12.5 and 12.6 (or the van der Waals theory of Section 12.2). Actually it goes much further, for it starts with an expression for the free energy of the system containing parameters $q_k, r_k, s_k, \ldots$, which represent the structure of the given system and the interactions operating in it, and goes on to show that, while the amplitudes of the various physical quantities near the critical point do depend on these parameters, the critical exponents do not! This *universality* (of critical exponents) suggests that we are dealing here with a class of systems which, despite their structural differences, display a critical behavior that is qualitatively the same for all members of the class. This leads to the concept of a *universality class* which, if Landau were right, would be a rather large one. The fact of the matter is that the concept of universality is very much overstated in Landau's theory; in reality, there are many different universality classes — each defined by the parameters d and n of Section 12.7 and by the range of the microscopic interactions — such that the critical exponents within a class are the same while they vary from one class to another. The way Landau's theory is set up, the parameter n is essentially equal to 1 (because the order parameter m_0 is treated as a scalar), the parameter d plays no role at all (though later on we shall see that the mean field exponents are, in fact, valid for all n if $d > 4$), while the microscopic interactions are implicitly long-ranged.[16]

An objection commonly raised against the Landau theory is that, knowing fully well that the thermodynamic functions of the given system are going to be *singular* at $t = 0$, a *Taylor-type expansion* of the free energy around $m = 0$ is patently a wrong start. While the objection is valid, it is worth noting how a regular function, (1) or (6), leads to an equation of state, (3) or (7), which yields different results for $t \to 0-$ from the ones for $t \to 0+$, the same being true of whether $h \to 0+$ or $0-$. The trick lies in the fact that we are not using equation (1) or (6) as such for *all* t; for $t < 0$, we use instead a modified form, as "corrected" by the Maxwell construction (see Figure 12.12). The spirit of the singularity is thereby captured, though the nature of the singularity, being closely tied with the nature of the original expansion, could not be any different from the mean-field type. The question now arises: how can the Landau theory be improved so that it may provide a more satisfactory picture of the critical phenomena? Pending exact analyses, one wonders if some generalization of the Landau approach, admitting more than one universality class, would provide a better picture than the one presented so far. It turns out that the scaling approach, initiated by Widom (1965), by Domb and Hunter (1965), and by Patashinskii and Pokrovskii (1966), provided the next step in the right direction.

[16]In certain systems such as superconductors, the effective interactions (which, for instance, lead to the formation of Cooper pairs of electrons) are, in fact, long-ranged. The critical exponents pertaining to such systems turn out to be the same as one gets from the mean field theory. For details, see Tilley and Tilley (1990).

12.10 Scaling hypothesis for thermodynamic functions

The scaling approach, which took the subject of phase transitions far beyond the mean field theory, emerged independently from three different sources — from Widom (1965), who searched for a generalization of the van der Waals equation of state that could accommodate nonclassical exponents; from Domb and Hunter (1965), who analyzed the behavior of the series expansions of higher derivatives of the free energy with respect to the magnetic field at the critical point of a magnetic system; and from Patashinskii and Pokrovskii (1966), who studied the behavior of multipoint correlation functions for the spins constituting a system. All three were led to the same form of a thermodynamic equation of state. Subsequently, Kadanoff (1966a) suggested a *scaling hypothesis* from which not only could this equation of state be derived but one could also obtain a number of relations among the critical exponents, which turned out to be equalities consistent with the findings of Section 12.8. This approach also made it clear why one needed only *two* independent numbers to describe the nature of the singularity in question; all other relevant numbers followed as consequences.

To set the stage for this development, we go back to the equation of state following from the Landau theory, namely (12.9.7), and write it in the asymptotic form

$$h(t,m) \approx 2r_1 tm + 4s_0 m^3. \tag{1}$$

In view of the relationship (12.9.4), we rewrite (1) in the form

$$h(t,m) \approx \frac{r_1^{3/2}}{s_0^{1/2}}|t|^{3/2}\left[2\,\mathrm{sgn}(t)\left(\frac{s_0^{1/2}}{r_1^{1/2}}\frac{m}{|t|^{1/2}}\right)+4\left(\frac{s_0^{1/2}}{r_1^{1/2}}\frac{m}{|t|^{1/2}}\right)^3\right]. \tag{2}$$

It follows that

$$m(t,h) \approx \frac{r_1^{1/2}}{s_0^{1/2}}|t|^{1/2} \times \text{ a function of } \left(\frac{s_0^{1/2}}{r_1^{3/2}}\frac{h}{|t|^{3/2}}\right) \tag{3}$$

and, within the context of the Landau theory, the function appearing here is *universal* for all systems conforming to this theory. In the same spirit, the relevant part of the free energy $\psi_h(t,m)$ — the part that determines the nature of the singularity — may be written in the form

$$\psi_h^{(s)}(t,m) \approx -hm + r_1 tm^2 + s_0 m^4 \tag{4}$$

$$= \frac{r_1^2}{s_0}t^2\left[-\left(\frac{s_0^{1/2}}{r_1^{3/2}}\frac{h}{|t|^{3/2}}\right)\left(\frac{s_0^{1/2}}{r_1^{1/2}}\frac{m}{|t|^{1/2}}\right)+\mathrm{sgn}\,(t)\left(\frac{s_0^{1/2}}{r_1^{1/2}}\frac{m}{|t|^{1/2}}\right)^2 + \left(\frac{s_0^{1/2}}{r_1^{1/2}}\frac{m}{|t|^{1/2}}\right)^4\right]. \tag{5}$$

Substituting (3) into (5), one gets

$$\psi^{(s)}(t,h) \approx \frac{r_1^2}{s_0} t^2 \times \text{a function of} \left(\frac{s_0^{1/2}}{r_1^{3/2}} \frac{h}{|t|^{3/2}} \right), \tag{6}$$

where, again, the function appearing here is *universal*. As a check, we see that differentiating (6) with respect to h we readily obtain (3).

The most notable feature of the equation of state, as expressed in (3), is that, instead of being the usual relationship among *three* variables m, h, and t, it is now a relationship among only *two* variables, namely $m/|t|^{1/2}$ and $h/|t|^{3/2}$. Thus, by scaling m with $|t|^{1/2}$ and h with $|t|^{3/2}$, we have effectively reduced the total number of variables by one. Similarly, we have replaced equation (4) by (6), which expresses the singular part of the free energy ψ scaled with t^2 as a function of the single variable h scaled with $|t|^{3/2}$. This reduction in the total number of effective variables may be regarded as the first important achievement of the scaling approach.

The next step consists of generalizing (6), to write

$$\psi^{(s)}(t,h) \approx F|t|^{2-\alpha} f(Gh/|t|^{\Delta}), \tag{7}$$

where α and Δ are *universal* numbers common to all systems in the given universality class, $f(x)$ is a *universal* function which is expected to have two different branches, f_+ for $t > 0$ and f_- for $t < 0$, while F and G (like r_1 and s_0) are *nonuniversal* parameters characteristic of the particular system under consideration. We expect α and Δ to determine all the critical exponents of the problem, while the amplitudes appearing in the various power laws will be determined by F, G, and the limiting values of the function $f(x)$ and its derivatives (as x tends to zero). Equation (7) constitutes the so-called *scaling hypothesis*, whose status will become much more respectable when it acquires legitimacy from the renormalization group theory; see Sections 14.1 and 14.3.

First of all it should be noted that the exponent of $|t|$, outside the function $f(x)$ in equation (7), has been chosen to be $(2-\alpha)$, rather than 2 of the corresponding mean field expression (6), so as to ensure that the specific heat singularity is correctly reproduced. Secondly, the fact that one must not encounter any singularities as one crosses the critical isotherm $(t = 0)$ at *nonzero* values of h or m requires that the exponents on the high-temperature side of the critical point be the same as on the low-temperature side, that is,

$$\alpha' = \alpha \quad \text{and} \quad \gamma' = \gamma. \tag{8}$$

From equation (7) it readily follows that

$$m(t,h) = -\left(\frac{\partial \psi^{(s)}}{\partial h} \right)_t \approx -FG|t|^{2-\alpha-\Delta} f'(Gh/|t|^{\Delta}) \tag{9}$$

and

$$\chi(t,h) = -\left(\frac{\partial^2 \psi^{(s)}}{\partial h^2}\right)_t \approx -FG^2|t|^{2-\alpha-2\Delta} f''(Gh/|t|^\Delta). \tag{10}$$

Letting $h \to 0$, we obtain for the spontaneous magnetization

$$m(t,0) \approx B|t|^\beta \quad (t \lesssim 0), \tag{11}$$

where

$$B = -FGf'_-(0), \quad \beta = 2-\alpha-\Delta, \tag{12a, b}$$

and for the low-field susceptibility

$$\chi(t,0) \approx |t|^{-\gamma} \begin{cases} C_+ & (t \gtrsim 0) & (13a) \\ C_- & (t \lesssim 0), & (13b) \end{cases}$$

where

$$C_\pm = -FG^2 f''_\pm(0), \quad \gamma = \alpha + 2\Delta - 2. \tag{14a, b}$$

Combining (12b) and (14b), we get

$$\Delta = \beta + \gamma = 2 - \alpha - \beta, \tag{15}$$

so that

$$\alpha + 2\beta + \gamma = 2. \tag{16}$$

To recover δ, we write the function $f'(x)$ of equation (9) as $x^{\beta/\Delta} g(x)$, so that

$$m(t,h) \approx -FG^{(1+\beta/\Delta)} h^{\beta/\Delta} g(Gh/|t|^\Delta). \tag{17}$$

Inverting (17), we can write

$$|t| \approx G^{1/\Delta} h^{1/\Delta} \times \text{a function of } (FG^{(1+\beta/\Delta)} h^{\beta/\Delta}/m). \tag{18}$$

It follows that, along the critical isotherm ($t = 0$), the argument of the function appearing in (18) would have a *universal* value (which makes the function vanish), with the result that

$$m \sim FG^{(1+\beta/\Delta)} h^{\beta/\Delta} \quad (t = 0). \tag{19}$$

Comparing (19) with (12.7.3), we infer that

$$\delta = \Delta/\beta. \tag{20}$$

Combining (20) with the previous relations, namely (12b) and (15), we get

$$\alpha + \beta(\delta + 1) = 2 \tag{21}$$

and

$$\gamma = \beta(\delta - 1). \tag{22}$$

Finally, combining (21) and (22), we have

$$\gamma = (2 - \alpha)(\delta - 1)/(\delta + 1). \tag{23}$$

For completeness, we write down for the specific heat C_h at $h = 0$

$$C_h^{(s)}(t,0) = -\left.\frac{\partial^2 \psi^{(s)}}{\partial t^2}\right|_{h \to 0} \approx -(2-\alpha)(1-\alpha)F|t|^{-\alpha} \begin{cases} f_+(0) & (t \gtrsim 0) \quad (24a) \\ f_-(0) & (t \lesssim 0). \quad (24b) \end{cases}$$

We thus see that the scaling hypothesis (7) leads to a number of relations among the critical exponents of the system, emphasizing the fact that only *two* of them are truly independent. Comparing these relations with the corresponding ones appearing in Section 12.8 — namely, (16), (21), (22), and (23) with (12.8.1), (12.8.8), (12.8.23), and (12.8.24) — we feel satisfied that they are mutually consistent, though the present ones are far more restrictive than the ones there. Besides exponent relations, we also obtain here relations among the various amplitudes of the problem; though individually these amplitudes are nonuniversal, certain combinations thereof turn out to be universal. For instance, the combination $(FC_\pm/B^2)$, which consists of coefficients appearing in equations (7), (11), and (13), is universal; see equations (12a) and (14a). The same is true of the ratio C_+/C_-. For further information on this question, see the original papers by Watson (1969) and a review by Privman, Hohenberg, and Aharony (1991).

We now pose the question: why do "universality classes" exist in the first place? In other words, what is the reason that a large variety of systems differing widely in their structures should belong to a single universality class and hence have common critical exponents and common scaling functions? The answer lies in the role played by the correlations among the microscopic constituents of the system which, as $T \to T_c$, become large enough to prevail over *macroscopic* distances in the system and in turn make structural details at the *local* level irrelevant. We now turn our attention to this important aspect of the problem.

12.11 The role of correlations and fluctuations

Much can be learned about criticality by scattering radiation — light, x-rays, neutrons, and so on — off the system of interest; see Section 10.7.A. In a standard scattering experiment, a well-collimated beam of light, or other radiation, with known wavelength λ is directed at the sample and one measures the intensity, $I(\theta)$, of the light scattered at an angle θ from the "forward" direction of the beam. The radiation undergoes a shift in the wavevector, **k**, which is related to the parameters θ and λ by

$$|\boldsymbol{k}| = \frac{4\pi}{\lambda} \sin \frac{1}{2}\theta. \tag{1}$$

Now, the scattered intensity $I(\theta)$ is determined by the *fluctuations* in the medium. If the medium were perfectly uniform (i.e., spatially homogenous), there would be no scattering at all! If one has in mind light scattering from a fluid, then the relevant fluctuations correspond to regions of different refractive index and, hence, of different particle density $n(\boldsymbol{r})$. For neutron scattering from a magnet, fluctuations in the spin or magnetization density are the relevant quantities, and so on. We need to study here the *normalized* scattering intensity $I(\theta; T, H)/I^{\text{ideal}}(\theta)$, where $I(\theta; T, H)$ is the actual scattering intensity observed at angle θ, which will normally depend on such factors as temperature, magnetic field, and so on, while $I^{\text{ideal}}(\theta)$ is the scattering that would take place if the individual particles (or spins) that cause the scattering could somehow be taken far apart so that they no longer interact and hence are quite uncorrelated with one another. Now, this normalized scattering intensity turns out to be essentially proportional to the quantity

$$\tilde{g}(\boldsymbol{k}) = \int g(\boldsymbol{r}) e^{i\boldsymbol{k}\cdot\boldsymbol{r}} d\boldsymbol{r}, \tag{2}$$

which represents the Fourier transform of the appropriate real-space correlation function $g(\boldsymbol{r})$, which will be defined shortly.

As the critical point of the system (say, a fluid) is approached, one observes an enormously enhanced level of scattering, especially at low angles which corresponds, via equations (1) and (2), to long wavelength density fluctuations in the fluid. In the critical region, the scattering is so large that it can be visible to the unaided eye, particularly through the phenomenon of *critical opalescence*. This behavior is, by no means, limited to fluids. Thus if, for example, one scatters neutrons from iron in the vicinity of the Curie point, one likewise sees a dramatic growth in the low-angle neutron scattering intensity, as sketched in Figure 12.13. One sees that for small-angle scattering there is a pronounced peak in $I(\theta; T)$ as a function of temperature, and this peak approaches closer and closer to T_c as the angle is decreased. Of course, one could never actually observe zero-angle scattering directly, since that would mean picking up the oncoming beam itself, but one *can* extrapolate to zero angle. When this is done, one finds that the zero-angle scattering $I(0; T)$ actually *diverges* at T_c. This is the most dramatic manifestation of the phenomenon of critical opalescence and is quite general, in that it is observed whenever appropriate scattering experiments can be performed. Empirically, one may write for small-angle scattering

$$I_{\max}(\theta) \sim k^{-\lambda_1}, \quad \{T_{\max}(\theta) - T_c\} \sim k^{\lambda_2}, \tag{3}$$

so that

$$I_{\max}(\theta)\{T_{\max}(\theta) - T_c\}^{\lambda_1/\lambda_2} = \text{const.} \tag{4}$$

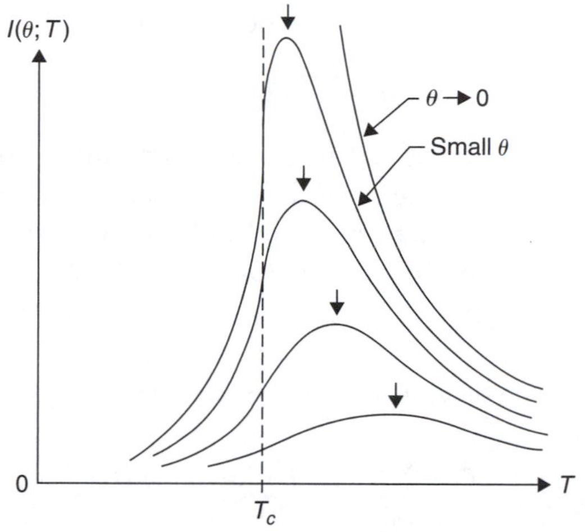

FIGURE 12.13 Schematic plot of the elastic scattering intensity of neutrons scattered at an angle θ from a magnetic system, such as iron, in the vicinity of the critical point T_c. The small arrows mark the smoothly rounded maxima (at fixed θ) which occur at a temperature $T_{\text{max}}(\theta)$ that approaches T_c as $\theta \to 0$.

Here, λ_1 and λ_2 are positive exponents (which, as will be seen later, are determined by the universality class to which the system belongs), while k, for a given θ, is determined by equation (1); note that, for small θ, k is essentially proportional to θ.

The first real insight into the problem of critical scattering in fluids was provided by Ornstein and Zernike (1914) and Zernike (1916) who emphasized the difference between the direct influence of the microscopic interactions among the atoms of the fluid, which are necessarily *short-ranged*, and the indirect (but more crucial) influence of the density–density correlations that become *long-ranged* as the critical temperature is approached; it is the latter that are truly responsible for the propagation of long-range order in the system and for practically everything else that goes with it. Unfortunately, the original work of Ornstein and Zernike makes difficult reading; moreover, it is based on the classical theory of van der Waals. Nevertheless, the subject has been neatly clarified in the review articles by Fisher (1964, 1983) and Domb (1985), to which the reader may turn for further details. Here, we shall stick to the language of the magnetic systems and work out the most essential parts of the theory in somewhat general terms.

We define the spin–spin *correlation function* $g(i,j)$, for the pair of spins at sites i and j, by the standard definition

$$g(i,j) = \overline{\sigma_i\sigma_j} - \overline{\sigma}_i\overline{\sigma}_j. \tag{5}$$

For $i = j$, expression (5) denotes the "mean-square fluctuation in the value of the variable σ at site i"; on the other hand, as the separation between the sites i and j increases indefinitely, the spins σ_i and σ_j get uncorrelated, so that $\overline{\sigma_i\sigma_j} \to \overline{\sigma}_i\overline{\sigma}_j$ and the function $g(i,j) \to 0$.

In view of the fact that expression (5) can also be written as

$$g(i,j) = \overline{(\sigma_i - \overline{\sigma}_i)(\sigma_j - \overline{\sigma}_j)}, \tag{6}$$

the function $g(i,j)$ may also be looked upon as a measure of the "correlation among the fluctuations in the order parameter of the system at sites i and j." This makes sense because σ_i may, quite appropriately, be regarded as the locally fluctuating order parameter linked to site i, just as $\overline{\sigma}$ is the order parameter for the whole system. We shall now establish connections between the function $g(i,j)$ and some important thermodynamic properties of the system.

We start with the partition function of the system, see equation (12.3.9),

$$Q_N(H,T) = \sum_{\{\sigma_i\}} \exp\left[\beta J \sum_{\text{n.n.}} \sigma_i\sigma_j + \beta\mu H \sum_i \sigma_i\right], \tag{7}$$

where the various symbols have their usual meanings. It follows that

$$\frac{\partial}{\partial H}(\ln Q_N) = \beta\mu \overline{\left(\sum_i \sigma_i\right)} = \beta\overline{M}, \tag{8}$$

where $M(= \mu\sum_i \sigma_i)$ denotes the net magnetization of the system. Next, since

$$\begin{aligned}\frac{\partial^2}{\partial H^2}(\ln Q_N) &= \frac{\partial}{\partial H}\left(\frac{1}{Q_N}\frac{\partial Q_N}{\partial H}\right) = \frac{1}{Q_N}\frac{\partial^2 Q_N}{\partial H^2} - \frac{1}{Q_N^2}\left(\frac{\partial Q_N}{\partial H}\right)^2 \\ &= \beta^2(\overline{M^2} - \overline{M}^2),\end{aligned} \tag{9}$$

we obtain for the magnetic susceptibility of the system

$$\chi \equiv \frac{\partial \overline{M}}{\partial H} = \beta(\overline{M^2} - \overline{M}^2) \tag{10a}$$

$$= \beta\mu^2\left\{\overline{\left(\sum_i \sigma_i\right)^2} - \overline{\left(\sum_i \sigma_i\right)}^2\right\} = \beta\mu^2 \sum_i\sum_j g(i,j). \tag{10b}$$

Equation (10a) is generally referred to as the *fluctuation–susceptibility* relation; it may be compared with the corresponding relation for fluids, namely (4.5.7), which connects isothermal compressibility κ_T with the density fluctuations in the system. Equations (10) and (4.5.7) represent the equilibrium limit of the *fluctuation–dissipation theorem* discussed later in Section 15.6. Equation (10b), on the other hand, relates χ to a summation of the correlation function $g(i,j)$ over all i and j; assuming homogeneity, this may be written as

$$\chi = N\beta\mu^2 \sum_{\boldsymbol{r}} g(\boldsymbol{r}) \quad (\boldsymbol{r} = \boldsymbol{r}_j - \boldsymbol{r}_i). \tag{11}$$

Treating $\boldsymbol{r}$ as a continuous variable, equation (11) may be written as

$$\chi = \frac{N\beta\mu^2}{a^d}\int g(\boldsymbol{r})d\boldsymbol{r}, \tag{12}$$

where a is a microscopic length, such as the lattice constant, so defined that $Na^d = V$, the volume of the system; for a similar result appropriate to fluids, see equation (10.7.14). Finally, introducing the Fourier transform of the function $g(\boldsymbol{r})$, through equation (2), we observe that

$$\chi = \frac{N\beta\mu^2}{a^d}\tilde{g}(0); \tag{13}$$

compare this result to equation (10.7.21) for fluids.

Our next task consists in determining the mathematical form of the functions $g(\boldsymbol{r})$ and $\tilde{g}(\boldsymbol{k})$. Pending exact calculations, let us see what the mean field theory has to offer in this regard. Following Kadanoff (1976b), we consider a magnetic system subject to an external field H which is *nonuniform*, that is, $H = \{H_i\}$, where H_i denotes the field at site i. Using mean field approximation, the thermal average of the variable σ_i is given by, see equation (12.5.10),

$$\overline{\sigma}_i = \tanh(\beta\mu H_{\text{eff}}), \tag{14}$$

where

$$H_{\text{eff}} = H_i + (J/\mu)\sum_{\text{n.n.}}\overline{\sigma}_j; \tag{15}$$

note that, in view of the nonuniformity of H, the product $(q\overline{\sigma})$ of equation (12.5.10) has been replaced by a sum of $\overline{\sigma}_j$ over all the nearest neighbors of spin i. If $\overline{\sigma}$ varies slowly in space, which means that the applied field is not too nonuniform, then (15) may be approximated as

$$H_{\text{eff}} \simeq H_i + (qJ/\mu)\overline{\sigma}_i + (cJa^2/\mu)\nabla^2\overline{\sigma}_i, \tag{16}$$

where c is a number of order unity whose actual value depends on the structure of the lattice, while a is an effective lattice constant; note that the term involving $\nabla\overline{\sigma}_i$ cancels on summation over the q nearest neighbors that are supposed to be positioned in some symmetrical fashion around the site i. At the same time, the function $\tanh x$, for small x, may be approximated by $x - x^3/3$. Retaining only essential terms, we get from (14) and (16)

$$\beta\mu H_i = (1 - q\beta J)\overline{\sigma}_i + \frac{1}{3}(q\beta J)^3\overline{\sigma}_i^3 - c\beta Ja^2\nabla^2\overline{\sigma}_i. \tag{17}$$

Now, the conditions for criticality are $\{H_i\} = 0$ and $q\beta_c J = 1$; see equation (12.5.13). So, near criticality, we may introduce our familiar variables

$$h_i = \beta\mu H_i, \quad t = (T - T_c)/T_c \simeq (\beta_c - \beta)/\beta_c; \tag{18}$$

equation (17) then reduces to

$$\left(t + \frac{1}{3}\overline{\sigma}_i^2 - c' a^2 \nabla^2\right)\overline{\sigma}_i = h_i, \tag{19}$$

where c' is another number of order unity. Equation (19) generalizes equation (12.10.1) of Landau's theory by taking into account the nonuniformity of $\overline{\sigma}$.

Differentiating (19) with respect to h_j, we get

$$\left(t + \overline{\sigma}_i^2 - c' a^2 \nabla^2\right)\frac{\partial \overline{\sigma}_i}{\partial h_j} = \delta_{i,j}. \tag{20}$$

The "response function" $\partial\overline{\sigma}_i/\partial h_j$ is identical with the correlation function $g(i,j)$;[17] equation (20) may, therefore, be written as

$$\left(t + \overline{\sigma}_i^2 - c' a^2 \nabla^2\right) g(i,j) = \delta_{i,j}. \tag{21}$$

For $t > 0$ and $\{h_i\} \to 0$, $\overline{\sigma}_i \to 0$; equation (21) then becomes

$$\left(t - c' a^2 \nabla^2\right) g(i,j) = \delta_{i,j}. \tag{22}$$

Assuming homogeneity, so that $g(i,j) = g(\boldsymbol{r})$ where $\boldsymbol{r} = \boldsymbol{r}_j - \boldsymbol{r}_i$, and introducing Fourier transforms, equation (22) gives the form

$$\left(t + c' a^2 k^2\right)\tilde{g}(\boldsymbol{k}) = \text{const.} \tag{23}$$

It follows that $\tilde{g}(\boldsymbol{k})$ is a function of the magnitude k only (which is not surprising in view of the assumed symmetry of the lattice). Thus

$$\tilde{g}(k) \sim \frac{1}{t + c' a^2 k^2}, \tag{24}$$

[17] Remembering that $\overline{M} = \mu \Sigma_i \overline{\sigma}_i$, we change the field $\{H_i\}$ to $\{H_i + \delta H_i\}$, with the result that

$$\delta\overline{M} = \mu \sum_i \left[\sum_j (\partial\overline{\sigma}_i/\partial H_j)\delta H_j\right].$$

Now, for simplicity, we let all δH_j be the same; then

$$(\delta\overline{M}/\delta H) = \mu \sum_i \sum_j (\partial\overline{\sigma}_i/\partial H_j).$$

Comparing this with (10b), we infer that $(\partial\overline{\sigma}_i/\partial H_j) = \beta\mu g(i,j)$ and hence $(\partial\overline{\sigma}_i/\partial h_j) = g(i,j)$.

which is the famous Ornstein–Zernike result derived originally for fluids.

Now, taking the inverse Fourier transform of $\tilde{g}(k)$, we obtain (disregarding numerical factors that are not so essential for the present argument)

$$g(r) \sim \int \frac{e^{-i\mathbf{k}\cdot\mathbf{r}}}{t + c'a^2k^2} d^d(ka) \tag{25a}$$

$$\sim \int_0^\infty \frac{a^d}{t + c'a^2k^2} \left(\frac{1}{kr}\right)^{(d-2)/2} J_{(d-2)/2}(kr)k^{d-1}dk; \tag{25b}$$

see equations (8) and (11) of Appendix C. The integral in (25b) is tabulated; see Gradshteyn and Ryzhik (1965, p. 686). We get

$$g(r) \sim \left(\frac{a^2}{\xi r}\right)^{(d-2)/2} K_{(d-2)/2}\left(\frac{r}{\xi}\right) \quad \{\xi = a(c'/t)^{1/2}\}, \tag{26}$$

$K_\mu(x)$ being a modified Bessel function. For $x \gg 1$, $K_\mu(x) \sim x^{-1/2}e^{-x}$; equation (26) then gives

$$g(r) \sim \frac{a^{d-2}}{\xi^{(d-3)/2} r^{(d-1)/2}} e^{-r/\xi} \quad (r \gg \xi). \tag{27}$$

On the other hand, for $x \ll 1$, $K_\mu(x)$, for $\mu > 0, \sim x^{-\mu}$; equation (26) then gives

$$g(r) \sim \frac{a^{d-2}}{r^{d-2}} \quad (r \ll \xi; d > 2). \tag{28}$$

In the special case $d = 2$, we obtain instead

$$g(r) \sim \ln(\xi/r) \quad (r \ll \xi; d = 2). \tag{29}$$

It is worth noting that equation (26) simplifies considerably when $d = 3$. Since $K_{1/2}(x)$ is exactly equal to $(\pi/2x)^{1/2}e^{-x}$ for *all* x,

$$g(r)|_{d=3} \sim \frac{a}{r} e^{-r/\xi} \tag{30}$$

for *all* r. Equation (30) is another important result of Ornstein and Zernike.

Clearly, the quantity ξ appearing here is a measure of the "distances over which the spin–spin (or density–density) correlations in the system extend" — hence the name *correlation length*. So long as T is significantly above T_c, $\xi = O(a)$; see (26). However, as T approaches T_c, ξ increases indefinitely — ultimately diverging at $T = T_c$. The resulting singularity is also of the power-law type:

$$\xi \sim at^{-1/2} \quad (t \gtrsim 0). \tag{31}$$

The divergence of ξ at $T = T_c$ is perhaps the most important clue we have for our general understanding of the critical phenomena. As $\xi \to \infty$, correlations extend *over* the *entire* system, paving the way for the propagation of long-range order (even though the microscopic interactions, which are at the root of the phenomenon, are themselves short-ranged). Moreover, since correlations extend over *macroscopic* distances in the system, any structural details that differentiate one system from another at the *microscopic* level lose significance, leading thereby to universal behavior!

Going back to equations (13) and (24), we see that the singularity in χ is indeed of the type expected in a mean field theory, namely

$$\chi \sim \tilde{g}(0) \sim t^{-1} \quad (t \gtrsim 0). \tag{32}$$

In view of the foregoing results, one may write

$$\frac{1}{\tilde{g}(k)} \sim \frac{1}{\chi}(1 + \xi^2 k^2). \tag{33}$$

In a so-called Ornstein–Zernike analysis, one plots $1/\tilde{g}(k)$ (or $1/I(k)$, where $I(k)$ is the intensity of the light scattered at angle θ, see equation (1)) in the critical region versus k^2. The data for small k ($ka \lesssim 0.1$), for which the above treatment holds, fall close to a straight line whose intercept with the vertical axis determines $\chi(t)$. As $t \to 0$, this intercept tends to zero but the successive isotherms remain more or less parallel to one another; the reduced slope evidently serves to determine $\xi(t)$. For $t \simeq 0$, these plots show a slight downward curvature, indicating departure from the k^2-law to one in which the power of k is somewhat less than 2. Finally, as regards the plot $I(\theta; T)$ of Figure 12.13 earlier in this section, the maximum in the curve, according to equation (24), should lie at $t = 0$ for all θ and the height of the maximum should be $\sim k^{-2}$ (i.e., essentially $\sim \theta^{-2}$); thus, according to the mean field expression for $\tilde{g}(k)$, the exponent λ_1 in equation (3) should be 2 while λ_2 should be 0.

12.12 The critical exponents ν and η

According to the mean field theory, the divergence of ξ at $T = T_c$ is governed by the power law (12.11.31), with a critical exponent $\frac{1}{2}$. We anticipate that the experimental data on actual systems may not conform to this law. We, therefore, introduce a new critical exponent, ν, such that

$$\xi \sim t^{-\nu} \quad (h \to 0, t \gtrsim 0). \tag{1}$$

In the spirit of the scaling hypothesis, see Section 12.10, the corresponding exponent ν' appropriate to $t \lesssim 0$ would be the same as ν.[18] Table 12.1 in Section 12.7 shows

[18] It turns out that the exponent ν' is relevant only for scalar models, for which $n = 1$; for vector models ($n \geq 2$), ξ is infinite at all $T \leq T_c$ and hence ν' is irrelevant.

experimental results for ν obtained from a variety of systems; we see that the observed values of ν, while varying very little from system to system, differ considerably from the mean field value.

As regards the correlation function, the situation for $t \gtrsim 0$ is described very well by a law of the type (12.11.27), namely

$$g(r) \sim e^{-r/\xi(t)} \times \text{some power of } r \quad (t \gtrsim 0), \tag{2}$$

where $\xi(t)$ is given by (1). The variation of $g(r)$ with r in this regime is governed primarily by the exponential, so $g(r)$ falls rapidly as r exceeds ξ (which is typically of the order of the lattice constant a). As $t \to 0$ and hence $\xi \to \infty$, the behavior of $g(r)$ would be expected to be like equation (12.11.28) or (12.11.29). A problem now arises: we have an *exact* expression for $g(r)$ at $T = T_c$ for a two-dimensional Ising model (see Section 13.4), according to which

$$g(r) \sim r^{-1/4} \quad (d = 2, n = 1, t = 0), \tag{3}$$

which is quite different from the mean field expression (12.11.29). We, therefore, generalize our classical result to

$$g(r) \sim r^{-(d-2+\eta)} \quad (t = 0), \tag{4}$$

which introduces another critical exponent, η. Clearly, η for the two-dimensional Ising model is $\frac{1}{4}$, which can even be confirmed by experiments on certain systems that are effectively two-dimensional. Table 12.1 shows experimental values of η for some systems in three dimensions; typically, η turns out to be a small number, which makes it rather difficult to measure reliably.

We shall now derive some scaling relations involving the exponents ν and η. First of all we write down the correlation function $g(r; t, h)$ and its Fourier transform $\tilde{g}(k; t, h)$ in a scaled form. For this, we note that, while h scales with t^{Δ}, the only natural variable with which r will scale is ξ; accordingly, r will scale with $t^{-\nu}$. We may, therefore, write

$$g(r; t, h) \approx \frac{G(rt^{\nu}, h/t^{\Delta})}{r^{d-2+\eta}}, \quad \tilde{g}(k; t, h) \approx \frac{\tilde{G}(k/t^{\nu}, h/t^{\Delta})}{k^{2-\eta}}, \tag{5a, b}$$

where the functions $G(x, y)$ and $\tilde{G}(z, y)$, like the exponents Δ, ν, and η, are *universal* for a given universality class; in expressions (5), for simplicity, we have suppressed nonuniversal parameters that vary from system to system within a class. In the absence of the field ($h = 0$), expressions (5) reduce to

$$g(r; t, 0) \approx \frac{G_0(rt^{\nu})}{r^{d-2+\eta}}, \quad \tilde{g}(k; t, 0) \approx \frac{\tilde{G}_0(k/t^{\nu})}{k^{2-\eta}}, \tag{6a, b}$$

where $G_0(x)$ and $\tilde{G}_0(z)$ are again universal. At the critical point $(h=0, t=0)$, we have simply

$$g_c(r) \sim \frac{1}{r^{d-2+\eta}}, \quad \tilde{g}_c(k) \sim \frac{1}{k^{2-\eta}}. \tag{7a, b}$$

We now recall equation (12.11.12), which relates χ to an integral of $g(\boldsymbol{r})$ over $d\boldsymbol{r}$, and substitute expression (6a) into it. We get, ignoring nonuniversal parameters as well as numerical factors,

$$\chi \sim \int \frac{G_0(rt^{\nu})}{r^{d-2+\eta}} r^{d-1}\, dr. \tag{8}$$

By a change of variables, this gives

$$\chi \sim t^{-(2-\eta)\nu}. \tag{9}$$

Invoking the standard behavior of χ, we obtain

$$\gamma = (2-\eta)\nu. \tag{10}$$

Note that the same scaling relation can also be obtained by appealing to equations (12.11.13) and (6b); the argument is that, in the limit $k \to 0$, the function $\tilde{G}_0(z)$ must be $\sim z^{2-\eta}$ (so that k is eliminated), leaving behind a result identical to (9). In passing, we note that in the critical region

$$\chi \sim \xi^{2-\eta}. \tag{11}$$

Relation (10) is consistent with the *Fisher inequality* (1969)

$$\gamma \le (2-\eta)\nu,$$

and is obviously satisfied by the mean field exponents $(\gamma = 1, \nu = \frac{1}{2}, \eta = 0)$; it also checks well with the experimental data given earlier in Table 12.1. In fact, this relation provides a much better method of extracting the elusive exponent η, from a knowledge of γ and ν, than determining it directly from experiment. Incidentally, the presence of η explains the slight downward curvature of the Ornstein–Zernike plot, $1/\tilde{g}(k)$ versus k^2, as $k \to 0$, for the appropriate expression for $1/\tilde{g}(k)$ now is

$$\frac{1}{\tilde{g}(k)} \sim \frac{1}{\chi}\left(1 + \xi^{2-\eta} k^{2-\eta}\right), \tag{12}$$

rather than (12.11.33).

We shall now derive another exponent relation involving ν, but first notice that all exponent relations derived so far have no *explicit* dependence on the dimensionality d of the system (though the actual values of the exponents do depend on d). There is, however, one important relationship that *does* involve d explicitly. For this, let us visualize what happens inside the system (say, a magnetic one) as $t \to 0$ from above. At some stage the correlation length ξ becomes significantly larger than the atomic spacings, with the result that magnetic domains, of alternating magnetization, begin to appear. The closer we get to the critical point, the larger the size of these domains; one may, in fact, say that the volume Ω of any such domain is $\sim \xi^d$. Now, the singular part of the free energy density of the system — or, for that matter, of any one of these domains — is given by, see equation (12.10.7) with $h = 0$,

$$f^{(s)}(t) \sim t^{2-\alpha}, \tag{13}$$

which vanishes as $t \to 0$. At the same time, the domain volume Ω diverges. It seems natural to expect that $f^{(s)}$, being a density, would vanish as $1/\Omega$, that is,

$$f^{(s)}(t) \sim \Omega^{-1} \sim \xi^{-d} \sim t^{d\nu}. \tag{14}$$

Comparing (13) and (14), we obtain the desired relationship

$$d\nu = 2 - \alpha, \tag{15}$$

which is generally referred to as a *hyperscaling relation* — to emphasize the fact that it goes beyond, and cannot be derived from, the ordinary scaling formulation of Section 12.10 without invoking something else, such as the domain volume Ω.

The relation in equation (15) is consistent with the *Josephson inequalities* (1967):

$$d\nu \geq 2 - \alpha, \quad d\nu' \geq 2 - \alpha', \tag{16a, b}$$

proved rigorously by Sokal (1981); of course, the scaling theory does not distinguish between exponents pertaining to $t > 0$ and their counterparts pertaining to $t < 0$. It is important to note that the classical exponents ($\nu = \frac{1}{2}, \alpha = 0$) satisfy (15) *only* for $d = 4$, which shows that the hyperscaling relations, (15) and any others that follow from it, have a rather different status than the other scaling relations (that do not involve d explicitly). The renormalization group theory, to be discussed in Chapter 14, shows why the hyperscaling relations are to be expected fairly generally, why typically they hold for $d < 4$ but break down for $d > 4$; see also Section 12.13. The reader may check that relation (15) is satisfied reasonably well by the experimental data of Table 12.1, with $d = 3$; it is also satisfied by the exponents derived theoretically by solving different models exactly, or almost exactly, as in Chapter 13.

Combining (15) with other scaling relations, see Section 12.10, we may write

$$d\nu = 2 - \alpha = 2\beta + \gamma = \beta(\delta + 1) = \gamma(\delta + 1)/(\delta - 1). \tag{17}$$

It follows that

$$2 - \eta = \gamma/\nu = d(\delta - 1)/(\delta + 1), \tag{18}$$

which is consistent with the *Buckingham–Gunton inequality* (1969)

$$2 - \eta \le d(\delta - 1)/(\delta + 1). \tag{19}$$

Notice that the experimental observation that, for magnetic systems in three dimensions, $\delta < 4.8$, implies that $\eta \ge 0.034$.

12.13 A final look at the mean field theory

We now return to the question: why does the mean field theory fail to represent the true nature of the critical phenomena in real systems? The short answer is — because it neglects fluctuations! As emphasized in Section 12.11, correlations among the microscopic constituents of a given system are at the very heart of the phenomenon of phase transitions, for it is through them that the system acquires a long-range order (even when the microscopic interactions are themselves short-ranged). At the same time, there is so direct a relationship between correlations and fluctuations, see equation (12.11.6), that they grow together and, as the critical point is approached, become a dominant feature of the system. Neglecting fluctuations is, therefore, a serious drawback of the mean field theory.

The question now arises: is mean field theory *ever* valid? In other words, can fluctuations *ever* be neglected? To answer this question, we recall the fluctuation–susceptibility relation (12.11.10a), namely

$$\chi = (\overline{M^2} - \overline{M}^2)/kT = \overline{(\Delta M)^2}/kT \tag{1}$$

and write it in the form

$$\overline{(\Delta M)^2}/\overline{M}^2 = kT\chi/\overline{M}^2. \tag{2}$$

Now, in order that the neglect of fluctuations be justified, we must have:

$$kT\chi \ll \overline{M}^2. \tag{3}$$

Requirement (3) is generally referred to as the *Ginzburg criterion* (1960); for a more detailed discussion of this criterion, along with physical illustrations, see Als-Nielsen and Birgeneau (1977).

We apply condition (3) to a domain, of volume $\Omega \sim \xi^d$, close to but below the critical point; we are assuming here a system of the Ising type ($n = 1$), so that ξ is finite for $t < 0$. Invoking the power-law behavior of χ and $\overline{M}$, we have

$$kT_c(A\xi^d|t|^{-\gamma}) \ll (B\xi^d|t|^{\beta})^2 \quad (t \lesssim 0), \tag{4}$$

where A and B are positive constants. Since $\xi \sim a|t|^{-\nu}$, we get

$$|t|^{d\nu - 2\beta - \gamma} \ll B^2 a^d / AkT_c. \tag{5}$$

In view of the scaling relation $\alpha + 2\beta + \gamma = 2$, we may as well write

$$|t|^{d\nu - (2-\alpha)} \ll D, \tag{6}$$

where D is a positive number of order unity.[19] For the mean field theory (with $\nu = \frac{1}{2}, \alpha = 0$) to be valid, condition (6) assumes the form

$$|t|^{(d-4)/2} \ll D. \tag{7}$$

Now, since $|t|$ can be made as small as we like, condition (7) will be violated unless $d > 4$. We, therefore, conclude that the mean field theory is valid for $d > 4$; by implication, it is inadequate for $d \leq 4$.

The preceding result has been established for scalar models $(n = 1)$ only. In Section 13.5, we shall see that in the case of the spherical model, which pertains to the limit $n \to \infty$, the mean field results do apply when $d > 4$. This means that, once again, fluctuations *can* be neglected if $d > 4$. Now fluctuations are supposed to decrease with decreasing n; the validity of the mean field theory for $d > 4$ should, therefore, hold for *all* n.

Ordinarily, when a system is undergoing a phase transition, expression (2), which is a measure of the relative fluctuations in the system, is expected to be of order unity. Condition (6) then suggests that the exponents ν and α obey the hyperscaling relation

$$d\nu = 2 - \alpha. \tag{8}$$

Experience shows that this relation is indeed obeyed when $d < 4$. At $d = 4$, the mean field theory begins to take over and thereafter, for all $d > 4$, the critical exponents are stuck at the mean field values (which are independent of both d and n). The dimensionality $d = 4$ is often referred to as the *upper critical dimension* for the kind of systems under study.

An alternative way of looking into the question posed at the beginning of this section is to examine the specific heat of the system which, according to the mean field theory, undergoes a jump discontinuity at the critical point whereas in real systems it shows a weak divergence. The question now arises: what is the source of this divergence that is missed by the mean field theory? The answer again lies in the "neglect of fluctuations." To see it more explicitly, we look at the internal energy of the system which, in the absence of the field, is given by

$$U = -J\overline{\left(\sum_{\text{n.n.}} \sigma_i \sigma_j\right)} = -J \sum_{\text{n.n.}} \overline{\sigma_i \sigma_j}. \tag{9}$$

In the mean field theory, one replaces $\overline{\sigma_i \sigma_j}$ by $\overline{\sigma_i}\,\overline{\sigma_j} (= \overline{\sigma}^2)$, see equation (12.5.5), which leads to the jump discontinuity in the specific heat of magnitude $\frac{3}{2}/Nk$; see equation (12.5.18).

[19]To see this, we note that $A \sim N\mu^2/\Omega k T_c$ while $B \sim N\mu/\Omega$, with the result that $D \sim Na^d/\Omega = O(1)$.

The *fluctuational* part of U, which is neglected in the mean field theory, may be written as

$$U_f = -J \sum_{\text{n.n.}} (\overline{\sigma_i \sigma_j} - \overline{\sigma}_i \overline{\sigma}_j) = -J \sum_{\text{n.n.}} g(r_{ij}), \tag{10}$$

where $g(r)$ is the spin–spin correlation function, for which we may use the mean field expression (12.11.26); thus, we will be using the mean field theory itself to predict its own shortcomings! Since the nearest-neighbor distances r_{ij} in (10) are all much smaller than ξ, one may be tempted to use for $g(r_{ij})$ the zeroth-order approximation (12.11.28). This, however, produces a temperature-independent term, which does not contribute to the specific heat of the system. We must, therefore, go to the next approximation, which can be obtained by using the asymptotic formulae

$$K_\mu(x)\big|_{x\ll 1} \approx \begin{cases} \frac{1}{2}\Gamma(\mu)\left(\frac{1}{2}x\right)^{-\mu} + \frac{1}{2}\Gamma(-\mu)\left(\frac{1}{2}x\right)^{\mu} & \text{for} \quad 0 < \mu < 1 & \text{(11a)} \\ x^{-1} + \left(\frac{1}{2}x\right)\ln\left(\frac{1}{2}x\right) & \text{for} \quad \mu = 1 & \text{(11b)} \\ \frac{1}{2}\Gamma(\mu)\left(\frac{1}{2}x\right)^{-\mu} - \frac{1}{2}\Gamma(\mu-1)\left(\frac{1}{2}x\right)^{2-\mu} & \text{for} \quad \mu > 1, & \text{(11c)} \end{cases}$$

with $\mu = (d-2)/2$ and $x = r_{ij}/\xi$. The temperature-dependent part of U_f comes from the second term(s) in (11); remembering that ξ here is $\sim at^{-1/2}$, we get[20]

$$(U_f/NJ)_{\text{thermal}} \sim \begin{cases} t^{(d-2)/2} & \text{for} \quad 2 < d < 4 & \text{(12a)} \\ t\ln(1/t) & \text{for} \quad d = 4 & \text{(12b)} \\ t & \text{for} \quad d > 4. & \text{(12c)} \end{cases}$$

The *fluctuational* part of the specific heat then turns out to be

$$C_f/Nk \sim \begin{cases} t^{(d-4)/2} & \text{for} \quad 2 < d < 4 & \text{(13a)} \\ \ln(1/t) & \text{for} \quad d = 4 & \text{(13b)} \\ \text{const.} & \text{for} \quad d > 4. & \text{(13c)} \end{cases}$$

It follows that the specific-heat singularity for $d > 4$ is indeed a "jump discontinuity," and hence the mean field theory remains applicable in this case. For $d = 4$, C_f shows a logarithmic divergence and for $d < 4$ a power-law divergence, making mean field theory invalid for $d \leq 4$.

It is rather instructive to see what part of the fluctuation–correlation spectrum, $\tilde{g}(k)$, contributes significantly to the divergence of the specific heat at $t = 0$. For this, we examine

[20]Note that the negative sign in (10) cancels the implicit negative sign of $\Gamma(-\mu)$ in (11a), that of $\ln(\frac{1}{2}x)$ in (11b), and the explicit negative sign in (11c).

the quantity $-\partial g(r_{ij})/\partial t$, which essentially determines the behavior of C_f near the critical point; see equation (10). Using equation (12.11.25a), we have

$$-\frac{\partial g(r_{ij})}{\partial t} \sim a^d \int \frac{e^{-i\boldsymbol{k}\cdot\boldsymbol{r}_{ij}}}{t^2(1+\xi^2 k^2)^2} k^{d-1} dk. \tag{14}$$

Now, the values of k that are much larger than ξ^{-1} contribute little to this integral; the only significant contributions come from the range $(0, k_{\max})$, where $k_{\max} = O(\xi^{-1})$; moreover, since $r_{ij} \ll \xi$, the exponential for these values of k is essentially equal to 1. Expression (14) may, therefore, be written as

$$-\frac{\partial g(r_{ij})}{\partial t} \sim a^d \int \frac{k^{d-1} dk}{t^2(1+\xi^2 k^2)^2} \tag{15}$$

which, for $d < 4$, scales as $(a/\xi)^d t^{-2} \sim t^{(d-4)/2}$; compare with (13a). We thus see that the most significant contribution to the criticality of the problem arises from fluctuations whose length scale, k^{-1}, is of order ξ or longer and hardly any contribution comes from fluctuations whose length scale is shorter. Now, it is only the latter that are likely to pick up the structural details of the system at the atomic level; since they do not play any significant role in bringing about the phenomenon, the precise nature of criticality remains independent of the structural details. This explains why a large variety of systems, differing so much in their structure at the macroscopic level, may, insofar as critical behavior is concerned, fall into a single universality class.

Problems

12.1. Assume that in the virial expansion

$$\frac{Pv}{kT} = 1 - \sum_{j=1}^{\infty} \frac{j}{j+1} \beta_j \left(\frac{\lambda^3}{v}\right)^j, \tag{10.4.22}$$

where β_j are the irreducible cluster integrals of the system, only terms with $j = 1$ and $j = 2$ are appreciable in the critical region. Determine the relationship between β_1 and β_2 at the critical point, and show that $kT_c/P_c v_c = 3$.

12.2. Assuming the *Dietrici* equation of state,

$$P(v-b) = kT \exp(-a/kTv),$$

evaluate the critical constants P_c, v_c, and T_c of the given system in terms of the parameters a and b, and show that the quantity $kT_c/P_c v_c = e^2/2 \simeq 3.695$.

Further show that the following statements hold in regard to the Dietrici equation of state:

(a) It yields the same expression for the second virial coefficient B_2 as the van der Waals equation does.

(b) For all values of P and for $T \geq T_c$, it yields a *unique* value of v.

(c) For $T < T_c$, there are *three* possible values of v for certain values of P and the critical volume v_c is always intermediate between the largest and the smallest of the three volumes.

(d) The Dietrici equation of state yields the same critical exponents as the van der Waals equation does.

12.3. Consider a nonideal gas obeying a modified van der Waals equation of state

$$(P + a/v^n)(v - b) = RT \quad (n > 1).$$

Examine how the critical constants P_c, v_c, and T_c, and the critical exponents β, γ, γ', and δ, of this system depend on the number n.

12.4. Following expressions (12.5.2), define

$$p = (1 + L)/2 \quad \text{and} \quad q = (1 - L)/2 \quad (-1 \le L \le 1) \tag{1}$$

as the probabilities that a spin chosen at random in a lattice composed of N spins is either "up" or "down." The partition function of the system may then be written as

$$Q(B, T) = \sum_L g(L) e^{\beta N(\frac{1}{2}qJL^2 + \mu BL)}, \tag{2}$$

where $g(L)$ is the multiplicity factor associated with a particular value of L, that is,

$$g(L) = N!/(Np)!(Nq)!; \tag{3}$$

note that in writing the Hamiltonian here we have made the assumption of *random mixing*, according to which

$$(N_{++} + N_{--} - N_{+-}) = \frac{1}{2}qN(p^2 + q^2 - 2pq) = \frac{1}{2}qNL^2.$$

(a) Determine the value, L^*, of L that *maximizes* the summand in (2). Check that L^* is identical to the mean value, $\bar{L}$, as given by equation (12.5.10).

(b) Write down the free energy A and the internal energy U of the system, and show that the entropy S conforms to the relation

$$S(B, T) = -Nk(p^* \ln p^* + q^* \ln q^*),$$

where $p^* = p(L^*)$ and $q^* = q(L^*)$.

12.5. Using the correspondence established in Section 12.4, apply the results of the preceding problem to the case of a lattice gas. Show, in particular, that the pressure, P, and the volume per particle, v, are given by

$$P = \mu B - \frac{1}{8}q\varepsilon_0\left(1 + \bar{L}^2\right) - \frac{1}{2}kT\ln\left(\frac{1 - \bar{L}^2}{4}\right)$$

and

$$v^{-1} = \frac{1}{2}(1 \pm \bar{L}).$$

Check that the critical constants of this system are: $T_c = q\varepsilon_0/4k$, $P_c = kT_c(\ln 2 - \frac{1}{2})$, and $v_c = 2$, so that the quantity $kT_c/P_cv_c = 1/(\ln 4 - 1) \simeq 2.589$.

12.6. Consider an Ising model with an *infinite-range* interaction such that each spin interacts equally strongly with all other spins:

$$H = -c\sum_{i<j}\sigma_i\sigma_j - \mu B\sum_i \sigma_i.$$

Express this Hamiltonian in terms of the parameter $L(= N^{-1}\Sigma_i\sigma_i)$ and show that, in the limit $N \to \infty$ and $c \to 0$, the mean field theory, with $J = Nc/q$, is exact for this model.

12.7. Study the Heisenberg model of a ferromagnet, based on the interaction (12.3.6), in the *mean field approximation* and show that this also leads to a phase transition of the kind met with in the Ising model. Show, in particular, that the transition temperature T_c and the Curie–Weiss constant C are given by

$$T_c = \frac{qJ}{k}\frac{2s(s+1)}{3} \quad \text{and} \quad C = \frac{N(g\mu_B)^2}{Vk}\frac{s(s+1)}{3}.$$

Note that the ratio $T_c/CV = 2qJ/N(g\mu_B)^2$ is the *molecular field constant* of the problem; compare to equation (12.5.8).

12.8. Study the spontaneous magnetization of the Heisenberg model in the *mean field approximation* and examine the dependence of $\overline{L}_0$ on T (i) in the neighborhood of the critical temperature where $(1 - T/T_c) \ll 1$, and (ii) at sufficiently low temperatures where $T/T_c \ll 1$. Compare these results with the corresponding ones, namely (12.5.14) and (12.5.15) for the Ising model.

[In this connection, it may be pointed out that, at very low temperatures, the experimental data do not agree with the theoretical formula derived here. We find instead a much better agreement with the formula $\overline{L}_0 = \{1 - A(kT/J)^{3/2}\}$, where A is a numerical constant (equal to 0.1174 in the case of a simple cubic lattice). This formula is known as Bloch's $T^{3/2}$-law and is derivable from the spin-wave theory of ferromagnetism; see Wannier (1966), Section 15.5.]

12.9. An antiferromagnet is characterized by the fact that the exchange integral J is negative, which tends to align neighboring spins antiparallel to one another. Assume that a given lattice structure is such that the whole lattice can be divided into two interpenetrating sublattices, a and b say, so that the spins belonging to each sublattice, a as well as b, tend to align themselves in the same direction, while the directions of alignment in the two sublattices are opposite to one another. Using the Ising as well as Heisenberg type of interaction, and working in the mean field approximation, evaluate the paramagnetic susceptibility of such a lattice at high temperatures.

12.10. The *Néel temperature* T_N of an antiferromagnet is defined as that temperature below which the sublattices a and b possess *nonzero* spontaneous magnetizations M_a and M_b, respectively. Determine T_N for the model described in the preceding problem.

12.11. Suppose that each atom of a crystal lattice can be in one of several *internal states* (which may be denoted by the symbol σ) and the interaction energy between an atom in state σ' and its nearest neighbor in state σ'' is denoted by $u(\sigma', \sigma'')\{= u(\sigma'', \sigma')\}$. Let $f(\sigma)$ be the probability of an atom being in a particular state σ, *independently* of the states in which its nearest neighbors are. The interaction energy and the entropy of the lattice may then be written as

$$E = \frac{1}{2}qN \sum_{\sigma', \sigma''} u(\sigma', \sigma'')f(\sigma')f(\sigma'')$$

and

$$S/Nk = -\sum_{\sigma} f(\sigma)\ln f(\sigma),$$

respectively. Minimizing the free energy $(E - TS)$, show that the equilibrium value of the function $f(\sigma)$ is determined by the equation

$$f(\sigma) = C\exp\{-(q/kT)\Sigma_{\sigma'} u(\sigma, \sigma')f(\sigma')\},$$

where C is the constant of normalization. Further show that, for the special case $u(\sigma', \sigma'') = -J\sigma'\sigma''$, where the σ can be either $+1$ or -1, this equation reduces to the Weiss equation (12.5.11), with $f(\sigma) = \frac{1}{2}(1 + \overline{L}_0\sigma)$.

12.12. Consider a binary alloy containing N_A atoms of type A and N_B atoms of type B, so that the relative concentrations of the two components are: $x_A = N_A/(N_A + N_B) \le \frac{1}{2}$ and $x_B = N_B/(N_A + N_B) \ge \frac{1}{2}$. The degree of *long-range order*, X, is such that

$$\begin{bmatrix} A \\ a \end{bmatrix} = \frac{1}{2}Nx_A(1 + X), \quad \begin{bmatrix} A \\ b \end{bmatrix} = \frac{1}{2}Nx_A(1 - X),$$

$$\begin{bmatrix} B \\ a \end{bmatrix} = \frac{1}{2}N(x_B - x_A X), \quad \begin{bmatrix} B \\ b \end{bmatrix} = \frac{1}{2}N(x_B + x_A X),$$

where $N = N_A + N_B$, while the symbol $\begin{bmatrix} A \\ a \end{bmatrix}$ denotes the number of atoms of type A occupying sites of the sublattice a, and so on. In the *Bragg–Williams approximation*, the number of nearest-neighbor pairs of different kinds can be written down straightaway; for instance,

$$\begin{bmatrix} AA \\ ab \end{bmatrix} = \frac{1}{2} qN \cdot x_A(1+X) \cdot x_A(1-X),$$

and so on. The configurational energy of the lattice then follows from equation (12.4.9). In the same approximation, the entropy of the lattice is given by $S = k \ln W$, where

$$W = \frac{(\frac{1}{2}N)!}{\begin{bmatrix} A \\ a \end{bmatrix}! \begin{bmatrix} B \\ a \end{bmatrix}!} \cdot \frac{(\frac{1}{2}N)!}{\begin{bmatrix} A \\ b \end{bmatrix}! \begin{bmatrix} B \\ b \end{bmatrix}!}.$$

Minimizing the free energy of the lattice, show that the equilibrium value of X is determined by the equation

$$\frac{X}{x_B + x_A X^2} = \tanh\left(\frac{2qx_A\varepsilon}{kT} X\right); \quad \varepsilon = \frac{1}{2}(\varepsilon_{11} + \varepsilon_{22}) - \varepsilon_{12} > 0.$$

Note that, in the special case of equal concentrations ($x_A = x_B = \frac{1}{2}$), this equation assumes the more familiar form

$$X = \tanh\left(\frac{q\varepsilon}{2kT} X\right).$$

Further show that the transition temperature of the system is given by

$$T_c = 4x_A(1-x_A)T_c^0,$$

where $T_c^0 (= q\varepsilon/2k)$ is the transition temperature in the case of equal concentrations.

[*Note*: In the Kirkwood approximation (see Kubo (1965), problem 5.19), T_c^0 turns out to be $(\varepsilon/k)\{1 - \surd[1-(4/q)]\}^{-1}$, which may be written as $(q\varepsilon/2k)(1 - 1/q + \cdots)$. To this order, the Bethe approximation also yields the same result.]

12.13. Consider a two-component solution of N_A atoms of type A and N_B atoms of type B, which are supposed to be randomly distributed over $N (= N_A + N_B)$ sites of a single lattice. Denoting the energies of the nearest-neighbor pairs AA, BB, and AB by $\varepsilon_{11}, \varepsilon_{22}$, and ε_{12}, respectively, write down the free energy of the system in the Bragg–Williams approximation and evaluate the chemical potentials μ_A and μ_B of the two components. Next, show that if $\varepsilon = (\varepsilon_{11} + \varepsilon_{22} - 2\varepsilon_{12}) < 0$, that is, if the atoms of the same species display greater affinity to be neighborly, then for temperatures below a critical temperature T_c, which is given by the expression $q|\varepsilon|/2k$, the solution separates out into two phases of unequal relative concentrations.

[*Note*: For a study of phase separation in an isotopic mixture of hard-sphere bosons and fermions, and for the relevance of this study to the actual behavior of He^3–He^4 solutions, see Cohen and van Leeuwen (1960, 1961).]

12.14. Modify the Bragg–Williams approximation (12.5.29) to include a *short-range order* parameter s, such that

$$\overline{N}_{++} = \frac{1}{2} qN\gamma \left(\frac{1+\overline{L}}{2}\right)^2 (1+s), \ \overline{N}_{--} = \frac{1}{2} qN\gamma \left(\frac{1-\overline{L}}{2}\right)^2 (1+s),$$

$$\overline{N}_{+-} = 2 \cdot \frac{1}{2} qN\gamma \left(\frac{1+\overline{L}}{2}\right)\left(\frac{1-\overline{L}}{2}\right)(1-s).$$

(a) Evaluate γ from the condition that the total number of nearest-neighbor pairs is $\frac{1}{2}qN$.

(b) Show that the critical temperature T_c of this model is $(1-s^2)qJ/k$.

(c) Determine the nature of the specific-heat singularity at $T = T_c$, and compare your result with both the Bragg–Williams approximation of Section 12.5 and the Bethe approximation of Section 12.6.

12.15. Show that in the *Bethe approximation* the entropy of the Ising lattice at $T = T_c$ is given by the expression

$$\frac{S_c}{Nk} = \ln 2 + \frac{q}{2}\ln\left(1-\frac{1}{q}\right) - \frac{q(q-2)}{4(q-1)}\ln\left(1-\frac{2}{q}\right).$$

Compare this result with the one following from the *Bragg–Williams approximation*, namely (12.5.20).

12.16. Examine the critical behavior of the low-field susceptibility, χ_0, of an Ising model in the Bethe approximation of Section 12.6, and compare your results with equations (12.5.22) of the Bragg–Williams approximation.

12.17. A function $f(x)$ is said to be *concave* over an interval (a,b) if it satisfies the property

$$f\{\lambda x_1 + (1-\lambda)x_2\} \geq \lambda f(x_1) + (1-\lambda)f(x_2),$$

where x_1 and x_2 are two arbitrary points in the interval (a,b) while λ is a positive number in the interval $(0,1)$. This means that the chord joining the points x_1 and x_2 lies *below* the curve $f(x)$. Show that this also means that the tangent to the curve $f(x)$ at any point x in the interval (a,b) lies *above* the curve $f(x)$ or, equivalently, that the second derivative $\partial^2 f/\partial x^2$ throughout this interval ≤ 0.

12.18. In view of the thermodynamic relationship

$$C_V = TV(\partial^2 P/\partial T^2)_V - TN(\partial^2 \mu/\partial T^2)_V$$

for a fluid, μ being the chemical potential of the system, Yang and Yang (1964) pointed out that, if C_V is singular at $T = T_c$, then either $(\partial^2 P/\partial T^2)_V$ or $(\partial^2 \mu/\partial T^2)_V$ or both will be singular. Define an exponent Θ by writing

$$(\partial^2 P/\partial T^2)_V \sim (T_c - T)^{-\Theta} \quad (T \lesssim T_c),$$

and show that (Griffiths, 1965b)

$$\Theta \leq \alpha' + \beta \quad \text{and} \quad \Theta \leq (2 + \alpha'\delta)/(\delta + 1).$$

12.19. Determine the numerical values of the coefficients r_1 and s_0 of equation (12.9.5) in (i) the Bragg–Williams approximation of Section 12.5 and (ii) the Bethe approximation of Section 12.6. Using these values of r_1 and s_0, verify that equations (12.9.4), (12.9.9), (12.9.10), (12.9.11), and (12.9.15) reproduce correctly the results obtained in the zeroth and the first approximation, respectively.

12.20. Consider a system with a modified expression for the Landau free energy, namely

$$\psi_h(t,m) = -hm + q(t) + r(t)m^2 + s(t)m^4 + u(t)m^6,$$

with $u(t)$ a fixed positive constant. Minimize ψ with respect to the variable m and examine the spontaneous magnetization m_0 as a function of the parameters r and s. In particular, show the following:[21]

(a) For $r > 0$ and $s > -(3ur)^{1/2}$, $m_0 = 0$ is the only real solution.

(b) For $r > 0$ and $-(4ur)^{1/2} < s \leq -(3ur)^{1/2}$, $m_0 = 0$ or $\pm m_1$, where $m_1^2 = \frac{\sqrt{(s^2-3ur)}-s}{3u}$. However, the minimum of ψ at $m_0 = 0$ is lower than the minima at $m_0 = \pm m_1$, so the ultimate equilibrium value of m_0 is 0.

(c) For $r > 0$ and $s = -(4ur)^{1/2}$, $m_0 = 0$ or $\pm(r/u)^{1/4}$. Now, the minimum of ψ at $m_0 = 0$ is of the same height as the ones at $m_0 = \pm(r/u)^{1/4}$, so a nonzero spontaneous magnetization is as likely to occur as the zero one.

[21] To fix ideas, it is helpful to use (r,s)-plane as our "parameter space."

(d) For $r > 0$ and $s < -(4ur)^{1/2}$, $m_0 = \pm m_1$ — which implies a *first-order* phase transition (because the two possible states available here differ by a *finite* amount in m). The line $s = -(4ur)^{1/2}$, with r positive, is generally referred to as a "line of first-order phase transitions."

(e) For $r = 0$ and $s < 0$, $m_0 = \pm(2|s|/3u)^{1/2}$.

(f) For $r < 0$, $m_0 = \pm m_1$ for all s. As $r \to 0$, $m_1 \to 0$ if s is positive.

(g) For $r = 0$ and $s > 0$, $m_0 = 0$ is only solution. Combining this result with (f), we conclude that the line $r = 0$, with s positive, is a "line of second-order phase transitions," for the two states available here differ by a *vanishing* amount in m.

The lines of first-order phase transitions and second-order phase transitions meet at the point $(r = 0, s = 0)$, which is commonly referred to as a *tricritical point* (Griffiths, 1970).

12.21. In the preceding problem, put $s = 0$ and approach the tricritical point along the r-axis, setting $r \approx r_1 t$. Show that the critical exponents pertaining to the tricritical point in this model are

$$\alpha = \frac{1}{2}, \beta = \frac{1}{4}, \gamma = 1, \text{ and } \delta = 5.$$

12.22. Consider a fluid near its critical point, with isotherms as sketched in Figure 12.3. Assume that the singular part of the Gibbs free energy of the fluid is of the form

$$G^{(s)}(T, P) \sim |t|^{2-\alpha} g(\pi/|t|^{\Delta}),$$

where $\pi = (P - P_c)/P_c$, $t = (T - T_c)/T_c$ while $g(x)$ is a *universal* function, with branches g_+ for $t > 0$ and g_- for $t < 0$; in the latter case, the function g_- has a point of *infinite* curvature at a value of π that varies smoothly with t, such that $\pi(0) = 0$ and $(\partial\pi/\partial t)_{t\to 0} = \text{const}$.

(a) Using the above expression for $G^{(s)}$, determine the manner in which the densities, ρ_l and ρ_g, of the two phases approach one another as $t \to 0$ from below.

(b) Also determine how $(P - P_c)$ varies with $(\rho - \rho_c)$ as the critical point is approached along the critical isotherm $(t = 0)$.

(c) Examine as well the critical behavior of the isothermal compressibility κ_T, the adiabatic compressibility κ_S, the specific heats C_P and C_V, the coefficient of volume expansion α_P, and the latent heat of vaporization l.

12.23. Consider a model equation of state which, near the critical point, can be written as

$$h \approx am(t + bm^2)^{\Theta} \quad (1 < \Theta < 2; a, b > 0).$$

Determine the critical exponents β, γ, and δ of this model, and check that they obey the scaling relation (12.10.22).

12.24. Assuming that the correlation function $g(\boldsymbol{r}_i, \boldsymbol{r}_j)$ is a function only of the distance $r = |\boldsymbol{r}_j - \boldsymbol{r}_i|$, show that $g(r)$ for $r \neq 0$ satisfies the differential equation

$$\frac{d^2 g}{dr^2} + \frac{d-1}{r}\frac{dg}{dr} - \frac{1}{\xi^2} g = 0.$$

Check that expression (12.11.27) for $g(r)$ satisfies this equation in the regime $r \gg \xi$, while expression (12.11.28) does so in the regime $r \ll \xi$.

12.25. Consider the correlation function $g(r; t, h)$ of Section 12.11 with $h > 0$.[22] Assume that this function has the following behavior:

$$g(r) \sim e^{-r/\xi(t,h)} \times \text{some power of } r \quad (t \gtrsim 0),$$

such that $\xi(0, h) \sim h^{-\nu^c}$. Show that $\nu^c = \nu/\Delta$.

Next, assume that the susceptibility $\chi(0, h) \sim h^{-\gamma^c}$. Show that $\gamma^c = \gamma/\Delta = (\delta - 1)/\delta$.

[22] For more details, see Tarko and Fisher (1975).

12.26. Liquid He^4 undergoes a superfluid transition at $T \simeq 2.17$ K. The order parameter in this case is a complex number Ψ, which is related to the Bose condensate density ρ_0 as

$$\rho_0 \sim |\Psi|^2 \sim |t|^{2\beta} \quad (t \lesssim 0).$$

The superfluid density ρ_s, on the other hand, behaves as

$$\rho_s \sim |t|^{\nu} \quad (t \lesssim 0).$$

Show that the ratio[23]

$$(\rho_0/\rho_s) \sim |t|^{\eta\nu} \quad (t \lesssim 0).$$

12.27. The surface tension, σ, of a liquid approaches zero as $T \to T_c$ from below. Define an exponent μ by writing

$$\sigma \sim |t|^{\mu} \quad (t \lesssim 0).$$

Identifying σ with the "free energy associated with a unit area of the liquid–vapor interface," argue that $\mu = (d-1)\nu = (2-\alpha)(d-1)/d$.

[*Note*: Analysis of the experimental data on surface tension yields: $\mu = 1.27 \pm 0.02$, which agrees with the fact that for most fluids $\alpha \simeq 0.1$.]

[23]The corresponding ratio for a magnetic system is M_0^2/Γ, where M_0 is the spontaneous magnetization and Γ the helicity modulus of the system; for details, see Fisher, Barber, and Jasnow (1973).

13

Phase Transitions: Exact (or Almost Exact) Results for Various Models

In the preceding chapter we saw that the onset of a phase transition in a given physico-chemical system is characterized by (singular) features whose qualitative nature is determined by the universality class to which the system belongs. In this chapter we propose to consider a variety of model systems belonging to different universality classes and analyze them theoretically to find out how these features arise and how they vary from class to class.

In this context, we recall that the parameters distinguishing one universality class from another are: (i) the space dimensionality d, (ii) the dimensionality of the order parameter, often referred to as the spin dimensionality, n, and (iii) the range of the microscopic interactions. As regards the latter, unless a statement is made to the contrary, we shall assume a short-range interaction which, in most cases, will be of the nearest-neighbor type; the only parameters open for selection will then be d and n.

We start our analysis with the properties of one-dimensional fluids, the one-dimensional Ising ($n = 1$) model, and the general n-vector models (again in one dimension). We then return to the Ising model — this time in two dimensions — and follow it with a study of two models in general d but with $n \to \infty$. These studies will give us a fairly good idea as to what to expect in the most practical, three-dimensional situations for which, unfortunately, we have no exact solutions — though a variety of mathematical techniques have been developed to obtain *almost exact* results in many cases of interest. For completeness, these and other results of physical importance will be revisited in the last section of this chapter.

13.1 One-dimensional fluid models

A system of interacting particles in one dimension can be solved analytically for several cases. In particular, a system of hard spheres in one dimension can be solved in the canonical ensemble (Tonks, 1936) while the one-dimensional isobaric ensemble allows the solution of a more general set of nearest-neighbor interactions. None of these models exhibits a transition to an ordered phase but they do display many of the short-range correlation properties characteristic of fluids.

© 2011 Elsevier Ltd. All rights reserved.

13.1.A Hard spheres on a ring

The partition function of a one-dimensional system of hard spheres was first evaluated by Tonks (1936). For this we consider the case of N hard spheres with diameter D on a ring of circumference L, obeying periodic boundary conditions. The free energy, the pressure, and the pair correlation function of this system can be determined exactly both for finite N and in the thermodynamic limit. The hard sphere pair interaction is given by

$$u(r) = \begin{cases} \infty & \text{for } r \le D \\ 0 & \text{for } r > D. \end{cases} \tag{1}$$

The configurational partition function can be written as an integral over the spatially ordered positions of the N spheres, with particle 1 set at $x_1 = 0$, while the other particles labeled $j = 2, 3, \ldots, N$ are restricted by the conditions $x_{j-1} + D < x_j < L - (N - j + 1)D$:

$$Z_N(L) = \frac{L}{N} \int_{D}^{L-(N-1)D} dx_2 \int_{x_2+D}^{L-(N-2)D} dx_3 \cdots \int_{x_{N-1}+D}^{L-D} dx_N = \frac{L(L-ND)^{N-1}}{N!}; \tag{2}$$

the prefactor L here comes from integrating x_1 over the circumference of the ring, while the factor $1/N$ delabels that final integral. The Helmholtz free energy in the thermodynamic limit turns out to be

$$A(N, L, T) = -NkT \ln\left(\frac{L-ND}{N\lambda}\right) - NkT, \tag{3}$$

where λ is the thermal deBroglie wavelength. The pressure is then given by

$$P = -\left(\frac{\partial A}{\partial L}\right)_{T,N} = \frac{nkT}{1-nD}, \tag{4}$$

where n $(= N/L)$ is the one-dimensional number density. This is equivalent to the pressure of an ideal gas of N particles in a free volume $L - ND$. The isothermal compressibility is

$$\kappa_T = \frac{1}{n}\left(\frac{\partial n}{\partial p}\right)_T = \frac{(1-nD)^2}{nkT} = \kappa_T^{\text{ideal}}(1-nD)^2 \,. \tag{5}$$

The pair correlation function for the particles on the ring can also be determined exactly (M. Foss-Feig, unpublished). This is accomplished by integrating over all configurations in which particle 1 is fixed at the origin and, in succession, each of the other particles is fixed at position x (if possible). This gives

$$g(x) = \sum_{j=1}^{N-1} g_j(x). \tag{6}$$

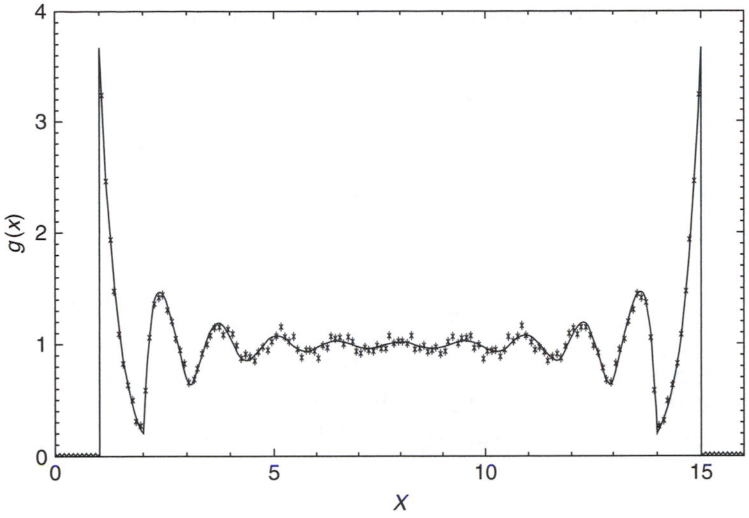

FIGURE 13.1 The pair correlation function for a system of 12 hard disks on a ring. The scaled number density $nD = 0.75$. The solid line represents the exact solution from equations (6) and (7), as compared to a Monte Carlo simulation.

where $g_j(x)$ is defined on the range $jD \leq x \leq L - (N-j)D$ by

$$g_j(x) = \left(\frac{L(N-1)!}{N(L-ND)^{N-1}}\right)\left(\frac{(x-jD)^{j-1}(L-x-(N-j)D)^{N-1-j}}{(j-1)!(N-1-j)!}\right); \tag{7}$$

see Figure 13.1. In the thermodynamic limit, the correlation function becomes

$$g(x) = \sum_{j=1}^{\infty} g_j(x), \tag{8}$$

where $g_j(x)$ is defined on the range $jD \leq x < \infty$ by

$$g_j(x) = \frac{(\beta P(x-jD))^{j-1}\exp\left(-\beta P(x-jD)\right)}{(1-nD)(j-1)!}, \tag{9}$$

where $\beta P = n/(1-nD)$; see Sells, Harris, and Guth (1953). Using equation (10.7.12) gives the correct virial equation of state pressure

$$\frac{P}{nkT} = 1 + nDg(D^+) = \frac{1}{1-nD}; \tag{10}$$

where $g(D^+)$ is the pair correlation function at contact; see Problem 13.28.

13.1.B Isobaric ensemble of a one-dimensional fluid

Takahashi (1942) has shown that a one-dimensional system of particles that interact with nearest neighbors via pair potential $u(r)$ can be analyzed analytically using an isobaric ensemble. The potential energy of the system can be written as

$$U = \sum_{j=1}^{N+1} u(x_j - x_{j-1}), \tag{11}$$

where the left and right walls are treated as particles fixed at $x_0 = 0$ and $x_{N+1} = L$. The isobaric partition function is given by

$$Y_N(P,T) = \frac{1}{\lambda} \int_0^\infty \exp(-\beta PL) Q_N(L,T)\, dL, \tag{12}$$

where λ is the thermal deBroglie wavelength. Equation (12) can be factorized in terms of integrals over the distances between nearest neighbors $y_i\,(= x_i - x_{i-1})$:

$$Y_N(P,T) = \left(\frac{1}{\lambda} \int_0^\infty \exp(-\beta P y - \beta u(y))\, dy \right)^{N+1} = Y_1(p,T)^{N+1}. \tag{13}$$

The bulk Gibbs free energy is then given by

$$G(N,P,T) = -NkT \ln\left(Y_1(P,T)\right), \tag{14}$$

and the average system size at pressure P by

$$\left(\frac{\partial G}{\partial P}\right)_{T,N} = L = N\langle y\rangle, \tag{15}$$

where

$$\langle y\rangle = \frac{\int_0^\infty y \exp[-\beta P y - \beta u(y)]\, dy}{\int_0^\infty \exp[-\beta P y - \beta u(y)]\, dy} \tag{16}$$

is the average nearest-neighbor distance between the particles. The isothermal compressibility,

$$\kappa_T = \frac{-1}{L}\left(\frac{\partial L}{\partial P}\right)_{T,N} = \frac{N}{kTL}\left(\langle y^2\rangle - \langle y\rangle^2\right), \tag{17}$$

is proportional to the variance of the nearest-neighbor distances. It is now easy to show that one-dimensional models cannot form a long-range ordered lattice. The average distance between two particles labeled by i and j is $\langle x_i - x_j\rangle = (i-j)\langle y\rangle$ but the variance is given by $\langle (x_i - x_j)^2\rangle = |i-j|(\langle y^2\rangle - \langle y\rangle^2)$. Therefore, if a chosen particle is located on a particular lattice site, then a particle m sites away will on average be separated from it by m lattice spacings, but the variance of this particle's position from that site location grows linearly with m.

In the thermodynamic limit, the structure factor $S(k)$, equation (10.7.18), can be written in the form

$$S(k) = \left\langle \sum_{j=-\infty}^{\infty} e^{ik(x_j - x_0)} \right\rangle. \tag{18}$$

Since $(x_j - x_0) = \sum_{i=1}^{j} y_i$, the structure factor can be summed exactly to give

$$S(k) = -1 + \sum_{j=0}^{\infty} z^j + \sum_{j=0}^{\infty} (z^*)^j = \frac{1 - |z|^2}{1 + |z|^2 - z - z^*}, \tag{19}$$

where

$$z = \left\langle e^{iky} \right\rangle = \frac{\int_0^\infty \exp\left[-\beta P y - \beta u(y) + iky\right] dy}{\int_0^\infty \exp\left[-\beta P y - \beta u(y)\right] dy}, \tag{20}$$

and z^* is the complex conjugate of z. The fluctuation-compressibility relation now gives: $S(k \to 0) = \kappa_T / \kappa_T^{\text{ideal}} = \left(\langle y^2 \rangle - \langle y \rangle^2\right) / \langle y \rangle^2$. For the particular case of hard spheres, the pressure is given by equation (4) and the structure factor is given by

$$S(k) = \frac{(kD)^2}{(kD)^2 + 2(\beta P D)^2 (1 - \cos(kD)) + 2(\beta P D)(kD)\sin(kD)}; \tag{21}$$

see Figure 13.2. Equation (10.7.20a), applied to the hard sphere pair correlation function in equation (9), also gives equation (21).

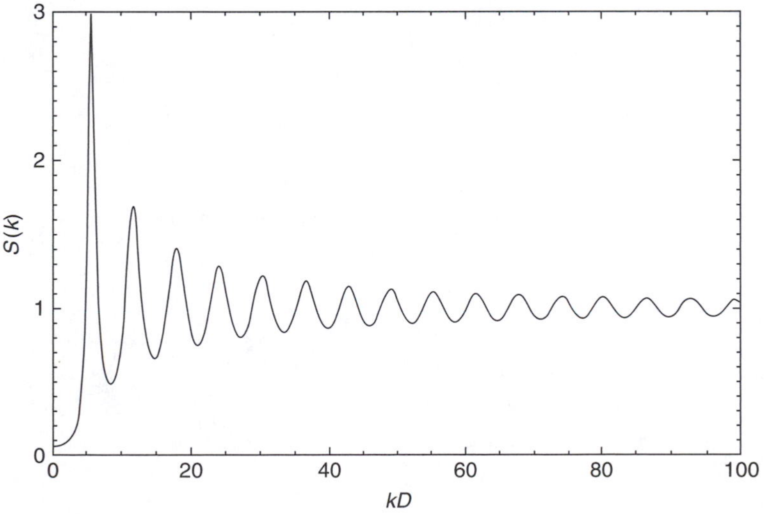

FIGURE 13.2 The structure factor $S(k)$ for a system of hard spheres on a line at density $nD = 0.75$. The structure factor at $k = 0$ is $S(0) = (1 - nD)^2 = \kappa_T / \kappa_T^{\text{ideal}}$.

13.2 The Ising model in one dimension

In this section we present an exact treatment of the Ising model in one dimension. This is important for several reasons. First of all, there do exist phenomena, such as adsorption on a linear polymer or on a protein chain, the elastic properties of fibrous proteins, and so on, that can be looked upon as one-dimensional nearest-neighbor problems. Secondly, it helps us evolve mathematical techniques for treating lattices in higher dimensions, which is essential for understanding the critical behavior of a variety of physical systems met with in nature. Thirdly, it enables us to estimate the status of the Bethe approximation as a "possible" theory of the Ising model, for it demonstrates mathematically that, at least in one dimension, this approximation leads to exact results.

In a short paper published in 1925, Ising himself gave an exact solution to this problem in one dimension. He employed a combinatorial approach that has by now been superseded by other approaches. Here we shall follow the *transfer matrix* method, first introduced by Kramers and Wannier (1941). In the one-dimensional case, this method worked with immediate success. Three years later, in 1944, it became, through Onsager's ingenuity, the first method to treat successfully the *field-free* Ising model in two dimensions. To apply this method, we replace the actual lattice by one having the topology of a closed, endless structure; thus, in the one-dimensional case we replace the straight, open chain by a curved one such that the Nth spin becomes a neighbor of the first (see Figure 13.3). This replacement eliminates the inconvenient end effects; it does not, however, alter the thermodynamic properties of the (infinitely long) chain. The important advantage of this replacement is that it enables us to write the Hamiltonian of the system,

$$H_N\{\sigma_i\} = -J\sum_{\text{n.n.}}\sigma_i\sigma_j - \mu B\sum_{i=1}^{N}\sigma_i, \tag{1}$$

in a symmetrical form, namely

$$H_N\{\sigma_i\} = -J\sum_{i=1}^{N}\sigma_i\sigma_{i+1} - \frac{1}{2}\mu B\sum_{i=1}^{N}(\sigma_i + \sigma_{i+1}), \tag{2}$$

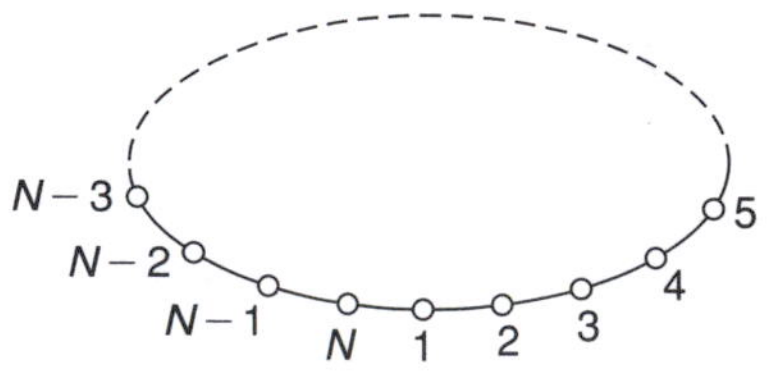

FIGURE 13.3 An Ising chain with a closed, endless structure.

because $\sigma_{N+1} \equiv \sigma_1$. The partition function of the system is then given by

$$Q_N(B,T) = \sum_{\sigma_1=\pm1} \cdots \sum_{\sigma_N=\pm1} \exp\left[\beta \sum_{i=1}^{N} \{J\sigma_i\sigma_{i+1} + \frac{1}{2}\mu B(\sigma_i + \sigma_{i+1})\}\right] \tag{3a}$$

$$= \sum_{\sigma_1=\pm1} \cdots \sum_{\sigma_N=\pm1} \langle\sigma_1|\boldsymbol{P}|\sigma_2\rangle\langle\sigma_2|\boldsymbol{P}|\sigma_3\rangle \cdots \langle\sigma_{N-1}|\boldsymbol{P}|\sigma_N\rangle\langle\sigma_N|\boldsymbol{P}|\sigma_1\rangle, \tag{3b}$$

where $\boldsymbol{P}$ denotes an operator with matrix elements

$$\langle\sigma_i|\boldsymbol{P}|\sigma_{i+1}\rangle = \exp\left[\beta\left\{J\sigma_i\sigma_{i+1} + \frac{1}{2}\mu B(\sigma_i + \sigma_{i+1})\right\}\right],$$

that is,

$$(\boldsymbol{P}) = \begin{pmatrix} e^{\beta(J+\mu B)} & e^{-\beta J} \\ e^{-\beta J} & e^{\beta(J-\mu B)} \end{pmatrix}. \tag{4}$$

According to the rules of matrix algebra, the summations over the various σ_i in equation (3b) lead to the simple result

$$Q_N(B,T) = \sum_{\sigma_1=\pm1} \langle\sigma_1|\boldsymbol{P}^N|\sigma_1\rangle = \text{Trace}\,(\boldsymbol{P}^N) = \lambda_1^N + \lambda_2^N, \tag{5}$$

where λ_1 and λ_2 are the eigenvalues of the matrix $\boldsymbol{P}$. These eigenvalues are given by the equation

$$\begin{vmatrix} e^{\beta(J+\mu B)} - \lambda & e^{-\beta J} \\ e^{-\beta J} & e^{\beta(J-\mu B)} - \lambda \end{vmatrix} = 0, \tag{6}$$

that is, by

$$\lambda^2 - 2\lambda e^{\beta J}\cosh(\beta\mu B) + 2\sinh(2\beta J) = 0. \tag{7}$$

One readily obtains

$$\begin{pmatrix} \lambda_1 \\ \lambda_2 \end{pmatrix} = e^{\beta J}\cosh(\beta\mu B) \pm \{e^{-2\beta J} + e^{2\beta J}\sinh^2(\beta\mu B)\}^{1/2}. \tag{8}$$

Quite generally, $\lambda_2 < \lambda_1$; so, $(\lambda_2/\lambda_1)^N \to 0$ as $N \to \infty$. Thus, it is only the *larger* eigenvalue, λ_1, that determines the major physical properties of the system in the thermodynamic limit; see equation (5). It follows that

$$\frac{1}{N}\ln Q_N(B,T) \approx \ln\lambda_1 \tag{9}$$

$$= \ln[e^{\beta J}\cosh(\beta\mu B) + \{e^{-2\beta J} + e^{2\beta J}\sinh^2(\beta\mu B)\}^{1/2}]. \tag{10}$$

The Helmholtz free energy then turns out to be

$$A(B,T) = -NJ - NkT\ln[\cosh(\beta\mu B) + \{e^{-4\beta J} + \sinh^2(\beta\mu B)\}^{1/2}]. \tag{11}$$

The various other properties of the system follow readily from equation (11). Thus,

$$U(B,T) \equiv -T^2 \frac{\partial}{\partial T}\left(\frac{A}{T}\right) = -NJ - \frac{N\mu B \sinh(\beta\mu B)}{\{e^{-4\beta J} + \sinh^2(\beta\mu B)\}^{1/2}}$$
$$+ \frac{2NJe^{-4\beta J}}{[\cosh(\beta\mu B) + \{e^{-4\beta J} + \sinh^2(\beta\mu B)\}^{1/2}]\{e^{-4\beta J} + \sinh^2(\beta\mu B)\}^{1/2}}, \tag{12}$$

from which the specific heat can be derived, and

$$\overline{M}(B,T) \equiv -\left(\frac{\partial A}{\partial B}\right)_T = \frac{N\mu \sinh(\beta\mu B)}{\{e^{-4\beta J} + \sinh^2(\beta\mu B)\}^{1/2}}, \tag{13}$$

from which the susceptibility can be derived.

Right away we note that, as $B \to 0, \overline{M}$ (for all finite β) $\to 0$. This rules out the possibility of spontaneous magnetization, and hence of a phase transition, at any finite temperature T. Of course, at $T = 0$, $\overline{M}$ (for any value of B) is equal to the saturation value $N\mu$, which implies perfect order in the system. This means that there is, after all, a phase transition at a critical temperature T_c, which coincides with absolute zero!

Figure 13.4 shows the degree of magnetization, $\overline{M}$, of the lattice as a function of the parameter $(\beta\mu B)$ for different values of (βJ). For $J = 0$, we have the *paramagnetic* result $\overline{M} = N\mu \tanh(\beta\mu B)$; compare to equation (3.9.27). A positive J enhances magnetization and, in turn, leads to a faster approach toward saturation. As $\beta J \to \infty$, the magnetization curve becomes a step function — indicative of a singularity at $T = 0$. The low-field susceptibility of the system is given by the initial slope of the magnetization curve; one obtains

$$\chi_0(T) = \frac{N\mu^2}{kT} e^{2J/kT}, \tag{14}$$

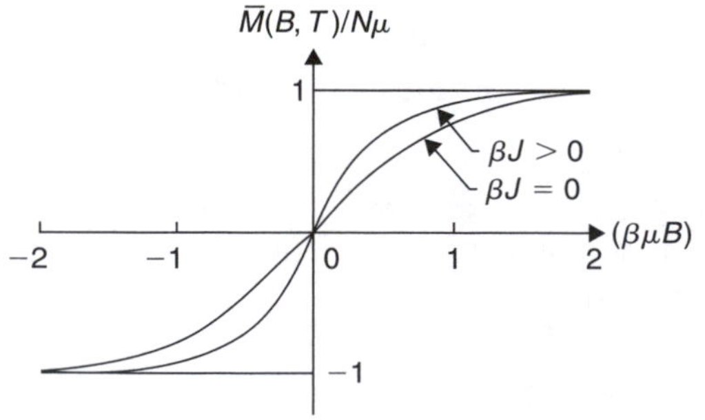

FIGURE 13.4 The degree of magnetization of an Ising chain as a function of the parameter $(\beta\mu B)$.

which diverges as $T \to 0$. It should be noted that the singularity here is *not* of the power-law type; it is exponential instead.

The zero-field energy and the zero-field specific heat of the system follow from equation (12); one gets

$$U_0(T) = -NJ\tanh(\beta J) \tag{15}$$

and

$$C_0(T) = Nk(\beta J)^2 \operatorname{sech}^2(\beta J). \tag{16}$$

Figure 13.5 shows the variation of the specific heat C_0 as a function of temperature. Although it passes through a maximum, C_0 is a smooth function of T, vanishing as $T \to 0$. Note that equations (15) and (16) are identical to the corresponding equations, (12.6.29) and (12.6.30), of the Bethe approximation, with coordination number 2, for which $T_c = 0$. It turns out that for a one-dimensional chain the Bethe approximation, in fact, yields *exact* results; for a fuller demonstration of this, see Problem 13.3.

At this stage it seems instructive to express the free energy of the system, near its critical point, in a *scaled* form, as in Section 12.10. Unfortunately, there is a problem here. Since $T_c = 0$, the conventional definition, $t = (T - T_c)/T_c$, does not work. A closer look at equations (11) and (14), however, suggests that we may adopt instead the definition

$$t = e^{-pJ/kT} \quad (p > 0) \tag{17}$$

so that, as $T \to T_c, t \to 0$ as desired while for temperatures close to T_c, t is much less than unity. The definition of h remains the same, namely $\mu B/kT$. The free energy function $(A + NJ)/NkT$ then takes the form

$$\psi^{(s)}(t,h) = -\ln[\cosh h + (t^{4/p} + \sinh^2 h)^{1/2}] \tag{18a}$$

$$\approx -(t^{4/p} + h^2)^{1/2} \quad (t, h \ll 1), \tag{18b}$$

which may be written in the scaled form

$$\psi^{(s)}(t,h) \approx t^{2/p} f(h/t^{2/p}). \tag{19}$$

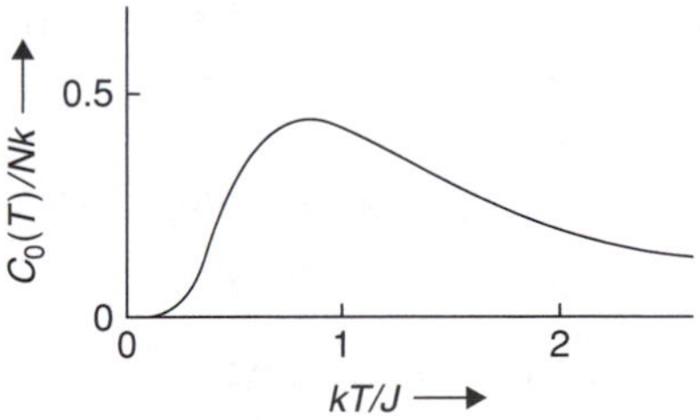

FIGURE 13.5 The zero-field specific heat of an Ising chain as a function of temperature.

At the same time, equation (14) becomes

$$\frac{\chi_0(T)}{(N\mu^2/kT)} \approx t^{-2/p}. \tag{20}$$

Comparing these results with the scaling formulation of Section 12.10, we infer that for this model

$$\alpha = 2 - 2/p, \quad \Delta = 2/p, \quad \text{and} \quad \gamma = 2/p, \tag{21}$$

in conformity with the exponent relation (12.10.14b). Note that the exponents β and δ for this model cannot be defined in the normal, conventional sense. One may, however, write equation (13) in the form

$$m = \sinh h/(t^{4/p} + \sinh^2 h)^{1/2} \tag{22a}$$

$$\approx h/(t^{4/p} + h^2)^{1/2} \quad (t, h \ll 1) \tag{22b}$$

$$= -t^0 f'(h/t^{2/p}), \tag{22c}$$

suggesting that β may formally be taken as zero. At the same time, since

$$m|_{t=0} = 1, \tag{23}$$

which is $\sim h^0$, the exponent δ may formally be taken as infinite.

We now study spin–spin correlations in the Ising chain. For this, we set $B = 0$ but at the same time generalize the interaction parameter J to become site-dependent (the reason for which will become clear soon). The partition function of the system is then given by

$$Q_N(T) = \sum_{\sigma_1=\pm1} \cdots \sum_{\sigma_N=\pm1} \prod_i e^{\beta J_i \sigma_i \sigma_{i+1}}; \tag{24}$$

compare to equation (3a). With $B = 0$, it is simpler to work with an *open* chain, which has only $(N - 1)$ nearest-neighbor pairs; the advantage of this choice is that in the summand of (24) the variables σ_1 and σ_N appear only once! A summation over either of these can be carried out easily; doing this over σ_N, we have

$$\sum_{\sigma_N=\pm1} e^{\beta J_{N-1}\sigma_{N-1}\sigma_N} = 2\cosh(\beta J_{N-1}\sigma_{N-1}) = 2\cosh(\beta J_{N-1}), \tag{25}$$

regardless of the sign of σ_{N-1}. We thus obtain the recurrence relation

$$Q_N(T; J_1, \ldots, J_{N-1}) = 2\cosh(\beta J_{N-1}) Q_{N-1}(T; J_1, \ldots, J_{N-2}). \tag{26}$$

By iteration, we get

$$Q_N(T) = \prod_{i=1}^{N-1} \{2\cosh(\beta J_i)\} \sum_{\sigma_1=\pm 1} 1 = 2^N \prod_{i=1}^{N-1} \cosh(\beta J_i), \tag{27}$$

so that

$$\frac{1}{N}\ln Q_N(T) = \ln 2 + \frac{1}{N}\sum_{i=1}^{N-1} \ln\cosh(\beta J_i), \tag{28}$$

which may be compared with equation (9) — remembering that, in the absence of the field, $\lambda_1 = 2\cosh(\beta J)$. We are now ready to calculate the correlation function, $g(r)$, of the Ising chain.

It is straightforward to see from equation (24) that

$$\overline{\sigma_k\sigma_{k+1}} = \frac{1}{Q_N}\left(\frac{1}{\beta}\frac{\partial}{\partial J_k}\right) Q_N = \left(\frac{1}{\beta}\frac{\partial}{\partial J_k}\right) \ln Q_N. \tag{29}$$

Substituting from equation (28), and remembering that $\overline{\sigma}_k = 0$ at all finite temperatures, we obtain for the *nearest-neighbor correlation function*

$$g_k(n.n.) = \overline{\sigma_k\sigma_{k+1}} = \tanh(\beta J_k). \tag{30}$$

For a pair of spins separated by r lattice constants, we get

$$\begin{aligned} g_k(r) = \overline{\sigma_k\sigma_{k+r}} &= \overline{(\sigma_k\sigma_{k+1})(\sigma_{k+1}\sigma_{k+2})\ldots(\sigma_{k+r-1}\sigma_{k+r})} \quad (\text{since all } \sigma_i^2 = 1) \\ &= \frac{1}{Q_N}\left(\frac{1}{\beta}\frac{\partial}{\partial J_k}\right)\left(\frac{1}{\beta}\frac{\partial}{\partial J_{k+1}}\right)\cdots\left(\frac{1}{\beta}\frac{\partial}{\partial J_{k+r-1}}\right) Q_N \\ &= \prod_{i=k}^{k+r-1} \tanh(\beta J_i). \end{aligned} \tag{31}$$

Reverting to a common J, we obtain the desired result

$$g(r) = \tanh^r(\beta J), \tag{32}$$

which may be written in the standard form

$$g(r) = e^{-r/\xi}, \quad \text{with} \quad \xi = \left[\ln\coth(\beta J)\right]^{-1}. \tag{33a, b}$$

For $\beta J \gg 1$,

$$\xi \approx \frac{1}{2}e^{2\beta J}, \tag{34}$$

which diverges as $T \to 0$. In terms of the variable t, as defined in equation (17),

$$\xi \sim t^{-2/p} \quad (t \ll 1), \tag{35}$$

giving $\nu = 2/p$. And since our $g(r)$ does not contain any power of r, we infer that $(d-2+\eta)=0$ — giving $\eta = 1$; one may check that the same result follows from equation (12.12.10) or (12.12.11). In passing, we note that, regardless of the choice of the number p in defining t, we have for this model

$$\gamma = \nu = 2 - \alpha. \tag{36}$$

We further note that, since $d = 1$ here, the hyperscaling relation, $d\nu = 2 - \alpha$, is also obeyed.

Finally, we observe that expression (33b) for ξ is in conformity with the general result

$$\xi^{-1} = \ln(\lambda_1/\lambda_2), \tag{37}$$

where λ_1 is the largest eigenvalue of the transfer matrix $\boldsymbol{P}$ of the problem and λ_2 the next largest; for a derivation of this result, see Section 5.3 of Yeomans (1992). In our case, $\lambda_1 = 2\cosh(\beta J)$ and $\lambda_2 = 2\sinh(\beta J)$, see equation (8) with $B = 0$, and hence expression (33b) for ξ.

13.3 The *n*-vector models in one dimension

We now consider a generalization of the Ising chain in which the spin variable $\boldsymbol{\sigma}_i$ is an n-dimensional vector of magnitude unity, whose components can vary *continuously* over the range -1 to $+1$; in contrast, the Ising spin σ_i could have only a *discrete* value, $+1$ or -1. We shall see that the vector models (with $n \geq 2$), while differing quantitatively from one another, differ rather qualitatively from the scalar models (for which $n=1$). While some of these qualitative differences will show up in the present study, more will become evident in higher dimensions. Here we follow a treatment due to Stanley (1969a,b) who first solved this problem for general n.

Once again we employ an *open* chain composed of N spins constituting $(N-1)$ nearest-neighbor pairs. The Hamiltonian of the system, in zero field, is given by

$$H_N\{\boldsymbol{\sigma}_i\} = -\sum_{i=1}^{N-1} J_i \boldsymbol{\sigma}_i \cdot \boldsymbol{\sigma}_{i+1}. \tag{1}$$

We assume our spins to be classical, so we do not have to worry about the commutation properties of their components. And since the components $\sigma_{i\alpha}$ $(\alpha = 1, \ldots, n)$ of each spin vector $\boldsymbol{\sigma}_i$ are now continuous variables, the partition function of the system will involve integrations, rather than summations, over these variables. Associating equal *a priori* probabilities with solid angles of equal magnitude in the n-dimensional spin-vector space, we may write

$$Q_N = \int \frac{d\Omega_1}{\Omega(n)} \cdots \frac{d\Omega_N}{\Omega(n)} \prod_{i=1}^{N-1} e^{\beta J_i \boldsymbol{\sigma}_i \cdot \boldsymbol{\sigma}_{i+1}}, \tag{2}$$

where $\Omega(n)$ is the total solid angle in an n-dimensional space; see equation (7b) of Appendix C, which gives

$$\Omega(n) = 2\pi^{n/2} / \Gamma(n/2). \tag{3}$$

We first carry out integration over $\boldsymbol{\sigma}_N$, keeping the other $\boldsymbol{\sigma}_i$ fixed. The relevant integral to do is

$$\frac{1}{\Omega(n)} \int e^{\beta J_{N-1} \boldsymbol{\sigma}_{N-1} \cdot \boldsymbol{\sigma}_N} d\Omega_N. \tag{4}$$

For $\boldsymbol{\sigma}_N$ we employ spherical polar coordinates, with polar axis in the direction of $\boldsymbol{\sigma}_{N-1}$, while for $d\Omega_N$ we use expression (9) of Appendix C. Integration over angles other than the polar angle θ yields a factor of

$$2\pi^{(n-1)/2} / \Gamma\{(n-1)/2\}. \tag{5}$$

The integral over the polar angle is

$$\int_0^{\pi} e^{\beta J_{N-1} \cos\theta} \sin^{n-2}\theta \, d\theta = \frac{\pi^{1/2} \Gamma\{(n-1)/2\}}{(\frac{1}{2}\beta J_{N-1})^{(n-2)/2}} I_{(n-2)/2}(\beta J_{N-1}), \tag{6}$$

where $I_\mu(x)$ is a modified Bessel function; see Abramowitz and Stegun (1964), formula 9.6.18. Combining (3), (5), and (6), we obtain for (4) the expression

$$\frac{\Gamma(n/2)}{(\frac{1}{2}\beta J_{N-1})^{(n-2)/2}} I_{(n-2)/2}(\beta J_{N-1}), \tag{7}$$

regardless of the direction of $\boldsymbol{\sigma}_{N-1}$. By iteration, we get

$$Q_N = \prod_{i=1}^{N-1} \frac{\Gamma(n/2)}{(\frac{1}{2}\beta J_i)^{(n-2)/2}} I_{(n-2)/2)}(\beta J_i); \tag{8}$$

the last integration, over $d\Omega_1$, gave simply a factor of unity.

Expression (8) is valid for *all* n — including $n = 1$, for which it gives: $Q_N = \prod_i \cosh(\beta J_i)$. This last result differs from expression (13.2.27) by a factor of 2^N; the reason for this difference lies in the fact that the Q_N of the present study is *normalized* to go to unity as the βJ_i go to zero [see equation (2)] whereas the Q_N of the preceding section, being a sum over 2^N discrete states [see equation (13.2.24)] goes to 2^N instead. This difference is important in the evaluation of the entropy of the system; it is of no consequence for the calculations that follow.

First of all we observe that the partition function Q_N is analytic at all β — except possibly at $\beta = \infty$ where the singularity of the problem is expected to lie. Thus, no long-range order is expected to appear at any finite temperature T — except at $T = 0$ where, of course, perfect order is supposed to prevail. In view of this, the correlation function for the nearest-neighbor pair $(\boldsymbol{\sigma}_k, \boldsymbol{\sigma}_{k+1})$ is simply $\overline{\boldsymbol{\sigma}_k \cdot \boldsymbol{\sigma}_{k+1}}$ and is given by, see equations (2) and (8),

$$g_k(n.n.) = \frac{1}{Q_N}\left(\frac{1}{\beta}\frac{\partial}{\partial J_k}\right) Q_N = \frac{I_{n/2}(\beta J_k)}{I_{(n-2)/2}(\beta J_k)}. \tag{9}$$

The internal energy of the system turns out to be

$$U_0 \equiv -\frac{\partial}{\partial \beta}(\ln Q_N) = -\sum_{i=1}^{N-1} J_i \frac{I_{n/2}(\beta J_i)}{I_{(n-2)/2}(\beta J_i)}; \tag{10}$$

not surprisingly, U_0 is simply a sum of the expectation values of the nearest-neighbor interaction terms $-J_i \boldsymbol{\sigma}_i \cdot \boldsymbol{\sigma}_{i+1}$, which is identical to a sum of the quantities $-J_i g_i(n.n.)$ over all nearest-neighbor pairs in the system.

The calculation of $g_k(r)$ is somewhat tricky because of the vector character of the spins, but things are simplified by the fact that we are dealing with a one-dimensional system only. Let us consider the trio of spins $\boldsymbol{\sigma}_k, \boldsymbol{\sigma}_{k+1}$ and $\boldsymbol{\sigma}_{k+2}$, and suppose for a moment that our spins are three-dimensional vectors; our aim is to evaluate $\overline{\boldsymbol{\sigma}_k \cdot \boldsymbol{\sigma}_{k+2}}$. We choose spherical polar coordinates with polar axis in the direction of $\boldsymbol{\sigma}_{k+1}$; let the direction of $\boldsymbol{\sigma}_k$ be defined by the angles (θ_0, ϕ_0) and that of $\boldsymbol{\sigma}_{k+2}$ by (θ_2, ϕ_2). Then

$$\overline{\boldsymbol{\sigma}_k \cdot \boldsymbol{\sigma}_{k+2}} = \overline{\cos\theta(k, k+2)} = \overline{\cos\theta_0 \cos\theta_2 + \sin\theta_0 \sin\theta_2 \cos(\phi_0 - \phi_2)}. \tag{11}$$

Now, with $\boldsymbol{\sigma}_{k+1}$ fixed, spins $\boldsymbol{\sigma}_k$ and $\boldsymbol{\sigma}_{k+2}$ will orient themselves *independently of one another* because, apart from $\boldsymbol{\sigma}_{k+1}$, there is no other channel of interaction between them. Thus, the pairs of angles (θ_0, ϕ_0) and (θ_2, ϕ_2) vary independently of one another; this makes $\overline{\cos(\phi_0 - \phi_2)} = 0$ and $\overline{\cos\theta_0 \cos\theta_2} = \overline{\cos(\theta_0)}\ \overline{\cos(\theta_2)}$. It follows that

$$\overline{\boldsymbol{\sigma}_k \cdot \boldsymbol{\sigma}_{k+2}} = \overline{\boldsymbol{\sigma}_k \cdot \boldsymbol{\sigma}_{k+1}}\ \overline{\boldsymbol{\sigma}_{k+1} \cdot \boldsymbol{\sigma}_{k+2}}. \tag{12}$$

Extending this argument to general n and to a segment of length r, we get

$$g_k(r) = \prod_{i=k}^{k+r-1} g_i(\text{n.n.}) = \prod_{i=k}^{k+r-1} I_{n/2}(\beta J_i)/I_{(n-2)/2}(\beta J_i). \tag{13}$$

With a common J, equations (9), (10), and (13) take the form

$$g(\text{n.n.}) = I_{n/2}(\beta J)/I_{(n-2)/2}(\beta J), \tag{14}$$

$$U_0 = -(N-1)J\, I_{n/2}(\beta J)/I_{(n-2)/2}(\beta J) \tag{15}$$

and

$$g(r) = \{I_{n/2}(\beta J)/I_{(n-2)/2}(\beta J)\}^r. \tag{16}$$

The last result here may be written in the standard form $e^{-r/\xi}$, with

$$\xi = [\ln\{I_{(n-2)/2}(\beta J)/I_{n/2}(\beta J)\}]^{-1}. \tag{17}$$

For $n = 1$, we have: $I_{1/2}(x)/I_{-1/2}(x) = \tanh x$; the results of the preceding section are then correctly recovered.

For a study of the low-temperature behavior, where $\beta J \gg 1$, we invoke the asymptotic expansion

$$I_\mu(x) = \frac{e^x}{\surd(2\pi x)}\left[1 - \frac{4\mu^2 - 1}{8x} + \cdots\right] \quad (x \gg 1), \tag{18}$$

with the result that

$$g(\text{n.n.}) \approx 1 - \frac{n-1}{2\beta J}, \tag{14a}$$

$$U_0 \approx -(N-1)J\left[1 - \frac{n-1}{2\beta J}\right] \tag{15a}$$

and

$$\xi \approx \frac{2\beta J}{n-1} \sim T^{-1}. \tag{17a}$$

Clearly, the foregoing results hold only for $n \geq 2$; for $n = 1$, the asymptotic expansion (18) is no good because it yields the same result for $\mu = \frac{1}{2}$ as for $\mu = -\frac{1}{2}$. In that case one is obliged to use the closed form result, $g(n.n.) = \tanh(\beta J)$, which for $\beta J \gg 1$ gives

$$g(\text{n.n.}) \approx 1 - 2e^{-2\beta J} \tag{14b}$$

$$U_0 \approx -(N-1)J[1 - 2e^{-2\beta J}] \tag{15b}$$

and

$$\xi \approx \frac{1}{2}e^{2\beta J}, \tag{17b}$$

in complete agreement with the results of the preceding section. For completeness, we have for the low-temperature specific heat of the system

$$C_0 \approx (N-1)\begin{cases} \frac{1}{2}(n-1)k & \text{for} \quad n \geq 2 \qquad (19a) \\ 4k(\beta J)^2 e^{-2\beta J} & \text{for} \quad n = 1. \qquad (19b) \end{cases}$$

The most obvious distinction between one-dimensional models with *continuous* symmetry ($n \geq 2$) and those with *discrete* symmetry ($n = 1$) is in relation to the nature of the singularity at $T = 0$. While in the case of the former it is a power-law singularity, with critical exponents[1]

$$\alpha = 1, \quad \nu = 1, \quad \eta = 1, \quad \text{and hence} \quad \gamma = 1, \tag{20}$$

in the case of the latter it is an exponential singularity. Nevertheless, by introducing the temperature parameter $t = e^{-p\beta J}$, see equation (13.2.17), we converted this exponential singularity in T into a power-law singularity in t, with

$$\alpha = 2 - 2/p, \quad \gamma = \nu = 2/p, \quad \eta = 1. \tag{21}$$

However, the inherent arbitrariness in the choice of the number p left an ambiguity in the values of these exponents; we now see that by choosing $p = 2$ we can bring exponents (21) in line with (20).

Next we observe that the critical exponents (20) for $n \geq 2$ turn out to be independent of n — a feature that seems peculiar to situations where $T_c = 0$. In higher dimensions, where T_c is finite, the critical exponents do vary with n; for details, see Section 13.7. In any case, the amplitudes always depend on n. In this connection we note that, since each of the N spins comprising the system has n components, the total number of degrees of freedom in this problem is Nn. It seems appropriate that the extensive quantities, such as U_0 and C_0, be divided by Nn, so that they are expressed as *per degree of freedom*. A look at equation (15a) then tells us that our parameter J must be of the form nJ', so that in the thermodynamic limit

$$\frac{U_0}{Nn} \approx -J' + \frac{n-1}{2n}kT \tag{22}$$

and, accordingly,

$$\frac{C_0}{Nn} \approx \frac{n-1}{2n}k. \tag{23}$$

[1]Note that, since $T_c = 0$ here, the assertion that the free energy function $\psi \equiv (A/NkT) \sim (T - T_c)^{2-\alpha}$ implies that, near $T = T_c$ in the present case, $A \sim T^{3-\alpha}$; accordingly, the specific heat $C_0 \sim T^{2-\alpha}$. Comparison with expression (19a) would be inappropriate because C_0, in the limit $T \to 0$, cannot be nonzero; the result quoted in (19a) is an artifice of the model considered, which must somehow be "subtracted away." The next approximation yields a result proportional to the *first* power in T, giving $\alpha = 1$. For a parallel situation, see equation (13.5.35) for the spherical model (which pertains to the case $n = \infty$).

Equation (17a) then becomes

$$\xi \approx \frac{2n}{n-1}\frac{J'}{kT}. \tag{24}$$

Note that the amplitudes appearing in equations (22) through (24) are such that the limit $n \to \infty$ exists; this limit pertains to the so-called *spherical model*, which will be studied in Section 13.5.

Figure 13.6 shows the normalized energy $u_0(= U_0/NnJ')$ as a function of temperature for several values of n, including the limiting case $n = 1$. We note that u_0 (which, in the thermodynamic limit, is equal and opposite to the nearest-neighbor correlation function g(n.n.)) increases monotonically with n — implying that g(n.n.), and hence $g(r)$, decrease monotonically as n increases. This is consistent with the fact that the correlation length ξ also decreases as n increases; see equation (24). The physical reason for this behavior is that an increase in the number of degrees of freedom available to each spin in the system effectively diminishes the degree of correlation among any two of them.

Another feature emerges here that distinguishes vector models ($n \geq 2$) from the scalar model ($n = 1$); this is related to the manner in which the quantity u_0 approaches its ground-state value -1. While for $n = 1$, the approach is quite slow — leading to a vanishing

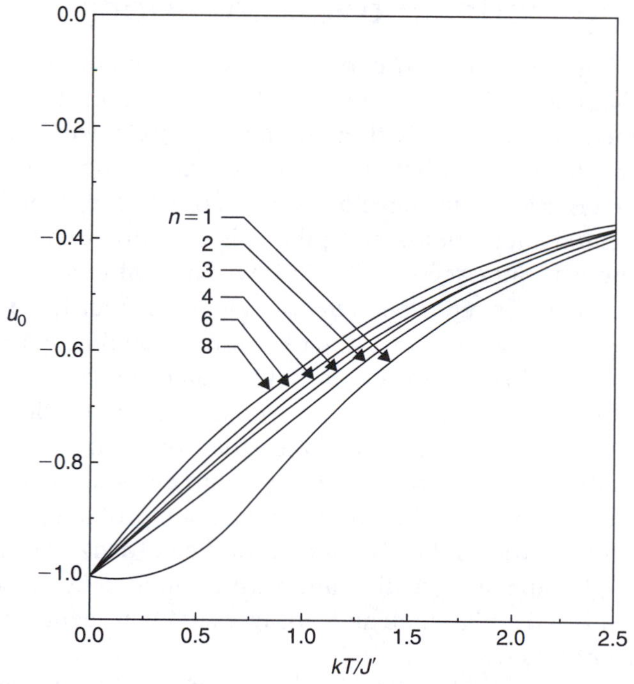

FIGURE 13.6 The normalized energy $u_0(= U_0/NnJ')$ of a one-dimensional chain as a function of the temperature parameter kT/J' for several values of n (after Stanley, 1969a,b). Note that for this *classical* model, the slopes as $T \to 0$ are given by the equipartition theorem for $n-1$ degrees of freedom per spin.

specific heat, see equations (15b) and (19b) — for $n \geq 2$, the approach is essentially linear in T, leading to a *finite* specific heat; see equations (15a) and (19a). This last result violates the third law of thermodynamics, according to which the specific heat of a real system must go to zero as $T \to 0$. The resolution of this dilemma lies in the fact that the low-lying states of a system with *continuous* symmetry ($n \geq 2$) are dominated by long-wavelength excitations, known as *Goldstone modes*, which in the case of a magnetic system assume the form of "spin waves," characterized by a particle-like spectrum: $\omega(k) \sim k^2$. The very low-temperature behavior of the system is primarily governed by these modes, and the thermal energy associated with them is given by

$$U_{\text{thermal}} \sim \int \frac{\hbar\omega}{\exp(\hbar\omega/k_B T) - 1} k^{d-1} dk \sim T^{(d+2)/2}; \tag{25}$$

this results in a specific heat $\sim T^{d/2}$, which indeed is consistent with the third law. For a general account of the Goldstone excitations, see Huang (1987); for their role as "spin waves" in a magnetic system, see Plischke and Bergersen (1989).

For further information on one-dimensional models, see Lieb and Mattis (1966) and Thompson (1972a,b).

13.4 The Ising model in two dimensions

As stated earlier, Ising (1925) himself carried out a combinatorial analysis of the one-dimensional model and found that there was no phase transition at a finite temperature T. This led him to conclude, erroneously though, that his model would not exhibit a phase transition in higher dimensions either. In fact, it was this "supposed failure" of the Ising model that motivated Heisenberg to develop, in 1928, the theory of ferromagnetism based on a more sophisticated interaction among the spins; compare the Heisenberg interaction (12.3.6) with the Ising interaction (12.3.7). It was only after some exploitation of the Heisenberg model that people returned to investigate the properties of the Ising model.

The first exact, quantitative result for the two-dimensional Ising model was obtained by Kramers and Wannier (1941) who successfully located the critical temperature of the system. They were followed by Onsager (1944) who derived an explicit expression for the free energy in zero field and thereby established the precise nature of the specific-heat singularity. These authors employed the *transfer matrix method* that was introduced in Section 13.2 to solve the corresponding one-dimensional problem; its application to the two-dimensional model, even in the absence of the field, turned out to be an extremely difficult task. Although some of these difficulties were softened by subsequent treatments due to Kaufman (1949) and to Kaufman and Onsager (1949), it seemed very natural to look for other simpler approaches.

One such approach was developed by Kac and Ward (1952), later refined by Potts and Ward (1955), in which combinatorial arguments were used to express the partition function of the system as the determinant of a certain matrix A. This method throws special light on the "topological conditions" that give rise to an exact solution in two dimensions but are clearly absent in three dimensions; a particularly lucid account of this method has been given by Baker (1990). In 1960 Hurst and Green introduced yet another approach to investigate the Ising problem; this involved the use of "triangular arrays of quantities closely related to antisymmetric determinants" and became rightly known as the method of Pfaffians. This method applies rather naturally to the study of "configurations of dimer molecules on a given lattice" which, in turn, is closely related to the Ising problem; for details, see Kasteleyn (1963), Montroll (1964), and Green and Hurst (1964). A pedagogical account of the approach through Pfaffians is given in Thompson (1972b), where a comprehensive treatment of the original, algebraic approach can also be found. Another combinatorial solution, which is generally regarded as the simplest, was obtained by Vdovichenko (1965) and by Glasser (1970), and is readily accessible in Stanley (1971). For an exhaustive account of the two-dimensional Ising model, see McCoy and Wu (1973).

We analyze this problem with the help of a combinatorial approach assisted, from time to time, by a graphical representation. The zero-field partition function of the system is given by the familiar expression

$$Q(N,T) = \sum_{\{\sigma_i\}} \prod_{\text{n.n.}} e^{K\sigma_i\sigma_j} \quad (K = J/kT). \tag{1}$$

Our first step consists of carrying out a high-temperature and a low-temperature expansion of the partition function and establishing an intimate relation between the two.

High-temperature expansion

Since the product $(\sigma_i\sigma_j)$ can only be $+1$ or -1, we may write

$$e^{K\sigma_i\sigma_j} = \cosh K + \sigma_i\sigma_j \sinh K = \cosh K(1+\sigma_i\sigma_j v), \quad v = \tanh K. \tag{2}$$

The product over all nearest-neighbor pairs then takes the form

$$\prod_{\text{n.n.}} e^{K\sigma_i\sigma_j} = (\cosh K)^{\mathscr{N}} \prod_{\text{n.n.}} (1+\sigma_i\sigma_j v), \tag{3}$$

$\mathscr{N}$ being the total number of nearest-neighbor pairs on the lattice; for a lattice with periodic boundary conditions, $\mathscr{N} = \frac{1}{2}qN$ where q is the coordination number. The partition function may then be written as

$$Q(N,T) = (\cosh K)^{\mathscr{N}} \sum_{\sigma_1=\pm1} \cdots \sum_{\sigma_N=\pm1} \left[1 + v\sum_{(i,j)} \sigma_i\sigma_j + v^2 \sum_{\substack{(i,j)\ (k,l) \\ (i,j)\neq(k,l)}} \sum \sigma_i\sigma_j\sigma_k\sigma_l + \ldots\right]. \tag{4}$$

Now we represent each product $(\sigma_i\sigma_j)$ appearing in (4) by a "bond connecting sites i and j on the given lattice"; then, each coefficient of v^r in the expansion would be represented by a "graph consisting of r different bonds on the lattice." Figure 13.7 shows all possible graphs, with $r = 1$ and 2, on a square lattice. Notice that in each case we have some of

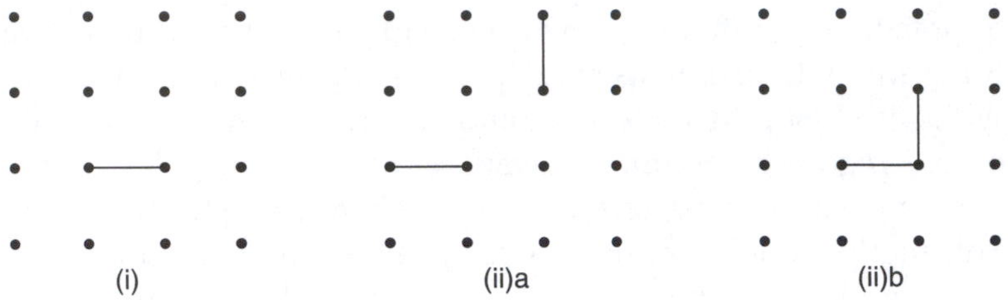

FIGURE 13.7 Graphs with $r = 1$ and $r = 2$ on a square lattice. Graph (i) is for $r = 1$. The $r = 2$ graphs are of two types, ones that do not include a common site as in (ii)a and ones that do include a common site as in (ii)b.

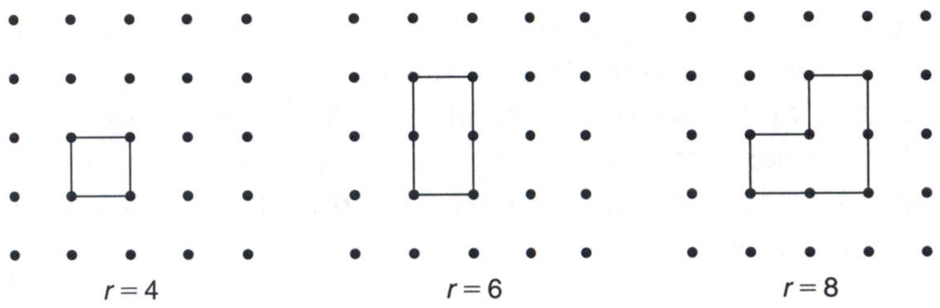

FIGURE 13.8 Examples of closed graphs with r bonds on a square lattice.

the σ_i appearing *only once* in the term, which makes all these terms vanish on summation over $\{\sigma_i\}$. The same is true for $r = 3$. Only when we reach $r = 4$ do we receive a nonvanishing contribution from terms of the type $(\sigma_i\sigma_j\sigma_j\sigma_k\sigma_k\sigma_l\sigma_l\sigma_i) \equiv 1$ which, on summation over $\{\sigma_i\}$, yield a contribution of 2^N each. It is obvious that a nonvanishing term corresponds to a graph in which each vertex is met by an *even* number of bonds — making the graph necessarily a closed one; see Figure 13.8, where some other closed graphs are also shown. In view of these observations, expression (4) may be written as

$$Q(N,T) = 2^N(\cosh K)^{\mathscr{N}} \sum_r n(r)v^r \quad [n(0) = 1], \tag{5}$$

where $n(r)$ is the number of graphs that can be drawn on the given lattice using r bonds such that each vertex of the graph is met by an even number of bonds. For simplicity, we shall refer to these graphs as *closed graphs*. Our problem thus reduces to one of enumerating such graphs on the given lattice.

Since $v = \tanh(J/kT)$, the higher the temperature the smaller the v. Expansion (5) is, therefore, particularly useful at higher temperatures (even though it is exact for all T). As an illustration, we apply this result to a one-dimensional Ising chain. In the case of an *open* chain, no closed graphs are possible, so all we get from (5) is the term with $r = 0$; with $\mathscr{N} = N - 1$, this gives

$$Q(N,T) = 2^N(\cosh K)^{N-1}, \tag{6}$$

which agrees with our previous result (13.2.27). In the case of a *closed* chain, we do have a closed graph — the one with $r = N$; we now get (with $\mathcal{N} = N$)

$$Q(N,T) = 2^N (\cosh K)^N [1 + v^N] = 2^N [(\cosh K)^N + (\sinh K)^N], \tag{7}$$

which agrees with expression (13.2.5), with $(\lambda_1)_{B=0} = 2\cosh K$ and $(\lambda_2)_{B=0} = 2\sinh K$.

Low-temperature expansion

We start with the observation that the ground state of the system consists of all spins aligned in the same direction, with the total energy $E_0 = -J\mathcal{N}$. As one spin is flipped, q *unlike* nearest-neighbor pairs are created at the expense of *like* ones, and the energy of the system increases by an amount $2qJ$. It seems appropriate, therefore, that the Hamiltonian of the system be written in terms of the number, N_{+-}, of unlike nearest-neighbor pairs in the lattice, that is,

$$H(N_{+-}) = -J(N_{++} + N_{--} - N_{+-}) = -J(\mathcal{N} - 2N_{+-}). \tag{8}$$

The partition function of the system may then be written as

$$Q(N,T) = e^{K\mathcal{N}} \sum_r m(r) e^{-2Kr} \quad [m(0) = 1], \tag{9}$$

where $m(r)$ denotes the "number of distinct ways in which the N spins of the lattice can be so arranged as to yield r unlike nearest-neighbor pairs." It is obvious that the first nonzero term in (9), after the one with $r = 0$, would be the one with $r = q$.

A graphical representation of the number $m(r)$ is straightforward. Referring to Figure 13.9, which pertains to a square lattice, we see that each term in expansion (9) can be associated with a closed graph that cordons off region(s) of "up" spins from those of "down" spins, the perimeter of the graph being precisely the number of unlike nearest-neighbor pairs in the lattice for that particular configuration. Our problem then reduces to one of enumerating closed graphs, of appropriate perimeters, that can be drawn on the given lattice.

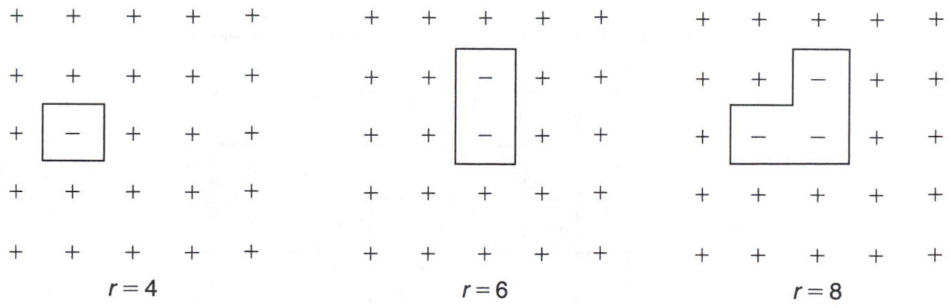

FIGURE 13.9 Graphs cordoning off regions of "up" spins from those of "down" spins, with r *unlike* nearest-neighbor pairs.

Now, since expansion (9) is a power series in the variable e^{-2K} that increases as T increases, this expansion is particularly useful at lower temperatures (even though it is exact for all T). We shall now establish an important relation between the coefficients appearing in expansion (5) and the ones appearing in (9).

The duality transformation

To establish the desired relation we construct a lattice *dual* to the one given. By definition, we draw right bisectors of all the bonds in the lattice, so that the points of intersection of these bisectors become the sites of the new lattice. The resulting lattice may not be similar in structure to the one we started with; for instance, while the dual of a square lattice is itself a square lattice, the dual of a triangular lattice ($q = 6$) is, in fact, a honeycomb lattice ($q = 3$), and vice versa; see Figures 13.10 and 13.11. The argument now runs as follows:

We start with the given lattice on which one of the $n(r)$ closed graphs, with r bonds, is drawn and construct the lattice dual to this one, placing spins of one sign on the sites inside this graph and spins of opposite sign on the sites outside. Then this graph represents precisely a configuration with r unlike nearest-neighbor pairs in the dual lattice and hence qualifies to be counted as one of the $m(r)$ graphs on the dual lattice. Conversely, if we start with one of the $m(r)$ graphs, of perimeter r, representing a configuration with r unlike nearest-neighbor pairs in the original lattice and go through the process of constructing the dual lattice, then this graph will qualify to be one of the $n(r)$ closed graphs, with r bonds, on the dual lattice. In fact, there is a *one-to-one* correspondence between graphs of one kind on the given lattice and graphs of the other kind on the dual lattice; compare Figure 13.8 with Figure 13.9. It follows that

$$n(r) = m_D(r) \quad \text{and} \quad m(r) = n_D(r), \tag{10a, b}$$

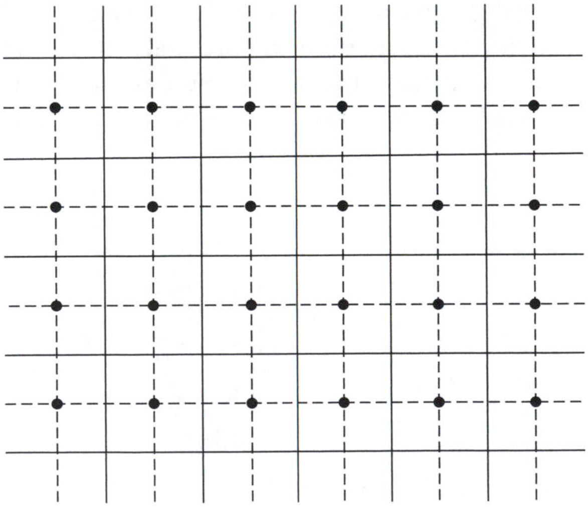

FIGURE 13.10 A square lattice and its dual (which is also square).

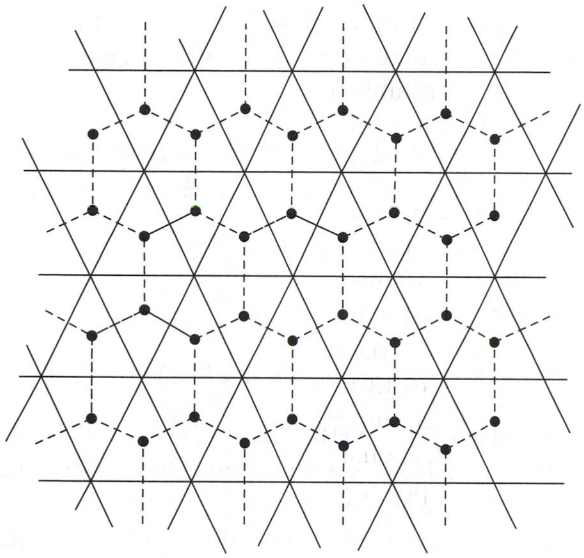

FIGURE 13.11 A honeycomb lattice ($q=3$) and its dual, which is triangular ($q=6$).

where the suffix D refers to the dual lattice.

With relations (10) established, we go back to equation (9) and introduce another temperature variable $K^*(=J/kT^*)$ such that

$$\tanh K^* = e^{-2K}; \tag{11}$$

note that equation (11) can also be written in the symmetrical form

$$\sinh(2K)\sinh(2K^*) = 1. \tag{12}$$

Substituting (10b) and (11) into (9), we get

$$Q(N,T) = e^{K\mathcal{N}} \sum_r n_D(r) v^{*r}, \quad v^* = \tanh K^*. \tag{13}$$

At the same time we apply equation (5) to the *dual* lattice at temperature T^*, to get

$$Q_D(N_D, T^*) = 2^{N_D} (\cosh K^*)^{\mathcal{N}} \sum_r n_D(r) v^{*r}, \tag{14}$$

where $N_D = qN/q_D$; see again Figure 13.11. Comparing (13) and (14), we arrive at the desired relation

$$Q(N,T) = 2^{-N_D} (\sinh K^* \cosh K^*)^{-\mathcal{N}/2} Q_D(N_D, T^*), \tag{15}$$

which relates the partition function of the given lattice at temperature T to that of the dual lattice at temperature T^*. Equation (15) constitutes the so-called *duality transformation.*

Location of the critical point

For a square lattice, which is self-dual, there should be no distinction between Q and Q_D. With $q = 4$, and hence $\mathscr{N} = 2N$, equation (15) becomes

$$Q(N,T) = [\sinh(2K^*)]^{-N} Q(N,T^*), \tag{16}$$

which may also be written as

$$[\sinh(2K)]^{-N/2} Q(N,T) = [\sinh(2K^*)]^{-N/2} Q(N,T^*). \tag{17}$$

It will be noted from equation (11) or (12) that as $T \to \infty$, $T^* \to 0$ and as $T \to 0$, $T^* \to \infty$; equation (17), therefore, establishes a one-to-one correspondence between the high-temperature and the low-temperature values of the partition function of the lattice. It then follows that if there exists a singularity in the partition function at a particular temperature T_c, there must exist an equivalent singularity at the corresponding temperature T_c^*. And in case we have only one singularity, as indeed follows from one of the theorems of Yang and Lee (1952), it must exist at a temperature T_c such that $T_c^* = T_c$. The critical temperature of the square lattice is, therefore, given by the equation, see formula (12),

$$\sinh(2K_c) = 1, \tag{18}$$

which gives

$$K_c = \frac{1}{2}\sinh^{-1} 1 = \frac{1}{2}\ln(\sqrt{2}+1) = \frac{1}{2}\ln\cot(\pi/8) \simeq 0.4407. \tag{19}$$

For comparison, we note that for the same lattice the Bragg–Williams approximation gave $K_c = 0.25$ while the Bethe approximation gave $K_c = \frac{1}{2}\ln 2 \simeq 0.3466$.

The situation for other lattices such as the triangular or the honeycomb, which are *not* self-dual, is complicated by the fact that the functions Q and Q_D in equation (15) are not the same. One then needs another trick — the so-called *star–triangle transformation* — which was first alluded to by Onsager (1944) in his famous paper on the solution of the square lattice problem but was written down explicitly by Wannier (1945); for details, see Baxter (1982). Unlike the duality transformation, this one establishes a relation between a *high*-temperature model on the triangular lattice and again a *high*-temperature model on the honeycomb lattice, and so on. Combining the two transformations, one can eliminate the dual lattice altogether and obtain a relation between a high-temperature and a low-temperature model on the same lattice. The location of the critical point is then straightforward; one obtains for the triangular lattice ($q = 6$)

$$K_c = \frac{1}{2}\sinh^{-1}\frac{1}{\sqrt{3}} \simeq 0.2747, \tag{20}$$

and for the honeycomb lattice ($q = 3$)

$$K_c = \frac{1}{2}\sinh^{-1}\sqrt{3} \simeq 0.6585. \tag{21}$$

The numerical values of K_c, given by equations (19) through (21), reinforce the fact that higher coordination numbers help propagate long-range order in the system more effectively and hence raise the critical temperature T_c.

The partition function and the specific-heat singularity

The partition function of the Ising model on a square lattice is given by, see references cited at the beginning of this section,

$$\frac{1}{N}\ln Q(T) = \ln\{2^{1/2}\cosh(2K)\} + \frac{1}{\pi}\int_0^{\pi/2} d\phi \ln\{1 + \sqrt{(1 - \kappa^2\sin^2\phi)}\}, \tag{22}$$

where

$$\kappa = 2\sinh(2K)/\cosh^2(2K). \tag{23}$$

Differentiating (22) with respect to $-\beta$, one obtains for the internal energy per spin

$$\frac{1}{N}U_0(T) = -2J\tanh(2K) + \frac{1}{\pi}\left(\kappa\frac{d\kappa}{d\beta}\right) \times \int_0^{\pi/2} d\phi \frac{\sin^2\phi}{\{1 + \sqrt{(1 - \kappa^2\sin^2\phi)}\}\sqrt{(1 - \kappa^2\sin^2\phi)}}. \tag{24}$$

Rationalizing the integrand, the integral in (24) can be written as

$$\frac{1}{\kappa^2}\left\{-\frac{\pi}{2} + K_1(\kappa)\right\}, \tag{25}$$

where $K_1(\kappa)$ is the complete elliptic integral of the *first* kind, κ being the modulus of the integral:

$$K_1(\kappa) = \int_0^{\pi/2} \frac{d\phi}{\sqrt{(1 - \kappa^2\sin^2\phi)}}. \tag{26}$$

Now, a logarithmic differentiation of (23) with respect to β gives

$$\frac{1}{\kappa}\frac{d\kappa}{d\beta} = 2J\{\coth(2K) - 2\tanh(2K)\}. \tag{27}$$

Substituting these results into (24), we obtain

$$\frac{1}{N}U_0(T) = -J\coth(2K)\left\{1+\frac{2\kappa'}{\pi}K_1(\kappa)\right\}, \tag{28}$$

where κ' is the *complementary* modulus:

$$\kappa' = 2\tanh^2(2K)-1 \quad (\kappa^2+\kappa'^2=1). \tag{29}$$

Figure 13.12 shows the variation of the moduli κ and κ' with the temperature parameter $(kT/J) = K^{-1}$. We note that, while κ is always positive, κ' can be positive or negative; actually, κ lies between 0 and 1 while κ' lies between -1 and 1. At the critical point, where $\sinh(2K_c) = 1$ and hence $K_c^{-1} \simeq 2.269$, the moduli κ and κ' are equal to 1 and 0, respectively.

To determine the specific heat of the lattice, we differentiate (28) with respect to temperature. In doing so, we make use of the following results:

$$\frac{d\kappa}{d\beta} = -\frac{\kappa'}{\kappa}\frac{d\kappa'}{d\beta}, \quad \frac{d\kappa'}{d\beta} = 8J\tanh(2K)\{1-\tanh^2(2K)\} \tag{30}$$

and

$$\frac{dK_1(\kappa)}{d\kappa} = \frac{1}{\kappa'^2\kappa}\{E_1(\kappa)-\kappa'^2K_1(\kappa)\}, \tag{31}$$

where $E_1(\kappa)$ is the complete elliptic integral of the *second* kind:

$$E_1(\kappa) = \int_0^{\pi/2}\sqrt{(1-\kappa^2\sin^2\phi)}\,d\phi. \tag{32}$$

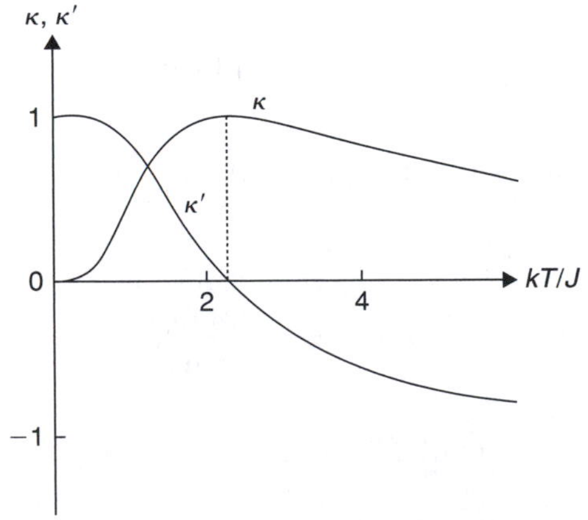

FIGURE 13.12 Variation of the moduli κ and κ' with (kT/J).

We finally obtain

$$\frac{1}{Nk}C_0(T) = \frac{2}{\pi}\{K\coth(2K)\}^2\left[2\{K_1(\kappa) - E_1(\kappa)\} - (1-\kappa')\left\{\frac{\pi}{2} + \kappa' K_1(\kappa)\right\}\right]. \tag{33}$$

Now, the elliptic integral $K_1(\kappa)$ has a singularity at $\kappa = 1$ (i.e., at $\kappa' = 0$), in the neighborhood of which

$$K_1(\kappa) \approx \ln\{4/|\kappa'|\} \quad \text{and} \quad E_1(\kappa) \approx 1. \tag{34}$$

Accordingly, the specific heat of the lattice displays a *logarithmic singularity* at a temperature T_c, given by the condition: $\kappa_c = 1$ (or $\kappa'_c = 0$), which is identical to (18). In the vicinity of the critical point, equation (33) reduces to

$$\frac{1}{Nk}C_0(T) \simeq \frac{8}{\pi}K_c^2\left[\ln\{4/|\kappa'|\} - \left(1 + \frac{\pi}{4}\right)\right]; \tag{35}$$

at the same time, the parameter κ' reduces to, see equation (30),

$$\kappa' \simeq 2\sqrt{2}K_c\left(1 - \frac{T}{T_c}\right). \tag{36}$$

The specific heat singularity is, therefore, given by

$$\begin{aligned}\frac{1}{Nk}C_0(T) &\simeq \frac{8}{\pi}K_c^2\left[-\ln\left|1 - \frac{T}{T_c}\right| + \left\{\ln\left(\frac{\sqrt{2}}{K_c}\right) - \left(1 + \frac{\pi}{4}\right)\right\}\right] \\ &\simeq -0.4945\ln\left|1 - \frac{T}{T_c}\right| + \text{const.},\end{aligned} \tag{37}$$

signaling a logarithmic divergence at $T = T_c$.

Figures 13.13 and 13.14 show the temperature dependence of the internal energy and the specific heat of the square lattice, as given by the Onsager expressions (28) and (33); for comparison, the results of the Bragg–Williams approximation and of the Bethe approximation (with $q = 4$) are also included. The specific-heat singularity, given correctly by the Onsager expression (37), is seen as a (logarithmic) peak in Figure 13.14, which differs markedly from the jump discontinuity predicted by the mean field theory. We conclude that the critical exponent $\alpha = \alpha' = 0(\log)$.

In passing, we note that the internal energy of the lattice is *continuous* at the critical point, having a value of $-\sqrt{2}J$ per spin and an infinite, positive slope; needless to say, the continuity of the internal energy implies that the transition takes place without any latent heat.

Other properties

We now consider the temperature dependence of the *order parameter* (i.e., the spontaneous magnetization), of the lattice. An exact expression for this quantity was first derived

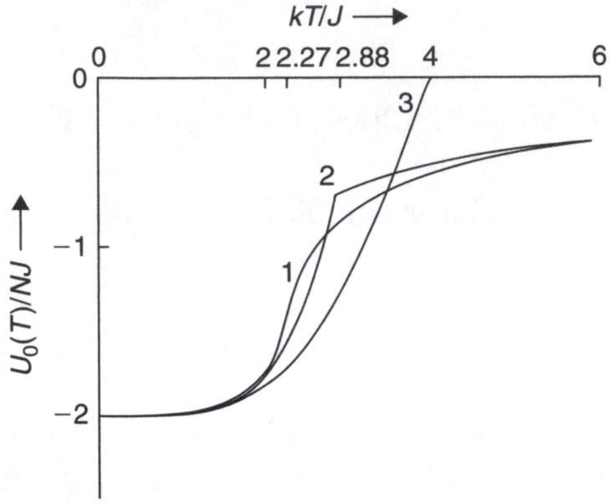

FIGURE 13.13 The internal energy of a square lattice ($q = 4$) according to (1) the Onsager solution, (2) the Bethe approximation, and (3) the Bragg–Williams approximation.

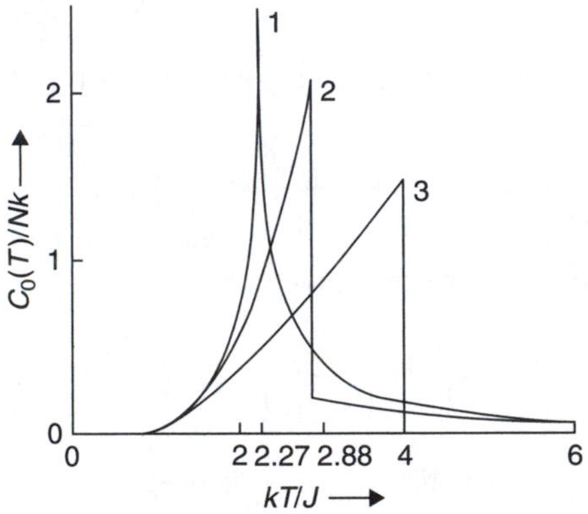

FIGURE 13.14 The specific heat of a square lattice ($q = 4$) according to (1) the Onsager solution, (2) the Bethe approximation, and (3) the Bragg–Williams approximation.

by Onsager (1949), though he never published the details of his derivation. The first published derivation is due to Yang (1952), who showed that

$$\overline{L}_0(T) \equiv \frac{1}{N\mu}\overline{M}(0,T) = \begin{cases} [1 - \{\sinh(2K)\}^{-4}]^{1/8} & \text{for} \quad T \le T_c \qquad (38a) \\ 0 & \text{for} \quad T \ge T_c, \qquad (38b) \end{cases}$$

where K, as usual, is J/kT. In the limit $T \to 0$,

$$\overline{L}_0(T) \simeq 1 - 2\exp(-8J/kT), \tag{39}$$

which implies a very slow variation with T. On the other hand, in the limit $T \to T_c-$,

$$\overline{L}_0(T) \approx \left\{8\sqrt{2}K_c\left(1 - \frac{T}{T_c}\right)\right\}^{1/8} \simeq 1.2224\left(1 - \frac{T}{T_c}\right)^{1/8}, \tag{40}$$

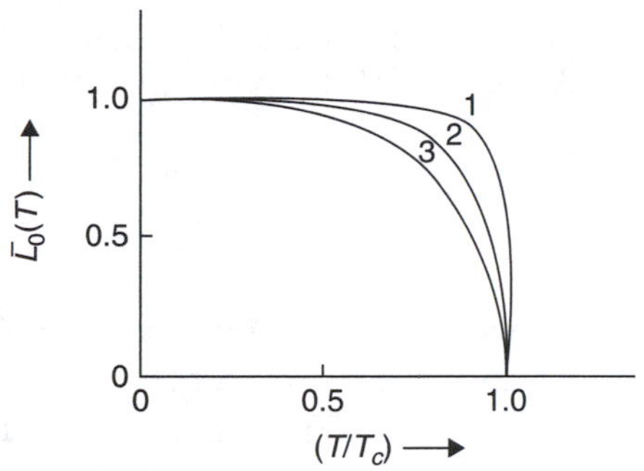

FIGURE 13.15 The spontaneous magnetization of a square lattice ($q = 4$) according to (1) the Onsager solution, (2) the Bethe approximation, and (3) the Bragg–Williams approximation.

which indicates a very fast variation with T. The detailed dependence of $\bar{L}_0$ on T is shown in Figure 13.15; for comparison, the results of the Bragg–Williams approximation and the Bethe approximation are also included. We infer that the critical exponent β for this model is $\frac{1}{8}$, which is very different from the mean field value of $\frac{1}{2}$.

Onsager also calculated the correlation length ξ of the lattice, which showed that the critical exponent $\nu = \nu' = 1$ — in sharp contrast to the classical value of $\frac{1}{2}$. Finally, he set up calculations for the correlation function $g(r)$ from which one could infer that the exponent $\eta = \frac{1}{4}$, again in disagreement with the classical value of zero. Precise asymptotic expressions for the correlation function in different regimes of temperature were derived by later authors (Fisher, 1959; Kadanoff, 1966a; Au-Yang and Perk, 1984):

$$g(r) \approx \frac{\{4(K_c - K)\}^{1/4}}{2^{3/8}(\pi r/\xi)^{1/2}} e^{-r/\xi}, \qquad \xi = \{4(K_c - K)\}^{-1} \tag{41}$$

for $T > T_c$,

$$g(r) \approx \frac{\{4(K - K_c)\}^{1/4}}{2^{21/8}\pi(r/\xi)^2} e^{-2r/\xi}, \qquad \xi = \{4(K - K_c)\}^{-1} \tag{42}$$

for $T < T_c$, and

$$g(r) \approx \frac{2^{1/12} \exp\{3\zeta'(-1)\}}{r^{1/4}} \tag{43}$$

at $T = T_c$; in the last expression, $\zeta'(x)$ denotes the derivative of the Riemann zeta function. We note from expressions (41) and (42) that the correlation length ξ, while diverging at $T = T_c$, is finite on *both* sides of the critical point. This feature is peculiar to the scalar model ($n = 1$) only, for in the case of vector models ($n \geq 2$), ξ turns out to be infinite at all $T \leq T_c$.

The zero-field susceptibility of this system has also been worked out (see Barouch et al., 1973; Tracy and McCoy, 1973; Wu et al., 1976); asymptotically, one finds that

$$\chi_0 \approx \frac{N\mu^2}{kT_c} \times \begin{cases} C_+ t^{-7/4} & \text{for} \quad t \gtrsim 0 \quad (44a) \\ C_- |t|^{-7/4} & \text{for} \quad t \lesssim 0, \quad (44b) \end{cases}$$

where t, as usual, is $(T - T_c)/T_c$ while the constants C_+ and C_- are about 0.96258 and 0.02554, respectively. We see that the critical exponent $\gamma = \gamma' = \frac{7}{4}$, as opposed to the mean field value of 1, and the ratio $C_+/C_- \simeq 37.69$, as opposed to the mean field value of 2. Assembling all the exponents in one place, we have for the two-dimensional Ising model

$$\alpha = \alpha' = 0(\log), \quad \beta = \frac{1}{8}, \quad \gamma = \gamma' = \frac{7}{4}, \quad \nu = \nu' = 1, \quad \eta = \frac{1}{4}. \tag{45}$$

Since this model has not yet been solved in the presence of a field, a direct evaluation of the exponent δ has not been possible. Assuming the validity of the scaling relations, however, we can safely conclude that $\delta = 15$ — again very different from the classical value of 3. All in all, the results of this section tell us very clearly, and loudly, how inadequate the mean field theory can be.

Before we close this section, a few remarks seem to be in order. First of all, it may be mentioned that for the model under consideration one can also calculate the *interfacial tension s*, which may be defined as the "free energy associated, per unit area, with the interfaces between the domains of *up* spins and those of *down* spins"; in our analogy with the gas–liquid systems, this corresponds to the conventional surface tension σ. The corresponding exponent μ, that determines the manner in which $s \to 0$ as $T \to T_c-$, turns out to be 1 for this model; see Baxter (1982). This indeed obeys the scaling relation $\mu = (d-1)\nu$, as stated in Problem 12.26. Second, we would like to point out that, while the solution of the two-dimensional Ising model was the first exact treatment that exposed the inadequacy of the mean field theory, it was also the first to disclose the underlying universality of the problem. As discovered by Onsager himself, if the spin–spin interactions were allowed to have different strengths, J and J', in the horizontal and vertical directions of the lattice, the specific-heat divergence at $T = T_c$ continued to be logarithmic — independently of the ratio J'/J — even though the value of T_c itself and of the various amplitudes appearing in the final expressions were modified. A similar result for the spontaneous magnetization was obtained by Chang (1952) who showed that, regardless of the value of J'/J, the exponent β continued to be $\frac{1}{8}$. Further corroborative evidence for universality came from the analysis of two-dimensional lattices other than the square one which, despite structural differences, led to the same critical exponents as the ones listed in equation (45).

13.4.A The two-dimensional Ising model on a finite lattice

Phase transitions, viewed as critical phenomena, cannot occur in a finite system since a statistical mechanical model with a finite number of degrees of freedom cannot have a nonanalytic partition function or free energy. Criticality occurs only in the thermodynamic limit. Since real physical systems are of finite size, the manner in which finite-size effects manifest themselves as the correlation length ξ approaches the system size is of considerable importance in understanding how critical singularities get rounded off in real systems. In this regard, the two-dimensional nearest-neighbor Ising model on a square lattice in zero field can be solved on a finite square lattice with periodic boundary conditions (Kaufman, 1949), which allows for a detailed exploration of finite-size effects, especially near the bulk critical point; see Ferdinand and Fisher (1969). Kaufman's solution is based on a determination of *all* the eigenvalues of the transfer matrix.

Onsager (1944) only required the largest eigenvalue since his solution was based on a strip geometry with the length of one side taken to infinity. We here consider the Ising model on a lattice with n rows and m columns with periodic boundary conditions; see Figure 13.16. Each column of n spins has 2^n possible configurations, so the transfer matrix $\boldsymbol{P}$ that couples nearest-neighbor columns is a $2^n \times 2^n$ matrix of Boltzmann factors with eigenvalues λ_α, with $\alpha = 1, 2, \ldots, 2^n$. Just as in the case of the one-dimensional Ising model studied in Section 13.2, the partition function of a system with n rows and m columns can be written as the trace of a transfer matrix $\boldsymbol{P}$:

$$Q_{n,m}(K) = \text{Trace}(\boldsymbol{P}^m) = \sum_{\alpha=1}^{2^n} \lambda_\alpha^m, \tag{46}$$

where the eigenvalues of the transfer matrix fall into two classes:

$$\lambda_\alpha = \begin{cases} (2\sinh(2K))^{n/2} \sum \exp\left(\frac{1}{2}\left(\pm\gamma_0 \pm \gamma_2 \pm \ldots \pm \gamma_{2n-2}\right)\right), \\ (2\sinh(2K))^{n/2} \sum \exp\left(\frac{1}{2}\left(\pm\gamma_1 \pm \gamma_3 \pm \ldots \pm \gamma_{2n-1}\right)\right). \end{cases} \tag{47}$$

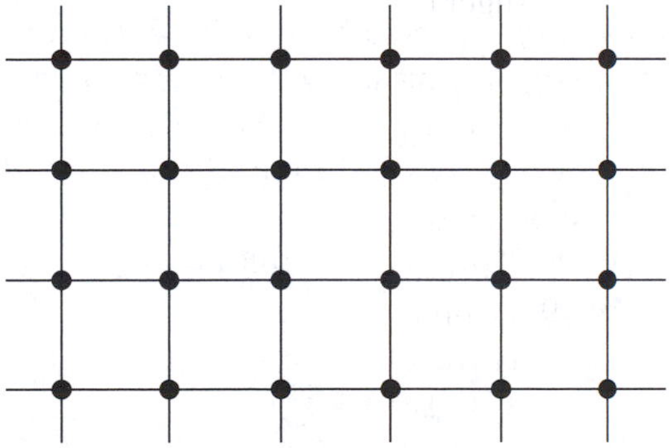

FIGURE 13.16 A finite square lattice with $n = 4$ rows and $m = 6$ columns. In view of the periodic boundary conditions, sites on the leftmost column interact with sites on the rightmost column and the bottom row interacts with the top row.

The quantity γ_q for $0 < q < 2n$ is the positive root of the equation

$$\cosh(\gamma_q) = \frac{\cosh^2(2K)}{\sinh(2K)} - \cos\left(\frac{\pi q}{n}\right), \tag{48}$$

while the $q = 0$ case is given by

$$e^{\gamma_0} = e^{2K}\tanh(K). \tag{49}$$

Only terms with an even number of minus signs inside the exponentials appear in the sums in equation (47), so the partition function can be written as

$$Q_{n,m}(K) = \frac{1}{2}(2\sinh(2K))^{nm/2}(Y_1 + Y_2 + Y_3 + Y_4), \tag{50}$$

where

$$Y_1 = \prod_{q=0}^{n-1}\left(2\cosh\left(\frac{m}{2}\gamma_{2q+1}\right)\right), \tag{51a}$$

$$Y_2 = \prod_{q=0}^{n-1}\left(2\sinh\left(\frac{m}{2}\gamma_{2q+1}\right)\right), \tag{51b}$$

$$Y_3 = \prod_{q=0}^{n-1}\left(2\cosh\left(\frac{m}{2}\gamma_{2q}\right)\right), \tag{51c}$$

$$Y_4 = \prod_{q=0}^{n-1}\left(2\sinh\left(\frac{m}{2}\gamma_{2q}\right)\right); \tag{51d}$$

see Kaufman (1949). This form of the partition function allows for an exact calculation of the free energy, internal energy, and specific heat on finite lattices; see Figure 13.17. The logarithmic singularity in the specific heat at the *bulk* critical point evolves from a specific heat peak that grows logarithmically with the system size, that is $C_{nm}(K_c)/nmk \approx (8K_c^2/\pi)\ln(n) \simeq 0.4945\ln(n)$; see Ferdinand and Fisher (1969). Note also that the coefficient of $\ln(n)$ here is the same as the coefficient of the $\ln(|1 - T/T_c|)$ term in the bulk specific heat, as given in equation (37).

The low-temperature series expansion for the partition function can be written as $Q_{n,m}(K) = e^{2nmK}\tilde{Q}_{n,m}(K)$, where

$$\tilde{Q}_{n,m}(K) = \sum_{q=0}^{nm} g_q x^{2q}, \tag{52}$$

$x = e^{-2K}$ is the Boltzmann factor for a single excitation, and the coefficients g_q denote the number of configurations with energy $4qJ$ above the ground state. The sum of the

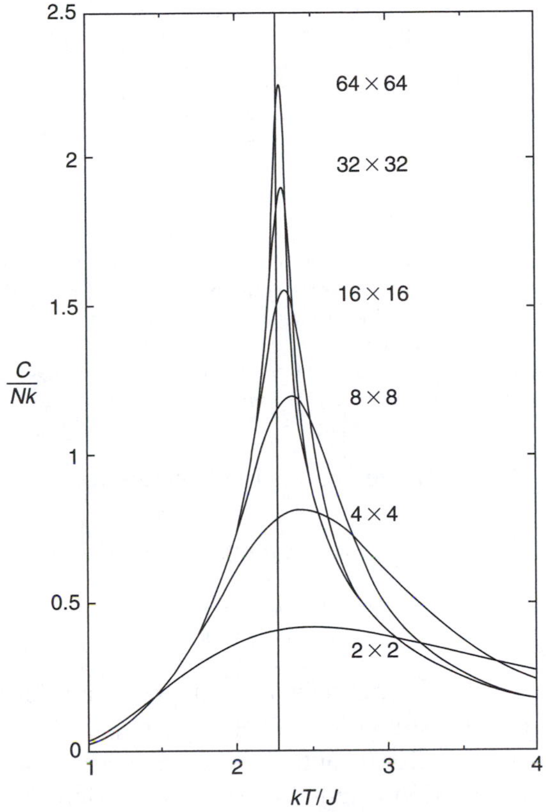

FIGURE 13.17 Specific heat of the two-dimensional Ising model for finite $2 \times 2, 4 \times 4, \ldots, 64 \times 64$ lattices. The specific heat is analytic for all finite lattices. The maximum value of the specific heat grows proportional to the logarithm of the linear dimension of the lattice and the location of the maximum approaches the bulk critical temperature (denoted by the vertical line) proportional to the inverse of the linear dimension of the lattice. From Ferdinand and Fisher (1969). Reprinted with permission; copyright © 1969, American Physical Society.

coefficients counts all the microstates in the system; therefore

$$\lim_{K \to 0} \tilde{Q}_{n,m}(K) = \sum_{q=0}^{nm} g_q = 2^{nm}.$$

The coefficients g_q represent the number of microstates pertaining to energy $(-2nmJ + 4qJ)$, with the corresponding entropy being $k \ln g_q$.

The first term in the series is $g_0 = 2$ since there are two degenerate ground states, namely all spins up or all spins down. It is straightforward to see that only even orders in x appear in this expansion. Examples of the low-order graphs that contribute to the series

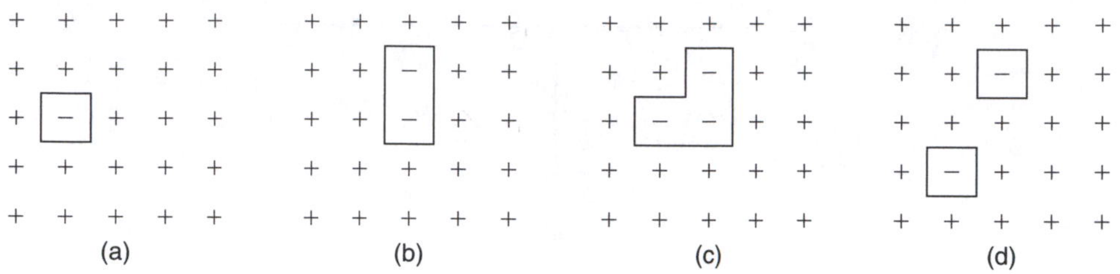

FIGURE 13.18 The lowest few excited states of the lattice. The $q=2$ states (a), have a single down spin in a sea of up spins or a single up spin in a sea of down spins; these states have energy $8J$ above the ground state and there are $g_2=2nm$ configurations. The $q=3$ states (b), have a pair of down spins in a sea of up spins, or vice versa; these states have energy $12J$ above the ground state and $g_3=4nm$ configurations. The $q=4$ states (c) and (d), can have a single grouping of opposite spins, or a pair of isolated flipped spins; these states have energy $16J$ above the ground state and the total number of configurations $g_4=(nm)^2+9nm$.

are shown in Figure 13.18. The first few terms in the series are

$$\tilde{Q}_{n,m}(K)=2+(2nm)x^4+(4nm)x^6+\left((nm)^2+9nm\right)x^8 + \left(4(nm)^2+24nm\right)x^{10}+\ldots. \tag{53}$$

If both n and m are even, the model's ferromagnetic/antiferromagnetic symmetry ($J\to -J$ and $s_i\to -s_i$ on one sublattice) gives: $g_q=g_{nm-q}$. Due to the self-duality of the two-dimensional square lattice, exactly the same coefficients g_q also appear in the high-temperature series expansion where the expansion variable is $\tanh K$.

The probability P_q of finding an equilibrium state with energy $4qJ$ above the ground state is given by

$$P_q=\frac{g_q x^{2q}}{\tilde{Q}_{n,m}(K)}, \tag{54}$$

and the internal energy and the heat capacity per spin are given by

$$\frac{U}{NJ}=-2+\frac{4}{N}\sum_{q=0}^{N} qP_q \quad (N=nm), \tag{55a}$$

$$\frac{C}{Nk}=\frac{16}{N}\left(\frac{J}{kT}\right)^2\left(\sum_{q=0}^{N} q^2P_q-\left(\sum_{q=0}^{N} qP_q\right)^2\right). \tag{55b}$$

One can cast Kaufman's solution, equation (50) and equation (51), in the form of a low-temperature expansion of the form shown in (52), thereby giving an exact determination of the partition function and the equilibrium energy distribution; see Beale (1996).

The low-temperature series (52) can be written as

$$\tilde{Q}_{n,m}(K) = \sum_{q=0}^{nm} g_q x^{2q} = (Z_1 + Z_2 + Z_3 + Z_4), \tag{56}$$

where if n is even, then

$$Z_1 = \frac{1}{2} \prod_{q=0}^{n/2-1} c_{2q+1}^2, \tag{57a}$$

$$Z_2 = \frac{1}{2} \prod_{q=0}^{n/2-1} s_{2q+1}^2, \tag{57b}$$

$$Z_3 = \frac{1}{2} c_0 c_n \prod_{q=1}^{n/2-1} c_{2q}^2, \tag{57c}$$

$$Z_4 = \frac{1}{2} s_0 s_n \prod_{q=1}^{n/2-1} s_{2q}^2; \tag{57d}$$

while if n is odd, then

$$Z_1 = \frac{1}{2} c_n \prod_{q=0}^{(n-3)/2} c_{2q+1}^2, \tag{58a}$$

$$Z_2 = \frac{1}{2} s_n \prod_{q=0}^{(n-3)/2} s_{2q+1}^2, \tag{58b}$$

$$Z_3 = \frac{1}{2} c_0 \prod_{q=1}^{(n-1)/2} c_{2q}^2, \tag{58c}$$

$$Z_4 = \frac{1}{2} s_0 \prod_{q=1}^{(n-1)/2} s_{2q}^2. \tag{58d}$$

The factors in equations (57) and (58) are

$$c_0 = (1-x)^m + (x(1+x))^m, \tag{59a}$$

$$s_0 = (1-x)^m - (x(1+x))^m, \tag{59b}$$

$$c_n = (1+x)^m + (x(1-x))^m, \tag{59c}$$

$$s_n = (1+x)^m - (x(1-x))^m, \tag{59d}$$

$$c_q^2 = \frac{1}{2^{m-1}} \left(\left(\sum_{j=0}^{\lfloor \frac{m}{2} \rfloor} \frac{m! \left(\alpha_q^2 - \beta^2\right)^j \alpha_q^{m-2j}}{(2j)!\,(m-2j)!} \right) + \beta^m \right), \tag{59e}$$

$$s_q^2 = \frac{1}{2^{m-1}}\left(\left(\sum_{j=0}^{\lfloor \frac{m}{2} \rfloor} \frac{m!\left(\alpha_q^2 - \beta^2\right)^j \alpha_q^{m-2j}}{(2j)!(m-2j)!}\right) - \beta^m\right), \tag{59f}$$

$$\beta = 2x(1-x^2), \tag{59g}$$

$$\alpha_q = (1+x^2)^2 - \beta\cos\left(\frac{\pi q}{n}\right). \tag{59h}$$

The function $\lfloor z \rfloor$ denotes the largest integer less than or equal to z. The quantities c_q^2 and s_q^2 were expanded using the binomial series in order to explicitly remove all square roots that would hide the polynomial nature of the final result. A symbolic programming language can be used to numerically expand the partition function as a polynomial in the variable x in the form (52). One must set the numerical precision in the calculation to somewhat more than $nm \ln 2/\ln 10$ decimal digits in order to determine the exact values of the integer coefficients $\{g_q\}$. The numerical calculation can be checked against the low-order result (53) or with an exact enumeration of energies on small lattices.

The low-temperature series for the Ising model on a 32×32 lattice is

$$\begin{aligned}\tilde{Q}_{32,32}(K) = {} & 2 + 2048x^4 + 4096x^6 + 1057792x^8 + 4218880x^{10} \\ & + 371621888x^{12} + 2191790080x^{14} \\ & + 100903637504x^{16} + 768629792768x^{18} \\ & + 22748079183872x^{20} + \cdots + 4096x^{2042} \\ & + 2048x^{2044} + 2x^{2048},\end{aligned} \tag{60}$$

where the largest coefficient is

$$\begin{aligned}g_{512} = {} & 6,342,873,169,001,916,568,766,443,273,025,000,331,593,063, \\ & 924,436,135,196,680,443,689,656,478,072,741,300,511,612, \\ & 123,900,652,711,596,311,283,701,724,071,226,144,241,851, \\ & 411,641,714,893,727,789,741,510,169,213,344,005,116,385, \\ & 197,594,692,089,556,614,547,788,150,860,200,720,413,211, \\ & 442,412,355,672,291,841,364,265,145,274,980,444,405,423, \\ & 129,672,679,584,959,498,234,944,801,613,246,300,853,599, \\ & 317,229,362,316,\end{aligned} \tag{61}$$

that is, there are about 6.342×10^{306} configurations with energy halfway between the ferromagnetic and antiferromagnetic ordered states. This single microstate comprises 3.5 percent of the 2^{1024} total configurations of the model. The exact results for the microcanonical entropy and the energy distribution for the 128×128 lattice are shown in Figures 13.19 and 13.20. These results provide excellent tests of Monte Carlo simulation methods, including broad histogram methods; see Beale (1996), Wang and Landau (2001), and Landau and Binder (2009).[2]

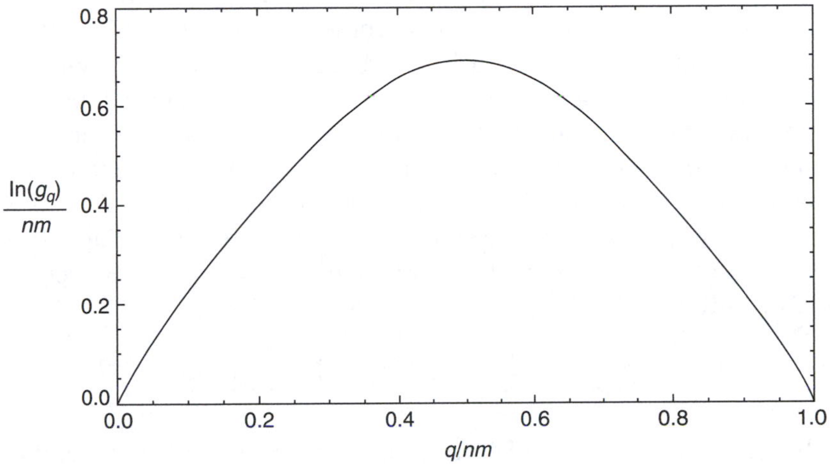

FIGURE 13.19 Microcanonical entropy per spin $S/Nk = \ln(g_q)/nm$ for the two dimensional Ising model on a 128×128 lattice as calculated from equations (56), (57), and (59). The slope of the curve is proportional to the inverse temperature, so the state with $q/nm = 1/2$ represents the infinite temperature state with energy halfway between the ordered ferromagnetic and antiferromagnetic states; the largest coefficient $g_{8192} \simeq 1.049 \times 10^{4930}$ is the number of configurations with $q = nm/2 = 8192$. Likewise, the states at $q = 0$ and $q = nm$ represent the ferromagnetic and antiferromagnetic ground states, so the slopes of the curve diverge logarithmically in the thermodynamic limit.

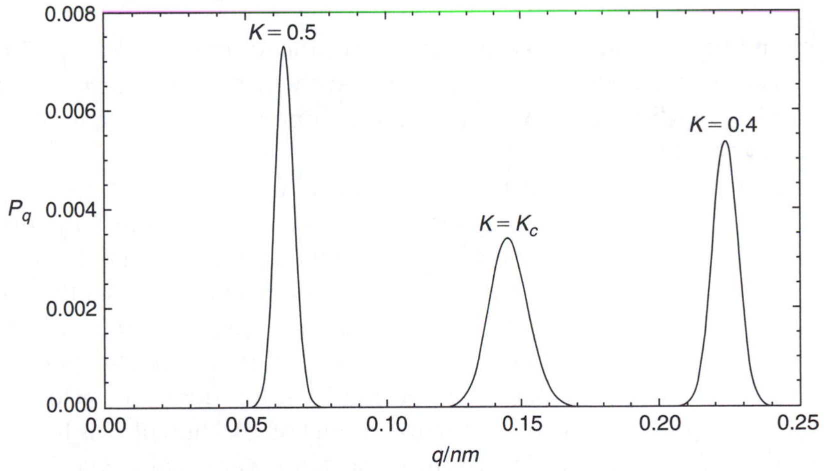

FIGURE 13.20 The exact energy distribution P_q for the two-dimensional Ising model on a 128×128 lattice for $K = 0.4$, $K = K_c \simeq 0.4407$, and $K = 0.5$. The variance of the energy distribution is proportional to the specific heat, so is largest near $K = K_c$. Refer to Figure 13.17.

[2]Mathematica code for calculating the low-temperature series coefficients for a two-dimensional Ising model, as well as the microcanonical entropies, internal energies, and specific heats for several lattice sizes can be found at www.elsevierdirect.com.

13.5 The spherical model in arbitrary dimensions

In the wake of Onsager's solution to the two-dimensional Ising problem in zero field, several attempts were made to go beyond Onsager — by solving either the three-dimensional problem in zero field or the two-dimensional problem with field. None of these attempts succeeded; the best one could accomplish was to rederive the Onsager solution by newer means. This led to the suggestion that one may instead consider certain "adaptations" of the Ising model, which may turn out to be mathematically tractable in more than two dimensions and hopefully throw some light on the problem of phase transitions in more realistic situations (where d is usually 3). One such adaptation was devised by Kac who, in 1947, considered a model in which the spin variable σ_i, instead of being restricted to the *discrete* choices -1 or $+1$, could vary *continuously*, from $-\infty$ to $+\infty$, subject to a Gaussian probability distribution law,

$$p(\sigma_i)d\sigma_i = (A/\pi)^{1/2}e^{-A\sigma_i^2}d\sigma_i \quad (i=1,\ldots,N), \tag{1}$$

so that σ_i^2, on an average, $=1/(2A)$. Clearly, for conformity with the standard practice, namely $\sigma_i^2 = 1$, the constant A here should be equal to $\frac{1}{2}$; we may, however, leave it arbitrary for the time being. The resulting model is generally referred to as the *Gaussian model*, and its partition function in the presence of the field is given by the multiple integral

$$Q_N = \int\limits_{-\infty}^{+\infty}\cdots\int\limits_{-\infty}^{+\infty}\left(\frac{A}{\pi}\right)^{N/2} e^{-A\sum\limits_i \sigma_i^2 + K\sum\limits_{\text{n.n.}}\sigma_i\sigma_j + h\sum\limits_i \sigma_i}\prod_i d\sigma_i \qquad (K=\beta J, h=\beta\mu B). \tag{2}$$

The exponent in the integrand is a symmetric, quadratic function in the σ_i; using standard techniques, it can be diagonalized. Integrations over the (transformed) σ_j can then be carried out straightforwardly — and in any number of dimensions; for details, see Berlin and Kac (1952) or Baker (1990).

One finds that for $d > 2$ the Gaussian model undergoes a phase transition at a *finite* temperature T_c which, for a simple hypercubic lattice, is determined by the condition $K_c = A/d$; note that, with $A=\frac{1}{2}$, this result is precisely the one predicted by the mean field theory (with $q = 2d$). There are differences, though. First of all, the present model does not exhibit a phase transition at a finite temperature for $d \le 2$. Secondly, the critical exponents for $2 < d < 4$ are nonclassical, in the sense that some of them are d-dependent, though for $d > 4$ they do become classical. More importantly, at temperatures below T_c, where K exceeds A/d, the integral in (2) diverges and the model breaks down! This led Kac to abandon this model and invent a new one in which the spins were again *continuous* variables but subject to an *overall* constraint,

$$\sum_i \sigma_i^2 = N, \tag{3}$$

rather than to individual constraints, $\sigma_i^2 = 1$ for each i, or to an arbitrary probability distribution law. Constraint (3) allows individual spins to vary over a rather wide range, $-N^{1/2}$ to $+N^{1/2}$, but restricts the super spin vector $\{\sigma_i\}$ to the "surface of an N-dimensional hypersphere of radius $N^{1/2}$"; in the Ising model, the same vector is restricted to the "corners of a hypercube inscribed within the above hypersphere." The resulting model is generally referred to as the *spherical model.*

Constraint (3) can be taken care of by inserting an appropriate delta function in the integrand of the partition function. Using the representation

$$\delta\left(N - \sum_i \sigma_i^2\right) = \frac{1}{2\pi i} \int_{x-i\infty}^{x+i\infty} e^{z(N - \sum_i \sigma_i^2)} dz, \tag{4}$$

the partition function of the spherical model is given by

$$Q_N = \frac{1}{2\pi i} \int_{x-i\infty}^{x+i\infty} dz\, e^{zN} \int_{-\infty}^{+\infty} \cdots \int_{-\infty}^{+\infty} e^{-z\sum_i \sigma_i^2 + K \sum_{\text{n.n.}} \sigma_i \sigma_j + h \sum_i \sigma_i} \prod_i d\sigma_i. \tag{5}$$

For a fixed z, the integral over the σ_i can be carried out in the same manner as in the Gaussian model; see equation (2). Let the result of that calculation be denoted by the symbol $Z_N(K, h; z)$. The partition function Q_N of the spherical model is then given by the complex integral

$$Q_N = \frac{1}{2\pi i} \int_{x-i\infty}^{x+i\infty} dz\, e^{zN} Z_N(K, h; z), \tag{6}$$

which can be evaluated by the *saddle-point* method — also known as the method of *steepest descent*; see Section 3.2. One finds that the saddle point of the integrand in (6) lies at the point $z = x_0$, where x_0 is determined by the condition

$$\left. \frac{\partial}{\partial z} \{zN + \ln Z_N(K, h; z)\} \right|_{z = x_0} = 0, \tag{7}$$

with the result that, asymptotically,

$$\frac{1}{N} \ln Q_N \approx x_0 + \frac{1}{N} \ln Z_N(K, h; x_0). \tag{8}$$

The thermodynamic properties of the system can then be worked out in detail.

It turned out that many physicists felt uncomfortable at the necessity of using the method of steepest descent, so a search for an alternative approach seemed desirable. In this connection Lewis and Wannier (1952) pointed out that while the ensemble underlying the model of Berlin and Kac was canonical in the variable E it was microcanonical in the variable $\sum_i \sigma_i^2$. They proposed that one consider instead an ensemble that is canonical in both E and $\sum_i \sigma_i^2$; the method of steepest descent could then be avoided. All one requires now is that the constraint (3) be obeyed only in the sense of an ensemble average,

$$\left\langle \sum_i \sigma_i^2 \right\rangle = N, \tag{9}$$

rather than in the original sense that was comparatively more rigid. The resulting model is referred to as the *mean spherical model*.

Constraint (9) can easily be taken care of by modifying the Hamiltonian of the system by including a term proportional to $\sum_i \sigma_i^2$, that is, by writing

$$H = -J\sum_{\text{n.n.}} \sigma_i \sigma_j - \mu B \sum_i \sigma_i + \lambda \sum_i \sigma_i^2, \tag{10}$$

where λ is the so-called *spherical field*, and requiring that

$$\langle (\partial H / \partial \lambda) \rangle = N. \tag{11}$$

The partition function of the revised model is thus given by

$$Q_N = \int_{-\infty}^{+\infty} \cdots \int_{-\infty}^{+\infty} e^{-\beta\lambda \sum_i \sigma_i^2 + K \sum_{\text{n.n.}} \sigma_i \sigma_j + h \sum_i \sigma_i} \prod_i d\sigma_i \tag{12a}$$

$$= Z_N(K, h; \beta\lambda), \tag{12b}$$

with the constraint

$$\frac{1}{\beta} \frac{\partial \ln Z_N(K, h; \beta\lambda)}{\partial \lambda} = -N. \tag{13}$$

Comparing (13) with (7), we readily see that the parameter x_0 of the spherical model is precisely equal to the parameter $\beta\lambda$ of the mean spherical model. The free energy resulting from (12), however, differs a little from the one given by equation (8), which is not surprising because the transition from a model that was microcanonical in the variable $\mathcal{S}^2 (\equiv \sum_i \sigma_i^2)$ to one that is canonical modifies the nature of the free energy — it goes from "being at constant $\mathcal{S}$" to "being at constant λ." The two free energies are connected by the Legendre transformation

$$A_\lambda = A_{\mathcal{S}} + \lambda \langle \mathcal{S}^2 \rangle = A_{\mathcal{S}} + \lambda N, \tag{14}$$

so that

$$\frac{1}{N} A_{\mathcal{S}} = \frac{1}{N} A_\lambda - \lambda. \tag{15}$$

This is precisely the difference that arises from the use of expression (8) or expression (12).

We now proceed to examine the thermodynamic properties of the mean spherical model, especially the nature of its critical behavior in arbitrary dimensions. The importance of these results will be discussed toward the end of this section.

The thermodynamic functions

We consider a simple hypercubic lattice, of dimensions $L_1 \times \cdots \times L_d$, subject to periodic boundary conditions. The partition function of the system, as given by equation (12a), then turns out to be (see Joyce, 1972; Barber and Fisher, 1973)

$$Z_N(K, h; \beta\lambda) = \prod_{\boldsymbol{k}} \left[\frac{\pi}{\beta(\lambda - \mu_{\boldsymbol{k}})} \right]^{1/2} e^{Nh^2/4\beta(\lambda-\mu_0)}, \tag{16}$$

where $\mu_{\boldsymbol{k}}$ are the eigenvalues of the problem,

$$\mu_{\boldsymbol{k}} = J \sum_{j=1}^{d} \cos(k_j a), \quad k_j = \frac{2\pi n_j}{L_j} \quad \{n_j = 0, 1, \ldots, (N_j - 1)\}, \tag{17a}$$

$$N_j = L_j/a, \quad N = \prod_j N_j, \tag{17b}$$

and a the lattice constant of the system. The free energy A_λ is then given by

$$A_\lambda = \frac{1}{2\beta} \sum_{\boldsymbol{k}} \ln \frac{\beta(\lambda - \mu_{\boldsymbol{k}})}{\pi} - \frac{N\mu^2 B^2}{4(\lambda - \mu_0)}, \tag{18}$$

while the parameter λ is determined by the *constraint equation*, see equation (13),

$$\frac{1}{2\beta} \sum_{\boldsymbol{k}} \frac{1}{(\lambda - \mu_{\boldsymbol{k}})} + \frac{N\mu^2 B^2}{4(\lambda - \mu_0)^2} = N. \tag{19}$$

The magnetization $\overline{M}$ and the low-field susceptibility χ_0 follow readily from equation (18):[3]

$$\overline{M} = \frac{N\mu^2 B}{2(\lambda - \mu_0)}, \quad \chi_0 = \frac{N\mu^2}{2(\lambda - \mu_0)}. \tag{20a, b}$$

Introducing the variable $m(\equiv \overline{M}/N\mu)$, the constraint equation (19) may be written in the form

$$\sum_{\boldsymbol{k}} \frac{1}{(\lambda - \mu_{\boldsymbol{k}})} = 2N\beta(1 - m^2). \tag{21}$$

[3]Note, however, that to calculate the *field-dependent* susceptibility, $(\partial\overline{M}/\partial B)_{T,\mathscr{S}}$, subject to the spherical constraint (19), one must keep in mind the field dependence of λ while differentiating (20a).

Next, the entropy of the system in zero field is given by[4]

$$S_0 = -(\partial A/\partial T)_{\mu_{\boldsymbol{k}},B=0} = \frac{1}{2}k_B \sum_{\boldsymbol{k}} [1 - \ln\{\beta(\lambda - \mu_{\boldsymbol{k}})\}] \tag{22}$$

and the corresponding specific heat by

$$C_0 = T\left(\frac{\partial S_0}{\partial T}\right) = \frac{1}{2}k_B \sum_{\boldsymbol{k}} \left[1 - \frac{T(\partial\lambda/\partial T)_0}{(\lambda - \mu_{\boldsymbol{k}})}\right] = N\left[\frac{1}{2}k_B - \left(\frac{\partial\lambda}{\partial T}\right)_0\right]; \tag{23}$$

here, use has been made of equation (19), with $B = 0$.

To make further progress we need to determine λ, as a function of B and T, from the constraint equation (19). But first note, from equation (18), that for the free energy of the system to be well-behaved, λ must be larger than the largest eigenvalue μ_0 — which, by (17a), is equal to Jd. At the same time, equation (20b) tells us that the singularity of the problem presumably lies at $\lambda = \mu_0$. We may thus infer that, as T decreases from higher values downward, λ also decreases and eventually reaches its lowest possible value, μ_0, at some critical temperature, T_c, where the system undergoes a phase transition. The condition for criticality, therefore, is

$$\lambda_c = \mu_0 = Jd, \tag{24}$$

which suggests that we may introduce a "reduced field," ϕ, by the definition

$$\phi = (\lambda - \lambda_c)/J = (\lambda/J) - d; \tag{25}$$

the condition for criticality then becomes

$$\phi_c = 0. \tag{26}$$

It follows that, as we approach the critical point from above, the parameter ϕ becomes much smaller than unity; ultimately, it becomes zero as T_c is reached and stays so for *all* $T < T_c$.

Now, substituting for the eigenvalues $\mu_{\boldsymbol{k}}$ into the sum appearing in equations (19) and (21), and making use of the representation

$$\frac{1}{z} = \int_0^\infty e^{-zx}dx, \tag{27}$$

[4]We denote Boltzmann's constant by the symbol k_B here so as to avoid confusion with the wavenumber k.

we have

$$\sum_{\boldsymbol{k}} \frac{1}{(\lambda - \mu_{\boldsymbol{k}})} = \frac{1}{J} \sum_{\{n_j\}} \int_0^\infty \exp\left\{ -\left[\phi + \sum_{j=1}^{d} \left\{ 1 - \cos\left(\frac{2\pi n_j}{N_j} \right) \right\} \right] x \right\} dx$$
$$= \frac{1}{J} \int_0^\infty e^{-\phi x} \prod_j \left[\sum_{n_j=0}^{N_j - 1} \exp\left\{ -x + x \cos\left(\frac{2\pi n_j}{N_j} \right) \right\} \right] dx. \tag{28}$$

For $N_j \gg 1$, the summation over n_j may be replaced by integration; writing $\theta_j = 2\pi n_j / N_j$, one gets

$$\sum_{n_j} \exp\{\cdots\} \approx \int_0^{2\pi} e^{-x + x\cos\theta_j} \frac{N_j}{2\pi} d\theta_j = N_j e^{-x} I_0(x), \tag{29}$$

where $I_0(x)$ is a modified Bessel function. Multiplying over j, one finally gets

$$\sum_{\boldsymbol{k}} \frac{1}{\lambda - \mu_{\boldsymbol{k}}} = \frac{N}{J} W_d(\phi), \tag{30}$$

where $W_d(\phi)$ is the so-called *Watson function*, defined by[5]

$$W_d(\phi) = \int_0^\infty e^{-\phi x} [e^{-x} I_0(x)]^d dx. \tag{31}$$

Equations (19) and (21) now take the form

$$W_d(\phi) = 2K - \frac{(\beta \mu B)^2}{2K\phi^2} \tag{32a}$$
$$= 2K(1 - m^2). \tag{32b}$$

The asymptotic behavior of the functions $W_d(\phi)$, for $\phi \ll 1$, is examined in Appendix G.

The critical behavior

We now analyze the various physical properties of the mean spherical model in different regimes of d.

(a) $d < 2$. For this regime we take expression (7a) of Appendix G and substitute it into equation (32a), with $B = 0$. We obtain

$$\phi|_{B=0} \approx \left[\frac{\Gamma\{(2-d)/2\}}{2(2\pi)^{d/2} K} \right]^{2/(2-d)} \sim \left(\frac{k_B T}{J} \right)^{2/(2-d)}. \tag{33}$$

[5]Note that our definition of the function $W_d(\phi)$ differs slightly from the one adopted by Barber and Fisher (1973); this difference arises from the fact that our J is *twice*, and our ϕ is *one-half*, of theirs.

We see that ϕ in this case goes to zero only as $T \to 0$. The phase transition, therefore, takes place at $T_c = 0$. Equations (20b), (23), (24), and (25) then give the following for the low-temperature susceptibility

$$\chi_0 = \frac{N\mu^2}{2J\phi} \sim \frac{N\mu^2}{J}\left(\frac{k_B T}{J}\right)^{-2/(2-d)} \tag{34}$$

and for the low-temperature specific heat

$$C_0 - \frac{1}{2}Nk_B = -NJ\left(\frac{\partial \phi}{\partial T}\right)_0 \sim -Nk_B\left(\frac{k_B T}{J}\right)^{d/(2-d)}. \tag{35}$$

(b) $d = 2$. We now use expression (7b) of Appendix G and obtain

$$\phi|_{B=0} \sim \exp(-4\pi J/k_B T), \tag{36}$$

so that once again $T_c = 0$ but now, at low temperatures,

$$\chi_0 \sim (N\mu^2/J)\exp(4\pi J/k_B T) \tag{37}$$

and

$$C_0 - \frac{1}{2}Nk_B \sim -Nk_B(J/k_B T)^2 \exp(-4\pi J/k_B T). \tag{38}$$

(c) $2 < d < 4$. We now substitute expression (7c) of Appendix G into equation (32a), with the result

$$W_d(0) - \frac{|\Gamma\{(2-d)/2\}|}{(2\pi)^{d/2}}\phi^{(d-2)/2} = 2K - \frac{(\beta\mu B)^2}{2K\phi^2}. \tag{39}$$

The critical point is now determined by setting $B = 0$ and letting $\phi \to 0$; the condition for criticality then turns out to be

$$K_c = \frac{1}{2}W_d(0). \tag{40}$$

The variation of ϕ with T as one approaches the critical point is given by

$$\phi|_{B=0} \approx \left[\frac{2(2\pi)^{d/2}(K_c - K)}{|\Gamma\{(2-d)/2\}|}\right]^{2/(d-2)} \quad (K \lesssim K_c). \tag{41a}$$

We also note that once ϕ becomes zero it stays so for all temperatures below, that is,

$$\phi|_{B=0} = 0 \quad (K \geq K_c). \tag{41b}$$

It then follows that

$$\chi_0 \sim (K_c - K)^{-2/(d-2)} \sim (T - T_c)^{-2/(d-2)} \quad (T \gtrsim T_c) \tag{42}$$

and is infinite for $T \le T_c$. At the same time

$$C_0 - \frac{1}{2}Nk_B \sim (T - T_c)^{(4-d)/(d-2)} \quad (T \gtrsim T_c) \tag{43}$$

and it vanishes for $T \le T_c$. The spontaneous magnetization is determined by equations (32b), (40), and (41b); we obtain a remarkably simple result

$$m_0 = (1 - K_c/K)^{1/2} = (1 - T/T_c)^{1/2} \quad (T \le T_c). \tag{44}$$

Finally, if in equation (39) we retain B but set $T = T_c$, we get

$$\phi_c \sim B^{4/(d+2)} \quad (T = T_c); \tag{45}$$

equation (20a) then gives

$$m_c = \frac{\mu B}{2J\phi_c} \sim B^{(d-2)/(d+2)} \quad (T = T_c). \tag{46}$$

The foregoing results give, for the critical exponents of the system,

$$\alpha = \frac{d-4}{d-2}, \quad \beta = \frac{1}{2}, \quad \gamma = \frac{2}{d-2}, \quad \delta = \frac{d+2}{d-2} \quad (2 < d < 4). \tag{47}$$

(d) $d > 4$. In this regime we employ expression (8) of Appendix G. The condition for criticality remains the same as in (40); the variation of ϕ with T as we approach the critical point is, however, different. We now have, for all $d > 4$,

$$\phi|_{B=0} \sim (K_c - K)^1 \quad (K \lesssim K_c). \tag{48}$$

The subsequent results are modified accordingly:

$$\chi_0 \sim (T - T_c)^{-1}, \quad C_0 - \frac{1}{2}Nk_B \sim (T - T_c)^0 \qquad (T \gtrsim T_c) \tag{49}$$

$$\phi_c \sim B^{2/3}, \quad m_c \sim B^{1/3} \qquad (T = T_c). \tag{50}$$

Equations (41b) and (44) continue to apply as such. We, therefore, conclude that

$$\alpha = 0, \quad \beta = \frac{1}{2}, \quad \gamma = 1, \quad \delta = 3 \quad (d > 4). \tag{51}$$

(e) $d = 4$. For this borderline case, we use expression (12) of Appendix G. Once again, the condition for criticality remains the same; however, the variation of ϕ with T as one approaches the critical point is now determined by the implicit relation

$$\phi \ln(1/\phi) \approx 8\pi^2(K_c - K) \quad (K \lesssim K_c). \tag{52}$$

Introducing the conventional parameter

$$t = (T - T_c)/T_c = (K_c - K)/K, \tag{53}$$

we get, to leading order in t,

$$\phi|_{B=0} \sim t/\ln(1/t) \quad (0 < t \ll 1). \tag{54}$$

It follows that

$$\chi_0 \sim t^{-1}\ln(1/t), \quad C_0 \sim 1/\ln(1/t) \quad (0 < t \ll 1) \tag{55}$$

$$\phi_c \sim B^{2/3}/\{\ln(1/B)\}^{1/3}, \quad m_c \sim \{B\ln(1/B)\}^{1/3} \quad (t = 0). \tag{56}$$

Spin–spin correlations

Following the procedure that led to equations (16) through (19), we obtain *in the absence of the field*

$$G(\boldsymbol{r}, \boldsymbol{r}') \equiv \overline{\sigma(\boldsymbol{r})\sigma(\boldsymbol{r}')} = \frac{1}{2N\beta}\sum_{\boldsymbol{k}} \frac{\exp\{i(\boldsymbol{k}\cdot\boldsymbol{R})\}}{\lambda - \mu_{\boldsymbol{k}}} \quad (\boldsymbol{R} = \boldsymbol{r} - \boldsymbol{r}'); \tag{57}$$

compare to equation (19), with $B = 0$. The summation over $\boldsymbol{k}$ in (57) can be handled in the same manner as was done in (28); however, the resulting summation over n_j now turns out to be

$$\begin{aligned} \sum_{n_j}\{\cdots\} &\approx \int_0^{2\pi} \exp\{iR_j\theta_j/a\} e^{-x + x\cos\theta_j} \frac{N_j}{2\pi} d\theta_j \\ &= N_j e^{-x} I_{R_j/a}(x); \end{aligned} \tag{58}$$

compare to (29). This leads to the result

$$G(\boldsymbol{R}) = \frac{1}{2K}\int_0^\infty e^{-\phi x}\prod_j [e^{-x} I_{R_j/a}(x)]dx; \tag{59}$$

compare to equation (32a), with $B = 0$. For the functions $I_n(x)$ we may use the asymptotic expression (see Singh and Pathria, 1985a)

$$I_n(x) \approx \frac{e^{x - n^2/2x}}{\sqrt{(2\pi x)}} \quad (x \gg 1), \tag{60}$$

so that, for $\phi \ll 1$,

$$G(\boldsymbol{R}) \approx \frac{1}{2(2\pi)^{d/2}K}\int_0^\infty e^{-\phi x - R^2/(2a^2x)}x^{-d/2}dx$$
$$= \frac{1}{(2\pi)^{d/2}K}\left(\frac{a^2}{\xi R}\right)^{(d-2)/2} K_{(d-2)/2}\left(\frac{R}{\xi}\right), \tag{61}$$

where $K_\mu(x)$ is the other modified Bessel function while

$$\xi = a/(2\phi)^{1/2}. \tag{62}$$

For $R \gg \xi$, we may use the asymptotic result $K_\mu(x) \approx (\pi/2x)^{1/2}e^{-x}$; equation (61) then becomes

$$G(\boldsymbol{R}) \approx \frac{a^{d-2}}{2K\xi^{(d-3)/2}(2\pi R)^{(d-1)/2}}e^{-R/\xi}, \tag{63}$$

which identifies ξ as the *correlation length* of the system.

Now, comparing equation (62) with equation (20b), we find that

$$\chi_0 = \frac{N\mu^2}{2J\phi} = \frac{N\mu^2}{Ja^2}\xi^2. \tag{64}$$

In view of the fact that $\xi \sim \chi_0^{1/2}$, we infer that, in *all* regimes of d, the exponent $\nu = \frac{1}{2}\gamma$ and hence, by relations (12.12.10) and (12.12.11), the exponent $\eta = 0$. To obtain this last result directly from equation (61), we observe that, as $T \to T_c$ from above, the parameter $\phi \to 0$ and hence $\xi \to \infty$. We may then use the approximation $R/\xi \ll 1$ and employ the formula

$$K_\mu(x) \approx \frac{1}{2}\Gamma(\mu)\left(\frac{1}{2}x\right)^{-\mu} \quad (\mu > 0), \tag{65}$$

to obtain

$$G(\boldsymbol{R})|_{T=T_c} \approx \frac{\Gamma\{(d-2)/2\}}{4\pi^{d/2}K_c}\frac{a^{d-2}}{R^{d-2}} \quad (d > 2). \tag{66}$$

The power of R appearing here clearly shows that $\eta = 0$. Finally, substituting equation (41a) into equation (62), we get

$$\xi \approx \frac{1}{2}a\left[\frac{|\Gamma\{(2-d)/2\}|}{4\pi^{d/2}(K_c - K)}\right]^{1/(d-2)} \quad (K \lesssim K_c), \tag{67}$$

which shows that for $2 < d < 4$ the critical exponent $\nu = 1/(d-2)$.

For $T < T_c$ we expect the function $G(\boldsymbol{R})$ to affirm the presence of *long-range order* in the system, that is, in the limit $R \to \infty$, it should tend to a limit, $\overline{\sigma}^2$, that is *nonzero*. To demonstrate this property of $G(\boldsymbol{R})$, we need to take a closer look at the derivations of this subsection that were carried out with the express purpose of obtaining results valid in the thermodynamic limit ($N \to \infty$). This resulted in "errors" that were negligible in the region $T \gtrsim T_c$ but are not so when $T < T_c$. The first such error crept in when we replaced the summations over $\{n_j\}$ in equation (28) by integrations; that suppressed contribution from the term with $\boldsymbol{n} = 0$. Equation (30), therefore, accounts for *only* the ($\boldsymbol{k} \neq 0$)-terms of the original sum in (28), and the missing term, $1/J\phi$, may be added to it *ad hoc*.[6] Equation (19), with $B = 0$, then becomes

$$\frac{1}{2\beta}\left[\frac{1}{J\phi} + \frac{N}{J}W_d(\phi)\right] = N. \tag{68}$$

Now, when ϕ becomes very very small, $W_d(\phi)$ may be approximated by $W_d(0)$, which is precisely equal to $2K_c$; equation (68) then gives

$$\phi \approx [2N(K - K_c)]^{-1} \quad (K > K_c), \tag{69}$$

rather than zero! The correlation length then turns out to be

$$\xi = a/(2\phi)^{1/2} \approx a[N(K - K_c)]^{1/2} \quad (K > K_c), \tag{70}$$

rather than infinite! Now, the same error was committed once again in going from equation (57) to (59); so, the primary result for $G(\boldsymbol{R})$, as given in equation (61), may be similarly amended by adding the missing term $1/(2N\beta J\phi)$ which, by (69), is exactly equal to $(1 - K_c/K)$. Thus, for $R \ll \xi$, we obtain, instead of (66),

$$G(\boldsymbol{R}) \approx \left(1 - \frac{K_c}{K}\right) + \frac{\Gamma\{(d-2)/2\}}{4\pi^{d/2}K}\frac{a^{d-2}}{R^{d-2}} \quad (K > K_c). \tag{71}$$

Now if we let $R \to \infty$, $G(\boldsymbol{R})$ does approach a *nonzero* value $\overline{\sigma}^2$, which is precisely the same as m_0^2 given by equation (44). It is remarkable, though, that in the present derivation the magnetic field B has not been introduced at any stage of the calculation, which underscores the fundamental role played by correlations in bringing about long-range order in the system.

In the preceding paragraph we outlined the essential argument that led to the desired expression, (71), for $G(\boldsymbol{R})$. For a more rigorous analysis of this problem, see Singh and Pathria (1985b, 1987a).

Physical significance of the spherical model

With a constraint as relaxed as in equation (3), or even more so in (9), one wonders how meaningful the spherical model is from a physical point of view. Relief comes from the fact, first established by Stanley (1968, 1969a,b), that the spherical model provides a correct

[6] This is reminiscent of a similar problem, and a similar *ad hoc* solution, encountered in the study of Bose–Einstein condensation in Section 7.1; see also Section 13.6.

representation of the $(n \to \infty)$-limit of an n-vector model with nearest-neighbor interactions; see also Kac and Thompson (1971). This connection arises from the very nature of the constraint imposed on the model, which introduces a super spin vector $\mathscr{S}$ with N degrees of freedom; it is not surprising that, in the limit $N \to \infty$, the model in some sense acquires the same sort of freedom that an n-vector model has in the limit $n \to \infty$.

In any case, this connection brings the spherical model in line with, and actually makes it a good guide for, all models with *continuous* symmetry, namely the ones with $n \geq 2$. And since it can be solved exactly in arbitrary dimensions, this model gives us some idea as to what to expect of models for which n is finite. For instance, we have seen that, for $d > 4$, the critical exponents of the spherical model are the same as the ones obtained from the mean field theory. Now, fluctuations are neglected in the mean field theory but, among the variety of models we are considering, fluctuations should be largest in the spherical model, for it has the largest number of degrees of freedom. If, for $d > 4$, fluctuations turn out to be negligible in the spherical model, they would be even more so in models with finite n. It thus follows that, regardless of the actual value of n, mean field theory should be valid for *all* these models when $d > 4$. See, in this connection, Section 14.4 as well.

For $d < 4$, the final results depend significantly on n. The spherical model now provides a starting point from which one may carry out the so-called $(1/n)$-expansions to determine how models with finite n might behave in this regime. Such an approach was initiated by Abe and collaborators (1972, 1973) and independently by Ma (1973); for a detailed account of this approach, along with the results following from it, see Ma (1976c). Finally, for a comprehensive discussion of the spherical model, including the one with long-range interactions, see the review article by Joyce (1972).

13.6 The ideal Bose gas in arbitrary dimensions

In this section we propose to examine the problem of Bose–Einstein condensation in an ideal Bose gas in *arbitrary* dimensions. As was first shown by Gunton and Buckingham (1968), the phenomenon of Bose–Einstein condensation falls in the same universality class as the phase transition in the spherical model; accordingly, the ideal Bose gas too corresponds to the $(n \to \infty)$-limit of an n-vector model. It must, however, be borne in mind that liquid He^4, whose transition from a normal to a superfluid state is often regarded as a manifestation of the "Bose–Einstein condensation in an interacting Bose liquid," actually pertains to the case $n = 2$. Now, just as the spherical model turns out to be a good guide for all models with continuous symmetry including the XY model (for which $n = 2$), in the same way the ideal Bose gas has also been a good guide for liquid He^4.

We consider a Bose gas composed of N noninteracting particles confined to a box of volume $V(= L_1 \times \cdots \times L_d)$ at temperature T. Following the procedure of Section 7.1, we obtain for the pressure P of the gas

$$P = -\frac{k_B T}{V} \sum_{\varepsilon} \ln(1 - z e^{-\beta\varepsilon}) = \frac{k_B T}{\lambda^d} g_{(d+2)/2}(z), \tag{1}$$

where $\lambda[= h/\sqrt{(2\pi m k_B T)}]$ is the mean thermal wavelength of the particles, z is the fugacity of the gas, which is related to the chemical potential μ through the formula

$$z = \exp(\beta\mu) < 1 \quad (\beta = 1/k_B T), \tag{2}$$

while $g_\nu(z)$ are Bose–Einstein functions whose main properties are discussed in Appendix D. The quantity z is determined by the equation

$$N = \sum_{\varepsilon} (z^{-1} e^{\beta\varepsilon} - 1)^{-1} = N_0 + N_e, \tag{3}$$

where N_0 is the mean number of particles in the ground state ($\varepsilon = 0$),

$$N_0 = z/(1-z), \tag{4}$$

while N_e is the mean number of particles in the excited states ($\varepsilon > 0$):

$$N_e = \frac{V}{\lambda^d} g_{d/2}(z). \tag{5}$$

At high temperatures, where z is significantly below the limiting value 1, N_0 is negligible in comparison with N; the quantity z is then determined by the simplified equation

$$N = \frac{V}{\lambda^d} g_{d/2}(z), \tag{6}$$

and the pressure P in turn is given by the expression

$$P = \frac{N k_B T}{V} \frac{g_{(d+2)/2}(z)}{g_{d/2}(z)}. \tag{7}$$

The internal energy of the gas may be obtained from the relationship

$$U = \frac{1}{2} d(PV); \tag{8}$$

see the corresponding derivation of equation (7.1.12) as well as of equation (6.4.4).

Now, making use of the recurrence relation (D.10) and remembering that the mean thermal wavelength $\lambda \propto T^{-1/2}$, we get from equation (6)

$$\frac{1}{z}\left(\frac{\partial z}{\partial T}\right)_v = -\frac{d}{2T} \frac{g_{d/2}(z)}{g_{(d-2)/2}(z)} \quad \left(v = \frac{V}{N}\right), \tag{9}$$

and from equation (1)

$$\frac{1}{z}\left(\frac{\partial z}{\partial T}\right)_P = -\frac{d+2}{2T} \frac{g_{(d+2)/2}(z)}{g_{d/2}(z)}. \tag{10}$$

It is now straightforward to show that the specific heats C_V and C_P of the gas are given by the formulae

$$\frac{C_V}{Nk_B} = \frac{d(d+2)}{4}\frac{g_{(d+2)/2}(z)}{g_{d/2}(z)} - \frac{d^2}{4}\frac{g_{d/2}(z)}{g_{(d-2)/2}(z)} \tag{11}$$

and

$$\frac{C_P}{Nk_B} = \frac{(d+2)^2}{4}\frac{\{g_{(d+2)/2}(z)\}^2 g_{(d-2)/2}(z)}{\{g_{d/2}(z)\}^3} - \frac{d(d+2)}{4}\frac{g_{(d+2)/2}(z)}{g_{d/2}(z)}, \tag{12}$$

respectively; it follows that the ratio

$$\frac{C_P}{C_V} = \frac{(d+2)}{d}\frac{g_{(d+2)/2}(z)g_{(d-2)/2}(z)}{\{g_{d/2}(z)\}^2}. \tag{13}$$

The isothermal compressibility κ_T and the adiabatic compressibility κ_S turn out to be

$$\kappa_T = -\frac{1}{v}\left(\frac{\partial v}{\partial P}\right)_T = -\frac{1}{v}\frac{(\partial v/\partial z)_T}{(\partial P/\partial z)_T} = \frac{g_{(d+2)/2}(z)g_{(d-2)/2}(z)}{\{g_{d/2}(z)\}^2}\frac{1}{P} \tag{14}$$

and

$$\kappa_S = -\frac{1}{v}\left(\frac{\partial v}{\partial P}\right)_S = -\frac{1}{v}\frac{(\partial v/\partial T)_z}{(\partial P/\partial T)_z} = \frac{d}{d+2}\frac{1}{P}, \tag{15}$$

respectively; note that the ratio κ_T/κ_S is precisely equal to the ratio C_P/C_V, as is expected thermodynamically.

As the temperature of the gas is reduced, keeping v constant, the fugacity z increases and ultimately approaches its limiting value 1 — marking the end of the regime where N_0 was negligible in comparison with N. Whether this limit is reached at a finite T or at $T=0$ depends entirely on the value of d; see equation (6), which tells us that if the function $g_{d/2}(z)$, as $z \to 1-$, is *bounded* then the limit in question will be reached at a finite T. On the other hand, if $g_{d/2}(z)$, as $z \to 1-$, is *unbounded* then the desired limit will be reached at $T=0$ instead. To settle this question, we refer to equations (D.9) and (D.11), which summarize the behavior of the function $g_\nu(z)$ as $z \to 1-$ (or as $\alpha \to 0+$, where $\alpha = -\ln z$); we thus have

$$g_{d/2}(e^{-\alpha}) \approx \begin{cases} \Gamma\left(\frac{2-d}{2}\right)\alpha^{-(2-d)/2} + \text{const.} & \text{for} \quad d<2 \quad (16a) \\ \ln(1/\alpha) + \frac{1}{2}\alpha & \text{for} \quad d=2 \quad (16b) \\ \zeta\left(\frac{d}{2}\right) - \left|\Gamma\left(\frac{2-d}{2}\right)\right|\alpha^{(d-2)/2} & \text{for} \quad 2<d<4 \quad (16c) \\ \zeta(2) - \{\ln(1/\alpha)+1\}\alpha & \text{for} \quad d=4 \quad (16d) \\ \zeta\left(\frac{d}{2}\right) - \zeta\left(\frac{d-2}{2}\right)\alpha & \text{for} \quad d>4, \quad (16e) \end{cases}$$

$\zeta(\nu)$ being the Riemann zeta function. Similarity of this system with the spherical model is quite transparent.

We readily see from equation (6) that, for $d > 2$, $\alpha \to 0$ at a *finite* temperature T_c, given by

$$\lambda_c^d = v\zeta(d/2), \tag{17}$$

with the result that

$$T_c = \frac{h^2}{2\pi m k_B}\left[\frac{1}{v\zeta(d/2)}\right]^{2/d}; \tag{18}$$

for $d \leq 2$, $\alpha \to 0$ only as $\lambda \to \infty$, so $T_c = 0$. For brevity, we confine our further discussion only to $d > 2$.

The critical behavior

As T approaches T_c from above, the manner in which $\alpha \to 0$ is determined by substituting the appropriate expression (16) into (6) and utilizing the criticality condition (17). For $2 < d < 4$, one gets asymptotically

$$\left|\Gamma\left(\frac{2-d}{2}\right)\right|\alpha^{(d-2)/2} \approx \frac{1}{v}(\lambda_c^d - \lambda^d) \simeq \frac{d}{2}\zeta\left(\frac{d}{2}\right)\left[\frac{T}{T_c} - 1\right]. \tag{19}$$

For $T \gtrsim T_c$, this gives

$$\alpha \sim t^{2/(d-2)} \quad [t = (T - T_c)/T_c, 0 < t \ll 1]. \tag{20}$$

As $T \to T_c$, the specific heat C_P and the isothermal compressibility κ_T diverge because the function $g_{(d-2)/2}(z)$ appearing in equations (12) and (14), being $\sim\alpha^{-(4-d)/2}$ [see equation (D.9)], becomes divergent; for small t,

$$C_P \sim \kappa_T \sim t^{-(4-d)/(d-2)}. \tag{21}$$

The specific heat C_V, on the other hand, approaches a finite value,

$$\left(\frac{C_V}{Nk_B}\right)_{T\to T_C+} = \frac{d(d+2)}{4}\frac{\zeta\{(d+2)/2\}}{\zeta(d/2)}, \tag{22}$$

with a derivative that, depending on the actual value of d, might diverge:

$$\frac{1}{Nk_B}\left(\frac{\partial C_V}{\partial T}\right)_V = \frac{1}{T}\left[\frac{d^2(d+2)}{8}\frac{g_{(d+2)/2}(z)}{g_{d/2}(z)} - \frac{d^2}{4}\frac{g_{d/2}(z)}{g_{(d-2)/2}(z)} - \frac{d^3}{8}\frac{\{g_{d/2}(z)\}^2 g_{(d-4)/2}(z)}{\{g_{(d-2)/2}(z)\}^3}\right] \tag{23a}$$

$$\sim -\alpha^{-(d-3)} \sim -t^{-2(d-3)/(d-2)} \quad (3 < d < 4). \tag{23b}$$

Equating the exponent appearing here with $(1+\alpha)$, we conclude[7] that the critical exponent α for this system is $(d-4)/(d-2)$; compare to equation (13.5.47). For a proper appreciation of the critical behavior of C_V, we must as well examine the region $T < T_c$, along with the limit $T \to T_c-$.

For $T < T_c$, the fugacity z is essentially equal to 1; equations (5) and (17) then give

$$N_e = \frac{V}{\lambda^d}\zeta\left(\frac{d}{2}\right) = N\left(\frac{\lambda_c}{\lambda}\right)^d = N\left(\frac{T}{T_c}\right)^{d/2}. \tag{24}$$

It follows that

$$N_0 = N - N_e = N\left[1 - \left(\frac{T}{T_c}\right)^{d/2}\right]. \tag{25}$$

Equation (4) then tells us that the precise value of z in this region is given by

$$z = N_0/(N_0+1) \simeq 1 - 1/N_0, \tag{26}$$

which gives

$$\alpha = -\ln z \simeq 1/N_0, \tag{27}$$

rather than zero. Disregarding this subtlety, equation (1) gives

$$P = \frac{k_B T}{\lambda^d}\zeta\left(\frac{d+2}{2}\right) \propto T^{(d+2)/2}. \tag{28}$$

Since P here is a function of T only, the quantities κ_T and C_P in this region are infinite; see, however, Problem 13.26. From equations (8) and (28), we get

$$U = \frac{1}{2}d\frac{k_B TV}{\lambda^d}\zeta\left(\frac{d+2}{2}\right), \tag{29}$$

which gives

$$\frac{C_V}{Nk_B} = \frac{d(d+2)}{4}\frac{v}{\lambda^d}\zeta\left(\frac{d+2}{2}\right) = \frac{d(d+2)}{4}\frac{\zeta\{(d+2)/2\}}{\zeta(d/2)}\left(\frac{T}{T_c}\right)^{d/2}. \tag{30}$$

As $T \to T_c-$, we obtain precisely the same limit as in (22) — showing that C_V is *continuous* at the critical point. Its derivative, however, turns out to be different from the one given in

[7]We equate this exponent with $(1+\alpha)$ because if $C_V \sim t^{-\alpha}$, then $\partial C_V/\partial t \sim t^{-\alpha-1}$. We hasten to add that the critical exponent α should not be confused with the physical quantity denoted by the same symbol, namely $\alpha(= -\ln z)$, which was introduced just before equations (16) and has been used throughout this section.

(23), for now

$$\frac{1}{Nk_B}\left(\frac{\partial C_V}{\partial T}\right)_V = \frac{1}{T_c}\frac{d^2(d+2)}{8}\frac{\zeta\{(d+2)/2\}}{\zeta(d/2)}\left(\frac{T}{T_c}\right)^{(d-2)/2} \tag{31a}$$

$$\rightarrow \frac{1}{T_c}\frac{d^2(d+2)}{8}\frac{\zeta\{(d+2)/2\}}{\zeta(d/2)} \tag{31b}$$

as $T \to T_c-$.

As for the condensate fraction, N_0/N, equation (25) gives

$$\frac{N_0}{N} = 1 - \left(\frac{T}{T_c}\right)^{d/2} \approx \frac{d}{2}|t| \quad [t < 0, |t| \ll 1]. \tag{32}$$

Now, the order parameter in the present problem is a complex number, Ψ_0, such that $|\Psi_0|^2 = N_0/V$, the condensate particle density in the system; see Gunton and Buckingham (1968). We therefore, expect that, for $|t| \ll 1, N_0$ would be $\sim t^{2\beta}$; equation (32) then tells us that the critical exponent β in this case has the classical value $\frac{1}{2}$ for *all* $d > 2$. To determine the exponents γ and δ, we must introduce a "complex Bose field, conjugate to the order parameter Ψ_0" and examine quantities such as the "Bose susceptibility" χ as well as the variation of Ψ_0 with the Bose field at $T = T_c$. Proceeding that way, one obtains: $\gamma = 2/(d-2)$ and $\delta = (d+2)/(d-2)$, just as for the spherical model.

The pair correlations

We now examine the pair correlation function of the ideal Bose gas

$$G(\boldsymbol{R}) = \frac{1}{V}\sum_{\boldsymbol{k}} \frac{e^{i\boldsymbol{k}\cdot\boldsymbol{R}}}{e^{\alpha+\beta\varepsilon(\boldsymbol{k})} - 1}. \tag{33}$$

As usual, we replace the summation over $\boldsymbol{k}$ by integration (mindful of the fact that this replacement suppresses the ($\boldsymbol{k} = 0$)-term which may, therefore, be kept aside). Making use of equation (C.11) in Appendix C, we get

$$G(\boldsymbol{R}) = \frac{N_0}{V} + \frac{1}{(2\pi)^d}\int \frac{e^{i\boldsymbol{k}\cdot\boldsymbol{R}}}{e^{\alpha+\beta\hbar^2k^2/2m} - 1} d^dk$$

$$= \frac{N_0}{V} + \frac{1}{(2\pi)^{d/2}R^{(d-2)/2}}\int_0^\infty \left[\sum_{j=1}^\infty e^{-j\alpha - j\beta\hbar^2k^2/2m}\right] J_{(d-2)/2}(kR)k^{d/2}dk$$

$$= \frac{N_0}{V} + \frac{1}{\lambda^d}\sum_{j=1}^\infty e^{-j\alpha - \pi R^2/j\lambda^2} j^{-d/2} \quad \left[\lambda = \hbar\left(\frac{2\pi\beta}{m}\right)^{1/2}\right]; \tag{34}$$

compare to equations (3) through (5), which pertain to the case $R = 0$. For $R > 0$, one may extend the summation over j from $j = 0$ to $j = \infty$, for the term so added is identically zero.

At the same time, the summation over j may be replaced by integration — committing errors $O(e^{-R/\lambda})$, which are negligible so long as $R \gg \lambda$; for details, see Zasada and Pathria (1976). We thus obtain

$$G(\mathbf{R}) = \frac{N_0}{V} + \frac{1}{\lambda^d}\int_0^\infty e^{-j\alpha - \pi R^2/j\lambda^2} j^{-d/2} dj$$

$$= \frac{N_0}{V} + \frac{2}{\lambda^2 (2\pi \xi R)^{(d-2)/2}} K_{(d-2)/2}\left(\frac{R}{\xi}\right), \tag{35}$$

where $K_\mu(x)$ is a modified Bessel function while

$$\xi = \lambda/(2\pi^{1/2}\alpha^{1/2}). \tag{36}$$

For $T \gtrsim T_c$, we may use expression (20) for α; equation (36) then gives

$$\xi \sim \lambda t^{-1/(d-2)} \; (0 < t \ll 1), \tag{37}$$

which means that $\xi \gg \lambda$. Now, if $R \gg \xi$, equation (35) reduces to

$$G(\mathbf{R}) \approx \frac{1}{\lambda^2 (2\pi\xi)^{(d-3)/2} R^{(d-1)/2}} e^{-R/\xi}, \tag{38}$$

which identifies ξ as the *correlation length* of the system. Equation (37) then tells us that for, $2 < d < 4$, the critical exponent ν of the ideal Bose gas is $1/(d-2)$. At $T = T_c, \xi$ is infinite; equation (35) now gives

$$G(\mathbf{R}) \approx \frac{\Gamma\{(d-2)/2\}}{\pi^{(d-2)/2} \lambda_c^2 R^{d-2}}, \tag{39}$$

which shows that the critical exponent $\eta = 0$. For $T < T_c$, ξ continues to be infinite but now the condensate fraction, which is a measure of the long-range order in the system, is *nonzero*. The correlation function then assumes the form

$$G(\mathbf{R}) = |\Psi_0|^2 + \frac{A(T)}{R^{d-2}}, \tag{40}$$

where

$$A(T) = \frac{\Gamma\{(d-2)/2\}}{\pi^{(d-2)/2} \lambda^2} \propto T; \tag{41}$$

compare this result to the corresponding equation (13.5.71) of the spherical model.

In the paper quoted earlier, Gunton and Buckingham also generalized the study of Bose–Einstein condensation to the single-particle energy spectrum $\varepsilon \sim k^\sigma$, where σ is a positive number not necessarily equal to 2. They found that the phase transition, at a finite temperature T_c, now took place for all $d > \sigma$, and the critical exponents in the regime $\sigma < d < 2\sigma$ turned out to be

$$\alpha = \frac{d-2\sigma}{d-\sigma}, \quad \beta = \frac{1}{2}, \quad \gamma = \frac{\sigma}{d-\sigma}, \quad \delta = \frac{d+\sigma}{d-\sigma}, \quad \nu = \frac{1}{d-\sigma}, \quad \eta = 2-\sigma. \tag{42}$$

While mathematically correct, these results left one with the awkward conclusion that the Bose gas in its extreme relativistic state ($\sigma = 1$) was in a *different* universality class than the one in the nonrelativistic state ($\sigma = 2$). It was shown later by Singh and Pandita (1983) that, if one employs the appropriate energy spectrum $\varepsilon = c\surd(m_0^2c^2 + \hbar^2k^2)$ and, at the same time, allows for the possibility of particle–antiparticle pair production in the system, as had been suggested earlier by Haber and Weldon (1981, 1982), then the relativistic Bose gas falls in the *same* universality class as the nonrelativistic one; see Problem (13.27).

13.7 Other models

In Section 13.4 we saw that a two-dimensional lattice model characterized by a discrete order parameter ($n = 1$) underwent a phase transition, accompanied by a spontaneous magnetization m_0, at a *finite* temperature T_c; naturally, one would expect the same if d were greater than 2. On the other hand, the spherical model, which is characterized by a continuous order parameter (with $n = \infty$), undergoes such a transition *only* if $d > 2$. The question then arises whether intermediate models, with $n = 2, 3, \ldots$, would undergo a phase transition at a finite T_c if d were equal to 2. The answer to this important question was provided by Mermin and Wagner (1966) who, making use of a well-known inequality due to Bogoliubov (1962), established the following theorem:[8]

> *Systems composed of spins with continuous symmetry* ($n \geq 2$) *and short-range interactions do not acquire spontaneous magnetization at any finite temperature T if the space dimensionality d* ≤ 2.

In this sense, systems with $n \geq 2$ behave in a manner similar to the spherical model — and quite unlike the Ising model!

The marginal case ($n = 2, d = 2$), however, deserves a special mention. Clearly, this refers to an *XY* model in two dimensions, which has a direct relevance to superfluid He^4 adsorbed on a substrate. As shown by Kosterlitz and Thouless (1972, 1973), this model exhibits a curious phase transition in that, while no long-range order develops at any finite temperature T, various physical quantities such as the susceptibility, the correlation length, the specific heat, and the superfluid density do become singular at a finite temperature T_c, whose precise value is determined by point defects, such as vortices or

[8]For a review of this theorem and other allied questions, see (Griffiths, 1972, pp. 84–89).

dislocations, in the system. The correlation length ξ, as $T \to T_c+$, displays an essential singularity,

$$\xi \sim e^{b'/(T-T_c)^{1/2}}, \tag{1}$$

where b' is a nonuniversal constant; of course, the critical temperature itself is nonuniversal and, for a square lattice, is given by

$$K_c = J/kT_c \simeq 1.12; \tag{2}$$

see Hasenbusch and Pinn (1997) and Dukovski, Machta, and Chayes (2002). The singular part of the specific heat shows a somewhat similar behavior, namely

$$c^{(s)} \sim \xi^{-2} \tag{3}$$

which, measured against a regular background, is rather indetectable since every derivative of the specific heat is finite at the critical point, yet the function is nonanalytic. The superfluid density behaves rather strangely; it approaches a finite value, as $T \to T_c-$, preceded by a square-root cusp. The correlation function is no different; at $T = T_c$, Kosterlitz (1974) found a logarithmic factor along with a power law, namely

$$g(r) \sim \frac{[\ln(r/a)]^{1/8}}{r^{1/4}}, \tag{4}$$

while for $T < T_c$ we encounter a temperature-dependent exponent η such that

$$g(r) \sim \frac{1}{r^{\eta(T)}}, \tag{5}$$

where $\eta(T) \approx kT/2\pi J$, for $kT \ll J$, as shown by Berezinskii (1970) and $\eta(T_c) = 1/4$, as shown by Kosterlitz (1974); see Berche, Sanchez, and Paredes (2002) for a numerical determination of $\eta(T)$. This phase, with a power-law decay of correlations, is said to display quasi-long-range order. For further details of this transition, see Kosterlitz and Thouless (1978) and Nelson (1983); for a pedagogical account, see Plischke and Bergersen (1989), Section 5.E.

For other exactly soluble models in two dimensions, see Baxter (1982), Wu (1982), Nienhuis (1987), and Cardy (1987), where references to other relevant literature on the subject can also be found.

We now proceed to consider the situation in *three* dimensions. Here, no exact solutions exist except for the extreme case $n = \infty$, which has been discussed in Section 13.5. However, an enormous amount of effort has been spent in obtaining approximate solutions which, over the years, have become accurate enough to be regarded as "almost exact." Irrespective of the model under study, the problem has generally been attacked along three different lines which, after some refinements, have led to almost identical results. In summary, these approaches may be described as follows.

The method of series expansions

In this approach, the partition function or other relevant properties of the system are expanded as a high-temperature series such as (13.4.5), with expansion parameter $v = \tanh(\beta J)$, or as a low-temperature series such as (13.4.9), with expansion parameter $w = \exp(-2\beta J)$; in the presence of an external field, one would have series with two expansion parameters instead of one. In either case, the first major task involves the numerical computation of coefficients, such as $n(r)$ and $m(r)$, on the basis of graph theory and other allied techniques, while the second major task involves analysis of these coefficients with a view to determining the location and the nature of the singularity governing the property in question. The latter task is generally accomplished by employing the *ratio method*, which examines the trend of the ratio of two consecutive coefficients, such as $n(r)$ and $n(r-1)$, as $r \to \infty$, or by constructing *Padé approximants* which, in a sense, provide a continuation of the known (finite) series beyond its normal range of validity up to its radius of convergence — thus locating and examining the nature of the relevant singularity. Since their inception (in the mid-1950s for the ratio method and the early 1960s for the Padé approximants), these techniques have been expanded, refined, and enriched in so many ways that it would be hopeless to attempt a proper review of them here. Suffice it to say that the reader may refer to Volume 3 of the Domb–Green series, which is devoted solely to the method of series expansions — in particular, to the articles by Gaunt and Guttmann (1974) on the asymptotic analysis of the various coefficients, by Domb (1974) on the Ising model, by Rushbrooke, Baker, and Wood (1974) on the Heisenberg model, by Stanley (1974) on the n-vector models, and by Betts (1974) on the XY model. For more recent reviews, see Guttmann (1989) and Baker (1990), where references to other relevant work on the subject are also available.

The renormalization group method

This method is based on the crucial observation (Wilson, 1971) that, as the critical point is approached, the correlation length of the system becomes exceedingly large — with the result that the sensitivity of the system to a *length transformation* (or renormalization) gets exceedingly diminished. At the critical point itself, the correlation length becomes infinite and, with it, the system becomes totally insensitive to such a transformation! The *fixed point* of the transformation is then identified with the critical point of the system, and the behavior of the system parameters such as K and h, see equation (13.5.2), in the neighborhood of the fixed point determines the critical exponents, and so on. Since very few systems could be solved exactly, approximation procedures had to be developed to handle most of the cases under study.

One such procedure starts with known results for the upper critical dimension $d = 4$ and carries out expansions in terms of the (small) parameter $\varepsilon = 4 - d$, while the other starts with known results for the spherical model ($n = \infty$) and carries out expansions in terms of the (small) parameter $1/n$; in the former case, the coefficients of the expansion would be n-dependent, while in the latter case they would be d-dependent.

Highly sophisticated manipulations enable one to obtain reliable results for ε that are as large as 1 (which corresponds to the most practical dimension $d = 3$) and for n as small as 1 (which corresponds to a large variety of systems that can be described through an order parameter that is scalar). We propose to discuss this method at length in Chapter 14; for the present, suffice it to say that the reader interested in further details may refer to Volume 6 of the Domb–Green series, which is devoted entirely to this topic.

Monte Carlo methods

As the name implies, these methods employ pseudorandom numbers to simulate statistical fluctuations for carrying out numerical integrations and computer simulations of systems with large numbers of degrees of freedom. Such methods have been in vogue for quite some time and, fortunately, have kept pace with developments along other lines of approach — so much so that they have adapted themselves to the ideas of the renormalization group as well (see Ma, 1976b). The interested reader may refer to Binder (1986, 1987, 1992), Frenkel and Smit (2002), Binder and Heermann (2002), and Landau and Binder (2009). We propose to discuss computer simulation methods further in Chapter 16.

As mentioned earlier, the aforementioned methods lead to results that are mutually compatible and, within the stated margins of error, essentially identical. Table 13.1 lists the generally accepted values of the critical exponents of a *three-dimensional* system with $n = 0, 1, 2, 3$, and ∞. It includes all the major exponents except α, which can be determined by using the scaling relation $\alpha + 2\beta + \gamma = 2$ (or the hyperscaling relation $d\nu = 2 - \alpha$); we thus obtain, for $n = 0, 1, 2, 3$, and ∞, $\alpha = 0.235, 0.111, -0.008, -0.114$, and -1, respectively — of course, with appropriate margins of error. The theoretical results assembled here may be compared with the corresponding experimental ones listed earlier in Table 12.1, remembering that, while all other entries there are Ising-like ($n = 1$), the case of superfluid He^4 pertains to $n = 2$.

Table 13.1 includes exponents for the case $n = 0$ as well. This relates to the fact that if the spin dimensionality n is treated as a continuously varying parameter then the limit $n \to 0$ corresponds to the statistical behavior of self-avoiding random walks and hence of polymers (de Gennes, 1972; des Cloizeaux, 1974). The role of t in that case is played by the parameter $1/N$, where N is the number of steps constituting the walk or the number of

Table 13.1 Theoretical Values of the Critical Exponents in Three Dimensions

	$n = 0$	$n = 1$	$n = 2$	$n = 3$	$n = \infty$
β	0.302 ± 0.004	0.324 ± 0.006	0.346 ± 0.009	0.362 ± 0.012	0.5
γ	1.161 ± 0.003	1.241 ± 0.004	1.316 ± 0.009	1.39 ± 0.01	2.0
δ	4.85 ± 0.08	4.82 ± 0.06	4.81 ± 0.08	4.82 ± 0.12	5.0
ν	0.588 ± 0.001	0.630 ± 0.002	0.669 ± 0.003	0.705 ± 0.005	1.0
η	0.026 ± 0.014	0.031 ± 0.011	0.032 ± 0.015	0.031 ± 0.022	0.0

Source: After Baker (1990).

monomers constituting the polymer chain; thus, the limit $N \to \infty$ corresponds to $t \to 0$, whereby one approaches the critical point of the system. Concepts such as the correlation function, the correlation length, the susceptibility, the free energy, and so on, can all be introduced systematically into the problem and one obtains a well-defined model that fits neatly with the rest of the family. For details, see de Gennes (1979) and des Cloizeaux (1982).[9]

Finally, we look at the situation with $d \geq 4$. As mentioned earlier, especially toward the end of Section 13.5, critical exponents for $d > 4$ are independent of n and have values as predicted by the mean field theory. To recapitulate, they are:

$$\alpha = 0, \quad \beta = \frac{1}{2}, \quad \gamma = 1, \quad \delta = 3, \quad \nu = \frac{1}{2}, \quad \eta = 0. \tag{6}$$

At the borderline dimensionality $d = 4$, two nonclassical features appear. First, the nature of the singularity is such that it cannot be represented by a power law alone; logarithmic factors are also present. Second, the dependence on n shows up in a striking fashion. In this context, we simply quote the results; for details, see Brézin et al. (1976):

$$c^{(s)} \sim |\ln t|^{(4-n)/(n+8)} \qquad (h = 0, t \gtrsim 0) \tag{7}$$

$$m_0 \sim |t|^{1/2} |\ln |t||^{3/(n+8)} \qquad (h = 0, t \lesssim 0) \tag{8}$$

$$\chi \sim t^{-1} |\ln t|^{(n+2)/(n+8)} \qquad (h = 0, t \gtrsim 0) \tag{9}$$

$$h \sim m^3 |\ln m|^{-1} \qquad (t = 0, h \gtrsim 0), \tag{10}$$

along with the fact that $\eta = 0$ and hence $\xi \sim \chi^{1/2}$. In the limit $n \to \infty$, these results go over to the ones pertaining to the spherical model; see Section 13.5.

Problems

13.1. Making use of expressions (12.3.17) through (12.3.19), (13.2.12), and (13.2.13), show that the expectation values of the numbers N_+, N_-, N_{++}, N_{--}, and N_{+-} in the case of an Ising chain are

$$\overline{N}_{\pm} = N \frac{P(\beta, B) \pm \sinh(\beta\mu B)}{2P(\beta, B)},$$

$$\overline{N}_{++} = \frac{N}{2D(\beta, B)} e^{\beta\mu B} [P(\beta, B) + \sinh(\beta\mu B)],$$

$$\overline{N}_{--} = \frac{N}{2D(\beta, B)} e^{-\beta\mu B} [P(\beta, B) - \sinh(\beta\mu B)]$$

[9]Values of n other than the ones appearing in Table 13.1 are sometimes encountered. For instance, certain antiferromagnetic order–disorder transitions require for their description an order parameter with $n = 4, 6, 8$, or 12; see Mukamel (1975), and Bak, Krinsky, and Mukamel (1976). Another example of this is provided by the superfluidity of liquid He^3, which seems to require an order parameter with $n = 18$; see, for instance, Anderson (1984), and Vollhardt and Wölfle (1990). Even negative values of n have been investigated; see Balian and Toulouse (1973) and Fisher (1973).

and

$$\overline{N}_{+-} = \frac{N}{D(\beta,B)} e^{-4\beta J},$$

where

$$P(\beta,B) = \{e^{-4\beta J} + \sinh^2(\beta\mu B)\}^{1/2}$$

and

$$D(\beta,B) = P(\beta,B)[P(\beta,B) + \cosh(\beta\mu B)].$$

Check that (i) the preceding expressions satisfy the requirement that $\overline{N}_{++} + \overline{N}_{--} + \overline{N}_{+-} = N$ and (ii) they agree with the quasichemical approximation (12.6.22), regardless of the value of B.

13.2. **(a)** Show that the partition function of an Ising lattice can be written as

$$Q_N(B,T) = \sum_{N_+,N_{+-}}{}' g_N(N_+,N_{+-}) \exp\{-\beta H_N(N_+,N_{+-})\},$$

where

$$H_N(N_+,N_{+-}) = -J\left(\frac{1}{2}qN - 2N_{+-}\right) - \mu B(2N_+ - N),$$

while other symbols have their usual meanings; compare these results to equations (12.3.19) and (12.3.20).

(b) Next, determine the combinatorial factor $g_N(N_+,N_{+-})$ for an Ising chain ($q = 2$) and show that, asymptotically,

$$\begin{aligned}\ln g_N(N_+,N_{+-}) \approx\; & N_+ \ln N_+ + (N - N_+)\ln(N - N_+) \\ & - \left(N_+ - \frac{1}{2}N_{+-}\right)\ln\left(N_+ - \frac{1}{2}N_{+-}\right) \\ & - \left(N - N_+ - \frac{1}{2}N_{+-}\right)\ln\left(N - N_+ - \frac{1}{2}N_{+-}\right) \\ & - 2\left(\frac{1}{2}N_{+-}\right)\ln\left(\frac{1}{2}N_{+-}\right).\end{aligned}$$

Now, assuming that $\ln Q_N \approx$ (the logarithm of the *largest* term in the sum $\sum'$), evaluate the Helmholtz free energy $A(B,T)$ of the system and show that this leads to precisely the same results as the ones quoted in the preceding problem as well as the ones obtained in Section 13.2.

13.3. Using the approximate expression, see Fowler and Guggenheim (1940),

$$g_N(N_1,N_{12}) \simeq \frac{\left(\frac{1}{2}qN\right)!}{N_{11}!N_{22}!\left[\left(\frac{1}{2}N_{12}\right)!\right]^2}\left(\frac{N_1!N_2!}{N!}\right)^{q-1},$$

for evaluating the partition function of an Ising lattice, show that one is led to the same results as the ones following from the Bethe approximation.

[Note that, for $q = 2$, the quantity $\ln g$ here is *asymptotically* exact; see Problem 13.2(b). No wonder that the Bethe approximation gives exact results in the case of an Ising chain.]

13.4. Making use of relation (13.2.37), along with expressions (13.2.8) for the eigenvalues λ_1 and λ_2 of the transfer matrix $\boldsymbol{P}$, determine the correlation length $\xi(B,T)$ of the Ising chain in the presence of a magnetic field. Evaluate the critical exponent ν^c, as defined in Problem 12.25, and check that $\nu^c = \nu/\Delta$, where ν and Δ are standard exponents defined in Sections 12.10 and 12.12.

13.5. Consider a one-dimensional Ising system in a *fluctuating* magnetic field B, so that

$$Q_N(s,T) \sim \int_{-\infty}^{\infty} dB \sum_{\{\sigma_i\}} \exp\left\{-\frac{\beta N B^2}{2s} + \sum_{i=1}^{N} [\beta\mu B\sigma_i + \beta J\sigma_i\sigma_{i+1}]\right\},$$

with $\sigma_{N+1} = \sigma_1$. Note that when the system is very large (i.e., $N \gg 1$) the typical value of B is very small; nevertheless, the presence of this small fluctuating field leads to an order–disorder transition in this one-dimensional system! Determine the critical temperature of this transition.

13.6. Solve exactly the problem of a field-free Ising chain with nearest-neighbor and next-nearest-neighbor interactions, so that

$$H\{\sigma_i\} = -J_1 \sum_i \sigma_i\sigma_{i+1} - J_2 \sum_i \sigma_i\sigma_{i+2},$$

and examine the various properties of interest of this model.

[*Hint*: Introduce a new variable $\tau_i = \sigma_i\sigma_{i+1} = \pm 1$, with the result that

$$H\{\tau_i\} = -J_1 \sum_i \tau_i - J_2 \sum_i \tau_i\tau_{i+1},$$

which is formally similar to expression (13.2.1)].

13.7. Consider a double Ising chain such that the nearest-neighbor coupling constant along either chain is J_1 while the one linking adjacent spins in the two chains is J_2. Then, in the absence of the field,

$$H\{\sigma_i, \sigma_i'\} = -J_1 \sum_i (\sigma_i\sigma_{i+1} + \sigma_i'\sigma_{i+1}') - J_2 \sum_i \sigma_i\sigma_i'.$$

Show that the partition function of this system is given by

$$\frac{1}{2N}\ln Q \approx \frac{1}{2}\ln[2\cosh K_2\{\cosh 2K_1 + \sqrt{(1+\sinh^2 2K_1 \tanh^2 K_2)}\}],$$

where $K_1 = \beta J_1$ and $K_2 = \beta J_2$. Examine the various thermodynamic properties of this system.

[*Hint*: Express the Hamiltonian H in a *symmetric* form by writing the last term as $-\frac{1}{2}J_2\Sigma_i(\sigma_i\sigma_i' + \sigma_{i+1}\sigma_{i+1}')$ and use the transfer matrix method.]

13.8. Write down the transfer matrix $\boldsymbol{P}$ for a one-dimensional spin-1 Ising model in zero field, described by the Hamiltonian

$$H_N\{\sigma_i\} = -J\sum_i \sigma_i\sigma_{i+1} \quad \sigma_i = -1, 0, +1.$$

Show that the free energy of this model is given by

$$\frac{1}{N}A(T) = -kT\ln\left\{\frac{1}{2}\left[(1+2\cosh\beta J) + \sqrt{\{8 + (2\cosh\beta J - 1)^2\}}\right]\right\}.$$

Examine the limiting behavior of this quantity in the limits $T \to 0$ and $T \to \infty$.

13.9. **(a)** Apply the theory of Section 13.2 to a one-dimensional lattice gas and show that the pressure P and the volume per particle v are given by

$$\frac{P}{kT} = \ln\left[\frac{1}{2}\left\{(y+1) + \sqrt{[(y-1)^2 + 4y\eta^2]}\right\}\right]$$

and

$$\frac{1}{v} = \frac{1}{2}\left[1 + \frac{y-1}{\sqrt{[(y-1)^2 + 4y\eta^2]}}\right],$$

where $y = z\exp(4\beta J)$ and $\eta = \exp(-2\beta J)$, z being the fugacity of the gas. Examine the high and the low temperature limits of these expressions.

(b) A *hard-core* lattice gas pertains to the limit $J \to -\infty$; this makes $y \to 0$ and $\eta \to \infty$ such that the quantity $y\eta^2$, which is equal to z, stays finite. Show that this leads to the equation of state

$$\frac{P}{kT} = \ln\left(\frac{1-\rho}{1-2\rho}\right) \quad \left(\rho = \frac{1}{v}\right).$$

13.10. For a one-dimensional system, such as the ones discussed in Sections 13.2 and 13.3, the correlation function $g(r)$ at all temperatures is of the form $\exp(-ra/\xi)$, where a is the lattice constant of the system. Using the fluctuation-susceptibility relation (12.11.11), show that the low-field susceptibility of such a system is given by

$$\chi_0 = N\beta\mu^2 \coth(a/2\xi).$$

Note that as $T \to 0$ and, along with it, $\xi \to \infty$, χ_0 becomes directly proportional to ξ — consistent with the fact that the critical exponent $\eta = 1$.

For an n-vector model (including the scalar case $n = 1$), ξ is given by equation (13.3.17), which leads to the result

$$\chi_0 = N\beta\mu^2 \frac{I_{(n-2)/2}(\beta J) + I_{n/2}(\beta J)}{I_{(n-2)/2}(\beta J) - I_{n/2}(\beta J)}.$$

Check that for the special case $n = 1$ this result reduces to equation (13.2.14).

13.11. Show that for a one-dimensional, field-free Ising model

$$\overline{\sigma_k\sigma_l\sigma_m\sigma_n} = \{\tanh\beta J\}^{n-m+l-k},$$

where $k \le l \le m \le n$.

13.12. Recall the symbol $n(r)$, of equation (13.4.5), which denotes the number of *closed graphs* that can be drawn on a given lattice using exactly r bonds. Show that for a square lattice wrapped on a torus (which is equivalent to imposing periodic boundary conditions)

$$n(4) = N, \quad n(6) = 2N, \quad n(8) = \frac{1}{2}N^2 + \frac{9}{2}N, \ldots.$$

Substituting these numbers into equation (13.4.5) and taking logs, one gets

$$\ln Q(N,T) = N\left\{\ln(2\cosh^2 K) + v^4 + 2v^6 + \frac{9}{2}v^8 + \cdots\right\}, v = \tanh K.$$

Note that the term in N^2 has disappeared — in fact, all higher powers of N do the same. Why?

13.13. According to Onsager, the field-free partition function of a rectangular lattice (with interaction parameters J and J' in the two perpendicular directions) is given by

$$\frac{1}{N}\ln Q(T) = \ln 2 + \frac{1}{2\pi^2}\int_0^\pi\int_0^\pi \ln\{\cosh(2\gamma)\cosh(2\gamma') - \sinh(2\gamma)\cos\omega - \sinh(2\gamma')\cos\omega'\}d\omega d\omega',$$

where $\gamma = J/kT$ and $\gamma' = J'/kT$. Show that if $J' = 0$, this leads to expression (13.2.9) for the linear chain with $B = 0$ while if $J' = J$, one is led to expression (13.4.22) for the square net. Locate the critical point of the rectangular lattice and study its thermodynamic behavior in the neighborhood of that point.

13.14. Write the elliptic integral $K_1(\kappa)$ in the form

$$K_1(\kappa) = \int_0^{\pi/2} \frac{1 - \kappa\sin\phi}{\sqrt{(1-\kappa^2\sin^2\phi)}}d\phi + \int_0^{\pi/2} \frac{\kappa\sin\phi}{\sqrt{(1-\kappa^2\sin^2\phi)}}d\phi,$$

and show that, as $\kappa \to 1-$, the first integral $\to \ln 2$ while the second $\approx \ln[2/(1-\kappa^2)^{1/2}]$. Hence $K_1(\kappa) \approx \ln(4/|\kappa'|)$, as in equation (13.4.34).

[*Hint*: In the second integral, substitute $\cos\phi = x$.]

13.15. Using equations (13.4.22) and (13.4.28) at $T = T_c$, show that the entropy of the two-dimensional Ising model on a square lattice at its critical point is given by

$$\frac{S_c}{Nk} = \frac{2G}{\pi} + \frac{1}{2}\ln 2 - \sqrt{2}K_c \simeq 0.3065;$$

here, G is Catalan's constant, which is approximately equal to 0.915966. Compare this result with the ones following from the Bragg–Williams approximation and from the Bethe approximation; see Problem 12.15.

13.16. The spontaneous magnetization of a two-dimensional Ising model on a square lattice at $T < T_c$ is given by equation (13.4.38a). Express this result in the form $B|t|^\beta\{1 + b|t| + \cdots\}$, where B and β are stated in equation (13.4.40) while $b = (1 - 9K_c/\sqrt{2})/8$. As usual, $t = (T - T_c)/T_c < 0$ and $|t| \ll 1$.

13.17. Apply the theory of Section 13.4 to a two-dimensional lattice gas and show that, at $T = T_c$, the quantity $kT_c/P_c v_c \simeq 10.35$.

13.18. Show that for the spherical model in one dimension the free energy *at constant* λ is given by

$$\frac{\beta A_\lambda}{N} = \frac{1}{2}\ln\left[\frac{\beta\{\lambda + \sqrt{(\lambda^2 - J^2)}\}}{2\pi}\right] - \frac{\beta\mu^2 B^2}{4(\lambda - J)},$$

while λ is determined by the constraint equation

$$\frac{1}{2\beta\sqrt{(\lambda^2 - J^2)}} + \frac{\mu^2 B^2}{4(\lambda - J)^2} = 1.$$

In the absence of the field ($B=0$), $\lambda = \sqrt{1 + 4\beta^2 J^2}/2\beta$; the free energy *at constant* $\mathcal{S}$ is then given by

$$\frac{\beta A_{\mathcal{S}}}{N} = \frac{1}{2}\ln\left[\frac{\sqrt{(1 + 4\beta^2 J^2)} + 1}{4\pi}\right] - \frac{1}{2}\sqrt{(1 + 4\beta^2 J^2)}.$$

13.19. Starting with expression (13.3.8) for the partition function of a one-dimensional n-vector model, with $J_i = nJ'$, show that

$$\operatorname*{Lim}_{n,N\to\infty} \frac{1}{nN}\ln Q_N(nK) = \frac{1}{2}\left[\sqrt{(4K^2+1)} - 1 - \ln\left\{\frac{\sqrt{(4K^2+1)}+1}{2}\right\}\right],$$

where $K = \beta J'$. Note that, apart from a constant term, this result is exactly the same as for the spherical model; the difference arises from the fact that the present result is normalized to give $Q_N = 1$ when $K = 0$.

[*Hint*: Use the asymptotic formulae (for $\nu \gg 1$)

$$\Gamma(\nu) \approx (2\pi/\nu)^{1/2}(\nu/e)^\nu$$

and

$$I_\nu(\nu z) \approx (2\pi\nu)^{-1/2}(z^2+1)^{-1/4}e^{\nu\eta},$$

where

$$\eta = \sqrt{(z^2+1)} - \ln[\{\sqrt{(z^2+1)}+1\}/z].]$$

13.20. Show that the low-field susceptibility, χ_0, of the spherical model at $T < T_c$ is given by the asymptotic expression

$$\chi_0 \approx (N\mu^2/k_B T)\cdot N m_0^2(T),$$

where $m_0(T)$ is the spontaneous magnetization of the system; note that in the thermodynamic limit the reduced susceptibility, $k_B T\chi_0/N\mu^2$, is infinite at all $T < T_c$. Compare to Problem 13.26.

13.21. In view of the fact that only those fluctuations whose length scale is large play a dominant role in determining the nature of a phase transition, the quantity $(\lambda - \mu_{\mathbf{k}})$ in the expression for the correlation function of the spherical model, see equation (13.5.57), may be replaced by

$$\lambda - \mu_{\mathbf{k}} = \lambda - J \sum_{j=1}^{d} \cos(k_j a) \simeq J \left[\phi + \frac{k^2 a^2}{2} \right],$$

where $\phi = (\lambda/J) - d$. Show that this approximation leads to the same result for $G(\boldsymbol{R})$ as we have in equation (13.5.61), with the same ξ as in equation (13.5.62).

For a similar reason, the quantity $[\exp(\alpha + \beta\varepsilon) - 1]$ in the correlation function (13.6.33) of the ideal Bose gas may be replaced by

$$e^{\alpha+\beta\varepsilon} - 1 \simeq \alpha + \beta\varepsilon = \alpha + (\beta\hbar^2/2m)k^2,$$

leading to the same $G(\boldsymbol{R})$ as in equation (13.6.35), with the same ξ as in equation (13.6.36).[10]

13.22. Consider a spherical model whose spins interact through a *long-range* potential varying as $(a/r)^{d+\sigma}$ $(\sigma > 0)$, r being the distance between two spins. This replaces the quantity $(\lambda - \mu_{\mathbf{k}})$ of equations (13.5.16) and (13.5.57) by an expression approximating $J[\phi + \frac{1}{2}(ka)^\sigma]$ for $\sigma < 2$ and $J[\phi + \frac{1}{2}(ka)^2]$ for $\sigma > 2$; note that the nearest-neighbor interaction corresponds to the limit $\sigma \to \infty$ and hence to the latter case.

Assuming σ to be less than 2, show that the above system undergoes a phase transition at a *finite* temperature T_c for all $d > \sigma$. Further show that the critical exponents for this model are

$$\alpha = \frac{d-2\sigma}{d-\sigma}, \quad \beta = \frac{1}{2}, \quad \gamma = \frac{\sigma}{d-\sigma}, \quad \delta = \frac{d+\sigma}{d-\sigma},$$
$$\nu = \frac{1}{d-\sigma}, \quad \eta = 2-\sigma$$

for $\sigma < d < 2\sigma$, and

$$\alpha = 0, \quad \beta = \frac{1}{2}, \quad \gamma = 1, \quad \delta = 3, \quad \nu = \frac{1}{\sigma}, \quad \eta = 2-\sigma$$

for $d > 2\sigma$.

13.23. Refer to Section 13.6 on the ideal Bose gas in d dimensions, and complete the steps leading to equations (13.6.9) through (13.6.15) and (13.6.23).

13.24. Show that for an ideal Bose gas in d dimensions and at $T > T_c$

$$V\left(\frac{\partial^2 P}{\partial T^2}\right)_v = \frac{Nk_B}{T}\left[\frac{d(d+2)}{4}\frac{g_{(d+2)/2}(z)}{g_{d/2}(z)} - \frac{d}{2}\frac{g_{d/2}(z)}{g_{(d-2)/2}(z)} - \frac{d^2}{4}\frac{\{g_{d/2}(z)\}^2 g_{(d-4)/2}(z)}{\{g_{(d-2)/2}(z)\}^3}\right]$$

and

$$\left(\frac{\partial^2 \mu}{\partial T^2}\right)_v = \frac{k_B}{T}\left[\frac{d(d-2)}{4}\frac{g_{d/2}(z)}{g_{(d-2)/2}(z)} - \frac{d^2}{4}\frac{\{g_{d/2}(z)\}^2 g_{(d-4)/2}(z)}{\{g_{(d-2)/2}(z)\}^3}\right],$$

where $\mu (= kT \ln z)$ is the chemical potential of the gas while other symbols have the same meanings as in Section 13.6. Note that these quantities satisfy the thermodynamic relationship

$$VT\left(\frac{\partial^2 P}{\partial T^2}\right)_v - NT\left(\frac{\partial^2 \mu}{\partial T^2}\right)_v = C_V.$$

[10] A comparison with the mean field results (12.11.25) and (12.11.26) brings out a close similarity that exists between these models and the mean field picture of a phase transition; for instance, they all share a common critical exponent η, which is zero. There are, however, significant differences; for one, the correlation length ξ for these models is characterized by a critical exponent ν which is *nonclassical* — in the sense that it is d-dependent whereas in the mean field case it is independent of d.

Also note that, since P here equals $(2U/dV)$, the quantities $(\partial P/\partial T)_v$ and $(\partial^2 P/\partial T^2)_v$ are directly proportional to C_V and $(\partial C_V/\partial T)_V$, respectively. Finally, examine the singular behavior of these quantities as $T \to T_c$ from above.

13.25. Show that for any given fluid

$$C_P = VT(\partial P/\partial T)_S(\partial P/\partial T)_V \kappa_T$$

and

$$C_V = VT(\partial P/\partial T)_S(\partial P/\partial T)_V \kappa_S,$$

where the various symbols have their usual meanings. In the two-phase region, these formulae take the form

$$C_P = VT(dP/dT)^2 \kappa_T \quad \text{and} \quad C_V = VT(dP/dT)^2 \kappa_S,$$

respectively. Using the last of these results, rederive equation (13.6.30) for C_V at $T < T_c$.

13.26. Show that for any given fluid

$$\kappa_T = \rho^{-2}(\partial\rho/\partial\mu)_T,$$

where $\rho(=N/V)$ is the particle density and μ the chemical potential of the fluid. For the ideal Bose gas at $T < T_c$,

$$\rho = \rho_0 + \rho_e \approx -\frac{k_B T}{V\mu} + \frac{\zeta(d/2)}{\lambda^d}.$$

Using these results, show that[11]

$$\kappa_T \approx (V/k_B T)(\rho_0/\rho)^2 \quad (T < T_c);$$

note that in the thermodynamic limit the reduced compressibility, $k_B T \kappa_T / v$, is infinite at all $T < T_c$. Compare Problem 13.20.

13.27. Consider an ideal relativistic Bose gas composed of N_1 particles and N_2 antiparticles, each of rest mass m_0, with occupation numbers

$$\frac{1}{\exp[\beta(\varepsilon - \mu_1)] - 1} \quad \text{and} \quad \frac{1}{\exp[\beta(\varepsilon - \mu_2)] - 1},$$

respectively, and the energy spectrum $\varepsilon = c\sqrt{(p^2 + m_0^2 c^2)}$. Since particles and antiparticles are supposed to be created in pairs, the system is constrained by the conservation of the number $Q(=N_1 - N_2)$, rather than of N_1 and N_2 separately; accordingly, $\mu_1 = -\mu_2 = \mu$, say.

Set up the thermodynamic functions of this system in three dimensions and examine the onset of Bose–Einstein condensation as T approaches a critical value, T_c, determined by the "number density" Q/V. Show that the nature of the singularity at $T = T_c$ is such that, regardless of the severity of the relativistic effects, this system falls in the *same* universality class as the nonrelativistic one.

13.28. Derive equation (13.1.9) for hard spheres in one dimension from equation (13.1.7). Plot the pair correlation function for $nD = 0.25$, 0.50, 0.75, and 0.90. Determine the structure factor $S(k)$ numerically and plot it for the same densities. Compare your results with equation (13.1.21).

13.29. Use the pair correlation function (13.1.8) and (13.1.9) to determine analytically the structure factor for hard spheres in one dimension. Show that $S(k)$ is given by equation (13.1.21). Plot $g(x)$ and $S(k)$ for $nD = 0.25$, 0.50, 0.75, and 0.90.

13.30. Use the Takahashi method of Section 13.1 for a system of point masses and harmonic springs of length a. Allow the particles to pass through each other, so that the partition function can be evaluated in a closed form. Show that the system is stable at zero pressure. Determine the average

[11] This remarkable relationship between the isothermal compressibility of a *finite-sized* Bose gas and the condensate density in the corresponding *bulk* system was first noticed by Singh and Pathria (1987b).

distance between particles that are far apart on the chain and the variance of that distance. Determine the structure factor and plot it for several values of the parameter $m\omega^2 a^2/kT$, where m is the mass and $m\omega^2$ is the spring constant. Show that the specific heat of this system is independent of temperature, as given by the equipartition theorem.

13.31. Confirm the first few coefficients in the low-temperature series for the two-dimensional Ising model in equation (13.4.53). Write a program to calculate the energies of all 2^{16} states for a 4×4 periodic lattice. Show that the coefficients here are

$$g_q = \{2, 0, 32, 64, 424, 1728, 6688, 13568, 20524, 13568, 6688, 1728, 424, 64, 32, 0, 2\}.$$

Extend your code to calculate the partition function for a 6×6 lattice, which has 2^{36} states.

13.32. Calculate the exact zero field partition function of the one-dimensional Ising model on a periodic chain of n spins using equation (13.2.5) and write $Q_N(0, T)$ in the form of equation (13.4.52). Show that for $x \to 1$, the partition function $\to 2^n$. Evaluate the microcanonical entropy $S(q)/k = \ln g_q$ and plot it for the case $n = 16$.

13.33. Use the code posted at www.elsevierdirect.com to evaluate equations (13.4.56) through (13.4.59) to determine the low-temperature series coefficients for the two-dimensional Ising model for an 8×8 lattice. Plot the internal energy and the specific heat as a function of temperature. Repeat your calculation for 16×16 and 32×32 lattices.

13.34. Use the data posted at www.elsevierdirect.com to evaluate equations (13.4.56) through (13.4.59) to plot the two-dimensional Ising model internal energy and specific heat as a function of temperature for an $L \times L$ lattice, where $L = 64$. Compare your results with the ones displayed in Figure 13.17.

14

Phase Transitions: The Renormalization Group Approach

In this chapter we propose to discuss what has turned out to be the most successful approach to the problem of phase transitions. This approach is based on ideas first propounded by Kadanoff (1966b) and subsequently developed by Wilson (1971) and others into a powerful calculational tool. The main point of Kadanoff's argument is that, as the critical point of a system is approached, its correlation length becomes exceedingly large — with the result that the sensitivity of the system to a *length transformation* (or a *change of scale,* as one may call it) gets exceedingly diminished. At the critical point itself, the correlation length becomes infinitely large and with it the system becomes totally insensitive to such a transformation. It is then conceivable that, if one is not too far from the critical point (i.e., $|t|, h \ll 1$), the given system (with lattice constant a) may bear a close resemblance to a transformed system (with lattice constant $a' = la$, where $l > 1$, and presumably modified parameters t' and h'), renormalized so that all distances in it are measured in terms of the new lattice constant a'; clearly, the rescaled correlation length ξ' (in units of a') would be one-lth of the original correlation length ξ (in units of a). This resemblance in respect of critical behavior is expected only if ξ' is also much larger than a', just as ξ was in comparison with a, which in turn requires that $l \ll (\xi/a)$; by implication, $|t'|$ and h' would also be $\ll 1$.

These considerations lead to a formulation similar to the one presented in Section 12.10, with the difference that, while there we had to rely on a scaling *hypothesis,* here we have a convincing argument based on the role played by correlations among the microscopic constituents of the system which, in the vicinity of the critical point, are so large-scale that they make all other length scales, including the one that determines the structure of the lattice, essentially irrelevant. Unfortunately, Kadanoff's approach did not provide a systematic means of deriving the critical exponents or of constructing the scaling functions that appear in the formulation. Those deficiencies were remedied by Wilson by introducing the concept of a renormalization group (RG) into the theory.

We propose to discuss Wilson's approach in Sections 14.3 and 14.4, but first we present a formulation of scaling along the lines indicated above and follow it with an exploration of simple examples of renormalization that pave way for establishing Wilson's theory. Finally, in Section 14.5, we outline the theory of finite-sizing scaling that too has benefited greatly from the RG approach.

© 2011 Elsevier Ltd. All rights reserved.

14.1 The conceptual basis of scaling

The scale change in a given system can be effected in several different ways, the earliest one being due to Kadanoff who suggested that, when large-scale correlations prevail, the individual spins in the system may be grouped into "blocks of spins," each block consisting of l^d original spins, and let each block play the role of a "single spin" in the transformed system; see Figure 14.1, where a block–spin transformation with $l = 2$ and $d = 2$ is shown. The spin variable of a block may be denoted by the symbol σ', which arises from the values of the individual spins in the block, but a rule has to be established so that σ' too is either $+1$ or -1, just as the original σ_i were.[1] The transformed system then consists of N' "spins," where

$$N' = l^{-d}N, \tag{1}$$

occupying a lattice structure with lattice constant $a' = la$. To preserve the spatial density of the degrees of freedom in the system, all spatial distances must be rescaled by the factor l, so that for any two "spins" in the transformed system

$$\boldsymbol{r}' = l^{-1}\boldsymbol{r}. \tag{2}$$

A second way of effecting a scale change is to write down the partition function of the system,

$$Q = \sum_{\{\sigma_i\}} \exp[-\beta H_N\{\sigma_i\}], \tag{3}$$

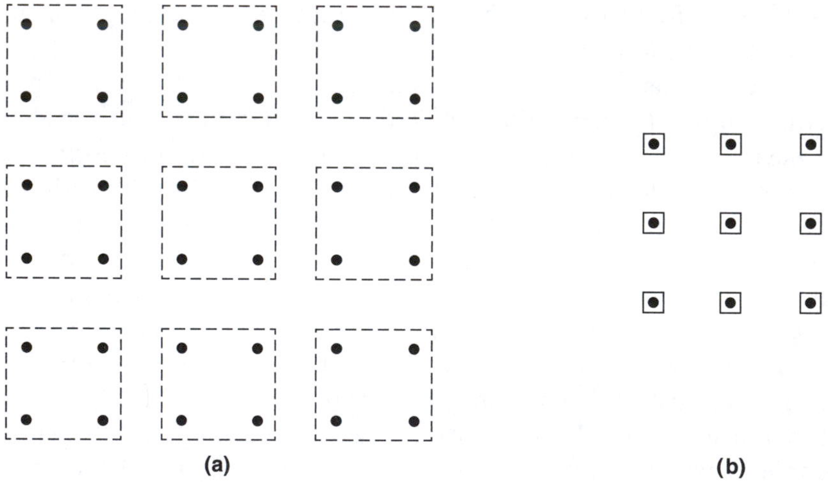

FIGURE 14.1 A block–spin transformation, with $l = 2$ and $d = 2$. The original lattice (a) has $N\ (= 36)$ spins, the transformed one (b) has $N'\ (= 9)$; after rescaling, the latter looks very much the same as the former, especially in the limit $N, N' \to \infty$.

[1]For simplicity, we employ the language of the *scalar* model here.

and carry out summation over a subset of $(N-N')$ spins, such that one is left with a summation, over the remaining N' spins, which can (hopefully) be expressed in a form similar to (3), namely

$$Q = \sum_{\{\sigma_i'\}} \exp[-\beta H_{N'}\{\sigma_i'\}]. \tag{4}$$

If the desired passage, from expression (3) to (4), can be accomplished with some degree of accuracy, we should expect a close resemblance between the critical behavior of the original system represented by equation (3) and the transformed one represented by (4); see Figure 14.2, where an example of this procedure with $l=\sqrt{2}$ and $d=2$ is shown. We note that this procedure forms the very backbone of the Wilson approach and is generally referred to as "decimation," although the fraction of the spins removed, $(N-N')/N$, is rarely equal to 1/10. Other ways of effecting a scale change will be mentioned later.

It is quite natural to expect that the free energy of the transformed system (or, at least, that part of it that determines the critical behavior) is the same as that of the original system. The singular parts of the free energy *per spin* of the two systems should, therefore, be related as

$$N'\psi^{(s)}(t',h') = N\psi^{(s)}(t,h), \tag{5}$$

so that

$$\psi^{(s)}(t,h) = l^{-d}\psi^{(s)}(t',h'). \tag{6}$$

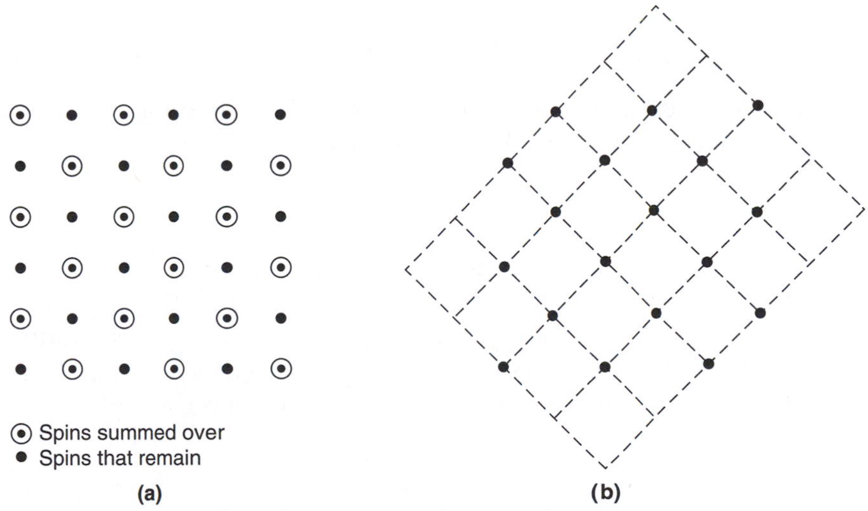

FIGURE 14.2 A scale transformation by "decimation," with $l=\sqrt{2}$ and $d=2$. The original lattice (a) has $N\ (=36)$ spins, the transformed one (b) has $N'\ (=18)$; the latter is yet to be rescaled (and rotated through an angle $\pi/4$) so that it looks very much the same as the former, especially in the limit $N, N' \to \infty$.

Now, since both t and t' are small in magnitude, one may assume that they are linearly related, that is,

$$t' = l^{y_t} t, \tag{7}$$

where y_t, as yet, is an unknown number. Similarly, one may assume that

$$h' = l^{y_h} h, \tag{8}$$

with the result that

$$\psi^{(s)}(t,h) = l^{-d}\psi^{(s)}(l^{y_t} t, l^{y_h} h); \tag{9}$$

like y_t, the number y_h is also unknown at this stage of the game.

We now assert that the function $\psi^{(s)}$, which governs much of the critical behavior of the system, is essentially insensitive to a change of scale; we should, therefore, be able to eliminate the scale factor l from expression (9). This essentially forces us to replace the variables t' and h' by a single, l-independent variable, namely

$$\frac{h'}{|t'|^{y_h/y_t}} = \frac{h}{|t|^{y_h/y_t}} = \frac{h}{|t|^{\Delta}}, \text{ say } \left(\Delta = \frac{y_h}{y_t}\right); \tag{10}$$

at the same time, it requires us to write

$$\psi^{(s)}(t',h') = |t'|^{d/y_t}\tilde{\psi}(h'/|t'|^{\Delta}), \tag{11}$$

leading to the identical result

$$\psi^{(s)}(t,h) = |t|^{d/y_t}\tilde{\psi}(h/|t|^{\Delta}); \tag{12}$$

note that, as of now, the function $\tilde{\psi}$ is also unknown.[2] Comparing (12) with equation (12.10.7), we readily identify the critical exponent α as

$$\alpha = 2 - (d/y_t); \tag{13}$$

more importantly, the present considerations have led to the same scaled form for the free energy density of the system as was hypothesized in Section 12.10. We have thus raised the status of expression (12.10.7) from being a mere hypothesis to being a well-founded result whose conceptual basis lies in the prevalence of large-scale correlations in the system. As in Section 12.10, we infer that the exponents β, γ, and δ are now given by

$$\beta = 2 - \alpha - \Delta = (d - y_h)/y_t, \tag{14}$$

$$\gamma = -(2 - \alpha - 2\Delta) = (2y_h - d)/y_t, \tag{15}$$

[2]Some authors derive equation (12) from (9) by choosing l to be t^{-1/y_t}. As will be seen shortly, see equation (19), this amounts to letting l be $O(\xi/a)$, which violates the requirement, $l \ll \xi/a$, mentioned earlier.

and

$$\delta = \Delta/\beta = y_h/(d - y_h). \tag{16}$$

As remarked earlier, the rescaled correlation length ξ' of the transformed system and the original correlation length ξ of the given system are related as

$$\xi' = l^{-1}\xi. \tag{17}$$

At the same time, we expect ξ' to be $\sim |t'|^{-\nu}$, just as $\xi \sim |t|^{-\nu}$. It follows that

$$\left(\frac{\xi'}{\xi}\right) = \left(\frac{t'}{t}\right)^{-\nu} = l^{-\nu y_t}. \tag{18}$$

Comparing (17) and (18), we conclude that

$$\nu = 1/y_t \tag{19}$$

and hence, by (13),

$$d\nu = 2 - \alpha. \tag{20}$$

We thus obtain not only a useful expression for the critical exponent ν but also the *hyper-scaling relation* (12.12.15) on a basis far sounder than the one employed in Section 12.12.

Finally we look at the correlation functions of the two systems. At the critical point we expect that for the transformed system

$$g(\boldsymbol{r}'_1, \boldsymbol{r}'_2) = \langle \sigma'(\boldsymbol{r}'_1)\sigma'(\boldsymbol{r}'_2)\rangle \sim (r')^{-(d-2+\eta)}, \tag{21}$$

just as for the original system

$$g(\boldsymbol{r}_1, \boldsymbol{r}_2) = \langle \sigma(\boldsymbol{r}_1)\sigma(\boldsymbol{r}_2)\rangle \sim r^{-(d-2+\eta)}. \tag{22}$$

In order that equations (21) and (22) be mutually compatible, we must rescale the spin variables such that

$$\sigma'(\boldsymbol{r}') = l^{(d-2+\eta)/2}\sigma(\boldsymbol{r}). \tag{23}$$

As for η, we may use the scaling relation $\gamma = (2 - \eta)\nu$, to get

$$\eta = d + 2 - 2y_h. \tag{24}$$

14.2 Some simple examples of renormalization

14.2.A The Ising model in one dimension

We start with the partition function (13.2.3a) of a closed Ising chain consisting of N spins, namely

$$Q_N(B,T) = \sum_{\{\sigma_i\}} \exp\left[\sum_{i=1}^{N}\left\{K_0 + K_1\sigma_i\sigma_{i+1} + \tfrac{1}{2}K_2(\sigma_i + \sigma_{i+1})\right\}\right]$$

$$(K_0 = 0, K_1 = \beta J, K_2 = \beta\mu B); \tag{1}$$

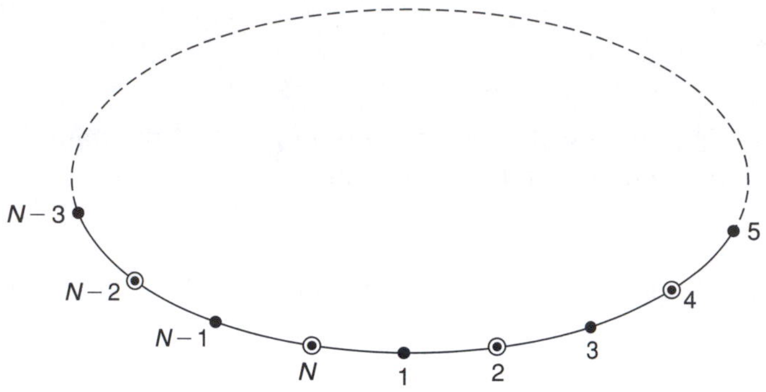

FIGURE 14.3 A closed Ising chain to be "decimated" by carrying out summations over $\sigma_2, \sigma_4, \ldots$.

the parameter K_0 has been introduced here for reasons that will become clear in the sequel. For simplicity, we assume N to be even and carry out summation in (1) over all σ_i for which i is even, that is, over $\sigma_2, \sigma_4, \ldots$; see Figure 14.3. Writing the summand in (1) as

$$\prod_{i=1}^{N} \exp\{K_0 + K_1\sigma_i\sigma_{i+1} + \tfrac{1}{2}K_2(\sigma_i + \sigma_{i+1})\} = \prod_{j=1}^{\frac{1}{2}N} \exp\{2K_0 + K_1(\sigma_{2j-1}\sigma_{2j} + \sigma_{2j}\sigma_{2j+1}) + \tfrac{1}{2}K_2(\sigma_{2j-1} + 2\sigma_{2j} + \sigma_{2j+1})\}, \tag{2}$$

the summations over σ_{2j} $(= +1 \text{ or } -1)$ can be carried out straightforwardly, with the result

$$\prod_{j=1}^{\frac{1}{2}N} \exp(2K_0) \cdot 2\cosh\{K_1(\sigma_{2j-1} + \sigma_{2j+1}) + K_2\} \cdot \exp\left\{\tfrac{1}{2}K_2(\sigma_{2j-1} + \sigma_{2j+1})\right\}. \tag{3}$$

Denoting σ_{2j-1} by σ'_j, the partition function Q_N assumes the form

$$Q_N(B,T) = \sum_{\{\sigma'_j\}} \prod_{j=1}^{\frac{1}{2}N} \exp(2K_0) \cdot 2\cosh\{K_1(\sigma'_j + \sigma'_{j+1}) + K_2\} \cdot \exp\left\{\frac{1}{2}K_2(\sigma'_j + \sigma'_{j+1})\right\}. \tag{4}$$

The crucial step now consists in expressing (4) in a form similar to (1), namely

$$Q_N(B,T) = \sum_{\{\sigma'_j\}} \exp\left[\sum_{j=1}^{N'} \left\{K'_0 + K'_1\sigma'_j\sigma'_{j+1} + \tfrac{1}{2}K'_2(\sigma'_j + \sigma'_{j+1})\right\}\right]. \tag{5}$$

This requires that, for *all* choices of the variables σ'_j and σ'_{j+1},

$$\exp\left\{K'_0 + K'_1\sigma'_j\sigma'_{j+1} + \tfrac{1}{2}K'_2(\sigma'_j + \sigma'_{j+1})\right\}$$
$$= \exp(2K_0) \cdot 2\cosh\{K_1(\sigma'_j + \sigma'_{j+1}) + K_2\} \cdot \exp\left\{\tfrac{1}{2}K_2(\sigma'_j + \sigma'_{j+1})\right\}. \tag{6}$$

The various choices being $\sigma'_j = \sigma'_{j+1} = +1, \sigma'_j = \sigma'_{j+1} = -1$ and $\sigma'_j = -\sigma'_{j+1} = \pm 1$, we obtain from (6)

$$\exp(K'_0 + K'_1 + K'_2) = \exp(2K_0 + K_2) \cdot 2\cosh(2K_1 + K_2), \tag{7a}$$

$$\exp(K'_0 + K'_1 - K'_2) = \exp(2K_0 - K_2) \cdot 2\cosh(2K_1 - K_2), \tag{7b}$$

and

$$\exp(K'_0 - K'_1) = \exp(2K_0) \cdot 2\cosh K_2. \tag{7c}$$

Solving for K'_0, K'_1, and K'_2, we get

$$e^{K'_0} = 2e^{2K_0}\{\cosh(2K_1 + K_2)\cosh(2K_1 - K_2)\cosh^2 K_2\}^{1/4}, \tag{8a}$$

$$e^{K'_1} = \{\cosh(2K_1 + K_2)\cosh(2K_1 - K_2)/\cosh^2 K_2\}^{1/4}, \tag{8b}$$

and

$$e^{K'_2} = e^{K_2}\{\cosh(2K_1 + K_2)/\cosh(2K_1 - K_2)\}^{1/2}. \tag{8c}$$

We may now remark on the need to have the parameter K_0 included in expression (1) and, accordingly, K'_0 in (5). Since we end up having three equations (7a), (7b), and (7c), to determine the parameters appropriate to the transformed system, the problem could not be solved with K_1 and K_2 only; thus, even if K_0 were set equal to zero, a $K'_0 \neq 0$ is essential for a proper representation of the transformed system. To highlight the role played by this parameter in determining the free energy of the given system, we observe on the basis of equations (1) and (5) that, with $K_0 = 0$,

$$Q_N(K_1, K_2) = e^{N'K'_0} Q_{N'}(K'_1, K'_2) \tag{9}$$

and hence for the free energy we have (in units of kT)

$$A_N(K_1, K_2) = -N'K'_0 + A_{N'}(K'_1, K'_2). \tag{10}$$

Since $N' = \frac{1}{2}N$, we obtain for the free energy *per spin* the recurrence relation

$$f(K_1, K_2) = -\tfrac{1}{2}K'_0 + \tfrac{1}{2}f(K'_1, K'_2), \tag{11}$$

which relates the free energy per spin of the given system with that of the transformed system; the role played by K_0' is clearly significant. For example, in the limit $T \to \infty$, when (K_1, K_2) and along with them (K_1', K_2') tend to zero, equations (8a) and (11) give

$$f(0,0) = -K_0' = -\ln 2, \tag{12}$$

which is indeed the correct result (arising from the limiting value of the entropy, $Nk\ln 2$, of the system).

We note that the parameter K_0 does not appear in equations (8b) and (8c), which determine K_1' and K_2' in terms of K_1 and K_2. As will be seen in Sections 14.3 and 14.4, it is these two relations that determine the singular part of the free energy of the system and hence its critical behavior; the parameters K_0 and K_0' affect only the regular part of the free energy and hence play no direct role in determining the critical behavior of the system. We might hasten to add that, while renormalization is generally used as a technique for studying the properties of a given system in the vicinity of its critical point, it can be useful over a much broader range of the variables K_1 and K_2. For instance, in the absence of the field ($K_2 = 0$), equations (8) and (11) give

$$K_0' = \ln\{2\surd[\cosh(2K_1)]\}, \quad K_1' = \ln\surd[\cosh(2K_1)], \quad K_2' = 0 \tag{13}$$

and hence

$$f(K_1, 0) = -\tfrac{1}{2}\ln\{2\surd[\cosh(2K_1)]\} + \tfrac{1}{2}f(\ln\surd[\cosh(2K_1)], 0); \tag{14}$$

the functional equation (14) has the solution

$$f(K_1, 0) = -\ln(2\cosh K_1), \tag{15}$$

valid at *all* K_1. On the other hand, in the paramagnetic case ($K_1 = 0$), we get

$$K_0' = \ln(2\cosh K_2), \quad K_1' = 0, \quad K_2' = K_2 \tag{16}$$

and hence

$$f(0, K_2) = -\tfrac{1}{2}\ln(2\cosh K_2) + \tfrac{1}{2}f(0, K_2), \tag{17}$$

with the solution

$$f(0, K_2) = -\ln(2\cosh K_2), \tag{18}$$

valid at *all* K_2. The case when both K_1 and K_2 are present is left as an exercise for the reader; see Problem 14.2. The critical behavior of this system will be studied in Section 14.4.

14.2.B The spherical model in one dimension

We adopt the same topology as in Figure 14.3 and write down the partition function of the one-dimensional spherical model consisting of N spins, see equation (13.5.12a),

$$Q_N = \int_{-\infty}^{\infty} \ldots \int_{-\infty}^{\infty} \exp\left[\sum_{i=1}^{N} \left\{K_0 + K_1\sigma_i\sigma_{i+1} + K_2\sigma_i - \Lambda\sigma_i^2\right\}\right] d\sigma_1 \ldots d\sigma_N$$

$$(K_0 = 0, K_1 = \beta J, K_2 = \beta\mu B, \Lambda = \beta\lambda), \tag{19}$$

where Λ is chosen so that

$$\left\langle \sum_{i=1}^{N} \sigma_i^2 \right\rangle = -\frac{\partial}{\partial\Lambda} \ln Q_N = N; \tag{20}$$

see equations (13.5.9) and (13.5.13). Assuming N to be even, we carry out integrations over $\sigma_2, \sigma_4, \ldots$. For this, we write our integrand as

$$\prod_{i=1}^{N} \exp\left\{K_0 + K_1\sigma_i\sigma_{i+1} + K_2\sigma_i - \Lambda\sigma_i^2\right\} = \prod_{j=1}^{\frac{1}{2}N} \exp\Big\{2K_0 + K_1(\sigma_{2j-1}\sigma_{2j} + \sigma_{2j}\sigma_{2j+1})$$
$$+ K_2(\sigma_{2j-1} + \sigma_{2j}) - \Lambda\left(\sigma_{2j-1}^2 + \sigma_{2j}^2\right)\Big\} \tag{21}$$

and integrate over σ_{2j}, using the formula

$$\int_{-\infty}^{\infty} e^{-px^2+qx} dx = \left(\frac{\pi}{p}\right)^{1/2} e^{q^2/4p} \quad (p > 0). \tag{22}$$

After simplification, we get

$$\prod_{j=1}^{\frac{1}{2}N} \left(\frac{\pi}{\Lambda}\right)^{1/2} \exp\left\{\left(2K_0 + \frac{K_2^2}{4\Lambda}\right) + \frac{K_1^2}{2\Lambda}\sigma_{2j-1}\sigma_{2j+1} + \left(K_2 + \frac{K_1K_2}{\Lambda}\right)\sigma_{2j-1} - \left(\Lambda - \frac{K_1^2}{2\Lambda}\right)\sigma_{2j-1}^2\right\}. \tag{23}$$

Denoting σ_{2j-1} by σ_j', expression (19) may now be written in the renormalized form

$$Q_N = \int_{-\infty}^{\infty} \ldots \int_{-\infty}^{\infty} \exp\left[\sum_{j=1}^{N'} \left\{K_0' + K_1'\sigma_j'\sigma_{j+1}' + K_2'\sigma_j' - \Lambda'\sigma_j'^2\right\}\right] d\sigma_1' \ldots d\sigma_{N'}', \tag{24}$$

where

$$K_0' = \frac{1}{2}\ln\left(\frac{\pi}{\Lambda}\right) + 2K_0 + \frac{K_2^2}{4\Lambda}, \quad K_1' = \frac{K_1^2}{2\Lambda}, \tag{25a, b}$$

$$K_2' = K_2\left(1 + \frac{K_1}{\Lambda}\right), \quad \Lambda' = \Lambda - \frac{K_1^2}{2\Lambda} \tag{25c, d}$$

and, of course, $N' = \frac{1}{2}N$. It follows that, with $K_0 = 0$,

$$Q_N(K_1, K_2, \Lambda) = e^{N'K_0'} Q_{N'}(K_1', K_2', \Lambda') \tag{26}$$

and hence for the free energy *per spin* (in units of kT) we have the recurrence relation

$$f(K_1, K_2, \Lambda) = -\tfrac{1}{2}K_0' + \tfrac{1}{2}f(K_1', K_2', \Lambda'). \tag{27}$$

The critical behavior of this system will be studied in Section 14.4. Presently we would like to demonstrate how the free energy of the system, over a broad range of the variables K_1 and K_2, can be determined by using the recurrence relation (27) along with the transformation equations (25).

First of all we identify two *invariants* of the transformation, namely

$$\Lambda'^2 - K_1'^2 = \Lambda^2 - K_1^2 \tag{28a}$$

and

$$(\Lambda' - K_1')/K_2' = (\Lambda - K_1)/K_2. \tag{28b}$$

It turns out that it is precisely these combinations that appear in the constraint equation of the system as well; see Problem 13.18. It follows that the constraint equation (20) is *RG-invariant*, that is, once it is satisfied in the original system, its counterpart

$$\left\langle \sum_{j=1}^{N'} \sigma_j'^2 \right\rangle = N' \tag{29}$$

is automatically satisfied in the transformed system — without any need to rescale the spin variables.[3] Now, in the absence of the field ($K_2 = 0$), equations (25) and (27) give

$$K_0' = \frac{1}{2}\ln\left(\frac{\pi}{\Lambda}\right), \quad K_1' = \frac{K_1^2}{2\Lambda}, \quad K_2' = 0, \quad \Lambda' = \Lambda - \frac{K_1^2}{2\Lambda} \tag{30}$$

and hence

$$f(K_1, \Lambda) = -\frac{1}{4}\ln\left(\frac{\pi}{\Lambda}\right) + \frac{1}{2}f\left(\frac{K_1^2}{2\Lambda}, \Lambda - \frac{K_1^2}{2\Lambda}\right). \tag{31}$$

[3]This is further related to the fact that the critical exponent η in this case is equal to 1; see equation (14.1.23) which, with $d = 1$, gives: $\sigma'(\boldsymbol{r}') = \sigma(\boldsymbol{r})$.

The functional equation (31) has the solution

$$f(K_1, \Lambda) = \frac{1}{2} \ln \left[\frac{\Lambda + \surd(\Lambda^2 - K_1^2)}{2\pi} \right], \tag{32}$$

valid at *all* K_1; the appropriate value of Λ is given by the constraint equation

$$\frac{\partial f}{\partial \Lambda} = \frac{1}{2\surd(\Lambda^2 - K_1^2)} = 1, \quad \text{that is} \quad \Lambda = \frac{1}{2}\surd(1 + 4K_1^2). \tag{33}$$

Note that the invariant (28a) in this case is equal to $\frac{1}{4}$. On the other hand, in the paramagnetic case ($K_1 = 0$), we get

$$K_0' = \frac{1}{2} \ln\left(\frac{\pi}{\Lambda}\right) + \frac{K_2^2}{4\Lambda}, \quad K_1' = 0, \quad K_2' = K_2, \quad \Lambda' = \Lambda \tag{34}$$

and hence

$$f(K_2, \Lambda) = -\frac{1}{4} \ln\left(\frac{\pi}{\Lambda}\right) - \frac{K_2^2}{8\Lambda} + \frac{1}{2} f(K_2, \Lambda), \tag{35}$$

with the solution

$$f(K_2, \Lambda) = -\frac{1}{2} \ln\left(\frac{\pi}{\Lambda}\right) - \frac{K_2^2}{4\Lambda}, \tag{36}$$

valid at *all* K_2; the appropriate value of Λ is now given by

$$\frac{\partial f}{\partial \Lambda} = \frac{1}{2\Lambda} + \frac{K_2^2}{4\Lambda^2} = 1, \quad \text{that is} \quad \Lambda = \frac{\surd(1 + 4K_2^2) + 1}{4}. \tag{37}$$

The case when both K_1 and K_2 are present is left as an exercise for the reader; see Problem 14.3.

14.2.C The Ising model in two dimensions

As our third example of renormalization, we consider an Ising model on a square lattice in two dimensions. In the field-free case, the partition function of this system is given by

$$Q_N(T) = \sum_{\{\sigma_i\}} \exp\left\{ \sum_{\text{n.n.}} K\sigma_i\sigma_j \right\} \quad (K = \beta J), \tag{38}$$

where the summation in the exponent goes over all nearest-neighbor pairs of spins in the lattice. Writing the summand in (38) as a product of factors pertaining to different pairs of

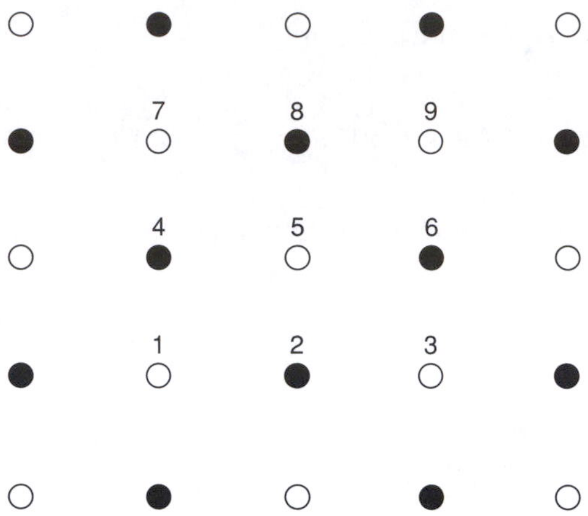

FIGURE 14.4 A section of the two-dimensional Ising lattice. The summed-over spins are denoted by open circles, the remaining ones by solid dots. To begin with, we concentrate on the summation over σ_5.

spins, we may highlight those factors that contain a particular spin, say σ_5, and carry out summation over this spin (see Figure 14.4):

$$\sum_{\sigma_5=\pm 1} \prod_{\text{n.n.}} \ldots e^{K\sigma_2\sigma_5} \cdot e^{K\sigma_4\sigma_5} \cdot e^{K\sigma_5\sigma_6} \cdot e^{K\sigma_5\sigma_8} \ldots \tag{39}$$

$$= \prod_{\text{n.n.}}{}' \ldots [2\cosh K(\sigma_2 + \sigma_4 + \sigma_6 + \sigma_8)] \ldots, \tag{40}$$

where the primed product goes over the remaining nearest-neighbor pairs in the lattice. This procedure of summation is supposed to be continued until one-half of the spins, shown as open circles in Figure 14.4, are all summed over. Clearly, this will generate a host of factors of the type shown in expression (40) but the real task now is to express these factors in a form similar, or at least as close as possible, to the factors appearing in the original expression (38); moreover, this mode of expression should be valid for all possible values of the remaining spins, namely $\sigma_2, \sigma_4, \ldots = \pm 1$.

For the factor explicitly displayed in (40), there are 16 possible values for the spins involved, of which only four turn out to be nonequivalent; they are

$$\text{(i) } \sigma_2 = \sigma_4 = \sigma_6 = \sigma_8, \tag{41a}$$

$$\text{(ii) } \sigma_2 = \sigma_4 = \sigma_6 = -\sigma_8, \tag{41b}$$

$$\text{(iii) } \sigma_2 = \sigma_4 = -\sigma_6 = -\sigma_8, \tag{41c}$$

$$\text{(iv) } \sigma_2 = -\sigma_4 = -\sigma_6 = \sigma_8. \tag{41d}$$

However, even four values are too many to accommodate by a factor of the form

$$\exp\{A + B(\sigma_2\sigma_4 + \sigma_2\sigma_6 + \sigma_4\sigma_8 + \sigma_6\sigma_8)\},$$

which contains only nearest-neighbor interactions in the transformed lattice and hence only two parameters to choose. Clearly, we need two more degrees of freedom, and it turns out that these are provided by the next-nearest-neighbor interactions and by interactions among a quartet of spins sitting on the corners of an elementary square in the new lattice. Thus, we are obliged to set

$$2\cosh K(\sigma_2+\sigma_4+\sigma_6+\sigma_8) = \exp\{K_0' + \tfrac{1}{2}K'(\sigma_2\sigma_4+\sigma_2\sigma_6+\sigma_4\sigma_8+\sigma_6\sigma_8) \\ + L'(\sigma_2\sigma_8+\sigma_4\sigma_6) + M'\sigma_2\sigma_4\sigma_6\sigma_8\}; \tag{42}$$

the reason why we have written $\frac{1}{2}K'$, rather than K', will become clear shortly. Now, the four possibilities listed above require that

$$2\cosh 4K = \exp(K_0' + 2K' + 2L' + M'), \tag{43a}$$

$$2\cosh 2K = \exp(K_0' - M'), \tag{43b}$$

$$2 = \exp(K_0' - 2L' + M'), \tag{43c}$$

$$2 = \exp(K_0' - 2K' + 2L' + M'), \tag{43d}$$

with the result that

$$K_0' = \ln 2 + \tfrac{1}{2}\ln\cosh 2K + \tfrac{1}{8}\ln\cosh 4K, \tag{44}$$

$$K' = \tfrac{1}{4}\ln\cosh 4K, \tag{45}$$

$$L' = \tfrac{1}{8}\ln\cosh 4K, \tag{46}$$

$$M' = \tfrac{1}{8}\ln\cosh 4K - \tfrac{1}{2}\ln\cosh 2K. \tag{47}$$

Continuing this process, we find that the factor exp $\left(\frac{1}{2}K'\sigma_2\sigma_4\right)$ appears once again when summation over σ_1 is carried out, the factor exp $\left(\frac{1}{2}K'\sigma_2\sigma_6\right)$ appears once again when summation over σ_3 is carried out, and so on; no further factors involving the products $\sigma_2\sigma_8, \sigma_4\sigma_6$ and $\sigma_2\sigma_4\sigma_6\sigma_8$ appear. The net result is that the partition function (38) assumes the form

$$Q_N = e^{N'K_0'}\sum_{\{\sigma_j'\}}\exp\left\{K'\sum_{\text{n.n.}}\sigma_j'\sigma_k' + L'\sum_{\text{n.n.n.}}\sigma_j'\sigma_k' + M'\sum_{\text{sq.}}\sigma_j'\sigma_k'\sigma_l'\sigma_m'\right\}, \tag{48}$$

where $N' = \frac{1}{2}N$.

Clearly, we have not been able to establish an *exact* correspondence between the transformed system and the original one (in which no interactions other than the nearest-neighbor ones were present). It seems more reasonable now that we redefine the original system as one having all the interactions appearing in expression (48), but with $L = M = 0$. We may then write

$$Q_N(K,0,0) = e^{N'K_0'}Q_{N'}(K',L',M') \tag{49}$$

and hence for the free energy *per spin* (in units of kT)

$$f(K,0,0) = -\tfrac{1}{2}K_0' + \tfrac{1}{2}f(K',L',M'), \tag{50}$$

where K_0', K', L', and M' are given by equations (44) through (47). In view of the appearance of new parameters, L' and M', in the recurrence relation (50), no further progress can be made without introducing some sort of an approximation, for which see Section 14.4. But one thing is clear: when renormalization is carried out in two dimensions or more, the connectivity of the lattice requires that the Hamiltonian of the decimated system contain some higher-order interactions not present in the original system, in order that the latter be represented correctly on transformation. It is obvious that further renormalizations would require more and more such interactions, and hence the need for more and more parameters would grow without limit. It may then be advisable that the Hamiltonian of the given system be regarded as a function of an "infinitely large number of parameters" (all but a few of which are zero to begin with), such that the number of parameters with a nonzero value grows indefinitely as renormalization transformations are carried out in succession and the number of degrees of freedom of the system steadily reduced.

We now present a formulation of the renormalization group approach to the study of critical phenomena.

14.3 The renormalization group: general formulation

We start with a system whose Hamiltonian depends on a large number of parameters $K_1, K_2, \ldots$ (all but a few of which are zero to begin with) and on the spin configuration $\{\sigma_i\}$ of the lattice. The free energy of the system is then given by

$$\exp(-\beta A) = \sum_{\{\sigma_i\}} \exp[-\beta H_{\{\sigma_i\}}(\{K_\alpha\})] \quad \alpha = 1,2,\ldots. \tag{1}$$

We now effect a "decimation" of the system, which reduces the number of degrees of freedom from N to N' and the correlation length from ξ to ξ', such that

$$N' = l^{-d}N, \quad \xi' = l^{-1}\xi \quad (l > 1). \tag{2a, b}$$

Expressing the Hamiltonian of the transformed system in a form similar to the one for the original system, except that we now have new parameters K_α', along with the additional K_0', and new spins σ_j', equation (1) takes the form

$$\exp(-\beta A) = \exp(N'K_0') \sum_{\{\sigma_j'\}} \exp\left[-\beta H_{\{\sigma_j'\}}(\{K_\alpha'\})\right], \tag{3}$$

so that the free energy *per spin* (in units of β^{-1}) is given by

$$f(\{K_\alpha\}) = l^{-d}[-K_0' + f(\{K_\alpha'\})]. \tag{4}$$

We now look closely at the transformation $\{K_\alpha\} \to \{K'_\alpha\}$ by introducing a *vector space* $\mathcal{K}$ in which the set of parameters K_α is represented by the tip of a position vector $\boldsymbol{K}$; on transformation, $\boldsymbol{K}$ changes to $\boldsymbol{K}'$, which may be looked upon as a kind of "flow from one position in the vector space to another." This flow may be represented by the transformation equation

$$\boldsymbol{K}' = \mathcal{R}_l(\boldsymbol{K}) \tag{5}$$

where $\mathcal{R}_l$ is the renormalization-group operator appropriate to the case under consideration. A repeated application of this process leads to a sequence of vectors $\boldsymbol{K}', \boldsymbol{K}'', \ldots$, such that

$$\boldsymbol{K}^{(n)} = \mathcal{R}_l(\boldsymbol{K}^{(n-1)}) = \ldots = \mathcal{R}_l^n(\boldsymbol{K}^{(0)}) \quad n = 0, 1, 2, \ldots, \tag{6}$$

where $\boldsymbol{K}^{(0)}$ denotes the original $\boldsymbol{K}$. At the end of this sequence, the correlation length ξ and the singular part of the free energy f_s are given by

$$\xi^{(n)} = l^{-n}\xi^{(0)}, \quad f_s^{(n)} = l^{nd} f_s^{(0)}; \tag{7a, b}$$

see equations (2b) and (4).

Now, the transformation (5) may have a *fixed point*, $\boldsymbol{K}^*$, so that

$$\mathcal{R}_l(\boldsymbol{K}^*) = \boldsymbol{K}^*. \tag{8}$$

Equation (2b) then tells us that $\xi(\boldsymbol{K}^*) = l^{-1}\xi(\boldsymbol{K}^*)$, which means that $\xi(\boldsymbol{K}^*)$ is either zero or infinite! The former possibility is of little interest to us, so let us dwell only on the latter (which makes the system with parameters $\boldsymbol{K} = \boldsymbol{K}^*$ critical); in simple situations, the fixed point, $\boldsymbol{K}^*$, will correspond to the critical point, $\boldsymbol{K}_c$, of the given system. Conceivably, an arbitrary point $\boldsymbol{K}$, on successive transformations such as (6), may end up at the fixed point $\boldsymbol{K}^*$. Since the correlation length ξ can only decrease on transformation, see equation (7a), and is infinite at the end of this sequence of transformations, it must be infinite at $\boldsymbol{K}$ as well (the same being true for *all* points intermediate between $\boldsymbol{K}$ and $\boldsymbol{K}^*$). The collection of all those points which, on successive transformations, flow into the fixed point, constitutes a surface over which ξ is infinite; this surface is generally referred to as the *critical surface*. All flow lines in this surface are directed toward, and terminate at, the fixed point, while points off this surface may initially move toward the fixed point but eventually their flow lines will veer away from it; see Figure 14.5. Reasons behind this pattern of flow will become clear soon.

For the analysis of the critical behavior we examine the pattern of flow in the neighborhood of the fixed point $\boldsymbol{K}^*$.[4] Setting

$$\boldsymbol{K} = \boldsymbol{K}^* + \boldsymbol{k}, \tag{9}$$

[4]In general, the vector $\boldsymbol{K}^*$ will contain components not present in the original problem. In such a case, one has to locate, on the critical surface, a point $\boldsymbol{K}_c$ that is free of these "unnecessary" components; since ξ is infinite at $\boldsymbol{K} = \boldsymbol{K}_c$ as well, the latter may be identified as the *critical point* of the given system. As will be seen in the sequel, the critical behavior of the system is still determined by the flow pattern in the neighborhood of the fixed point.

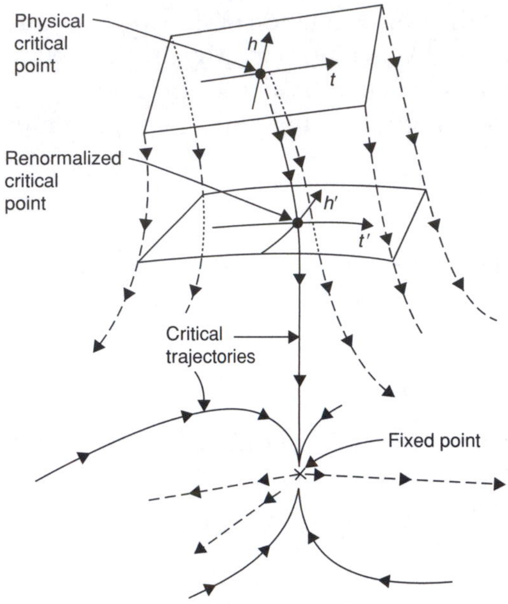

FIGURE 14.5 The parameter space of a physical system, showing critical trajectories (solid lines) flowing into the fixed point. The subspace of the relevant variables, t and h, is everywhere "orthogonal" to these trajectories (on all of which $t = h = 0$); the critical trajectories differ from one another only in respect of irrelevant variables that vanish as the fixed point is approached. The dashed lines depict that part of the flow in which the relevant variables play a decisive role.

we have by equations (5) and (8)

$$\boldsymbol{K}' = \boldsymbol{K}^* + \boldsymbol{k}' = \mathcal{R}_l(\boldsymbol{K}^* + \boldsymbol{k}) \tag{10}$$

so that

$$\boldsymbol{k}' = \mathcal{R}_l\left(\boldsymbol{K}^* + \boldsymbol{k}\right) - \boldsymbol{K}^* \tag{11}$$

Assuming $\{k_\alpha\}$, and hence $\{k'_\alpha\}$, to be small, we may linearize equation (11) to write

$$\boldsymbol{k}' \approx \left.\frac{d\mathcal{R}_l}{d\boldsymbol{K}}\right|_{\boldsymbol{K}=\boldsymbol{K}^*} \boldsymbol{k} = \mathcal{A}_l^* \boldsymbol{k} \tag{12}$$

where $\mathcal{A}_l^*$ is a matrix arising from the linearization of the operator $\mathcal{R}_l$ around the fixed point $\boldsymbol{K}^*$. The eigenvalues λ_i and the eigenvectors $\boldsymbol{\phi}_i$ of the matrix $\mathcal{A}_l^*$ play a vital role in determining the critical behavior of the system.

If the vectors $\boldsymbol{\phi}_i$ form a complete set, we may expand $\boldsymbol{k}$ and $\boldsymbol{k}'$ in terms of $\boldsymbol{\phi}_i$,

$$\boldsymbol{k} = \sum_i u_i \boldsymbol{\phi}_i, \quad \boldsymbol{k}' = \sum_i u'_i \boldsymbol{\phi}_i, \tag{13a, b}$$

with the result that

$$u'_i = \lambda_i u_i \quad i = 1, 2, \ldots; \tag{14}$$

the coefficients u_i appearing here are generally referred to as the *scaling fields* of the problem. Under successive transformations (all in the neighborhood of the fixed point), these fields are given by

$$u_i^{(n)} = \lambda_i^n u_i^{(0)}. \tag{15}$$

It is obvious that the fields u_i are certain linear combinations of the original parameters k_α; they may, therefore, be looked upon as the "generalized coordinates" of the problem. The relative influence of these coordinates in determining the critical behavior of the system depends crucially on the respective eigenvalues λ_i. With $u_i^{(n)}$ given by (15), we have three possible courses for a given coordinate u_i.

(a) If $\lambda_i > 1$, the coordinate u_i grows with n and, with successive transformations, becomes more and more significant. Clearly, u_i in this case is a *relevant variable* which, by itself, drives the system away from the fixed point — thus making the fixed point unstable. By experience, we know that the temperature parameter $t\,[= (T - T_c)/T_c]$ and the magnetic field parameter $h\,[= \mu B/kT_c]$ are two basic quantities that vanish at the critical point and are clearly relevant to the problem of phase transitions. We, therefore, expect that our analysis will produce at least two relevant variables, u_1 and u_2 say, which could be identified with t and h, respectively, so that

$$u_1 = at + O(t^2), \quad u_2 = bh + O(h^2), \tag{16, 17}$$

with both λ_1 and λ_2 greater than unity.

(b) If $\lambda_i < 1$, the coordinate u_i decays with n and, with successive transformations, becomes less and less significant. Clearly, u_i in this case is an *irrelevant variable* which, by itself, drives the system toward the fixed point. Now, if all the relevant variables are set at zero, then successive transformations (by virtue of the irrelevant variables) *will* drive the system to the fixed point. We must then be cruising on the critical surface itself (where all trajectories are known to flow into the fixed point). It follows that on the critical surface all relevant variables of the problem are zero; furthermore, the divergence of the correlation length is also tied to the same fact.

(c) If $\lambda_i = 1$, the coordinate u_i, in the linear approximation, stays constant; it neither grows nor decays very rapidly unless one enters the nonlinear, beyond-scaling, regime of the variable u_i. Quite appropriately, u_i in this case is termed a *marginal variable*; it does not affect the critical behavior of the system as significantly as a relevant variable does, but it may throw in logarithmic factors along with the conventional power laws. The ability to identify such variables and to predict the consequent departures from simple power-law scaling is one of the added virtues of the RG approach.

The above considerations enable us to understand the pattern of flow shown in Figure 14.5. While the points *on* the critical surface flow into the fixed point, those *off* this surface flow toward the fixed point by virtue of the irrelevant variables and, at the same time, away from it by virtue of the relevant variables; the net result is an initial approach toward but a final recession away from the fixed point. Points close to the fixed point and in directions "orthogonal" to the critical surface have only relevant variables to contend with, so right away they move away from the fixed point. It is this part of the flow that determines the critical behavior of the given system.

Disregarding the irrelevant variables, we now examine the manner in which the correlation length, ξ, and the (singular part of the) free energy, f_s, of the system are affected by the transformation (15). In view of equations (7a,b), we have

$$\xi(u_1, u_2, \ldots) = l^n \xi(\lambda_1^n u_1, \lambda_2^n u_2, \ldots) \tag{18}$$

and

$$f_s(u_1, u_2, \ldots) = l^{-nd} f_s(\lambda_1^n u_1, \lambda_2^n u_2, \ldots). \tag{19}$$

Identifying u_1 with t, as in (16), and remembering the definition of the critical exponent ν, we obtain from (18)

$$u_1^{-\nu} = l^n (\lambda_1^n u_1)^{-\nu}, \tag{20}$$

with the result that

$$\nu = \ln l / \ln \lambda_1. \tag{21}$$

At first sight one might wonder why ν should depend on l. In fact, it doesn't because of the following argument. On physical grounds we expect that two successive transformations with scale factors l_1 and l_2 would be equivalent to a single transformation with scale factor $l_1 l_2$, that is,[5]

$$\mathcal{R}_{l_1}^* \mathcal{R}_{l_2}^* = \mathcal{R}_{l_1 l_2}^*. \tag{22}$$

This forces the eigenvalues λ_i to be of the form l^{y_i}, for

$$l_1^{y_i} l_2^{y_i} = (l_1 l_2)^{y_i}. \tag{23}$$

Relation (21) then becomes

$$\nu = 1/y_1, \tag{24}$$

manifestly independent of l.

Equation (19) may now be written as

$$f_s(t, h, \ldots) = l^{-nd} f_s(l^{ny_1} t, l^{ny_2} h, \ldots). \tag{25}$$

[5]This requirement makes the set of operators $\mathcal{R}_l$ a *semigroup* — *not* a *group* because the inverse of $\mathcal{R}_l$ does not exist. The reason for the nonexistence of $\mathcal{R}_l^{-1}$ is that once a number of degrees of freedom of the system are summed over there is no definitive way of recreating them.

To ensure that the above relationship is independent of the choice of l, we follow the same line of argument as in Section 14.1, after equation (14.1.9), with the result

$$f_s(t,h,\ldots) = |t|^{d\nu} \tilde{f}_s(h/|t|^{\Delta},\ldots), \tag{26}$$

where

$$\Delta = y_2/y_1. \tag{27}$$

Equation (26) is formally the same as the scaling relationship postulated in Section 12.10 and, one might say, argued out in Section 14.1. The big difference here is that not only has this relationship been derived on a firmer basis but now we also have a recipe for evaluating the critical exponents ν, Δ, and so on, from first principles. What one has to do here is to determine the RG operator $\mathcal{R}_l$ for the given problem, linearize it around the appropriate fixed point $\boldsymbol{K}^*$, determine the eigenvalues $\lambda_i\,(= l^{y_i})$ and use equations (24) and (27) to evaluate ν and Δ. At the same time, recalling the definition of the critical exponent α, we infer from (26) that

$$2 - \alpha = d\nu; \tag{28}$$

the remaining exponents follow with the help of the scaling relations

$$\beta = (2-\alpha) - \Delta, \quad \gamma = 2\Delta - (2-\alpha), \quad \delta = \Delta/\beta, \quad \eta = 2 - (\gamma/\nu). \tag{29}$$

We find that the hyperscaling relation (28) is an integral part of the RG formulation; it comes out naturally — with no external imposition whatsoever. It is, however, disconcerting that, according to the above argument, this relation should hold for *all* d — notwithstanding the fact that, for $d > 4$, all critical exponents are "stuck" at the mean field values and relation (28) gets replaced by

$$2 - \alpha = 4\nu \quad (d > 4), \tag{30}$$

with $\alpha = 0$ and $\nu = \frac{1}{2}$. The reason for this peculiar behavior is somewhat subtle; it arises from the fact that in certain situations an "irrelevant variable" may well raise its "dangerous" head and affect the outcome of the calculation in a rather significant manner.

To see how this happens, we may consider a *continuous spin model*, very much like the one considered in Section 13.5, with the probability distribution law

$$p(\sigma_i)d\sigma_i = \text{const.}\, e^{-\frac{1}{2}\sigma_i^2 - \tilde{u}\sigma_i^4}\, d\sigma_i \quad (\tilde{u} > 0); \tag{31}$$

compared with equation (13.5.1). The free energy of the system then depends on the parameter $\tilde{u}$ as well as on t and h, and we obtain instead of (25)

$$f_s(t,h,\tilde{u}) = l^{-nd} f_s\left(l^{ny_1} t, l^{ny_2} h, l^{ny_3} \tilde{u},\ldots\right). \tag{32}$$

Now, using the RG approach, one finds (see Appendix D of Fisher, 1983) that, for $d > 4$,

$$y_1 = 2, \quad y_2 = \frac{1}{2}d + 1, \quad y_3 = 4 - d, \tag{33}$$

showing very clearly that, for $d > 4$, the parameter $\tilde{u}$ is an irrelevant variable. One is, therefore, tempted to ignore $\tilde{u}$ and arrive at equation (26), with

$$\nu = \frac{1}{2} \quad \text{and} \quad \Delta = (d+2)/4. \tag{34}$$

The very fact that Δ turns out to be d-dependent shows that there is something wrong here. It turns out that though, on successive transformations, the variable $\tilde{u}$ does tend to zero, its influence on the function f_s does not. It may, therefore, be prudent to write

$$f_s(t, h, \tilde{u}) = |t|^{d\nu} \tilde{f}_s(h/|t|^{\Delta}, \tilde{u}/|t|^{\phi}), \tag{35}$$

where

$$\phi = \frac{y_3}{y_1} = \frac{4-d}{2}. \tag{36}$$

Now, by analysis, one finds that

$$\operatorname*{Lim}_{w \to 0} \tilde{f}_s(v, w) \approx \frac{1}{w} F(vw^{1/2}), \tag{37}$$

where $F(vw^{1/2})$ is a perfectly well-behaved function. The singularity of $\tilde{f}_s$ in w changes the picture altogether, and we get in the desired limit

$$f_s(t, h, \tilde{u}) \approx |t|^{d\nu + \phi} F\left(h/|t|^{\Delta + \frac{1}{2}\phi}\right). \tag{38}$$

The "revised" value of Δ now is

$$\Delta_{\text{rev}} = \frac{d+2}{4} + \frac{4-d}{4} = \frac{3}{2}, \tag{39}$$

which is indeed independent of d and agrees with the corresponding mean field value. At the same time, the "revised" form of the hyperscaling relation now is

$$2 - \alpha = \frac{d}{2} + \frac{4-d}{2} = 2, \tag{40}$$

as stated in (30).

The lesson to be learnt here is that the standard derivations of the scaling relations rest on certain assumptions, often left unstated, about the nonsingular or nonvanishing behavior of various scaling functions and their arguments. In many cases these assumptions are valid and may even be confirmed by explicit calculations or otherwise; in certain circumstances, however, they fail — in which case a scaling relation may change its form. Fortunately, such circumstances are not that common.

14.4 Applications of the renormalization group

We start our considerations with the models examined in Section 14.2.

14.4.A The Ising model in one dimension

The renormalization group transformation in this case is given by equations (14.2.8b and c), namely

$$K_1' = \tfrac{1}{4}\ln[\cosh(2K_1+K_2)\cosh(2K_1-K_2)] - \tfrac{1}{2}\ln\cosh K_2 \tag{1a}$$

and

$$K_2' = K_2 + \tfrac{1}{2}\ln[\cosh(2K_1+K_2)/\cosh(2K_1-K_2)], \tag{1b}$$

where $K_1 = J/kT$ and $K_2 = \mu B/kT$. It is straightforward to see that this transformation has a "line of *trivial* fixed points," with $K_1 = 0$ and K_2 arbitrary. These fixed points pertain to either $J = 0$ or $T = \infty$; in either case, one has a correlation length that vanishes. There is also a *nontrivial* fixed point at $K_1 = \infty$ and $K_2 = 0$, which may be realized by first setting $B = 0$ and then letting $T \to 0$; the correlation length at this fixed point will be infinite. In the vicinity of this point, we have

$$K_1' \simeq K_1 - \tfrac{1}{2}\ln 2, \quad K_2' \simeq 2K_2. \tag{2a, b}$$

Now, since $K_1^* = \infty$, K_1 is not an appropriate variable to carry out an expansion around the fixed point. We may adopt instead a new variable, see equation (13.2.17), namely

$$t = \exp(-pK_1) \quad (p > 0), \tag{3}$$

so that $t^* = 0$; now, in the vicinity of the fixed point, we have

$$t' \simeq 2^{p/2}t. \tag{4}$$

Identifying K_2 as the variable h, and remembering that the scale factor l here is 2, we readily obtain from equations (2b) and (4)

$$y_1 = p/2, \quad y_2 = 1. \tag{5}$$

The critical exponents of the model now follow straightforwardly from (5); we get

$$\nu = 2/p, \quad \Delta = 2/p, \tag{6}$$

from which, by equations (14.3.28) and (14.3.29),

$$\alpha = 2 - 2/p, \quad \beta = 0, \quad \gamma = 2/p, \quad \delta = \infty, \quad \eta = 1, \tag{7}$$

in complete agreement with the results found in Section 13.2. As for the choice of p, see remarks following equation (13.3.21).

14.4.B The spherical model in one dimension

The RG transformation in this case is given by equations (14.2.25b, c, and d), namely

$$K_1' = \frac{K_1^2}{2\Lambda}, \quad K_2' = K_2\left(1 + \frac{K_1}{\Lambda}\right), \quad \Lambda' = \Lambda - \frac{K_1^2}{2\Lambda}, \tag{8a, b, c}$$

where $K_1 = J/kT, K_2 = \mu B/kT$, and $\Lambda = \lambda/kT, \lambda$ being the "spherical field" that was employed to take care of the constraint on the model. The nontrivial fixed point is again at $T = 0$, where $\lambda = J$ [see equation (13.5.24), with $d = 1$] and hence $\Lambda = K_1$. Equations (8) then reduce to the linearized form (valid for small T)

$$K_1' \simeq \tfrac{1}{2}K_1, \quad K_2' \simeq 2K_2, \quad \Lambda' \simeq \tfrac{1}{2}\Lambda. \tag{9a, b, c}$$

Equations (9a) and (9c) contain essentially the same information, namely $T' \simeq 2T$. Clearly, T itself is a good variable for expansion in this case — giving $y_1 = 1$. Equation (9b), just like (2b), gives $y_2 = 1$, and we obtain

$$\nu = 1, \quad \Delta = 1, \tag{10}$$

whereby

$$\alpha = 1, \quad \beta = 0, \quad \gamma = 1, \quad \delta = \infty, \quad \eta = 1, \tag{11}$$

in complete agreement with the results for one-dimensional models with $n \geq 2$, as quoted in equation (13.3.20).

14.4.C The Ising model in two dimensions

The RG transformation in this case is given by equations (14.2.45) through (14.2.47), namely

$$K' = \frac{1}{4}\ln\cosh 4K, \tag{12}$$

$$L' = \frac{1}{8}\ln\cosh 4K, \tag{13}$$

and

$$M' = \frac{1}{8}\ln\cosh 4K - \frac{1}{2}\ln\cosh 2K. \tag{14}$$

It will be recalled that, while effecting this transformation, we started only with nearest-neighbor interactions (characterized by a single parameter $K = \beta J$) but, due to the connectivity of the lattice, ended up with more — namely, the next-nearest-neighbor interactions (characterized by L') and the four-spin interactions (characterized by M') — in addition to the nearest-neighbor interactions (characterized by K'). On subsequent transformations, still higher-order interactions come into play and the problem becomes formidable unless some approximations are introduced.

In one such approximation, due originally to Wilson (1975), we discard all interactions other than the ones represented by the parameters K and L, and at the same time assume K and L to be small enough so that equations (12) and (13) reduce to

$$K' = 2K^2, \quad L' = K^2. \tag{15a, b}$$

Now, if the parameter L had been introduced right in the beginning, the transformation equations, in this very approximation, would have been

$$K' = 2K^2 + L, \quad L' = K^2. \tag{16a, b}$$

We shall treat equations (16) as if they were the *exact* transformation equations of the problem and see what they lead to.

It is straightforward to see that the transformation (16) has a nontrivial fixed point at

$$K^* = \frac{1}{3}, \quad L^* = \frac{1}{9}. \tag{17}$$

Linearizing around this fixed point, we get

$$k_1' = \frac{4}{3}k_1 + k_2, \quad k_2' = \frac{2}{3}k_1, \tag{18}$$

where k_1 and k_2 represent deviations of the parameters K and L from the fixed-point values K^* and L^*, respectively. The transformation matrix $\mathcal{A}_l^*$ of equation (14.3.12) is then given by

$$\mathcal{A}^*_{\surd 2} = \begin{pmatrix} \frac{4}{3} & 1 \\ \frac{2}{3} & 0 \end{pmatrix}, \tag{19}$$

whose eigenvalues are

$$\lambda_1 = \frac{1}{3}(2 + \surd 10), \quad \lambda_2 = \frac{1}{3}(2 - \surd 10) \tag{20a, b}$$

and whose eigenvectors are

$$\phi_1 \sim \begin{pmatrix} 2 + \surd 10 \\ 2 \end{pmatrix}, \quad \phi_2 \sim \begin{pmatrix} 2 - \surd 10 \\ 2 \end{pmatrix}. \tag{21a, b}$$

The scaling fields u_i are then determined by equation (14.3.13a) which, on inversion, gives

$$u_1 \sim \{2k_1 + (\surd 10 - 2)k_2\}, \quad u_2 \sim \{2k_1 - (\surd 10 + 2)k_2\}. \tag{22a, b}$$

Clearly, the field u_1, with $\lambda_1 > 1$, is the relevant variable of the problem, while the field u_2 is irrelevant. The "critical curve" in the (K, L)-plane is determined by the condition $u_1 = 0$, while the linear part of this curve, in the vicinity of the fixed point ($\boldsymbol{u} = 0$), is mapped by the relation (22a). In terms of the variables k_1 and k_2, this segment of the critical curve is given by the equation

$$k_2 \approx -\frac{\sqrt{10}+2}{3} k_1, \tag{23}$$

which represents a straight line of slope -1.7208; see Figure 14.6.

To determine the *physical* critical point of this model, we have to locate on the critical curve a point with $L = 0$, for the original problem had no interactions other than the one represented by the parameter K; the corresponding value of K would be our K_c.[6] This requires a mapping of the critical curve right up to the K-axis. While this has been done numerically (Wilson, 1975), a crude estimate of K_c can be made by simply extending the straight-line segment (23) down to the desired limit. One thus obtains[7]

$$K_c = \frac{1}{3} + \frac{1}{9}\frac{3}{\sqrt{10}+2} = \frac{4+\sqrt{10}}{18} = 0.3979, \tag{24}$$

which may be compared with the exact result found in Section 13.4, namely 0.4407.

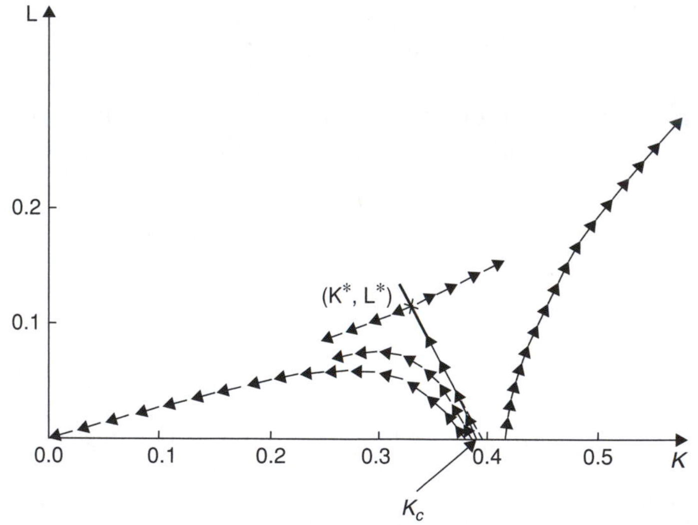

FIGURE 14.6 A section of the critical curve for the two-dimensional Ising model near the nontrivial fixed point $(K^* = \frac{1}{3}, L^* = \frac{1}{9})$. Points on the critical curve flow into the fixed point, while those off it flow away toward the trivial fixed point $(K^* = L^* = 0)$ or $(K^* = L^* = \infty)$.

[6]Remember that at each and every point on the critical surface — in this case, the critical curve — the correlation length is infinite; accordingly, each and every such point is qualified to be a critical point. The *physical* critical point is the one that is free of unnecessary parameters.

[7]The result obtained through numerical analysis was 0.3921.

Another quantity that can be estimated here is the critical exponent ν. From equations (14.3.21) and (14.3.20a),one obtains

$$\nu = \frac{\ln l}{\ln \lambda_1} = \frac{\ln \sqrt{2}}{\ln[(2+\sqrt{10})/3]} = 0.6385, \tag{25}$$

which may be compared with the exact value 1. Even though a comparison of the results obtained here with the ones following from exact analysis is not very flattering, the basic merits of the RG approach are quite obvious.

One important aspect of critical phenomena, namely their *universality* over a large class of systems, is manifest even in this simple example. Imagine, for instance, that in the case of the given system a next-nearest-neighbor interaction L_0 were indeed present. Our approximate treatment would then lead to the same fixed point and the same critical curve as above, but our *physical* critical point would now be given by that "point on the critical curve whose L-value is L_0"; we may denote this critical point by $K_c(L_0)$. As for the critical behavior, it will still be determined by an expansion around the fixed point, for that is where the "relevant part" of the flow is; see again Figure 14.6. Clearly, the critical behavior of the given system, insofar as exponents are concerned, will be the same, regardless of the actual value of L_0. And, by extension, the same will be true of any two systems which have the same basic topology but differ only in the details of the spin–spin interactions.

As for the accuracy of the results obtained here, improvements are needed in several important respects. First of all, the exclusion of all interaction parameters other than K and L constitutes a rather inadequate approximation; one should at least include the four-spin interaction, represented by the parameter M, and may possibly ignore the ones that appear on successive transformations. Next, the assumption that the parameters K and L are small is also unjustified, especially for K; this makes a numerical approach to the problem rather essential. Thirdly, we disregarded the renormalization of the spins, from the original $\sigma(\boldsymbol{r})$ to $\sigma'(\boldsymbol{r}')$, as required by equation (14.1.23); in the present problem, this would amount to introducing a factor of $(\sqrt{2})^{\eta/2}$, that is, $2^{1/16}$, for η here is $\frac{1}{4}$. In the actual treatment, one may have to introduce an unknown parameter, ρ, and determine its "true" value by theoretical analysis (see Wilson, 1975). Highly sophisticated procedures have been developed over the years to accommodate (or circumvent) these problems, leading to very accurate — in fact, almost exact — results for the model under considerations. For details, see the review article by Niemeijer and van Leeuwen (1976), where references to other pertinent literature on the subject can also be found.[8]

14.4.D The ε-expansion

Application of the RG approach to systems in higher dimensions, namely with $d > 2$, presents serious mathematical difficulties. One is then forced to resort to approximation procedures such as the ε-expansion, first introduced by Wilson (1972); see also Wilson

[8] In this reference one can also find a systematic method of constructing the scaling function $f_s(u_1, u_2, \ldots)$ from a knowledge of the regular function $f(K_0')$ of equation (14.3.4).

and Fisher (1972). This procedure was inspired by the observation that the field-theoretic calculations of the RG formulation become especially simple as the upper critical dimension, $d=4$, is approached; it, therefore, seemed desirable to introduce a parameter $\varepsilon(=4-d)$ and carry out expansions of the various quantities of interest around $\varepsilon=0$. The model adopted for these calculations was the same as the one referred to in Section 14.3, namely the *continuous, n-vector spin model,* with the probability distribution given by equation (14.3.31).[9] An important advantage of using continuous spins $\boldsymbol{\sigma}(\boldsymbol{r})\{=\sigma^{(\mu)}(\boldsymbol{r}), \mu=1,\ldots,n\}$, with $-\infty<\sigma^{(\mu)}<\infty$, is that one can introduce Fourier transforms $\boldsymbol{\sigma}(\boldsymbol{q})$ and make use of the "momentum shell integration" technique of Wilson (1971). The parameters of interest now are (see Fisher, 1983)

$$r=\frac{T-T_0}{T_0R_0^2}=\frac{t_0}{R_0^2},\quad u=\tilde{u}\frac{T^2a^d}{T_0^2R_0^4} \tag{26a, b}$$

and, of course, the magnetic field parameter h; here, T_0 denotes the mean-field critical temperature qJ/k (q being the coordination number), R_0 is a measure of the range of interactions, a is the lattice constant, whereas $\tilde{u}$ is the real-space parameter appearing in equation (14.3.31). The transformation equations, with a scale factor l, turn out to be

$$r'=l^2r+4(l^2-1)c(n+2)u-l^2\ln l(n+2)(2\pi^2)^{-1}ru, \tag{27a}$$

$$u'=(1+\varepsilon\ln l)u-(n+8)\ln l(2\pi^2)^{-1}u^2, \tag{27b}$$

and

$$h'=l^3\left(1-\frac{1}{2}\varepsilon\ln l\right)h, \tag{27c}$$

correct to the orders displayed; the parameter c in equation (27a) is related to a cutoff in the momentum space which, in turn, is a reflection of the underlying lattice structure.

Transformation (27) has two fixed points — the so-called *Gaussian* fixed point, with

$$r^*=u^*=h^*=0, \tag{28}$$

and a *non-Gaussian* fixed point, with

$$r^*=-\frac{8\pi^2c(n+2)}{(n+8)}\varepsilon,\quad u^*=\frac{2\pi^2}{(n+8)}\varepsilon,\quad h^*=0. \tag{29}$$

We now examine two distinct situations.

[9]It can be shown that, by a suitable transformation, the Ising model ($n=1$), which is a *discrete* (rather than a *continuous*) model, can also be rendered "continuous" with a probability distribution similar to (14.3.31). For details, see Appendix A in Fisher (1983).

Dimension $d \gtrsim 4$, so that ε is a small negative number

One readily sees from equation (27b) that the parameter u in this case decreases on transformation, so on successive transformations it will tend to zero. Clearly, only the Gaussian fixed point is the one appropriate to this case. Linearizing around this point, we obtain for the transformation matrix $\mathcal{A}_l^*$

$$\mathcal{A}_l^* = \begin{pmatrix} l^2 & 4(l^2-1)c(n+2) \\ 0 & 1+\varepsilon \ln l \end{pmatrix}, \tag{30}$$

with eigenvalues

$$\lambda_r = l^2, \quad \lambda_u = (1+\varepsilon \ln l) < 1 \tag{31}$$

and, of course,

$$\lambda_h = l^3\left(1 - \frac{1}{2}\varepsilon \ln l\right). \tag{32}$$

It follows that

$$y_1 = 2, \quad y_2 \approx 3 - \frac{1}{2}\varepsilon, \quad y_3 \approx \varepsilon, \tag{33}$$

as in equation (14.3.33). Note that the parameter u in this case is an irrelevant variable but, as discussed at the end of Section 14.3, it is a *dangerously irrelevant variable* that does eventually affect the results of the calculation in hand.

Dimension $d \lesssim 4$, so that ε is a small positive number

The parameter u now behaves very differently. If it is already zero, it stays so; otherwise, it moves away from that value, carrying the system to some other fixed point — possibly the non-Gaussian one, with coordinates given in (29). The resulting pattern of flow in the (r, u)-plane is shown in Figure 14.7; clearly, the Gaussian fixed point is no longer appropriate and the problem now revolves around the non-Gaussian fixed point instead. Linearizing around the latter, we obtain

$$\mathcal{A}_l^* = \begin{pmatrix} l^2\left\{1 - \dfrac{n+2}{n+8}\varepsilon \ln l\right\} & 4(l^2-1)c(n+2) \\ 0 & 1-\varepsilon \ln l \end{pmatrix}, \tag{34}$$

with eigenvalues

$$l^2\left\{1 - \frac{n+2}{n+8}\varepsilon \ln l\right\} \quad \text{and} \quad (1-\varepsilon \ln l) < 1. \tag{35}$$

We note that of the "generalized coordinates" u_1 and u_2, which are certain linear combinations of the parameters $\Delta r (= r - r^*)$ and $\Delta u (= u - u^*)$, only u_1 is a relevant variable of

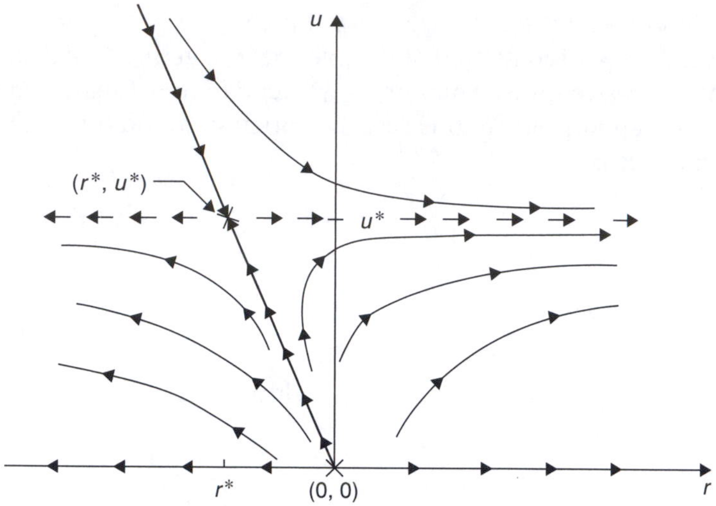

FIGURE 14.7 A section of the critical curve and a sketch of the RG flows in the (r, u)-plane for $0 < \varepsilon \ll 1$. Note that the critical curve is straight only to order ε.

the problem.[10] Identifying u_1 with the temperature parameter t, we obtain

$$y_t \approx 2 - \frac{n+2}{n+8}\varepsilon. \tag{36}$$

Combining this with expression (33) for y_h, namely

$$y_h \approx 3 - \frac{1}{2}\varepsilon, \tag{37}$$

we obtain, see equations (14.3.24), (14.3.27), (14.3.28), and (14.3.29),

$$\nu \approx \frac{1}{2} + \frac{n+2}{4(n+8)}\varepsilon, \quad \Delta \approx \frac{3}{2} + \frac{n-1}{2(n+8)}\varepsilon, \tag{38}$$

which gives

$$\alpha \approx \frac{4-n}{2(n+8)}\varepsilon, \quad \beta \approx \frac{1}{2} - \frac{3}{2(n+8)}\varepsilon, \quad \gamma \approx 1 + \frac{n+2}{2(n+8)}\varepsilon, \tag{39}$$

$$\delta \approx 3 + \varepsilon, \quad \eta \approx 0, \tag{40}$$

correct to the first power in ε.

For obvious reasons the value of ε of greatest interest to us is $\varepsilon = 1$, for which the above results are totally inadequate; they do show the correct trends, though. For better numerical accuracy it is essential to extend these calculations to higher orders in ε. Considerable

[10]It can be seen quite easily that the generalized coordinate u_2 is directly proportional to Δu, making u an irrelevant variable of the problem; see Problem 14.6, with $a_{21} = 0$.

progress has been made in this direction, so that we now have expressions available that include terms up to ε^3 and, in some cases, even ε^4; for details, see Wallace (1976). One wonders if that degree of extension would be good enough for obtaining reliable results for ε as large as 1. The answer is yes, and we will illustrate it with an example.

For the spherical model we know exact values of the various critical exponents which may, for the purpose of illustration, be expressed as power series in ε. Thus, for instance,

$$\gamma = \frac{2}{d-2} = \left(1 - \frac{1}{2}\varepsilon\right)^{-1} = 1 + \frac{1}{2}\varepsilon + \frac{1}{4}\varepsilon^2 + \frac{1}{8}\varepsilon^3 + \frac{1}{16}\varepsilon^4 + \cdots. \tag{41}$$

Since the radius of convergence of this series is 2, the value 1 of ε is not really as large as it seems. In fact, the terms displayed in (41) already give, for $\varepsilon = 1$, $\gamma = 1.9375$, as opposed to the correct value 2. The situation is clearly encouraging and, with better methods of summing up diagrams, the convergence of the ε-expansions can be improved greatly. In fact, some of the entries in Table 13.1 were originally obtained (or at least rechecked) with the help of this method.

Before closing this subsection we would like to point out a somewhat unusual piece of information contained in the first-order results obtained above. This refers to the exponent α, for which we note the prediction that for large n it is negative and hence the (singular part of the) specific heat vanishes at $T = T_c$ (which we know to be the case with the spherical model) whereas for small n it is positive and hence the specific heat diverges (which we know to be the case with Ising-like systems). The inversion, from one case to the other, takes place at $n = 4$ where α, according to the first-order expression (39), vanishes. The inclusion of the second-order term in ε upholds this prediction qualitatively but changes it quantitatively. We now have

$$\alpha \approx \frac{4-n}{2(n+8)}\varepsilon - \frac{(n+2)^2(n+28)}{4(n+8)^3}\varepsilon^2, \tag{42}$$

so that, with $\varepsilon = 1$, the inversion takes place between $n = 1$ and $n = 2$ — in agreement with the more accurate results quoted in Section 13.7.

14.4.E The 1/n expansion

Another approach to the problem of determining critical exponents, as functions of d and n, is to adopt the limiting case $n = \infty$ as the starting point and carry out expansions in powers of the small quantity $1/n$. Clearly, the leading terms in these expansions would pertain to the spherical model, which has been studied in Section 13.5, and the correction terms would enable us to get some useful information on models with finite n. We quote some first-order results here:[11]

$$\eta = \frac{4(4-d)S_d}{d}\frac{1}{n} + O\left(\frac{1}{n^2}\right), \tag{43}$$

$$\gamma = \frac{2}{d-2}\left\{1 - \frac{6S_d}{n} + O\left(\frac{1}{n^2}\right)\right\}, \tag{44}$$

and

$$\alpha = -\frac{4-d}{d-2}\left\{1 - \frac{8(d-1)S_d}{4-d}\frac{1}{n} + O\left(\frac{1}{n^2}\right)\right\}, \tag{45}$$

where

$$S_d = \frac{\sin\{\pi(d-2)/2\}\Gamma(d-1)}{2\pi\{\Gamma(d/2)\}^2} \quad (2 < d < 4). \tag{46}$$

We note that the coefficients of expansion in this approach are functions of d just as the coefficients of expansion in the preceding approach were functions of n; in this sense, the two expansions are complementary to one another. Unfortunately, there has not been much progress in the evaluation of further terms of these expansions (except for the one mentioned in the note); accordingly, the usefulness of this approach has been rather limited.

14.4.F Other topics

As mentioned earlier, the renormalization group approach has provided a very clear explanation of the concept of *universality*, in that it arises when several physical systems, despite their microscopic structural differences, are governed by a common fixed point and hence display a common critical behavior. Typically, this behavior is linked to the dimensionality d of the physical space, the dimensionality n of the spin vector $\boldsymbol{\sigma}$ and the range of the spin–spin interaction. Now, depending on the precise nature of the Hamiltonian and the relative importance of the various parameters therein, it is quite possible that under certain circumstances the critical behavior of the system may "cross over" from being characteristic of one fixed point to being characteristic of another fixed point. For instance, we may write for the spin–spin interaction in the lattice

$$H_{\text{int}} = -\frac{1}{2}\sum_{\boldsymbol{r},\boldsymbol{r}'}\sum_{\alpha,\beta} J^{\alpha\beta}(\boldsymbol{r}-\boldsymbol{r}')\sigma^{\alpha}(\boldsymbol{r})\sigma^{\beta}(\boldsymbol{r}') \quad \alpha,\beta = 1,\ldots,n. \tag{47}$$

[11] In the special case $d = 3$, the expansion for η is known to a higher order, namely

$$\eta = \frac{8}{3\pi^2 n} - \left(\frac{8}{3}\right)^3 \frac{1}{\pi^4 n^2} + O\left(\frac{1}{n^3}\right).$$

If the given interaction is isotropic in the physical space but anisotropic in the spin components (assumed three in number), so that $J^{\alpha\beta} = J^{\alpha}\delta_{\alpha\beta}$, then the system is ordinarily supposed to be a Heisenberg ferromagnet; however, the anisotropy of the interaction may finally drive the system toward an Ising fixed point (if one of the J^{α} dominates over the other two) or toward an *XY* fixed point (if two of the J^{α} are equally strong and dominate over the third one). In either case, we encounter what is generally referred to as a *crossover phenomenon*.

Similarly, anisotropy in the physical space, $J(\boldsymbol{R}) = J(R_i)\delta_{ij}$ or $K(R)R_iR_j$, may result in a crossover from a d-dimensional behavior to a d'-dimensional behavior (where $d' < d$). In the same vein, one may consider a long-range interaction, $J(\boldsymbol{R}) \sim R^{-d-\sigma}\delta_{ij}$, leading to a critical behavior which, for $\sigma < 2$, is quite seriously σ-dependent; see, for instance, Problem 13.22. However, as σ goes over from the value $2-$ to $2+$, the system crosses over to the universality class characterized by a short-range interaction and remains in that class for all $\sigma > 2$. Crossover phenomena constitute a very fascinating topic in the subject of phase transitions but we cannot pursue them here any more; the interested reader may refer to an excellent review by Aharony (1976).

Another topic of considerable interest deals with the so-called *interfacial phase transitions* in both magnets and fluids. In his seminal paper of 1944, Onsager included in his model a row of "mismatched spins," calculated the boundary tension (or what is more commonly referred to as the interfacial free energy) of this row and examined how this quantity vanished as T approached T_c. In the case of a fluid system, this corresponds to the disappearance of the meniscus between the liquid and the vapor and hence to the vanishing of the conventional surface tension as $T \to T_c-$; see, in this connection, Problem 12.27. A theoretical study of such interfacial layers involves consideration of the free energy of an *inhomogeneous* system, which has been a subject of considerable research for quite some time. We refer the interested reader to two review articles — by Abraham (1986) and by Jasnow (1986) — for further reading on this topic.

A major ingredient employed by the RG approach is the fact that the critical behavior of a system is invariant under a *scale transformation*. It did not take very long to realize that an important connection exists between this transformation and the well-known *conformal transformation* in a complex plane, for the latter too is, roughly speaking, a scale transformation in which the scale factor l varies continuously with position. Though, in principle, this connection could be relevant in all dimensions, the most fruitful applications have been in the realm of two dimensions (where the conformal group consists of analytic functions of a complex variable). Among the important results emerging from the conformal transformation approach, one may mention the form of the many-point correlation functions, the critical behavior of finite-sized strips of different sizes and shapes, and the nature of the surface critical effects. For details, see the review article by Cardy (1987).

Another area of interest pertains to the so-called *multicritical points*, for which reference may be made to Lawrie and Sarbach (1984) for theoretical studies and to Knobler and Scott (1984) for experimental results.

14.5 Finite-size scaling

In our study of phase transitions so far, we generally worked in the *thermodynamic limit*, that is, we started with a lattice of size $L_1 \times \ldots \times L_d$, containing $N_1 \times \ldots \times N_d$ spins (where $N_j = L_j/a$, a being the lattice constant), but at some appropriate stage of the calculation resorted to the limit $L_j \to \infty$. This limiting process is crucial in some important respects; while it simplifies subsequent calculations, it also generates singularities which, as we know, are a hallmark of systems undergoing phase transitions. It is of considerable interest, both theoretically and experimentally, to find out what happens (or does not happen) if some of the L_j are allowed to stay finite. The resulting analysis is quite complicated, but considerable progress has been made in this direction during the last 25 years or so. Accordingly, a whole new subject entitled "finite-size scaling" has emerged, of which only a brief summary will be presented here. The reader interested in further details may refer to Barber (1983), Cardy (1988), and Privman (1990).

To fix ideas, we start with a d-dimensional *bulk* system ("bulk" in the sense that it is infinite in all its dimensions) that undergoes a phase transition at a *finite* critical temperature $T_c(\infty)$; clearly, the dimensionality d must be greater than the "lower critical dimension" $d_<$. We also assume that d is less than the "upper critical dimension" $d_>$, so that the critical exponents of the system are d-dependent and obey the hyperscaling relation

$$d\nu = 2 - \alpha = 2\beta + \gamma. \tag{1}$$

We now consider a similar system that is infinite in only d' dimensions, where $d' < d$, and finite in the remaining dimensions; the geometry of this system may be expressed as $L^{d-d'} \times \infty^{d'}$, where $L \gg a$ and, for simplicity, is taken to be the same in all finite dimensions. We may expect this system to be critical at a finite temperature $T_c(L)$, not very different from $T_c(\infty)$. In reality, this is so only if d' too is greater than $d_<$; otherwise, the system continues to be regular at all finite temperatures and the criticality sets in only at $T = 0$.[12] The cases $d' > d_<$ and $d' \leq d_<$, therefore, merit separate treatments.

Our primary goal here is to determine the L-dependence of the various physical quantities pertaining to the system when the system is undergoing a phase transition. We attain this goal by setting up a *finite-size scaling law* that generalizes equation (12.10.7) or equation (14.3.26) to systems with a finite L. Now, since the only relevant length in the region of a phase transition is the correlation length ξ of the system, it is natural that we scale L with ξ — leading to the combination

$$(L/\xi) \sim Lt^{\nu} = (L^{1/\nu}t)^{\nu}. \tag{2}$$

At the same time, the combination (h/t^{Δ}) appearing in the bulk scaling law may be written as

$$(h/t^{\Delta}) = (hL^{\Delta/\nu})/(L^{1/\nu}t)^{\Delta}. \tag{3}$$

[12] For the special case $d' = 0$, when the system is fully finite, this point has already been emphasized in Section 12.1. Here we assert that, even when some of the system dimensions are infinite (and hence the total number of spins is infinite), a *finite-temperature singularity* does not arise unless the number of those infinite dimensions exceeds $d_<$.

The appropriate combinations of L with t and h, therefore, are $L^{1/\nu}t$ and $L^{\Delta/\nu}h$, respectively. The "singular" part of the free energy density of the system may then be written in the form (see Privman and Fisher, 1984)

$$f^{(s)}(t,h;L) \equiv \frac{A^{(s)}}{VkT} \approx L^{-d} Y(x_1,x_2), \tag{4}$$

where x_1 and x_2 are the scaled variables of the system, namely

$$x_1 = C_1 L^{1/\nu} t, \quad x_2 = C_2 L^{\Delta/\nu} h, \tag{5a, b}$$

with

$$t = \frac{T - T_c(\infty)}{T_c(\infty)}, \quad h = \frac{\mu B}{kT} \quad |t|, h \ll 1, \tag{6a, b}$$

while C_1 and C_2 are certain *nonuniversal* scale factors peculiar to the system under study. Expressed in terms of the variables x_1 and x_2, the function Y is expected to be a *universal* function — common to all systems in the same universality class as the system under study. Of course, the definition of the universality class will now include (apart from the conventional parameters d, n, and the range of the spin–spin interaction) the parameter d' as well as the nature of the boundary conditions imposed on the system (which, unless stated otherwise, will be assumed to be *periodic*).

We note that, in the limit $L \to \infty$, expression (4) reduces to equation (12.10.7), provided that the function Y has the asymptotic form

$$Y(x_1,x_2) \approx |x_1|^{d\nu} f_\pm(x_2/|x_1|^\Delta) \quad |x_1|, x_2 \gg 1, \tag{7}$$

thus identifying the nonuniversal parameters F and G with $C_1^{d\nu}$ and C_2/C_1^Δ, respectively. This enables us to write C_1 and C_2 in terms of F and G, namely

$$C_1 \sim F^{1/(2-\alpha)}, \quad C_2 \sim F^{(\beta+\gamma)/(2-\alpha)} G, \tag{8a, b}$$

which provides a means of determining the nonuniversal parameters C_1 and C_2 from a knowledge of the bulk parameters F and G; any other factors appearing in (8) would be universal. Once C_1 and C_2 are known, no more nonuniversal amplitudes are needed to describe the behavior of the system — regardless of whether it is finite-sized or infinite in extent. We are now in a position to examine the consequences of the scaling law (4).

With appropriate differentiations, we obtain from equation (4) the following expressions for the zero-field susceptibility per unit volume and the "singular" part of the zero-field specific heat per unit volume:

$$\begin{aligned} \chi_0(t;L) &= -\frac{1}{V}\left(\frac{\partial^2 A^{(s)}}{\partial B^2}\right)_{B=0} \approx -\frac{kT\mu^2 C_2^2 L^{2\Delta/\nu - d}}{(kT)^2}\left(\frac{\partial^2 Y(x_1,x_2)}{\partial x_2^2}\right)_{x_2=0} \\ &= \frac{\mu^2 C_2^2 L^{\gamma/\nu}}{kT} Y_\chi(x_1), \end{aligned} \tag{9}$$

and

$$c_0^{(s)}(t;L) = -\frac{T}{V}\left(\frac{\partial^2 A^{(s)}}{\partial T^2}\right)_{B=0} \approx -\frac{kT^2C_1^2L^{2/\nu-d}}{T_c^2(\infty)}\left(\frac{\partial^2 Y(x_1,x_2)}{\partial x_1^2}\right)_{x_2=0}$$
$$= \frac{kT^2C_1^2L^{\alpha/\nu}}{T_c^2(\infty)}Y_c(x_1), \tag{10}$$

where $Y_\chi(x_1)$ and $Y_c(x_1)$ are appropriate derivatives of the universal function $Y(x_1,x_2)$ and, hence, are themselves universal. We may, for further analysis, supplement the above results with the corresponding ones for the correlation length of the finite-sized system, namely

$$\xi(t,h;L) = LS(x_1,x_2) \tag{11}$$

and

$$\xi_0(t;L) = LS(x_1), \tag{12}$$

where $S(x_1) = S(x_1,0)$; note that the functions $S(x_1,x_2)$ and $S(x_1)$ are also universal. We shall now focus our attention on equations (9), (10), and (12), to see what messages they deliver in different regimes of the variables T and L.

Case A: $T \gtrsim T_c(\infty)$

With $t > 0$ and $L \gg a$, the variable x_1 in this regime would be positive and much greater than unity. The functions Y_χ, Y_c, and S are then expected to assume the form

$$Y_\chi(x_1) \approx \Gamma x_1^{-\gamma}, \quad Y_c(x_1) \approx A x_1^{-\alpha}, \quad S(x_1) \approx N x_1^{-\nu}, \tag{13a, b, c}$$

so that we recover the standard bulk results

$$\chi_0 \approx \frac{\mu^2 \Gamma C_1^{-\gamma} C_2^2}{kT_c(\infty)} t^{-\gamma}, \quad c_0^{(s)} \approx kAC_1^{2-\alpha} t^{-\alpha}, \quad \xi_0 \approx NC_1^{-\nu} t^{-\nu}, \tag{14a, b, c}$$

complete with nonuniversal amplitudes and universal factors. The effect of L in this regime appears only as a *correction* to the bulk results; under periodic boundary conditions, such *correction* terms turn out to be exponentially small, that is, $O(e^{-L/\xi_0})$ where $\xi_0 \sim a$.[13]

Case B: $T \simeq T_c(\infty)$

This case refers to the "core region" where $|x_1|$ is of order unity and hence $|t|$ is of order $L^{-1/\nu}$; the bulk critical point, $t = 0$, is at the heart of this region. Equations (9), (10), and (12) now yield the first significant results of finite-size scaling, namely

$$\chi_0 \sim \frac{\mu^2 C_2^2}{kT_c(\infty)} L^{\gamma/\nu}, \quad c_0^{(s)} \sim kC_1^2 L^{\alpha/\nu}, \quad \xi_0 \sim L. \tag{15a, b, c}$$

[13] See, for instance, Luck (1985), and Singh and Pathria (1985b, 1987a).

Case C: $T < T_c(\infty)$

Here we must distinguish between the cases $d' > d_<$ and $d' \le d_<$. In the first case, the system becomes critical at a temperature $T_c(L)$ that is not too far removed from $T_c(\infty)$; in the second, the system remains regular at all finite temperatures and becomes critical only at $T = 0$.

(i) $d' > d_<$

In view of the fact that the system is now singular at $T = T_c(L)$ rather than at $T_c(\infty)$, it seems natural to define a *shifted temperature variable* $\dot{t}$ such that

$$\dot{t} = \frac{T - T_c(L)}{T_c(\infty)}; \tag{16}$$

compare this with (6a). Thus, for any temperature T,

$$\dot{t} = t - \tau; \quad \tau = [T_c(L) - T_c(\infty)]/T_c(\infty), \tag{17}$$

which prompts us to define a *shifted scaled variable*

$$\dot{x}_1 = C_1 L^{1/\nu}\dot{t} = x_1 - X; \quad X = C_1 L^{1/\nu}\tau. \tag{18}$$

Clearly, the scaling functions governing the system would now be singular at $\dot{x}_1 = 0$, that is, at $x_1 = X$. With no other arguments present, we presume that $|X|$ will be of order unity; the shift in T_c is thus given by

$$|\tau| = |X| C_1^{-1} L^{-1/\nu} = O(L^{-1/\nu}). \tag{19}$$

Now, as $T \to T_c(L)$, the correlation length of the system approaches infinity — with the result that, insofar as the qualitative nature of the critical behavior is concerned, the finite variable L, however large, becomes essentially unimportant. The behavior of the system, in the immediate neighborhood of $T_c(L)$, would, therefore, be characteristic of a d'-dimensional bulk system rather than of a d-dimensional one; accordingly, it would be governed by the critical exponents $\dot{\alpha}, \dot{\beta}, \ldots$ pertaining to d' dimensions rather than by the exponents $\alpha, \beta, \ldots$ pertaining to d dimensions. We, therefore, expect that, as $\dot{x}_1 \to 0$, the functions Y_χ, Y_c, and S of equations (9), (10), and (12) assume the form

$$Y_\chi(x_1) \approx \dot{\Gamma}\dot{x}_1^{-\dot{\gamma}}, \quad Y_c(x_1) \approx \dot{A}\dot{x}_1^{-\dot{\alpha}}, \quad S(x_1) \approx \dot{N}\dot{x}_1^{-\dot{\nu}}, \tag{20a, b, c}$$

with the result that

$$\chi_0 \approx [\mu^2/kT_c(\infty)]\dot{\Gamma} C_1^{-\dot{\gamma}} C_2^2 L^{(\gamma-\dot{\gamma})/\nu}\dot{t}^{-\dot{\gamma}}, \tag{21a}$$

$$c_0^{(s)} \approx k\dot{A} C_1^{2-\dot{\alpha}} L^{(\alpha-\dot{\alpha})/\nu}\dot{t}^{-\dot{\alpha}}, \tag{21b}$$

and

$$\xi_0 \approx \dot{N} C_1^{-\dot{\nu}} L^{(\nu-\dot{\nu})/\nu} \dot{t}^{-\dot{\nu}}. \tag{21c}$$

It is obvious that, for $\dot{t} < 0$ but such that $|\dot{t}| \ll 1$, the same results would hold — except that $\dot{t}$ would be replaced by $|\dot{t}|$.

The contents of equations (21), insofar as the dependence on L and $\dot{t}$ is concerned, have been verified by direct calculation on a variety of systems over the years; for details, see the review articles by Barber (1983) and Privman (1990) cited earlier. More recently, Allen and Pathria (1989) have verified the nonuniversal amplitudes as well by carrying out an explicit calculation for the spherical model ($n = \infty$) in the general geometry $L^{d-d'} \times \infty^{d'}$, with both d and d' greater than $d_<$. Remarkably enough, they found that, just as the critical exponents $\dot{\alpha}, \dot{\beta}, \ldots$ are the same functions of d' as the exponents $\alpha, \beta, \ldots$ are of d, the universal coefficients $\dot{\Gamma}, \dot{A}, \ldots$ too are the same functions of d' as the coefficients $\Gamma, A, \ldots$ are of d; the same is true of the coefficients appearing in the presence of a magnetic field (see Allen and Pathria, 1991). One wonders if this would also be the case for general n!

(ii) $d' \le d_<$

In this case the singularity of the problem lies at $T = 0$, with the result that at *all* finite temperatures the system is regular and hence expressible by smooth, analytic functions. We may, therefore, generalize the scaling law (4) to apply at all temperatures down to $T = 0$ by simply allowing the scale factors C_1 and C_2 to become T-dependent and writing (after Singh and Pathria, 1985b, 1986a)

$$x_1 = \tilde{C}_1(T) L^{1/\nu} t, \quad x_2 = \tilde{C}_2(T) L^{\Delta/\nu} h, \tag{22a, b}$$

leaving t and h unchanged; the quantities $\tilde{C}_1$ and $\tilde{C}_2$ must be such that, as $T \to T_c(\infty)$ from below, they approach the quantities C_1 and C_2 of equations (5). Expressions (9) and (10) now take the form

$$\chi_0(t; L) = \frac{\mu^2 \tilde{C}_2^2 L^{\gamma/\nu}}{kT} Y_\chi(x_1) \tag{23}$$

and

$$c_0^{(s)}(t; L) = kT^2 \left[\frac{\partial}{\partial T} (\tilde{C}_1 t) \right]^2 L^{\alpha/\nu} Y_c(x_1), \tag{24}$$

respectively, while expression (12) remains formally the same.

Now, as we approach the critical temperature T_c (which is zero here), we again expect the quantities χ_0, $c_0^{(s)}$, and ξ_0 to behave in a manner characteristic of a d'-dimensional bulk system. So, let us assume that, in limit $d \to 0$, our scale factors behave as

$$\tilde{C}_1 \sim T^r, \quad \tilde{C}_2 \sim T^s \quad (T \to 0), \tag{25a, b}$$

and our universal functions behave as

$$Y_\chi(x_1) \sim |x_1|^\theta, \quad Y_c(x_1) \sim |x_1|^\phi, \quad S(x_1) \sim |x_1|^\sigma \quad (x_1 \to -\infty). \tag{26a, b, c}$$

The resulting T-dependence of $\chi_0, c_0^{(s)}$, and ξ_0 then is

$$\chi_0 \sim T^{2s-1+\theta r}, \quad c_0^{(s)} \sim T^{(2+\phi)r}, \quad \xi_0 \sim T^{\sigma r}. \tag{27a, b, c}$$

The corresponding results for an n-vector, d'-dimensional model (with $n \geq 2$ and $d' < d_<$, where $d_< = 2$) are[14]

$$\chi_0 \sim T^{-2/(2-d')}, \quad c_0^{(s)} \sim T^{d'/(2-d')}, \quad \xi_0 \sim T^{-1/(2-d')}. \tag{28a, b, c}$$

Comparing (27) with (28), we infer that

$$\theta = \frac{1}{r}\left[1 - 2s - \frac{2}{2-d'}\right], \quad \phi = \frac{d'}{r(2-d')} - 2, \quad \sigma = \frac{-1}{r(2-d')}. \tag{29a, b, c}$$

Very shortly we shall find, see equations (44), that

$$r = -1/\nu(d-2), \quad s = \beta/\nu(d-2), \tag{30a, b}$$

with the results

$$\theta = 2\beta + \frac{\nu d'(d-2)}{(2-d')}, \quad \phi = -\frac{\nu d'(d-2)}{(2-d')} - 2, \quad \sigma = \frac{\nu(d-2)}{(2-d')}. \tag{31a, b, c}$$

The L-dependence of the various quantities now turns out to be

$$\chi_0 \sim L^{(\gamma+\theta)/\nu} \sim L^{2(d-d')/(2-d')} \tag{32a}$$

$$c_0^{(s)} \sim L^{(\alpha+\phi)/\nu} \sim L^{-2(d-d')/(2-d')} \tag{32b}$$

and

$$\xi_0 \sim L^{1+\sigma/\nu} \sim L^{(d-d')/(2-d')}. \tag{32c}$$

It is remarkable that in these last expressions the critical exponents pertaining to d dimensions have disappeared altogether and the resulting powers of L depend *entirely* on the geometry of the system! Expression (32a) agrees with the earlier results for χ_0 pertaining to a "block" geometry ($d' = 0$) and to a "cylindrical" geometry ($d' = 1$), namely

$$\chi_0 \sim \begin{cases} L^d & \text{for} \quad d' = 0 \quad (33a) \\ L^{2(d-1)} & \text{for} \quad d' = 1; \quad (33b) \end{cases}$$

[14]For the spherical model ($n = \infty$), these results appear in Section 13.5; see equations (13.5.34), (13.5.35), and (13.5.64). Since the criticality in this case occurs at absolute zero, these results hold for all models with *continuous* symmetry, that is, with $n \geq 2$. See, for instance, Section 13.3, where n is general but $d' = 1$.

see Fisher and Privman (1985). For $d' = 2$, the L-dependence of the various quantities studied here becomes exponential instead of a power law.

To obtain results valid for *all* T in the range $0 < T < T_c(\infty)$, we need to know the full T-dependence of the scale factors $\tilde{C}_1$ and $\tilde{C}_2$. It turns out that this too can be determined from the properties of the corresponding bulk system — in particular, from the field-free bulk correlation function $G(\boldsymbol{R}, T)$, which is known to possess the following forms:

$$G(\boldsymbol{R}, T) \sim R^{-(d-2+\eta)} \quad T = T_c(\infty) \tag{34}$$

and

$$G(\boldsymbol{R}, T) = m_0^2(T) + A(T)R^{-(d-2)} \quad T < T_c(\infty), \tag{35}$$

where $m_0(T)$ is the order parameter of the bulk system and $A(T)$ another system-dependent parameter.[15] Now, the correlation function of the finite-sized system may be written in the scaled form

$$G(\boldsymbol{R}, t, h; L) \approx \tilde{D}(T)R^{-(d-2+\eta)} Z(x_1, x_2, \boldsymbol{x}_3), \tag{36}$$

where the scaled variables x_1 and x_2 are the same given by equations (22a, b) while $\boldsymbol{x}_3 = \boldsymbol{R}/L$; as usual, the scale factor $\tilde{D}(T)$ is nonuniversal while the function $Z(x_1, x_2, \boldsymbol{x}_3)$ is universal.

Expression (36) already conforms to (34), with $x_1 = x_2 = x_3 = 0$. For conformity with (35), the function Z must possess the following asymptotic form:

$$Z(x_1, x_2, \boldsymbol{x}_3) \approx Z_1(|x_1|^\nu x_3)^{d-2+\eta} + Z_2(|x_1|^\nu x_3)^\eta \quad x_1 \to -\infty,\ x_2 = 0,\ x_3 \to 0, \tag{37}$$

with

$$Z_1 = \frac{m_0^2(T)}{\tilde{D}(T)[\tilde{C}_1(T)|t|]^{\nu(d-2+\eta)}}, \quad Z_2 = \frac{A(T)}{\tilde{D}(T)[\tilde{C}_1(T)|t|]^{\nu\eta}}. \tag{38a, b}$$

It follows that

$$\tilde{C}_1(T)|t| \sim \left[\frac{m_0^2(T)}{A(T)}\right]^{1/\nu(d-2)}, \quad \tilde{D}(T) \sim \left[\frac{A^\beta(T)}{m_0^{\nu\eta}(T)}\right]^{2/\nu(d-2)}; \tag{39a, b}$$

here, use has been made of the fact that

$$\nu(d-2+\eta) = (2-\alpha) - \gamma = 2\beta. \tag{40}$$

We shall now establish a relationship between the scale factors $\tilde{C}_2$ and $\tilde{D}$. For this, we utilize the fluctuation–susceptibility relation (12.11.12) which, with the help of

[15]Note that the exponent η appears only in equation (34) and not in (35); for details, see Schultz et al. (1964) and Fisher et al. (1973).

expression (36), gives for the zero-field susceptibility per unit volume

$$\chi_0(t;L) = \frac{\mu^2\tilde{D}(T)}{a^{2d}kT}\int \frac{Z(x_1,0,\boldsymbol{R}/L)}{R^{d-2+\eta}}d^dR$$
$$= \frac{\mu^2\tilde{D}(T)}{a^{2d}kT}L^{2-\eta}Z_\chi(x_1), \tag{41}$$

where Z_χ is another universal function. Comparing (41) with (23), we get

$$\tilde{D}(T) \sim a^{2d}\tilde{C}_2^2(T). \tag{42}$$

Equation (39b) now gives

$$\tilde{C}_2(T) \sim a^{-d}[A^\beta(T)/m_0^{\nu\eta}(T)]^{1/\nu(d-2)}. \tag{43}$$

Equations (39) and (43) give us the full T-dependence of the scale factors $\tilde{C}_1, \tilde{C}_2$, and $\tilde{D}$ for *all* T in the range $0 < T < T_c(\infty)$; these equations were first derived by Singh and Pathria (1987a).

Before utilizing these results we note that since, in the limit $T \to 0$, $m_0(T)$ approaches a constant value while $A(T) \sim T$, expressions (39a) and (43) yield

$$\tilde{C}_1|t| \sim T^{-1/\nu(d-2)}, \quad \tilde{C}_2 \sim T^{\beta/\nu(d-2)}, \tag{44a, b}$$

exactly as stipulated in equations (25) and (30). As for the T-dependence of the quantities χ_0, $c_0^{(s)}$, and ξ_0, we observe that, regardless of whether we keep L fixed and let $T \to 0$ *or* keep T fixed and let $L \to \infty$, in either case $x_1 \to -\infty$; the asymptotic forms (26) of the universal functions Y_χ, Y_c, and S, therefore, apply throughout the region under study. Now, with θ, ϕ, and σ given by equations (31), our final results for χ_0, $c_0^{(s)}$, and ξ_0 turn out to be

$$\chi_0 \sim \frac{\mu^2A(T)}{a^{2d}kT}\left[\frac{m_0^2(T)}{A(T)}\right]^{2/(2-d')} L^{2(d-d')/(2-d')}, \tag{45}$$

$$c_0^{(s)} \sim k\left\{T\frac{\partial}{\partial T}\left[\frac{m_0^2(T)}{A(T)}\right]\right\}^2\left[\frac{m_0^2(T)}{A(T)}\right]^{-(4-d')/(2-d')} L^{-2(d-d')/(2-d')}, \tag{46}$$

and

$$\xi_0 \sim \left[\frac{m_0^2(T)}{A(T)}\right]^{1/(2-d')} L^{(d-d')/(2-d')}, \tag{47}$$

complete with nonuniversal amplitudes. Comparing (45) with (47), we find that, in the regime under study,

$$\chi_0/\xi_0^2 \sim \mu^2A(T)/a^{2d}kT, \tag{48}$$

a function of T only. In the case of the spherical model, since $A(T)$ is proportional to T at *all* $T < T_c(\infty)$, see equation (13.5.71), the quantity χ_0/ξ_0^2 is a constant — independent of *both* T and L; see also equation (13.5.64).

It is important to note that the above formulation ties very neatly with the one provided by the scale factors C_1 and C_2 of equations (5) that covered cases A and B pertaining to the regions $T \gtrsim T_c(\infty)$ and $T \simeq T_c(\infty)$, respectively. To see this, we observe that, as $T \to T_c(\infty)$ from below, $m_0(T)$ becomes $\sim |t|^{\beta}$ and $A(T) \sim |t|^{\nu\eta}$; expressions (39a) and (43) then assume the form

$$\tilde{C}_1(T)|t| \sim |t|^{(2\beta-\nu\eta)/\nu(d-2)} \sim |t|^1 \tag{49a}$$

and

$$\tilde{C}_2(T) \sim |t|^0. \tag{49b}$$

Clearly, $\tilde{C}_1(T)$ and $\tilde{C}_2(T)$ now assume some constant values that may be identified with C_1 and C_2 — thus providing a *unified* formulation through the *same* universal functions $Y(x_1,x_2)$, $S(x_1,x_2)$, and $Z(x_1,x_2,\boldsymbol{x}_3)$ covering the regions of *both* first-order and second-order phase transitions! Remarkable though it is, this finding is not really surprising because, with L finite and $d' \le d_<$, the system is critical only at $T = 0$ and analytic everywhere else; so, its properties should indeed be expressible by a single set of functions throughout. Of course, as $L \to \infty$, the criticality spreads all the way from $T = 0$ to $T = T_c(\infty)$.

As regards the spin dimensionality n, our results for cases A, B, and C(i) were quite general; only in case C(ii) did we specialize to systems with *continuous* symmetry ($n \ge 2$). With a slight modification, the case of discrete symmetry ($n = 1$) can also be taken care of. The net result essentially is the replacement of the number 2 in equations (28) and henceforth by the "lower critical dimension," $d_<$, of the system — leading to results such as[16]

$$\chi_0 \sim L^{\zeta}, \quad c_0^{(s)} \sim L^{-\zeta}, \quad \xi_0 \sim L^{\zeta/d_<} \quad T < T_c(\infty), \tag{50a, b, c}$$

with $\zeta = d_<(d-d')/(d_< - d')$; compared with equations (32). Once again, the L-dependence of the various quantities of interest follows a power law, which changes to an exponential when $d' = d_<$; in the case of scalar models, this happens at $d' = 1$.

Throughout this discussion we have assumed that the total dimensionality d of the system is less than the "upper critical dimension" $d_>$. The case $d \ge d_>$ presents some special problems but the net result is that, while the situation in the region $T < T_c(\infty)$ is described by the same set of expressions as above, in the region $T \simeq T_c(\infty)$ it is considerably modified. For instance, one now gets in the region $T \simeq T_c(\infty)$, for $d > d_>$ and $d' < d_<$,

$$\chi_0 \sim L^{2(d-d')/(d_>-d')}, \quad c_0^{(s)} \sim L^0, \quad \xi_0 \sim L^{(d-d')/(d_>-d')}, \tag{51a, b, c}$$

[16] See, for instance, Singh and Pathria (1986b).

which may be compared with the corresponding results, (15a, b, c), for $d < d_>$. Furthermore, if $d = d_>$ and/or $d' = d_<$, factors containing $\ln L$ appear along with the power laws displayed in (51). For details, see Singh and Pathria (1986b, 1988, 1992).

Finally we would like to emphasize the fact that finite-size effects in any given system are quite sensitive to the choice of the boundary conditions imposed on the system. For simplicity, we assumed the boundary conditions to be *periodic*. In real situations, there may be reasons to adopt different boundary conditions such as *antiperiodic, free*, and so on. This, in general, changes the mathematical character of the finite-size effects and the finite-size corrections in not only the singular part(s) of the various quantities studied but in their regular part(s) as well. For comparison between theory and experiment, this aspect of the problem is of vital importance and deserves a close scrutiny. For lack of space we cannot go into this matter any further here; the interested reader may refer to a review article by Privman (1990), where other references on this topic can also be found. An allied subject in this regard is the "critical behavior of surfaces," for which reference may be made to Binder (1983) and Diehl (1986).

Problems

14.1. Show that the decimation transformation of a one-dimensional Ising model, with $l = 2$, can be written in terms of the transfer matrix $\boldsymbol{P}$ as

$$\boldsymbol{P}'\{\boldsymbol{K}'\} = \boldsymbol{P}^2\{\boldsymbol{K}\}, \tag{1}$$

where $\boldsymbol{K}$ and $\boldsymbol{K}'$ are the coupling constants of the original and the decimated lattice, respectively. Next show that, with $\boldsymbol{P}$ given by

$$(\boldsymbol{P}\{\boldsymbol{K}\}) = e^{K_0}\begin{pmatrix} e^{K_1+K_2} & e^{-K_1} \\ e^{-K_1} & e^{K_1-K_2} \end{pmatrix}, \tag{2}$$

see equation (13.2.4), relation (1) leads to the same transformation equations among $\boldsymbol{K}$ and $\boldsymbol{K}'$ as (14.2.8a, b, and c).

14.2. Verify that expression (15) of Section 14.2 indeed satisfies the functional equation (14) for the field-free Ising model in one dimension. Next show (or at least verify) that, with the field present, the functional equation (11), with $\boldsymbol{K}'$ given by (8), is satisfied by the more general expression

$$f(K_1, K_2) = -\ln\left[e^{K_1}\cosh K_2 + \left\{e^{-2K_1} + e^{2K_1}\sinh^2 K_2\right\}^{1/2}\right].$$

14.3. Verify that expression (32) of Section 14.2 indeed satisfies the functional equation (31) for the field-free spherical model in one dimension. Next show (or at least verify) that, with the field present, the functional equation (27), with $\boldsymbol{K}'$ given by (25), is satisfied by the more general expression

$$f(K_1, K_2, \Lambda) = \frac{1}{2}\ln\left[\frac{\Lambda + \sqrt{\Lambda^2 - K_1^2}}{2\pi}\right] - \frac{K_2^2}{4(\Lambda - K_1)},$$

where Λ is determined by the constraint equation

$$\frac{\partial f}{\partial \Lambda} = \frac{1}{2\sqrt{\Lambda^2 - K_1^2}} + \frac{K_2^2}{4(\Lambda - K_1)^2} = 1.$$

14.4. Consider the field-free spherical model in one dimension whose partition function is given by equation (14.2.24) as well as by (14.2.19), with $K_2' = K_2 = 0$. Substituting $\sigma_j' = (2\Lambda/K_1)^{1/2} s_j'$ in the former and comparing the resulting expression with the latter, show that

$$Q_N(K_1, \Lambda) = \left(\frac{2\pi}{K_1}\right)^{N'/2} Q_{N'}(K_1, \Lambda''),$$

where $N' = \frac{1}{2}N$ and $\Lambda'' = (2\Lambda^2/K_1) - K_1$. This leads to the functional relation

$$f(K_1, \Lambda) = -\frac{1}{4}\ln\left(\frac{2\pi}{K_1}\right) + \frac{1}{2} f(K_1, \Lambda'').$$

Check that expression (14.2.32) satisfies this relation.

14.5. An approximate way of implementing an RG transformation on a square lattice is provided by the so-called Migdal–Kadanoff transformation[17] shown in Figure 14.8. It consists of two essential steps:

(i) First, one-half of the bonds in the lattice are simply removed, so as to change the length scale of the lattice by a factor of 2; to compensate for this, the coupling strength of the remaining bonds is changed from J to $2J$. This takes us from Figure 14.8(a) to Figure 14.8(b).

(ii) Next, the sites marked by crosses in Figure 14.8(b) are eliminated by a set of one-dimensional decimation transformations, leading to Figure 14.8(c) with coupling strength J'.

(a) Show that the recursion relation for a spin-$\frac{1}{2}$ Ising model on a square lattice, according to the above transformation, is

$$x' = 2x^2/(1 + x^4),$$

where $x = \exp(-2K)$ and $x' = \exp(-2K')$. Disregarding the trivial fixed points $x^* = 0$ and $x^* = 1$, show that a nontrivial fixed point of this transformation is

$$x^* = \frac{1}{3}\left[-1 + 2\sqrt{2}\sinh\left\{\frac{1}{3}\sinh^{-1}\frac{17}{2\sqrt{2}}\right\}\right] \simeq 0.5437;$$

compare this with the actual value of x_c, which is $(\sqrt{2} - 1) \simeq 0.4142$.

(b) Linearizing around this nontrivial fixed point, show that the eigenvalue λ of this transformation is

$$\lambda = 2(1 - x^*)/x^* \simeq 1.6785$$

and hence the exponent $\nu = \ln 2/\ln\lambda \simeq 1.338$; compare this with the actual value of ν, which is 1.

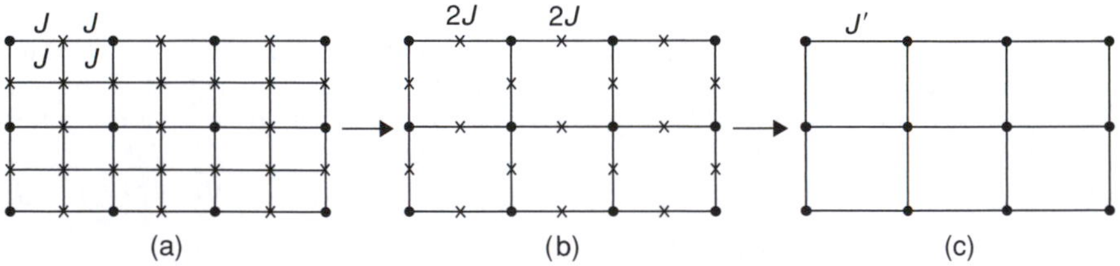

FIGURE 14.8 Migdal–Kadanoff transformation on a square lattice.

[17] See Kadanoff (1976a).

14.6. Consider the linearized RG transformation (14.3.12), with

$$\mathcal{A}_l^* = \begin{pmatrix} a_{11} & a_{12} \\ a_{21} & a_{22} \end{pmatrix}, \tag{3}$$

such that $(a_{11}a_{22} - a_{12}a_{21}) \neq 0$. We now introduce the "generalized coordinates" u_1 and u_2 through equations (14.3.13); clearly, u_1 and u_2 are certain linear combinations of the system parameters k_1 and k_2.

(a) Show that the slopes of the lines $u_1 = 0$ and $u_2 = 0$, in the (k_1, k_2)-plane, are

$$m_1 = \frac{a_{21}}{\lambda_2 - a_{22}} = \frac{\lambda_2 - a_{11}}{a_{12}} \quad \text{and} \quad m_2 = \frac{a_{21}}{\lambda_1 - a_{22}} = \frac{\lambda_1 - a_{11}}{a_{12}},$$

respectively; here, λ_1 and λ_2 are the eigenvalues of the matrix $\mathcal{A}_l^*$. Verify that the product $m_1 m_2 = -a_{21}/a_{12}$ and hence the two lines are mutually perpendicular if and only if $a_{12} = a_{21}$.

(b) Check that, in the special case when $a_{12} = 0$ but $a_{21} \neq 0$, the above slopes assume the simple form

$$m_1 = \infty \quad \text{and} \quad m_2 = a_{21}/(a_{11} - a_{22})$$

whereas, in the special case when $a_{21} = 0$ but $a_{12} \neq 0$, they become

$$m_1 = (a_{22} - a_{11})/a_{12} \quad \text{and} \quad m_2 = 0;$$

note that Figure 14.7 pertains to the latter case.

(c) Examine as well the cases for which either a_{11} or a_{22} is zero; Figure 14.6 pertains to the latter of these cases.

14.7. Check that the critical exponents (14.4.38) through (14.4.40), in the limit $n \to \infty$, agree with the corresponding exponents for the spherical model of Section 13.5 with $d \lesssim 4$.

14.8. Show, from equations (14.4.43) through (14.4.46), that for $d \lesssim 4$

$$\eta \simeq \frac{1}{2n}\varepsilon^2, \quad \gamma \simeq 1 + \frac{1}{2}\left(1 - \frac{6}{n}\right)\varepsilon, \quad \alpha \simeq -\frac{1}{2}\left(1 - \frac{12}{n}\right)\varepsilon,$$

where $\varepsilon = (4 - d) \ll 1$. Check that these results agree with the ones following from equations (14.4.38) through (14.4.40) for $n \gg 1$.

14.9. Using the various scaling relations, derive from equations (14.4.43) through (14.4.45) comparable expressions for the remaining exponents β, δ, and ν. Repeat for these exponents the exercise suggested in the preceding problem.

15

Fluctuations and Nonequilibrium Statistical Mechanics

In this course we have been mostly concerned with the evaluation of *statistical averages* of the various physical quantities; these averages represent, with a high degree of accuracy, the results expected from relevant measurements on the given system *in equilibrium*. Nevertheless, there do occur *deviations* from, or *fluctuations* about, these mean values. Even though they are generally small, their study is of great physical interest for several reasons.

First, such a study enables us to develop a mathematical scheme with the help of which the magnitude of the relevant fluctuations, under a variety of physical situations, can be estimated. Not surprisingly, we find that while in a single-phase system the fluctuations are thermodynamically negligible but they can assume considerable importance in multiphase systems, especially in the neighborhood of a critical point. In the latter case, we obtain a rather high degree of *spatial correlation* among the molecules of the system which, in turn, gives rise to phenomena such as *critical opalescence*.

Second, it provides a natural framework for understanding a class of phenomena that come under the heading "Brownian motion"; these phenomena relate properties such as the mobility of a fluid system, its coefficient of diffusion, and so on, with temperature through the so-called *Einstein relations*. The mechanism of Brownian motion is vital in formulating, and in a certain sense answering, questions as to how "a given physical system, which is not in a state of equilibrium, finally approaches such a state" while "a physical system, which is already in a state of equilibrium, persists to stay in that state."

Third, the study of fluctuations, as a function of time, leads to the concept of certain "correlation functions" that play a vital role in relating the dissipative properties of a system, such as the viscous resistance of a fluid or the electrical resistance of a conductor, with the microscopic properties of the system in a state of equilibrium; this relationship (between irreversible processes on one hand and equilibrium properties on the other) manifests itself in the so-called *fluctuation–dissipation theorem*. At the same time, a study of the "frequency spectrum" of fluctuations, which is related to the time-dependent correlation function through the fundamental theorem of Wiener and Khintchine, is of considerable value in assessing the "noise" met with in electrical circuits as well as in the transmission of electromagnetic signals.

© 2011 Elsevier Ltd. All rights reserved.

15.1 Equilibrium thermodynamic fluctuations

We begin by deriving a *probability distribution law* for the fluctuations of certain basic thermodynamic quantities pertaining to a given physical system; the *mean square fluctuations* can then be evaluated, in a straightforward manner, with the help of this law. We assume that the given system, which may be referred to as 1, is embedded in a reservoir, which may be referred to as 2, such that a mutual exchange of energy, and of volume, can take place between the two; of course, the overall energy E and the overall volume V are supposed to be fixed. For convenience, we do not envisage an exchange of particles here, so the numbers N_1 and N_2 remain individually constant. The equilibrium division of E into $\overline{E}_1$ and $\overline{E}_2$, and of V into $\overline{V}_1$ and $\overline{V}_2$, must be such that parts 1 and 2 of the composite system $(1+2)$ have a *common* temperature T^* and a *common* pressure P^*; see Sections 1.2 and 1.3, especially equations (1.3.6). Of course, the entropy of the composite system will have its largest value in the equilibrium state; in any other state, such as the one characterized by a fluctuation, it must have a lower value. If ΔS denotes the deviation in the entropy of the composite system from its equilibrium value S_0, then

$$\Delta S \equiv S - S_0 = k \ln \Omega_f - k \ln \Omega_0, \tag{1}$$

where Ω_f (or Ω_0) denotes the number of distinct microstates of the system $(1+2)$ in the presence (or in the absence) of a fluctuation from the equilibrium state; see equation (1.2.6). The probability that the proposed fluctuation may indeed occur is then given by

$$p \propto \Omega_f \propto \exp(\Delta S/k); \tag{2}$$

see Section 3.1, especially equation (3.1.3). In terms of other thermodynamic quantities, we may write

$$\Delta S = \Delta S_1 + \Delta S_2 = \Delta S_1 + \int_0^f \frac{dE_2 + P_2 dV_2}{T_2}; \tag{3}$$

note that the pressure P_2 and the temperature T_2 of the reservoir may, in principle, vary during the build-up of the fluctuation! Now, even if the fluctuation is sizable from the point of view of system 1, it will be small from the point of view of 2. The "variables" P_2 and T_2 may, therefore, be replaced by the constants P^* and T^*, respectively; at the same time, the increments dE_2 and dV_2 may be replaced by $-dE_1$ and $-dV_1$, respectively. Equation (3) then becomes

$$\Delta S = \Delta S_1 - (\Delta E_1 + P^* \Delta V_1)/T^*. \tag{4}$$

Accordingly, formula (2) takes the form

$$p \propto \exp\{-(\Delta E_1 - T^* \Delta S_1 + P^* \Delta V_1)/kT^*\}. \tag{5}$$

Clearly, the probability distribution law (5) does not depend, in any manner, on the peculiarities of the reservoir in which the given system was supposedly embedded. Formula (5), therefore, applies equally well to a system that attained equilibrium in a statistical ensemble (or, for that matter, to any macroscopic part of a given system itself). Consequently, we may drop the suffix 1 from the symbols $\Delta E_1, \Delta S_1$, and ΔV_1, and the star from the symbols P^* and T^*, and write

$$p \propto \exp\{-(\Delta E - T\Delta S + P\Delta V)/kT\}. \tag{6}$$

In most cases, the fluctuations are exceedingly small in magnitude; the quantity ΔE may, therefore, be expanded as a Taylor series about the equilibrium value $(\Delta E)_0 = 0$, with the result

$$\Delta E = \left(\frac{\partial E}{\partial S}\right)_0 \Delta S + \left(\frac{\partial E}{\partial V}\right)_0 \Delta V + \frac{1}{2}\left[\left(\frac{\partial^2 E}{\partial S^2}\right)_0 (\Delta S)^2 + 2\left(\frac{\partial^2 E}{\partial S \partial V}\right)_0 \Delta S \Delta V + \left(\frac{\partial^2 E}{\partial V^2}\right)_0 (\Delta V)^2\right] + \cdots \tag{7}$$

Substituting (7) into (6) and retaining terms up to second order only, we obtain

$$p \propto \exp\{-(\Delta T \Delta S - \Delta P \Delta V)/2kT\}; \tag{8}$$

here, use has been made of the relations

$$\left(\frac{\partial E}{\partial S}\right)_0 = T, \quad \left(\frac{\partial E}{\partial V}\right)_0 = -P, \tag{9}$$

and of the fact that the expression within the square brackets in (7) is equivalent to

$$\Delta\left(\frac{\partial E}{\partial S}\right)_0 \Delta S + \Delta\left(\frac{\partial E}{\partial V}\right)_0 \Delta V = \Delta T \Delta S - \Delta P \Delta V. \tag{10}$$

With the help of (8), the mean square fluctuations of various physical quantities and the statistical correlations among different fluctuations can be readily calculated. We note, however, that of the four Δ terms appearing in this formula only two can be chosen independently; the other two must assume the role of "derived quantities." For instance, if we choose ΔT and ΔV to be the *independent variables*, then ΔS and ΔP can be written as

$$\Delta S = \left(\frac{\partial S}{\partial T}\right)_V \Delta T + \left(\frac{\partial S}{\partial V}\right)_T \Delta V = \frac{C_V}{T}\Delta T + \left(\frac{\partial P}{\partial T}\right)_V \Delta V \tag{11}$$

and

$$\Delta P = \left(\frac{\partial P}{\partial T}\right)_V \Delta T + \left(\frac{\partial P}{\partial V}\right)_T \Delta V = \left(\frac{\partial P}{\partial T}\right)_V \Delta T - \frac{1}{\kappa_T V}\Delta V, \tag{12}$$

κ_T being the *isothermal compressibility* of the system. Substituting (11) and (12) into (8), we get

$$p \propto \exp\left\{-\frac{C_V}{2kT^2}(\Delta T)^2 - \frac{1}{2kT\kappa_T V}(\Delta V)^2\right\}, \tag{13}$$

which shows that the fluctuations in T and V are *statistically independent, Gaussian variables*! A quick glance at (13) yields the results

$$\overline{(\Delta T)^2} = \frac{kT^2}{C_V}, \quad \overline{(\Delta V)^2} = kT\kappa_T V, \tag{14a}$$

while

$$\overline{(\Delta T \Delta V)} = 0. \tag{14b}$$

Similarly, if we choose ΔS and ΔP as our *independent variables*, we are led to the distribution law

$$p \propto \exp\left\{-\frac{1}{2kC_P}(\Delta S)^2 - \frac{\kappa_S V}{2kT}(\Delta P)^2\right\}, \tag{15}$$

which gives

$$\overline{(\Delta S)^2} = kC_P, \quad \overline{(\Delta P)^2} = \frac{kT}{\kappa_S V}, \tag{16a}$$

while

$$\overline{(\Delta S \Delta P)} = 0; \tag{16b}$$

here, κ_S denotes the *adiabatic compressibility* of the system.

We note that, in general, the mean square fluctuation of an extensive quantity is directly proportional to the size of the system while that of an intensive quantity is inversely proportional to the same; in either case, the *relative, root-mean-square fluctuation* of any quantity is inversely proportional to the square root of the size of the system. Thus, except for situations such as the ones met with in a critical region, normal fluctuations are thermodynamically negligible. This does not mean that fluctuations are altogether irrelevant to the physical phenomena taking place in the system; in fact, as will be seen in the sequel, the very presence of fluctuations at the *microscopic* level is of fundamental importance to several properties of the system displayed at the *macroscopic* level!

With the help of the foregoing results, we may evaluate the mean square fluctuation in the energy of the system. With T and V as independent variables, we have

$$\Delta E = \left(\frac{\partial E}{\partial T}\right)_V \Delta T + \left(\frac{\partial E}{\partial V}\right)_T \Delta V. \tag{17}$$

Squaring this expression and taking averages, keeping in mind equations (14), we get

$$\begin{aligned}\overline{(\Delta E)^2} &= kT^2C_V + kT\kappa_T V\left\{\left(\frac{\partial E}{\partial V}\right)_T\right\}^2 \\ &= kT^2C_V + kT\kappa_T\left(\frac{N^2}{V}\right)\left\{\left(\frac{\partial E}{\partial N}\right)_T\right\}^2. \end{aligned} \tag{18}$$

Now, the results derived in the preceding paragraphs determine the fluctuations of the various physical quantities pertaining to *any macroscopic subsystem* of a given system, provided that the number of particles in the subsystem remains fixed. The expression (14b) for $\overline{(\Delta V)^2}$ may, therefore, be used to derive an expression for the mean square fluctuation of the variable v (the volume per particle) and the variable n (the particle density) of the subsystem. We readily obtain

$$\overline{(\Delta v)^2} = kT\kappa_T V/N^2, \quad \overline{(\Delta n)^2} = \frac{1}{v^4}\overline{(\Delta v)^2} = kT\kappa_T N^2/V^3; \tag{19}$$

note that the last result obtained here is in complete agreement with equation (4.5.7), which was derived on the basis of the grand canonical ensemble. A little reflection shows that this result applies equally well to a subsystem with a fixed volume V and a fluctuating number of particles N. The mean square fluctuation in N is then given by

$$\overline{(\Delta N)^2} = V^2\overline{(\Delta n)^2} = kT\kappa_T N^2/V. \tag{20}$$

Substituting (20) into (18), we obtain once again the grand canonical result for $\overline{(\Delta E)^2}$, namely

$$\overline{(\Delta E)^2} = kT^2C_V + \overline{(\Delta N)^2}\{(\partial E/\partial N)_T\}^2, \tag{21}$$

as in equation (4.5.14).

In passing, we note that the first part of expression (21) denotes the mean square fluctuation in the energy E of a subsystem for which both N and V are fixed, just as we have in the canonical ensemble (N, V, T). Conversely, if we assume the energy E to be fixed, then the temperature of the subsystem will fluctuate, and the mean square value of the quantity ΔT will be given by (kT^2C_V) divided by the square of the thermal capacity of the subsystem. The net result will, therefore, be (kT^2/C_V), which is the same as in (14a).

15.2 The Einstein–Smoluchowski theory of the Brownian motion

The term "Brownian motion" derives its name from the botanist Robert Brown who, in 1828, made careful observations on the tiny pollen grains of a plant under a microscope. In his own words: "While examining the form of the particles immersed in water, I observed many of them very evidently in motion. These motions were such as to satisfy me ... that they arose neither from currents in the fluid nor from its gradual evaporation, but belonged to the particle itself." We now know that the real source of this motion lies in the incessant, and more or less random, bombardment of the *Brownian particles*, as these grains (or, for that matter, any colloidal suspensions) are usually referred to, by the *molecules* of the surrounding fluid. It was Einstein who, in a number of papers (beginning in 1905), first provided a sound theoretical analysis of the Brownian motion on the basis of the so-called "random walk problem" and thereby established a far-reaching relationship between the irreversible nature of this phenomenon and the mechanism of molecular fluctuations.

To illustrate the essential theme of Einstein's approach, we first consider the problem in *one* dimension. Let $x(t)$ denote the position of the Brownian particle at time t, given that its position coincided with the point $x = 0$ at time $t = 0$. To simplify matters, we assume that each molecular impact (which, on an average, takes place after a time τ^*) causes the particle to jump a (small) distance l — of *constant* magnitude — in either a positive or negative direction along the x-axis. It seems natural to regard the possibilities $\Delta x = +l$ and $\Delta x = -l$ to be *equally likely*; though somewhat less natural, we may also regard the successive impacts on, and hence the successive jumps of, the Brownian particle to be *mutually uncorrelated*. The probability that the particle is found at the point x at time t is then equal to the probability that, in a series of $n (= t/\tau^*)$ successive jumps, the particle makes $m (= x/l)$ more jumps in the positive direction of the x-axis than in the negative, that is, it makes $\frac{1}{2}(n+m)$ jumps in the positive direction and $\frac{1}{2}(n-m)$ in the negative.[1] The desired probability is then given by the binomial expression

$$p_n(m) = \frac{n!}{\left\{\frac{1}{2}(n+m)\right\}!\left\{\frac{1}{2}(n-m)\right\}!}\left(\frac{1}{2}\right)^n, \tag{1}$$

with the result that

$$\overline{m} = 0 \quad \text{and} \quad \overline{m^2} = n. \tag{2}$$

Thus, for $t \gg \tau^*$, we have for the net displacement of the particle

$$\overline{x(t)} = 0 \quad \text{and} \quad \overline{x^2(t)} = l^2 \frac{t}{\tau^*} \propto t^1. \tag{3}$$

Accordingly, the root-mean-square displacement of the particle is proportional to the square root of the time elapsed:

$$x_{\text{r.m.s.}} = \sqrt{\left(\overline{x^2(t)}\right)} = l\sqrt{(t/\tau^*)} \propto t^{1/2}. \tag{4}$$

It should be noted that the proportionality of the *net* overall displacement of the Brownian particle to the *square root* of the total number of elementary steps is a typical consequence

[1] Since the quantities x and t are *macroscopic* in nature while l and τ^* are microscopic, the numbers n and m are much larger than unity; consequently, it is safe to assume that they are *integral* as well.

of the random nature of the steps and it manifests itself in a large variety of phenomena in nature. In contrast, if the successive steps were fully coherent (or else if the motion were completely predictable and reversible over the time interval t),[2] then the net displacement of the Brownian particle would have been proportional to t^1.

Smoluchowski's approach to the problem of Brownian motion, which appeared in 1906, was essentially the same as that of Einstein; the difference lay primarily in the mathematical procedure. Smoluchowski introduced the *probability function* $p_n(x_0|x)$, which denotes the "probability that, after a series of n steps, the Brownian particle, initially at the point x_0, reaches the point x"; the number x here denotes the distance traveled by the Brownian particle in terms of the length of the elementary step. Clearly,

$$p_n(x_0|x) = \sum_{z=-\infty}^{\infty} p_{n-1}(x_0|z)p_1(z|x) \quad (n \geq 1); \tag{5}$$

moreover, since a single step is equally likely to take the particle to the right or to the left,

$$p_1(z|x) = \frac{1}{2}\delta_{z,x-1} + \frac{1}{2}\delta_{z,x+1}, \tag{6}$$

while

$$p_0(z|x) = \delta_{z,x}. \tag{7}$$

Equation (5) is known as the *Smoluchowski equation*. To solve it, we introduce a *generating function* $Q_n(\xi)$, namely

$$Q_n(\xi) = \sum_{x=-\infty}^{\infty} p_n(x_0|x)\xi^{x-x_0}, \tag{8}$$

from which it follows that

$$Q_0(\xi) = \sum_{x=-\infty}^{\infty} p_0(x_0|x)\xi^{x-x_0} = \sum_{x=-\infty}^{\infty} \delta_{x_0,x}\xi^{x-x_0} = 1. \tag{9}$$

Substituting (6) into (5), we obtain

$$p_n(x_0|x) = \frac{1}{2}p_{n-1}(x_0|x-1) + \frac{1}{2}p_{n-1}(x_0|x+1). \tag{10}$$

[2]The term "reversible" here is related to the fact that the Newtonian equations of motion, which govern this class of phenomena, preserve their form if the direction of time is reversed (i.e., if we change t to $-t$, etc.); alternatively, one would expect that if at any instant of time we reverse the velocities of the particles in a given mechanical system, the system would "retrace" its path exactly. This is *not* true of equations describing "irreversible" phenomena, such as the *diffusion equation* (19), with which the phenomenon of Brownian motion is intimately related.

Multiplying (10) by ξ^{x-x_0} and adding over all x, we obtain the recurrence relation

$$Q_n(\xi) = \frac{1}{2}[\xi + (1/\xi)]Q_{n-1}(\xi), \tag{11}$$

so that, by iteration,

$$Q_n(\xi) = \left\{\frac{1}{2}[\xi + (1/\xi)]\right\}^n Q_0(\xi) = (1/2)^n[\xi + (1/\xi)]^n. \tag{12}$$

Expanding this expression binomially and comparing the result with (8), we get

$$p_n(x_0|x) = \begin{cases} \left(\dfrac{1}{2}\right)^n \dfrac{n!}{\{\frac{1}{2}(n+x-x_0)\}!\{\frac{1}{2}(n-x+x_0)\}!} & \text{for } |x-x_0| \le n \\ 0 & \text{for } |x-x_0| > n. \end{cases} \tag{13}$$

Identifying $(x - x_0)$ with m, we find this result to be in complete agreement with our previous result (1).[3] Accordingly, any conclusions drawn from the Smoluchowski approach will be the same as the ones drawn from the Einstein approach.

To obtain an asymptotic form of the function $p_n(m)$, we apply Stirling's formula, $n! \approx (2\pi n)^{1/2}(n/e)^n$, to the factorials appearing in (1), with the result

$$\ln p_n(m) \approx \left(n + \frac{1}{2}\right)\ln n - \frac{1}{2}(n+m+1)\ln\left\{\frac{1}{2}(n+m)\right\} - \frac{1}{2}(n-m+1)\ln\left\{\frac{1}{2}(n-m)\right\} - n\ln 2 - \frac{1}{2}\ln(2\pi).$$

For $m \ll n$ (which is generally true because $\overline{m} = 0$ and $m_{\text{r.m.s.}} = n^{1/2}$, while $n \gg 1$), we obtain

$$p_n(m) \approx \frac{2}{\surd(2\pi n)}\exp(-m^2/2n). \tag{14}$$

Taking x to be a continuous variable (and remembering that $p_n(m) \equiv 0$ either for even values of m or for odd values of m, so that in the distribution (14), $\Delta m = 2$ and not 1), we may write this result in the *Gaussian* form:

$$p(x)dx = \frac{dx}{\surd(4\pi Dt)}\exp\left(-\frac{x^2}{4Dt}\right), \tag{15}$$

where

$$D = l^2/2\tau^*. \tag{16}$$

[3]It is easy to recognize the additional fact that if n is even, then $p_n(m) \equiv 0$ for odd m, and if n is odd, then $p_n(m) \equiv 0$ for even m.

Later on, we shall see that the quantity D introduced here is identical to the *diffusion coefficient* of the given system; equation (16) connects this quantity with the microscopic quantities l and τ^*. To appreciate this connection, one has simply to note that the problem of Brownian motion can also be looked on as a problem of "diffusion" of Brownian particles through the medium of the fluid; this point of view is also due to Einstein. However, before we embark on these considerations, we would like to present here the results of an actual observation made on the Brownian motion of a spherical particle immersed in water; see Lee, Sears, and Turcotte (1963). It was found that the 403 values of the net displacement Δx of the particle, observed after successive intervals of 2 seconds each, were distributed as follows:

Displacement Δx, *in units of* $\mu (= 10^{-4}\text{cm})$	*Frequency of occurrence n*
less than -5.5	0
between -5.5 and -4.5	1
between -4.5 and -3.5	2
between -3.5 and -2.5	15
between -2.5 and -1.5	32
between -1.5 and -0.5	95
between -0.5 and $+0.5$	111
between $+0.5$ and $+1.5$	87
between $+1.5$ and $+2.5$	47
between $+2.5$ and $+3.5$	8
between $+3.5$ and $+4.5$	5
greater than $+4.5$	0

The mean square value of the displacement here turns out to be: $\overline{(\Delta x)^2} = 2.09 \times 10^{-8}\text{cm}^2$. The observed frequency distribution has been plotted as a "block diagram" in Figure 15.1. We have included, in this figure, a Gaussian curve based on the observed value of the mean square displacement; we find that the experimental data fit the theoretical curve fairly well. We can also derive here an experimental value for the diffusion coefficient of the medium; we obtain: $D = \overline{(\Delta x)^2}/2t = 5.22 \times 10^{-9}\text{cm}^2/\text{s}$.[4]

We now turn to the study of the Brownian motion from the point of view of diffusion. We denote the number density of the Brownian particles in the fluid by the symbol $n(\boldsymbol{r}, t)$ and their current density by $\boldsymbol{j}(\boldsymbol{r}, t)\{= n(\boldsymbol{r}, t)\boldsymbol{v}(\boldsymbol{r}, t)\}$; then, according to Fick's law,

$$\boldsymbol{j}(\boldsymbol{r}, t) = -D\nabla n(\boldsymbol{r}, t), \tag{17}$$

[4]In the next section we shall see that, for a *spherical* particle, $D = kT/6\pi\eta a$ where η is the coefficient of viscosity of the medium and a the radius of the Brownian particle. In the case under study, $T \simeq 300\,\text{K}$, $\eta \simeq 10^{-2}$ poise, and $a \simeq 4 \times 10^{-5}$ cm. Substituting these values, we obtain for the Boltzmann constant: $k \simeq 1.3 \times 10^{-16}\text{erg/K}$.

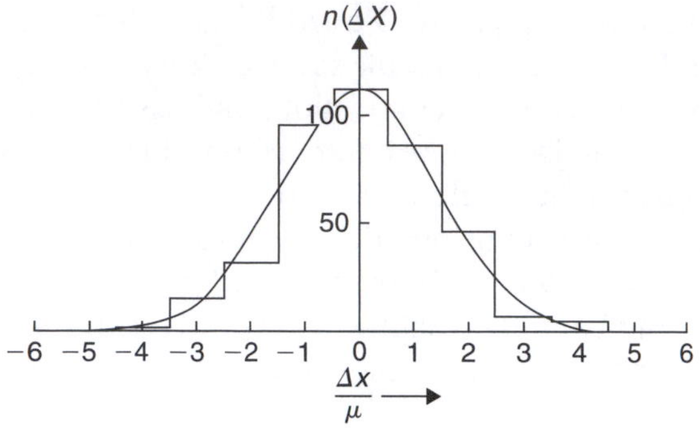

FIGURE 15.1 The statistical distribution of the successive displacements, Δx, of a Brownian particle immersed in water: $(\Delta x)_{\text{r.m.s.}} \simeq 1.45\,\mu$.

where D denotes for the *diffusion coefficient* of the medium. We also have here the equation of continuity, namely

$$\nabla \cdot \boldsymbol{j}(\mathbf{r}, t) + \frac{\partial n(\mathbf{r}, t)}{\partial t} = 0. \tag{18}$$

Substituting (17) into (18), we obtain the diffusion equation

$$\nabla^2 n(\mathbf{r}, t) - \frac{1}{D}\frac{\partial n(\mathbf{r}, t)}{\partial t} = 0. \tag{19}$$

Of the various possible solutions of this equation, the one relevant to the present situation is

$$n(\mathbf{r}, t) = \frac{N}{(4\pi D t)^{3/2}} \exp\left(-\frac{r^2}{4Dt}\right), \tag{20}$$

which is a spherically symmetric solution and is already normalized:

$$\int_0^\infty n(\mathbf{r}, t) 4\pi r^2 dr = N, \tag{21}$$

N being the total number of (Brownian) particles immersed in the fluid. A comparison of the (three-dimensional) result (20) with the (one-dimensional) result (15) brings out most vividly the relationship between the random walk problem on one hand and the phenomenon of diffusion on the other.

It is clear that in the last approach we have considered the motion of an "ensemble" of N Brownian particles placed under "equivalent" physical conditions, rather than considering the motion of a single particle over a length of time (as was done in the random walk approach). Accordingly, the averages of the various physical quantities obtained here will be in the nature of "ensemble averages"; they must, of course, agree with the long-time averages of the same quantities obtained earlier.

Now, by virtue of the distribution (20), we obtain

$$\langle \boldsymbol{r}(t) \rangle = 0; \quad \langle r^2(t) \rangle = \frac{1}{N} \int_0^\infty n(\boldsymbol{r}, t) 4\pi r^4 dr = 6Dt \propto t^1, \tag{22}$$

in complete agreement with our earlier results, namely

$$\overline{x(t)} = 0; \quad \overline{x^2(t)} = l^2 t/\tau^* = 2Dt \propto t^1. \tag{23}$$

Thus, the "ensemble" of the Brownian particles, initially concentrated at the origin, "diffuses out" as time increases, the nature and the extent of its spread at any time t being given by equations (20) and (22), respectively. The diffusion process, which is clearly *irreversible*, gives us a fairly good picture of the statistical behavior of a single particle in the ensemble. However, the important thing to bear in mind is that, whether we focus our attention on a single particle in the ensemble or look at the ensemble as a whole, the ultimate source of the phenomenon lies in the incessant, and more or less random, impacts received by the Brownian particles from the molecules of the fluid. In other words, the irreversible character of the phenomenon ultimately arises from the random, fluctuating forces exerted by the fluid molecules on the Brownian particles. This leads us to another systematic theory of the Brownian motion, namely the theory of Langevin (1908). For a detailed analysis of the problem, see Uhlenbeck and Ornstein (1930), Chandrasekhar (1943, 1949), MacDonald (1948–1949), and Wax (1954).

15.3 The Langevin theory of the Brownian motion

We consider the simplest case of a "free" Brownian particle, surrounded by a fluid environment; the particle is assumed to be free in the sense that it is not acted on by any other force except the one arising from the molecular bombardment. The equation of motion of the particle will then be

$$M \frac{d\boldsymbol{v}}{dt} = \boldsymbol{\mathcal{F}}(t), \tag{1}$$

where M is the particle mass, $\boldsymbol{v}(t)$ the particle velocity, and $\boldsymbol{\mathcal{F}}(t)$ the force acting on the particle by virtue of the impacts received from the fluid molecules. Langevin suggested that the force $\boldsymbol{\mathcal{F}}(t)$ may be written as a sum of two parts: (i) an "averaged-out" part, which represents the *viscous drag*, $-\boldsymbol{v}/B$, experienced by the particle (accordingly, B is the *mobility* of the system, that is, the drift velocity acquired by the particle by virtue of a unit "external" force)[5] and (ii) a "rapidly fluctuating" part $\boldsymbol{F}(t)$ which, over long intervals of

[5]If Stokes's law is applicable, then $B = 1/(6\pi\eta a)$, where η is the coefficient of viscosity of the fluid and a the radius of the particle (assumed spherical).

time (as compared to the characteristic time τ^*), averages out to zero; thus, we may write

$$M\frac{d\boldsymbol{v}}{dt} = -\frac{\boldsymbol{v}}{B} + \boldsymbol{F}(t); \quad \overline{\boldsymbol{F}(t)} = 0. \tag{2}$$

Taking the ensemble average of (2), we obtain[6]

$$M\frac{d}{dt}\langle\boldsymbol{v}\rangle = -\frac{1}{B}\langle\boldsymbol{v}\rangle, \tag{3}$$

which gives

$$\langle\boldsymbol{v}(t)\rangle = \boldsymbol{v}(0)\exp(-t/\tau) \quad (\tau = MB). \tag{4}$$

Thus, the mean drift velocity of the particle decays, at a rate determined by the *relaxation time* τ, to the ultimate value zero. We note that this result is typical of the phenomena governed by *dissipative* properties such as the viscosity of the fluid; the *irreversible* nature of the result is also evident.

Dividing (2) by the mass of the particle, we obtain an equation for the *instantaneous* acceleration, namely

$$\frac{d\boldsymbol{v}}{dt} = -\frac{\boldsymbol{v}}{\tau} + A(t); \quad \overline{\boldsymbol{A}(t)} = 0. \tag{5}$$

We now construct the scalar product of (5) with the *instantaneous* position $\boldsymbol{r}$ of the particle and take the ensemble average of the product. In doing so, we make use of the facts that (i) $\boldsymbol{r}\cdot\boldsymbol{v} = \frac{1}{2}(dr^2/dt)$, (ii) $\boldsymbol{r}\cdot(d\boldsymbol{v}/dt) = \frac{1}{2}(d^2r^2/dt^2) - v^2$, and (iii) $\langle\boldsymbol{r}\cdot\boldsymbol{A}\rangle = 0$.[7] We obtain

$$\frac{d^2}{dt^2}\langle r^2\rangle + \frac{1}{\tau}\frac{d}{dt}\langle r^2\rangle = 2\langle v^2\rangle. \tag{6}$$

If the Brownian particle has already attained thermal equilibrium with the molecules of the fluid, then the quantity $\langle v^2\rangle$ in this equation may be replaced by its *equipartition value* $3kT/M$. The equation is then readily integrated, with the result

$$\langle r^2\rangle = \frac{6kT\tau^2}{M}\left\{\frac{t}{\tau} - (1 - e^{-t/\tau})\right\}, \tag{7}$$

[6]The process of "averaging over an ensemble" implies that we are imagining a large number of systems similar to the one originally under consideration and are taking an average over this collection at *any* time t. By the very nature of the function $\boldsymbol{F}(t)$, the ensemble average $\langle\boldsymbol{F}(t)\rangle$ must be zero at all times.

[7]This is so because we have no reason to expect a statistical correlation between the position $\boldsymbol{r}(t)$ of the Brownian particle and the force $\boldsymbol{F}(t)$ exerted on it by the molecules of the fluid; see, however, Manoliu and Kittel (1979). Of course, we do expect a correlation between the variables $\boldsymbol{v}(t)$ and $\boldsymbol{F}(t)$; consequently, $\langle\boldsymbol{v}\cdot F\rangle \neq 0$ (see Problem 15.7).

where the constants of integration have been so chosen that at, $t = 0$, both $\langle r^2 \rangle$ and its first time-derivative vanish. We observe that, for $t \ll \tau$,

$$\langle r^2 \rangle \simeq \frac{3kT}{M} t^2 = \langle v^2 \rangle t^2, \tag{8}$$ [8]

which is consistent with the reversible equations of motion whereby one would simply have

$$\boldsymbol{r} = \boldsymbol{v}t. \tag{9}$$

On the other hand, for $t \gg \tau$,

$$\langle r^2 \rangle \simeq \frac{6kT\tau}{M} t = (6BkT)t, \tag{10}$$ [9]

which is essentially the same as the Einstein–Smoluchowski result (15.2.22); incidentally, we obtain here a simple, but important, relationship between the coefficient of diffusion D and the mobility B, namely

$$D = BkT, \tag{11}$$

which is generally referred to as the *Einstein relation*.

The irreversible character of equation (10) is self-evident; it is also clear that it arises essentially from the viscosity of the medium. Moreover, the Einstein relation (11), which connects the coefficient of diffusion D with the mobility B of the system, tells us that the ultimate source of the viscosity of the medium (as well as of diffusion) lies in the random, fluctuating forces arising from the incessant motion of the fluid molecules; see also the fluctuation–dissipation theorem of Section 15.6.

In this context, if we consider a particle of charge e and mass M moving in a viscous fluid under the influence of an external electric field of intensity E, then the "coarse-grained" motion of the particle will be determined by the equation

$$M\frac{d}{dt}\langle \boldsymbol{v} \rangle = -\frac{1}{B}\langle \boldsymbol{v} \rangle + e\boldsymbol{E}; \tag{12}$$

compare this to equation (3). The "terminal" drift velocity of the particle would now be given by the expression $(eB)\boldsymbol{E}$, which prompts one to define (eB) as the "mobility" of the system and denote it by the symbol μ. Consequently, one obtains, instead of (11),

$$D = \frac{kT}{e}\mu, \tag{13}$$

which, in fact, is the original version of the Einstein relation; sometimes this is also referred to as the *Nernst relation*.

[8]Note that the limiting solution (8) corresponds to "dropping out" the second term on the left side of equation (6).
[9]Note that the limiting solution (10) corresponds to "dropping out" the first term on the left side of equation (6).

So far we have not felt any *direct* influence of the rapidly fluctuating term $\boldsymbol{A}(t)$ that appears in the equation of motion (5) of the Brownian particle. For this, let us try to evaluate the quantity $\langle v^2(t)\rangle$ which, in the preceding analysis, was assumed to have already attained its "limiting" value $3kT/M$. For this evaluation we replace the variable t in equation (5) by u, multiply both sides of the equation by $\exp(u/\tau)$, rearrange and integrate over du between the limits $u = 0$ and $u = t$; we thus obtain the formal solution

$$\boldsymbol{v}(t) = \boldsymbol{v}(0)e^{-t/\tau} + e^{-t/\tau}\int_0^t e^{u/\tau}\boldsymbol{A}(u)du. \tag{14}$$

Thus, the drift velocity $\boldsymbol{v}(t)$ of the particle is also a fluctuating function of time; of course, since $\langle \boldsymbol{A}(u)\rangle = 0$ for all u, the *average* drift velocity is given by the first term alone, namely

$$\langle \boldsymbol{v}(t)\rangle = \boldsymbol{v}(0)e^{-t/\tau}, \tag{15}$$

which is the same as our earlier result (4). For the mean square velocity $\langle v^2(t)\rangle$, we now obtain from (14)

$$\langle v^2(t)\rangle = v^2(0)e^{-2t/\tau} + 2e^{-2t/\tau}\left[\boldsymbol{v}(0)\cdot\int_0^t e^{u/\tau}\langle \boldsymbol{A}(u)\rangle du\right] + e^{-2t/\tau}\int_0^t\int_0^t e^{(u_1+u_2)/\tau}\langle \boldsymbol{A}(u_1)\cdot\boldsymbol{A}(u_2)\rangle du_1 du_2. \tag{16}$$

The second term on the right side of this equation is identically zero, because $\langle \boldsymbol{A}(u)\rangle$ vanishes for all u. In the third term, we have the quantity $\langle \boldsymbol{A}(u_1)\cdot\boldsymbol{A}(u_2)\rangle$, which is a measure of the "statistical correlation between the value of the fluctuating variable $\boldsymbol{A}$ at time u_1 and its value at time u_2"; we call it the *autocorrelation function* of the variable $\boldsymbol{A}$ and denote it by the symbol $K_A(u_1, u_2)$ or simply by $K(u_1, u_2)$. Before proceeding with (16) any further, we place on record some of the important properties of the function $K(u_1, u_2)$.

(i) In a stationary ensemble (i.e., one in which the overall macroscopic behavior of the systems does not change with time), the function $K(u_1, u_2)$ depends only on the time interval $(u_2 - u_1)$. Denoting this interval by the symbol s, we have

$$K(u_1, u_1 + s) \equiv \langle \boldsymbol{A}(u_1)\cdot\boldsymbol{A}(u_1 + s)\rangle = K(s), \text{ independently of } u_1. \tag{17}$$

(ii) The quantity $K(0)$, which is identically equal to the mean square value of the variable $\boldsymbol{A}$ at time u_1, must be *positive definite*. In a stationary ensemble, it would be a constant, independent of u_1:

$$K(0) = \text{ const.} > 0. \tag{18}$$

(iii) For any value of s, the magnitude of the function $K(s)$ cannot exceed $K(0)$.

Proof: Since

$$\begin{aligned}\langle |\boldsymbol{A}(u_1) \pm \boldsymbol{A}(u_2)|^2\rangle &= \langle A^2(u_1)\rangle + \langle A^2(u_2)\rangle \pm 2\langle \boldsymbol{A}(u_1)\cdot \boldsymbol{A}(u_2)\rangle \\ &= 2\{K(0) \pm K(s)\} \geq 0,\end{aligned}$$

the function $K(s)$ cannot go outside the limits $-K(0)$ and $+K(0)$; consequently,

$$|K(s)| \leq K(0) \quad \text{for all } s. \tag{19}$$

(iv) The function $K(s)$ is symmetric about the value $s = 0$, that is,

$$K(-s) = K(s) = K(|s|). \tag{20}$$

Proof:

$$\begin{aligned}K(s) &\equiv \langle \boldsymbol{A}(u_1)\cdot \boldsymbol{A}(u_1+s)\rangle = \langle \boldsymbol{A}(u_1-s)\cdot \boldsymbol{A}(u_1)\rangle^{10} \\ &= \langle \boldsymbol{A}(u_1)\cdot \boldsymbol{A}(u_1-s)\rangle \equiv K(-s).\end{aligned}$$

(v) As s becomes large in comparison with the characteristic time τ^*, the values $\boldsymbol{A}(u_1)$ and $\boldsymbol{A}(u_1+s)$ become *uncorrelated*, that is

$$K(s) \equiv \langle \boldsymbol{A}(u_1)\cdot \boldsymbol{A}(u_1+s)\rangle \xrightarrow[s\gg\tau^*]{} \langle \boldsymbol{A}(u_1)\rangle \cdot \langle \boldsymbol{A}(u_1+s)\rangle = 0. \tag{21}$$

In other words, the "memory" of the molecular impacts received during a given interval of time, say between u_1 and $u_1 + du_1$, is "completely lost" after a lapse of time large in comparison with τ^*. It follows that the magnitude of the function $K(s)$ is significant only so long as the variable s is of the same order of magnitude as τ^*.

Figures 15.7 through 15.9 later in this chapter show the s-dependence of certain typical correlation functions $K(s)$; they fully conform to the properties listed here.

We now evaluate the double integral appearing in (16):

$$I = \int_0^t \int_0^t e^{(u_1+u_2)/\tau} K(u_2 - u_1)\, du_1 du_2. \tag{22}$$

Changing over to the variables

$$S = \frac{1}{2}(u_1 + u_2) \quad \text{and} \quad s = (u_2 - u_1), \tag{23}$$

the integrand becomes $\exp(2S/\tau)K(s)$, the element $(du_1 du_2)$ gets replaced by the corresponding element $(dSds)$ while the limits of integration, in terms of the variables S and s,

[10] This is the only crucial step in the proof. It involves a "shift," by an amount s, in both instants of the measurement process; the equality results from the fact that the ensemble is supposed to be stationary.

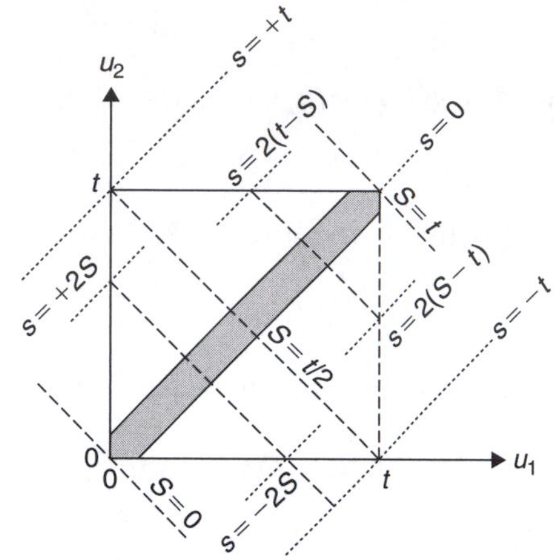

FIGURE 15.2 Limits of integration, of the double integral I, in terms of the variables S and s.

can be read from Figure 15.2; we find that, for $0 \le S \le t/2$, s goes from $-2S$ to $+2S$ while, for $t/2 \le S \le t$, it goes from $-2(t-S)$ to $+2(t-S)$. Accordingly,

$$I = \int_0^{t/2} e^{2S/\tau} dS \int_{-2S}^{+2S} K(s) ds + \int_{t/2}^{t} e^{2S/\tau} dS \int_{-2(t-S)}^{+2(t-S)} K(s) ds. \tag{24}$$

In view of property (v) of the function $K(s)$, see equation (21), the integrals over s draw significant contribution only from a very narrow region, of the order of τ^*, around the value $s = 0$ (i.e., from the shaded region in Figure 15.2); contributions from regions with larger values of $|s|$ are negligible. Thus, if $t \gg \tau^*$, the limits of integration for s may be replaced by $-\infty$ and $+\infty$, with the result

$$I \simeq C \int_0^t e^{2S/\tau} dS = C\frac{\tau}{2}(e^{2t/\tau} - 1), \tag{25}$$

where

$$C = \int_{-\infty}^{\infty} K(s) ds. \tag{26}$$

Substituting (25) into (16), we obtain

$$\langle v^2(t) \rangle = v^2(0) e^{-2t/\tau} + C\frac{\tau}{2}(1 - e^{-2t/\tau}). \tag{27}$$

Now, as $t \to \infty$, $\langle v^2(t) \rangle$ must tend to the equipartition value $3kT/M$; therefore,

$$C = 6kT/M\tau \tag{28}$$

and hence

$$\langle v^2(t)\rangle = v^2(0) + \left\{\frac{3kT}{M} - v^2(0)\right\}(1 - e^{-2t/\tau}). \tag{29}$$ [11]

We note that if $v^2(0)$ were itself equal to the equipartition value $3kT/M$, then $\langle v^2(t)\rangle$ would always remain the same, which shows that statistical equilibrium, once attained, has a natural tendency to persist.

Substituting (29) into the right side of (6), we obtain a more representative description of the manner in which the quantity $\langle r^2\rangle$ varies with t; we thus have

$$\frac{d^2}{dt^2}\langle r^2\rangle + \frac{1}{\tau}\frac{d}{dt}\langle r^2\rangle = 2v^2(0)e^{-2t/\tau} + \frac{6kT}{M}(1 - e^{-2t/\tau}), \tag{30}$$

with the solution

$$\langle r^2\rangle = v^2(0)\tau^2(1 - e^{-t/\tau})^2 - \frac{3kT}{M}\tau^2(1 - e^{-t/\tau})(3 - e^{-t/\tau}) + \frac{6kT\tau}{M}t. \tag{31}$$

Solution (31) satisfies the initial conditions that both $\langle r^2\rangle$ and its first time-derivative vanish at $t = 0$; moreover, if we put $v^2(0) = 3kT/M$, it reduces to solution (7) obtained earlier. Once again, we note the *reversible* nature of the motion for $t \ll \tau$, with $\langle r^2\rangle \simeq v^2(0)t^2$, and its *irreversible* nature for $t \gg \tau$, with $\langle r^2\rangle \simeq (6BkT)t$.

Figures 15.3 and 15.4 show the variation, with time, of the ensemble averages $\langle v^2(t)\rangle$ and $\langle r^2(t)\rangle$ of a Brownian particle, as given by equations (29) and (31), respectively. All important features of our results are manifestly evident in these plots.

Brownian motion continues to be a topic of contemporary research nearly 200 years after Brown's discovery and over 100 years after Einstein and Smoluchowski's analysis and early measurements by Perrin. The renewed interest is due to the growth in the technological importance of colloids across a wide range of fields and the development of digital video and computer image analysis. An interesting example is the detailed observation and analysis of rotational and two-dimensional translational Brownian motion of ellipsoidal particles by Han et al. (2006) in a thin microscope slide. The case of rotational Brownian motion was first analyzed by Einstein (1906b) and first measured by Perrin (1934, 1936). Both rotational and translational modes diffuse according to Langevin dynamics but the translational diffusion is coupled to the rotational diffusion since the translational diffusion constant parallel to the longer axis is larger than the diffusion constant perpendicular

[11] One may check that

$$\frac{d}{dt}\langle v^2(t)\rangle = \frac{2}{\tau}\left[v^2(\infty) - \langle v^2(t)\rangle\right] = -\frac{2}{\tau}\Delta\langle v^2(t)\rangle,$$

where $v^2(\infty) = 3kT/M$ and $\Delta\langle v^2(t)\rangle$ is the "deviation of the quantity concerned from its equilibrium value." In this form of the equation, we have a typical example of a "relaxation phenomenon," with *relaxation time* $\tau/2$.

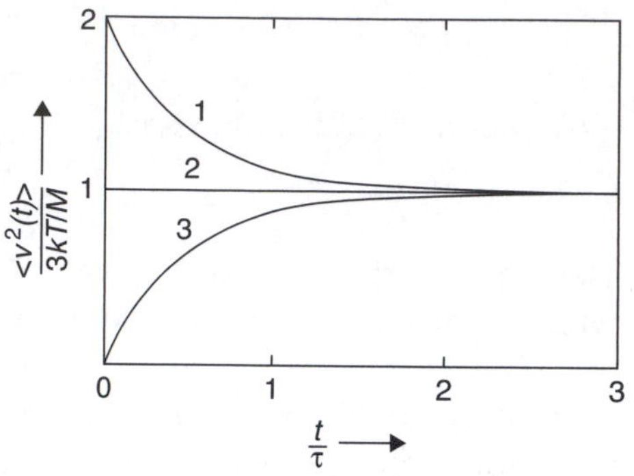

FIGURE 15.3 The mean square velocity of a Brownian particle as a function of time. Curves 1, 2, and 3 correspond, respectively, to the initial conditions $v^2(0) = 6kT/M, 3kT/M$, and 0.

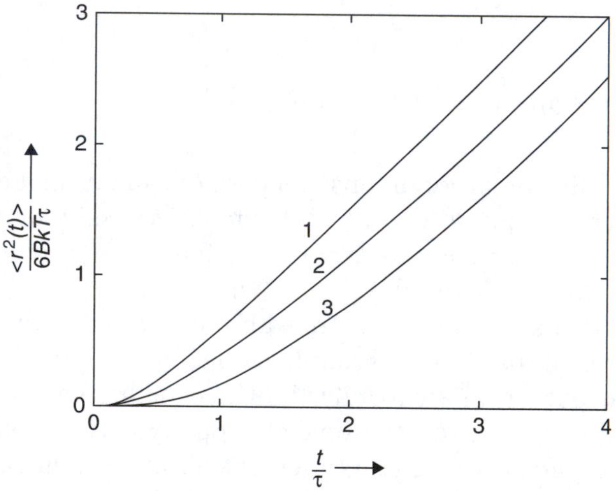

FIGURE 15.4 The mean square displacement of a Brownian particle as a function of time. Curves 1, 2, and 3 correspond, respectively, to the initial conditions $v^2(0) = 6kT/M$, $3kT/M$, and 0.

to that axis. The rotational diffusion and both long-axis (a) and short-axis (b) body-frame diffusions are all Gaussian:

$$p_\theta(\Delta\theta, t) = \frac{1}{\sqrt{4\pi D_\theta t}} \exp\left(-\frac{(\Delta\theta)^2}{4D_\theta t}\right), \tag{32a}$$

$$p_{\rm a}(\Delta x_{\rm a}, t) = \frac{1}{\sqrt{4\pi D_{\rm a} t}} \exp\left(-\frac{(\Delta x_{\rm a})^2}{4D_{\rm a} t}\right), \tag{32b}$$

$$p_{\rm b}(\Delta x_{\rm b}, t) = \frac{1}{\sqrt{4\pi D_{\rm b} t}} \exp\left(-\frac{(\Delta x_{\rm b})^2}{4D_{\rm b} t}\right), \tag{32c}$$

with diffusion constants D_θ, D_a, and D_b. Experiments have observed the complex two-dimensional spatial diffusion at short times ($t \lesssim \tau_\theta = 1/(2D_\theta)$), as predicted by the Langevin theory. The long-time ($t \gg \tau_\theta$) spatial diffusion is isotropic with diffusion constant $\overline{D} = (D_a + D_b)/2$.

15.3.A Brownian motion of a harmonic oscillator

An analysis similar to the one for a diffusing Brownian particle can also be performed for a particle in a harmonic oscillator potential that prevents the particle from diffusing away from the origin and allows a more general analysis of the relationship between the position and velocity response functions and the power spectra of the fluctuations; see Kappler (1938) and Chandrasekhar (1943). The one-dimensional equation of motion for a Brownian particle of mass M in a harmonic oscillator potential with spring constant $M\omega_0^2$ is

$$\frac{d^2x}{dt^2} + \gamma\frac{dx}{dt} + \omega_0^2 x = \frac{F(t)}{M}, \tag{33}$$

where $\gamma\,(= 6\pi\eta a/M)$ is the damping coefficient of a spherical particle in a fluid with viscosity η. Just as in the case of diffusive Brownian motion, the force $F(t)$ can be a time-dependent external force designed to explore the response function or a time-dependent random force due to collisions with molecules in the fluid to analyze the equilibrium fluctuations. Assuming the system was in equilibrium in the distant past, the position at time t is given by

$$x(t) = \int_{-\infty}^{t} \chi_{xx}(t-t')F(t')dt', \tag{34}$$

where

$$\chi_{xx}(s) = \frac{1}{M\omega_1} e^{-\frac{\gamma s}{2}} \sin(\omega_1 s) \tag{35}$$

is the xx response function and $\omega_1 = \sqrt{\omega_0^2 - \frac{\gamma^2}{4}}$.[12] The velocity response is given by

$$v(t) = \int_{-\infty}^{t} \chi_{vx}(t-t')F(t')dt', \tag{36}$$

[12]This form of the response function assumes that the oscillator is underdamped. The notation χ_{xx} refers to the notation used in Section 15.6.A in which the response of the position coordinate x depends on the applied field F that couples to the Hamiltonian via a term $-F(t)x(t)$.

where

$$\chi_{vx}(s) = \frac{1}{M} e^{-\frac{\gamma s}{2}} \left(\cos(\omega_1 s) - \frac{\gamma}{2\omega_1} \sin(\omega_1 s) \right). \tag{37}$$

The response of the system can be decomposed into a sum of independent terms involving a sinusoidal applied force $\hat{F}(\omega)e^{i\omega t}$. This takes the form

$$\hat{x}(\omega) = \tilde{\chi}_{xx}(\omega)\hat{F}(\omega), \tag{38}$$

where the frequency-dependent response function can be decomposed into real and imaginary parts $\hat{\chi}'_{xx}(\omega)$ and $\hat{\chi}''_{xx}(\omega)$:

$$\tilde{\chi}_{xx}(\omega) = \int_0^\infty \chi_{xx}(s) e^{i\omega s} ds = \hat{\chi}'_{xx}(\omega) + i\hat{\chi}''_{xx}(\omega), \tag{39a}$$

$$\hat{\chi}'_{xx}(\omega) = \frac{\omega_0^2 - \omega^2}{M[(\omega_0^2 - \omega^2)^2 + \gamma^2\omega^2]}, \tag{39b}$$

$$\hat{\chi}''_{xx}(\omega) = \frac{\gamma\omega}{M[(\omega_0^2 - \omega^2)^2 + \gamma^2\omega^2]}. \tag{39c}$$

The real part here describes the dispersion and the imaginary part describes the dissipation, that is, it sets the average rate of energy dissipation due to the sinusoidal external force.

Now let's consider the natural fluctuations of the position and the velocity of the particle *in equilibrium* due to the random collisions with the atoms in the fluid. We will use the same Langevin formalism as was used earlier with Brownian motion of a free particle. The random force averages to zero and is assumed to be delta-function correlated in time:

$$\langle F \rangle = 0, \tag{40a}$$

$$\langle F(t)F(t') \rangle = \Gamma\delta(t - t'), \tag{40b}$$

where $\Gamma = 2\gamma MkT$. With this choice, the long-time average position and velocity of the particle are both zero,

$$\langle x(t) \rangle = \langle v(t) \rangle = 0, \tag{41}$$

and the average of the squares of the position and velocity both obey the equipartition theorem:

$$\langle x^2(t) \rangle = \frac{kT}{M\omega_0^2}, \quad \langle v^2(t) \rangle = \frac{kT}{M}. \tag{42a,b}$$

The xx correlation function is given by

$$\begin{aligned} G_{xx}(t-t') &= \langle x(t)x(t')\rangle \\ &= \frac{kT}{M\omega_0^2}\exp\left(\frac{-\gamma|t-t'|}{2}\right)\left(\cos\left(\omega_1|t-t'|\right)+\frac{\gamma}{2\omega_1}\sin\left(\omega_1|t-t'|\right)\right), \end{aligned} \tag{43}$$

and the xx power spectrum by

$$S_{xx}(\omega) = \int_{-\infty}^{\infty} G_{xx}(s)e^{i\omega s}ds = \frac{2\gamma kT}{M}\frac{1}{\left(\omega_0^2-\omega^2\right)^2+\gamma^2\omega^2}. \tag{44}$$

Note that the imaginary part of the response function, $\hat{\chi}''_{xx}(\omega)$, in equation (39c) is proportional to the power spectrum $S_{xx}(\omega)$:

$$\hat{\chi}''_{xx}(\omega) = \frac{\omega}{2kT}S_{xx}(\omega). \tag{45}$$

This result indicates that the dissipation that results from driving a system out of equilibrium by an external force is proportional to the power spectrum of the natural fluctuations that occur in equilibrium. While this result was derived here for a very specific model, it constitutes an example of the very general fluctuation–dissipation theorem we will derive in Section 15.6.A.

15.4 Approach to equilibrium: the Fokker–Planck equation

In our analysis of the Brownian motion we have considered the behavior of a dynamical variable, such as the position $\boldsymbol{r}(t)$ or the velocity $\boldsymbol{v}(t)$ of a Brownian particle, from the point of view of fluctuations in the value of the variable. To determine the average behavior of such a variable, we sometimes invoked an "ensemble" of Brownian particles immersed in identical environments and undergoing diffusion. A treatment along these lines was carried out toward the end of Section 15.2, and the most important results of that treatment are summarized in equation (15.2.20) for the density function $n(\boldsymbol{r},t)$ and in equation (15.2.22) for the mean square displacement $\langle r^2(t)\rangle$.

A more generalized way of looking at "the manner in which, and the rate at which, a given distribution of Brownian particles approaches a state of thermal equilibrium" is provided by the so-called *Master Equation*, a simplified version of which is known as the Fokker–Planck equation. For illustration, we examine the displacement, $x(t)$, of the given set of particles along the x-axis. At any time t, let $f(x,t)dx$ be the probability that an arbitrary particle in the ensemble may have a displacement between x and $x+dx$. The function

$f(x,t)$ must satisfy the normalization condition

$$\int_{-\infty}^{\infty} f(x,t)dx = 1. \tag{1}$$

The Master Equation then reads:

$$\frac{\partial f(x,t)}{\partial t} = \int_{-\infty}^{\infty} \{-f(x,t)W(x,x') + f(x',t)W(x',x)\}dx', \tag{2}$$

where $W(x,x')dx'\delta t$ denotes the probability that, in a short interval of time δt, a particle having displacement x makes a "transition" to having a displacement between x' and $x' + dx'$.[13]

The first part of the integral in equation (2) corresponds to all those transitions that remove particles from the displacement x at time t to some other displacement x' and, hence, represent a net *loss* to the function $f(x,t)$; similarly, the second part of the integral corresponds to all those transitions that bring particles from some other displacement x' at time t to the displacement x and, hence, represent a net *gain* to the function $f(x,t)$.[14]

The structure of the Master Equation is thus founded on very simple and straightforward premises. Of course, under certain conditions, this equation, or any generalization thereof (such as the one including velocity, or momentum, coordinates in the argument of f), can be reduced to the simple form

$$\frac{\partial f}{\partial t} = -\frac{f - f_0}{\tau}, \tag{3}$$

which has proved to be a very useful first approximation for studying problems related to *transport phenomena*. Here, f_0 denotes the *equilibrium distribution function* (for $\partial f/\partial t = 0$ when $f = f_0$), while τ is the *relaxation time* that determines the rate at which the fluctuations in the system drive it to a state of equilibrium.

In studying Brownian motion on the basis of equation (2), we can safely assume that it is only transitions between "closely neighboring" states x and x' that have an appreciable probability of occurring; in other words, the transition probability function $W(x,x')$ is sharply peaked around the value $x' = x$ and falls rapidly to zero away from x. Denoting the interval $(x' - x)$ by ξ, we may write

$$W(x,x') \to W(x;\xi), \; W(x',x) \to W(x';-\xi) \tag{4}$$

[13]We are tacitly assuming here a "Markovian" situation where the *transition probability function* $W(x,x')$ depends *only* on the present position x (and, of course, the subsequent position x') of the particle but *not* on the previous history of the particle.

[14]In the case of fermions, an account must be taken of the *Pauli exclusion principle*, which controls the "occupation of single-particle states in the system"; for instance, we cannot, in that case, consider a transition that tends to transfer a particle to a state that is already occupied. This requires an appropriate modification of the Master Equation.

where $W(x;\xi)$ and $W(x';-\xi)$ have sharp peaks around the value $\xi = 0$ and fall rapidly to zero elsewhere.[15] This enables us to expand the right side of (2) as a Taylor series around $\xi = 0$. Retaining terms up to second order only, we obtain

$$\frac{\partial f(x,t)}{\partial t} = -\frac{\partial}{\partial x}\{\mu_1(x)f(x,t)\} + \frac{1}{2}\frac{\partial^2}{\partial x^2}\{\mu_2(x)f(x,t)\}, \tag{5}$$

where

$$\mu_1(x) = \int_{-\infty}^{\infty} \xi W(x;\xi)d\xi = \frac{\langle\delta x\rangle_{\delta t}}{\delta t} = \langle v_x\rangle \tag{6}$$

and

$$\mu_2(x) = \int_{-\infty}^{\infty} \xi^2 W(x;\xi)d\xi = \frac{\langle(\delta x)^2\rangle_{\delta t}}{\delta t}. \tag{7}$$

Equation (5) is the so-called Fokker–Planck equation, which occupies a classic place in the field of Brownian motion and fluctuations.

We now consider a specific system of Brownian particles (of negligible mass), each particle being acted on by a linear restoring force, $F_x = -\lambda x$, and having mobility B in the surrounding medium; the assumption of negligible mass implies that the relaxation time $\tau(= MB)$ of equation (15.3.4) is very small, so the time t here may be regarded as very large in comparison with that τ. The mean viscous force, $-\langle v_x\rangle/B$, is then balanced by the linear restoring force, with the result that

$$-\frac{\langle v_x\rangle}{B} + F_x = 0 \tag{8}$$

and hence

$$\langle v_x\rangle \equiv \mu_1(x) = -\lambda Bx. \tag{9}$$

Next, in view of equation (15.3.10), we have

$$\frac{\langle(\delta x)^2\rangle}{\delta t} \equiv \mu_2(x) = 2BkT; \tag{10}$$

it will be noted that the influence of λ on this quantity is being neglected here. Substituting (9) and (10) into (5), we obtain

$$\frac{\partial f}{\partial t} = \lambda B\frac{\partial}{\partial x}(xf) + BkT\frac{\partial^2 f}{\partial x^2}. \tag{11}$$

[15]Clearly, this assumption limits our analysis to what may be called the "Brownian motion approximation," in which the object under consideration is presumed to be on a very different scale of magnitude than the molecules constituting the environment. It is obvious that if one tries to apply this sort of analysis to "understand" the behavior of molecules *themselves*, one cannot hope for anything but a "crude, semiquantitative" outcome.

Now we apply equation (11) to an "ensemble" of Brownian particles, initially concentrated at the point $x = x_0$. To begin with, we note that, in the absence of the restoring force ($\lambda = 0$), equation (11) reduces to the one-dimensional *diffusion equation*

$$\frac{\partial f}{\partial t} = D\frac{\partial^2 f}{\partial x^2} \quad (D = BkT), \tag{12}$$

which conforms to our earlier results (15.2.19) and (15.3.11). The present derivation shows that the process of diffusion is essentially a "random walk, at the molecular level." In view of equation (15.2.20), the function $f(x,t)$ here would be

$$f(x,t) = \frac{1}{(4\pi Dt)^{1/2}} \exp\left\{-\frac{(x-x_0)^2}{4Dt}\right\}, \tag{13}$$

with

$$\overline{x} = x_0 \quad \text{and} \quad \overline{x^2} = x_0^2 + 2Dt; \tag{14}$$

the last result shows that the mean square distance traversed by the particle(s) increases *linearly* with time, without any upper limit on its value. The restoring force, however, puts a check on the diffusive tendency of the particles. For instance, in the presence of such a force ($\lambda \neq 0$), the terminal distribution f_∞ (for which $\partial f/\partial t = 0$) is determined by the equation

$$\frac{\partial}{\partial x}(xf_\infty) + \frac{kT}{\lambda}\frac{\partial^2 f_\infty}{\partial x^2} = 0, \tag{15}$$

which gives

$$f_\infty(x) = \left(\frac{\lambda}{2\pi kT}\right)^{1/2} \exp\left(-\frac{\lambda x^2}{2kT}\right), \tag{16}$$

with

$$\overline{x} = 0 \quad \text{and} \quad \overline{x^2} = kT/\lambda. \tag{17}$$

The last result agrees with the fact that the mean square value of x must ultimately comply with the *equipartition theorem*, namely $\overline{(\frac{1}{2}\lambda x^2)}_\infty = \frac{1}{2}kT$. From the point of view of equilibrium statistical mechanics, if we regard Brownian particles with kinetic energy $p_x^2/2m$ and potential energy $\frac{1}{2}\lambda x^2$ as *loosely coupled* to a thermal environment at temperature T, then we may directly write

$$f_{eq}(x,p_x)dxdp_x \propto e^{-(p_x^2/2m+\lambda x^2/2)/kT}dxdp_x. \tag{18}$$

On integration over p_x, expression (18) leads directly to the distribution function (16).

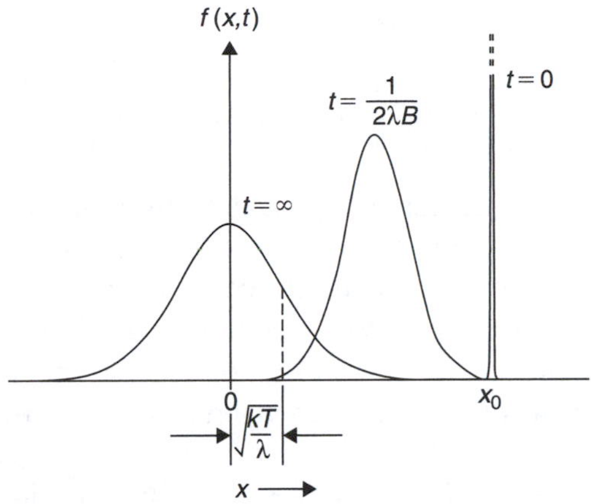

FIGURE 15.5 The distribution function (19) at times $t = 0$, $t = 1/(2\lambda B)$, and $t = \infty$.

The general solution of equation (11), relevant to the ensemble under consideration, is given by

$$f(x,t) = \left\{\frac{\lambda}{2\pi kT(1-e^{-2\lambda Bt})}\right\}^{1/2} \exp\left\{-\frac{\lambda(x - x_0 e^{-\lambda Bt})^2}{2kT(1-e^{-2\lambda Bt})}\right\}, \tag{19}$$

with

$$\overline{x} = x_0 e^{-\lambda Bt} \quad \text{and} \quad \overline{x^2} = x_0^2 e^{-2\lambda Bt} + \frac{kT}{\lambda}(1 - e^{-2\lambda Bt}); \tag{20}$$

in the limit $\lambda \to 0$, we recover the purely "diffusive" situation, as described by equations (13) and (14), while for $t \gg (\lambda B)^{-1}$, we approach the "terminal" situation, as described by equations (16) and (17). Figure 15.5 shows the manner in which an ensemble of Brownian particles approaches a state of equilibrium under the combined influence of the restoring force and the molecular bombardment; clearly, the relaxation time of the present process is $\sim (\lambda B)^{-1}$.

A physical system to which the foregoing theory is readily applicable is provided by the oscillating component of a moving-coil galvanometer. Here, we have a coil of wire and a mirror that are suspended by a fine fiber, so they can rotate about a vertical axis. Random, incessant collisions of air molecules with the suspended system produce a succession of torques of *fluctuating* intensity; as a result, the angular position θ of the system continually fluctuates and the system exhibits an *unsteady* zero. This is clearly another example of the Brownian motion! The role of the viscous force in this case is played by the mechanism of air damping (or, else, electromagnetic damping) of the galvanometer, while the restoring torque, $N_\theta = -c\theta$, arises from the torsional properties of the fiber. In equilibrium, we expect that

$$\overline{\left(\frac{1}{2}c\theta^2\right)} = \frac{1}{2}kT, \quad \text{that is,} \quad \overline{\theta^2} = \frac{kT}{c}; \tag{21}$$

compare this to equation (17). An experimental determination of the mean square deflection, $\overline{\theta^2}$, of such a system was made by Kappler (1931) who, in turn, applied his results to derive, with the help of equation (21), an empirical value for the Boltzmann constant k (or, for that matter, the Avogadro number N_A). The system used by Kappler had a moment of inertia $I = 4.552 \times 10^{-4}\,\text{g cm}^2$ and a time period of oscillation $\tau = 1379\,\text{s}$; accordingly, the constant c of the restoring torque had a value given by the formula $\tau = 2\pi(I/c)^{1/2}$, so that

$$c = 4\pi^2(I/\tau^2) = 9.443 \times 10^{-9}\,\text{g cm}^2\text{s}^{-2}/\text{rad}.$$

The observed value of $\overline{\theta^2}$, at a temperature of 287.1 K, was 4.178×10^{-6}. Substituting these numbers in (21), Kappler obtained: $k = 1.374 \times 10^{-16}$ erg K^{-1}. And, since the gas constant R is equal to 8.31×10^7 erg $\text{K}^{-1}\text{mole}^{-1}$, he obtained for the Avogadro number: $N_A = R/k = 6.06 \times 10^{23}\,\text{mole}^{-1}$.

One might expect that by suspending the mirror system in an "evacuated" casing the fluctuations caused by the collisions of the air molecules could be severely reduced. This is not true because even at the lowest possible pressures there still remain a tremendously large number of molecules in the system that keep the Brownian motion "alive." The interesting part of the story, however, is that the mean square deflection of the system, caused by molecular bombardment, is not at all affected by the density of the molecules; for a system *in equilibrium*, it is determined solely by the temperature. This situation is depicted, rather dramatically, in Figure 15.6 where we have two traces of oscillations of the mirror system, the upper one having been taken at the atmospheric pressure and the lower one at a pressure of 10^{-4} mm of mercury. The root-mean-square deviation is very nearly the same in the two cases! Nevertheless, one does note a difference of "quality" between the two traces that relates to the "frequency spectrum" of the fluctuations and arises for the following reason. When the density of the surrounding gas is relatively high, the molecular impulses come in rapid succession, with the result that the *individual* deflections of the system are large in number but small in magnitude. As the pressure is lowered, the time intervals between successive impulses become longer, making the *individual* deflections smaller in number but larger in magnitude. However, the overall deflection, observed over a long interval of time, remains essentially the same.

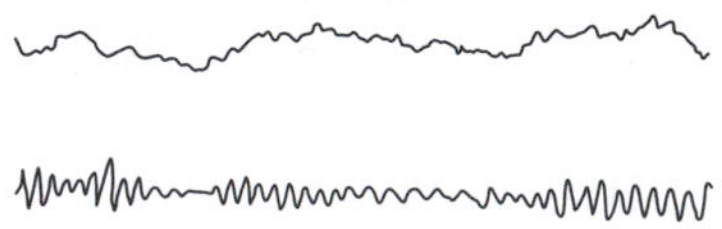

FIGURE 15.6 Two traces of the thermal oscillations of a mirror system suspended in air; the upper trace was taken at the atmospheric pressure, the lower one at a pressure of 10^{-4} mm of mercury.

15.5 Spectral analysis of fluctuations: the Wiener–Khintchine theorem

We have already made reference to the (spectral) quality of a fluctuation pattern. Referring once again to the patterns shown in Figure 15.6, we note that, even though the mean square fluctuation of the variable θ is the same in the two cases, the second pattern is far more "jagged" than the first; in other words, the high-frequency components are far more prominent in the second pattern. At the same time, there is a lot more "predictability" in the first pattern (insofar as it is represented by a much smoother curve); in other words, the *correlation function*, or the *memory function*, $K(s)$ of the first pattern extends over much larger values of s. In fact, these two aspects of a fluctuation process, namely its time-dependence and its frequency spectrum, are very closely related to one another. And the most natural course for studying this relationship is to carry out a Fourier analysis of the given process.

For this study we consider only those variables, $y(t)$, whose *mean square value*, $\langle y^2(t)\rangle$, has already attained an equilibrium, or stationary, value:

$$\langle y^2(t)\rangle = \text{const.} \tag{1}$$

Such a variable is said to be *statistically stationary*. As an example of such a variable, we may recall the velocity $v(t)$ of a "free" Brownian particle at times t much larger than the relaxation time τ, see equation (15.3.29), or the displacement $x(t)$ of a Brownian particle moving under the influence of a restoring force ($F_x = -\lambda x$) at times t much larger than $(\lambda B)^{-1}$, see equation (15.4.20). Now, if the variable $y(t)$ were *strictly periodic* (and hence *completely predictable*), with a time period $T = 1/f_0$, then we could write

$$y(t) = a_0 + \sum_{n=1}^{\infty} a_n \cos(2\pi n f_0 t) + \sum_{n=1}^{\infty} b_n \sin(2\pi n f_0 t), \tag{2}$$

where

$$a_0 = \frac{1}{T}\int_0^T y(t)dt, \tag{3}$$

$$a_n = \frac{2}{T}\int_0^T y(t)\cos(2\pi n f_0 t)dt, \tag{4}$$

and

$$b_n = \frac{2}{T}\int_0^T y(t)\sin(2\pi n f_0 t)dt; \tag{5}$$

in this case, the coefficients a and b would be completely known and would define, with no uncertainty, the frequency spectrum of the variable $y(t)$. If, on the other hand, the given variable is *more* or *less* a random function of time, then the coefficients a and b would themselves be statistical in nature. To apply the concept of periodicity to such a function, we must take the "time interval of repetition" to be infinitely large, that is, we let $f_0 \to 0$.

In the proposed limit, equation (3) would read

$$a_0 = \operatorname*{Lim}_{T\to\infty} \frac{1}{T}\int_0^T y(t)dt \equiv \langle y(t)\rangle; \tag{6}$$

thus, the coefficient a_0, which represents the mean (or d.c.) value of the variable y, may be determined *either* by taking a time average (over a sufficiently long interval) of the variable or by taking an ensemble average (at any instant of time t). For convenience, and without loss of generality, we take $a_0 = 0$; in other words, we assume that from the actual values of the variable $y(t)$ its mean value, $\langle y(t)\rangle$, has already been subtracted.[16]

Taking the ensemble average of equations (4) and (5), we obtain, for all n,

$$\langle a_n\rangle = \langle b_n\rangle = 0. \tag{7}$$

However, by taking the ensemble average of equation (2) *squared*, we obtain

$$\begin{aligned}\langle y^2(t)\rangle &= \sum_n \frac{1}{2}\langle a_n^2\rangle + \sum_n \frac{1}{2}\langle b_n^2\rangle \\ &= \sum_n \frac{1}{2}\left\{\langle a_n^2\rangle + \langle b_n^2\rangle\right\} = \text{const.}\end{aligned} \tag{8}$$

The term $\frac{1}{2}\{\langle a_n^2\rangle + \langle b_n^2\rangle\}$ represents the respective "share," belonging to the frequency nf_0, in the total, time-independent value of the quantity $\langle y^2(t)\rangle$. Now, in view of the randomness of the phases of the various components, we have, for all n, $\langle a_n^2\rangle = \langle b_n^2\rangle$; consequently, equation (8) may be written as

$$\langle y^2\rangle = \sum_n \langle a_n^2\rangle \simeq \int_0^\infty w(f)df, \tag{9}$$

where

$$\langle a_n^2\rangle = w(nf_0)\Delta(nf_0), \quad \text{that is,} \quad w(nf_0) = \frac{1}{f_0}\langle a_n^2\rangle; \tag{10}$$

the function $w(f)$ defines the *power spectrum* of the variable $y(t)$.

We shall now show that the power spectrum $w(f)$ of the fluctuating variable $y(t)$ is completely determined by its autocorrelation function $K(s)$. For this, we make use of

[16]Obviously, this does not affect the spectral quality of the fluctuations, except that now we do not have a component with frequency zero. To represent the actual situation, one may have to add, to the resulting spectrum, a suitably weighted $\delta(f)$-term.

equation (4), which gives

$$\langle a_n^2 \rangle = 4f_0^2 \int_0^{1/f_0} \int_0^{1/f_0} \langle y(t_1)y(t_2) \rangle \cos(2\pi n f_0 t_1) \cos(2\pi n f_0 t_2) dt_1 dt_2. \tag{11}$$

Changing over to the variables

$$S = \frac{1}{2}(t_1 + t_2) \quad \text{and} \quad s = (t_2 - t_1),$$

and remembering that the interval T over which the integrations extend is much larger than the duration over which the "memory" of the variable y lasts, we obtain

$$\langle a_n^2 \rangle \simeq 2f_0^2 \int_{S=0}^{1/f_0} \int_{s=-\infty}^{\infty} K(s)\{\cos(2\pi n f_0 s) + \cos(4\pi n f_0 S)\} dS ds; \tag{12}$$

compare this to the steps that led us from equations (15.3.22) to (15.3.25) and (15.3.26). The second part of the integral in (12) vanishes on integration over S; the first part then gives

$$\langle a_n^2 \rangle = 4f_0 \int_0^{\infty} K(s) \cos(2\pi n f_0 s) ds. \tag{13}$$

Comparing (13) with (10), we obtain the desired formula

$$w(f) = 4 \int_0^{\infty} K(s) \cos(2\pi f s) ds. \tag{14}$$

Taking the inverse of (14), we obtain

$$K(s) = \int_0^{\infty} w(f) \cos(2\pi f s) df. \tag{15}$$

For $s = 0$, formula (15) yields the important relationship

$$K(0) = \int_0^{\infty} w(f) df = \langle y^2 \rangle; \tag{16}$$

see equation (9) as well as the definition of the autocorrelation function of the variable y, namely $K(s) = \langle y(t_1) y(t_1 + s) \rangle$. Equations (14) and (15), which connect the complementary functions $w(f)$ and $K(s)$, constitute a theorem that goes after the names of Wiener (1930) and Khintchine (1934).

We shall now look at some special cases of the variable $y(t)$ to illustrate the use of the *Wiener–Khintchine theorem.*

Case 1

If the given variable $y(t)$ is extremely irregular, and hence unpredictable, then its correlation function $K(s)$ would extend over a negligibly small range of the time interval s.[17] We may then write

$$K(s) = c\delta(s). \tag{17a}$$

Equation (14) then gives

$$w(f) = 2c \quad \text{for all } f. \tag{17b}$$

A spectrum in which the distribution (of power) over different frequencies is uniform is known as a "flat" or a "white" spectrum. We note, however, that if the uniformity of distribution were literally true for all frequencies, from 0 to ∞, then the integral in (16), which is identically equal to $\langle y^2 \rangle$, would diverge! We, therefore, expect that, in any realistic situation, the correlation function $K(s)$ will not be as sharply peaked as in (17a). Typically, $K(s)$ will extend over a small range, $O(\sigma)$, of the variable s, which in turn will define a "frequency zone," with $f = O(1/\sigma)$, such that the function $w(f)$ would undergo a change of character as f passes through this zone; toward lower frequencies, $w(f) \to \text{const.} \neq 0$, while toward higher frequencies, $w(f) \to \text{const.} = 0$. One possible representation of this situation is shown in Figure 15.7 where we have taken, rather arbitrarily,

$$K(s) = K(0)\frac{\sin(as)}{as} \quad (a > 0), \tag{18a}$$

for which

$$w(f) = \begin{cases} \dfrac{2\pi}{a}K(0) & \text{for } f < \dfrac{a}{2\pi} \\[2ex] 0 & \text{for } f > \dfrac{a}{2\pi}. \end{cases} \tag{18b}$$

In the limit $a \to \infty$, equations (18) reduce to (17), with $c = \pi a^{-1} K(0)$.

Case 2

On the other hand, if the variable $y(t)$ is extremely regular, and hence predictable, then its correlation function would extend over large values of s; its power spectrum would then appear in the form of "peaks," located at certain "characteristic frequencies" of the variable. In the simplest case of a *monochromatic* variable, with characteristic frequency f^*, the correlation function would be

$$K(s) = K(0)\cos(2\pi f^* s), \tag{19a}$$

[17] This is essentially true of the rapidly fluctuating force $\boldsymbol{F}(t)$ experienced by a Brownian particle due to the incessant molecular impulses received by it.

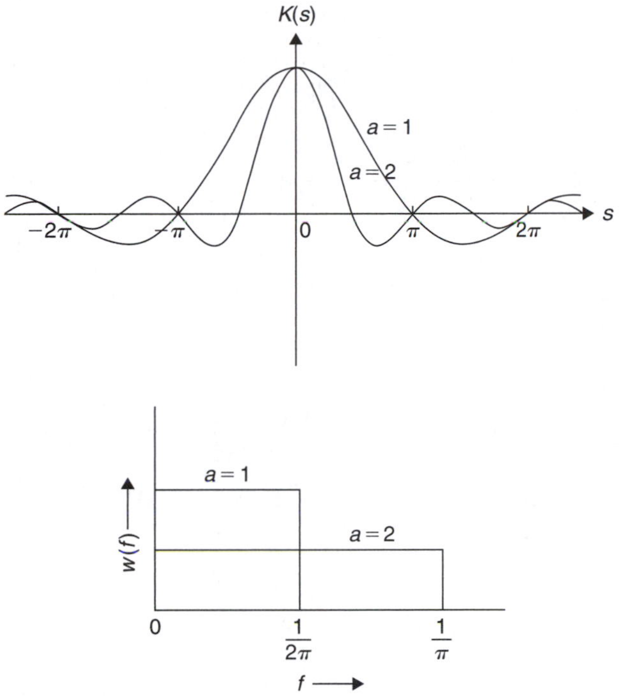

FIGURE 15.7 The autocorrelation function $K(s)$ and the power distribution function $w(f)$ of a given variable $y(t)$; the parameter a appears in terms of an arbitrary unit of (time)$^{-1}$.

for which

$$w(f) = K(0)\delta(f - f^*); \tag{19b}$$

see Figure 15.8. A very special case arises when $f^* = 0$; then, both $y(t)$ and $K(s)$ are constant in value, and the function $w(f)$ is peaked at the d.c. frequency $f = 0$.

Case 3

If the variable $y(t)$ represents a signal that arises from, or has been filtered through, a lightly damped tuned circuit (a "narrowband" filter), then its power will be distributed over a "hump" around the mean frequency f^*. The function $K(s)$ will then appear in the nature of an "attenuated" function whose time scale, σ, is determined by the width, Δf, of the hump in the power spectrum. A situation of this kind is shown in Figure 15.9.

The relevance of spectral analysis to the problem of the actual observation of a fluctuating variable is best brought out by examining the power spectrum of the velocity $\boldsymbol{v}(t)$ of a Brownian particle. Considering the x-component alone, the autocorrelation function $K_{v_x}(s)$, or simply $K(s)$, in this case is given by

$$K(s) = \frac{kT}{M} e^{-|s|/\tau} \quad (\tau = MB); \tag{20}$$

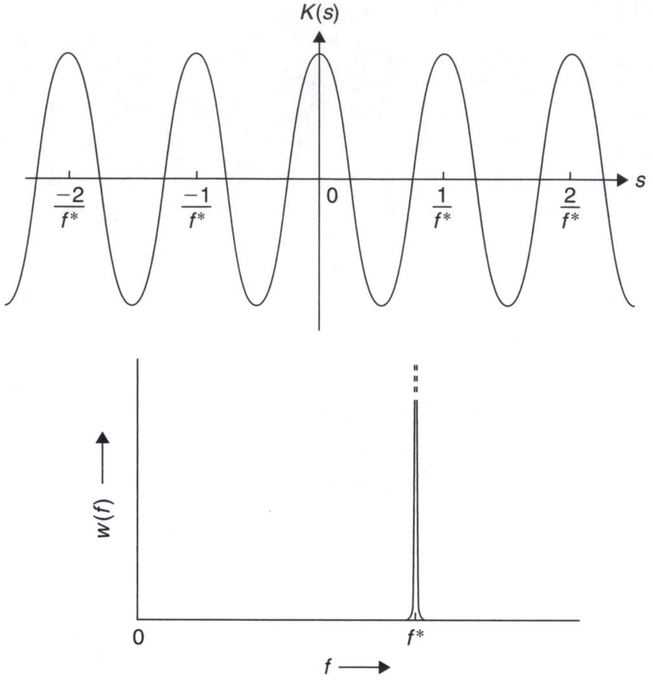

FIGURE 15.8 The autocorrelation function $K(s)$ and the power distribution function $w(f)$ of a monochromatic variable $y(t)$, with characteristic frequency f^*.

see equation (15.6.10). The power spectrum $w(f)$ is then given by the expression

$$w(f) = \frac{4kT}{M}\int_0^\infty e^{-s/\tau}\cos(2\pi fs)ds = \frac{4kT\tau}{M}\frac{1}{1+(2\pi f\tau)^2}, \tag{21}$$

which indeed satisfies the relationship

$$\begin{aligned}\int_0^\infty w(f)df &= \frac{2kT}{\pi M}\tan^{-1}(2\pi f\tau)\Big|_0^\infty \\ &= \frac{kT}{M} = \langle v_x^2\rangle,\end{aligned} \tag{22}$$

in agreement with the *equipartition theorem* (as applied to a single component of the velocity $\boldsymbol{v}$). For $f \ll \tau^{-1}$, the power distribution is practically independent of f, which implies a practically "white" spectrum, with

$$w(f) \simeq \frac{4kT\tau}{M} = 4BkT. \tag{23}$$

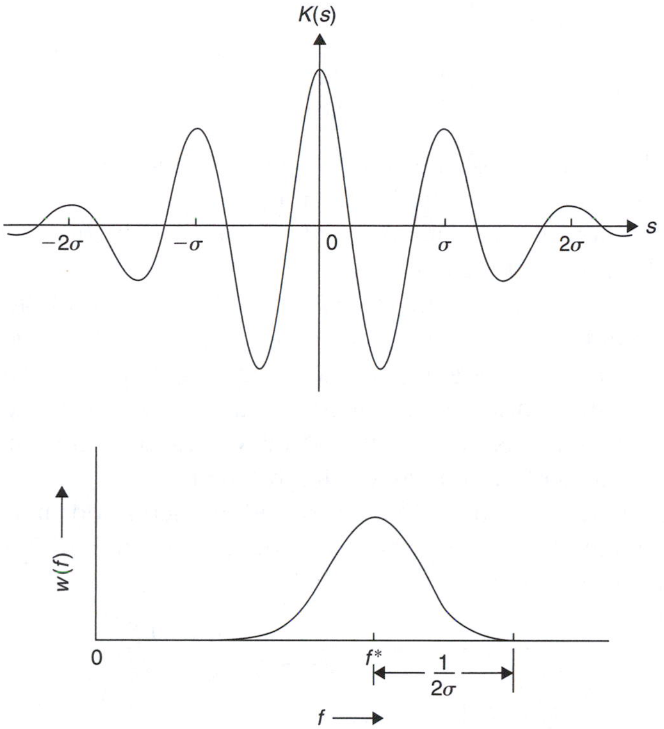

FIGURE 15.9 The autocorrelation function $K(s)$ and the power distribution function $w(f)$ of a variable that has been filtered through a lightly damped tuned circuit, with mean frequency f^* and width $\Delta f \sim (1/\sigma)$.

We can then write for the velocity fluctuations in the frequency range $(f, f + \Delta f)$, with $f \ll \tau^{-1}$,

$$\langle \Delta v_x^2 \rangle_{(f, f+\Delta f)} \simeq w(f)\Delta f \simeq (4BkT)\Delta f. \tag{24}$$

In general, our measuring instrument (or the eye, in the case of a visual examination of the particle) has a *finite* response time τ_0, as a consequence of which it is unable to respond to frequencies larger than, say, τ_0^{-1}. The *observed* fluctuation is then given by the "pruned" expression

$$\langle v_x^2 \rangle_{\text{obs}} \simeq \int_0^{1/\tau_0} w(f) df = \frac{2kT}{\pi M} \tan^{-1}\left(2\pi \frac{\tau}{\tau_0}\right), \tag{25}$$

instead of the "full" expression (22). In a typical case, the mass of the Brownian particle $M \sim 10^{-12}$g, its diameter $2a \sim 10^{-4}$ cm, and the coefficient of viscosity of the fluid $\eta \sim 10^{-2}$ poise, so that the relaxation time $\tau = M/(6\pi\eta a) \sim 10^{-7}$ seconds. However, the response time τ_0, in the case of visual observation, is of the order of 10^{-1} s; clearly, $\tau/\tau_0 \sim 10^{-6} \ll 1$.

Equation (25) then reduces to[18]

$$\langle v_x^2 \rangle_{\text{obs}} \simeq \frac{4kT\tau}{M\tau_0} \ll \frac{kT}{M}; \tag{26}$$

thus, in view of the *finiteness* of the response time τ_0, the observed root-mean-square velocity of the Brownian particle will be down by a factor of $2(\tau/\tau_0)^{1/2} \sim 10^{-3}$; numerically, this takes us down from a root-mean-square velocity which, at room temperatures, is $\sim 10^{-1}$ cm/s to a value $\sim 10^{-4}$ cm/s.

It is gratifying to note that the outcome of actual observations of Brownian particles is in complete agreement with the latter result; for a more detailed analysis of this question, see MacDonald (1950). The foregoing discussion highlights the fact that, in the process of observing a fluctuating variable, our measuring instrument picks up signals over only a *limited* range of frequencies (as determined by the response time of the instrument); signals belonging to higher frequencies are simply left out.

The theory of this section can be readily applied to fluctuations in the motion of electrons in an (L, R) circuit. Corresponding to equations (21) through (24), we now have for fluctuations in the electric current I

$$w(f) = \frac{4kT\tau'}{L}\frac{1}{1+(2\pi f\tau')^2} \quad \left(\tau' = \frac{L}{R}\right), \tag{27}$$

so that

$$\int_0^\infty w(f)df = \frac{kT}{L} = \langle I^2 \rangle, \tag{28}$$

in agreement with the *equipartition* theorem: $\langle \frac{1}{2}LI^2 \rangle = \frac{1}{2}kT$. For $f \ll 1/\tau'$, equation (27) reduces to

$$w(f) \simeq \frac{4kT}{R}, \tag{29}$$

which, once again, implies "white" noise; accordingly, for low frequencies,

$$\langle \Delta I^2 \rangle_{(f, f+\Delta f)} \simeq w(f)\Delta f \simeq \frac{4kT}{R}\Delta f. \tag{30}$$

Equivalently, we obtain for fluctuations in the voltage

$$\langle \Delta V^2 \rangle_{(f, f+\Delta f)} \simeq (4RkT)\Delta f. \tag{31}$$

Equation (31) constitutes the so-called *Nyquist theorem*, which was first discovered empirically by Johnson (1927a,b; 1928) and was later derived by Nyquist (1927–1928) on the basis

[18]The fluctuations constituting this result belong *entirely* to the region of the "white" noise, with $\Delta f = 1/\tau_0$; see equation (24), with $B = \tau/M$.

of an argument involving the second law of thermodynamics and the exchange of energy between two resistances in thermal equilibrium.[19]

15.6 The fluctuation–dissipation theorem

In Section 15.3 we obtained a result of considerable importance, namely

$$\frac{1}{B} \equiv \frac{M}{\tau} = \frac{M^2}{6kT}C = \frac{M^2}{6kT}\int_{-\infty}^{\infty} K_{\boldsymbol{A}}(s)ds$$

$$= \frac{1}{6kT}\int_{-\infty}^{\infty} K_{\boldsymbol{F}}(s)ds; \tag{1}$$

see equations (15.3.4), (15.3.26), and (15.3.28). Here, $K_{\boldsymbol{A}}(s)$ and $K_{\boldsymbol{F}}(s)$ are, respectively, the autocorrelation functions of the fluctuating acceleration $\boldsymbol{A}(t)$ and the fluctuating force $\boldsymbol{F}(t)$ experienced by the Brownian particle:

$$K_{\boldsymbol{A}}(s) = \langle \boldsymbol{A}(0) \cdot \boldsymbol{A}(s) \rangle = \frac{1}{M^2}\langle \boldsymbol{F}(0) \cdot \boldsymbol{F}(s) \rangle = \frac{1}{M^2}K_{\boldsymbol{F}}(s). \tag{2}$$[20]

Equation (1) establishes a fundamental relationship between the coefficient, $1/B$, of the "averaged-out" part of the total force $\mathcal{F}(t)$ experienced by the Brownian particle due to the impacts of the fluid molecules and the statistical character of the "fluctuating" part, $F(t)$, of that force; see Langevin's equation (15.3.2). In other words, it relates the coefficient of viscosity of the fluid, which represents *dissipative* forces operating in the system, with the temporal character of the molecular *fluctuations*; the content of equation (1) is, therefore, referred to as a *fluctuation–dissipation theorem*.

The most striking feature of this theorem is that it relates, in a fundamental manner, the fluctuations of a physical quantity pertaining to the *equilibrium state* of a given system to a dissipative process which, in practice, is realized only when the system is subject to an external force that drives it *away from equilibrium*. Consequently, it enables us to determine the *nonequilibrium properties* of the given system on the basis of a knowledge of the thermal fluctuations occurring in the system when the system is in one of its *equilibrium*

[19]We note that the foregoing results are essentially equivalent to Einstein's original result for charge fluctuations in a conductor, namely

$$\langle \delta q^2 \rangle_t = \tfrac{2kT}{R}t;$$

compare, as well, the Brownian-particle result: $\langle x^2 \rangle_t = 2BkTt$.

[20]We note that the functions $K_{\boldsymbol{A}}(s)$ and $K_{\boldsymbol{F}}(s)$, which are nonzero only for $s = O(\tau^*)$, see equation (15.3.21), may, for certain purposes, be written as

$$K_{\boldsymbol{A}}(s) = \tfrac{6kT}{M^2B}\delta(s) \quad \text{and} \quad K_{\boldsymbol{F}}(s) = \tfrac{6kT}{B}\delta(s).$$

In this form, the functions are nonzero only for $s = 0$.

states! For an expository account of the fluctuation–dissipation theorem, the reader may refer to Kubo (1966).

At this stage we recall that in equation (15.3.11) we obtained a relationship between the *diffusion coefficient* D and the *mobility* B, namely $D = BkT$. Combining this with equation (1), we get

$$\frac{1}{D} = \frac{1}{6(kT)^2} \int_{-\infty}^{\infty} K_F(s) ds. \tag{3}$$

Now, the diffusion coefficient D can be related directly to the autocorrelation function $K_{\boldsymbol{v}}(s)$ of the fluctuating variable $\boldsymbol{v}(t)$. For this, one starts with the observation that, by definition,

$$\boldsymbol{r}(t) = \int_0^t \boldsymbol{v}(u) du, \tag{4}$$

which gives

$$\langle r^2(t) \rangle = \int_0^t \int_0^t \langle \boldsymbol{v}(u_1) \cdot \boldsymbol{v}(u_2) \rangle du_1 du_2. \tag{5}$$

Proceeding in the same manner as for the integral in equation (15.3.22), one obtains

$$\langle r^2(t) \rangle = \int_0^{t/2} dS \int_{-2S}^{+2S} K_{\boldsymbol{v}}(s) ds + \int_{t/2}^{t} dS \int_{-2(t-S)}^{+2(t-S)} K_{\boldsymbol{v}}(s) ds; \tag{6}$$

compare this to equation (15.3.24).

The function $K_{\boldsymbol{v}}(s)$ can be determined by making use of expression (15.3.14) for $\boldsymbol{v}(t)$ and following exactly the same procedure as for determining the quantity $\langle v^2(t) \rangle$, which is nothing but the maximal value, $K_{\boldsymbol{v}}(0)$, of the desired function. Thus, one obtains

$$K_{\boldsymbol{v}}(s) = \begin{cases} v^2(0) e^{-(2t+s)/\tau} + \dfrac{3kT}{M} e^{-s/\tau}(1 - e^{-2t/\tau}) & \text{for} \quad s > 0 \quad (7) \\ v^2(0) e^{-(2t+s)/\tau} + \dfrac{3kT}{M} e^{s/\tau}(1 - e^{-2(t+s)/\tau}) & \text{for} \quad s < 0; \quad (8) \end{cases}$$

compare these results to equation (15.3.27). It is easily seen that formulae (7)and (8) can be combined into a single one, namely

$$K_{\boldsymbol{v}}(s) = v^2(0) e^{-|s|/\tau} + \left\{ \frac{3kT}{M} - v^2(0) \right\} (e^{-|s|/\tau} - e^{-(2t+s)/\tau}) \quad \text{for all } s; \tag{9}$$

compare this to equation (15.3.29). In the case of a "stationary ensemble,"

$$K_v(s) = \frac{3kT}{M} e^{-|s|/\tau}, \tag{10}$$

which is consistent with property (15.3.20). It should be noted that the time scale for the correlation function $K_v(s)$ is provided by the *relaxation time* τ of the Brownian motion, which is many orders of magnitude larger than the *characteristic time* τ^* that provides the time scale for the correlation functions $K_A(s)$ and $K_F(s)$.

It is now instructive to verify that the substitution of expression (10) into (6) leads to formula (15.3.7) for $\langle r^2 \rangle$, while the substitution of the more general expression (9) leads to formula (15.3.31); see Problem 15.17. In either case,

$$\langle r^2 \rangle \xrightarrow[t \gg \tau]{} 6Dt. \tag{11}$$

In the same limit, equation (6) reduces to

$$\langle r^2 \rangle \simeq \int_0^t dS \int_{-\infty}^{\infty} K_v(s) ds = t \int_{-\infty}^{\infty} K_v(s) ds. \tag{12}$$

Comparing the two results, we obtain the desired relationship:

$$D = \frac{1}{6} \int_{-\infty}^{\infty} K_v(s) ds. \tag{13}$$

In passing, we note, from equations (3) and (13), that

$$\int_{-\infty}^{\infty} K_v(s) ds \int_{-\infty}^{\infty} K_F(s) ds = (6kT)^2; \tag{14}$$

see also Problem 15.7.

It is not surprising that the equations describing a fluctuation–dissipation theorem can be adapted to any situation that involves a dissipative mechanism. For instance, fluctuations in the motion of electrons in an electric resistor give rise to a "spontaneous" thermal e.m.f., which may be denoted as $\boldsymbol{\mathcal{B}}(t)$. In the spirit of the Langevin theory, this e.m.f. may be split into two parts: (i) an "averaged-out" part, $-\boldsymbol{RI}(t)$, which represents the resistive (or dissipative) aspect of the situation, and (ii) a "rapidly fluctuating" part, $\boldsymbol{V}(t)$, which, over long intervals of time, averages out to zero. The "spontaneous" current in the resistor is then given by the equation

$$L\frac{d\boldsymbol{I}}{dt} = -R\boldsymbol{I} + \boldsymbol{V}(t); \quad \langle \boldsymbol{V}(t) \rangle = 0. \tag{15}$$

Comparing this with the *Langevin equation* (15.3.2) and pushing the analogy further, we infer that there exists a direct relationship between the resistance R and the temporal character of the fluctuations in the variable $\boldsymbol{V}(t)$. In view of equations (1) and (13), this relationship would be

$$R = \frac{1}{6kT}\int_{-\infty}^{\infty}\langle \boldsymbol{V}(0)\cdot\boldsymbol{V}(s)\rangle ds \tag{16}$$

or, equivalently,

$$\frac{1}{R} = \frac{1}{6kT}\int_{-\infty}^{\infty}\langle \boldsymbol{I}(0)\cdot\boldsymbol{I}(s)\rangle ds. \tag{17}$$

A generalization of the foregoing result has been given by Kubo (1957, 1959); see, for instance, Problem 6.19 in Kubo (1965), or Section 23.2 of Wannier (1966). On generalization, the *electric current density* $\boldsymbol{j}(t)$ is given by the expression

$$j_i(t) = \sum_l \int_{-\infty}^{t} E_l(t')\Phi_{li}(t-t')dt' \quad (i,l = x,y,z); \tag{18}$$

here, $\boldsymbol{E}(t)$ denotes the applied electric field while

$$\Phi_{li}(s) = \frac{1}{kT}\langle j_l(0)j_i(s)\rangle. \tag{19}$$

Clearly, the quantities $kT\Phi_{li}(s)$ are the components of the *autocorrelation tensor* of the fluctuating vector $\boldsymbol{j}(t)$. In particular, if we consider the static case $\boldsymbol{E} = (E,0,0)$, we obtain for the *conductivity* of the system

$$\begin{aligned}\sigma_{xx} \equiv \frac{j_x}{E} &= \int_{-\infty}^{t}\Phi_{xx}(t-t')dt' = \int_{0}^{\infty}\Phi_{xx}(s)ds \\ &= \frac{1}{2kT}\int_{-\infty}^{\infty}\langle j_x(0)j_x(s)\rangle ds,\end{aligned} \tag{20}$$

which may be compared with equation (17). If, on the other hand, we take $\boldsymbol{E} = (E\cos\omega t, 0, 0)$, we obtain instead

$$\sigma_{xx}(\omega) = \frac{1}{2kT}\int_{-\infty}^{\infty}\langle j_x(0)j_x(s)\rangle e^{-i\omega s}ds. \tag{21}$$

Taking the inverse of (21), we get

$$\langle j_x(0) j_x(s)\rangle = \frac{kT}{\pi}\int_{-\infty}^{\infty} \sigma_{xx}(\omega) e^{i\omega s} d\omega. \tag{22}$$

If we now assume that $\sigma_{xx}(\omega)$ does not depend on ω (and may, therefore, be denoted by the simpler symbol σ), then

$$\langle j_x(0) j_x(s)\rangle = (2kT\sigma)\delta(s); \tag{23}$$

see footnote 20. A reference to equations (15.5.17) shows that, in the present approximation, thermal fluctuations in the electric current are charaterized by a "white" noise.

15.6.A Derivation of the fluctuation–dissipation theorem from linear response theory

In this section we will show that the nonequilibrium response of a thermodynamic system to a small driving force is very generally related to the time-dependence of equilibrium fluctuations. In hindsight, this is not too surprising since natural fluctuations about the equilibrium state also induce small deviations of observables from their average values. The response of the system to these natural fluctuations should be the same as the response of the system to deviations from the equilibrium state as caused by small driving forces; see Martin (1968), Forster (1975), and Mazenko (2006).

Let us compute the time-dependent changes to an observable A caused by a small time-dependent external applied field $h(t)$ that couples linearly to some observable B. The Hamiltonian for the system then becomes

$$H(t) = H_0 - h(t)B, \tag{24}$$

where H_0 is the unperturbed Hamiltonian in the equilibrium state. Remarkably, the calculation for determining the nonequilibrium response to the driving field is easiest using the quantum-mechanical density matrix approach developed in Section 5.1. The equilibrium density matrix is given by

$$\hat{\rho}_{\text{eq}} = \frac{\exp(-\beta H_0)}{\text{Tr}\left(\exp(-\beta H_0)\right)}, \tag{25}$$

where equilibrium averages involve traces over the density matrix:

$$\langle A\rangle_{\text{eq}} = \text{Tr}\left(A\hat{\rho}_{\text{eq}}\right). \tag{26}$$

When the Hamiltonian includes a small time-dependent field $h(t)$, then this additional term drives the system slightly out of equilibrium. We will assume that the field was zero in the distant past so the system was initially in the equilibrium state defined by the

Hamiltonian H_0. We then turn on the field and measure the time-dependent deviations of the observable A from its equilibrium value. The small time-dependent applied field $h(t)$ induces a small change to the density matrix

$$\hat{\rho}(t) = \hat{\rho}_{\text{eq}} + \delta\hat{\rho}(t). \tag{27}$$

The equation of motion of the density matrix in equation (5.1.10) gives

$$\frac{\partial\hat{\rho}}{\partial t} = \frac{\partial\delta\hat{\rho}}{\partial t} = \frac{1}{i\hbar}\left[H,\hat{\rho}(t)\right] \approx \frac{1}{i\hbar}\left(\left[H_0,\delta\hat{\rho}\right] - h(t)\left[B,\hat{\rho}_{\text{eq}}\right]\right), \tag{28}$$

where $[,]$ denotes the commutator. Since we are considering only the linear response of the system to the applied field, we have ignored the higher-order term proportional to $h(t)\left[B,\delta\hat{\rho}\right]$. Solving (28) for the time-dependent change to the density matrix, we get

$$\delta\hat{\rho}(t) = \frac{i}{\hbar}\int_{-\infty}^{t} h(t')\exp\left(\frac{-iH_0(t-t')}{\hbar}\right)\left[B,\hat{\rho}_{\text{eq}}\right]\exp\left(\frac{iH_0(t-t')}{\hbar}\right)dt'. \tag{29}$$

This form uses the interaction representation in which operators evolve in time due to the unperturbed Hamiltonian H_0. We can use the change in the density matrix at time t to calculate the change in the observable A compared to its equilibrium value, namely

$$\langle\delta A(t)\rangle = \langle A(t)\rangle - \langle A\rangle_{\text{eq}} = \text{Tr}\left(A\hat{\rho}(t)\right) - \text{Tr}\left(A\hat{\rho}_{\text{eq}}\right) = \text{Tr}\left(A\delta\hat{\rho}(t)\right). \tag{30}$$

Using the cyclic property of traces, $\text{Tr}\,(QRS) = \text{Tr}\,(SQR)$, we find that $\langle\delta A(t)\rangle$ is the convolution of a response function with the applied field:

$$\langle\delta A(t)\rangle = \frac{i}{\hbar}\int_{-\infty}^{t}\left\langle\left[A(t),B(t')\right]\right\rangle_{\text{eq}} h(t')dt'. \tag{31}$$

Note that this nonequilibrium response function of the system to the driving force depends on the *equilibrium* average of the commutator of the observables A and B at different times. The effect of the field on the observable A is causal since $\langle\delta A(t)\rangle$ depends only on the applied field at earlier times. Since the relation is linear, time-translationally invariant, and causal, the Fourier spectra of $\langle\delta A\rangle$ and h, namely

$$\delta\hat{A}(\omega) = \int_{-\infty}^{\infty}\langle\delta A(t)\rangle\, e^{i\omega t}dt, \text{ and} \tag{32a}$$

$$\hat{h}(\omega) = \int_{-\infty}^{\infty} h(t)e^{i\omega t}dt, \tag{32b}$$

are related by

$$\delta\hat{A}(\omega) = \left(\hat{\chi}'_{AB}(\omega) + i\hat{\chi}''_{AB}(\omega)\right)\hat{h}(\omega), \tag{33}$$

where $\hat{\chi}''_{AB}(\omega)$ is given by

$$\hat{\chi}''_{AB}(\omega) = \frac{1}{2\hbar}\int_{-\infty}^{\infty} \langle [A(t), B(0)] \rangle_{\text{eq}}\, e^{i\omega t} dt. \tag{34}$$

The quantity $\hat{\chi}'_{AB}(\omega)$ is given by the Kramers–Kronig relation using the principal part P of an infinite integral over $\hat{\chi}''_{AB}(\omega)$:

$$\hat{\chi}'_{AB}(\omega) = P\int_{-\infty}^{\infty} \frac{\hat{\chi}''_{AB}(\omega')}{\omega' - \omega}\frac{d\omega'}{\pi}. \tag{35}$$

If A and B are the same operator, then $\hat{\chi}'_{AA}(\omega)$ and $\hat{\chi}''_{AA}(\omega)$ are, respectively, the real and imaginary parts of the response function. If A and B have the same symmetry under time-reversal, $\omega\hat{\chi}''_{AB}(\omega)$ is real and is an even function of ω. For a set of observables A_i, the set of response functions $\omega\hat{\chi}''_{ij}(\omega)$ form a symmetric positive matrix that gives the rate of energy dissipation due to the external driving forces. See Jackson (1999) for a general causality discussion and Martin (1968), Forster (1975), or Mazenko (2006) for the details of this calculation.

Now we consider the *equilibrium* temporal correlations between the fluctuations of the observables A and B, namely

$$G_{AB}(t - t') = \left\langle \delta A(t)\delta B(t')\right\rangle_{\text{eq}}. \tag{36}$$

At equal times, this measures the AB equilibrium fluctuations described in Section 15.1, that is, $G_{AB}(0) = \langle \delta A \delta B\rangle_{\text{eq}}$. The power spectrum of the AB equilibrium fluctuations is defined by

$$S_{AB}(\omega) = \int_{-\infty}^{\infty} G_{AB}(t) e^{i\omega t} dt = \int_{-\infty}^{\infty} \langle \delta A(t)\delta B(0)\rangle_{\text{eq}}\, e^{i\omega t} dt. \tag{37}$$

The similarity between the forms of $S_{AB}(\omega)$ and $\hat{\chi}''_{AB}(\omega)$ in equations (34) and (37) leads to an important relation between the power spectrum and the linear response function, namely the *fluctuation–dissipation theorem*:

$$\hat{\chi}''_{AB}(\omega) = \frac{1}{2\hbar}\left(1 - e^{-\beta\hbar\omega}\right) S_{AB}(\omega). \tag{38}$$

The power spectrum $S_{AB}(\omega)$ measures equilibrium fluctuations whereas $\hat{\chi}''_{AB}(\omega)$ is proportional to the average rate of power dissipation that results from the time-varying applied field. The classical limit of the fluctuation–dissipation theorem is obtained by letting $\hbar\omega/kT \to 0$ with the result

$$\hat{\chi}''_{AB}(\omega) = \frac{\omega}{2kT} S_{AB}(\omega); \tag{39}$$

compare this to equation (15.3.45). More complete discussions of the fluctuation–dissipation theorem can be found in Martin (1968), Forster (1975), and Mazenko (2006).

15.6.B Inelastic scattering

Inelastic scattering is an important experimental technique for measuring the dynamical behavior of materials. By measuring the intensity of radiation scattered from a sample as a function of wavevector transfer and frequency change relative to the incident monochromatic radiation source, one can measure the spatio-temporal correlations in the material. This technique is now commonly applied to the scattering of neutrons, electrons, light, and x-rays. The frequency changes in the scattered wave are caused by inelastic scattering from quantum excitations in the sample; see Forster (1975), Squires (1997), and Mazenko (2006).

The frequency-dependent scattering intensity is directly proportional to the dynamical structure factor

$$S(\boldsymbol{k},\omega) = \frac{1}{N}\int_{-\infty}^{\infty}\left\langle \sum_{i,j} e^{-i\boldsymbol{k}\cdot(\boldsymbol{r}_i(t)-\boldsymbol{r}_j(0))}\right\rangle e^{i\omega t}\,dt, \tag{40}$$

where $\boldsymbol{k}$ represents the wavevector transfer of the scattering process and ω represents the frequency difference from that of the incident beam. A positive value of ω corresponds to a detected frequency that is less than the frequency of the incident beam. The dynamical structure factor $S(\boldsymbol{k},\omega)$ encodes both the spatial and temporal equilibrium correlations of fluctuations in the material and can be decomposed into two terms, one that represents scattering from a single particle at different times and another that represents scattering from different particles:

$$S(\boldsymbol{k},\omega) = S_{\text{self}}(\boldsymbol{k},\omega) + S_{\text{coherent}}(\boldsymbol{k},\omega), \tag{41a}$$

$$S_{\text{self}}(\boldsymbol{k},\omega) = \frac{1}{N}\int_{-\infty}^{\infty}\left\langle \sum_{i} e^{-i\boldsymbol{k}\cdot(\boldsymbol{r}_i(t)-\boldsymbol{r}_i(0))}\right\rangle e^{i\omega t}\,dt, \tag{41b}$$

$$S_{\text{coherent}}(\boldsymbol{k},\omega) = \frac{1}{N}\int_{-\infty}^{\infty}\left\langle \sum_{i\neq j} e^{-i\boldsymbol{k}\cdot(\boldsymbol{r}_i(t)-\boldsymbol{r}_j(0))}\right\rangle e^{i\omega t}\,dt. \tag{41c}$$

The dynamical structure factor $S(\boldsymbol{k},\omega)$ can also be written in terms of the spatio-temporal Fourier transforms of the time-dependent density $n(\boldsymbol{r},t)$ to connect it to the power spectrum as defined in Section 15.6.A:

$$S(\boldsymbol{k},\omega) = S_{AB}(\omega) = \int_{-\infty}^{\infty} \langle \delta A(t)\delta B(0)\rangle\, e^{i\omega t}\,dt, \tag{42a}$$

$$\delta A(t) = \frac{1}{\sqrt{N}}\int e^{-i\boldsymbol{k}\cdot\boldsymbol{r}}\delta n(\boldsymbol{r},t)d\boldsymbol{r} = \frac{1}{\sqrt{N}}\hat{n}_{-\boldsymbol{k}}(t), \tag{42b}$$

$$\delta B(0) = \frac{1}{\sqrt{N}}\int e^{i\boldsymbol{k}\cdot\boldsymbol{r}}\delta n(\boldsymbol{r},0)d\boldsymbol{r} = \frac{1}{\sqrt{N}}\hat{n}_{\boldsymbol{k}}(0). \tag{42c}$$

Positive frequency changes $\omega > 0$ represent scattering events that create quantum excitations in the material with energy $\hbar\omega$ and are referred to as Stokes scattering. Negative frequency changes are called anti-Stokes scattering and correspond to scattering events that destroy an excitation in the material with energy $\hbar\omega$. Since an excitation must first exist in order for it to be destroyed, the anti-Stokes scattering rate in equilibrium is lower relative to the Stokes scattering rate by the Boltzmann factor for the excitation:

$$S(\boldsymbol{k}, -\omega) = e^{-\beta\hbar\omega} S(\boldsymbol{k}, \omega). \tag{43}$$

This relation is sometimes used in Raman scattering to measure the temperature of the sample. The heights of Stokes and anti-Stokes peaks are symmetric if the excitation energies are small compared to thermal energies, that is $\hbar\omega \ll kT$. The static structure factor $S(\boldsymbol{k})$ in equation (10.7.18) is obtained by integrating $S(\boldsymbol{k}, \omega)$ over all ω:

$$S(\boldsymbol{k}) = \frac{1}{2\pi} \int_{-\infty}^{\infty} S(\boldsymbol{k}, \omega) d\omega. \tag{44}$$

This singles out equal-time scattering events and corresponds to quasielastic scattering measurements that are unable to resolve the energy changes due to the excitations in the material.

Three commonly measured types of inelastic laser scattering are: Raman scattering, Brillouin scattering, and Rayleigh scattering. Raman scattering measures electronic, vibrational, and rotational excitations of atoms and molecules, electronic band structure, and optical phonon modes. Brillouin scattering measures long-wavelength acoustic sound modes. The widths of Raman and Brillouin scattering peaks are determined by the lifetimes of their respective modes. Rayleigh scattering measures the heat diffusion mode centered at $\omega = 0$ with width proportional to the thermal diffusivity. The wavelength of visible light is large compared to atomic scales, so the wavevector transfers possible with light scattering are very small compared to the size of the Brillouin zone. This limitation is removed for inelastic x-ray and neutron scattering where experiments can probe wavevectors away from the center of the Brillouin zone.

The dynamical scattering of a laser beam from a liquid includes three peaks: a Rayleigh peak centered at $\omega = 0$ due to scattering from the thermal fluctuations in the liquid and the Stokes and anti-Stokes Brillouin peaks at $\omega_k \approx \pm ck$ due to scattering from acoustic phonons with sound speed c. For scattering in this wavevector and frequency range, the dynamical structure factor is symmetric in ω since $\hbar\omega \ll kT$. An early Brillouin scatttering measurement of the dynamical structure factor of a liquid is shown in Figure 15.10.

The fluctuation–dissipation theorem enables one to develop a theory of the dynamical structure factor based on the hydrodynamic response of a system that is weakly perturbed from equilibrium. In the case of a fluid, small perturbations in pressure and temperature result in propagating sound waves and thermal diffusion. This results in the following

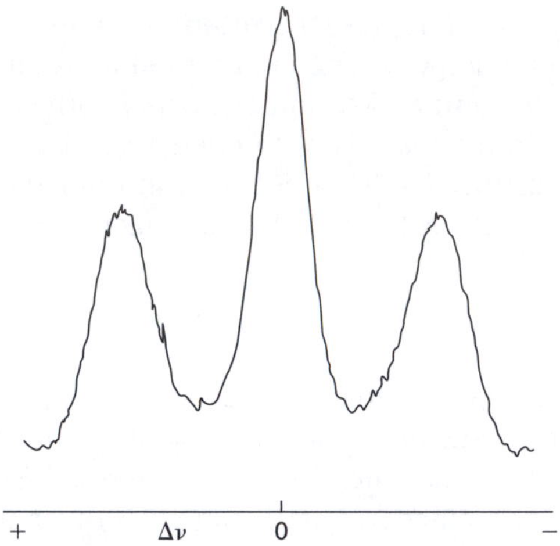

FIGURE 15.10 Dynamical structure factor for carbon tetrachloride using 632.8 nm He-Ne laser with a 90° scattering angle. The structure factor is symmetric since $\hbar\omega \ll kT$. The Landau-Placzek ratio of the integrated intensity under the Rayleigh peak to the integrated intensities under the two Brillouin peaks is related to ratio of the constant-pressure heat capacity to the constant-volume heat capacity: $I_R/(2I_B) = C_P/C_V - 1 = 0.72 \pm 0.03$, Landau and Placzek (1934). The locations of the Brillouin peaks give the sound speed and the widths of the peaks measure the thermal diffusivity and sound attenuation coefficient of the liquid; see equation (45). From Cummins and Gammon (1966), reprinted with permission; copyright © 1966, American Institute of Physics.

theoretical form for the dynamical structure factor:

$$S(k,\omega) = S(k)\left[\left(\frac{\gamma-1}{\gamma}\right)\frac{2D_T k^2}{\omega^2 + (D_T k^2)^2} + \frac{1}{\gamma}\left(\frac{\Gamma k^2}{(\omega^2 - c^2k^2)^2 + (\Gamma k^2)^2} + \frac{\Gamma k^2}{(\omega^2 + c^2k^2)^2 + (\Gamma k^2)^2}\right)\right]. \tag{45}$$

The parameters in equation (45) are the thermal diffusivity D_T, the sound speed c, and the sound attenuation coefficient Γ, while $\gamma = C_P/C_V$ is the ratio of the constant-pressure and constant-volume heat capacities; see Forster (1975) and Hansen and McDonald (1986).

15.7 The Onsager relations

Most physical phenomena exhibit a kind of symmetry, sometimes referred to as *reciprocity*, that arises from certain basic properties of the microscopic processes that operate behind the (observable) macroscopic situations. A notable example of this is met with in the *thermodynamics of irreversible processes* where one deals with a variety of flow processes, such as heat flow, electric current, mass transfer, and so on. These flows (or "currents") are driven by "forces," such as a temperature difference, a potential difference, a pressure difference, and so on, which come into play because of a natural tendency among physical systems which happen to be out of equilibrium to approach a state of equilibrium. If the given state of the system is not too far removed from a state of equilibrium, then one might assume a *linear* relationship between the forces X_i and the currents $\dot{x}_i$:

$$\dot{x}_i = \gamma_{ij} X_j, \tag{1}$$

where γ_{ij} are the *kinetic coefficients* of the system.[21] Simple examples of such coefficients are thermal conductivity, electrical conductivity, diffusion coefficient, and so on. There are, however, *nondiagonal elements*, $\gamma_{ij}(i \neq j)$, as well that may or may not vanish; they are responsible for the so-called *cross effects*. It is the symmetry properties of the matrix (γ_{ij}) that form the subject matter of this section.

The most obvious way to approach this problem is to consider the entropy, $S(x_i)$, of the system in the *disturbed state* relative to its maximal value, $S(\tilde{x}_i)$, in the relevant *state of equilibrium*. It is the natural tendency of the function $S(x_i)$ to approach its maximal value $S(\tilde{x}_i)$ that brings into play the driving forces X_i; these forces give rise to currents $\dot{x}_i$, which take the "coordinates" x_i toward their equilibrium values $\tilde{x}_i$. If the deviations $(x_i - \tilde{x}_i)$ are small, then the function $S(x_i)$ may be expressed as a Taylor series around the values $x_i = \tilde{x}_i$; retaining terms up to the second order only, we have

$$\begin{aligned} S(x_i) = S(\tilde{x}_i) &+ \left(\frac{\partial S}{\partial x_i}\right)_{x_i = \tilde{x}_i} (x_i - \tilde{x}_i) \\ &+ \frac{1}{2}\left(\frac{\partial^2 S}{\partial x_i \partial x_j}\right)_{x_{i,j} = \tilde{x}_{i,j}} (x_i - \tilde{x}_i)(x_j - \tilde{x}_j). \end{aligned} \tag{2}$$

In view of the fact that the function $S(x_i)$ is *maximum* at $x_i = \tilde{x}_i$, its first derivatives vanish; we may, therefore, write

$$\Delta S \equiv S(x_i) - S(\tilde{x}_i) = -\frac{1}{2}\beta_{ij}(x_i - \tilde{x}_i)(x_j - \tilde{x}_j), \tag{3}$$

where

$$\beta_{ij} = -\left(\frac{\partial^2 S}{\partial x_i \partial x_j}\right)_{x_{i,j} = \tilde{x}_{i,j}} = \beta_{ji}. \tag{4}$$

The driving forces X_i may be defined in the spirit of the second law of thermodynamics, that is,

$$X_i = \left(\frac{\partial S}{\partial x_i}\right) = -\beta_{ij}(x_j - \tilde{x}_j). \tag{5}$$

[21]In writing equation (1), and other subsequent equations, we follow the summation convention that implies *an automatic summation over a repeated index.*

We note that in the present approximation the forces X_i depend linearly on the displacements $(x_i - \tilde{x}_i)$; in the state of equilibrium, they vanish. Now, in view of equation (15.1.2), the *ensemble average* of the product $x_i X_j$ is given by

$$\langle x_i X_j \rangle = \frac{\int_{-\infty}^{\infty} (x_i X_j) \exp\left\{-\frac{1}{2k}\beta_{ij}(x_i - \tilde{x}_i)(x_j - \tilde{x}_j)\right\} \prod_i dx_i}{\int_{-\infty}^{\infty} \exp\left\{-\frac{1}{2k}\beta_{ij}(x_i - \tilde{x}_j)(x_j - \tilde{x}_j)\right\} \prod_i dx_i}; \tag{6}$$

the limits of integration in (6) have been extended to $-\infty$ and $+\infty$ because the integrals here do not draw any significant contribution from large values of the variables involved. In the same way,

$$\langle x_i \rangle = \frac{\int_{-\infty}^{\infty} x_i \exp\left\{-\frac{1}{2k}\beta_{ij}(x_i - \tilde{x}_i)(x_j - \tilde{x}_j)\right\} \prod_i dx_i}{\int_{-\infty}^{\infty} \exp\left\{-\frac{1}{2k}\beta_{ij}(x_i - \tilde{x}_i)(x_j - \tilde{x}_j)\right\} \prod_i dx_i} = \tilde{x}_i. \tag{7}$$

Differentiating (7) with respect to $\tilde{x}_j$ (and remembering that the integral in the denominator is a constant, independent of the actual values of the quantities $\tilde{x}_i$), and comparing the resulting expression with (6), we obtain the remarkable result

$$\langle x_i X_j \rangle = -k\delta_{ij}. \tag{8}$$

We now proceed toward the key point of the argument. First of all, we note that, though equations (1) are concerned with *irreversible* phenomena, the microscopic processes underlying these phenomena obey *time reversal*, which means that the temporal correlations of the relevant variables are the same whether measured forward or backward in time. Thus,

$$\langle x_i(0) x_j(s) \rangle = \langle x_i(0) x_j(-s) \rangle; \tag{9}$$

also, by a shift in the zero of time,

$$\langle x_i(0) x_j(-s) \rangle = \langle x_i(s) x_j(0) \rangle. \tag{10}$$

Combining (9) and (10), we get

$$\langle x_i(0) x_j(s) \rangle = \langle x_i(s) x_j(0) \rangle. \tag{11}$$

If we now subtract, from both sides of this equation, the quantity $\langle x_i(0) x_j(0) \rangle$, divide the resulting equation by s and let $s \to 0$, we obtain

$$\langle x_i(0) \dot{x}_j(0) \rangle = \langle \dot{x}_i(0) x_j(0) \rangle. \tag{12}$$

Substituting from (1) and making use of (8), we obtain on the left side of (12)

$$\langle x_i(0)\gamma_{jl}X_l(0)\rangle = -k\gamma_{jl}\delta_{il} = -k\gamma_{ji}$$

and on its right side

$$\langle \gamma_{il}X_l(0)x_j(0)\rangle = -k\gamma_{il}\delta_{jl} = -k\gamma_{ij}.$$

It follows that

$$\gamma_{ij} = \gamma_{ji}. \tag{13}$$

Equations (13) constitute the *Onsager reciprocity relations*; they were first derived by Onsager in 1931 and have become an essential part of the thermodynamics of irreversible phenomena.

In view of equations (1) and (13), the currents $\dot{x}_i$ may be written as

$$\dot{x}_i = \frac{\partial f}{\partial X_i}, \tag{14}$$

where the *generating function* f is a quadratic function of the forces X_i:

$$f = \frac{1}{2}\gamma_{ij}X_iX_j. \tag{15}$$

The function f is especially important in that it determines directly the rate at which the entropy of the system changes with time:

$$\dot{S} = \frac{\partial S}{\partial x_i}\dot{x}_i = X_i\dot{x}_i = X_i\frac{\partial f}{\partial X_i} = 2f. \tag{16}$$

As the system approaches the state of equilibrium, its entropy must *increase* toward the equilibrium value $S(\tilde{x}_i)$. The function f must, therefore, be *positive definite*, which places certain restrictions on the coefficients γ_{ij}.

Analogous to equation (1), we could also write

$$\dot{X}_i = \zeta_{ij}(x_j - \tilde{x}_j), \tag{17}$$

the quantities ζ_{ij} being another set of coefficients pertaining to the system. From equations (1) and (5), on the other hand, we obtain

$$\begin{aligned}\dot{X}_i &= -\beta_{ij}\dot{x}_j = -\beta_{ij}(\gamma_{jl}X_l) = -\beta_{ij}\gamma_{jl}\{-\beta_{lm}(x_m - \tilde{x}_m)\}\\ &= \beta_{ij}\gamma_{jl}\beta_{lm}(x_m - \tilde{x}_m).\end{aligned} \tag{18}$$

Comparing (17) and (18), we obtain a relationship between the new coefficients ζ_{ij} and the kinetic coefficients γ_{ij}:

$$\zeta_{im} = \beta_{ij}\gamma_{jl}\beta_{lm}. \tag{19}$$

Now, in view of the symmetry properties of the matrices β and γ, we get

$$\zeta_{im} = \zeta_{mi}; \tag{20}$$

thus, the coefficients ζ_{ij}, introduced through the phenomenological equations (17), also obey the reciprocity relations. It then follows that the quantities $\dot{X}_i$, see equation (14), may be written as,

$$\dot{X}_i = \frac{\partial f'}{\partial x_i}, \tag{21}$$

where

$$f' = \frac{1}{2}\zeta_{ij}(x_i - \tilde{x}_i)(x_j - \tilde{x}_j). \tag{22}$$

The entropy change dS may now be written as

$$\begin{aligned} dS &= \frac{\partial S}{\partial x_j}dx_j = X_j dx_j = -\beta_{ji}(x_i - \tilde{x}_i)dx_j \\ &= (x_i - \tilde{x}_i)d\{-\beta_{ij}(x_j - \tilde{x}_j)\} = (x_i - \tilde{x}_i)dX_i, \end{aligned} \tag{23}$$

so that

$$\frac{\partial S}{\partial X_i} = (x_i - \tilde{x}_i); \tag{24}$$

clearly, the entropy S is now regarded as an *explicit* function of the forces X_i (rather than of the coordinates x_i). The time derivative of S now takes the form

$$\dot{S} = \frac{\partial S}{\partial X_i}\dot{X}_i = (x_i - \tilde{x}_i)\frac{\partial f'}{\partial x_i} = 2f'. \tag{25}$$

Comparing (16) and (25), we conclude that the functions f and f' are, in fact, the same; they are only expressed in terms of two different sets of variables.

It seems important to mention here that Onsager's reciprocity relations have an intimate connection with the fluctuation–dissipation theorem of the preceding section. Following equations (15.6.18) and (15.6.19), and adopting the summation convention, we have in the present context

$$\dot{x}_i(t) = \frac{1}{kT}\int_{-\infty}^{t} E_l(t')\langle \dot{x}_l(t')\dot{x}_i(t)\rangle dt' \tag{26}$$

or, setting $(t - t') = s$,

$$\dot{x}_i(t) = \frac{1}{kT}\int_{0}^{\infty} E_l(t-s)\langle \dot{x}_l(t-s)\langle \dot{x}_i(t)\rangle ds; \tag{27}$$

compare this to equation (1). Interchanging the indices i and l, we obtain

$$\dot{x}_l(t) = \frac{1}{kT}\int_0^\infty E_i(t-s)\langle \dot{x}_i(t-s)\dot{x}_l(t)\rangle ds. \tag{28}$$

The crucial point now is that the correlation functions appearing in equations (27) and (28) are identical in value, for

$$\langle \dot{x}_l(t-s)\dot{x}_i(t)\rangle = \langle \dot{x}_l(0)\dot{x}_i(s)\rangle = \langle \dot{x}_l(0)\dot{x}_i(-s)\rangle = \langle \dot{x}_l(t)\dot{x}_i(t-s)\rangle; \tag{29}$$

in establishing (29), the first and third steps followed from "a shift in time" while the second step followed from the "principle of *dynamical reversibility* of microscopic processes." The equivalence depicted in equation (29) is, in essence, the content of Onsager's reciprocity relations. In particular, if the correlation functions appearing in (27) and (28) are sharply peaked at the value $s = 0$, then these equations reduce to the phenomenological equations (1), and equation (29) becomes synonymous with the Onsager relations (13).

In the end, we make some further remarks concerning relations (13). We recall that, in arriving at these relations, we had to make an appeal to the invariance of the microscopic processes under time reversal. The situation is somewhat different in the case of a "system in rotation" (or a "system in an external magnetic field"), for then the invariance under time reversal holds only if there is also a simultaneous change of sign of the angular velocity $\boldsymbol{\Omega}$ (or of the magnetic field $\boldsymbol{B}$). The kinetic coefficients, which in this case might depend on the parameter $\boldsymbol{\Omega}$ (or $\boldsymbol{B}$), will now satisfy the relations

$$\gamma_{ij}(\boldsymbol{\Omega}) = \gamma_{ji}(-\boldsymbol{\Omega}) \tag{13a}$$

and

$$\gamma_{ij}(\boldsymbol{B}) = \gamma_{ji}(-\boldsymbol{B}). \tag{13b}$$

In addition, our proof here rested on the *implicit* assumption that the quantities x_i themselves do not change under time reversal. If, for some reason, these quantities are proportional to the velocities of a certain macroscopic motion, then they will also change their sign under time reversal. Now, if both x_i and x_j belong to this category, then equation (12), which is crucial to our proof, would remain unaltered; consequently, the coefficients γ_{ij} and γ_{ji} would still be equal. However, if only one of them belongs to this category while the other one does not, then equation (12) would change to

$$\langle x_i(0)\dot{x}_j(0)\rangle = -\langle \dot{x}_i(0)x_j(0)\rangle; \tag{12$'$}$$

the coefficients γ_{ij} and γ_{ji} would then obey the relations

$$\gamma_{ij} = -\gamma_{ji}. \tag{13$'$}$$

For the application of Onsager's relations to different physical problems, reference may be made to the monographs by de Groot (1951), de Groot and Mazur (1962), and Prigogine (1967).

Problems

15.1. Making use of expressions (15.1.11) and (15.1.12) for ΔS and ΔP, and expressions (15.1.14) for $\overline{(\Delta T)^2}, \overline{(\Delta V)^2}$, and $\overline{(\Delta T \Delta V)}$, show that

(a) $\overline{(\Delta T \Delta S)} = kT$; **(b)** $\overline{(\Delta P \Delta V)} = -kT$;

(c) $\overline{(\Delta S \Delta V)} = kT(\partial V/\partial T)_P$; **(d)** $\overline{(\Delta P \Delta T)} = kT^2 C_V^{-1}(\partial P/\partial T)_V$.

[Note that results (a) and (b) give: $\overline{(\Delta T \Delta S - \Delta P \Delta V)} = 2kT$, which follows directly from the probability distribution function (15.1.8).]

15.2. Establish the probability distribution (15.1.15), which leads to the expressions in (15.1.16) for $\overline{(\Delta S)^2}$, $\overline{(\Delta P)^2}$, and $\overline{(\Delta S \Delta P)}$. Show that these results can also be obtained by following the procedure of the preceding problem.

15.3. If we choose the quantities E and V as "independent" variables, then the probability distribution function (15.1.8) does not reduce to a form as simple as (15.1.13) or (15.1.15); it is marked instead by the presence, in the exponent, of a cross term proportional to the product $\Delta E \Delta V$. Consequently, the variables E and V are not *statistically independent*: $\overline{(\Delta E \Delta V)} \neq 0$.

Show that

$$\overline{(\Delta E \Delta V)} = kT\left\{T\left(\frac{\partial V}{\partial T}\right)_P + P\left(\frac{\partial V}{\partial P}\right)_T\right\};$$

verify as well expressions (15.1.14) and (15.1.18) for $\overline{(\Delta V)^2}$ and $\overline{(\Delta E)^2}$.

[Note that in the case of a two-dimensional normal distribution, namely

$$p(x,y) \propto \exp\left\{-\frac{1}{2}(ax^2 + 2bxy + cy^2)\right\},$$

the quantities $\langle x^2 \rangle$, $\langle xy \rangle$, and $\langle y^2 \rangle$ can be obtained in a straightforward manner by carrying out a logarithmic differentiation of the formula

$$\int_{-\infty}^{\infty}\int_{-\infty}^{\infty} \exp\left\{-\frac{1}{2}(ax^2 + 2bxy + cy^2\right\} dx\, dy = \frac{2\pi}{\sqrt{(ac - b^2)}}$$

with respect to the parameters a, b, and c. This leads to the *covariance matrix* of the distribution, namely

$$\begin{pmatrix} \langle x^2 \rangle & \langle xy \rangle \\ \langle yx \rangle & \langle y^2 \rangle \end{pmatrix} = \frac{1}{(ac - b^2)}\begin{pmatrix} c & -b \\ -b & a \end{pmatrix}.$$

If $b = 0$, then

$$\langle x^2 \rangle = 1/a, \quad \langle xy \rangle = 0, \quad \langle y^2 \rangle = 1/c.]^{22}$$

15.4. A string of length l is stretched, under a constant tension F, between two fixed points A and B. Show that the mean square (fluctuational) displacement $y(x)$ at point P, distant x from A, is given by

$$\overline{\{y(x)\}^2} = \frac{kT}{Fl}x(l - x).$$

[22] For the covariance matrix of an n-dimensional normal distribution, see Landau and Lifshitz (1958), Section 110.

Further show that, for $x_2 \geq x_1$,

$$\overline{y(x_1)y(x_2)} = \frac{kT}{Fl} x_1(l - x_2).$$

[*Hint*: Calculate the energy, Φ, associated with the fluctuation in question; the desired probability distribution is then given by $p \propto \exp(-\Phi/kT)$, from which the required averages can be readily evaluated.]

15.5. How small must the volume, V_A, of a gaseous subsystem (at normal temperature and pressure) be, so that the root-mean-square deviation in the number, N_A, of particles occupying this volume be 1 percent of the mean value $\overline{N_A}$?

15.6. Pospišil (1927) observed the Brownian motion of soot particles, of radii 0.4×10^{-4} cm, immersed in a water–glycerine solution, of viscosity 0.0278 poise at a temperature of 18.8°C. The observed value of $\overline{x^2}$, in a 10-second time interval, was $3.3 \times 10^{-8}\text{cm}^2$. Making use of these data, determine the Boltzmann constant k.

15.7. In the notation of Section 15.3, show that for a Brownian particle

$$\langle \boldsymbol{v}(t) \cdot \boldsymbol{F}(t) \rangle = 3kT/\tau, \quad \text{while } \langle \boldsymbol{v}(t) \cdot \boldsymbol{\mathcal{F}}(t) \rangle = 0.$$

On the other hand,

$$\langle \boldsymbol{r}(t) \cdot \boldsymbol{\mathcal{F}}(t) \rangle = -3kT, \quad \text{while } \langle \boldsymbol{r}(t) \cdot \boldsymbol{F}(t) \rangle = 0.$$

15.8. Integrate equation (15.3.14) to obtain

$$\boldsymbol{r}(t) = \boldsymbol{v}(0)\tau(1 - e^{-t/\tau}) + \tau \int_0^t \{1 - e^{(u-t)/\tau}\} \boldsymbol{A}(u) du,$$

so that $\boldsymbol{r}(0) = 0$. Taking the square of this expression and making use of the autocorrelation function $K_A(s)$, derive formula (15.3.31) for $\langle r^2(t) \rangle$.

15.9. While detecting a very feeble current with the help of a moving-coil galvanometer, one must ensure that an observed deflection is not just a stray kick arising from the Brownian motion of the suspended system. If we agree that a deflection θ, whose magnitude exceeds $4\theta_{\text{r.m.s.}} [= 4(kT/c)^{1/2}]$, is highly unlikely to be due to the Brownian motion, we obtain a *lower* limit to the magnitude of the current that can be reliably recorded with the help of the given galvanometer. Express this limiting current in terms of the time period τ and the critical damping resistance R_c of the galvanometer.

15.10. **(a)** Integrate Langevin's equation (15.3.5) for the velocity component v_x over a *small* interval of time δt, and show that

$$\frac{\langle \delta v_x \rangle}{\delta t} = -\frac{v_x}{\tau} \quad \text{and} \quad \frac{\langle (\delta v_x)^2 \rangle}{\delta t} = \frac{2kT}{M\tau}.$$

(b) Now, set up the Fokker–Planck equation for the *distribution function* $f(v_x, t)$ and, making use of the foregoing results for $\mu_1(v_x)$ and $\mu_2(v_x)$, derive an explicit expression for this function. Study the various cases of interest, especially the one for which $t \gg \tau$.

15.11. Generalize the analysis of the Langevin theory of a harmonic oscillator, as given by equation (15.3.33), to the case of an oscillator starting at time $t = 0$ with the initial position $x(0)$ and the initial velocity $v(0)$. Derive, for this system, the quantities $\langle x^2(t) \rangle$ and $\langle v^2(t) \rangle$ and show that, in the limit $\omega_0 \to 0$, these expressions reproduce equations (15.3.29) and (15.3.31) while, in the limit $M \to 0$, they lead to the relevant results of Section 15.4.

15.12. Generalize the Fokker–Planck equation to the case of a particle executing Brownian motion in *three* dimensions. Determine the general solution of this equation and study its important features.

15.13. The autocorrelation function $K(s)$ of a certain *statistically stationary* variable $y(t)$ is given by

(a) $K(s) = K(0)e^{-\alpha s^2} \cos(2\pi f^* s)$

or by

(b) $K(s) = K(0)e^{-\alpha|s|} \cos(2\pi f^* s)$,

where $\alpha > 0$. Determine, and discuss the nature of, the power spectrum $w(f)$ in each of these cases and investigate its behavior in the limits (a) $\alpha \to 0$, (b) $f^* \to 0$, and (c) both α and $f^* \to 0$.

15.14. Show that if the autocorrelation function $K(s)$ of a certain *statistically stationary* variable $y(t)$ is given by

$$K(s) = K(0)\frac{\sin(as)}{as}\frac{\sin(bs)}{bs} \qquad (a > b > 0),$$

then the power spectrum $w(f)$ of that variable is given by

$$w(f) = \begin{cases} \dfrac{2\pi}{a}K(0) & \text{for } 0 < f \leq \dfrac{a-b}{2\pi}, \\[2ex] \dfrac{2\pi}{ab}K(0)\left\{\dfrac{a+b}{2} - \pi f\right\} & \text{for } \dfrac{a-b}{2\pi} \leq f \leq \dfrac{a+b}{2\pi}, \\[2ex] 0 & \text{for } \dfrac{a+b}{2\pi} \leq f < \infty. \end{cases}$$

Verify that the function $w(f)$ satisfies the requirement (15.5.16).

[Note that, in the limit $b \to 0$, we recover the situation pertaining to equations (15.5.18).]

15.15. **(a)** Show that the mean square value of the variable $Y(t)$, defined by the formula

$$Y(t) = \int_u^{u+t} y(u)du,$$

where $y(u)$ is a *statistically stationary* variable with power spectrum $w(f)$, is given by

$$\langle Y^2(t)\rangle = \frac{1}{2\pi^2}\int_0^\infty \frac{w(f)}{f^2}\{1 - \cos(2\pi ft)\}\,df;$$

and, accordingly,

$$w(f) = 4\pi f \int_0^\infty \frac{\partial}{\partial t}\langle Y^2(t)\rangle \sin(2\pi ft)dt$$

$$= 2\int_0^\infty \frac{\partial^2}{\partial t^2}\langle Y^2(t)\rangle \cos(2\pi ft)dt.$$

For details, see MacDonald (1962), Section 2.2.1. A comparison of the last result with equation (15.5.14) suggests that

$$K_y(s) = \frac{1}{2}\frac{\partial^2}{\partial s^2}\langle Y^2(s)\rangle.$$

(b) Apply the foregoing analysis to the motion of a Brownian particle, taking y to be the velocity of the particle and Y its displacement.

15.16. Show that the power spectra $w_{\boldsymbol{v}}(f)$ and $w_{\boldsymbol{A}}(f)$ of the fluctuating variables $\boldsymbol{v}(t)$ and $\boldsymbol{A}(t)$ that appear in the Langevin equation (15.3.5) are connected by the relation

$$w_{\boldsymbol{v}}(f) = w_{\boldsymbol{A}}(f)\frac{\tau^2}{1 + (2\pi f\tau)^2},$$

τ being the relaxation time of the problem. Hence, by equation (15.5.21) $w_{\boldsymbol{A}}(f) = 12kT/M\tau$.

15.17. **(a)** Verify equations (15.6.7) through (15.6.9).
(b) Substituting expression (15.6.9) for $\boldsymbol{K}_v(s)$ into equation (15.6.6), derive formula (15.3.31) for $\langle r^2(t)\rangle$.

15.18. Determine $\hat{\chi}''_{vx}(\omega)$ and $S_{vx}(\omega)$ for a Brownian particle in a harmonic oscillator potential. Show that the response function and the power spectrum for this case are related by the classical limit of the fluctuation–dissipation theorem.

15.19. Derive the linear response density matrix (15.6.29) from the equation of motion (15.6.28).

15.20. Show that $G_{AB}(t) = G_{BA}(t - i\beta\hbar)$ and use the cyclic property of the traces to derive the fluctuation–dissipation theorem $\hat{\chi}''_{AB}(\omega) = \frac{1}{2\hbar}\left(1 - e^{-\beta\hbar\omega}\right)S_{AB}(\omega)$.

15.21. Show that $G_{AB}(t) = G_{BA}(t - i\beta\hbar)$. Use this result to show that, in the classical limit, $\hat{\chi}''_{AB}(t)$ becomes $\left\langle \frac{dA(t)}{dt}B(0)\right\rangle$. Further show that this leads to equation (15.6.39).

15.22. Determine the self-diffusion term in the dynamical structure factor $S_{\text{self}}(\boldsymbol{k},\omega)$ in equation (15.6.41b) for the case of a single particle that diffuses according to the diffusion equation. Assume the process to be Gaussian for which $\langle e^f\rangle = \exp\left(\langle f^2/2\rangle\right)$.

15.23. Determine the angular frequency ω_k for the Brillouin peaks in water for 90° laser scattering, using a He-Ne laser with $\lambda = 632.8\,\text{nm}$. Determine the width of the Rayleigh peak and show that the Brillouin peaks are well-separated from the Rayleigh peak. The thermal diffusivity of water is $D_T = 1.4\times10^{-7}\,\text{m}^2/\text{s}$.

15.24. Describe the dynamical structure factor for Raman scattering for a He-Ne laser with $\lambda = 632.8\,\text{nm}$. The energy level responsible for this scattering has an energy of 0.05 eV and the lifetime of this state is 1 picosecond. Are the Stokes/anti-Stokes scatterings symmetric as $\omega \to -\omega$ at room temperature?

16 Computer Simulations

Computer simulations play an important role in modern statistical mechanics. The history of the use of computer simulations in science parallels the history of early digital computing. The people and places involved centered around Los Alamos and other U.S. national laboratories where the first digital computers became available for use by scientists after World War II. Early leaders in the development of computer simulation methods included Fermi, Ulam, von Neumann, Teller, Metropolis, Rosenbluth, and others who were also involved in the Manhattan Project (Metropolis, 1987).

Computer simulations in statistical mechanics fall into two broad classes: Monte Carlo (MC) and molecular dynamics (MD), although variants span the range between the two. Both methods involve numerically evolving simple models of materials through a set of microstates in order to determine the thermodynamic averages of measurable quantities. Computer simulations provide a means to study physical systems that is complementary to both experiment and theory. The following are a few of the advantages of computer simulations:

- Computer simulations can provide insight into the equilibrium and nonequilibrium behavior of model systems for ranges of parameters where theoretical approximations are invalid or untested.
- Computer simulations provide a means to test the range of validity of theoretical approximations against specific model systems.
- Computer simulations allow visualization of physical processes that can provide new insights into complex phenomena.
- Computer simulations allow detailed examination of behaviors that might not be accessible experimentally.
- Computer simulations can be used to examine fundamental physical processes that can be used to guide theory.
- Computer simulations can be used to model systems that do not exist in nature to provide assistance in understanding existing materials and engineering new ones.

16.1 Introduction and statistics

While certain critical aspects of computer simulation theory should be followed rigorously, much of computer simulation development and use is an art form. There are many possible simulation approaches for any given problem and some choices will be more effective

© 2011 Elsevier Ltd. All rights reserved.

at elucidating important physical properties than others. This brief chapter concentrates on equilibrium simulations but computer simulations are also widely used to model dynamical and nonequilibrium processes. The task of determining equilibrium thermodynamic averages of model systems is accomplished by generating a sequence of microstates that are chosen from the equilibrium ensemble of the model. For example, an MD simulation might be used to integrate Newton's equations of motion for generating a time-series of states in phase space as the system explores the constant-energy hypersurface of the Hamiltonian. By comparison, an MC simulation of the same model might generate a sequence of states chosen by a random walk among the configurational microstates of the canonical ensemble. Both methods are examples of *importance sampling*, which focuses computational effort on generating microstates that are representative of the equilibrium ensemble rather than sampling *all* of the phase space. It is this huge improvement in efficiency that makes computer simulations of statistical mechanical models feasible. The sequence of states produced by either method can be used to estimate equilibrium averages. Allen and Tildesley (1990), Binder and Heermann (2002), Frenkel and Smit (2002), and Landau and Binder (2009) provide more detailed discussions of computer simulations and their applications in statistical physics.

Let q represent a microstate of the system and $A(q)$ a thermodynamic observable that is a function of the microstate. In an MC simulation, q might represent the positions of all the particles in the system while in an MD simulation q might represent the positions and momenta of all the particles. The observable $A(q)$ might represent the potential energy, virial contribution to the pressure, pair correlation function, and so on. The initial microstate chosen to start a simulation will generally not be typical of the set of microstates that make up the equilibrium ensemble, but the goal of a simulation is to evolve the microstate through a large enough subset of the microstates of the equilibrium ensemble so that averages of observables approach their equilibrium values. After a simulation has run long enough for the system to approach equilibrium, the simulation then generates a sequence of M configurations, $\{q_j\}_{j=1}^{M}$, chosen from the set of microstates in the equilibrium ensemble and stores a sequence of values, $\{A(q_j)\}_{j=1}^{M}$, for each of the thermodynamic variables one wants to measure.[1] Since the microstates are chosen from the equilibrium ensemble, the equilibrium average of A is approximated by a simple average of the set of values $\{A(q_j)\}_{j=1}^{M}$. Of course, a simulation can only provide a finite sequence of states, so a statistical analysis of the uncertainty of the results is a crucial part of any simulation.

The equilibrium average of the variable A is given by

$$\langle A\rangle = \langle A\rangle_M \pm \sigma_M \tag{1}$$

[1]Alternatively, one can store the full configuration set of statistically independent microstates for later analysis. This tactic requires a large amount of storage space but is useful if there is a large computational cost of generating statistically independent configurations and one needs to calculate averages of many different observables at a later time using the stored configurations. This is sometimes done in large-scale lattice quantum chromodynamics simulations.

where the simulation average $\langle A \rangle_M$ and uncertainty σ_M are determined by

$$\langle A \rangle_M = \frac{1}{M} \sum_{j=1}^{M} A(q_j), \tag{2a}$$

$$\sigma_M = \frac{\sqrt{\langle A^2 \rangle_M - \langle A \rangle_M^2}}{\sqrt{M/(2\tau+1)}}, \tag{2b}$$

$$\langle A^2 \rangle_M - \langle A \rangle_M^2 = \frac{1}{M} \sum_{j=1}^{M} [A(q_j) - \langle A \rangle_M]^2. \tag{2c}$$

The "correlation time" τ is defined as follows. Since the states q_j are generated sequentially by the simulation, each new state q_{j+1} is guaranteed to be close to the previous state q_j, so the values $A(q_j)$ in the sequence are highly correlated. The correlations in the values of $A(q_j)$ decrease with the "correlation time" τ which can be calculated from the correlation function $\phi_{AA}(t)$, namely

$$\phi_{AA}(t) = \frac{\langle A(t)A(0) \rangle - \langle A(t) \rangle \langle A(0) \rangle}{\langle A^2 \rangle - \langle A \rangle^2}, \tag{3a}$$

$$\tau = \sum_{t>0} \phi_{AA}(t). \tag{3b}$$

The variable t is a measure of the separation between pairs of configurations in the ordered sequence. In the case of molecular dynamics simulations, τ represents a physical time for the system to move far enough along its trajectory on the energy surface to result in decorrelated values of A. Monte Carlo simulations explore equilibrium microstates in a random walk, so τ does not correspond to physical time but rather the average number of Monte Carlo sweeps needed to give statistically independent values for A. The quantity $M/(2\tau+1)$ represents the number of statistically independent configurations in the sequence of M values.[2]

[2]The correlations $\phi_{AA}(t)$ in equation (16.1.3a) can be measured using subsequences of the M configurations:

$$\langle A(t)A(0) \rangle \approx \frac{1}{M'} \sum_{j=1}^{M'} A(q_{j+t})A(q_j),$$

and

$$\langle A(t) \rangle \approx \frac{1}{M'} \sum_{j=1}^{M'} A(q_{j+t}).$$

By definition, the correlation function $\phi_{AA}(0)$ is unity and the correlations decay to zero as $t \to \infty$. Once one can place a reliable upper bound on the size of the correlation time τ for a given system from a knowledge of its equilibrium correlations, one can simply skip more than τ configurations between storing values of $A(q_j)$ to ensure that the numbers in the sequence are now approximately statistically independent.

16.2 Monte Carlo simulations

The term *Monte Carlo method*, named for the gambling casinos in Monaco, was coined by Nicholas Metropolis (1987) – "a suggestion not unrelated to the fact that Stan [Ulam] had an uncle who would borrow money from relatives because he 'just had to go to Monte Carlo'." The goal of a Monte Carlo simulation in equilibrium statistical mechanics is to use pseudorandom numbers to draw a representative sample of microstates $\{q\}$ from the equilibrium probability distribution

$$P_{\text{eq}}(q) = \frac{\exp(-\beta E(q))}{\sum_{q'} \exp[-\beta E(q')]}. \tag{1}$$

This means that "instead of choosing configurations randomly, then weighting them with $\exp(-E/kT)$, we choose configurations with a probability $\exp(-E/kT)$ and weight them evenly" (Metropolis et al., 1953). If a simulation can accomplish this, then thermodynamic averages can be calculated using the simple averages in equation (16.1.2). This importance sampling of the states provides a huge computational advantage over normal random sampling.

The following algorithm accomplishes the goal of randomly selecting microstates q from the set of all microstates with a probability distribution that approaches the equilibrium distribution (1) (Metropolis et al., 1953; Kalos and Whitlock, 1986; Allen and Tildesley, 1990; Frenkel and Smit, 2002; Binder and Heermann, 2002; Landau and Binder, 2009). Consider an ensemble of microstates that has some initial distribution of probabilities $P(q,0)$ and let the distribution evolve according to the discrete stochastic rate equation

$$P(q,t+1) = P(q,t) + \sum_{q'} P(q',t)W(q' \to q) - P(q,t)\sum_{q'} W(q \to q'), \tag{2}$$

where $W(q \to q')$ is the transition rate from state q to state q'. If the transition rate obeys the balance condition

$$\sum_{q'} P_{\text{eq}}(q')W(q' \to q) = P_{\text{eq}}(q)\sum_{q'} W(q \to q'), \tag{3}$$

and the random process in equation (2) can reach every microstate from every other microstate in a finite number of steps, then the ensemble probability will approach the equilibrium distribution:

$$\lim_{t\to\infty} P(q,t) = P_{\text{eq}}(q). \tag{4}$$

In practice, equation (3) is usually implemented using the *detailed* balance condition

$$P_{\text{eq}}(q)W(q \to q') = P_{\text{eq}}(q')W(q' \to q). \tag{5}$$

Evaluating $P_{\text{eq}}(q)$ requires summing over all states to determine the partition function, but the ratio $P_{\text{eq}}(q')/P_{\text{eq}}(q)$ depends only on the energy difference $\Delta E = E(q') - E(q)$.

Therefore, the transition rates are related by

$$W(q \to q') = \exp(-\beta \Delta E)\, W(q' \to q). \tag{6}$$

This guarantees that the sequence of states generated by this stochastic process, beginning from any starting configuration, asymptotically becomes equivalent to selecting states by a random walk among the microstates of the equilibrium ensemble. This can be implemented in a computer code, as first proposed by Metropolis et al. (1953), using the transition rates

$$\begin{aligned} W(q \to q') &= 1 && \text{if } \Delta E \le 0, \\ W(q \to q') &= \exp(-\beta \Delta E) && \text{if } \Delta E > 0. \end{aligned} \tag{7}$$

Other choices for the transition rates are possible but this form, named after Metropolis, is one of the most commonly used.

16.2.A Metropolis Monte Carlo algorithm

The Metropolis method can be implemented in a computer program by using a pseudorandom number generator rand() that returns pseudorandom numbers that are uniformly distributed on the open unit interval (0.0,1.0); see Appendix I for a discussion of how pseudorandom numbers are generated. First, initialize the system by choosing a starting state q_0 from the set of all microstates of the model. It is helpful if q_0 is not atypical of the states in the equilibrium ensemble. This reduces the number of steps needed for the system to equilibrate. For example, a disordered liquid-like state would not be the best starting point for a simulation of a crystalline solid.

The Metropolis algorithm is defined by the following steps:

1. Generate a random trial state q_{trial} that is "nearby" the current state q_j of the system. "Nearby" here means that the trial state should be almost identical to the current state except for a small random change made, usually, to a single particle or spin. For example, one can create a trial state of a particle simulation by randomly moving one particle to a nearby location

$$x_i^{\text{trial}} = x_i + \Delta x(\text{rand}() - 0.5), \tag{8}$$

with two more calls to rand( ) to generate y_i^{trial} and z_i^{trial}. The trial state of a spin system usually involves a spin flip or a random rotation of a single spin.[3]

[3]There are Monte Carlo algorithms for spin systems that flip spins in large correlated clusters rather than one spin at a time; see Swendsen and Wang (1987) and Wolff (1989). These methods are very effective for simulations of some particular models. Also, one can attempt spin flips of all the spins on noninteracting sublattices at one time since the acceptance of each flip is independent of the other flipped spins. For example, a chessboard pattern update of a spin model in which spins only interact with nearest neighbors of a square lattice can be more efficient for some computer architectures or programming environments.

2. Determine the change in the energy of the trial state compared to the previous state, namely $\Delta E = E(q_{\text{trial}}) - E(q_j)$. If $\Delta E \leq 0$, accept the trial state, that is, set $q_{j+1} = q_{\text{trial}}$. If $\Delta E > 0$, then accept the trial state with probability $\exp(-\beta \Delta E)$. This is accomplished by using an additional call to the pseudorandom number generator. If $\text{rand}() < \exp(-\beta \Delta E)$, then accept the trial state. If the interactions are short-ranged, the calculation of the energy change will only involve interactions with a few nearby particles or spins. If the trial state is illegal in some way, that is, it is not an allowed state in the set of all configurations, then the state should be rejected. This is equivalent to setting the energy change at $+\infty$. If the trial state is rejected for either reason, then set the new state of the system equal to the previous state $q_{j+1} = q_j$, that is, leave the state at the old value q_j, throw away the trial state, and move on.
3. Perform steps 1 and 2 once for each particle or spin in the system. This is often done randomly to ensure detailed balance.[4] Steps 1 through 3 define one Monte Carlo sweep.
4. Repeat steps 1 through 3 for M_{eq} Monte Carlo sweeps to let the system equilibrate. The proper choice of M_{eq} is not obvious *a priori*. At the very least, all the measures $A(q)$ studied in the simulation should no longer have any obvious monotonic drift by the end of equilibration. This does not guarantee that the system has reached equilibrium since the system could well be trapped in the vicinity of a long-lived metastable state.
5. Repeat steps 1 through 3 for M Monte Carlo sweeps while keeping track of all the thermodynamic variables one wants to measure, namely $\{A(q_j)\}_{j=1}^{M}$. Use equations (16.1.1) and (16.1.2) to determine the equilibrium averages and uncertainties.

To determine averages at a different set of parameters (temperature, density, etc.), change the parameters by a small amount and repeat steps 1 through 5, including the equilibration step 4.[5] Using the last configuration of the previous run as the first configuration of the next run can often reduce the equilibration time. Figure 16.1 shows a Monte Carlo calculation of the specific heat of the two-dimensional Ising model on a 128×128 square lattice, as compared to the exact solution presented in Section 13.4.A.

[4]Sequential and other update methods that violate detailed balance are sometimes used for efficiency but special care should be taken to ensure that detailed balance is maintained on average.

[5]Histogram reweighting methods can sometimes be used to reduce the number of temperatures and fields that need to be simulated (Ferrenberg and Swendsen, 1988). For example, if a spin simulation at coupling K and field h collects a histogram that samples the joint energy-magnetization distribution $P_{K,h}(E,M)$, the distribution at nearby temperatures and fields is given by

$$P_{K+\Delta K,h+\Delta h}(E,M) = \frac{P_{K,h}(E,M)e^{\Delta K E + \Delta h M}}{\sum_{E,M} P_{K,h}(E,M)e^{\Delta K E + \Delta h M}}.$$

Other methods that are now widely used are: parallel tempering, multicanonical Monte Carlo, and "broad histogram" methods. These are particularly effective for studying systems with strongly first-order phase transitions. Monte Carlo renormalization group methods are very powerful for studying critical points. For a survey, see Landau and Binder (2009).

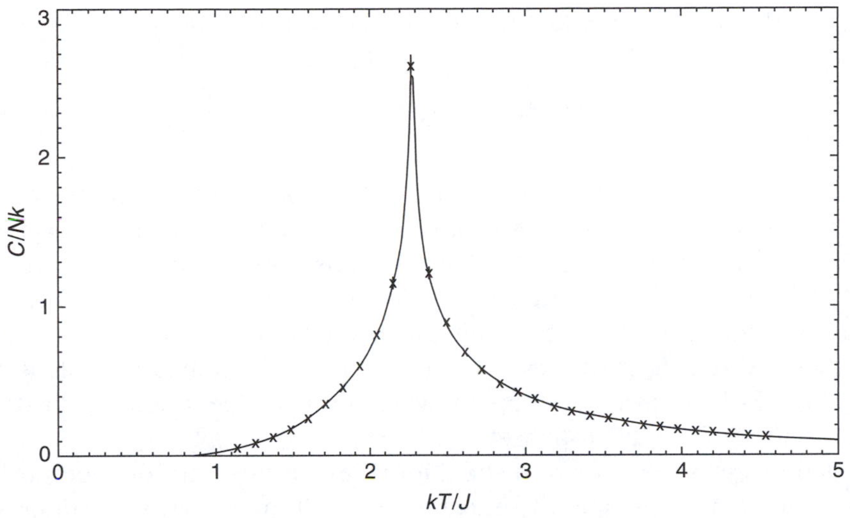

FIGURE 16.1 Monte Carlo specific heat (×'s) of the two-dimensional Ising model on a 128 × 128 lattice, as compared to the exact solution (solid line) from Section 13.4.A; see Kaufman (1948), Ferdinand and Fisher (1967), and Beale (1996). The MC error bars are smaller than the symbols used, except near the bulk critical temperature $T_c(\infty)$. Each data point represents an average using 10^5 Monte Carlo sweeps, except at the bulk critical point where 10^6 Monte Carlo sweeps were used to mitigate critical slowing down.

16.3 Molecular dynamics

The purpose of a molecular dynamics simulation is to integrate Newton's equations of motion for the set of particles in the given system. One advantage of MD over MC is that it approximates the time evolution of the equations of motion of the system, so MD can be used to study a host of dynamical properties. MD is usually more efficient at simulating systems with long-range interactions since all the particles are updated together. MD is sometimes easier to implement than MC for complex systems since appropriate MC moves are sometimes difficult to derive. There are MD variants that allow simulations of other ensembles, but the simplest case simulates a microcanonical ensemble in which the microstate of the given system explores its energy surface in the phase space; see Allen and Tildesley (1990) and Frenkel and Smit (2002) for details.

The equations of motion here are

$$\frac{d^2\boldsymbol{r}_i}{dt^2} = \frac{1}{m_i}\boldsymbol{F}_i = -\frac{1}{m_i}\nabla_i U(\boldsymbol{r}_1, \boldsymbol{r}_2, .., \boldsymbol{r}_N), \tag{1}$$

where $\boldsymbol{F}_i$ is the force on particle i arising from the N-particle potential energy function U. The MD simulation moves the system forward in time by discrete steps Δt. The most commonly used integration method in this context is due to Verlet (1967):

$$\boldsymbol{r}_i(t+\Delta t) = 2\boldsymbol{r}_i(t) - \boldsymbol{r}_i(t-\Delta t) + \frac{(\Delta t)^2}{m_i}\boldsymbol{F}_i(t). \tag{2}$$

This is equivalent to the leap-frog and velocity Verlet algorithms that update both positions and velocities of the particles; see Frenkel and Smit (2002). The Verlet method preserves the time-reversal symmetry of the Hamiltonian equations of motion and has an error per step of order $(\Delta t)^4$, while only requiring a single determination of the force on each particle, which is usually the most computationally time-consuming part of the simulation. Most importantly, the Verlet algorithm is symplectic, so the integration is equivalent to an exact solution of a "nearby" ghost Hamiltonian, which results in good long-term stability and good conservation of energy properties.

A simulation starts the system in some initial microstate with defined positions and velocities of all the particles, and the integration algorithm steps the positions and velocities of the particles forward in time. The simplest forms of the approximations employed for the velocities and the energy are

$$\boldsymbol{v}_i(t) = \frac{\boldsymbol{r}_i(t+\Delta t) - \boldsymbol{r}_i(t-\Delta t)}{2\Delta t}, \tag{3a}$$

$$E = \sum_{i=1}^{N} \frac{m_i}{2}\left(\frac{\boldsymbol{r}_i(t+\Delta t) - \boldsymbol{r}_i(t-\Delta t)}{2\Delta t}\right)^2 + U[\boldsymbol{r}_1(t), \boldsymbol{r}_2(t), .., \boldsymbol{r}_N(t)]. \tag{3b}$$

The time-step Δt is chosen to be small as compared to the shortest fundamental time scale in the Hamiltonian, while not so small as to limit the efficiency of the program. The numerical integration approximates a member of the microcanonical ensemble moving along the constant-energy hypersurface in the phase space. Calculating equilibrium averages properly depends on the Hamiltonian being ergodic;[6] this allows the system to sample all regions of the constant-energy hypersurface, so the MD time-averages are equivalent to averages over the microcanonical ensemble. If the total energy drifts more than some predetermined amount during the course of the simulation, then all the velocities can be rescaled to shift the total energy back to its initial value. Alternatively, one can use a thermostat to maintain the temperature at a desired value; see Frenkel and Smit (2002).

[6]Since an MD simulation creates a time evolution of the model system, one needs some assurance that the system is ergodic, that is, the time averages and the ensemble averages are the same. For example, a system of harmonic oscillators is not ergodic. A system of N particles in d dimensions has a $6N$-dimensional phase space, so the constant-energy hypersurface has $6N-1$ dimensions. The normal mode solution for N coupled oscillators has $3N$ constants of the motion, so the system explores only a $3N$-dimensional hypersurface. Even making the couplings between particles anharmonic does not eliminate the problem as first shown by Fermi, Pasta, and Ulam (1955) and explored theoretically by Kolmogorov (1954), Arnold (1963), and Moser (1962). MD simulations of equilibrium systems presume that the system is ergodic. There are only a few systems that are provably ergodic but, fortunately, most systems with realistic pair potentials appear to behave ergodically in two or more dimensions. In view of this, MD simulations of one-dimensional systems should be treated as suspect from this perspective.

A commonly used pair interaction for monatomic fluids such as neon and argon is the Lennard-Jones interaction

$$u(r) = 4\varepsilon\left(\left(\frac{D}{r}\right)^{12} - \left(\frac{D}{r}\right)^{6}\right), \tag{4}$$

where D is the molecular diameter and ε is the depth of the attractive well. The Lennard-Jones potential is attractive at long distances, and decays as $1/r^6$ to model the van der Waals attraction; at short distances, it diverges as $1/r^{12}$ to model the Pauli repulsion that prevents overlap of the electronic wavefunctions. Simulations are best carried out using dimensionless parameters. In the case of a fluid with Lennard-Jones interactions, all lengths can be measured in units of D, all energies (including kT) in units of ε, all forces in units of ε/D, all pressures in units of ε/D^3, all times in units of $\sqrt{mD^2/\varepsilon}$, and so on. Simulations then need to be conducted only for single values of reduced temperature kT/ε, the reduced density nD^3, and so on, while measuring observables in reduced units. Comparisons between simulations and experimental results can then be made using experimental values of D, m, and ε.[7] In dimensionless units, the Lennard-Jones force between a pair of particles is

$$\boldsymbol{F} = \pm\frac{\boldsymbol{r}}{r^2}\left(\frac{48}{r^{12}} - \frac{24}{r^6}\right). \tag{5}$$

Newton's third law of motion can be used to reduce the number of force calculations by a factor of two. The Lennard-Jones model was first studied in an MC simulation by Wood and Parker (1957) and in an MD simulation by Rahman (1964) and Verlet (1967).

16.3.A Molecular dynamics algorithm

First, start the system by choosing an initial state by setting the initial positions and velocities of all the particles. The initial velocities are usually set by choosing each component of the velocity vector of each particle from the Maxwell distribution. In reduced units, this is

$$P_{\text{Maxwell}}(v_x(0)) = \frac{1}{\sqrt{2\pi T}}\exp\left(-\frac{v_x^2(0)}{2T}\right); \tag{6}$$

[7]For example, the Lennard-Jones parameters appropriate for argon are $\varepsilon/k = 119.8\,\text{K}$ and $D = 0.3405\,\text{nm}$ (Levelt, 1960; Rowley, Nicholson, and Parsonage, 1975). Interaction potentials are almost always cut off at a finite distance between molecules to reduce the number of interactions that need to be considered at each time step. For the Lennard-Jones interaction, this is most commonly done at $r_{\max} = 2.5D$. If the potential is set to zero for distances greater than $r_{\max}$, then this would leave a small discontinuity in the potential. To eliminate this, the potential is often shifted upward by $-u(2.5D) \simeq 0.0163\varepsilon$, so that the potential is zero at $r_{\max}$. This allows a direct comparison between MC and MD simulations. If the shift is not made in the potential, one could not directly compare the results from MC and MD simulations because the discontinuity in the potential would result in a delta-function force that affects the motion in the MD simulation but not the configurations in the MC simulation. Comparisons of MC and MD results with experiments need to include perturbations from the shift and the missing tails of the pair potentials.

see Appendix I to see how to use a uniform pseudorandom number generator to select from a Gaussian distribution. The initial velocities can then be used to set the positions of the particles after the first time-step, namely

$$\boldsymbol{r}_i(\Delta t) = \boldsymbol{r}_i(0) + \boldsymbol{v}_i(0)\Delta t + \frac{1}{2}\frac{(\Delta t)^2}{m_i}\boldsymbol{F}_i(0). \tag{7}$$

1. Next, use equation (2) to move the system forward in time through $M_{\text{eq}} = \tau_{\text{eq}}/\Delta t$ time steps. The equilibration time τ_{eq} must be chosen large enough for the system to equilibrate; see the Monte Carlo discussion in Section 16.2. A thermostat is often used to evolve the system to a state with the desired temperature; see Frenkel and Smit (2002).
2. Now, use equation (2) to move the system forward in time through $M = \tau_{\text{avg}}/\Delta t$ time steps while keeping track of all the thermodynamic variables $\{A(q_j)\}$ one wants to measure. Finally, use equation (16.1.2) to determine the equilibrium averages and uncertainties.

To determine averages at a different set of parameters (temperature, density, etc.), change the parameters by a small amount and repeat steps 1 and 2. Using the last configuration of the previous run as the first configuration of the new run can often reduce the equilibration time.

16.4 Particle simulations

Fluids can be modeled by both MC and MD simulations by placing N particles in a periodic box with volume V interacting via a pair potential. Hansen and McDonald (1986), Allen and Tildesley (1990), and Frenkel and Smit (2002) provide excellent surveys of this topic. Calculating energy changes of trial moves in MC or forces in MD only involves pairs of particles whose closest periodic copies are within the cutoff distance of each other. MC simulations typically sample the canonical ensemble,[8] so they control the temperature and density, and measure the energy, pressure, and so on. MD simulations typically sample the microcanonical ensemble,[9] so they control the energy and density, and they measure the temperature, pressure, and so on. The equipartition theorem gives for the temperature

[8]Monte Carlo simulations of isobaric or grand canonical ensembles are also widely used by adding PV or μN terms to the Hamiltonian (Frenkel and Smit, 2002).

[9]Molecular dynamics simulations of other ensembles are possible. For example, one can include extra dynamical variables that allow the total energy or volume to fluctuate in order to approximate a canonical or isobaric ensemble; see Frenkel and Smit (2002). Variants of MC and MD simulations that span the range between the two include hybrid Monte Carlo methods that mix MC and MD methods into one code to take advantage of the strengths of both methods. Alternatively, one can include Langevin random force terms and damping in an MD simulation to create coupling to a heat bath.

T in a d-dimensional system

$$kT = \frac{1}{Nd}\left\langle \sum_{i=1}^{N} m_i v_i^2 \right\rangle. \tag{1}$$

The virial equation of state in equations (3.7.15) and (10.7.11) can be used to determine the pressure P in either type of simulation:

$$\frac{P}{nkT} = 1 + \frac{1}{NdkT}\left\langle \sum_{i<j} \boldsymbol{F}(\boldsymbol{r}_{ij}) \cdot \boldsymbol{r}_{ij} \right\rangle = 1 - \frac{n}{2dkT}\int \frac{du}{dr} r g(r) d\boldsymbol{r}. \tag{2}$$

As discussed in Section 10.7, the pair correlation function of a fluid can be used to measure a variety of thermodynamic properties including the pressure, the isothermal compressibility κ_T, and the scattering structure factor $S(k)$; see equations (10.7.18) through (10.7.21). The pair correlation function $g(r)$ defined in equation (10.7.5) can be determined by collecting a histogram of the distances between all pairs of particles periodically during the simulation, accounting for the periodic boundary conditions, and scaling the histogram by an amount proportional to the volume of shells of radius r and thickness Δr. In three dimensions, the pair correlation function is given by

$$g(r) = \frac{2V}{N^2\left(\frac{4\pi}{3}\right)\left[(r+\Delta r)^3 - r^3\right]}\left\langle \sum_{i<j}^{N} \Delta_{\Delta r}(r_{ij} - r) \right\rangle, \tag{3}$$

where the step function $\Delta_{\Delta r}(\xi)$ is unity for $0 < \xi < \Delta r$ and zero otherwise. This expression is the ratio of the number of events in each bin in the histogram compared to the average number that would be expected for an ideal gas with the same density.

16.4.A Simulations of hard spheres

The system of hard spheres has been studied extensively in both MC and MD simulations, and was the first model studied using either method (Metropolis et al., 1953; Adler and Wainwright, 1957, 1959). The pair potential for hard spheres is

$$u(r) = \begin{cases} 0 & \text{for } r > D, \\ \infty & \text{for } r \le D. \end{cases} \tag{4}$$

Temperature is an irrelevant parameter for the spatial configurations sampled by this model since the pair potential does not have a finite energy scale. A full exploration of the phase diagram involves only varying the *reduced* number density nD^d. All thermodynamic properties are either independent of temperature, or scale with temperature in a trivial way. For example, the scaled pressure for a system of hard spheres P/nkT is a function only of the reduced number density nD^d. The hard sphere density is often expressed in terms of the packing fraction η, the fraction of the volume of the system actually occupied by the spheres. In three dimensions, the volume fraction is given by $\eta = \pi n D^3/6$. Since the pair potential is singular, the pressure cannot be calculated using the virial equation (2) but the pressure can be determined using the virial equation of state for hard spheres, namely (10.7.12).

An MC code for hard spheres is relatively simple since the energy change in a trial move is either zero or infinity. A trial displacement of a particle is rejected if the trial position of the particle is within a distance D of any other particle, and is accepted otherwise. This was the first statistical physics model ever studied in a computer simulation (Metropolis et al., 1953).

Implementing MD for hard spheres requires a different approach from the standard MD. Finite-difference integration methods will not work here since the potential is not differentiable. Instead, one can exploit the exact solution to the equations of motion. Each particle travels in a straight line at a constant velocity except at the instants when pairs of particles collide, that is when they are a distance D apart. Due to the singular nature of the potential, the collisions can be uniquely time-ordered. Each collision conserves both kinetic energy and momentum, so the velocities after each collision can be determined analytically from the velocities and the displacement vector between the centers of the two particles at the moment of the collision. The changes in the velocities of the two colliding particles are

$$\Delta \boldsymbol{v}_i = -\,\Delta \boldsymbol{v}_j = \left.\frac{-(\boldsymbol{r}_{ij}\cdot \boldsymbol{v}_{ij})\,\boldsymbol{r}_{ij}}{D^2}\right|_{|\boldsymbol{r}_{ij}|=D}, \tag{5}$$

where $\boldsymbol{r}_{ij}$ and $\boldsymbol{v}_{ij}$ are, respectively, the relative positions and relative velocities of the two particles. The simulation moves the particles forward in time from collision to collision and changes the velocities of the pairs of particles involved in the collisions, as given by equation (5). This was the first implementation of the MD method in statistical physics (Alder and Wainwright 1957, 1959).

The pair correlation function and the structure factor in the fluid phase are shown in Figure 16.2 and the phase diagram for hard spheres in Figure 16.3. At low densities, the equilibrium phase is a short-range ordered fluid. At high densities, the equilibrium phase of the model is a long-range ordered, face-centered cubic solid. It is worthwhile to note that an attractive interaction is *not* required for a model to have a crystalline phase. An attractive interaction *is*, however, required for the formation of a liquid–vapor coexistence line and a critical point. For this reason, the low-density phase of the hard sphere model is often referred to as a fluid phase rather than a liquid phase since the model does not have a liquid–vapor coexistence line. The liquid and solid volume fractions at the liquid–solid coexistence line are $\eta_l \simeq 0.491 \pm 0.002$ and $\eta_s \simeq 0.543 \pm 0.002$, respectively. The liquid–solid coexistence pressure is given by $P^*_{ls} = P_{ls}D^3/kT \simeq 11.55 \pm 0.11$; see Speedy (1997). In the low-density fluid phase for $\eta < \eta_l$, the reduced pressure is accurately modeled by the Carnahan–Starling equation of state (10.3.25)

$$\frac{P}{nkT} = \frac{1+\eta+\eta^2-\eta^3}{(1-\eta)^3}, \tag{6}$$

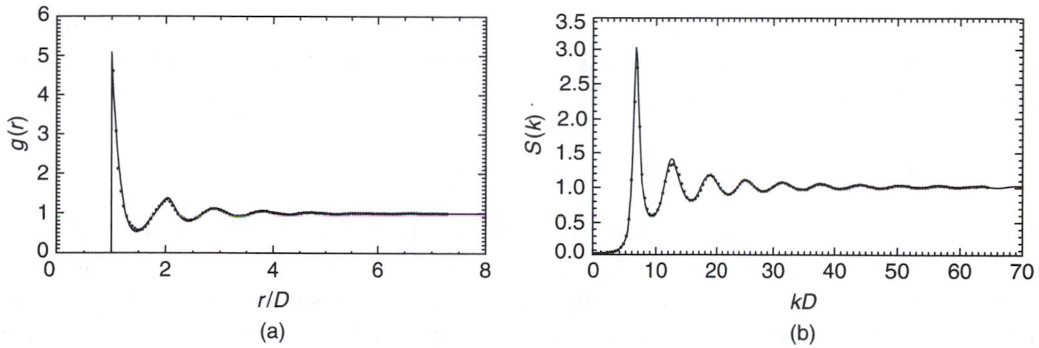

FIGURE 16.2 (a) The pair correlation function $g(r)$ and (b) the static structure factor $S(k)$ for a three-dimensional system of hard spheres at volume fraction $\eta = 0.49$ from a Monte Carlo simulation of 2916 particles (M. Glaser, unpublished). This value of the volume fraction is in the liquid phase close to the solid–liquid coexistence line. The solid lines depict the pair correlation function and the static structure factor from the Percus–Yevick approximation; see Percus and Yevick (1958), Wertheim (1963), and Hansen and McDonald (1986).

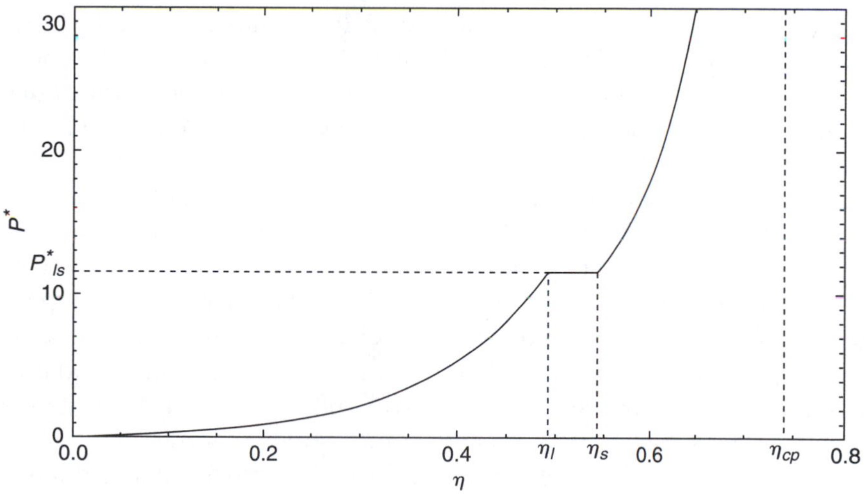

FIGURE 16.3 Sketch of the equilibrium phase diagram for hard spheres in three dimensions. The horizontal axis is the volume fraction $\eta = \pi n D^3/6$ and the vertical axis is the scaled pressure $P^* = PD^3/kT$. There are two equilibrium phases: a low-density fluid phase for $0 < \eta < \eta_l$ and a high-density solid phase for $\eta_s < \eta < \eta_{cp}$.

although there are other good parametizations as well. In the solid phase the pressure is approximately given by

$$\frac{P}{nkT} = \frac{3}{1-\eta^*} - 0.5921\frac{\eta^* - 0.7072}{\eta^* - 0.601}, \tag{7}$$

where $\eta^*(= \eta/\eta_{cp})$ is the ratio of the actual packing fraction to the maximum close-packed value η_{cp}, namely $\pi\sqrt{2}/6 \simeq 0.7405$ (Speedy, 1997; Frenkel and Smit, 2002). The pressure in the solid phase diverges as the density approaches the close-packed density. The model also exhibits a metastable disordered phase for densities between η_l and the random close-packed volume fraction $\eta_{rcp} \simeq 0.644 \pm 0.005$; see Rintoul and Torquato (1996). For a survey of hard sphere results, see Mulero et al. (2008).

16.5 Computer simulation caveats

Computer simulations are widely used in statistical physics and have played an important role in our understanding of many physical systems. Simulations complement theory and experiment and provide many advantages for the study of systems that are not amenable to exact or approximate theoretical analysis. However, it is important to understand the inherent limitation of this technique.

- Computer simulations necessarily involve a limited number of degrees of freedom, typically hundreds to thousands of particles or spins. This is not nearly large enough to display many of the behaviors that occur in thermodynamically large systems. For example, a model of a dense system of 1,000 particles in a three-dimensional cubic box will only have about 10 particles along each linear dimension of the box, so correlations beyond about five particle diameters are affected by the periodic boundary conditions. Extraction of accurate thermodynamic behavior from such a study will often involve an analysis of the finite-size scaling behavior of the model for a sequence of systems with different sizes.
- Computer simulations necessarily involve a limited time-scale of simulation. Typical molecular time scales are of the order of $t_0 \approx a/v$, where a a microscopic length scale and v a molecular velocity. For atomic scales near room temperature, $a \approx 0.1\,\text{nm}$ and $v \approx 100\,\text{m/s}$, which gives $t_0 \approx 10^{-12}\,\text{s}$. In an MD simulation, each time step moves the system forward in time by an amount Δt, which must be much less than t_0 in order to aptly integrate the equations of motion, say $\Delta t \approx 10^{-14}\,\text{s}$. A simulation that moves the system forward through 10^6 time steps will sample a physical time of only 10 ns, which may not be sufficient to reach many important time-scales of interest in the problem. This is especially problematic when the system has inherently slow time-scales, such as the critical slowing down near second-order phase transitions and the hysteresis near first-order phase transitions. Monte Carlo simulations are similarly hampered in that the simulation must run long enough for the model to explore a sufficiently large region of the phase space to capture the equilibrium behavior. In favorable cases, this and the previous issue can be mitigated by special simulation methods such as coarse graining, cluster update methods, parallel tempering, multicanonical Monte Carlo methods, Monte Carlo renormalization group, and so on; see footnote 5.
- Interactions between particles are usually highly simplified for computational efficiency and the interaction range is usually cut off. Long-range interactions (Couloub, dipole, van der Waals, etc.) need to be resummed or treated via perturbation theory to try to account for their effects.

- MC and MD simulations do not directly measure the number of microstates available to the system, so one cannot directly calculate the entropy or the free energy in the same way as other observables. If, for example, a determination of the free energy is necessary to locate a phase transition, then it can be determined by a thermodynamic integration to a state with a theoretically known free energy; see Frenkel and Smit (2002).
- MD simulations depend on the ergodicity of the Hamiltonian, so one-dimensional models, models that are weakly perturbed from mechanical equilibrium, and other nearly integrable models may get trapped in low-dimensional orbits that do not fully explore the constant-energy hypersurface of the Hamiltonian; see footnote 6.
- All pseudorandom number generators produce some level of correlation in their sequences. Even subtle correlations in pseudorandom number sequences can produce erroneous results in Monte Carlo simulations. Different classes of generators have different weaknesses, so switching to a generator based on a different algorithm will sometimes cure a problem caused by correlations produced by a particular generator. Testing a generator before using it in a simulation is always a good idea; see Appendix I.
- It is extremely important to confirm the validity of MC and MD simulation codes. This process is rather different from verifying a theoretical calculation. Some code evaluation and verification procedures include: testing the code initially on small systems with known properties, testing the code whenever possible against models with exact solutions or models that have been widely studied in the literature, examining results as a function of system size and run length, and retesting carefully whenever new interactions or code modules are added. In this connection, Frenkel and Smit (2002), Parker (2008), and Landau and Binder (2009) provide lists of good strategies.

Problems

16.1. Write a code to test a uniform pseudorandom number generator. If you do not have a canned generator available, write a generator based on L'Ecuyer's recommended generator in Appendix I. Apply the following tests: average $\langle x \rangle = 1/2$, variance $\langle x^2 \rangle - \langle x \rangle^2 = 1/12$, and the pair correlations test $\langle x_{i+k} x_i \rangle = 1/4$ for $k \neq 0$. Generate a histogram of pairs of numbers on a two-dimensional unit square and test that the distribution is statistically uniform.

16.2. Write a code to test a Gaussian pseudorandom number generator. If you do not have a canned generator available, write a generator based on the Box-Muller algorithm in Appendix I. Apply the following tests: average $\langle x \rangle = 0$, variance $\langle x^2 \rangle = 1$, and the pair correlations test $\langle x_{i+k} x_i \rangle = 0$ for $k \neq 0$. Generate a histogram of pairs of numbers in two dimensions and test that the distribution is statistically Gaussian.

16.3. Define a sequence of correlated random numbers

$$s_k = \alpha s_{k-1} + (1 - \alpha) r_k,$$

where r_k is a unit-variance, uncorrelated, Gaussian pseudorandom number while $0 < \alpha < 1$ defines the range of the correlations. Show that this sequence is Gaussian distributed, with a zero mean. Determine the variance in terms of α and compare your result with equation (16.1.2b). Write a code to determine the correlation function (16.1.3). Plot your measured correlation function and compare it to the exact correlation function.

16.4. Write a Monte Carlo code for a system of N hard spheres of diameter D on a one-dimensional ring of length L with periodic boundary conditions. Calculate the pair correlation function and compare it to equations (13.1.6) and (13.1.7). The pressure of the system is given by $P/nkT = 1 + nDg(D^+)$; see equation (10.7.12). Compare your pressure to the one obtained for the exact configurational partition function $Z_N = L(L - ND)^{N-1}/N!$; see equation (13.1.2).

16.5. Write a Monte Carlo code for a fluid of N hard spheres in a two-dimensional $L \times L$ square box with periodic boundary conditions in each direction. Calculate the pair correlation function and determine the scaled pressure using equation (10.7.12), namely $P/nkT = 1 + 2\eta g(D^+)$. Compare this pressure to the approximate form $P/nkT = (1 + \eta/8)/(1 - \eta)^2$.

16.6. Write a Monte Carlo code for a fluid of N hard spheres in a two-dimensional $L \times L$ square box and include a one-body gravity term $\sum_{i=1}^{N} mgy_i$ in the algorithm, that is, accept otherwise legal configurations with probability $\exp(-\beta mg\Delta y)$. You will need to use hard-wall boundary conditions on the top and bottom walls. Determine the average number density as a function of the vertical position in the box.

16.7. Write a molecular dynamics code for N Lennard-Jones particles in a two-dimensional $L \times L$ square box. Apply periodic boundary conditions in each direction. Determine the scaled pressure using the virial equation (16.4.2). Calculate and plot the pair correlation function of the system.

16.8. Write a molecular dynamics code for N Lennard-Jones particles in a two-dimensional $L \times L$ square box, and include a one-body gravity term in the energy: $\sum_{i=1}^{N} mgy_i$. Apply periodic boundary conditions in the x-direction but a repulsive WCA (Weeks, Chandler, and Andersen, 1971) potential on the top and bottom walls. The WCA potential is the repulsive part of a Lennard-Jones potential for $r/D < (2)^{1/6}$, with the potential shifted up by ε. Show that the average kinetic energy per particle is independent of the height y in the box but the average scaled density nD^2 depends on the vertical position in the box.

16.9. Write an MC code to simulate the one-dimensional Ising model on a periodic lattice of length L. Calculate the internal energy and specific heat of the model and compare them to equations (13.2.15) and (13.2.16). Calculate the correlation function $G(n) = \langle s_{i+n} s_i \rangle$ and it compare to equation (13.2.32).

16.10. Write an MC code to simulate the two-dimensional nearest-neighbor Ising model on a periodic $L \times L$ lattice in zero field. Calculate the internal energy and the specific heat of the system as functions of temperature and compare them to the exact results in section 13.4.A. See exact results for the two-dimensional Ising model for various lattice sizes at www.elsevierdirect.com.

16.11. Write an MC code to simulate the two-dimensional nearest-neighbor Ising model on a periodic $L \times L$ lattice in zero field. Calculate the energy distribution $P(E)$ over a range of temperatures including the critical point. Use this distribution to calculate the internal energy and the specific heat as functions of temperature. See exact results for the two-dimensional Ising model for various lattice sizes at www.elsevierdirect.com.

16.12. Write an MC code to simulate the one-dimensional XY model. Calculate the internal energy, the specific heat, the isothermal susceptibility, and the pair correlation function, and compare your results to the analytical results for the $n = 2$ case in Section 13.2.

16.13. Write an MC code to simulate the two-dimensional XY model. Calculate the internal energy, specific heat, isothermal susceptibility and the pair correlation function, and compare your results to the theoretical results given in Section 13.7.

Appendices

A Influence of boundary conditions on the distribution of quantum states

In this appendix we examine, under different boundary conditions, the asymptotic distribution of single-particle states in a bounded continuum. For simplicity, we consider a cuboidal enclosure of sides a, b, and c. The admissible solutions of the free-particle Schrödinger equation

$$\nabla^2\psi + k^2\psi = 0 \quad \left(\boldsymbol{k} = \boldsymbol{p}\hbar^{-1}\right), \tag{1}$$

which satisfy Dirichlet boundary conditions (namely, $\psi = 0$ everywhere at the boundary), are then given by

$$\psi_{lmn}(\boldsymbol{r}) \propto \sin\left(\frac{l\pi x}{a}\right)\sin\left(\frac{m\pi y}{b}\right)\sin\left(\frac{n\pi z}{c}\right), \tag{2}$$

with

$$k = \pi\left(\frac{l^2}{a^2} + \frac{m^2}{b^2} + \frac{n^2}{c^2}\right)^{1/2}; \quad l, m, n = 1, 2, 3, \ldots. \tag{3}$$

Note that in this case none of the quantum numbers l, m, or n can be zero, for that would make the wavefunction identically vanish. If, on the other hand, we impose Neumann boundary conditions (namely, $\partial\psi/\partial n = 0$ everywhere at the boundary), the desired solutions turn out to be

$$\psi_{lmn}(\boldsymbol{r}) \propto \cos\left(\frac{l\pi x}{a}\right)\cos\left(\frac{m\pi y}{b}\right)\cos\left(\frac{n\pi z}{c}\right), \tag{4}$$

with

$$l, m, n = 0, 1, 2, \ldots; \tag{5}$$

clearly, the value zero of the quantum numbers is now allowed! In each case, however, the negative-integral values of the quantum numbers do not lead to any new wavefunctions.

The total number $g(K)$ of distinct wavefunctions ψ, with wave number k not exceeding a given value K, may be written as

$$g(K) = {\sum_{l,m,n}}' f(l, m, n), \tag{6}$$

where $f(l,m,n) = 1$ for the numbers (l,m,n) belonging to the set (3) or (5), as the case may be; the summation $\sum'$ in each case is restricted by the condition

$$\left(\frac{l^2}{a^2} + \frac{m^2}{b^2} + \frac{n^2}{c^2}\right) \leq \frac{K^2}{\pi^2}. \tag{7}$$

We now define a sum

$$G(K) = \sum_{l,m,n}{}' f^*(l,m,n), \tag{8}$$

where $f^*(l,m,n) = 1$ for *all* integral values of l, m, and n (positive, negative, or zero), the restriction on the numbers (l,m,n) being the same as stated in (7). One can then show, by setting up correspondence of terms, that

$$\sum_{l,m,n}{}' f(l,m,n) = \frac{1}{8}\left[\sum_{l,m,n}{}' f^*(l,m,n) \mp \left\{\sum_{l,m}{}' f^*(l,m,0) + \sum_{l,n}{}' f^*(l,0,n) + \sum_{m,n}{}' f^*(0,m,n)\right\}\right.$$
$$\left. + \left\{\sum_{l}{}' f^*(l,0,0) + \sum_{m}{}' f^*(0,m,0) + \sum_{n}{}' f^*(0,0,n)\right\} \mp 1\right]; \tag{9}$$

the upper and the lower signs here correspond, respectively, to the Dirichlet and the Neumann boundary conditions.

Clearly, the first sum on the right side of (9) denotes the number of lattice points in the ellipsoid[1] $(X^2/a^2 + Y^2/b^2 + Z^2/c^2) = K^2/\pi^2$, the next three sums denote the numbers of lattice points in the ellipses, which are cross-sections of this ellipsoid with the Z-, Y- and X-planes, while the last three sums denote the numbers of lattice points on the principal axes of the ellipsoid. Now, if a, b, and c are sufficiently large in comparison with π/K, one may replace these numbers by the corresponding volume, areas, and lengths, respectively, with the result

$$g(K) = \frac{K^3}{6\pi^2}(abc) \mp \frac{K^2}{8\pi}(ab + ca + bc) + \frac{K}{4\pi}(a + b + c) \mp \frac{1}{8} + E(K); \tag{10}$$

the term $E(K)$ here denotes the net error committed in making the aforementioned replacements. We thus find that the main term of our result is directly proportional to the volume of the enclosure while the first correction term is proportional to its surface area (and, hence, represents a "surface effect"); the next-order term(s) appear in the nature of an "edge effect" and a "corner effect."

Now, a reference to the literature dealing with the determination of the "number of lattice points in a given domain" reveals that the error term $E(K)$ in equation (10) is $O(K^\alpha)$, where $1 < \alpha < 1.4$; hence, expression (10) for $g(K)$ is reliable only up to the surface term. In view of this, we may write

$$g(K) = \frac{K^3}{6\pi^2}V \mp \frac{K^2}{16\pi}S + \text{a lower-order remainder}; \tag{11}$$

[1] By the term "in the ellipsoid" we mean "not external to the ellipsoid," that is, the lattice points "on the ellipsoid" are also included. Other such expressions in the sequel carry a similar meaning.

in terms of ε^*, where

$$\varepsilon^* = \frac{8mL^2}{h^2}\varepsilon = \frac{4L^2}{h^2}P^2 = \frac{L^2}{\pi^2}K^2, \tag{12}$$

equation (11) reduces to equations (1.4.15) and (1.4.16) of the text.

In the case of periodic boundary conditions, namely

$$\psi(x,y,z) = \psi(x+a,y,z) = \psi(x,y+b,z) = \psi(x,y,z+c), \tag{13}$$

the appropriate wavefunctions are

$$\psi_{lmn}(\boldsymbol{r}) \propto \exp\{i(\boldsymbol{k}\cdot\boldsymbol{r})\}, \tag{14}$$

with

$$\boldsymbol{k} = 2\pi\left(\frac{l}{a}, \frac{m}{b}, \frac{n}{c}\right); \quad l,m,n = 0, \pm1, \pm2, \ldots. \tag{15}$$

The number of free-particle states $g(K)$ is now given by

$$g(K) = \sum_{l,m,n}{}' f^*(l,m,n), \tag{16}$$

such that

$$(l^2/a^2 + m^2/b^2 + n^2/c^2) \le K^2/(4\pi^2). \tag{17}$$

This is precisely the number of lattice points in the ellipsoid with semiaxes $Ka/2\pi$, $Kb/2\pi$, and $Kc/2\pi$, which, allowing for the approximation made in the earlier cases, is just equal to the volume term in (11). Thus, in the case of periodic boundary conditions, we do not obtain a surface term in the expression for the density of states.

For further information on this topic, see Fedosov (1963, 1964), Pathria (1966), Chaba and Pathria (1973), and Baltes and Hilf (1976).

B Certain mathematical functions

In this appendix we outline the main properties of certain mathematical functions that are of special importance to the subject matter of this text.

We start with the *gamma function* $\Gamma(\nu)$, which is identical with the *factorial function* $(\nu-1)!$ and is defined by the integral

$$\Gamma(\nu) \equiv (\nu-1)! = \int_0^\infty e^{-x}x^{\nu-1}dx; \quad \nu > 0. \tag{1}$$

First of all, we note that

$$\Gamma(1) \equiv 0! = 1. \tag{2}$$

Next, integrating by parts, we obtain the recurrence formula

$$\Gamma(\nu) = \frac{1}{\nu}\Gamma(\nu+1), \tag{3}$$

from which it follows that

$$\Gamma(\nu+1) = \nu(\nu-1)\cdots(1+p)p\Gamma(p), \quad 0 < p \le 1, \tag{4}$$

p being the fractional part of ν. For integral values of ν ($\nu = n$, say), we have the familiar representation

$$\Gamma(n+1) \equiv n! = n(n-1)\cdots 2\cdot 1; \tag{5}$$

on the other hand, if ν is a half-odd integral ($\nu = m + \frac{1}{2}$, say), then

$$\Gamma\left(m+\tfrac{1}{2}\right) \equiv \left(m-\tfrac{1}{2}\right)! = \left(m-\tfrac{1}{2}\right)\left(m-\tfrac{3}{2}\right)\cdots\tfrac{3}{2}\cdot\left(\tfrac{1}{2}\right)\Gamma\left(\tfrac{1}{2}\right) = \frac{(2m-1)(2m-3)\cdots 3\cdot 1}{2^m}\pi^{1/2}, \tag{6}$$

where use has been made of equation (21), whereby

$$\Gamma\left(\tfrac{1}{2}\right) \equiv \left(-\tfrac{1}{2}\right)! = \pi^{1/2}. \tag{7}$$

By repeated application of the recurrence formula (3), the definition of the function $\Gamma(\nu)$ can be extended to all ν, except for $\nu = 0, -1, -2, \ldots$ where the singularities of the function lie. The behavior of $\Gamma(\nu)$ in the neighborhood of a singularity can be determined by setting $\nu = -n + \varepsilon$, where $n = 0, 1, 2, \ldots$ and $|\varepsilon| \ll 1$, and using formula (3) $n+1$ times; we get

$$\begin{aligned}\Gamma(-n+\varepsilon) &= \frac{1}{(-n+\varepsilon)(-n+1+\varepsilon)\cdots(-1+\varepsilon)\varepsilon}\Gamma(1+\varepsilon)\\ &\approx \frac{(-1)^n}{n!\,\varepsilon}.\end{aligned} \tag{8}$$

Replacing x by αy^2, equation (1) takes the form

$$\Gamma(\nu) = 2\alpha^\nu \int_0^\infty e^{-\alpha y^2} y^{2\nu-1} dy, \quad \nu > 0. \tag{9}$$

We thus obtain another closely related integral, namely

$$I_{2\nu-1} \equiv \int_0^\infty e^{-\alpha y^2} y^{2\nu-1} dy = \frac{1}{2\alpha^\nu}\Gamma(\nu), \quad \nu > 0; \tag{10}$$

by a change of notation, this can be written as

$$I_\nu \equiv \int_0^\infty e^{-\alpha y^2} y^\nu dy = \frac{1}{2\alpha^{(\nu+1)/2}}\Gamma\left(\frac{\nu+1}{2}\right), \quad \nu > -1. \tag{11}$$

One can easily see that the foregoing integral satisfies the relationship

$$I_{\nu+2} = -\frac{\partial}{\partial \alpha} I_\nu. \tag{12}$$

The integrals I_ν appear so frequently in our study that we write down the values of some of them explicitly:

$$I_0 = \frac{1}{2}\left(\frac{\pi}{\alpha}\right)^{1/2}, \quad I_2 = \frac{1}{4}\left(\frac{\pi}{\alpha^3}\right)^{1/2}, \quad I_4 = \frac{3}{8}\left(\frac{\pi}{\alpha^5}\right)^{1/2} \cdots, \tag{13a}$$

while

$$I_1 = \frac{1}{2\alpha}, \quad I_3 = \frac{1}{2\alpha^2}, \quad I_5 = \frac{1}{\alpha^3}, \cdots. \tag{13b}$$

In connection with these integrals, it may as well be noted that

$$\int_{-\infty}^{\infty} e^{-\alpha y^2} y^\nu dy = 0 \qquad \text{if } \nu \text{ is an odd integer}$$

$$= 2I_\nu \quad \text{if } \nu \text{ is an even integer.} \tag{14}$$

Next, we consider the product of two gamma functions, say $\Gamma(\mu)$ and $\Gamma(\nu)$. Using representation (9), with $\alpha = 1$, we have

$$\Gamma(\mu)\Gamma(\nu) = 4\int_0^\infty \int_0^\infty e^{-(x^2+y^2)} x^{2\mu-1} y^{2\nu-1} dx\, dy, \quad \mu > 0, \nu > 0. \tag{15}$$

Changing over to the polar coordinates (r,θ), equation (15) becomes

$$\Gamma(\mu)\Gamma(\nu) = 4\int_0^\infty e^{-r^2} r^{2(\mu+\nu)-1} dr \int_0^{\pi/2} \cos^{2\mu-1}\theta \sin^{2\nu-1}\theta d\theta = 2\Gamma(\mu+\nu)\int_0^{\pi/2} \cos^{2\mu-1}\theta \sin^{2\nu-1}\theta d\theta. \tag{16}$$

Now, defining the *beta function* $B(\mu,\nu)$ by the integral

$$B(\mu,\nu) = 2\int_0^{\pi/2} \cos^{2\mu-1}\theta \sin^{2\nu-1}\theta d\theta, \quad \mu > 0, \nu > 0, \tag{17}$$

we obtain an important relationship:

$$B(\mu,\nu) = \frac{\Gamma(\mu)\Gamma(\nu)}{\Gamma(\mu+\nu)} = B(\nu,\mu). \tag{18}$$

Substituting $\cos^2\theta = \eta$, equation (17) takes the standard form

$$B(\mu,\nu) = \int_0^1 \eta^{\mu-1}(1-\eta)^{\nu-1} d\eta, \quad \mu > 0, \nu > 0, \tag{19}$$

while the special case $\mu = \nu = \frac{1}{2}$ gives

$$B\left(\tfrac{1}{2}, \tfrac{1}{2}\right) = 2 \int_0^{\pi/2} d\theta = \pi. \tag{20}$$

Coupled with equations (2) and (18), equation (20) yields

$$\Gamma\left(\tfrac{1}{2}\right) = \pi^{1/2}. \tag{21}$$

Stirling's formula for $\nu!$

We now derive an asymptotic expression for the factorial function

$$\nu! = \int_0^{\infty} e^{-x} x^{\nu} dx \tag{22}$$

for $\nu \gg 1$. It is not difficult to see that, for $\nu \gg 1$, the major contribution to this integral comes from the region of x that lies around the point $x = \nu$ and has a width of order $\surd\nu$. In view of this, we invoke the substitution

$$x = \nu + (\surd\nu)u, \tag{23}$$

whereby equation (22) takes the form

$$\nu! = \surd\nu \left(\frac{\nu}{e}\right)^{\nu} \int_{-\surd\nu}^{\infty} e^{-(\surd\nu)u} \left(1 + \frac{u}{\surd\nu}\right)^{\nu} du. \tag{24}$$

The integrand in (24) attains its maximum value, unity, at $u = 0$ and on both sides of the maximum it falls rapidly to zero, which suggests that it may be approximated by a Gaussian. We, therefore, expand the logarithm of the integrand around $u = 0$ and then reconstruct the integrand by taking the exponential of the resulting expression; this gives

$$\nu! = \surd\nu \left(\frac{\nu}{e}\right)^{\nu} \int_{-\surd\nu}^{\infty} \exp\left\{-\frac{u^2}{2} + \frac{u^3}{3\surd\nu} - \frac{u^4}{4\nu} + \cdots\right\} du. \tag{25}$$

If ν is sufficiently large, the integrand in (25) may be replaced by the single factor $\exp(-u^2/2)$; moreover, since the major contribution to this integral comes only from that range of u for which $|u|$ is of order unity, the lower limit of integration may be replaced by $-\infty$. We thus obtain the *Stirling formula*

$$\nu! \approx \surd(2\pi\nu)(\nu/e)^{\nu}, \quad \nu \gg 1. \tag{26}$$

A more detailed analysis leads to the Stirling series

$$\nu! = \surd(2\pi\nu) \left(\frac{\nu}{e}\right)^{\nu} \left[1 + \frac{1}{12\nu} + \frac{1}{288\nu^2} - \frac{139}{51840\nu^3} - \frac{571}{2488320\nu^4} + \cdots\right]. \tag{27}$$

Next, we consider the function $\ln(\nu!)$. Corresponding to formula (27), we have, for large ν,

$$\ln(\nu!) = \left(\nu + \frac{1}{2}\right)\ln\nu - \nu + \frac{1}{2}\ln(2\pi) + \left[\frac{1}{12\nu} - \frac{1}{360\nu^3} + \frac{1}{1260\nu^5} - \frac{1}{1680\nu^7} + \cdots\right]. \tag{28}$$

For most practical purposes, we may write

$$\ln(\nu!) \approx (\nu\ln\nu - \nu). \tag{29}$$

We note that formula (29) can be obtained very simply by an application of the Euler–Maclaurin formula. Since ν is large, we may consider its integral values only; then, by definition,

$$\ln(n!) = \sum_{i=1}^{n}(\ln i).$$

Replacing summation by integration, we obtain

$$\ln(n!) \simeq \int_1^n (\ln x)dx = (x\ln x - x)\Big|_{x=1}^{x=n}$$
$$\approx (n\ln n - n),$$

which is identical to (29).

We must, however, be warned that, whereas approximation (29) is fine as it is, it would be wrong to take its exponential and write $\nu! \approx (\nu/e)^\nu$, for that would affect the evaluation of $\nu!$ by a factor $O(\nu^{1/2})$, which can be considerably large; see (26). In the expression for $\ln(\nu!)$, the corresponding term is indeed negligible.

The Dirac δ-function

We start with the Gaussian distribution function

$$p(x, x_0, \sigma) = \frac{1}{\sqrt{(2\pi)}\sigma} e^{-(x-x_0)^2/2\sigma^2}, \tag{30}$$

which satisfies the normalization condition

$$\int_{-\infty}^{\infty} p(x, x_0, \sigma)dx = 1. \tag{31}$$

The function $p(x)$ is symmetric about the value x_0 where it has a maximum; the height of this maximum is inversely proportional to the parameter σ while its width is directly proportional to σ, the total area under the curve being a constant. As σ becomes smaller and smaller, the function $p(x)$ becomes narrower and narrower in width and grows higher and higher at the central point x_0, condition (31) being satisfied at all σ; see Figure B.1.

In the limit $\sigma \to 0$, we obtain a function whose value at $x = x_0$ is infinitely large while at $x \neq x_0$ it is vanishingly small, the area under the curve being still equal to unity. This limiting form of the

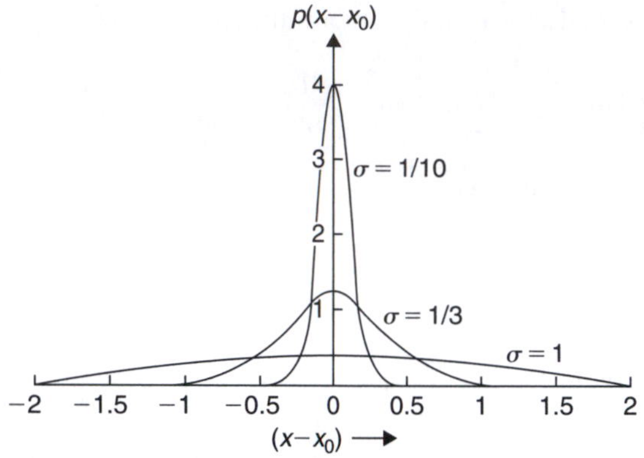

FIGURE B.1 Gaussian distribution function (30) for different values of σ.

function $p(x, x_0, \sigma)$ is, in fact, the δ-function of Dirac. Thus, we may define this function as the one satisfying the following properties:

$$\text{(i)} \quad \delta(x - x_0) = 0 \quad \text{for all } x \neq x_0, \tag{32}$$

$$\text{(ii)} \quad \int_{-\infty}^{\infty} \delta(x - x_0)dx = 1. \tag{33}$$

Conditions (32) and (33) inherently imply that, at $x = x_0$, $\delta(x - x_0) = \infty$ and that the range of integration in (33) need not extend all the way from $-\infty$ to $+\infty$; in fact, any range that includes the point $x = x_0$ would suffice. Thus, we may rewrite (33) as

$$\int_A^B \delta(x - x_0)dx = 1 \quad \text{if } A < x_0 < B. \tag{34}$$

It follows that, for any well-behaved function $f(x)$,

$$\int_A^B f(x)\delta(x - x_0)dx = f(x_0) \quad \text{if } A < x_0 < B. \tag{35}$$

Another limiting process frequently employed to represent the δ-function is the following:

$$\delta(x - x_0) = \operatorname*{Lim}_{\gamma \to 0} \frac{\gamma}{\pi\{(x - x_0)^2 + \gamma^2\}}. \tag{36}$$

To see the appropriateness of this representation, we note that, for $x \neq x_0$, this function vanishes like γ while, for $x = x_0$, it diverges like γ^{-1}; moreover, for all γ,

$$\int_{-\infty}^{\infty} \frac{\gamma}{\pi\{(x-x_0)^2+\gamma^2\}} dx = \frac{1}{\pi}\left[\tan^{-1}\frac{(x-x_0)}{\gamma}\right]_{-\infty}^{\infty} = 1. \tag{37}$$

An integral representation of the δ-function is

$$\delta(x-x_0) = \frac{1}{2\pi}\int_{-\infty}^{\infty} e^{ik(x-x_0)}\, dk, \tag{38}$$

which means that the δ-function is the "Fourier transform of a constant." We note that, for $x = x_0$, the integrand in (38) is unity throughout, so the function diverges. On the other hand, for $x \neq x_0$, the oscillatory character of the integrand is such that it makes the integral vanish. And, finally, the integration of this function, over a range of x that includes the point $x = x_0$, gives

$$\frac{1}{2\pi}\int_{-\infty}^{\infty}\left[\int_{x_0-L}^{x_0+L} e^{ik(x-x_0)}\, dx\right] dk = \int_{-\infty}^{\infty} \frac{e^{ikL}-e^{-ikL}}{2\pi(ik)} dk = \int_{-\infty}^{\infty} \frac{\sin(kL)}{\pi k} dk = 1, \tag{39}$$

independently of the choice of L.

It is instructive to see how the integral representation of the δ-function is related to its previous representations. For this, we introduce into the integrand of (38) a convergence factor $\exp(-\gamma k^2)$, where γ is a small, positive number. The resulting function, in the limit $\gamma \to 0$, should reproduce the δ-function; we thus expect that

$$\delta(x-x_0) = \frac{1}{2\pi}\operatorname*{Lim}_{\gamma\to 0}\int_{-\infty}^{\infty} e^{ik(x-x_0)-\gamma k^2}\, dk. \tag{40}$$

The integral in (40) is easy to evaluate if we recall that

$$\int_{-\infty}^{\infty} \cos(kx)e^{-\gamma k^2} dk = 2\int_{0}^{\infty} \cos(kx)e^{-\gamma k^2} dk = \sqrt{\left(\frac{\pi}{\gamma}\right)}\, e^{-x^2/4\gamma}, \tag{41}$$

while

$$\int_{-\infty}^{\infty} \sin(kx)e^{-\gamma k^2} dk = 0. \tag{42}$$

Accordingly, equation (40) becomes

$$\delta(x-x_0) = \operatorname*{Lim}_{\gamma\to 0}\frac{1}{\sqrt{(4\pi\gamma)}}e^{-(x-x_0)^2/4\gamma}, \tag{43}$$

which is precisely the representation we started with.[2]

Finally, the notation of the δ-function can be readily extended to spaces with more than one dimension. For instance, in n dimensions,

$$\delta(\mathbf{r}) = \delta(x_1)\cdots\delta(x_n), \tag{44}$$

so that

$$\text{(i)} \quad \delta(\mathbf{r}) = 0 \quad \text{for all } r \neq 0, \tag{45}$$

$$\text{(ii)} \quad \int_{-\infty}^{\infty}\cdots\int_{-\infty}^{\infty} \delta(\mathbf{r})dx_1\cdots dx_n = 1. \tag{46}$$

The integral representation of $\delta(\mathbf{r})$ is

$$\delta(\mathbf{r}) = \frac{1}{(2\pi)^n}\int_{-\infty}^{\infty}\cdots\int_{-\infty}^{\infty} e^{i(\mathbf{k}\cdot\mathbf{r})}d^nk. \tag{47}$$

Once again, we may write

$$\delta(\mathbf{r}) = \frac{1}{(2\pi)^n}\operatorname*{Lim}_{\gamma\to 0}\int_{-\infty}^{\infty}\cdots\int_{-\infty}^{\infty} e^{i(\mathbf{k}\cdot\mathbf{r})-\gamma k^2}d^nk \tag{48}$$

$$= \operatorname*{Lim}_{\gamma\to 0}\left(\frac{1}{4\pi\gamma}\right)^{n/2} e^{-r^2/4\gamma}. \tag{49}$$

C "Volume" and "surface area" of an *n*-dimensional sphere of radius *R*

Consider an n-dimensional space in which the position of a point is denoted by the vector $\mathbf{r}$, with Cartesian components $(x_1,\ldots,x_n)$. The "volume element" dV_n in this space would be

$$d^nr = \prod_{i=1}^{n}(dx_i);$$

accordingly, the "volume" V_n of a sphere of radius R would be given by

$$V_n(R) = \int\cdots\int_{0\le\sum_{i=1}^{n}x_i^2\le R^2} \prod_{i=1}^{n}(dx_i). \tag{1}$$

[2]The reader may check that the introduction into (38) of a convergence factor $\exp(-\gamma|k|)$, rather than $\exp(-\gamma k^2)$, leads to the representation (36).

Obviously, V_n will be proportional to R^n, so let us write it as

$$V_n(R) = C_n R^n, \tag{2}$$

where C_n is a constant that depends only on the dimensionality of the space. Clearly, the "volume element" dV_n can also be written as

$$dV_n = S_n(R)dR = nC_nR^{n-1}dR, \tag{3}$$

where $S_n(R)$ denotes the "surface area" of the sphere.

To evaluate C_n, we make use of the formula

$$\int_{-\infty}^{\infty} \exp(-x^2)dx = \pi^{1/2}. \tag{4}$$

Multiplying n such integrals, one for each x_i, we obtain

$$\pi^{n/2} = \int_{x_i=-\infty}^{x_i=\infty} \cdots \int \exp\left(-\sum_{i=1}^{n} x_i^2\right) \prod_{i=1}^{n} (dx_i) = \int_0^{\infty} \exp(-R^2) nC_nR^{n-1}dR = nC_n \cdot \frac{1}{2}\Gamma\left(\frac{n}{2}\right) = \left(\frac{n}{2}\right)! C_n; \tag{5}$$

here, use has been made of formula (B.11), with $\alpha = 1$. Thus,

$$C_n = \pi^{n/2} \Big/ \left(\frac{n}{2}\right)!, \tag{6}$$

so that

$$V_n(R) = \frac{\pi^{n/2}}{(n/2)!}R^n \quad \text{and} \quad S_n(R) = \frac{2\pi^{n/2}}{\Gamma(n/2)}R^{n-1}, \tag{7a,b}$$

which are the desired results.

Alternatively, one may prefer to use spherical polar coordinates right from the beginning — as, for instance, in the evaluation of the Fourier transform

$$I(\boldsymbol{k}) = \int f(r)e^{i\boldsymbol{k}\cdot\boldsymbol{r}}d^n r. \tag{8}$$

In that case,

$$d^n r = r^{n-1}(\sin\theta_1)^{n-2} \cdots (\sin\theta_{n-2})^1 dr\, d\theta_1 \cdots d\theta_{n-2} d\phi, \tag{9}$$

where the θ_i range from 0 to π while ϕ ranges from 0 to 2π. Choosing our polar axis to be in the direction of $\boldsymbol{k}$, equation (8) takes the form

$$I(\boldsymbol{k}) = \int f(r)e^{ikr\cos\theta_1} r^{n-1}(\sin\theta_1)^{n-2} \cdots (\sin\theta_{n-2})^1 dr\, d\theta_1 \cdots d\theta_{n-2} d\phi. \tag{10}$$

Integration over the angular coordinates $\theta_1, \theta_2, \theta_3, \ldots, \theta_{n-2}$ and ϕ yields factors

$$\pi^{1/2}\Gamma\left(\frac{n-1}{2}\right)\left(\frac{2}{kr}\right)^{(n-2)/2} J_{(n-2)/2}(kr) \times B\left(\frac{n-2}{2}, \frac{1}{2}\right) \cdot B\left(\frac{n-3}{2}, \frac{1}{2}\right) \cdots B\left(1, \frac{1}{2}\right) \cdot 2\pi,$$

where $J_\nu(x)$ is the ordinary Bessel function while $B(\mu, \nu)$ is the beta function; see equations (B.17) and (B.18). Equation (10) now becomes

$$I(\boldsymbol{k}) = (2\pi)^{n/2} \int_0^\infty f(r) \left(\frac{1}{kr}\right)^{(n-2)/2} J_{(n-2)/2}(kr) r^{n-1} dr, \tag{11}$$

which is our main result.

In the limit $k \to 0$, $J_\nu(kr) \to \left(\frac{1}{2}kr\right)^\nu / \Gamma(\nu + 1)$, so that

$$I(0) = \frac{2\pi^{n/2}}{\Gamma(n/2)} \int_0^\infty f(r) r^{n-1} dr, \tag{12}$$

consistent with (3) and (7b). On the other hand, if we take $f(r)$ to be a constant, say $1/(2\pi)^n$, we should obtain another representation of the Dirac δ-function in n dimensions; see equations (8) and (B.47). We thus have, from (11),

$$\delta(\boldsymbol{k}) = \frac{1}{(2\pi)^{n/2}} \int_0^\infty \left(\frac{1}{kr}\right)^{(n-2)/2} J_{(n-2)/2}(kr) r^{n-1} dr. \tag{13}$$

As a check, we introduce a factor $\exp(-\alpha r^2)$ in the integrand of (13) and obtain

$$\delta(\boldsymbol{k}) = \operatorname*{Lim}_{\alpha \to 0} \frac{1}{(2\pi)^{n/2}} \int_0^\infty e^{-\alpha r^2} \left(\frac{1}{kr}\right)^{(n-2)/2} J_{(n-2)/2}(kr) r^{n-1} dr = \operatorname*{Lim}_{\alpha \to 0} \left(\frac{1}{4\pi\alpha}\right)^{n/2} e^{-k^2/4\alpha}, \tag{14}$$

in complete agreement with (B.49). If, on the other hand, we use the factor $\exp(-\alpha r)$ rather than $\exp(-\alpha r^2)$, we get

$$\delta(\boldsymbol{k}) = \operatorname*{Lim}_{\alpha \to 0} \Gamma\left(\frac{n+1}{2}\right) \frac{\alpha}{\{\pi(k^2 + \alpha^2)\}^{(n+1)/2}}, \tag{15}$$

which generalizes (B.36).

D On Bose–Einstein functions

In the theory of Bose–Einstein systems we come across integrals of the type

$$G_\nu(z) = \int_0^\infty \frac{x^{\nu-1} dx}{z^{-1} e^x - 1} \quad (0 \le z < 1, \nu > 0; z = 1, \nu > 1). \tag{1}$$

In this appendix we study the behavior of $G_\nu(z)$ over the stated range[3] of the parameter z. First of all, we note that

$$\operatorname*{Lim}_{z\to 0} G_\nu(z) = \int_0^\infty z e^{-x} x^{\nu-1} dx = z\Gamma(\nu). \tag{2}$$

Hence, it appears useful to introduce another function, $g_\nu(z)$, such that

$$g_\nu(z) \equiv \frac{1}{\Gamma(\nu)} G_\nu(z) = \frac{1}{\Gamma(\nu)} \int_0^\infty \frac{x^{\nu-1} dx}{z^{-1} e^x - 1}. \tag{3}$$

For small z, the integrand in (3) may be expanded in powers of z, with the result

$$g_\nu(z) = \frac{1}{\Gamma(\nu)} \int_0^\infty x^{\nu-1} \sum_{l=1}^\infty (ze^{-x})^l dx = \sum_{l=1}^\infty \frac{z^l}{l^\nu} = z + \frac{z^2}{2^\nu} + \frac{z^3}{3^\nu} + \cdots; \tag{4}$$

thus, for $z \ll 1$, the function $g_\nu(z)$, *for all* ν, behaves like z itself. Moreover, $g_\nu(z)$ is a monotonically increasing function of z whose largest value in the physical range of interest obtains when $z \to 1$; then, for $\nu > 1, g_\nu(z)$ approaches the Riemann *zeta function* $\zeta(\nu)$:

$$g_\nu(1) = \sum_{l=1}^\infty \frac{1}{l^\nu} = \zeta(\nu) \quad (\nu > 1). \tag{5}$$

The numerical values of some of the $\zeta(\nu)$ are

$$\zeta(2) = \frac{\pi^2}{6} \simeq 1.64493, \quad \zeta(4) = \frac{\pi^4}{90} \simeq 1.08232, \quad \zeta(6) = \frac{\pi^6}{945} \simeq 1.01734,$$

$$\zeta\left(\tfrac{3}{2}\right) \simeq 2.61238, \quad \zeta\left(\tfrac{5}{2}\right) \simeq 1.34149, \quad \zeta\left(\tfrac{7}{2}\right) \simeq 1.12673,$$

and, finally,

$$\zeta(3) \simeq 1.20206, \quad \zeta(5) \simeq 1.03693, \quad \zeta(7) \simeq 1.00835.$$

For $\nu \le 1$, the *function* $g_\nu(z)$ diverges as $z \to 1$. The case $\nu = 1$ is rather simple, for the function $g_\nu(z)$ now assumes a closed form:

$$g_1(z) = \int_0^\infty \frac{dx}{z^{-1} e^x - 1} = \ln(1 - ze^{-x}) \Big|_0^\infty = -\ln(1 - z). \tag{6}$$

As $z \to 1$, $g_1(z)$ diverges logarithmically. Setting $z = e^{-\alpha}$, we have

$$g_1(e^{-\alpha}) = -\ln(1 - e^{-\alpha}) \xrightarrow[\alpha \to 0]{} \ln(1/\alpha). \tag{7}$$

[3]The behavior of $G_\nu(z)$ for $z > 1$ has been discussed by Clunie (1954).

For $\nu < 1$, the behavior of $g_\nu(e^{-\alpha})$, as $\alpha \to 0$, can be determined as follows:

$$g_\nu(e^{-\alpha}) = \frac{1}{\Gamma(\nu)} \int_0^\infty \frac{x^{\nu-1}dx}{e^{\alpha+x}-1} \approx \frac{1}{\Gamma(\nu)} \int_0^\infty \frac{x^{\nu-1}dx}{\alpha+x}.$$

Setting $x = \alpha \tan^2\theta$ and making use of equation (B.17), we obtain

$$g_\nu(e^{-\alpha}) \approx \frac{\Gamma(1-\nu)}{\alpha^{1-\nu}} \quad (0 < \nu < 1). \tag{8}$$

Expression (8) isolates the singularity of the function $g_\nu(e^{-\alpha})$ at $\alpha = 0$; the remainder of the function can be expanded in powers of α, with the result (see Robinson, 1951)

$$g_\nu(e^{-\alpha}) = \frac{\Gamma(1-\nu)}{\alpha^{1-\nu}} + \sum_{i=0}^{\infty} \frac{(-1)^i}{i!} \zeta(\nu - i)\alpha^i, \tag{9}$$

$\zeta(s)$ being the Riemann zeta function analytically continued to all $s \neq 1$.

A simple differentiation of $g_\nu(z)$ brings out the recurrence relation

$$z\frac{\partial}{\partial z}[g_\nu(z)] \equiv \frac{\partial}{\partial(\ln z)} g_\nu(z) = g_{\nu-1}(z). \tag{10}$$

This relation follows readily from the series expansion (4) but can also be derived from the defining integral (3). We thus have

$$z\frac{\partial}{\partial z}[g_\nu(z)] = \frac{z}{\Gamma(\nu)} \int_0^\infty \frac{e^x x^{\nu-1}dx}{(e^x - z)^2}.$$

Integrating by parts, we get

$$z\frac{\partial}{\partial z}[g_\nu(z)] = \frac{z}{\Gamma(\nu)} \left[-\frac{x^{\nu-1}}{e^x - z}\Big|_0^\infty + (\nu - 1)\int_0^\infty \frac{x^{\nu-2}dx}{e^x - z} \right].$$

The integrated part vanishes at both limits (provided that $\nu > 1$), while the part yet to be integrated yields precisely $g_{\nu-1}(z)$. The validity of the recurrence relation (10) is thus established for all $\nu > 1$.

Adopting (10) as a part of the definition of the function $g_\nu(z)$, the notion of this function may be extended to *all* ν, including $\nu \le 0$. Proceeding in this manner, Robinson showed that equation (9) applied to all $\nu < 1$ and to all *nonintegral* $\nu > 1$. For $\nu = m$, a positive integer, we have instead

$$g_m(e^{-\alpha}) = \frac{(-1)^{m-1}}{(m-1)!} \left[\sum_{i=1}^{m-1} \frac{1}{i} - \ln\alpha \right] \alpha^{m-1} + \sum_{\substack{i=0 \\ i \neq m-1}}^{\infty} \frac{(-1)^i}{i!} \zeta(m-i)\alpha^i. \tag{11}$$

Equations (9) and (11) together provide a complete description of the function $g_\nu(e^{-\alpha})$ for small α; it may be checked that both these expressions conform to the recurrence relation

$$\frac{\partial}{\partial\alpha} g_\nu(e^{-\alpha}) = -g_{\nu-1}(e^{-\alpha}). \tag{12}$$

For the special cases $\nu = \frac{5}{2}, \frac{3}{2}$, and $\frac{1}{2}$ we obtain from (9)

$$g_{5/2}(\alpha) = 2.36\alpha^{3/2} + 1.34 - 2.61\alpha - 0.730\alpha^2 + 0.0347\alpha^3 + \cdots, \tag{13a}$$

$$g_{3/2}(\alpha) = -3.54\alpha^{1/2} + 2.61 + 1.46\alpha - 0.104\alpha^2 + 0.00425\alpha^3 + \cdots, \tag{13b}$$

$$g_{1/2}(\alpha) = 1.77\alpha^{-1/2} - 1.46 + 0.208\alpha - 0.0128\alpha^2 + \cdots. \tag{13c}$$

The terms quoted here are sufficient to yield a better than 1 percent accuracy for all $\alpha \le 1$. The numerical values of these functions have been tabulated by London (1954) over the range $0 \le \alpha \le 2$.

The values of several important integrals involving relativistic bosons in Chapters 7 and 9 are:

$$\int_0^\infty x^2 \ln(1 - e^{-x})dx = -2\zeta(4) = -\frac{\pi^4}{45}, \tag{14a}$$

$$\int_0^\infty \frac{x^2}{e^x - 1}dx = 2\zeta(3) \simeq 2.40411, \tag{14b}$$

$$\int_0^\infty \frac{x^3}{e^x - 1}dx = 6\zeta(4) = \frac{\pi^4}{15}. \tag{14c}$$

E On Fermi–Dirac functions

In the theory of Fermi–Dirac systems we come across integrals of the type

$$F_\nu(z) = \int_0^\infty \frac{x^{\nu-1}dx}{z^{-1}e^x + 1} \quad (0 \le z < \infty, \nu > 0). \tag{1}$$

In this appendix we study the behavior of $F_\nu(z)$ over the entire range of the parameter z. For the same reason as in the case of Bose–Einstein integrals, we introduce here another function, $f_\nu(z)$, such that

$$f_\nu(z) \equiv \frac{1}{\Gamma(\nu)}F_\nu(z) = \frac{1}{\Gamma(\nu)}\int_0^\infty \frac{x^{\nu-1}dx}{z^{-1}e^x + 1}. \tag{2}$$

For small z, the integrand in (2) may be expanded in powers of z, with the result

$$f_\nu(z) = \frac{1}{\Gamma(\nu)}\int_0^\infty x^{\nu-1}\sum_{l=1}^\infty (-1)^{l-1}(ze^{-x})^l dx = \sum_{l=1}^\infty (-1)^{l-1}\frac{z^l}{l^\nu} = z - \frac{z^2}{2^\nu} + \frac{z^3}{3^\nu} - \cdots; \tag{3}$$

thus, for $z \ll 1$, the function $f_\nu(z)$, *for all* ν, behaves like z itself.

The functions $f_\nu(z)$ are related to the Bose–Einstein functions $g_\nu(z)$ as follows:

$$f_\nu(z) = g_\nu(z) - 2^{1-\nu}g_\nu(z^2) \quad (0 \le z < 1,\ \nu > 0;\ z = 1,\ \nu > 1). \tag{4}$$

This is useful for determining the values of relativistic Fermi–Dirac integrals needed in Chapter 9:

$$\int_0^\infty x^2 \ln(1+e^{-x})dx = \frac{7\pi^4}{360}, \tag{5a}$$

$$\int_0^\infty \frac{x^2}{e^x+1}dx = \frac{3\zeta(3)}{2} \simeq 1.80309, \tag{5b}$$

$$\int_0^\infty \frac{x^3}{e^x+1}dx = \frac{7\pi^4}{120}. \tag{5c}$$

The functions $f_\nu(z)$ and $f_{\nu-1}(z)$ are connected through the recurrence relation

$$z\frac{\partial}{\partial z}[f_\nu(z)] \equiv \frac{\partial}{\partial(\ln z)}f_\nu(z) = f_{\nu-1}(z); \tag{6}$$

this relation follows readily from the series expansion (3) but can also be derived from the defining integral (2).

To study the behavior of Fermi–Dirac integrals for large z, we introduce the variable

$$\xi = \ln z, \tag{7}$$

so that

$$F_\nu(e^\xi) \equiv \Gamma(\nu)f_\nu(e^\xi) = \int_0^\infty \frac{x^{\nu-1}dx}{e^{x-\xi}+1}. \tag{8}$$

For large ξ, the situation in (8) is primarily controlled by the factor $(e^{x-\xi}+1)^{-1}$, whose departure from its limiting values — namely, zero (as $x \to \infty$) and almost unity (as $x \to 0$) — is significant only in the neighborhood of the point $x = \xi$; see Figure E.1. The width of this "region of significance" is $O(1)$ and hence *much smaller* than the total, effective range of integration, which is $O(\xi)$. Therefore, in the lowest approximation, we may replace the actual curve of Figure E.1 by a step function, as

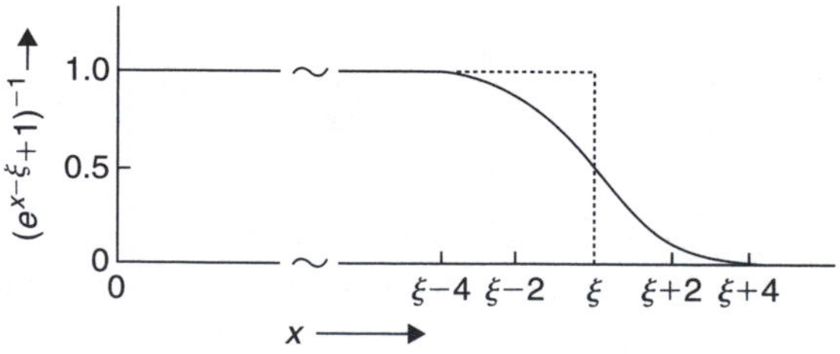

FIGURE E.1

shown by the dotted line. Equation (8) then reduces to

$$F_\nu(e^\xi) \approx \int_0^\xi x^{\nu-1} dx = \frac{\xi^\nu}{\nu} \tag{9}$$

and, accordingly,

$$f_\nu(e^\xi) \approx \frac{\xi^\nu}{\Gamma(\nu+1)}. \tag{10}$$

For a better approximation, we rewrite (8) as

$$F_\nu(e^\xi) = \int_0^\xi x^{\nu-1}\left[1 - \frac{1}{e^{\xi-x}+1}\right]dx + \int_\xi^\infty x^{\nu-1}\frac{1}{e^{x-\xi}+1}dx \tag{11}$$

and substitute in the respective integrals

$$x = \xi - \eta_1 \quad \text{and} \quad x = \xi + \eta_2, \tag{12}$$

with the result

$$F_\nu(e^\xi) = \frac{\xi^\nu}{\nu} - \int_0^\xi \frac{(\xi-\eta_1)^{\nu-1}d\eta_1}{e^{\eta_1}+1} + \int_0^\infty \frac{(\xi+\eta_2)^{\nu-1}d\eta_2}{e^{\eta_2}+1}. \tag{13}$$

Since $\xi \gg 1$ while our integrands are significant only for η of order unity, the upper limit in the first integral may be replaced by ∞. Moreover, one may use the same variable η in both the integrals, with the result

$$F_\nu(e^\xi) \approx \frac{\xi^\nu}{\nu} + \int_0^\infty \frac{(\xi+\eta)^{\nu-1} - (\xi-\eta)^{\nu-1}}{e^\eta+1}d\eta \tag{14}$$

$$= \frac{\xi^\nu}{\nu} + 2\sum_{j=1,3,5,\ldots}\binom{\nu-1}{j}\left[\xi^{\nu-1-j}\int_0^\infty \frac{\eta^j}{e^\eta+1}d\eta\right]; \tag{15}$$

in the last step the numerator in the integrand of (14) has been expanded in powers of η. Now,

$$\frac{1}{\Gamma(j+1)}\int_0^\infty \frac{\eta^j}{e^\eta+1}d\eta = 1 - \frac{1}{2^{j+1}} + \frac{1}{3^{j+1}} - \cdots = \left(1 - \frac{1}{2^j}\right)\zeta(j+1); \tag{16}$$

see equations (2) and (3), with $\nu = j+1$ and $z = 1$. Substituting (16) into (15), we obtain

$$f_\nu(e^\xi) = \frac{\xi^\nu}{\Gamma(\nu+1)}\left[1 + 2\nu\sum_{j=1,3,5\ldots}\left\{(\nu-1)\cdots(\nu-j)\left(1-\frac{1}{2^j}\right)\frac{\zeta(j+1)}{\xi^{j+1}}\right\}\right]$$

$$= \frac{\xi^\nu}{\Gamma(\nu+1)}\left[1 + \nu(\nu-1)\frac{\pi^2}{6}\frac{1}{\xi^2} + \nu(\nu-1)(\nu-2)(\nu-3)\frac{7\pi^4}{360}\frac{1}{\xi^4} + \cdots\right], \tag{17}$$

which is the desired asymptotic formula — commonly known as Sommerfeld's lemma (see Sommerfeld, 1928).[4]

By the same procedure, one can derive the following asymptotic result, which is clearly a generalization of (17):

$$\int_0^\infty \frac{\phi(x)dx}{e^{x-\xi}+1} = \int_0^\xi \phi(x)dx + \frac{\pi^2}{6}\left(\frac{d\phi}{dx}\right)_{x=\xi} + \frac{7\pi^4}{360}\left(\frac{d^3\phi}{dx^3}\right)_{x=\xi} + \frac{31\pi^6}{15120}\left(\frac{d^5\phi}{dx^5}\right)_{x=\xi} + \cdots, \qquad (18)$$

where $\phi(x)$ is any well-behaved function of x. It may be noted that the numerical coefficients in this expansion approach the limiting value 2.

Blakemore (1962) has tabulated numerical values of the function $f_\nu(e^\xi)$ in the range $-4 \le \xi \le +10$; his tables cover all integral orders from 0 to +5 and all half-odd integral orders from $-\frac{1}{2}$ to $+\frac{9}{2}$.

F A rigorous analysis of the ideal Bose gas and the onset of Bose–Einstein condensation

In this appendix we study the problem of the ideal Bose gas without arbitrarily extracting the condensate term ($\varepsilon = 0$) from the original sum for N in equation (7.2.1) and approximating the remainder by an integral ranging from $\varepsilon = 0$ to $\varepsilon = \infty$. We will instead evaluate the original sum as it is with the help of certain mathematical identities dating back to Poisson and Jacobi in the early nineteenth century. Luckily, these identities obviate the necessity of approximating sums by integrals and yield results valid for arbitrary values of N (though, for all practical purposes, we may assume that $N \gg 1$). For pertinent details of this procedure, see Pathria (1983) and the references quoted therein.

We consider an ideal Bose gas consisting of particles of mass m confined to the cubic geometry $L \times L \times L$ and subject to periodic boundary conditions so that the single-particle energy eigenvalues are given by

$$\varepsilon = \frac{h^2}{2mL^2}\left(n_1^2 + n_2^2 + n_3^2\right), \quad (n_i = 0, \pm 1, \pm 2, \ldots). \qquad (1)$$

[4]A more careful analysis carried out by Rhodes (1950), and followed by Dingle (1956), shows that the passage from equation (13) to (14) omits a term which, for large ξ, is of order $e^{-\xi}$. This term turns out to be $\cos\{(\nu-1)\pi\}\, F_\nu(e^{-\xi}) \equiv \cos\{(\nu-1)\pi\} F_\nu(1/z)$. For large z, this would be very nearly equal to $\cos\{(\nu-1)\pi\}/z$ and hence negligible in comparison with any of the terms appearing in (17). Of course, for $\nu = \frac{1}{2}, \frac{3}{2}, \frac{5}{2}, \ldots$, which are the values occurring in most of the important applications of Fermi–Dirac statistics, the missing term is identically zero.

For $\nu = 2$, the inclusion of the missing term leads to the identity

$$f_2(e^\xi) + f_2(e^{-\xi}) = \tfrac{1}{2}\xi^2 + \tfrac{\pi^2}{6},$$

which is relevant to the contents of Section 8.3.B.

The sum (7.1.2) then takes the form

$$N = \sum_{\varepsilon}\left(e^{\alpha+\beta\varepsilon} - 1\right)^{-1} = \sum_{\varepsilon}\sum_{j=1}^{\infty} e^{-j(\alpha+\beta\varepsilon)}$$

$$= \sum_{j=1}^{\infty} e^{-j\alpha}\left[\sum_{n_1} e^{-jwn_1^2}\sum_{n_2} e^{-jwn_2^2}\sum_{n_3} e^{-jwn_3^2}\right], \tag{2}$$

where $\alpha = -\mu/kT$, $\beta = 1/kT$, and

$$w = \frac{\beta h^2}{2mL^2} = \pi\left(\frac{\lambda}{L}\right)^2, \tag{3}$$

$\lambda\ (= h/\sqrt{2\pi mkT})$ being the mean thermal wavelength of the particles.

To evaluate the sums in (2), we make use of the Poisson summation formula (see Schwartz, 1966):

$$\sum_{n=-\infty}^{\infty} f(n) = \sum_{q=-\infty}^{\infty} F(q); \quad F(q) = \int_{-\infty}^{\infty} f(x)e^{2\pi iqx}dx. \tag{4}$$

The function $F(q)$ is, of course, the Fourier transform of the original function $f(n)$. Choosing $f(n) = e^{-jwn^2}$, we obtain the remarkable identity

$$\sum_{n=-\infty}^{\infty} e^{-jwn^2} = \sqrt{\frac{\pi}{jw}}\sum_{q=-\infty}^{\infty} e^{-\pi^2q^2/jw}. \tag{5}$$

It is instructive to note that the $q = 0$ term in (5) is precisely the result one would obtain if the summation over n were replaced by integration, as is customarily done in the treatment of this problem. Terms with $q \neq 0$, therefore, represent corrections that arise from the discreteness of the single-particle states. Using equation (5) for each of the three summations in (2), we obtain

$$N = \sum_{j=1}^{\infty} e^{-j\alpha}\prod_{i=1}^{3}\left[\sqrt{\frac{\pi}{jw}}\sum_{q_i} e^{-\pi^2q_i^2/jw}\right]$$

$$= \left(\frac{\pi}{w}\right)^{3/2}\sum_{\boldsymbol{q}}\sum_{j=1}^{\infty}\frac{e^{-j\alpha}}{j^{3/2}}\exp\left[-\frac{\pi^2}{jw}\left(q_1^2+q_2^2+q_3^2\right)\right]$$

$$= \frac{L^3}{\lambda^3}\sum_{j=1}^{\infty}\left[\frac{e^{-j\alpha}}{j^{3/2}} + \sum_{\boldsymbol{q}}{}'\frac{e^{-j\alpha}}{j^{3/2}}\exp\left(-\frac{\pi^2}{jw}\left(q_1^2+q_2^2+q_3^2\right)\right)\right], \tag{6}$$

where the primed summation in the second set of terms implies that the term with $\boldsymbol{q} = 0$ has been taken out of this sum.

The $\boldsymbol{q} = 0$ term in (6) is precisely the bulk result for the total number of particles in the excited states of the system, namely $Vg_{3/2}(e^{-\alpha})/\lambda^3$. In the second set of terms, the summation over j may be carried out with the help of a straightforward generalization of identity (4), namely

$$\sum_{n=a}^{b} f(n) = \frac{1}{2} f(a) + \frac{1}{2} f(b) + \sum_{l=-\infty}^{\infty} F_{a,b}(l); \quad F_{a,b}(l) = \int_{a}^{b} f(x) e^{2\pi i l x} dx, \tag{7}$$

where a and b are integers such that $b > a$. Applying (7) to the primed sum in (6), we obtain

$$\begin{aligned} \sum_{j=1}^{\infty} j^{-3/2} e^{-j\alpha} e^{-\gamma(\boldsymbol{q})/j} &= \sum_{j=0}^{\infty} j^{-3/2} e^{-j\alpha} e^{-\gamma(\boldsymbol{q})/j} \\ &= \sqrt{\frac{\pi}{\gamma(\boldsymbol{q})}} \sum_{l=-\infty}^{\infty} \exp\left[-2\sqrt{\gamma(\boldsymbol{q})}\,(\alpha + 2\pi i l)^{1/2}\right], \end{aligned} \tag{8}$$

where

$$\gamma(\boldsymbol{q}) = \frac{\pi^2 q^2}{w} = \frac{\pi L^2 q^2}{\lambda^2} > 0.$$

We readily note that, whatever the value of α, terms with $l \neq 0$ are at most of order $\exp(-L/\lambda)$ which, for $L \gg \lambda$, are altogether negligible. As a consequence, no errors of order $(\lambda/L)^n$ are committed if we retain only the term with $l = 0$. We thus obtain

$$N \approx \frac{L^3}{\lambda^3}\left[g_{3/2}\left(e^{-\alpha}\right) + \pi^{1/2}\alpha^{1/2} S(y)\right], \tag{9}$$

where

$$S(y) = \sum_{\boldsymbol{q}}{}' \frac{e^{-2R(\boldsymbol{q})}}{R(\boldsymbol{q})}; \quad R(\boldsymbol{q}) = y\sqrt{q_1^2 + q_2^2 + q_3^2}, \tag{10}$$

while y is given by

$$y = \pi^{1/2}\alpha^{1/2}\frac{L}{\lambda}. \tag{11}$$

In view of equation (13.6.36), the parameter y is a measure of the lateral dimension L of the system in terms of its correlation length ξ — to be precise, $y = L/2\xi$. We, therefore, expect that, as we lower the temperature of the system and enter the region of phase transition, this parameter will go from very large values to very small values over an infinitesimally small range of temperatures around the transition point. We'll examine this aspect of the problem a little later.

At this point, it is worthwhile to note that if the summation over $\boldsymbol{q}$ that appears in equation (10) is replaced by integration, which is justifiable only in the limit $y \to 0$, we obtain from equation (9)

$$N \approx \frac{L^3}{\lambda^3}\left[g_{3/2}\left(e^{-\alpha}\right) + \pi^{1/2}\alpha^{1/2}\left(\frac{\pi}{y^3}\right)\right] = \frac{L^3}{\lambda^3} g_{3/2}\left(e^{-\alpha}\right) + \frac{1}{\alpha}, \tag{12}$$

in perfect agreement with the bulk result obtained in Section 7.1, with $\alpha \ll 1$. To have an idea of the "degree of error" committed in making this replacement, we must evaluate this sum more accurately. For this, we make use of another mathematical identity, first established by Chaba and Pathria (1975), namely

$$\sum_{\boldsymbol{q}}{}' \left[\frac{y^2}{q^2 \left(y^2 + \pi^2 q^2\right)} + \frac{\pi}{q} e^{-2yq} \right] = 2\pi y + \frac{\pi^2}{y^2} + C_3, \tag{13}$$

where

$$C_3 = \pi \lim_{y \to 0} \left(\sum_{\boldsymbol{q}}{}' \frac{e^{-2yq}}{q} - \int_{\text{all } \boldsymbol{q}} \frac{e^{-2yq}}{q} d\boldsymbol{q} \right) \simeq -8.9136. \tag{14}$$

It is important to note that the constant C_3 is directly related to the Madelung constant of a simple cubic lattice; see, for instance, Harris and Monkhorst (1970). Now, since the second part of the sum appearing on the left side of (13) is directly proportional to $S(y)$, we can rewrite (9) in the form

$$N \approx \frac{L^3}{\lambda^3} g_{3/2}\left(e^{-\alpha}\right) + \frac{L^2}{\lambda^2} \frac{\pi}{y^2} + \frac{L^2}{\lambda^2} \left(\frac{C_3}{\pi} + 2y - \frac{y^2}{\pi} \sum_{\boldsymbol{q}}{}' \frac{1}{q^2 \left(y^2 + \pi^2 q^2\right)} \right). \tag{15}$$

We observe that the second term on the right side of (15) is equal to $1/\alpha$ — which is precisely N_0 when $\alpha \ll 1$. The condensate, therefore, emerges naturally in our analysis and does not have to be extracted prematurely, as is done in the customary treatment.

Now, in view of the fact that, for small α,

$$g_{3/2}\left(e^{-\alpha}\right) \approx \zeta\left(\frac{3}{2}\right) - 2\pi^{1/2}\alpha^{1/2}, \tag{16}$$

equation (15) is further simplified to

$$N \approx \frac{L^3}{\lambda^3} \zeta(3/2) + N_0 + \frac{L^2}{\lambda^2} \left(\frac{C_3}{\pi} - \frac{y^2}{\pi} \sum_{\boldsymbol{q}}{}' \frac{1}{q^2 \left(y^2 + \pi^2 q^2\right)} \right). \tag{17}$$

Introducing the bulk critical temperature $T_c(\infty)$, as defined in equation (7.1.24),

$$T_c(\infty) = \frac{h^2}{2\pi mk} \left[\frac{N}{L^3 \zeta\left(\frac{3}{2}\right)} \right]^{2/3} = T \frac{\lambda^2}{L^2} \left[\frac{N}{\zeta\left(\frac{3}{2}\right)} \right]^{2/3}; \tag{18}$$

we obtain from (17) the desired result, namely

$$N_0 = N \left[1 - \left(\frac{T}{T_c(\infty)} \right)^{3/2} \right] + \frac{L^2}{\lambda^2} \left(-\frac{C_3}{\pi} + \frac{y^2}{\pi} \sum_{\boldsymbol{q}}{}' \frac{1}{q^2 \left(y^2 + \pi^2 q^2\right)} \right). \tag{19}$$

The first part of this expression is the standard bulk result for N_0, while the second part represents the "finite-size correction" to this quantity. More important, while the main term here is of order N, the correction term is of order $N^{2/3}$. In the thermodynamic limit, the correction term loses its importance altogether and we are left with the conventional result following from the customary treatment.

Finally, we study the variation of the *scaling parameter* y as a function of T; this will also enable us to examine the manner in which the correlation length ξ and the condensate fraction f $(= N_0/N)$ build up as we move from temperatures above $T_c(\infty)$ to those below $T_c(\infty)$. For this, we introduce a *scaled temperature*, defined by $t = [T - T_c(\infty)]/T_c(\infty)$, and study the problem in three distinct regimes:

(a) For $t > 0$, such that $1 \gg t \gg N^{-1/3}$, we make use of the result for α, as stated in Problem 7.3. Combining this result with equation (11), we get

$$y \approx \frac{3}{4}\left[\zeta\left(\frac{3}{2}\right)\right]^{2/3} N^{1/3} t, \tag{20a}$$

$$\xi \approx \frac{2}{3}\left[\zeta\left(\frac{3}{2}\right)\right]^{-2/3} \ell t^{-1}, \tag{20b}$$

$$f \approx \frac{16\pi}{9}\left[\zeta\left(\frac{3}{2}\right)\right]^{-2} N^{-1} t^{-2}, \tag{20c}$$

where ℓ $(= L/N^{1/3})$ is the mean interatomic distance in the system.

(b) For $|t| = O(N^{-1/3})$, the parameter $y = O(1)$ and its value has to be determined numerically. At $t = 0$, this value is determined by the equation $S(y_0) = 2$; see equations (9) and (16). We thus get: $y_0 \simeq 0.973$. The correlation length ξ in this regime is $O(L)$ and the condensate fraction is $O(N^{-1/3})$.

(c) For $t < 0$, such that $1 \gg |t| \gg N^{-1/3}$, we get

$$y \approx \sqrt{\frac{2\pi}{3}}\left[\zeta\left(\frac{3}{2}\right)\right]^{-1/3} N^{-1/6} |t|^{-1/2}, \tag{21a}$$

$$\xi \approx \sqrt{\frac{3}{8\pi}}\left[\zeta\left(\frac{3}{2}\right)\right]^{1/3} \frac{L^{3/2}}{\ell^{1/2}} |t|^{1/2}, \tag{21b}$$

$$f \approx \frac{3}{2}|t|. \tag{21c}$$

We thus see how, over an infinitesimally small range of temperatures $O(N^{-1/3})$ around $t = 0$, the parameter y descends from values $O(N^{1/3})$ to values $O(N^{-1/6})$ while the correlation length ξ grows from values $O(\ell)$ to values $O(L^{3/2}/\ell^{1/2})$, and the condensate fraction f grows from values $O(N^{-1})$ to values $O(1)$. As $N \to \infty$, the transition region collapses onto a *singular* point $t = 0$ and the phenomenon of Bose–Einstein condensation becomes a *critical* one.

G On Watson functions

In this appendix we examine the asymptotic behavior of the functions

$$W_d(\phi) = \int_0^\infty e^{-\phi x}\left[e^{-x} I_0(x)\right]^d dx \tag{1}$$

for $0 \leq \phi \ll 1$. First of all, we note that if we set $\phi = 0$ the resulting integral converges only if $d > 2$. To see this, we observe that, with $\phi = 0$, convergence problems may arise in the limit of large x where the integrand

$$\left[e^{-x} I_0(x)\right]^d \approx (2\pi x)^{-d/2} \quad (x \gg 1). \tag{2}$$

Clearly, the integral will converge if $d > 2$; otherwise, it will diverge. We, therefore, conclude that

$$W_d(0) = \int_0^\infty [e^{-x} I_0(x)]^d\, dx \tag{3}$$

exists for $d > 2$.

Next we look at the derivative

$$W_d'(\phi) = -\int_0^\infty e^{-\phi x}[e^{-x} I_0(x)]^d x\, dx. \tag{4}$$

By the same argument as above, we conclude that

$$W_d'(0) = -\int_0^\infty \left[e^{-x} I_0(x)\right]^d x\, dx \tag{5}$$

exists for $d > 4$. The manner in which $W_d'(\phi)$ diverges for $d < 4$, as $\phi \to 0$, can be seen as follows:

$$\begin{aligned} W_d'(\phi) &= -\int_0^\infty e^{-y}\left[e^{-y/\phi} I_0(y/\phi)\right]^d \frac{1}{\phi^2} y\, dy \\ &\approx -\frac{1}{(2\pi)^{d/2}\phi^{(4-d)/2}} \int_0^\infty e^{-y} y^{(2-d)/2} dy \quad (\phi \ll 1) \\ &= -\frac{\Gamma\{(4-d)/2\}}{(2\pi)^{d/2}\phi^{(4-d)/2}}. \end{aligned} \tag{6}$$

Integrating (6) with respect to ϕ, and remembering the comments made earlier about $W_d(0)$, we obtain the desired results:

$$W_d(\phi) \approx \begin{cases} (2\pi)^{-d/2}\Gamma\{(2-d)/2\}\phi^{-(2-d)/2} + \text{const.} & \text{for} \quad d < 2 \qquad (7a) \\ (2\pi)^{-1}\ln(1/\phi) + \text{const.} & \text{for} \quad d = 2 \qquad (7b) \\ W_d(0) - (2\pi)^{-d/2}|\Gamma\{(2-d)/2\}|\phi^{(d-2)/2} & \text{for} \quad 2 < d < 4. \qquad (7c) \end{cases}$$

For $d > 4$, we have a simpler result:

$$W_d(\phi) \approx W_d(0) - |W_d'(0)|\phi, \tag{8}$$

for, in this case, both $W_d(0)$ and $W_d'(0)$ exist.

The borderline case $d = 4$ presents some problems that can be simplified by splitting the integral in (4) into two parts:

$$\int_0^\infty = \int_0^1 + \int_1^\infty. \tag{9}$$

The first part is clearly finite; the divergence of the function $W_4'(\phi)$, as $\phi \to 0$, arises from the second part which, for $\phi \ll 1$, can be written as

$$\approx \int_1^\infty e^{-\phi x}(2\pi x)^{-2} x\,dx = \frac{1}{4\pi^2}E_1(\phi), \tag{10}$$

where $E_1(\phi)$ is the exponential integral; see Abramowitz and Stegun (1964), Chapter 5. Since $E_1(\phi) \approx -\ln\phi$ for $\phi \ll 1$, we conclude that

$$W_4'(\phi) \approx \frac{1}{4\pi^2}\ln\phi. \tag{11}$$

Integrating (11) with respect to ϕ, we obtain

$$W_4(\phi) \approx W_4(0) - \frac{1}{4\pi^2}\phi\ln(1/\phi). \tag{12}$$

Equations (7), (8), and (12) constitute the main results of this appendix.

For the record, we quote a couple of numbers:

$$W_3(0) = 0.50546, \quad W_4(0) = 0.30987. \tag{13}$$

H Thermodynamic relationships

The following four equations relating partial derivatives are sometimes known as the *Four Famous Formulae*. They make it easy to derive thermodynamic relations in any one assembly or to convert relations from one assembly to another; by assembly we mean the set of thermodynamic parameters on which a system depends. If the quantities x, y, z are mutually related, then

$$\left(\frac{\partial x}{\partial y}\right)_z = 1 \Big/ \left(\frac{\partial y}{\partial x}\right)_z, \tag{1a}$$

$$\left(\frac{\partial}{\partial z}\left(\frac{\partial y}{\partial x}\right)_z\right)_x = \left(\frac{\partial}{\partial x}\left(\frac{\partial y}{\partial z}\right)_x\right)_z, \tag{1b}$$

$$\left(\frac{\partial x}{\partial y}\right)_z \left(\frac{\partial y}{\partial z}\right)_x \left(\frac{\partial z}{\partial x}\right)_y = -1, \tag{1c}$$

$$\left(\frac{\partial x}{\partial y}\right)_w = \left(\frac{\partial x}{\partial y}\right)_z + \left(\frac{\partial x}{\partial z}\right)_y \left(\frac{\partial z}{\partial y}\right)_w. \tag{1d}$$

Entropy $S(N, V, U)$ and the microcanonical ensemble

The entropy describes a closed, isochoric, adiabatic assembly, so it is a function of internal energy, volume, and number of molecules:

$$dS = \frac{1}{T}dU + \frac{P}{T}dV - \frac{\mu}{T}dN, \tag{2a}$$

$$\frac{1}{T} = \left(\frac{\partial S}{\partial U}\right)_{V,N}, \tag{2b}$$

$$\frac{P}{T} = \left(\frac{\partial S}{\partial V}\right)_{U,N}, \tag{2c}$$

$$\frac{\mu}{T} = -\left(\frac{\partial S}{\partial N}\right)_{U,V}. \tag{2d}$$

The Maxwell relations for the entropy are

$$\left(\frac{\partial(1/T)}{\partial V}\right)_{U,N} = \left(\frac{\partial(P/T)}{\partial U}\right)_{V,N}, \tag{3a}$$

$$\left(\frac{\partial(1/T)}{\partial N}\right)_{U,V} = -\left(\frac{\partial(\mu/T)}{\partial U}\right)_{V,N}, \tag{3b}$$

$$\left(\frac{\partial(P/T)}{\partial N}\right)_{U,V} = -\left(\frac{\partial(\mu/T)}{\partial V}\right)_{U,N}. \tag{3c}$$

The entropy is determined from the number of microstates in the microcanonical ensemble

$$\Gamma(N, V, U; \Delta U) = \mathrm{Tr}\big(\Delta_{\Delta U}(\mathcal{H} - U)\big), \tag{4}$$

where $\mathcal{H}$ is the Hamiltonian of the system and $\Delta_{\Delta U}(x)$ is the step function that is unity in the range 0 to ΔU and zero otherwise. The quantity $\Gamma(N, V, U; \Delta U)$ here denotes the number of discrete quantum states in the energy range between U and $U + \Delta U$. The entropy is then given by

$$S(N, V, U) = k \ln \Gamma(N, V, U; \Delta U). \tag{5}$$

The bulk value of the entropy does not depend on the value chosen for ΔU.

Helmholtz free energy $A(N, V, T) = U - TS$ and the canonical ensemble

The Helmholtz free energy describes a closed, isochoric, isothermal assembly, so it is a function of temperature, volume, and number of molecules:

$$dA = -SdT - PdV + \mu dN, \tag{6a}$$

$$S = -\left(\frac{\partial A}{\partial T}\right)_{V,N}, \tag{6b}$$

$$P = -\left(\frac{\partial A}{\partial V}\right)_{T,N}, \tag{6c}$$

$$\mu = \left(\frac{\partial A}{\partial N}\right)_{T,V}. \tag{6d}$$

The Maxwell relations for the Helmholtz free energy are

$$\left(\frac{\partial S}{\partial V}\right)_{T,N} = \left(\frac{\partial P}{\partial T}\right)_{V,N}, \tag{7a}$$

$$\left(\frac{\partial S}{\partial N}\right)_{T,V} = -\left(\frac{\partial \mu}{\partial T}\right)_{V,N}, \tag{7b}$$

$$\left(\frac{\partial P}{\partial N}\right)_{T,V} = -\left(\frac{\partial \mu}{\partial V}\right)_{T,N}. \tag{7c}$$

The Helmholtz free energy is determined from the canonical partition function

$$Q_N(V, T) = \mathrm{Tr}\left(\exp(-\beta \mathcal{H})\right) = \int e^{-\beta U} \left\{ \frac{1}{\Delta U} \Gamma(N, V, U; \Delta U) \right\} dU. \tag{8}$$

The Helmholtz free energy is given by

$$A(N, V, T) = -kT \ln Q_N(V, T). \tag{9}$$

Thermodynamic potential $\Pi(\mu, V, T) = -A + \mu N = PV$ and the grand canonical ensemble

The thermodynamic potential describes an open, isochoric, isothermal assembly, so it is a function of temperature, volume, and chemical potential:

$$d\Pi = SdT + PdV + Nd\mu\,, \tag{10a}$$

$$S = \left(\frac{\partial \Pi}{\partial T}\right)_{V,\mu}, \tag{10b}$$

$$P = \left(\frac{\partial \Pi}{\partial V}\right)_{T,\mu}, \tag{10c}$$

$$N = \left(\frac{\partial \Pi}{\partial \mu}\right)_{T,V}. \tag{10d}$$

The Maxwell relations for the thermodynamic potential are

$$\left(\frac{\partial S}{\partial V}\right)_{T,\mu} = \left(\frac{\partial P}{\partial T}\right)_{V,\mu}, \tag{11a}$$

$$\left(\frac{\partial S}{\partial \mu}\right)_{T,V} = \left(\frac{\partial N}{\partial T}\right)_{V,\mu}, \tag{11b}$$

$$\left(\frac{\partial P}{\partial \mu}\right)_{T,V} = \left(\frac{\partial N}{\partial V}\right)_{T,\mu}. \tag{11c}$$

The thermodynamic potential is a function only of a single extensive quantity, V, so $\Pi(\mu, V, T) = P(\mu, T)V$, that is, the pressure is the thermodynamic potential per unit volume. Therefore, we can write simpler thermodynamic relations in terms of pressure P, entropy density $s = S/V$, and number density $n = n/V$:

$$dP = sdT + nd\mu\,, \tag{12a}$$

$$s = \left(\frac{\partial P}{\partial T}\right)_{\mu}, \tag{12b}$$

$$n = \left(\frac{\partial P}{\partial \mu}\right)_{T}, \tag{12c}$$

$$\left(\frac{\partial s}{\partial \mu}\right)_{T} = \left(\frac{\partial n}{\partial T}\right)_{\mu}. \tag{12d}$$

The thermodynamic potential and pressure are determined from the grand canonical partition function

$$\mathcal{Q}(\mu, V, T) = \mathrm{Tr}\left(\exp(-\beta\mathcal{H} + \beta\mu N)\right) = \sum_{N=0}^{\infty} e^{\beta\mu N} Q_N(V, T), \tag{13}$$

with the result

$$P(\mu, T) = \frac{\Pi(\mu, V, T)}{V} = \frac{kT}{V} \ln \mathcal{Q}(\mu, V, T). \tag{14}$$

Gibbs free energy $G(N, P, T) = A + PV = U - TS + PV = \mu N$ and the isobaric ensemble

The Gibbs free energy describes a closed, isobaric, isothermal assembly, so it is a function of temperature, pressure, and number of molecules:

$$dG = -SdT + VdP + \mu dN, \tag{15a}$$

$$S = -\left(\frac{\partial G}{\partial T}\right)_{P,N}, \tag{15b}$$

$$V = \left(\frac{\partial G}{\partial P}\right)_{T,N}, \tag{15c}$$

$$\mu = \left(\frac{\partial G}{\partial N}\right)_{T,P}. \tag{15d}$$

The Maxwell relations for the Gibbs free energy are

$$\left(\frac{\partial S}{\partial P}\right)_{T,N} = -\left(\frac{\partial V}{\partial T}\right)_{P,N}, \tag{16a}$$

$$\left(\frac{\partial S}{\partial N}\right)_{T,P} = -\left(\frac{\partial \mu}{\partial T}\right)_{P,N}, \tag{16b}$$

$$\left(\frac{\partial V}{\partial N}\right)_{T,P} = \left(\frac{\partial \mu}{\partial P}\right)_{T,N}. \tag{16c}$$

The Gibbs free energy is a function only of a single extensive quantity, N, so $G(N, P, T) = N\mu(P, T)$, that is, the chemical potential is the Gibbs free energy per particle. Therefore, we can write simpler thermodynamic relations in terms of the pressure P, entropy per particle $s = S/N$, and specific volume $v = V/N$:

$$d\mu = -sdT + vdP, \tag{17a}$$

$$s = -\left(\frac{\partial \mu}{\partial T}\right)_P, \tag{17b}$$

$$v = \left(\frac{\partial \mu}{\partial P}\right)_T, \tag{17c}$$

$$\left(\frac{\partial s}{\partial P}\right)_T = -\left(\frac{\partial v}{\partial T}\right)_P. \tag{17d}$$

The Gibbs free energy and the chemical potential are determined from the isobaric partition function

$$Y_N(P, T) = \mathrm{Tr}\left(\exp(-\beta \mathcal{H} - \beta PV)\right) = \frac{1}{\lambda^3} \int_0^\infty e^{-\beta PV} Q_N(V, T) dV; \tag{18}$$

the cube of the thermal deBroglie wavelength is employed here to make the partition function dimensionless and is irrelevant in the classical thermodynamic limit. This leads us to the result

$$G(N,P,T) = N\mu(P,T) = -kT\ln Y_N(P,T). \tag{19}$$

The isobaric ensemble is often used in computer simulations to avoid two-phase regions that are present at first-order phase transitions.

Internal Energy $U(N,V,S)$

The internal energy describes a closed, isochoric, adiabatic assembly, so it is a function of entropy, volume, and number of molecules:

$$dU = TdS - PdV + \mu dN, \tag{20a}$$

$$T = \left(\frac{\partial U}{\partial S}\right)_{V,N}, \tag{20b}$$

$$P = -\left(\frac{\partial U}{\partial V}\right)_{S,N}, \tag{20c}$$

$$\mu = \left(\frac{\partial U}{\partial N}\right)_{S,V}. \tag{20d}$$

Maxwell relations for the internal energy are

$$\left(\frac{\partial T}{\partial V}\right)_{S,N} = -\left(\frac{\partial P}{\partial S}\right)_{V,N}, \tag{21a}$$

$$\left(\frac{\partial T}{\partial N}\right)_{S,V} = \left(\frac{\partial \mu}{\partial S}\right)_{V,N}, \tag{21b}$$

$$\left(\frac{\partial P}{\partial N}\right)_{S,V} = -\left(\frac{\partial \mu}{\partial V}\right)_{S,N}. \tag{21c}$$

Enthalpy $H(N,P,S) = U + PV$

The enthalpy describes a closed, isobaric, adiabatic assembly, so it is a function of entropy, pressure, and number of molecules.

$$dH = TdS + VdP + \mu dN, \tag{22a}$$

$$T = \left(\frac{\partial H}{\partial S}\right)_{P,N}, \tag{22b}$$

$$V = \left(\frac{\partial H}{\partial P}\right)_{S,N}, \tag{22c}$$

$$\mu = \left(\frac{\partial H}{\partial N}\right)_{S,P}. \tag{22d}$$

The Maxwell relations for the enthalpy are

$$\left(\frac{\partial T}{\partial P}\right)_{S,N} = \left(\frac{\partial V}{\partial S}\right)_{P,N}, \tag{23a}$$

$$\left(\frac{\partial T}{\partial N}\right)_{S,P} = \left(\frac{\partial \mu}{\partial S}\right)_{P,N}, \tag{23b}$$

$$\left(\frac{\partial V}{\partial N}\right)_{S,P} = \left(\frac{\partial \mu}{\partial P}\right)_{S,N}. \tag{23c}$$

This assembly is often used by chemists to describe chemical reactions that take place rapidly in the laboratory at fixed pressure where the speed of the reaction is too fast to allow a substantial heat exchange with the environment.

Magnetic free energy $F(T,H) = U(S,M) - TS - HM$

The magnetic free energy describes a system with a fixed number of magnetic dipoles, at temperature T and magnetic field H. The thermodynamic relations and Maxwell relations are

$$dF = -SdT - MdH, \tag{24a}$$

$$S = -\left(\frac{\partial F}{\partial T}\right)_H, \tag{24b}$$

$$M = -\left(\frac{\partial F}{\partial H}\right)_T, \tag{24c}$$

$$\left(\frac{\partial S}{\partial H}\right)_T = \left(\frac{\partial M}{\partial T}\right)_H. \tag{24d}$$

This assembly is defined by the fixed-field canonical ensemble where the canonical partition function for spins $\{\mathbf{s}_i\}$ with dipole moment μ and exchange energies J_{ij} is

$$Q_N(T,H) = \mathrm{Tr}\left(\exp(-\beta\mathcal{H})\right) = \sum_{\{\mathbf{s}\}} \exp\left(\sum_{(i,j)} K_{ij}\mathbf{s}_i \cdot \mathbf{s}_j + h\sum_i s_{z,i}\right), \tag{25}$$

where the coupling constants are $K_{ij} = J_{ij}/kT$ and the field coupling is $h = \mu H/kT$. The magnetic free energy is then given by

$$F(T,H) = -kT \ln Q_N(T,H). \tag{26}$$

Convexity and variances

The convexity of the entropy S that follows from the second law of thermodynamics requires that all second derivatives of free energies have a unique sign. These derivatives are all proportional to variances of different statistical quantities. A few important examples are the heat capacity, the isothermal compressibility, and the magnetic susceptibility:

$$C_V = T\left(\frac{\partial S}{\partial T}\right)_{V,N} = -\left(\frac{\partial^2 A}{\partial T^2}\right)_{V,N} = \frac{\langle \mathcal{H}^2\rangle - \langle \mathcal{H}\rangle^2}{kT^2} \geq 0, \tag{27}$$

$$\kappa_T = \frac{1}{n^2}\left(\frac{\partial n}{\partial \mu}\right)_T = \frac{1}{n^2}\left(\frac{\partial^2 P}{\partial \mu^2}\right)_T = \left(\frac{1}{nkT}\right)\frac{\langle N^2\rangle - \langle N\rangle^2}{\langle N\rangle} \geq 0, \tag{28}$$

$$\chi_T = \left(\frac{\partial M}{\partial H}\right)_T = -\left(\frac{\partial^2 F}{\partial H^2}\right)_T = \frac{\langle M^2\rangle - \langle M\rangle^2}{kT} \geq 0. \tag{29}$$

I Pseudorandom numbers

The random numbers used in computer simulations are not truly random but rather pseudorandom. They are intended to display as many of the characteristics of a random sequence as possible but are generated using simple integer algorithms. "Anyone who considers arithmetical methods of producing random digits is, of course, in a state of sin. For, as has been pointed out several times, there is no such thing as a random number — there are only methods to produce random numbers, and a strict arithmetic procedure of course is not such a method" (von Neumann, 1951). von Neumann's quip was not intended to dissuade the use of pseudorandom numbers but rather to encourage users to understand how such numbers are generated, their statistical properties and potential weaknesses. Pseudorandom number generators are essential in computer simulations in many fields but a generator should not blindly be trusted unless it has been thoroughly tested. There is an extensive literature on theoretical methods for creating and empirically testing pseudorandom number generators; see Knuth (1997), L'Ecuyer (1988, 1999), and Press et al. (2007).

The most commonly used and tested classes of pseudorandom number generators are based on integer arithmetic modulo large, usually prime, numbers. The simplest class of generators is the prime modulus linear congruential method that generates pseudorandom numbers in the range $[1, 2, \ldots, m-1]$, where m is a prime number. Each new number is based on the previous number according to the formula

$$r_j = a\, r_{j-1} \bmod m, \tag{1}$$

where the multiplier a is a carefully chosen and empirically tested integer less than m. If a is chosen properly, the generator will produce a sequence of integers that includes every integer in the range $[1, 2, \ldots, m-1]$ exactly once before the sequence begins to repeat after $m-1$ calls to the generator. Pseudorandom floating-point numbers in the open range $(0, 1)$ are produced by a floating-point multiplication by m^{-1}. This produces a uniformly distributed set of floating-point numbers on a comb of $m-1$ values between zero and one. The floating-point numbers 0.0 and 1.0 will *not* appear in the sequence. Often, in implementations, the modulus chosen is less than 2^{31} to allow for the integers to be represented by four-byte words. Generators of this form will repeat after about two billion calls. This may be fine for some applications but is often inadequate for a scientific code. Even a modest simulation can exhaust such a simple generator. However, linear congruential generators with different moduli and multipliers can be combined to give much longer periods and less statistical correlation (L'Ecuyer, 1988). The following algorithm to generate a uniformly distributed pseudorandom floating-point number r using two linear congruential generators gives a much longer sequence before repeating than does a single generator.

ALGORITHM

$q_j = a_1 q_{j-1} \bmod m_1$
$s_j = a_2 s_{j-1} \bmod m_2$
$z = (q_j - s_j)$
if $z \leq 0$, then $z = z + m_1 - 1$
$r = z m_1^{-1}$

The length of the period of the combined generator is equal to the product of the non-common prime factors of $m_1 - 1$ and $m_2 - 1$. L'Ecuyer (1988) recommended the following multiplers and moduli: $m_1 = 2147483563 = 2^{31} - 85$, $a_1 = 40014$, $m_2 = 2147483399 = 2^{31} - 249$, and $a_2 = 40692$. The two individual generators do well on the spectral test and other tests of correlations (Knuth, 1997) and $(m_1 - 1)/2$ and $(m_2 - 1)/2$ are relatively prime, so the period of the generator is 2.3×10^{18}. This can be implemented very easily in high-level languages using IEEE double precision floating-point arithmetic since $a_i m_i < 2^{53}$. L'Ecuyer (1988) and Press et al. (2007) also showed how the generators can be implemented using four-byte integer arithmetic. The exact order of the pseudorandom numbers can also be shuffled to improve statistics (Bays and Durham, 1976).

There are several other classes of pseudorandom number algorithms and many good generators available that have been extensively tested in the computing literature and are in wide use in the scientific literature; for summaries, see Newman and Barkema (1999), Gentle (2003), and Landau and Binder (2009). However, note that subtle correlations in a pseudorandom sequence can produce very large deviations compared to exact results, so care is warranted; see Ferrenberg, Landau, and Wong (1992), Beale (1996), and Figure I.1. Different classes of generators have different correlation properties, so substituting a generator based on a different algorithm can sometimes be a useful strategy for testing a computer code.

Gaussian distributed pseudorandom numbers

One often needs pseudorandom numbers that are drawn from a Gaussian distribution centered at the origin with unit variance: $P(g) = \exp\left(-g^2/2\right)/\sqrt{2\pi}$. The following algorithm for generating Gaussian-distributed pseudorandom numbers from pairs of uniformly distributed pseudorandom numbers is based on an algorithm by Box and Muller (1958).

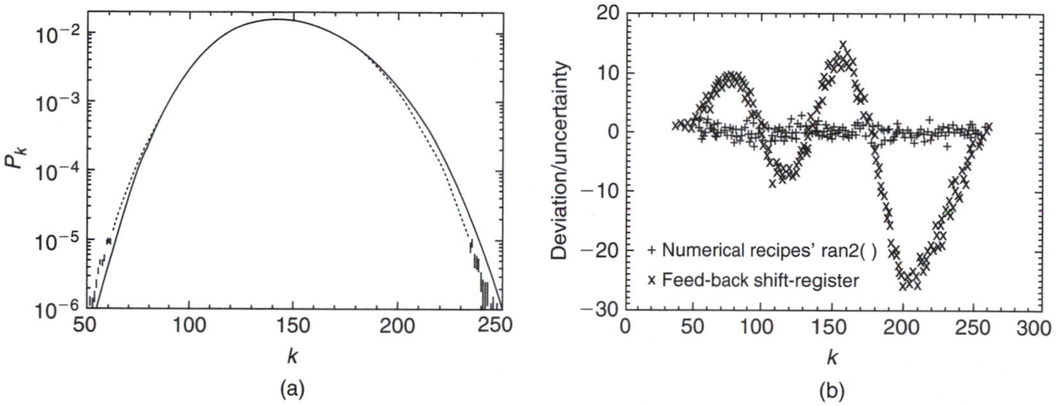

FIGURE I.1 (a) The exact energy distribution for a 32×32 Ising lattice at the critical temperature (solid line) and the distribution calculated from 10^7 configurations using the Wolff Monte Carlo algorithm with the R250 feedback shift-register pseudorandom number generator (error bars).[5] The distribution calculated using the Wolff algorithm with Numerical Recipes ran2( ) (Press et al., 2007) is also shown, but is almost indistinguishable from the exact solid curve on this scale. (b) Deviation of the Monte Carlo results from the exact distribution in units of the statistical uncertainty of each point. The +'s indicate the ran2( ) results and the $\times$'s indicate the feedback shift-register results. The χ^2-value for the two cases yields $\chi^2 = 190$ for 210 nonzero points and $\chi^2 = 2.8 \times 10^4$ for 217 nonzero points, giving deviations of -0.95σ and 1300σ, respectively; see Section 13.4.A and Beale (1996).

ALGORITHM

To generate pairs of Gaussian pseudorandom numbers g_1 and g_2 from uniformly distributed pseudorandom numbers x and y.

REPEAT

$x = 2\,\text{rand}() - 1.0$

$y = 2\,\text{rand}() - 1.0$

$s = x^2 + y^2$

UNTIL $s < 1$

$w = \sqrt{\frac{-2\ln s}{s}}$

$g_1 = xw$

$g_2 = yw$

[5]An example of a widely used and trusted pseudorandom number generator whose subtle correlations have been shown to adversely affect the results of some MC simulations is the R250 feedback shift-register generator. R250 generates a pseudorandom positive four-byte integer $0 \le r_i < 2^{31}$ by taking an exclusive or of two previous random integers kept in a 250 element table: $r_i = r_{i-103} \wedge r_{j-250}$. As with other generators, the floating-point numbers on $[0, 1]$ are produced by dividing the pseudorandom integer by 2^{31}. This particular feedback shift-register algorithm is very fast and produces fairly uncorrelated random pairs $\langle r_k r_{k-i} \rangle = 1/4$ for $i \neq 0$. Triplets are also uncorrelated $\langle r_k r_{k-i} r_{k-j} \rangle = 1/8$, except for *one* particular triplet correlation, $\langle r_k r_{k-103} r_{k-250} \rangle$, which can be shown to give $(6/7)(1/8)$ rather than the correct value of $1/8$; see Heuer, Dunweg, and Ferrenberg (1997). All higher order correlations that involve these same triplets are similarly affected. This triplet correlation being off by 14% can greatly affect Monte Carlo results (Ferrenberg, Landau, and Wong, 1992; Beale, 1996).

Bibliography

Abe, R. (1972). *Prog. Theor. Phys. Japan* **48**, 1414.

Abe, R. (1973). *Prog. Theor. Phys. Japan* **49**, 113, 1851.

Abe, R., and Hikami, S. (1973). *Prog. Theor. Phys. Japan* **49**, 442.

Abraham, D. B. (1986). In *Phase Transitions and Critical Phenomena*, eds. C. Domb and J. L. Lebowitz (Academic Press, London), Vol. 10, pp. 1–74.

Abramowitz, M., and Stegun, I. A. (eds.) (1964). *Handbook of Mathematical Functions* (National Bureau of Standards, Washington, DC).

Abrikosov, A. A., Gorkov, L. P., and Dzyaloshinskii, I. Y. (1965). *Quantum Field Theoretical Methods in Statistical Physics* (Pergamon Press, Oxford).

Abrikosov, A. A., and Khalatnikov, I. M. (1957). *J. Exptl. Theoret. Phys. USSR* **33**, 1154; English transl. *Soviet Phys. JETP* **6**, 888 (1958).

Adare, A., et al. (2010). *Phys. Rev. Lett.* **104**, 132301.

Aharony, A. (1976). In *Phase Transitions and Critical Phenomena*, eds. C. Domb and M. S. Green (Academic Press, London), Vol. 6, pp. 357–424.

Ahlers, G. (1980). *Rev. Mod. Phys.* **52**, 489.

Alder, B. J., and Wainwright, T. E. (1957). *J. Chem. Phys.* **27**, 1208.

Alder, B. J., and Wainwright, T. E. (1959). *J. Chem. Phys.* **31**, 459.

Allen, M. B., and Tildesley, D. J. (1990). *Computer Simulations of Liquids*, 2nd ed. (Oxford University Press, Oxford).

Allen, S., and Pathria, R. K. (1989). *Can. J. Phys.* **67**, 952.

Allen, S., and Pathria, R. K. (1991). *Can. J. Phys.* **69**, 753.

Alpher, R. A., Bethe, H. A., and Gamow, G. (1948). *Phys. Rev.* **73**, 803.

Alpher, R. A., and Herman, R. C. (1948). *Nature* **162**, 774.

Alpher, R. A., and Herman, R. C. (1950). *Rev. Mod. Phys.* **22**, 153.

Alpher, R. A., and Herman, R. C. (2001). *Genesis of the Big Bang* (Oxford University Press, New York).

Als-Nielsen, J. (1976). In *Phase Transitions and Critical Phenomena*, eds. C. Domb and M. S. Green (Academic Press, London), Vol. 5a, pp. 87–164.

Als-Nielsen, J., and Birgeneau, R. J. (1977). *Am. J. Phys.* **45**, 554.

Anderson, M. H., Ensher, J. R., Matthews, M. R., Wieman, C. E., and Cornell, E. E. (1995). *Science* **269**, 198.

Anderson, P. W. (1984). *Basic Notions of Condensed Matter Physics* (Benjamin/Cummings, Menlo Park, CA).

Anderson, W. (1929). *Z. Phys.* **56**, 851.

Arnold, V. I. (1963). *Russian Math. Survey* **18**, 13.

Ashcroft, N. W., and Mermin, N. D. (1976). *Sold State Physics* (Holt, Rinehart, & Winston, New York).

Auluck, F. C., and Kothari, D. S. (1946). *Proc. Camb. Phil. Soc.* **42**, 272.

Au-Yang, H., and Perk, J. H. H. (1984). *Phys. Lett. A* **104**, 131.

Bagnato, V., Pritchard, D. E., and Kleppner, D. (1987). *Phys. Rev. A* **35**, 4354.

Bahm, C. J., and Pethick, C. J. (1996). *Phys. Rev. Lett.* **76**, 6.

Bak, P., Krinsky, S., and Mukamel, D. (1976). *Phys. Rev. Lett.* **36**, 52.

Baker, G. A., Jr. (1990). *Quantitative Theory of Critical Phenomena* (Academic Press, New York).

Balian, R., and Toulouse, G. (1973). *Phys. Rev. Lett.* **30**, 544.

Baltes, H. P., and Hilf, E. R. (1976). *Spectra of Finite Systems* (Bibliographisches Institut AG, Zürich).

Band, W. (1955). *An Introduction to Quantum Statistics* (Van Nostrand, Princeton).

Barber, M. N. (1983). In *Phase Transitions and Critical Phenomena*, eds. C. Domb and J. L. Lebowitz (Academic Press, London), Vol. 8, pp. 145–266.

Barber, M. N., and Fisher, M. E. (1973). *Ann. Phys. (N.Y.)* **77**, 1.

Bardeen, J., Cooper, L. N., and Schrieffer, J. R. (1957). *Phys. Rev.* **108**, 1175.

Barnett, S. J. (1944). *Proc. Am. Acad. Arts & Sci.* **75**, 109.

Barouch, E., McCoy, B. M., and Wu, T. T. (1973). *Phys. Rev. Lett.* **31**, 1409.

Baxter, R. J. (1982). *Exactly Solved Models in Statistical Mechanics* (Academic Press, London).

Bays, C., and Durham, S. D. (1976). *ACM Trans. Math. Software* **2**, 59.

Beale, P. D. (1996). *Phys. Rev. Lett.* **76**, 79.

Belinfante, F. J. (1939). *Physica* **6**, 849, 870.

Berche, B., Sanchez, A. I. F., and Paredes, V. (2002). *Europhys. Lett.* **60**, 539.

Berezinskii, V. (1970). *Zh. Eksp. Teor. Fiz.* **59**, 907; (1971) Engl. transl. *Sov. Phys., JETP* **32**, 493.

Berlin, T. H., and Kac, M. (1952). *Phys. Rev.* **86**, 821.

Bernoulli, D. (1738). *Hydrodynamica* (Argentorati).

Beth, E., and Uhlenbeck, G. E. (1937). *Physica* **4**, 915.

Bethe, H. A. (1935). *Proc. R. Soc. Lond. A* **150**, 552.

Betts, D. D. (1974). In *Phase Transitions and Critical Phenomena*, eds. C. Domb and M. S. Green (Academic Press, London), Vol. 3, pp. 569–652.

Betts, D. D., and Lee, M. H. (1968). *Phys. Rev. Lett.* **20**, 1507.

Betts, D. D. et al. (1969). *Phys. Lett.* **29A**, 150; (1970) *ibid.* **32A**, 152.

Betts, D. D. et al. (1970). *Can. J. Phys.* **48**, 1566; (1971) *ibid.* **49**, 1327; (1973) *ibid.* **51**, 2249.

Betts, D. S. (1969). *Contemp. Phys.* **10**, 241.

Bijl, A. (1940). *Physica* **7**, 869.

Binder, K. (1983). In *Phase Transitions and Critical Phenomena*, eds. C. Domb and J. L. Lebowitz (Academic Press, London), Vol. 8, pp. 1–144.

Binder, K. (ed.) (1986). *Monte Carlo Methods in Statistical Physics*, 2nd ed., Topics in Current Physics (Springer-Verlag, Berlin), Vol. 7.

Binder, K. (ed.) (1987). *Applications of the Monte Carlo Method in Statistical Physics*, 2nd ed., Topics in Current Physics (Springer-Verlag, Berlin), Vol. 36.

Binder, K. (ed.) (1992). *The Monte Carlo Method in Condensed Matter Physics*, Topics in Applied Physics (Springer-Verlag, Berlin), Vol. 71.

Binder, K., and Heermann, D. W. (2002). *Monte Carlo Simulation in Statistical Physics*, 4th ed., (Springer-Verlag, Berlin).

Blakemore, J. S. (1962). *Semiconductor Statistics* (Pergamon Press, Oxford).

Bloch, F. (1928). *Z. Phys.* **52**, 555.

Bogoliubov, N. N. (1947). *J. Phys. USSR* **11**, 23.

Bogoliubov, N. N. (1962). *Phys. Abhandl. S.U.* **6**, 113.

Bogoliubov, N. N. (1963). *Lectures on Quantum Statistics* (Gordon & Breach, New York).

Boltzmann, L. (1868). *Wien. Ber.* **58**, 517.

Boltzmann, L. (1871). *Wien. Ber.* **63**, 397, 679, 712.

Boltzmann, L. (1872). *Wien. Ber.* **66**, 275.

Boltzmann, L. (1875). *Wien. Ber.* **72**, 427.

Boltzmann, L. (1876). *Wien. Ber.* **74**, 503.

Boltzmann, L. (1877). *Wien. Ber.* **76**, 373.

Boltzmann, L. (1879). *Wien. Ber.* **78**, 7.

Boltzmann, L. (1884). *Ann. Phys.* **22**, 31, 291, 616.

Boltzmann, L. (1896, 1898). *Vorlesungen über Gastheorie*, 2 vols. (J. A. Barth, Leipzig). English transl. (1964) *Lectures on Gas Theory* (University of California Press, Berkeley, CA).

Boltzmann, L. (1899). *Amsterdam Ber.* 477; see also (1909) *Wiss. Abhandl. von L. Boltzmann III*, 658.

Borner, G. (2003). *The Early Universe* (Springer, New York).

Bose, S. N. (1924). *Z. Physik* **26**, 178.

Box, G. E. P., and Muller, M. E. (1958). *Ann. Math. Stat.* **29**, 610.

Bragg, W. L., and Williams, E. J. (1934). *Proc. R. Soc. Lond. A* **145**, 699.

Bragg, W. L., and Williams, E. J. (1935). *Proc. R. Soc. Lond. A* **151**, 540; **152**, 231.

Brézin, E., le Guillou, J. C., and Zinn-Justin, J. (1976). In *Phase Transitions and Critical Phenomena*, eds. C. Domb and M. S. Green (Academic Press, London), Vol. 6, pp. 125–247.

Brueckner, K. A., and Gammel, J. L. (1958). *Phys. Rev.* **109**, 1040.

Brueckner, K. A., and Sawada, K. (1957). *Phys. Rev.* **106**, 1117, 1128.

Brush, S. G. (1957). *Ann. Sci.* **13**, 188, 273.

Brush, S. G. (1958). *Ann. Sci.* **14**, 185, 243.

Brush, S. G. (1961a). *Am. J. Phys.* **29**, 593.

Brush, S. G. (1961b). *Am. Scientist* **49**, 202.

Brush, S. G. (1965–66). *Kinetic Theory*, 3 vols. (Pergamon Press, Oxford).

Brush, S. G. (1967). *Rev. Mod. Phys.* **39**, 883.

Buckingham, M. J., and Gunton, J. D. (1969). *Phys. Rev.* **178**, 848.

Bush, V., and Caldwell, S. H. (1931). *Phys. Rev.* **38**, 1898.

Callen, H. B., and Welrton, T. R. (1951). *Phys. Rev.* **83**, 34.

Cardy, J. L. (1987). In *Phase Transitions and Critical Phenomena*, eds. C. Domb and J. L. Lebowitz (Academic Press, London), Vol. 11, pp. 55–126.

Cardy, J. L. (ed.) (1988). *Finite-Size Scaling* (North-Holland, Amsterdam).

Carnahan, N. F., and Starling, J. (1969). *J. Chem. Phys.* **51**, 635.

Chaba, A. N., and Pathria, R. K. (1973). *Phys. Rev. A* **8**, 3264.

Chaba, A. N., and Pathria, R. K. (1975). *Phys. Rev. B* **12**, 3697.

Chandrasekhar, S. (1939). *Introduction to the Study of Stellar Structure* (University of Chicago Press, Chicago); now available in a Dover edition.

Chandrasekhar, S. (1943). *Rev. Mod. Phys.* **15**, 1.

Chandrasekhar, S. (1949). *Rev. Mod. Phys.* **21**, 383.

Chang, C. H. (1952). *Phys. Rev.* **88**, 1422.

Chang, T. S. (1939). *Proc. Camb. Phil. Soc.* **35**, 265.

Chapman, S. (1916). *Trans. R. Soc. Lond. A* **216**, 279.

Chapman, S. (1917). *Trans. R. Soc. Lond. A* **217**, 115.

Chapman, S., and Cowling, T. G. (1939). *The Mathematical Theory of Non-uniform Gases* (Cambridge University Press, Cambridge, UK).

Chisholm, J. S. R., and Borde, A. H. (1958). *An Introduction to Statistical Mechanics* (Pergamon Press, New York).

Chrétien, M., Gross, E. P., and Deser, S. (eds.) (1968). *Statistical Physics, Phase Transitions and Superfluidity*, 2 vols. (Gordon & Breach, New York).

Clausius, R. (1857). *Ann. Phys.* **100**, 353; also published in (1857) *Phil. Mag.* **14**, 108.

Clausius, R. (1859). *Ann. Phys.* **105**, 239; also published in (1859) *Phil. Mag.* **17**, 81.

Clausius, R. (1870). *Ann. Phys.* **141**, 124; also published in (1870) *Phil. Mag.* **40**, 122.

Clunie, J. (1954). *Proc. Phys. Soc. A* **67**, 632.

Cohen, E. G. D., and van Leeuwen, J. M. J. (1960). *Physica* **26**, 1171.

Cohen, E. G. D., and van Leeuwen, J. M. J. (1961). *Physica* **27**, 1157.

Cohen, E. G. D. (ed.) (1962). *Fundamental Problems in Statistical Mechanics* (John Wiley, New York).

Cohen, M., and Feynman R. P. (1957). *Phys. Rev.* **107**, 13.

Compton, A. H. (1923a). *Phys. Rev.* **21**, 207, 483.

Compton, A. H. (1923b). *Phil. Mag.* **46**, 897.

Cooper, L. N. (1956). *Phys. Rev.* **104**, 1189.

Cooper, L. N. (1960). *Am. J. Phys.* **28**, 91.

Copi, C. J., Schramm, D. N., and Turner, M. S. (1997). *Phys. Rev. D.* **55**, 3389.

Corak, W. S., Garfunkel, M. P., Satterthwaite, C. B., and Wexler, A. (1955). *Phys. Rev.* **98**, 1699.

Cummins, H. Z., and Gammon, R. W. (1966). *J. Chem. Phys.* **44**, 2785.

Darwin, C. G., and Fowler, R. H. (1922a). *Phil. Mag.* **44**, 450, 823.

Darwin, C. G., and Fowler, R. H. (1922b). *Proc. Camb. Phil. Soc.* **21**, 262.

Darwin, C. G., and Fowler, R. H. (1923). *Proc. Camb. Phil. Soc.* **21**, 391, 730; see also Fowler, R. H. (1923) *Phil. Mag.* **45**, 1, 497; (1925) *Proc. Camb. Phil. Soc.* **22**, 861; (1926) *Phil. Mag.* **1**, 845; (1926) *Proc. R. Soc. Lond. A* **113**, 432.

Davis, K. B., Mewes, M.-O., Andrews, M. R., van Druten, N. J., Durfee, D. S., Kurn, D. M., and Ketterle, W. (1995). *Phys. Rev. Lett.* **75**, 3969.

Debye, P. (1912). *Ann. Phys.* **39**, 789.

de Gennes, P.-G. (1972). *Phys. Lett.* **38A**, 339.

de Gennes, P.-G. (1979). *Scaling Concepts in Polymer Physics* (Cornell University Press, Ithaca, NY).

de Groot, S. R. (1951). *Thermodynamics of Irreversible Processes* (North-Holland, Amsterdam).

de Groot, S. R., and Mazur, P. (1962). *Non-equilibrium Thermodynamics* (North-Holland, Amsterdam).

de Groot, S. R., Hooyman, G. J., and ten Seldam, C. A. (1950). *Proc. R. Soc. London A*, 203, 266.

de Klerk, D., Hudson, R. P., and Pellam, J. R. (1953). *Phys. Rev.* **89**, 326, 662.

DeMarco, B., and Jin, D. (1999). *Science* **285**, 1703.

Dennison, D. M. (1927). *Proc. R. Soc. Lond. A* **115**, 483.

des Cloizeaux, J. (1974). *Phys. Rev. A* **10**, 1665.

des Cloizeaux, J. (1982). In *Phase Transitions: Cargese 1980*, eds. M. Levy et al. (Plenum Press, New York), pp. 371–394.

Diehl, H. W. (1986). In *Phase Transitions and Critical Phenomena*, eds. C. Domb and J. L. Lebowitz (Academic Press, London), Vol. 10, pp. 75–267.

Dingle, R. B. (1956). *J. App. Res. B* **6**, 225.

Dirac, P. A. M. (1926). *Proc. R. Soc. Lond. A* **112**, 661, 671.

Dirac, P. A. M. (1929). *Proc. Camb. Phil. Soc.* **25**, 62.

Dirac, P. A. M. (1930). *Proc. Camb. Phil. Soc.* **26**, 361, 376.

Dirac, P. A. M. (1931). *Proc. Camb. Phil. Soc.* **27**, 240.

Domb, C. (1960). *Advan. Phys.* **9**, 150.

Domb, C. (1974). In *Phase Transitions and Critical Phenomena*, eds. C. Domb and M. S. Green (Academic Press, London), Vol. 3, pp. 357–484.

Domb, C. (1985). *Contemp. Phys.* **26**, 49.

Domb, C., and Green, M. S. (eds.) (1972–1976). *Phase Transitions and Critical Phenomena* (Academic Press, London), Vols. 1–6.

Domb, C., and Hunter, D. L. (1965). *Proc. Phys. Soc.* **86**, 1147.

Domb, C., and Lebowitz, J. L. (eds.) (1983–1992). *Phase Transitions and Critical Phenomena* (Academic Press, London), Vols. 7–15.

Drude, P. (1900). *Ann. Phys.* **1**, 566; **3**, 369.

Dukovski, I., Machta, J., and Chayes, L. (2002). *Phys. Rev. E* **65**, 026702.

Ehrenfest, P. (1905). *Wiener Ber.* **114**, 1301.

Ehrenfest, P. (1906). *Phys. Zeits.* **7**, 528.

Ehrenfest, P., and Ehrenfest, T. (1912). *Enzyklopädie der mathematischen Wissenschaften*, Vol. IV (Teubner, Leipzig). English transl. (1959) *The Conceptual Foundations of the Statistical Approach in Mechanics* (Cornell University Press, Ithaca, NY).

Einstein, A. (1902). *Ann. Phys.* **9**, 417.

Einstein, A. (1903). *Ann. Phys.* **11**, 170.

Einstein, A. (1905a). *Ann. Phys.* **17**, 132; see also (1906b) *ibid.* **20**, 199.

Einstein, A. (1905b). *Ann. Phys.* **17**, 549; see also (1906a) *ibid.* **19**, 289, 371.

Einstein, A. (1905c). *Ann. Phys.* **17**, 891; see also (1905d) *ibid.* **18**, 639 and (1906c) *ibid.* **20**, 627.

Einstein, A. (1907). *Jb. Radioakt.* **4**, 411.

Einstein, A. (1909). *Physik Z.* **10**, 185.

Einstein, A. (1910). *Ann. Phys.* **33**, 1275.

Einstein, A. (1924). *Berliner Ber.* **22**, 261.

Einstein, A. (1925). *Berliner Ber.* **1**, 3.

Eisenschitz, R. (1958). *Statistical Theory of Irreversible Processes* (Oxford University Press, Oxford).

Elcock, E. W., and Landsberg, P. T. (1957). *Proc. Phys. Soc.* **70**, 161.

Ensher, J. R., Jin, D. S., Matthews, M. R., Wieman, C. E., and Cornell, E. A. (1996). *Phys. Rev. Lett.* **77**, 4984.

Enskog, D. (1917). *Kinetische Theorie der Vorgänge in mässig verdünnten Gasen*, dissertation (Almquist & Wiksell, Uppsala).

Eyring, H., Henderson, D., Stover, B. J., and Eyring, E. M. (1963). *Statistical Mechanics and Dynamics* (John Wiley, New York).

Ezawa, Z. F. (2000). *Quantum Hall Effects: Field Theorectical Approach and Related Topics* (World Scientific, Singapore).

Farquhar, I. E. (1964). *Ergodic Theory in Statistical Mechanics* (Interscience Publishers, New York).

Fay, J. A. (1965). *Molecular Thermodynamics* (Addison-Wesley, Reading, MA).

Fedosov, B. V. (1963). *Dokl. Akad. Nauk SSSR* **151**, 786; English transl. 1963: *Sov. Math.* (Doklady) **4**, 1092.

Fedosov, B. V. (1964). *Dokl. Akad. Nauk SSSR* **157**, 536; English transl. 1964: *Sov. Math.* (Doklady) **5**, 988.

Ferdinand, A. E., and Fisher, M. E. (1969). *Phys. Rev.* **185**, 832.

Fermi, E. (1926). *Z. Physik* **36**, 902.

Fermi, E. (1928). *Zeit. für Phys.* **48**, 73; see also (1927) *Ace. Lencei* **6**, 602.

Fermi, E. (1936). *Thermodynamics* (Dover, New York).

Fermi, E., Pasta, J. G., and Ulam, S. M. (1955). LASL Report LA-1940.

Ferrenberg, A. M., Landau, D. P., and Wong, Y. J. (1992). *Phys. Rev. Lett.* **69**, 3382.

Ferrenberg, A. M., and Swendsen, R. H. (1988). *Phys. Rev. Lett.* **61**, 2635.

Fetter, A. L. (1963). *Phys. Rev. Lett.* **10**, 507.

Fetter, A. L. (1965). *Phys. Rev.* **138**, A 429.

Feynman, R. P. (1953). *Phys. Rev.* **91**, 1291, 1301.

Feynman, R. P. (1954). *Phys. Rev.* **94**, 262.

Feynman, R. P. (1955). *Progress in Low Temperature Physics*, ed. C. J. Gorter (North-Holland, Amsterdam), Vol. 1, p. 17

Feynman, R. P., and Cohen, M. (1956). *Phys. Rev.* **102**, 1189.

Fisher, M. E. (1959). *Physica* **25**, 521.

Fisher, M. E. (1964). *J. Math. Phys.* **5**, 944.

Fisher, M. E. (1967). *Rep. Prog. Phys.* **30**, 615.

Fisher, M. E. (1969). *Phys. Rev.* **180**, 594.

Fisher, M. E. (1972). In *Essays in Physics*, eds. G. K. T. Conn and G. N. Fowler (Academic Press, London), Vol. 4, pp. 43–89.

Fisher, M. E. (1973). *Phys. Rev. Lett.* **30**, 679.

Fisher, M. E. (1983). In *Critical Phenomena*, ed. F. J. W. Hahne (Springer-Verlag, Berlin), pp. 1–139.

Fisher, M. E., Barber, M. N., and Jasnow, D. (1973). *Phys. Rev. A* **8**, 1111.

Fisher, M. E., and Privman, V. (1985). *Phys. Rev. B* **32**, 447.

Fixsen, D. J., Cheng, E. S., Gales, J. M., Mather, J. C., Shafer, R. A., and Wright, E. L. (1996). *Astrophys. J.* **473**, 576.

Fokker, A. D. (1914). *Ann. Phys.* **43**, 812.

Forster, D. (1975). *Hydrodynamic Fluctuations, Broken Symmetry, and Correlation Functions* (W. A. Benjamin, Reading, MA).

Fowler, R. H. (1926). *Mon. Not. R. Astron. Soc.* **87**, 114.

Fowler, R. H. (1955). *Statistical Mechanics*, 2nd ed. (Cambridge University Press, Cambridge, UK).

Fowler, R. H., and Guggenheim, E. A. (1940). *Proc. R. Soc. Lond. A* **174**, 189.

Fowler, R. H., and Guggenheim, E. A. (1960). *Statistical Thermodynamics* (Cambridge University Press, Cambridge, UK).

Fowler, R. H., and Nordheim, L. (1928). *Proc. R. Soc. Lond. A* **119**, 173.

Freedman, W. L., Madore, B. F., Gibson, B. K., Ferrarese, L., Kelson, D. D., Sakai, S., Mould, J. R., Kennicutt, R. C., Jr., Ford, H. C., Graham, J. A., Huchra, J. P., Hughes, S. M. G., Illingworth, G. D., Macri, L. M., and Stetson, P. B. (2001). *Astrophys. J.* **553**, 47.

Frenkel, D., and Smit, B. (2002). *Molecular Simulation* (Academic, New York).

Friedmann, A. (1922). *Z. Phys.* **16**, 377.

Friedmann, A. (1924). *Z. Phys.* **21**, 326.

Fujiwara, I., ter Haar, D., and Wergeland, H. (1970). *J. Stat. Phys.* **2**, 329.

Galitskii, V. M. (1958). *J. Exptl. Theor. Phys. USSR* **34**, 151; English transl. (1958) *Soviet Physics JETP-USSR* **7**, 104–112.

Gaunt, D. S., and Guttmann, A. J. (1974). In *Phase Transitions and Critical Phenomena*, eds. C. Domb and M. S. Green (Academic Press, London), Vol. 3, pp. 181–243.

Gelmini, G. B. (2005). *Phys. Scr.* **T121**, 131.

Gentle, J. E. (2003). *Random Number Generation and Monte Carlo Methods* (Springer-Verlag, New York).

Giardeau, M. (1960). *J. Math. Phys.* **1**, 516.

Gibbs, J. W. (1902). *Elementary Principles in Statistical Mechanics* (Yale University Press, New Haven); reprinted by Dover Publications, New York (1960). See also *A Commentary on the Scientific Writings of J. Willard Gibbs*, ed. A. Haas, Yale University Press, New Haven (1936), Vol. II.

Ginzburg, V. L. (1960). *Sov. Phys. Solid State* **2**, 1824.

Glasser, M. L. (1970). *Ann. J. Phys.* **38**, 1033.

Goldman, I. I., Krivchenkov, V. D., Kogan V. I., and Galitskii V. M. (1960). *Problems in Quantum Mechanics*, translated by D. ter Haar (Academic Press, New York).

Gombás, P. (1949). *Die statistische Theorie des Atoms und ihre Anwendungen* (Springer, Vienna).

Gombás, P. (1952). *Ann. Phys.* **10**, 253.

Gradshteyn, I. S., and Ryzhik, I. M. (1965). *Table of Integrals, Series and Products* (Academic Press, New York).

Green, H. S., and Hurst, C. A. (1964). *Order–Disorder Phenomena* (Interscience Publishers, New York).

Greenspoon, S., and Pathria, R. K. (1974). *Phys. Rev. A* **9**, 2103.

Greiner, M., Regal, C. A., and Jin, D. S. (2003). *Nature* **426**, 537.

Griffiths, R. B. (1965a). *Phys. Rev. Lett.* **14**, 623.

Griffiths, R. B. (1965b). *J. Chem. Phys.* **43**, 1958.

Griffiths, R. B. (1970). *Phys. Rev. Lett.* **24**, 715.

Griffiths, R. B. (1972). In *Phase Transitions and Critical Phenomena*, eds. C. Domb and M. S. Green (Academic Press, London), Vol. 1, pp. 7–109.

Gross, E. P. (1961). *Nuovo Cim.* **20**, 454.

Gross, E. P. (1963). *J. Math. Phys.* **4**, 195.

Guggenheim, E. A. (1935). *Proc. R. Soc. Lond. A* **148**, 304.

Guggenheim, E. A. (1938). *Proc. R. Soc. Lond. A* **169**, 134.

Guggenheim, E. A. (1959). *Boltzmann's Distribution Law* (North-Holland, Amsterdam).

Guggenheim, E. A. (1960). *Elements of the Kinetic Theory of Gases* (Pergamon Press, Oxford).

Gunton, J. D., and Buckingham, M. J. (1968). *Phys. Rev.* **166**, 152.

Gupta, H. (1947). *Proc. Nat. Ins. Sci. India* **13**, 35.

Gurarie, V., and Radzihovsky, L. (2007). *Ann. Phys.* **322**, 2.

Guttmann, A. J. (1989). In *Phase Transitions and Critical Phenomena*, eds. C. Domb and J. L. Lebowitz (Academic Press, London), Vol. 13, pp. 1–234.

Haber, H. E., and Weldon, H. A. (1981). *Phys. Rev. Lett.* **46**, 1497.

Haber, H. E., and Weldon, H. A. (1982). *Phys. Rev. D* **25**, 502.

Halperin, B. I., and Nelson, D. R. (1978). *Phys Rev. Lett.* **41**, 121.

Han, Y., Alsayed, A. M., Nobili, M., Zhang, J., Lubensky, T. C., and Yodh, A. G. (2006). *Science* **314**, 626.

Hanbury Brown, R., and Twiss, R. Q. (1956). *Nature* **177**, 27; **178**, 1447.

Hanbury Brown, R., and Twiss, R. Q. (1957). *Proc. R. Soc. Lond. A* **242**, 300; **243**, 291.

Hanbury Brown, R., and Twiss, R. Q. (1958). *Proc. R. Soc. Lond. A* **248**, 199, 222.

Hansen, J. P., and McDonald, I. R. (1986). *Theory of Simple Liquids* (Academic, New York).

Harris, F. E., and Monkhorst, H. K. (1970). *Phys. Rev. B* **2**, 4400.

Harrison, S. F., and Mayer, J. E. (1938). *J. Chem. Phys.* **6**, 101.

Hasenbusch, M., and Pinn, K. (1997). *J. Phys. A* **30**, 63.

Haugerud, H., Haugest, T., and Ravndal, F. (1997). *Phys. Lett. A* **225**, 18.

Heisenberg, W. (1928). *Z. Physik* **49**, 619.

Heller, P. (1967). *Rep. Prog. Phys.* **30**, 731.

Henry, W. E. (1952). *Phys. Rev.* **88**, 561.

Henshaw, D. G. (1960). *Phys. Rev.* **119**, 9.

Henshaw, D. G., and Woods, A. D. B. (1961). *Phys. Rev.* **121**, 1266.

Herapath, J. (1821). *Ann. Philos.* **1**, 273, 340, 401.

Heuer, A., Dunweg, B., and Ferrenberg, A. M. (1997). *Comp. Phys. Comm.* **103**, 1.

Hill, T. L. (1953). *J. Phys. Chem.* **57**, 324.

Hill, T. L. (1956). *Statistical Mechanics* (McGraw-Hill, New York).

Hill, T. L. (1960). *Introduction to Statistical Thermodynamics* (Addison-Wesley, Reading, MA).

Hillebrandt, W., and Niemeyer, J. C. (2000). *Annu. Rev. Astron. Astrophys.* **38**, 191.

Hirschfelder, J. O., Curtiss, C. F., and Bird, R. B. (1954). *Molecular Theory of Gases and Liquids* (John Wiley, New York).

Hohenberg, P. C., and Halperin, B. I. (1977). *Rev. Mod. Phys.* **49**, 435.

Holland, M., and Cooper, J. (1996). *Phys. Rev. A* **53**, R1954.

Holland, M. J., Jin, D. S., Chiofalo, M. L., and Cooper, J. (1997). *Phys. Rev. Lett.* **78**, 3801.

Holliday, D., and Sage, M. L. (1964). *Ann. Phys.* **29**, 125.

Huang, K. (1959). *Phys. Rev.* **115**, 765.

Huang, K. (1960). *Phys. Rev.* **119**, 1129.

Huang, K. (1963). *Statistical Mechanics*, 1st ed. (John Wiley, New York).

Huang, K. (1987). *Statistical Mechanics*, 2nd ed. (John Wiley, New York).

Huang, K., and Klein, A. (1964). *Ann. Phys.* **30**, 203.

Huang, K., and Yang, C. N. (1957). *Phys. Rev.* **105**, 767.

Huang, K., Yang, C. N., and Luttinger, J. M. (1957). *Phys. Rev.* **105**, 776.

Hupse, J. C. (1942). *Physica* **9**, 633.

Hurst, C. A., and Green, H. S. (1960). *J. Chem. Phys.* **33**, 1059.

Ising, E. (1925). *Z. Physik* **31**, 253.

Jackson, H. W., and Feenberg, E. (1962). *Rev. Mod. Phys.* **34**, 686.

Jackson, J. D. (1999). *Electrodynamics*, 3rd ed. (John Wiley, New York).

Jasnow, D. (1986). In *Phase Transitions and Critical Phenomena*, eds. C. Domb and J. L. Lebowitz (Academic Press, London), Vol. 10, pp. 269–363.

Jeans, J. H. (1905). *Phil. Mag.* **10**, 91.

Jin, D. (2002). *Phys. World* **15**, 27.

Jochim, S., Bartenstein, M., Altmeyer, A., Hendl, G., Riedl, S., Chin, C., Denschlag, J. H., and Grimm, R. (2003). *Science* **302**, 2101.

Johnson, J. B. (1927a). *Nature* **119**, 50.

Johnson, J. B. (1927b). *Phys. Rev.* **29**, 367.

Johnson, J. B. (1928). *Phys. Rev.* **32**, 97.

Josephson, B. D. (1967). *Proc. Phys. Soc.* **92**, 269, 276.

Joule, J. P. (1851). *Mem. Proc. Manchester Lit. Phil. Soc.* **9**, 107; also published in 1857, *Phil. Mag.* **14**, 211.

Joyce, G. S. (1972). In *Phase Transitions and Critical Phenomena*, eds. C. Domb and M. S. Green (Academic Press, London), Vol. 2, pp. 375–442.

Kac, M., and Thompson, C. J. (1971). *Phys. Norveg.* **5**, 163.

Kac, M., Uhlenbeck, G. E., and Hemmer, P. C. (1963). *J. Math. Phys.* **4**, 216, 229; see also (1964) *ibid.* **5**, 60.

Kac, M., and Ward, J. C. (1952). *Phys. Rev.* **88**, 1332.

Kadanoff, L. P. (1966a). *Nuovo Cim.* **44**, 276.

Kadanoff, L. P. (1966b). *Physics* **2**, 263.

Kadanoff, L. P. (1976a). *Ann. Phys.* **100**, 359.

Kadanoff, L. P. (1976b). In *Phase Transitions and Critical Phenomena*, eds. C. Domb and M. S. Green (Academic Press, London), Vol. 5a, pp. 1–34.

Kadanoff, L. P. et al. (1967). *Rev. Mod. Phys.* **39**, 395.

Kahn, B. (1938). *On the Theory of the Equation of State*, Thesis, Utrecht; English translation appears in de Boer, J., and Uhlenbeck, G. E. (eds.) (1965) *Studies in Statistical Mechanics*, Vol. III (North-Holland, Amsterdam).

Kahn, B., and Uhlenbeck, G. E. (1938). *Physica* **5**, 399.

Kalos, M. H., and Whitlock, P. A. (1986). *Monte Carlo Methods* (Wiley, New York).

Kappler, E. (1931). *Ann. Phys.* **11**, 233.

Kappler, E. (1938). *Ann. Phys.* **31**, 377, 619.

Kasteleyn, P. W. (1963). *J. Math. Phys.* **4**, 287.

Katsura, S. (1959). *Phys. Rev.* **115**, 1417.

Kaufman, B. (1949). *Phys. Rev.* **76**, 1232.

Kaufman, B., and Onsager, L. (1949). *Phys. Rev.* **76**, 1244.

Kawatra, M. P., and Pathria, R. K. (1966). *Phys. Rev.* **151**, 132.

Keesom, P. H., and Pearlman, N. (1953). *Phys. Rev.* **91**, 1354.

Khalatnikov, I. M. (1965). *Introduction to the Theory of Superfluidity* (W. A. Benjamin, New York).

Khintchine, A. (1934). *Math. Ann.* **109**, 604.

Kiess, E. (1987). *Am. J. Phys.* **55**, 1006.

Kilpatrick, J. E. (1953). *J. Chem. Phys.* **21**, 274.

Kilpatrick, J. E., and Ford, D. I. (1969). *Am. J. Phys.* **37**, 881.

Kim, E., and Chan, M. H. W. (2004). *Nature* **427**, 225; (2004b) *Science*, **305**, 1941.

Kirkwood, J. G. (1965). *Quantum Statistics and Cooperative Phenomena* (Gordon & Breach, New York).

Kirsten, K., and Toms, D. J. (1996) *Phys. Rev. A* **54**, 4188.

Kittel, C. (1958). *Elementary Statistical Physics* (John Wiley, New York).

Kittel, C. (1969). *Thermal Physics* (John Wiley, New York).

Klein, M. J., and Tisza, L. (1949). *Phys. Rev.* **76**, 1861.

Knobler, C. M., and Scott, R. L. (1984). In *Phase Transitions and Critical Phenomena*, eds. C. Domb and J. L. Lebowitz (Academic Press, London), Vol. 9, pp. 163–231.

Knuth, D. E. (1997). *The Art of Computer Programming, Volume 2, Seminumerical Algorithms*, 3rd ed. (Addison-Wesley, Reading, MA).

Kolmogorov, A. N. (1954). *Dokl. Akad. Nauk SSSR* **98**, 527.

Komatsu, E., et al. (2010). *Astrophys. J.* (submitted): doi: arXiv.org/abs/1001.4538v3.

Kosterlitz, J. M. (1974). *J. Phys. C* **7**, 1046.

Kosterlitz, J. M., and Thouless, D. J. (1972). *J. Phys. C* **5**, L124.

Kosterlitz, J. M., and Thouless, D. J. (1973). *J. Phys. C* **6**, 1181.

Kosterlitz, J. M., and Thouless, D. J. (1978). *Prog. Low Temp. Phys.* **78**, 371.

Kothari, D. S., and Auluck, F. C. (1957). *Curr. Sci.* **26**, 169.

Kothari, D. S., and Singh, B. N. (1941). *Proc. R. Soc. Lond. A* **178**, 135.

Kothari, D. S., and Singh, B. N. (1942). *Proc. R. Soc. Lond. A* **180**, 414.

Kramers, H. A. (1938). *Proc. Kon. Ned. Akad. Wet. (Amsterdam)* **41**, 10.

Kramers, H. A., and Wannier, G. H. (1941). *Phys. Rev.* **60**, 252, 263.

Krönig, A. (1856). *Ann. Phys.* **99**, 315.

Kubo, R. (1956). *Can. J. Phys.* **34**, 1274.

Kubo, R. (1957). *Proc. Phys. Soc. Japan* **12**, 570.

Kubo, R. (1959). *Some Aspects of the Statistical Mechanical Theory of Irreversible Processes*, University of Colorado *Lectures in Theoretical Physics* (Interscience Publishers, New York), Vol. 1, p. 120.

Kubo, R. (1965). *Statistical Mechanics* (Interscience Publishers, New York).

Kubo, R. (1966). *Rep. Prog. Phys.* **29**, 255.

Landau, D. P., and Binder, K. (2009). *Monte-Carlo Simulations in Statistical Physics*, 3rd ed. (Cambridge University Press, New York).

Landau, L. D. (1927). *Z. Phys.* **45**, 430; reprinted in the *Collected Papers of L. D. Landau*, ed. D. ter Haar (Pergamon Press, Oxford), p. 8.

Landau, L. D. (1930). *Z. Phys.* **64**, 629.

Landau, L. D. (1937). *Phys. Z. Sowjetunion* **11**, 26.

Landau, L. D. (1941). *J. Phys. USSR* **5**, 71.

Landau, L. D. (1947). *J. Phys. USSR* **11**, 91.

Landau, L. D. (1956). *J. Exptl. Theoret. Phys. USSR* **30**, 1058; English transl. *Soviet Physics JETP* **3**, 920.

Landau, L. D., and Lifshitz, E. M. (1958). *Statistical Physics* (Pergamon Press, Oxford).

Landau, L. D., and Placzek, G. (1934). *Phys. Z. Sowjetunion* **5**, 172.

Landsberg, P. T. (1954a). *Proc. Natl. Acad. Sci. (U.S.A.)* **40**, 149.

Landsberg, P. T. (1954b). *Proc. Camb. Phil. Soc.* **50**, 65.

Landsberg, P. T. (1961). *Thermodynamics with Quantum Statistical Illustrations* (Interscience Publishers, New York).

Landsberg, P. T., and Dunning-Davies, J. (1965). *Phys. Rev.* **138**, A1049; see also their article in the *Statistical Mechanics of Equilibrium and Nonequilibrium* (1965), ed. J. Meixner (North-Holland, Amsterdam).

Langevin, P. (1905a). *J. Phys.* **4**, 678.

Langevin, P. (1905b). *Ann. Chim. et Phys.* **5**, 70.

Langevin, P. (1908). *Comptes Rend. Acad. Sci. Paris* **146**, 530.

Lawrie, I. D., and Sarbach, S. (1984). In *Phase Transitions and Critical Phenomena*, eds. C. Domb and J. L. Lebowitz (Academic Press, London), Vol. 9, pp. 1–161.

L'Ecuyer, P. (1988). *Comm. ACM* **31**, 742.

L'Ecuyer, P. (1999). *Math. Comput.* **68**, 249.

Lee, J. F., Sears, F. W., and Turcotte, D. L. (1963). *Statistical Thermodynamics* (Addison-Wesley, Reading, MA).

Lee, T. D., Huang, K., and Yang, C. N. (1957). *Phys. Rev.* **106**, 1135.

Lee, T. D., and Yang, C. N. (1952). *Phys. Rev.* **87**, 410; see also *ibid.* 404.

Lee, T. D., and Yang, C. N. (1957). *Phys. Rev.* **105**, 1119.

Lee, T. D., and Yang, C. N. (1958). *Phys. Rev.* **112**, 1419.

Lee, T. D., and Yang, C. N. (1959a,b). *Phys. Rev.* **113**, 1165; **116**, 25.

Lee, T. D., and Yang, C. N. (1960a,b,c). *Phys. Rev.* **117**, 12, 22, 897.

Leggett, A. J. (2006). *Quantum Liquids* (Oxford University Press, Oxford).

Leib, E., and Liniger, W. (1963). *Phys. Rev.* **130**, 1605.

Lennard-Jones, J. E. (1924). *Proc. R. Soc. Lond. A* **106**, 463.

Lenz, W. (1920). *Z. Physik* **21**, 613.

Lenz, W. (1929). *Z. Physik* **56**, 778.

Levelt, J. M. H. (1960). *Physica* **26**, 361.

Lewis, H. W., and Wannier, G. H. (1952). *Phys. Rev.* **88**, 682.

Lieb, E. H., and Mattis, D. C. (1966). *Mathematical Physics in One Dimension* (Academic Press, New York).

Lines, M. E., and Glass, A. M. (1977). *Principles and Applications of Ferroelectrics and Related Materials* (Clarendon Press, Oxford).

Liouville, J. (1838). *J. de Math.* **3**, 348.

London, F. (1938a). *Nature* **141**, 643.

London, F. (1938b). *Phys. Rev.* **54**, 947.

London, F. (1954). *Superfluids*, Vols. 1 and 2 (John Wiley, New York); reprinted by Dover Publications, New York, 1964.

Lorentz, H. A. (1904–1905). *Proc. Kon. Ned Akad. Wet. Amsterdam* **7**, 438, 585, 684. See also (1909) *The Theory of Electrons* (Teubner, Leipzig), pp. 47–50; reprinted by Dover Publications, New York (1952).

Loschmidt, J. (1876). *Wien. Ber.* **73**, 139.

Loschmidt, J. (1877). *Wien. Ber.* **75**, 67.

Luck, J. M. (1985). *Phys. Rev. B* **31**, 3069.

Lüders, G., and Zumino, B. (1958). *Phys. Rev.* **110**, 1450.

Ma, S.-K. (1973). *Phys. Rev. A* **7**, 2712.

Ma, S.-K. (1976a). *Modern Theory of Critical Phenomena* (Benjamin/Cummings, Reading, MA).

Ma, S.-K. (1976b). *Phys. Rev. Lett.* **37**, 461.

Ma, S.-K. (1976c). In *Phase Transitions and Critical Phenomena*, eds. C. Domb and M. S. Green (Academic Press, London), Vol. 6, pp. 249–292.

MacDonald, D. K. C. (1948–49). *Rep. Prog. Phys.* **12**, 56.

MacDonald, D. K. C. (1950). *Phil. Mag.* **41**, 814.

MacDonald, D. K. C. (1962). *Noise and Fluctuations: An Introduction* (John Wiley, New York).

MacDonald, D. K. C. (1963). *Introductory Statistical Mechanics for Physicists* (John Wiley, New York).

Majumdar, R. (1929). *Bull. Calcutta Math. Soc.* **21**, 107.

Malijevsky, A., and Kolafa, J. (2008). In *Theory and Simulation of Hard-Sphere Fluids and Related Systems*, ed. A. Mulero (Springer, Berlin), pp. 27–36.

Mandel, L., Sudarshan, E. C. G., and Wolf, E. (1964). *Proc. Phys. Soc. Lond.* **84**, 435.

Manoliu, A., and Kittel, C. (1979). *Am. J. Phys.* **47**, 678.

March, N. H. (1957). *Adv. Phys.* **6**, 1.

Martin, P. C. (1968). *Measurements and Correlations* (Gordon & Breach, New York).

Martin, P. C., and De Dominicis, C. (1957). *Phys. Rev.* **105**, 1417.

Mather, J. C. et al. (1994). *Astrophys. J.* **420**, 439.

Mather, J. C. et al. (1999). *Astrophys. J.* **512**, 511.

Matsubara, T., and Matsuda, H. (1956). *Prog. Theor. Phys. Japan* **16**, 416.

Maxwell, J. C. (1860). *Phil. Mag.* **19**, 19; **20**, 1.

Maxwell, J. C. (1867). *Trans. R. Soc. Lond.* **157**, 49; also published in 1868 in *Phil. Mag.* **35**, 129, 185.

Maxwell, J. C. (1879). *Camb. Phil. Soc. Trans.* **12**, 547.

Mayer, J. E., and Ackermann, P. G. (1937). *J. Chem. Phys.* **5**, 74.

Mayer, J. E., and Harrison, S. F. (1938). *J. Chem. Phys.* **6**, 87.

Mayer, J. E., and Mayer, M. G. (1940). *Statistical Mechanics* (John Wiley, New York).

Mayer, J. E. (1937). *J. Chem. Phys.* **5**, 67.

Mayer, J. E. (1942). *J. Chem. Phys.* **10**, 629.

Mayer, J. E. (1951). *J. Chem. Phys.* **19**, 1024.

Mazenko, G. F. (2006). *Nonequilibrium Statistical Mechanics* (John Wiley, New York).

McCoy, B. M., and Wu, T. T. (1973). *The Two-Dimensional Ising Model* (Harvard University Press, Cambridge, MA).

Mehra, J. M., and Pathria, R. K. (1994). In *The Beat of a Different Drum: The Life and Science of Richard Feynman* by J. Mehra (Clarendon Press, Oxford), Chap. 17, pp. 348–391.

Mermin, N. D. (1968). *Phys. Rev.* **176**, 250.

Mermin, N. D., and Wagner, H. (1966). *Phys. Rev. Lett.* **17**, 1133, 1307.

Metropolis, N. (1987). *Los Alamos Sci.* **15**, 125.

Metropolis, N., Rosenbluth, A. W., Rosenbluth, M. N., Teller, A. N., and Teller, E. (1953). *J. Chem. Phys.* **21**, 1087, 1092.

Milne, E. (1927). *Proc. Camb. Phil. Soc.* **23**, 794.

Minguzzi, A., Conti, S., and Tosi, M. P. (1997). *J. Phys.: Condens. Matter* **9**, L33.

Mohling, F. (1961). *Phys. Rev.* **122**, 1062.

Montroll, E. (1963). *The Many-Body Problem*, ed. J. K. Percus (Interscience Publishers, New York), pp. 525–533.

Montroll, E. (1964). *Applied Combinatorial Mathematics*, ed. E. F. Beckenbach (John Wiley, New York).

Morse, P. M. (1969). *Thermal Physics*, 2nd ed. (W. A. Benjamin, New York).

Moser, J. K. (1962). *Nach. Akad. Wiss. Gttingen, Math. Phys. Kl. II*, **1**, 1.

Mukamel, D. (1975). *Phys. Rev. Lett.* **34**, 481.

Mulero, A., Galan, C. A., Parra, M. I., and Cuadros, F. (2008). In *Theory and Simulation of Hard-Sphere Fluids and Related Systems*, ed. A. Mulero (Springer, Berlin), pp. 111–132.

Nelson, D. R. (1983). In *Phase Transitions and Critical Phenomena*, eds. C. Domb and J. L. Lebowitz (Academic Press, London), Vol. 7, pp. 1–99.

Newman, M. E. J., and Barkema, G. T. (1999). *Monte Carlo Methods in Statistical Physics* (Oxford University Press, Oxford).

New Worlds, New Horizons in Astronomy and Astrophysics. (2010). (National Academies Press, Washington). See www.nap.edu.

Niemeijer, Th., and van Leeuwen, J. M. J. (1976). In *Phase Transitions and Critical Phenomena*, eds. C. Domb and M. S. Green (Academic Press, London), Vol. 6, pp. 425–505.

Nienhuis, B. (1987). In *Phase Transitions and Critical Phenomena*, eds. C. Domb and J. L. Lebowitz (Academic Press, London), Vol. 11, pp. 1–53.

Nozières, P. (1964). *Theory of Interacting Fermi Systems* (W. A. Benjamin, New York).

Nyquist, H. (1927). *Phys. Rev.* **29**, 614.

Nyquist, H. (1928). *Phys. Rev.* **32**, 110.

Ono, S. (1951). *J. Chem. Phys.* **19**, 504.

Onsager, L. (1931). *Phys. Rev.* **37**, 405; **38**, 2265.

Onsager, L. (1944). *Phys. Rev.* **65**, 117.

Onsager, L. (1949). *Nuovo Cim.* **6** (Suppl. 2), 249, 261.

Ornstein, L. S., and Zernike, F. (1914). *Proc. Akad. Sci. Amsterdam* **17**, 793; reproduced in *The Equilibrium Theory of Classical Fluids*, eds. A. L. Frisch and J. L. Lebowitz (W. A. Benjamin, New York, 1964).

Pais, A. M., and Uhlenbeck, G. E. (1959). *Phys. Rev.* **116**, 250.

Paredes, B., Widera, A., Murg, V., Mandel, O., Folling, S., Cirac, I., Shlyapnikov, G. V., and Hansch, T. W. (2004). *Nature*, **429**, 277.

Parker, W. S. (2008). *Int. Stud. Philos. Sci.* **22**, 165.

Patashinskii, A. Z., and Pokrovskii, V. L. (1966). *Sov. Phys. JETP* **23**, 292.

Pathria, R. K. (1966). *Nuovo Cim. (Supp.), Ser. I*, **4**, 276.

Pathria, R. K. (1972). *Statistical Mechanics*, 1st ed. (Pergamon Press, Oxford).

Pathria, R. K. (1974). *The Theory of Relativity*, 2nd ed. (Pergamon Press, Oxford).

Pathria, R. K. (1983). *Can. J. Phys.* **61**, 228.

Pathria, R. K. (1998). *Phys. Rev. A* **58**, 1490.

Pathria, R. K., and Kawatra, M. P. (1962). *Prog. Theor. Phys. Japan* **27**, 638, 1085.

Pauli, W. (1925). *Z. Physik* **31**, 776.

Pauli, W. (1927). *Z. Physik* **41**, 81.

Pauli, W. (1940). *Phys. Rev.* **58**, 716.

Pauli, W., and Belinfante, F. J. (1940). *Physica* **7**, 177.

Peierls, R. E. (1935). *Ann. Inst. Henri Poincare* **5**, 177.

Peierls, R. E. (1936). *Proc. Camb. Phil. Soc.* **32**, 471, 477.

Penrose, O., and Onsager, L. (1956). *Phys. Rev.* **104**, 576.

Penzias, A. A., and Wilson, R. W. (1965). *Astrophys. J.* **142**, 419.

Percus, J. K. (1963). *The Many-Body Problem* (Interscience Publishers, New York).

Percus, J. K., and Yevick, G. J. (1958). *Phys. Rev.* **110**, 1.

Perrin, F. J. (1934). *Phys. Radium* **V**, 497.

Perrin, F. J. (1936). *Phys. Radium* **VII**, 1.

Pethick, C. J., and Smith, H. (2008). *Bose–Einstein Condensation in Dilute Gases* (Cambridge University Press, New York).

Pines, D. (1962). *The Many-Body Problem* (W. A. Benjamin, New York).

Pitaevskii, L. P. (1959). *Sov. Phys. JETP* **9**, 830.

Pitaevskii, L. P. (1961). *Sov. Phys. JETP* **13**, 451.

Pitaevskii, L. P., and Stringari, S. (2003). *Bose–Einstein Condensation* (Oxford University Press, Oxford).

Planck, M. (1900). *Verhandl. Deut. Phys. Ges.* **2**, 202, 237.

Planck, M. (1908). *Ann. Phys.* **26**, 1.

Planck, M. (1917). *Sitz. der Preuss. Akad.* 324.

Plischke, M., and Bergersen, B. (1989). *Equilibrium Statistical Physics* (Prentice-Hall, Englewood Cliffs, NJ).

Pospišil, W. (1927). *Ann. Phys.* **83**, 735.

Potts, R. B., and Ward, J. C. (1955). *Prog. Theor. Phys. Japan* **13**, 38.

Press, W. H., Teukolsky, S. A., Vetterling, W. T., and Flannery, B. P. (2007). *Numerical Recipes: The Art of Scientific Computing*, 3rd ed. (Cambridge University Press, New York).

Prigogine, I. (1967). *Introduction to Thermodynamics of Irreversible Processes*, 3rd ed. (John Wiley, New York).

Privman, V. (ed.) (1990). *Finite Size Scaling and Numerical Simulation of Statistical Systems* (World Scientific, Singapore).

Privman, V., and Fisher, M. E. (1984). *Phys. Rev. B* **30**, 322.

Privman, V., Hohenberg, P. C., and Aharony, A. (1991). In *Phase Transitions and Critical Phenomena*, eds. C. Domb and J. L. Lebowitz (Academic Press, London), Vol. 14, pp. 1–134, 364–367.

Purcell, E. M. (1956). *Nature* **178**, 1449.

Purcell, E. M., and Pound, R. V. (1951). *Phys. Rev.* **81**, 279.

Rahman, A. (1964). *Phys. Rev.* **136**, A405.

Ramsey, N. F. (1956). *Phys. Rev.* **103**, 20.

Rayleigh, L. (1900). *Phil. Mag.* **49**, 539.

Rayfield, G. W., and Reif, F. (1964). *Phys. Rev.* **136**, A1194; see also (1963) *Phys. Rev. Lett.* **11**, 305.

Ree, F. H., and Hoover, W. G. (1964). *J. Chem. Phys.* **40**, 939.

Regal, C. A., Greiner, M., and Jin, D. S. (2004). *Phys. Rev. Lett.* **92**, 040403.

Reif, F. (1965). *Fundamentals of Statistical and Thermal Physics* (McGraw-Hill, New York).

Rhodes, P. (1950). *Proc. R. Soc. Lond. A* **204**, 396.

Riecke, E. (1898). *Ann. Phys.* **66**, 353, 545.

Riecke, E. (1900). *Ann. Phys.* **2**, 835.

Riess, A. G., Macri, L., Casertano, S., Sosey, M., Lampeitl, H., Ferguson, H. C., Filippenko, A. V., Jha, S. W., Li, W., Chornock, R., and Sarkar, D. (2009). *Astrophys. J.* **699**, 539.

Rintoul, M. D., and Torquato, S. (1996). *J. Chem. Phys.* **105**, 9258.

Roberts, J. L., Claussen, N. R., Cornish, S. L., Donley, E. A., Cornell, E. A., and Wieman, C. E. (2001). *Phys. Rev. Lett.* **86**, 4211.

Roberts, T. R., and Sydoriak, S. G. (1955). *Phys. Rev.* **98**, 1672.

Robinson, J. E. (1951). *Phys. Rev.* **83**, 678.

Rowley, L. A., Nicholson, D., and Parsonage, N. G. (1975). *J. Comp. Phys.* **17**, 401.

Rowlinson, J. S., and Swinton, F. L. (1982). *Liquids Liquid Mixtures*, 3rd ed. (Butterworth Scientific, London).

Ruprecht, P. A., Holland, M. J., Burnett, K., and Edwards, M. (1995). *Phys. Rev. A* **51**, 4704.

Rushbrooke, G. S. (1938). *Proc. R. Soc. Lond. A* **166**, 296.

Rushbrooke, G. S. (1955). *Introduction to Statistical Mechanics* (Clarendon Press, Oxford).

Rushbrooke, G. S. (1963). *J. Chem. Phys.* **39**, 842.

Rushbrooke, G. S., Baker, G. A., Jr., and Wood, P. J. (1974). In *Phase Transitions and Critical Phenomena*, eds. C. Domb and M. S. Green (Academic Press, London), Vol. 3, pp. 245–356.

Sackur, O. (1911). *Ann. Phys.* **36**, 958.

Sackur, O. (1912). *Ann. Phys.* **40**, 67.

Sakharov, A. D. (1967). *JETP Lett.* **5**, 24.

Schiff, L. I. (1968). *Quantum Mechanics*, 3rd ed. (McGraw-Hill, New York).

Schramm, D. N., and Turner, M. S. (1998). *Rev. Mod. Phys.* **70**, 303.

Schrödinger, E. (1960). *Statistical Thermodynamics* (Cambridge University Press, Cambridge, UK).

Schultz, T. D., Lieb, E. H., and Mattis, D. C. (1964). *Rev. Mod. Phys.* **36**, 856.

Schwartz, L. (1966). *Mathematics of the Physical Sciences* (Addison-Wesley, Reading, MA).

Scott, G. G. (1951). *Phys. Rev.* **82**, 542.

Scott, G. G. (1952). *Phys. Rev.* **87**, 697.

Sells, R. L., Harris, C. W., and Guth, E. (1953). *J. Chem. Phys.* **21**, 1422.

Shannon, C. E. (1948). *Bell Syst. Tech. J.* **27**, 379, 623.

Shannon, C. E. (1949). *The Mathematical Theory of Communication* (University of Illinois Press, Urbana, IL).

Simon, F. (1930). *Ergeb. Exakt. Naturwiss.* **9**, 222.

Singh, S. (2005). *The Big Bang: The Origin of the Universe* (Fourth Estate, New York).

Singh, S., and Pandita, P. N. (1983). *Phys. Rev. A* **28**, 1752.

Singh, S., and Pathria, R. K. (1985a). *Phys. Rev. B* **31**, 4483.

Singh, S., and Pathria, R. K. (1985b). *Phys. Rev. Lett.* **55**, 347.

Singh, S., and Pathria, R. K. (1986a). *Phys. Rev. Lett.* **56**, 2226.

Singh, S., and Pathria, R. K. (1986b). *Phys. Rev. B* **34**, 2045.

Singh, S., and Pathria, R. K. (1987a). *Phys. Rev. B* **36**, 3769.

Singh, S., and Pathria, R. K. (1987b). *J. Phys. A* **20**, 6357.

Singh, S., and Pathria, R. K. (1988). *Phys. Rev. B* **38**, 2740.

Singh, S., and Pathria, R. K. (1992). *Phys. Rev. B* **45**, 9759.

Smith, B. L. (1969). *Contemp. Phys.* **10**, 305.

Smoluchowski, M. V. (1906). *Ann. Phys.* **21**, 756.

Smoluchowski, M. V. (1908). *Ann. Phys.* **25**, 205.

Sokal, A. D. (1981). *J. Stat. Phys.* **25**, 25, 51.

Sommerfeld, A. (1928). *Z. Physik* **47**, 1; see also Sommerfeld and Frank (1931) *Rev. Mod. Phys.* **3**, 1.

Sommerfeld, A. (1932). *Z. Physik* **78**, 283.

Sommerfeld, A. (1956). *Thermodynamics and Statistical Mechanics* (Academic Press, New York).

Speedy, R. J. (1997). *J. Phys.: Condens. Matter* **9**, 8591.

Squires, G. L. (1997). *Introduction to the Theory of Thermal Neutron Scattering* (Cambridge University Press, New York).

Stanley, H. E. (1968). *Phys. Rev.* **176**, 718.

Stanley, H. E. (1969a). *J. Phys. Soc. Japan* **26S**, 102.

Stanley, H. E. (1969b). *Phys. Rev.* **179**, 570.

Stanley, H. E. (1971). *Introduction to Phase Transitions and Critical Phenomena* (Oxford University Press, Oxford).

Stanley, H. E. (1974). In *Phase Transitions and Critical Phenomena*, eds. C. Domb and M. S. Green (Academic Press, London), Vol. 3, pp. 485–567.

Stefan, J. (1879). *Wien. Ber.* **79**, 391.

Steigman, G. (2006). *Int. J. Mod. Phys* **E15**, 1, doi: arXiv:astro-ph/0511534v1.

Stierstadt, K. et al. (1990). *Physics Data: Experimental Values of Critical Exponents and Amplitude Ratios of Magnetic Phase Transitions* (Fachinformationszentrum, Karlsruhe, Germany).

Stoner, E. C. (1929). *Phil. Mag.* **7**, 63.

Stoner, E. C. (1930). *Phil. Mag.* **9**, 944.

Swendsen, R. H., and Wang, J.-S. (1987). *Phys. Rev. Lett.* **58**, 86.

Takahashi, H. (1942). *Proc. Phys. Math. Soc. (Japan)* **24**, 60.

Tarko, H. B., and Fisher, M. E. (1975). *Phys. Rev. B* **11**, 1217.

ter Haar, D. (1954). *Elements of Statistical Mechanics* (Rinehart, New York).

ter Haar, D. (1955). *Rev. Mod. Phys.* **27**, 289.

ter Haar, D. (1961). *Rep. Prog. Phys.* **24**, 304.

ter Haar, D. (1966). *Elements of Thermostatistics* (Holt, Rinehart & Winston, New York).

ter Haar, D. (1967). *The Old Quantum Theory* (Pergamon Press, Oxford).

ter Haar, D. (1968). *On the History of Photon Statistics, Course 42 of the International School of Physics "Enrico Fermi"* (Academic Press, New York).

ter Haar, D., and Wergeland, H. (1966). *Elements of Thermodynamics* (Addison-Wesley, Reading, MA).

Tetrode, H. (1912). *Ann. Phys.* **38**, 434; corrections to this paper appear in (1912) *ibid.* **39**, 255.

Tetrode, H. (1915). *Proc. Kon. Ned. Akad. Amsterdam* **17**, 1167.

Thomas, L. H. (1927). *Proc. Camb. Phil. Soc.* **23**, 542.

Thompson, C. J. (1972a). In *Phase Transitions and Critical Phenomena*, eds. C. Domb and M. S. Green (Academic Press, London), Vol. 1, pp. 177–226.

Thompson, C. J. (1972b). *Mathematical Statistical Mechanics* (Princeton University Press, Princeton, NJ).

Thompson, C. J. (1988). *Classical Equilibrium Statistical Mechanics* (Clarendon Press, Oxford).

Tilley, D. R., and Tilley, J. (1990). *Superfluidity and Superconductivity*, 3rd ed. (Adam Hilger, Bristol).

Tinkham, M. (1996). *Introduction to Superconductivity* (McGraw-Hill, New York).

Tisza, L. (1938a). *Nature* **141**, 913.

Tisza, L. (1938b). *Compt. Rend. Paris* **207**, 1035, 1186.

Tolman, R. C. (1934). *Relativity, Thermodynamics and Cosmology* (Clarendon Press, Oxford).

Tolman, R. C. (1938). *The Principles of Statistical Mechanics* (Oxford University Press, Oxford).

Tonks, L. (1936). *Phys. Rev.* **50**, 955.

Tracy, C. A., and McCoy, B. M. (1973). *Phys. Rev. Lett.* **31**, 1500.

Tuttle, E. R., and Mohling, F. (1966). *Ann. Phys.* **38**, 510.

Uhlenbeck, G. E. (1966). In *Critical Phenomena*, eds. M. S. Green and J. V. Sengers (National Bureau of Standards, Washington, DC), pp. 3–6.

Uhlenbeck, G. E., and Beth, E. (1936). *Physica* **3**, 729.

Uhlenbeck, G. E., and Ford, G. W. (1963). *Lectures in Statistical Mechanics* (American Mathematical Society, Providence, RI).

Uhlenbeck, G. E., and Gropper, L. (1932). *Phys. Rev.* **41**, 79.

Uhlenbeck, G. E., and Ornstein, L. S. (1930). *Phys. Rev.* **36**, 823.

Ursell, H. D. (1927). *Proc. Camb. Phil. Soc.* **23**, 685.

van der Waals, J. D. (1873). *Over de Continuiteit van den Gas-en Vloeistoftoestand*, Thesis, Leiden.

van Hove, L. (1949). *Physica*, **15**, 951.

Vdovichenko, N. V. (1965). *Sov. Phys. JETP* **20**, 477; **21**, 350.

Verlet, L. (1967). *Phys. Rev.* **159**, 98.

Vinen, W. F. (1961). *Proc. R. Soc. Lond. A* **260**, 218; see also *Progress in Low Temperature Physics*, Vol. III (1961), ed. C. J. Gorter (North-Holland, Amsterdam).

Vinen, W. F. (1968). *Rep. Prog. Phys.* **31**, 61.

Vollhardt, D., and Wölfle, P. (1990). *The Superfluid Phases of Helium 3* (Taylor and Francis, London).

von Neumann, J. (1927). *Gottinger Nachr.* **1**, 24, 273.

von Neumann, J. (1951). In *Monte Carlo Method*, eds. A. S. Householder, G. E. Forsythe, and H. H. Germond (NBS-Appl. Math. Ser, U. S. Government Printing Office, Washington, DC).

Voronel, A. V. (1976). In *Phase Transitions and Critical Phenomena*, eds. C. Domb and M. S. Green (Academic Press, London), Vol. 5b, pp. 343–394.

Walker, C. B. (1956). *Phys. Rev.* **103**, 547.

Wallace, D. J. (1976). In *Phase Transitions and Critical Phenomena*, eds. C. Domb and M. S. Green (Academic Press, London), Vol. 6, pp. 293–356.

Walton, A. J. (1969). *Contemp. Phys.* **10**, 181.

Wang, F., and Landau, D. P. (2001). *Phys. Rev. Lett.* **86**, 2050; *Phys. Rev. E* **64**, 056101.

Wannier, G. H. (1945). *Rev. Mod. Phys.* **17**, 50.

Wannier, G. H. (1966). *Statistical Physics* (John Wiley, New York).

Waterston, J. J. (1892). *Philos. Trans. R. Soc. Lond. A* **183**, 5, 79; reprinted in his collected papers (1928), ed. J. S. Haldane (Oliver & Boyd, Edinburgh), pp. 207, 318. An abstract of Waterston's work did appear earlier; see *Proc. R. Soc. Lond. A* **5**, 604 (1846).

Watson, P. G. (1969). *J. Phys. C* **2**, 1883, 2158.

Wax, N. (ed.). (1954). *Selected Papers on Noise and Stochastic Processes* (Dover Publications, New York).

Weeks, J. D., Chandler, D., and Andersen, H. C. (1971). *Chem. Phys.* **54**, 5237.

Weinberg, S. (1993). *The First Three Minutes*, 2nd ed. (Basic, New York).

Weinberg, S. (2008). *Cosmology* (Oxford University Press, New York).

Weinstock, R. (1969). *Am. J. Phys.* **37**, 1273.

Weller, W. (1963). *Z. Naturforsch.* **18A**, 79.

Wergeland, H. (1969). *Lettere al Nuovo Cim.* **1**, 49.

Wertheim, M. S. (1963). *Phys. Rev. Lett.* **10**, 321.

Whitmore, S. C., and Zimmermann, W. (1965). *Phys. Rev. Lett.* **15**, 389.

Widom, B. (1965). *J. Chem. Phys.* **43**, 3892, 3898.

Wiebes, N.-H., and Kramers (1957). *Physica* **23**, 625.

Wien, W. (1896). *Ann. Phys.* **58**, 662.

Wiener, N. (1930). *Act. Math. Stockholm* **55**, 117.

Wilczek, F. (1990). *Fractional Statistics and Anyon Superconductivity* (World Scientific, Singapore).

Wilks, J. (1961). *The Third Law of Thermodynamics* (Oxford University Press, Oxford).

Wilson, A. H. (1960). *Thermodynamics and Statistical Mechanics* (Cambridge University Press, Cambridge, UK).

Wilson, K. G. (1971). *Phys. Rev. B* **4**, 3174, 3184.

Wilson, K. G. (1972). *Phys. Rev. Lett.* **28**, 548.

Wilson, K. G. (1975). *Rev. Mod. Phys.* **47**, 773.

Wilson, K. G., and Fisher, M. E. (1972). *Phys. Rev. Lett.* **28**, 240.

Wolff, U. (1989). *Phys. Rev. Lett.* **62**, 361.

Wood, W. W., and Parker, F. R. (1957). *J. Chem. Phys.* **27**, 720.

Woods, A. D. B. (1966). *Quantum Fluids*, ed. D. F. Brewer (North-Holland, Amsterdam), p. 239.

Wright, E. L. et al. (1994). *Astrophys. J.* **420**, 450.

Wu, F. Y. (1982). *Rev. Mod. Phys.* **54**, 235.

Wu, T. T. (1959). *Phys. Rev.* **115**, 1390.

Wu, T. T., McCoy, B. M., Tracy, C. A., and Barouch, E. (1976). *Phys. Rev. B* **13**, 316.

Yang, C. N. (1952). *Phys. Rev.* **85**, 808.

Yang, C. N., and Lee, T. D. (1952). *Phys. Rev.* **87**, 404; see also *ibid.*, 410.

Yang, C. N., and Yang, C. P. (1964). *Phys. Rev. Lett.* **13**, 303.

Yang, C. P. (1962). *J. Math. Phys.* **3**, 797.

Yarnell, J. L. et al. (1959). *Phys. Rev.* **113**, 1379, 1386.

Yarnell, J. L., Katz, M. J., Wenzel, R. G., and Koenig, S. H. (1973). *Phys. Rev. A* **7**, 2130.

Yeomans, J. M. (1992). *Statistical Mechanics of Phase Transitions* (Clarendon Press, Oxford).

Young, A. P. (1979). *Phys. Rev. B* **19**, 1855.

Zasada, C. S., and Pathria, R. K. (1976). *Phys. Rev. A* **14**, 1269.

Zermelo, E. (1896). *Ann. Phys.* **57**, 485; **59**, 793.

Zernike, F. (1916). *Proc. Akad. Sci. Amsterdam* **18**, 1520; reproduced in *The Equilibrium Theory of Classical Fluids*, eds. A. L. Frisch and J. L. Lebowitz (W. A. Benjamin, New York, 1964).

Zwierlein, M. W., Stan, C. A., Schunck, C. H., Raupach, S. M. F., Gupta, S., Hadzibabic, Z. M., and Ketterle, W. (2003). *Phys. Rev. Lett.* **91**, 250401.

Index

D